Telecommunication Networks: Protocols, Modeling and Analysis

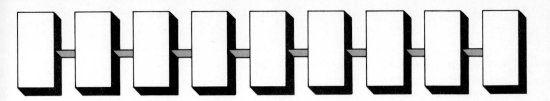

Telecommunication Networks: Protocols, Modeling and Analysis

MISCHA SCHWARTZ
Department of Electrical Engineering and
Center for Telecommunications Research
Columbia University

Addison-Wesley Publishing Company

Reading, Massachusetts
Menlo Park, California · Don Mills, Ontario
Wokingham, England · Amsterdam
Sydney · Singapore · Tokyo · Madrid
Bogotá · Santiago · San Juan

This book is in the **Addison-Wesley Series in Electrical and Computer Engineering**

Sponsoring Editor • *Tom Robbins*
Production Supervisor • *Bette J. Aaronson*
Copy Editor • *Stephanie Kaylin*
Text Designer • *Herb Caswell*
Illustrator • *George Nichols*
Technical Art Consultant • *Joseph Vetere*
Production Coordinator • *Ezra C. Holston*
Manufacturing Supervisor • *Hugh Crawford*
Cover Designer • *Marshall Henrichs*

Library of Congress Cataloging-in-Publication Data

Schwartz, Mischa.
    Telecommunication networks.

    Bibliography: p.
    Includes index.
    1. Data transmission systems.   2. Packet switching
(Data transmission)   3. Telecommunication — Switching
systems.   I. Title.
TK5105.S385   1987   004.6   85-30639
ISBN 0-201-16423-X

Reprinted with corrections May, 1987

To Charlotte

Preface

This book attempts to present the dramatic changes of the past two decades in the field of telecommunications. It covers both packet switching, used to improve data communications, and circuit switching, used in telephone networks. Both technologies may now be considered vehicles for the integrated transmission of voice, data, video, and other traffic. It is expected that elements of both types of switching will be used in the integrated networks of the future.

Organization

The book is organized around the layered architecture of the OSI Model. Chapter 3 provides the necessary introduction to this model. The chapter also provides a brief introduction to IBM's Systems Network Architecture (SNA) and to X.25, the international interface recommendation, so that the reader may compare them. SNA and X.25 are discussed in more detail in Chapter 5.

Chapters 4–9 are devoted to packet switching; Chapters 10, 11, and a portion of Chapter 12 cover circuit switching. The remainder of Chapter 12 introduces integrated networks.

The stress throughout the book is on the quantitative performance evaluation of telecommunication networks and systems. The vehicle for much of this analysis, in packet switching and in networks integrating aspects of both technologies, is queueing theory. To make the book as self-contained as possible, Chapter 2 provides an introduction to this subject. The only prerequisite for this material is knowledge of probability theory.

The book has been used, in note form, for formal classroom courses and for short courses given to practicing engineers, programmers, systems analysts, and other technical specialists. Although the stress, as noted above, is on quantitative performance evaluation, there is much qualitative material presented. The book can thus be used to present a comprehensive in-depth course on the subject of telecommunication networks, geared to the needs of engineers, analysts, and computer scientists interested in the field. It can also be used to provide a more qualitative introduction to basic concepts of the field, with the quantitative material following afterwards, if desired.

At Columbia, the material is largely covered in two semesters. The first semester, computer-communication networks, covers Chapters 1 – 6, 8, 9. The second semester, devoted to circuit switching, covers Chapters 10 – 12. Chapter 7, on the OSI Transport Layer, is presented in selected topics courses. Students taking the sequence have come from electrical engineering, computer science, operations research, and the Business School, among other areas. An introductory overview course geared to those having little or no knowledge of probability theory would use Chapters 1, 3, 4, 6, 7 and portions of Chapters 8 and 9. Some of the qualitative portions of Chapter 5, describing congestion control in X.25 and SNA, for example, could also be presented in such an introductory course. The following chart illustrates various options for use depending on time, interest, and prerequisites.

Type of Course	Chapters
One semester, computer communications, probability required	1 – 6, 8, 9, plus selected portions of 7, time permitting
One semester, computer communications, probability not required	1, 3, 4 (selected portions), 6, 8, 9, plus Section 5 – 1, introduction to 5 – 3, 5 – 3 – 2, and selected portions of 7
One semester, digital circuit switching, probability required *Note:* This course would be more advanced than the one on packet switching	1, 2, 10 – 12
Full year, two semesters, probability required	1 – 6, 8 – 12 plus selected portions of 7, time permitting

Acknowledgements

I have profited immensely from discussions with students, colleagues, and friends at Columbia University, at other academic institutions, and in industry. There is no substitute for teaching this material. I am particularly thankful to the many students at Columbia and in industrial short courses who, with their often insightful and penetrating questions, forced me time and again to rethink and understand better a particular concept, algorithm, or operation of a system. Carrying out research in the area also helped develop the intuition and insight necessary for writing this book. Here, too, I am indebted to graduate students, colleagues, and friends. There are too many such persons to thank individually. However, I would like to acknowledge the individual help of Profs. Thomas E. Stern and Aurel Lazar, who have participated in the teaching of the course material and in carrying out joint research in many of the topics covered in this book. Other members of the Electrical Engineering Department and our Center for Telecommunication Research played important roles in bringing this book to life. Mrs. Betty Lim, the technical typist for the Department of Electrical Engineering, must be particularly singled out. She typed and retyped the notes through a number of uses in class (what would she have done without the word processor?), prepared all the figures from rough handwritten form, and even saw to the printing of the notes and distribution to students. Thank you Betty! I couldn't have done it without your uncomplaining help.

New York   M. S.

Contents

7 Transport Layer 331

8 Polling and Random Access in Data Networks 403

9 Local Area Networks 451

10 Introduction to Circuit Switching 483

11 Call Processing in Digital Circuit-switching Systems   567

12 The Evolution Toward Integrated Networks   627

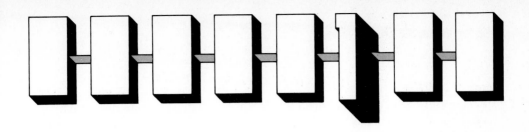

Introduction and Overview

1-1 Circuit and Packet Switching — A Brief Introduction

This book focuses on data communication networks, thousands of which are deployed worldwide to enhance the ability of data users to communicate with one another. These systems range from small networks interconnecting data terminals and computers within a single building or campuslike complex, to large geographically distributed networks covering entire countries or, in some cases, spanning the globe. Some networks are privately operated; others are public — available for a fee to all who want to use them.

Some networks use *packet-switched* technology, in which blocks of data called packets are transmitted from a source to a destination. Source and destination can be user terminals, computers, printers, or any other types of data-communicating and/or data-handling devices. In this technology, which is described in detail in this book, packets from multiple users share the same distribution and transmission facilities.

Other networks are of the *circuit-switched* type, most commonly portions of the ubiquitous telephone networks to which we are all accustomed. In these networks, which generally transmit voice or data, a private transmission path is established between any pair or group of users attempting to communicate and is held as long as transmission is required. Integrated networks, combining aspects of both packet- and circuit-switched technology, are now beginning to be deployed; they are expected, in one form or another, to dominate the field in the 1990s or later.

The most common types of traffic handled are interactive data, generally transmitted in short bursts of a few characters, to as many as 400–1000 characters, between terminals or between terminals and computers; file transfer, involving the transmission of up to millions of characters (or bytes) between computers, or between mass storage systems; and, increasingly, digital voice. Facsimiles, images, and other types of traffic are being considered for transmission as well.

The increasing transmission of digital voice needs some explanation. Voice transmission is still the most common mode of communication worldwide. It involves by far the largest investment in installed plant. The telephone networks developed to handle voice cover every part of the globe. All projections indicate that voice will continue to be the heaviest user of communication facilities worldwide.

The bulk of the telephone plant throughout the world, however, is still analog. It has been tailormade and designed to handle voice. Although telephone networks are used extensively for data transmission, data signals must normally be converted to voice-type analog signals using devices called *modems* [SCHW 1980a]. This limits the rate of transmission of data to at most 14.4 kbps (9600-bps transmission is more common as an upper limit), and then only using private, specially conditioned transmission facilities. More typically, over the usual dialed public telephone network, data bit rates of 1200 to 2400 bps are accommodated.

Increasingly, however, digital transmission and switching facilities are being introduced into telephone networks. In the early 1960s American Telephone & Telegraph (AT&T) in the United States pioneered the first commercial introduction of a digital carrier system. Most other countries followed soon thereafter. This has led to worldwide deployment of two digital carrier systems: the T1 system, transmitting at 1.544 Mbps, designed to handle 24 voice channels at 64 kbps each (soon to be changed to 32-kbps voice channels), and a 30-voice channel system operating at 2.048 Mbps. The former is used extensively in the United States, Canada, and Japan; the latter, as a CCITT-recom-

---

[SCHW 1980a] M. Schwartz, *Information Transmission, Modulation, and Noise*, 3d ed., McGraw-Hill, New York, 1980.

mended standard,* in the rest of the world. (The specific formats used are described briefly in Chapter 10. See [SCHW 1980a] for more details.) This type of digital carrier transmission is used extensively for so-called short-haul communication, up to 25 miles. Higher bit rates are used for longer-range transmission. Digital switching systems, handling the routing of calls between transmission facilities, began to be introduced in the mid-1970s. Telephone administrations have also begun to adopt them rapidly. The advent of long-haul fiber optic transmission systems will expedite the transition to all-digital telephone networks.

Once telephone networks become all digital, *any* kind of data, whether of the interactive data type or whether due to computers communicating with one another, digital voice, or digital images, presumably could traverse a network. Each type might be handled differently, or some could be combined (multiplexed). At present, however, because telephone networks are part analog and part digital, all-digital transmission throughout a network is not always possible. Nonetheless many telephone carriers now provide an all-digital transmission capability, but only over selected portions of a network that are fully digital from end to end.

Real-time telephone voice communication is universally of the circuit-switched type, although there is great interest in the possibility of transmitting voice in real time, in the form of packets, and studies of this type of voice communication are being carried out. A number of circuit-switched data networks have been developed in various parts of the world. We thus provide a detailed discussion of circuit-switched technology in this book, covering digital circuit-switched systems in detail. In these networks a dedicated channel or transmission path, end to end, is set up for users who desire to communicate.

A major portion of the book is devoted to packet-switched data networks. The history of these is more recent than that of circuit switching, beginning with some initial studies in the mid-1960s, intensive interest and development in the late 1960s and early to mid-1970s, and full development from that time on. The genesis of packet switching is tied directly to the revolutionary development of large mainframe ("host") computers in the 1960s. Data networking got its start in an attempt to share the cost, in both hardware and software development, of these very expensive systems.

Three major parallel activities fueled the accelerating development of packet switching as we know it today. Time-sharing service companies were set up in the 1960s to enable terminal users, at a relatively low cost, to access host

---

* The CCITT (the International Telegraph and Telephone Consultative Committee of the International Telecommunications Union) and other international standard-making bodies are described later in this book.

computers owned by these companies. They had to develop a networking capability in order to allow users anywhere in a country (ultimately, throughout the world) to communicate with the appropriate host. A prime example is General Electric (GE) Information Services, which currently operates one of the largest networks in the world devoted to data transmission. This network provides access to GE's privately owned hosts [SCHW 1977]. Another important example is Tymshare, Inc., which developed a network called TYMNET [SCHW 1977] to handle access to its host computers deployed throughout the United States. The TYMNET developers pioneered in a number of significant innovations in commercial packet switching. TYMNET has since expanded to become one of the two major public packet-switched networks in the United States (GTE Telenet is the other example). TYMNET is probably the largest packet-switched network in the world in terms of switching nodes; currently over 1000 are interconnected to form the network. TYMNET is discussed in detail later in this book.

The second major set of developments behind packet switching were provided by the computer manufacturers, who contributed in a number of ways. They developed special-purpose computers called communication processors to offload the communication-handling tasks from the large (host) computers. They developed large software packages to give the hosts specific communication access. Finally, and most importantly, they pioneered in introducing and developing the concept of layered communication architectures. This enabled any number of devices — including terminals, communication processors, computers, and application programs within these computers — to communicate with one another. In particular, IBM pioneered with the development of its Systems Network Architecture (SNA) beginning in the late 1960s. The first formal announcement of the use of SNA, and products based on it, came in 1974. The architecture has continued to evolve since then. Other computer manufacturers have developed their own architectures as well. SNA, Digital Equipment Corporation's DNA, and Burrough's BNA are described later in this book.

The proliferation of various communication architectures, enhancing the ability of members of a family of intelligent systems to communicate but creating possible barriers between systems of different manufacturers, led the International Organization for Standardization (ISO) to launch an intensive effort to develop a worldwide communication architecture standard that would allow systems to communicate openly. This effort, begun in 1978, culminated just two years later, in 1980, in a proposal for a Reference Model for Open Systems

[SCHW 1977] M. Schwartz, *Computer-Communication Network Design and Analysis*, Prentice-Hall, Englewood Cliffs, N.J., 1977.

Interconnection (OSI). The proposal, incorporating the concept of a seven-layered communication architecture much along the lines of IBM's SNA and architectures of other computer manufacturers, was finally approved as an international standard in May 1983 [IEEE 1983b]. Studies of each layer, leading eventually to detailed specifications for each, have been going on in parallel, with a number completed and adopted as standards. The OSI Reference Model is described in detail in this book, beginning with Chapter 3. It is important to note that, although the developing packet-switched technology was one of the prime movers behind this international standards effort, the architectural concept is not limited to packet switching. Actual network implementations could, and in some cases do, use circuit switching as a basis for communications. Except for possible differences in performance and performance objectives offered to the users, the users communicating would see a "transparent pipe." (Packet switching introduces time delay during transmission. Circuit switching usually provides a blocking mode of operation, with users denied access to the network — i.e., they receive a network busy signal if resources are not available.)

The third major development to drive packet-switched technology came out of work initiated and funded by the Advanced Research Projects Agency (ARPA) of the U.S. Department of Defense. This pioneering effort, which established a data communication network called ARPAnet, was also begun in the late 1960s and continued through the 1970s. The basic concept behind the work was the "computer utility": Large hosts, each with a specialized capability, would be interconnected. Users anywhere in the United States could then access any host and use its specialized software or "number-crunching" capability. An underlying, or backbone, communication network had to be developed to interconnect the hosts, and this led to a great deal of research on packet switching worldwide. At the network level, communication processors or concentrators capable of both interfacing with hosts and routing packets along the network had to be developed. Routing and flow control algorithms were devised. The very word packet was coined by ARPA investigators to distinguish between longer messages introduced into the network by hosts and the multiple smaller blocks of data into which these messages were split or segmented to improve network performance. This led to the use of the term packet switch as a label for the communication processor.

A layered architecture was developed for ARPAnet as well, involving so-called higher-level protocols that enable dissimilar hosts to communicate and lower-level protocols that ensure a satisfactory packet delivery service end to

[IEEE 1983b] "Open Systems Interconnection (OSI) — New International Standards Architecture and Protocols for Distributed Information Systems," special issue, *Proc. IEEE*, H. C. Folts and R. des Jardins, eds., vol. 71, no. 12, Dec. 1983.

end. The time-shared computing industry and the computer manufacturers, noted earlier, were also wrestling with these problems.

The ARPAnet experiment spawned many models worldwide. The U.S. public packet-switching carrier, GTE Telenet, grew out of this experiment. The Canadian Datapac (the first public data network in the world, which began operations in 1976), the French Transpac, Japanese networks, and many others around the world developed out of this experience, as well as out of the experiences of the time-shared companies and computer manufacturers mentioned earlier. Some discussion of ARPAnet appears later in this book. A detailed description of the development of ARPAnet and its configuration as of 1976, with accompanying citations to the literature, appears in [SCHW 1977].

This work in data transmission via packet-switched networks fostered standards development activity in the CCITT. (The standards activities of both ISO and CCITT are described later in this book.) This body is made up of official government representatives and observers from telephone administrations worldwide. In 1976 it came out with a recommendation for a three-layered interface architecture for packet switching called X.25. X.25 has since been adopted by almost every packet-switched network and vendor. This architecture is described in detail in Chapters 3 and 5.

1–2 The Need for Networks

The previous introductory section has provided a brief overview of the data networking field, with mention made of the circuit-switched telephone network we all use so often and the newer breed of packet-switched networks. What we mean by network, however, is not obvious. Why deploy networks in the first place?

Conceivably one could interconnect a small number of users directly. But consider the problem of directly connecting hundreds, thousands, or millions of users who desire to communicate with one another. This is obviously an insurmountable task. One might not even want to make direct connections if one could, since many of them might be used only infrequently. In addition, any new pair of users desiring to communicate would require a new direct connection to be made. The concept of a network thus arises quite naturally. A network consists essentially of network switches, or nodes, interconnected by transmission links. These links can be wire, cable, radio, satellite, or fiber optics facilities. A simple example appears in Fig. 1–1.

Note that a network such as this can accommodate a large number of users, up to the limit a switch can handle. Users who desire to communicate must thus have their respective traffic routed or switched, by the network nodes, along an

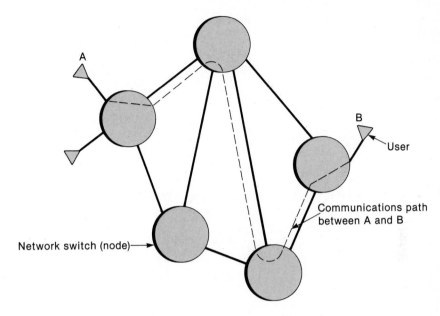

Figure 1-1 Example of a network

appropriate routing path. This network could be dedicated to voice (the telephone network), to data, or to both. The abstract diagram of Fig. 1-1 would apply in both cases. A network of this type allows many users to be interconnected in a cost-effective way. It can also provide added services and features that individual users might not be able to duplicate cost-effectively.

A specific example of a commercial packet-switched network appears in Fig. 1-2. This is the SITA network as of July 1983. The network is owned and operated by SITA (Société Internationale de Telecommunications Aéronautiques), a nonprofit cooperative organization serving 248 airlines in 154 countries. This network provides international airline reservations service, interconnecting visual display terminals at agents' sets throughout the world to airline reservation computers. SITA also provides a telecommunication service for telegraph-type messages concerning flight operations, administrative matters, and commercial activities [SITA].

The mesh network shown in Fig. 1-2 is the data transport portion of the

[SITA] *A Pocket Guide to SITA,* Information and Public Relations Department, SITA, Neuilly-sur-Seine, Fr., July 1983.

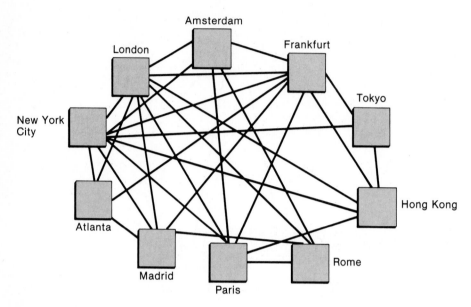

Figure 1–2 Backbone network, SITA, 1983 (from [SITA], Société Internationale de Telecommunications Aéronautiques, with permission)

---

network only. It is essentially a backbone network, to which many other switching centers, not shown in the figure, are connected. The switching centers
shown are interconnected by 9600- and 14,400-bps circuits. Many of the links
shown are satellite circuits. The reservations traffic, called *type A* traffic, is of the
interactive type and consists of short packets approximately 80 characters long.
The average response time, from an agent's terminal to a remote control computer and back, is 3 sec [SITA]. In a typical year over 5 billion type A messages
are transmitted. The telegraph-type traffic, called *type B*, consists of longer
messages, about 200 characters long. Over 400 million such messages are transmitted in a typical year. Stringent time requirements are not set on these non-
conversational messages.

In 1983 the total SITA network consisted of 181 switching centers, interconnecting 10,400 reservation terminals, 11,500 teleprinter terminals, and 54
airline reservation systems. (See [SITA] for a complete map.) The network is
obviously very sizable and requires large resources to run efficiently and cost-effectively. Its worldwide activities indicate the role international telecommunications plays in the world today. (A more detailed, technical description of an
earlier version of the SITA network appears in [SCHW 1977].)

We return now to a more abstract consideration of the concept of network, as exemplified by the network of Fig. 1–1. Consider the "users" shown connected to this network. We have been using the word user very informally thus far. In the case of SITA the users are terminals and host computers. In the case of a voice network they are most often human beings at either end who want to talk with one another. (With the advent of automated voice-response systems, of course, this is not so clear any more!) In the case of data networks in general, the word user has a much more complex meaning. It could represent a person sitting at the keyboard of a terminal or personal computer, it could represent a remote computer-controlled video terminal or printer, it could represent a computer, or it could represent an application program in a computer. This point was made indirectly in the previous section. A better term might be communicating entity, but, for the sake of conciseness and, one hopes, clarity, we shall use the word user in this much broader sense throughout this book.

A number of design questions arise in the context of the simple network model of Fig. 1–1. First is the access problem. How does one connect users to the network? Does each user get a hard-wired, dedicated port into a switch? Or do users share an access point? Can users dial in to a particular access port? The answer, of course, is that all these solutions, and others as well, have been adopted in network design. For the public telephone network, and for many terminal-oriented public data networks, dial-in access is commonly used. In some data networks, users share an access link and wait to transmit when polled. This is typical of many terminals in the SITA network [SCHW 1977]. The problem becomes more complex when users first access a private small network and *then* require connection to a large, geographically distributed network or set of networks. Here access is carried out in a variety of ways depending on the design of the small network. The "network" could actually be a digital PBX (private branch exchange) with dialed access, or it could be a local area network (LAN) that might operate in a token-passing (polling) mode or a (random) contention mode. Other access mechanisms are also possible. Packet-switching access mechanisms are described in detail in Chapters 8 and 9. Circuit-switched dialed access is described in Chapters 10 and 11.

Once the mode of access is decided on, another set of questions arises, relating to the design of the overall network. Is the network of the circuit-switched or packet-switched type? If the former, a complete, *dedicated* path end to end through the network must be established before communication can begin, as noted earlier. This is the path shown in Fig. 1–1 as an example. It requires the use of separate dedicated channels, commonly called trunks, on each of the links connecting the network switches. These channels might be separate frequency bands. More commonly, particularly with digital switches, the channels are time slots in a time-multiplexed repetitive series of frames in time that propagate between adjacent switches. The network switches must be organized to first set up a connection through the switch to a dedicated time slot (trunk or channel) on the appropriate outgoing link. The network must then repeat the process at each subsequent node along the path. If no free path consisting of a concatenation of free trunks exists end to end, the user can either be blocked (a "busy" signal is then returned to the user), or its traffic queued for later transmission. Details of this mode of communication appear in Chapters 10 through 12.

In the case of packet switching, the abstract model of Fig. 1–1 applies. However, the nodes now have a somewhat different function. Two modes of packet-switched data transmission are commonly distinguished. In one case, that of *virtual circuit* transmission, a path is first set up end to end through the network. User packets then traverse the network following the path chosen and arrive at the destination node in the sequence in which they were transmitted. They share link and switch facilities along the way, however, being stored at each intermediate node until ready to be read out, or forwarded, along the appropriate outgoing link. This method of transmission is also called *connection-oriented* transmission. The second mode of transmission of packets is a *connectionless* one, with individual *datagrams* moving between source and destination nodes. No initial connection is set up in this case. Datagrams are forwarded through the network on an individual basis. Routing at intermediate nodes is commonly based on the destination address of the datagram, which must be carried by each datagram. Datagrams are not necessarily guaranteed to arrive at the destination in the order of their transmission.

In both modes of packet-switched transmission — virtual circuit (connection oriented) and datagram (connectionless) — packets are queued at intermediate points along the route between source and destination. This introduces a time delay during transmission that does not normally appear in circuit-switched transmission. (The corresponding delay there is one of setting up the complete path end to end.) This delay, plus propagation and processing delay, accounts for the 3-sec response time of the SITA network, for example. Blocking of packets is a much rarer occurrence than blocking in circuit-switched networks. Packet switching of either type, datagram or virtual circuit, is discussed in detail in Chapters 5 and 6.

It is apparent from this brief discussion that the functions of and the requirements for a nodal switch are quite different in the two cases of packet switching and circuit switching. However, with the move toward integrated networks, which handle voice, data, and other types of traffic, and which use packet-switched technology, circuit-switched technology, and hybrid technologies under development, the functions of and requirements for a nodal switch will become more similar. Modern digital-circuit switches, for example, are being designed to handle packets as well as circuit-switched calls. Similarly, packet-switch designs are being studied that enable these switches to incorporate voice-handling integrated capability in addition. One possibility is to handle voice on a packet basis. Another is to continue to provide circuits for voice calls.

1-2-1 Interconnection of Networks

The abstract model of a network portrayed in Fig. 1-1, topologically in the form of a mesh, is characteristic of many networks worldwide. The SITA network of Fig. 1-2 provides one example. As already noted, whether networks are circuit switched or packet switched, they have the same general appearance in an abstract sense—they consist of nodal switches interconnected by transmission facilities. Many geographically distributed networks are considerably larger, consisting of hundreds and even thousands of nodes, but the usual network graph form is that of the mesh shown in Fig. 1-1.

More and more often, however, data traffic may have to traverse multiple networks. This is particularly true if access is through a local area network, as already noted. Communication via public data networks between different countries also requires traversing a number of networks. Large corporations may have multiple private networks that are interconnected. Interconnection of networks thus plays, and will continue to play, an increasingly significant role in the study of data networks.

Figure 1-3 is an example of a communication system that incorporates both the concept of interconnected networks and the integration of a variety of data sources. Three geographically distributed networks, A, B, and C, are shown, interconnected via so-called *gateways* (indicated by the letter G). These gateways provide the necessary protocol translation and interfacing between disparate networks of possibly different bit rates (bandwidth) and packet-handling capabilities, as well as different architectural constructs. The gateways could be separate intelligent systems (nodes) or could be embedded within network switches. Examples of both types are shown in the figure.

Access to a network could be via a common access point at a network node, a local area network (LAN), or a digital PBX. All three examples appear in Fig. 1-3. The local area networks indicated there are of the two most common types—bus and ring. These are discussed in detail in Chapter 9. LANs normally run at much higher bit rates than distributed networks. Their transmission capacity can range from 1-10 Mbps on coaxial cable facilities, with fiber optic LANs capable of running at bit rates of hundreds and even thousands of Mbps. The distributed networks, on the other hand, if of the packet-switched type, may be operated at speeds in the range of 2400 bps to up to 56 kbps. (Note that the SITA network is operated at 9600 bps and 14,400 bps.) Gateways are

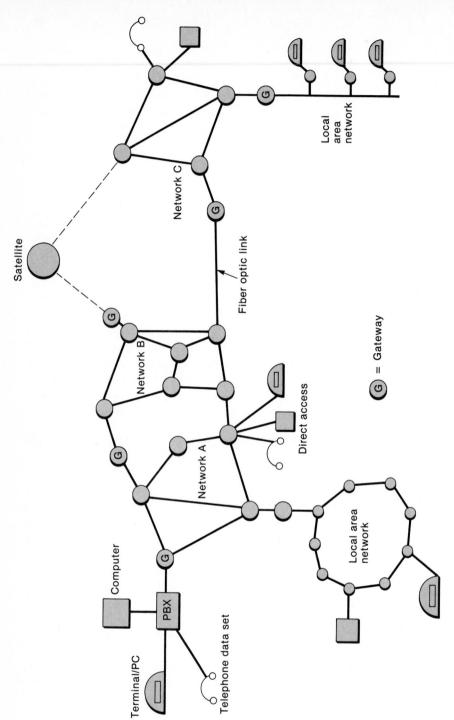

Figure 1–3 Interconnected, integrated networks

obviously required to provide the appropriate speed matching and flow control.

Much work remains to be done on problems arising from the interconnection of networks, as well as the integration of disparate traffic types in a single network or group of networks. Some of the issues involved are discussed in Chapters 7 and 12.

1-3 Layered Communication Architectures

The stress thus far in this chapter has been on communication over networks. Nothing has really been said about the data that must be transmitted or about the functional requirements at the user stations at either end that must be met to ensure that the data is delivered in the form required by the user. For example, the SITA network serves host computers of many varieties and from different manufacturers. So does the ARPA network. Each computer, and each application program in the computer, may require a different communication access method and protocol (i.e., a standard convention for communicating intelligently) for a viable "conversation" to proceed. This sets very specific requirements on either or both parties to the conversation. In essence the data must be presented to the end users in a form that they can recognize and manipulate. This may involve protocol conversion to accommodate an end user's format, code, and language syntax.

As another example, consider a videotex service provided by a supplier to multiple customers. In this service, the supplier maintains a database or multiple databases on a variety of subjects of potential interest to customers: travel information, plane/hotel reservation information, shopping information, and so forth. Much of this information is in visual form and is delivered to the customer's terminal in a prescribed graphics format.

The two parties to a "conversation" or session, the supplier and the customer, must first agree to set up a session through a network (or series of networks). They must agree on the format of the data ultimately delivered to the screen of the customer's terminal. (If the sequence of bits delivered to the terminal in a packet makes no sense to that terminal, even if the packet is delivered correctly, nonsense will appear on the screen.) The terminal must be able to regulate the rate of delivery of data (the rate of arrival of packets); otherwise the source (host) computer that controls the source database could overwhelm the terminal. If connectionless transmission is used, packets (datagrams) may arrive out of order. The recipient terminal must be in a position to resequence these packets. All of these tasks and others like them have nothing to do with the operation of the network. The network may be delivering packets correctly and to the right place, and yet the overall system may not be performing properly. These added controls, required at either end of the network within the host computer or terminal or other intelligent communicating system, as the case may be, are generally incorporated in software at each of the systems or devices. It has become common to carry out a sequence of required tasks in an organized fashion, giving rise to the concept of layered communication architectures.

IBM's Systems Network Architecture (SNA), mentioned in the introduction to this chapter, was one of the first layered architectures developed. The ISO Reference Model for Open Systems Interconnection (OSI), also mentioned in the introduction, is rapidly becoming an international standard for layered architectures. These architectures and others like them recognize that there are essentially two parts to the complete communications problem, that of ensuring timely, correct, and recognizable delivery of data to end users engaged in a conversation or session over a network or series of networks. The first part of the problem involves the communication network: Data delivered by an end user to a network must arrive at the destination correctly and in timely fashion. The second part of the problem is to ensure that the data ultimately delivered to the end user at the destination is recognizable and in the proper form for its correct use.

A number of *network protocols* have been developed to handle the first part of the problem. The second part is solved by introducing *higher-level protocols*. A complete end-user-oriented architecture encompasses both kinds of protocols. Figure 1 – 4 portrays communication between end users A and B of Fig. 1 – 1 as an example in terms of this characterization. An intermediate node in the network is shown as well. This node could also have end users connected to it; they would also have higher-level protocols associated with them. But as indicated in Fig. 1 – 4, the purpose of the intermediate node, so far as other network users such as A and B are concerned, is only to provide the appropriate network services.

The two groups — protocols providing network services and higher-level protocols — are typically broken down further into a series of levels or layers (both words have been used to describe the layering concept). Each layer is chosen to provide a particular service in terms of the basic problem just described: delivering data correctly, on time, and in recognizable form. More precisely, through the development of the OSI Reference Model, the concept has developed of having each layer provide a service to the layer above it.

To be specific, consider the Reference Model. It consists of the seven layers shown in Fig. 1 – 5. The bottom three layers shown provide the network services. Protocols implementing these layers must appear in every network node. The upper four layers provide services to the end users themselves and are thus associated with the end users, *not* with the networks.

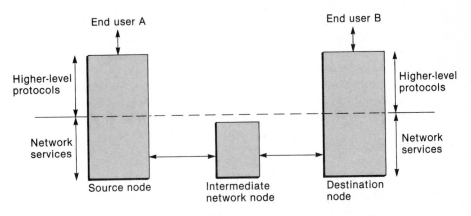

Figure 1–4 Functions of layered communication architectures

The data link layer and the physical layer below it provide an error-free communication link between two nodes in a network. The function of the physical layer is to ensure that a bit entering the physical medium at one end of a link gets through to the other end. Using this underlying bit transport service, the purpose of the data link protocol is to ensure that blocks of data are transferred reliably (error free) across a link. Such blocks are often called frames.

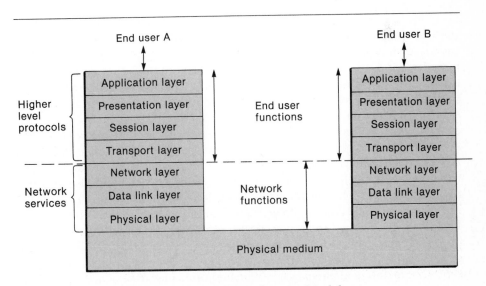

Figure 1–5 OSI Reference Model

This procedure generally requires synchronizing on the first bit in a block, recognizing when the block ends, detecting bit errors when they occur, and providing for their correction in some manner. (Usually this is done by requesting retransmission of a block found to have one or more bits in error.) Although the link in question may be (and often is) a point-to-point physical connection such as the links shown in Fig. 1–1, it could represent the connection between two nodes on either end of a local area network such as those shown in Fig. 1–3. Other examples of "virtual" links are also possible.

The function of the network layer of Fig. 1–5 is to route the data through the network, or through multiple networks if necessary, from source to destination nodes. This layer also provides for flow or congestion control, to prevent network resources (nodal buffers and transmission links) from filling up, possibly leading to deadlock situations. In carrying out these functions, the network layer uses the services of the data link layer below to ensure that a block of data inputted at one end of a link along a route through the network arrives there error free.

In packet-switched networks the units of data routed through the network from one end to the other are the packets we mentioned earlier. The blocks or frames of data transmitted over a link along a path in the network consist of the packets plus control information in the form of headers and trailers added to a packet just before it leaves a node. This control information enables the receiving node at the other end of a link to carry out the required synchronization and error detection. The control information is stripped off at each receiving node and reinserted when that node in turn transmits to its next neighbor across a link.

This concept of adding control information to the data has been enlarged in the OSI architecture to include the possibility of adding control information at each layer in the architecture. One thus arrives at the data unit picture portrayed in Fig. 1–6. (In the case of the data link layer a trailer may also be added, but the concept of encapsulating the data unit from above with control information still holds.) Each layer receives a block of data from the layer above, adds control information to the data, and passes it on to the layer below. Only the header (control information) is used by the receiving layer at the same level in the architecture. A given layer does *not*, in this concept, "look" at the data unit it receives from the layer above. Layers are therefore self-contained and isolated from one another. (In some older architectures layers sometimes use information carried in data from one or more layers above. This tends to destroy the isolation.) This leads to a valuable property of the layered architecture concept: Layers can be removed and replaced with new ones. Implementations (software products embodying the layered architecture) can be changed. The result is transparent to the layer above, providing the interface signals passed between layers are maintained unchanged. (The end users may experience a change in

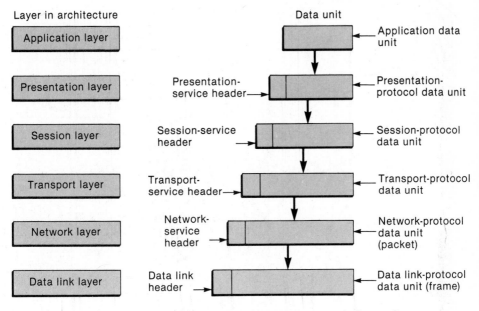

Figure 1-6 Data units involved in OSI network formation

performance because throughput, delay, and blocking performance may depend on the specific implementation of the architecture.)

A packet transmitted through the network from one end user to another thus generally consists of the actual information sent out by a user plus control information added at the various layers that will have to be stripped off as the packet arrives at the destination and starts working its way up through the layers. (We ignore in this introductory discussion the possibility of segmenting or blocking—i.e., combining—data units, which may be carried out at any level as well.) To the end users the network looks like a "transparent pipe" whose prime function, as already noted, is to route data units from source to destination, having them delivered at the desired end in timely fashion. The function of the upper layers, then, is to actually deliver the data correctly and in recognizable form. These upper layers do not know of the existence of the network. They care only that it provide the service required. This idea is embodied in diagrammatic form in Fig. 1-7.

The lowest of the higher-level OSI layers, the transport layer, is the one that ensures reliable, sequenced exchange of data between the two end users. The transport layer uses the services of the network layer for this purpose. In the connectionless (datagram-type) mode of operation, packets (datagrams) are not

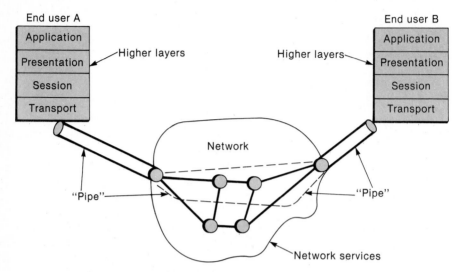

Figure 1–7 "Transparent pipe" service provided by network layer

required to arrive in order. The transport layer must thus sequence them properly for delivery to the layers above. It also controls the flow to ensure orderly reception of data units. (Because different end systems may transmit data units at different rates, a slower system conceivably could be overwhelmed by the faster one unless flow control is exerted.)

The existence of a session or conversation between two users implies the establishment and disconnection of the session. This is done by the session layer. This layer also manages the conversation, if necessary, to ensure an orderly exchange of data. Finally, the presentation layer manages and transforms the syntax of the data units being exchanged between the end users, while the application layer protocols provide the appropriate semantics or meaning to the information being exchanged [IEEE 1983b]. More details of the functions of the upper layers appear in Chapter 3.

1–4 Outline of the Book

With this basic introduction to data networks concluded, we are now in a position to outline the remaining chapters in the book. Although much qualitative material on data networks appears throughout, the stress where possible is on performance modeling and analysis. Much of the analysis will involve queueing theory and its application to both packet and circuit switching. In the following chapter we provide a self-contained introduction to queueing theory. Added material appears later in the book when needed.

The rest of the book is essentially divided into two parts. Chapters 3–9 cover topics in packet switching, following as much as possible the layered architecture approach. Chapters 10–12 cover digital circuit switching, with Chapter 12 also providing an introduction to integrated networks.

Chapter 3 is a more detailed introduction to the layered architecture concept, with stress placed on the OSI standard architecture and protocols, as well as on IBM's SNA. The CCITT interface standard X.25 is also described briefly, but most discussion on X.25 is deferred to Chapter 5.

Chapters 4–7 treat the layers one by one, as quantitatively as possible. Chapter 4 focuses on data link controls, describing first two generic types of data link control: the *stop-and-wait* protocol and the *go-back-N* protocols. Calculations are made of the throughput performance of idealized models of these protocols. The international standard data link control, HDLC, is then discussed, and the throughput performance of the go-back-N version of this protocol is calculated and compared with the idealized model. The chapter concludes with a brief discussion and analysis of an improved procedure, the *selective-repeat* protocol.

Chapters 5 and 6 move up to the network layer. Chapter 5 focuses on congestion control at this layer. Both virtual circuit and datagram controls are considered. The CCITT interface standard X.25 is described in detail, and an analysis is made of its (virtual circuit) congestion control mechanism. This mechanism uses a so-called sliding window control. This is followed with a discussion of SNA's network layer, called *path control* in that architecture. Its congestion control mechanism, called *virtual route pacing*, is analyzed and compared with the sliding window control. Models for both control procedures lead naturally to queueing network models. The chapter thus includes a discussion of the analysis of queueing networks. Input-buffer-limiting control mechanisms, used in DEC's DNA and other datagram-type architectures and networks, are also analyzed.

In Chapter 6 we discuss routing in packet-switched networks. Two algorithms are described, one appropriate for centralized routing, the other for decentralized routing, although both have been used for both types of routing. Routing in a variety of architectures and networks is then discussed, covering, for example, TYMNET, DEC's DNA, Burrough's BNA, GTE Telenet, and ARPAnet. Discussion of these networks and their architectures is included where appropriate. We also provide a simple performance analysis of the decentralized routing algorithm just noted, which is used in many datagram networks.

Chapter 7 takes us to the transport layer, in which we describe the transport protocols prescribed for the OSI architecture and the transmission control standard developed for the ARPAnet. Chapters 8 and 9 are devoted to access mechanisms mentioned earlier in this chapter. Chapter 8 discusses the two basic ways of handling access to a common, shared medium: controlled access (polling) and contention (random access). Chapter 9 applies these two techniques to the two main classes of local area networks: token passing (polling) and carrier sense multiple access (CSMA), a contention-type access mechanism.

Chapter 10 introduces the principles of circuit switching by first carrying out a simple but detailed comparative analysis of packet and circuit switching in a queued (nonblocked) mode of operation. We then turn to circuit switching exclusively, introducing the principles of traffic engineering used in circuit-switched networks. This leads to a discussion of nodal switches in circuit-switched nodes (switching offices or exchanges), with emphasis on digital switching systems. An analysis of the blocking performance of "essentially nonblocking" switches is included, and examples of commercial digital exchanges are described.

Chapter 11 covers computer-controlled call processing in digital switching exchanges. A model of the processing of calls is developed, and a queueing analysis of call processing in a hypothetical exchange is carried out. The chapter concludes with a performance analysis of overload (congestion) controls for circuit switches. In Chapter 12 we continue a discussion of circuit switching by describing routing in circuit-switched networks and the CCITT No. 7 packet-switch-oriented signaling system for handling routing messages in a circuit-switched system. We conclude the book with a brief introduction to hybrid or integrated multiplexing, with packet- and circuit-switched calls combined at a nodal access point.

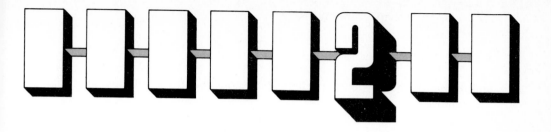

Introduction to Queueing Theory

This book focuses on the *performance* analysis of data networks. Although a great deal of qualitative material, describing real networks and network architectures, appears, the emphasis where possible is on quantitative considerations. These considerations involve the interplay among various performance parameters and how they relate to network resources that are to be controlled.

As noted in Chapter 1, two generic types of networks are considered: packet-switched and circuit-switched networks. In the packet-switched case, packets—blocks of data of varying length—are transmitted over a network from source to destination, following some routing path prescribed as part of the network design. The transmission facilities are shared by packets as they traverse the network. In the circuit-switched case, a transmission path end to end is set up for a pair of users who desire to establish a call. (The data flowing could equally well be voice or data messages.) The number and length of packets entering or traversing a network at any time, the number of calls arriving at a network entrance point in a given time, the length (the holding time of these calls)—all of these parameters generally vary statistically. In order to come up with quantitative measures of performance, therefore, probabilistic concepts must be used to study their interaction with a network. Queueing theory plays a key role in the analysis of networks, and this chapter covers the basic principles of queueing in order to prepare the reader for the quantitative material that follows.

Queueing arises very naturally in a study of packet-switched networks: Packets, arriving either at an entry point to the network or at an intermediate node on the way to the destination, are buffered, processed to determine the appropriate outgoing transmission link connected to the next node along the path, and then read out over that link when their time for transmission comes up. The time spent in the buffer waiting for transmission is a significant measure of performance for such a network, since end-to-end time delay, of which this wait time is a component, is one element of the network performance experienced by a user. The wait time depends, of course, on nodal processing time and packet length. It also depends on the transmission link capacity, in packets/sec capable of being transmitted, on the traffic rate in packets/sec arriving at the node in question, and on the service discipline used to handle the packets. Our queueing theory formulation will in fact consider most of these items. In our quantitative study of packet-switched networks we shall neglect nodal processing time for the most part, for the sake of simplicity. In our study of circuit-switched networks in Chapter 11, however, we shall consider the processing of calls at a node.

Queueing theory also arises naturally in the study of circuit-switched networks, not only in studying the processing of calls but in analyzing the relation at a node or switching exchange between trunks available (each capable of handling one call) and the probability that a call to be set up will be blocked or queued for service at a later time. In fact, historically, much of modern queueing theory developed out of telephone traffic studies at the beginning of the twentieth century. Integrated networks, which combine aspects of packet switching and circuit switching, will be considered later in this book, and the discussion there will of necessity involve the use of queueing concepts.

Consider the simplest model of a queue, as depicted in Fig. 2–1. To keep the discussion concrete, the queue in this case is shown handling packets of data. These packets could also be calls queueing up for service in a circuit-switched system. More generally, in the queueing literature jargon, they would be "cus-

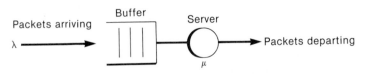

Figure 2–1 Model of a single-server queue

tomers" queueing up for service. The packets arrive randomly, at an average rate of λ packets/time (we shall use units of packets/sec most often). They queue up for service in the buffer shown and are then served, following some specified service discipline, at an average rate of μ packets/time. In the example of Fig. 2–1, only a single server is shown. More generally, multiple servers may be available, in which case more than one packet may be in service at any one time.

The concept of a server is of course well known to all of us from innumerable waits at the supermarket, bank, movie house, automobile toll booth, and so forth. In the context of a data network, the server is the transmission facility — an outgoing link, line, or trunk (all three terms will be used interchangeably in the material following) that transmits data at a prescribed digital rate of C data units/time. Most frequently, the data units are given in terms of bits or characters, and one talks of a transmission rate or link capacity C in units of bits/sec or characters/sec. A transmission link handling 1000-bit packets and transmitting at a rate of $C = 2400$ bps, for example, would be capable of transmitting at a rate of $\mu = 2.4$ packets/sec. More generally, if the average packet length in bits is $1/\mu'$ bits, and is given in units of bits/packet, $\mu = \mu'C$ packets/sec is the transmission capacity in units of packets/sec. (For circuit-switched calls, the "customer" would be a call; λ arrivals/sec represents the average *call* arrival rate, or the number of calls/sec handled on the average. The parameter $1/\mu$, in units of sec/call, is called the average call holding time.)

It is apparent that as the packet arrival rate λ approaches the packet rate capacity μ, the queue will start to build up. For a finite buffer (the situation in real life), the queue will eventually saturate at its maximum value as λ exceeds μ and continues to increase. If the buffer is filled, all further packets (customers) are blocked on arrival. We shall demonstrate this phenomenon quantitatively later in this chapter. If for simplicity the buffer is assumed to be infinite (an assumption we shall make often to simplify analysis), the queue becomes unstable as $\lambda \to \mu$. We will show that $\lambda < \mu$ to ensure stability in this case of a single-server queue. In particular, we shall find the parameter $\rho \equiv \lambda/\mu$ playing a critical role in queueing analysis. This parameter is often called the link *utilization* or *traffic intensity*. Note that it is defined as the ratio of load on the system to capacity of the system. For a single-server queue, as ρ approaches and exceeds one, the region of congestion is encountered, time delays begin to increase rapidly, and packets arriving are blocked more often.

To quantify the discussion of time delay, blocking performance, and packet throughput (the actual number of packets/time that get through the system), and their connection with both μ (the packet rate capacity) and the size of the buffer in Fig. 2–1, one needs a more detailed model of the queueing system. Specifically, those performance parameters among others will be shown to depend on the probabilities of state of the queue. The state is in turn defined to be the number of packets on the queue (including the one in service if the queue is nonempty). To calculate the probabilities of state, one must have knowledge of

1. The packet arrival process (the arrival statistics of the incoming packets),

2. The packet length distribution (this is called the service-time distribution in queueing theory), and

3. The service discipline (for example, FCFS — first come – first served — or FIFO service; LCFS — last come – first served; or some form of priority discipline).

For multiple-server queues, the state probabilities depend on the number of servers as well. (The servers represent the trunks or outgoing links simultaneously transmitting packets, or handling calls in the case of a circuit-switched system.)

In most of our work in this book we model the packet- and call-arrival processes as *Poisson processes.* The Poisson process is the most frequently used arrival process in queueing theory. For this reason we devote the next section to a brief discussion of this process and show how it is related intimately to exponential statistics. The simplest queueing system to analyze, the so-called M/M/1 queue, is one with Poisson arrivals and an exponential service-time distribution. It is easy to obtain the probabilities of state of this queueing system for both the finite and the infinite queue cases, as shown in Section 2–2. We then derive a simple but general relation between average time delay and average number of customers (packets or calls) in a queue, called Little's formula (Section 2–3). This relation will be found useful throughout the remainder of this book.

Continuing our introduction to queueing theory in this chapter, we then present two sections that generalize the M/M/1 queue analysis. In Section 2–4 we show how one can analyze state-dependent queues. (This material will be found particularly useful in the discussion of circuit switching in the latter part of the book.) In Section 2–5 we consider a queue with a general service time distribution and Poisson arrivals, the so-called M/G/1 queue. This enables us, as a special case, to determine the effect of fixed length packets. More generally, we derive a very interesting and extremely useful expression for the average time delay of a queue with general service (packet length or call-holding time) statistics. The result appears as a simple modification of the M/M/1 (exponential service time) time delay result. A brief discussion of priority queueing for the single-server case follows the M/G/1 analysis.

2–1 Poisson Process

As noted in the previous section, the Poisson process is the arrival process used most frequently to model the behavior of queues. It has been used extensively in

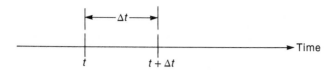

Figure 2–2 Time interval used in defining Poisson process

telephone traffic as well as in evaluating the performance of telephone switching systems and computer networks. Some of these analyses will be presented later in this book. In addition, the Poisson process has been used to model both photon generation and photodetector statistics, to represent shot noise processes, and to study semiconductor electron-hole generation phenomena, among other applications.

Three basic statements are used to define the Poisson arrival process. Consider a small time interval Δt ($\Delta t \to 0$), separating times t and $t + \Delta t$, as shown in Fig. 2–2. Then,

1. The probability of one arrival in the interval Δt is defined to be $\lambda \Delta t + o(\Delta t)$,* $\lambda \Delta t \ll 1$, and λ a specified proportionality constant.

2. The probability of zero arrivals in Δt is $1 - \lambda \Delta t + o(\Delta t)$.

3. Arrivals are memoryless: An arrival (event) in one time interval of length Δt is independent of events in previous or future intervals.

With this last definition the Poisson process is seen to be a special case of a *Markov* process, one in which the probability of an event at time $t + \Delta t$ depends on the probability at time t only [PAPO], [COX]. Note that with defining relations 1 and 2, more than one arrival or occurrence of an event in the interval Δt ($\Delta t \to 0$) is ruled out, at least to $0(\Delta t)$.

If one now takes a larger finite interval T, one finds the probability $p(k)$ of k arrivals in T to be given by

$$p(k) = (\lambda T)^k e^{-\lambda T}/k! \qquad k = 0, 1, 2, \ldots \qquad (2\text{–}1)$$

---

* $o(\Delta t)$ implies that other terms are higher order in Δt and that they go to zero more rapidly than Δt as $\Delta t \to 0$.

[PAPO] A. Papoulis, *Probability, Random Variables, and Stochastic Processes,* McGraw-Hill, New York, 2d ed., 1984.

[COX] D. R. Cox and H. D. Miller, *The Theory of Stochastic Processes,* Methuen, London, 1965.

Figure 2-3 Derivation of Poisson distribution

This is called the *Poisson distribution*. It is left to the reader to show that this distribution is properly normalized ($\sum_{k=0}^{\infty} p(k) = 1$) and that the mean or expected value is given by

$$E(k) = \sum_{k=0}^{\infty} kp(k) = \lambda T \qquad (2-2)$$

The variance $\sigma_k^2 \equiv E[k - E(k)]^2 = E(k^2) - E^2(k)$ turns out to be given by

$$\sigma_k^2 = E(k) = \lambda T \qquad (2-3)$$

This is also left as an exercise to the reader. The parameter λ, defined originally as a proportionality constant (see defining relation 1 for the Poisson process), turns out to be a rate parameter:

$$\lambda = E(k)/T$$

from Eq. (2-2). It thus represents the average rate of Poisson arrivals, as implied in our discussion in the previous section (see Fig. 2-1).

From Eqs. (2-2) and (2-3) it is apparent that the standard deviation σ_k of the distribution, normalized to the average value $E(k)$, tends to zero as λT increases: $\sigma_k/E(k) = 1/\sqrt{\lambda T}$. This implies that for large λT the distribution is closely packed about the average value λT. If one thus actually measures the (random) number of arrivals n in a large interval T ("large" implies $\lambda T \gg 1$, or $T \gg 1/\lambda$), n/T should be a good estimate of λ. Note also that $p(0) = e^{-\lambda T}$. As λT increases, with the distribution peaking eventually about $E(k) = \lambda T$, the probability of *no* arrivals in the interval T approaches zero exponentially with T.

The Poisson distribution of Eq. (2-1) is easily derived using the three defining relations of the Poisson process. Referring to Fig. 2-3, consider a sequence of m small intervals, each Δt units long. Let the probability of one event (arrival) in any interval Δt be $p = \lambda \Delta t$, while the probability of 0 events is $q = 1 - \lambda \Delta t$. Using the memoryless (independent) relation, it is then apparent that the probability of k events (arrivals) in the interval $T = m\Delta t$ is given by the *binomial* distribution

$$p(k) = \binom{m}{k} p^k q^{m-k} \qquad (2-4)$$

with

$$\binom{m}{k} \equiv m!/(m-k)!k!$$

Now let $\Delta t \to 0$, but with $T = m\Delta t$ fixed. Using the defining equation for the exponential,

$$\lim_{t \to 0} (1 + at)^{k/t} = e^{ak}$$

and approximating factorial terms by the Stirling approximation, one gets Eq. (2–1). Details are left to the reader.

Now consider a large time interval, and mark off the times at which a Poisson event (arrival) occurs. One gets a random sequence of points as shown in Fig. 2–4. The time between successive arrivals is represented by the symbol τ. It is apparent that τ is a continuously distributed positive random variable. For Poisson statistics, it turns out that τ is an *exponentially distributed* random variable; i.e., its probability density function $f_\tau(\tau)$ is given by

$$f_\tau(\tau) = \lambda e^{-\lambda \tau} \qquad \tau \geq 0 \tag{2–5}$$

This exponential interarrival distribution is sketched in Fig. 2–5. For Poisson arrivals the time between arrrivals is thus more likely to be small than large, the probability between two successive events decreasing exponentially with the time τ between them.

A simple calculation indicates that the mean value $E(\tau)$ of this exponential distribution is

$$E(\tau) = \int_0^\infty \tau f_\tau(\tau) d\tau = 1/\lambda \tag{2–6}$$

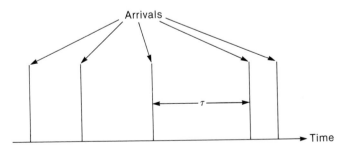

Figure 2–4 Poisson arrivals

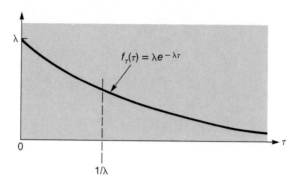

Figure 2–5 Exponential interarrival distribution

while its variance is given by

$$\sigma_\tau^2 = 1/\lambda^2 \qquad (2-7)$$

The average time between arrivals is as expected, for if the rate of arrivals is λ, the time between arrivals should be $1/\lambda$.

The fact that Poisson arrival statistics give rise to an exponential interarrival distribution is easily deduced from the Poisson distribution of Eq. (2–1).

Consider the time diagram of Fig. 2–6. Let τ be the random variable representing the time to the first arrival after some arbitrary time origin, as shown. Take any value x. *No* arrivals occur in the interval $(0, x)$ if and only if $\tau > x$. The probability that $\tau > x$ is just the probability that no arrivals occur in

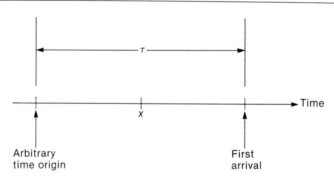

Figure 2–6 Derivation of exponential distribution

$(0, x)$; i.e.,

$$P(\tau > x) = \text{prob. (number of arrivals in } (0, x) = 0)$$
$$= e^{-\lambda x}$$

from Eq. (2–1). Then the probability that $\tau \le x$ is

$$P(\tau \le x) = 1 - e^{-\lambda x}$$

But this is just the cumulative probability distribution $F_\tau(x)$ of the random variable τ. Hence we have

$$F_\tau(x) = 1 - e^{-\lambda x} \tag{2–8}$$

from which the probability density function $f_\tau(x) = dF_\tau(x)/dx = \lambda e^{-\lambda x}$ follows immediately.

The close connection between the Poisson arrival process and the exponential interarrival time can be exploited immediately in discussing properties of the *exponential service-time distribution*. Thus consider a queue with a number of customers (packets or calls) waiting for service. Focus attention on the output of the queue, and mark the time at which a customer completes service. This is shown schematically in Fig. 2–7. Let the random variable representing time between completions be r, as shown. This must also be the service time if the next customer is served as soon as the one in service leaves the system. In particular, take the case where r is exponentially distributed in time, with an average value $E(r) = 1/\mu$. Thus

$$f_r(r) = \mu e^{-\mu r} \qquad r \ge 0 \tag{2–9}$$

But comparing Fig. 2–7 with Fig. 2–4, it is apparent that if r, the time between completions, is exponential, the completion times themselves must represent a Poisson process! The service process is the complete analog of the arrival process. On this basis, the probability of a service completion in the small time

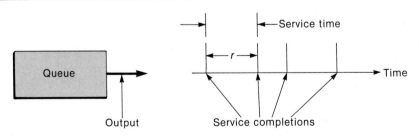

Figure 2–7 Service completions at output of a queue

interval $(t, t + \Delta t)$ is just $\mu \Delta t + 0(\Delta t)$, whereas the probability of *no* completion in $(t, t + \Delta t)$ is $1 - \mu \Delta t + 0(\Delta t)$, independent of past or future completions. The exponential model for service carries with it the memoryless property used as one of the defining relations of the Poisson process.

Before going on to our study of queueing, we introduce an additional property of the Poisson process. Say that m independent Poisson streams, of arbitrary rates $\lambda_1, \lambda_2, \ldots, \lambda_m$, respectively, are merged. Then the composite stream is itself Poisson, with rate parameter $\lambda = \sum_{i=1}^{n} \lambda_i$. This is an extremely useful property, and is one of the reasons that Poisson arrivals are often used to model arrival processes. In the context of packet-switched and circuit-switched networks, this situation occurs when combining statistically packets or calls from a number of data sources (terminals or telephones), each of which generates packets or calls (as the case may be) at a Poisson rate. A simple proof is as follows: Let $N^{(i)}(t, t + \Delta t)$ be the number of events in Poisson process $i, i = 1, 2, \ldots, m$ in the interval $(t, t + \Delta t)$. Let $N(t, t + \Delta t)$ be the total number of events from the composite stream. Then

$$\text{prob. } [N(t, t + \Delta t) = 0] = \prod_{i=1}^{n} \text{prob. } [N^{(i)}(t, t + \Delta t) = 0]$$

$$= \prod_{i=1}^{m} [1 - \lambda_i \Delta t + 0(\Delta t)] = 1 - \lambda \Delta t + 0(\Delta t), \qquad (2-10)$$

$\lambda = \sum_{i=1}^{m} \lambda_i$, since the individual processes are independent. A similar calculation shows that

$$\text{prob. } [N(t, t + \Delta t) = 1] = \lambda \Delta t + 0(\Delta t) \qquad (2-11)$$

This proves the desired relation.

Sums of Poisson processes are thus distribution conserving: They retain the Poisson property. This property will be used implicitly in a number of places in this book in working out examples of queueing and buffering calculations.

2-2 The M/M/1 Queue

We now use the material of the previous section, on the Poisson process, to determine the properties of the simplest model of a queue, the M/M/1 queue. This is a queue of the single-server type, with Poisson arrivals, exponential service-time statistics, and FIFO service. The notation M/M/1 used is due to British statistician D. G. Kendall. The Kendall notation for a general queueing system is of the form A/B/C. The symbol A represents the arrival distribution, B represents the service distribution, and C denotes the number of servers used.

The symbol M in particular, from the Markov process, is used to denote the Poisson process or the equivalent exponential distribution. An M/M/m queue is thus one with Poisson arrivals, exponential service statistics, and m servers. An M/G/1 queue has Poisson arrivals, *general* service distribution, and a single server. A special case is the M/D/1 queue, with D used to represent fixed (deterministic) or *constant* service time. We shall have more to say about these other queueing structures in later sections.

As noted in the introduction to this chapter, the statistical properties of the M/M/1 queue, the average queue occupancy, the probability of blocking for a finite queue, average throughput, and so forth, are readily determined once we find the probabilities of state p_n at the queue. By definition, p_n is the probability that there are n customers (packets or calls) in the queue, including the one in service. By implication the system is operating at steady state so that these probabilities do not vary with time. Starting from some initial defined values (for example, an empty queue state), one expects these probabilities to approach steady-state, stationary, non-time-varying values as time goes on, if the arrival and service-time distributions are invariant with time. We shall show later on that these stationary probabilities are readily determined from simple flow balance arguments. At this point we use a more general, time-dependent argument.

Specifically, let the arrival process to the single-server queue of Fig. 2-8 be Poisson, with parameter λ. Let the service-time process (packet length or call holding time) be exponential, with parameter μ, as shown. Then the probability $p_n(t + \Delta t)$ that there are n customers (packets or calls) in the queue at time $(t + \Delta t)$ may readily be found in terms of corresponding probabilities at time t. Referring to the state-time diagram of Fig. 2-9, it is apparent that if the queue is in state n at time $t + \Delta t$, it could only have been in states $n-1$, n, or $n+1$ at time

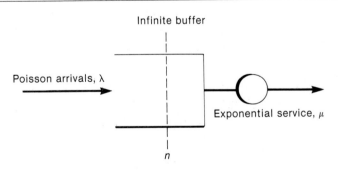

Infinite buffer

Poisson arrivals, λ

Exponential service, μ

n

Figure 2-8 M/M/1 queue

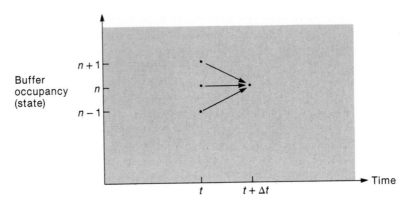

Figure 2–9 M/M/1 state-time diagram

t. (We assume here for simplicity that $n \geq 1$.) The probability $p_n(t + \Delta t)$ that the queue is in state n at time $t + \Delta t$ must be the sum of the (mutually exclusive) probabilities that the queue was in states $n - 1$, n, or $n + 1$ at time t, each multiplied by the (independent) probability of arriving at state n in the intervening Δt units of time. We thus have, as the generating equation for $p_n(t + \Delta t)$,

$$p_n(t + \Delta t) = p_n(t)[(1 - \lambda \Delta t)(1 - \mu \Delta t) + \mu \Delta t \cdot \lambda \Delta t + 0(\Delta t)]$$
$$+ p_{n-1}(t)[\lambda \Delta t(1 - \mu \Delta t) + 0(\Delta t)] \qquad (2-12)$$
$$+ p_{n+1}(t)[\mu \Delta t(1 - \lambda \Delta t) + 0(\Delta t)]$$

The transition probabilities of moving from one state to another have been obtained by considering the ways in which one could move between the two states and calculating the respective probabilities, using the properties of the arrival and service-time distributions. As an example, if the system remains in state n, $n \geq 1$, there could have been either one departure and one arrival, with probability $\mu \Delta t \cdot \lambda \Delta t$, or no departure and no arrival, with probability $(1 - \mu \Delta t)(1 - \lambda \Delta t)$, as shown. The other terms in Eq. (2–12) are obtained similarly.

Since $0(\Delta t)$ includes terms of order $(\Delta t)^2$ and higher, the terms involving $(\Delta t)^2$ in Eq. (2–12) should be incorporated in $0(\Delta t)$. (They were retained and shown explicitly to help the reader understand Eq. (2–12).) Simplifying Eq. (2–12) in this way, and dropping $0(\Delta t)$ terms altogether, one gets

$$p_n(t + \Delta t) = [1 - (\lambda + \mu)\Delta t]p_n(t) + \lambda \Delta t \, p_{n-1}(t) + \mu \Delta t p_{n+1}(t) \quad (2-12a)$$

Eq. (2–12a) can be used to study the time-dependent (transient) behavior of the M/M/1 queue given that the queue is started at time $t = 0$ in some known state or set of states. Alternatively, a differential-difference equation governing the time variation of $p_n(t)$ may be found by expanding $p_n(t + \Delta t)$ in a Taylor series about t and retaining the first two terms only:

$$p_n(t + \Delta t) \doteq p_n(t) + \frac{dp_n(t)}{dt} \Delta t \qquad (2-13)$$

Using Eq. (2-13) in Eq. (2-12a) and simplifying, one readily obtains the following equation:

$$\frac{dp_n(t)}{dt} = -(\lambda + \mu)p_n(t) + \lambda p_{n-1}(t) + \mu p_{n+1}(t) \qquad (2-14)$$

This is the differential-difference equation to be solved to find the time variation of $p_n(t)$ explicitly.

As stated earlier, $p_n(t)$ should approach a constant, stationary value p_n as time goes on. Assuming that this is the case (it will be shown later that the condition $\lambda < \mu$ ensures this in the case of the infinite queue), we must have $dp_n(t)/dt = 0$ at the stationary value of p_n as well. Equation (2-14), for the case of stationary, non-time-varying probabilities, then simplifies to the following equation involving the stationary state probabilities p_n of the M/M/1 queue:

$$(\lambda + \mu)p_n = \lambda p_{n-1} + \mu p_{n+1} \qquad n \geq 1 \qquad (2-15)$$

This equation can be given a physical interpretation that enables us to write it down directly, by inspection, without going through the lengthy process of deriving it from Eq. (2-12). More important, using the approach to be described, we shall be able to write similar equations down, by inspection, for more general state-dependent queues later in this chapter. More complex queueing systems arising in the study of both packet- and circuit-switched networks will be analyzed in a similar manner in later chapters.

Consider the state diagram of Fig. 2-10, which represents the M/M/1 queue. Because of the Poisson arrival and departure processes assumed, transitions between adjacent states only can take place with the rates shown. There is a rate λ of moving up one state due to arrivals in the system, whereas there is a rate μ of moving down due to service completions or departures. Alternatively, if one multiples the rates by Δt, one has the probability $\lambda \Delta t$ of moving up one state due to an arrival, or the probability $\mu \Delta t$ of dropping down one state due to a service completion (departure). (If the system is in state 0, i.e., it is empty, it can only move up to state 1 due to an arrival.)

The form of Eq. (2-15) indicates that there is a stationary balance principle at work: The left-hand side of Eq. (2-15) represents the rate of *leaving* state n, given the system was in state n with probability p_n. The right-hand side represents the rate of *entering* state n, from either state $n - 1$ or state $n + 1$. In order for stationary state probabilities to exist, the two rates must be equal.

Figure 2–10 State diagram, M/M/1 queue

Balance equations play a key role in the study of queueing systems. We shall encounter similar equations later in this chapter in studying state-dependent queues; in Chapter 5, in studying queueing networks; and in later chapters, in studying more complex queueing systems arising as models of data networks. The relation of balance equations to time-invariant or equilibrium state probabilities is explored rigorously and at length in the seminal book by Kelly [KELL].

The solution of Eq. (2–15) for the state probabilities can be carried out in a number of ways. The simplest way is to again apply balance arguments. Consider Fig. 2–11, which represents the state diagram of the M/M/1 queue again drawn as in Fig. 2–10, but with two closed "surfaces," 1 and 2, sketched as shown. If one calculates the total "probability flux" crossing surface 1, and equates the flux leaving (rate of leaving state n) to the flux entering (rate of entering state n), one gets Eq. (2–15). Now focus on surface 2, which encloses the entire set of points from 0 to n. The flux entering the surface is μp_{n+1}; the flux leaving is λp_n. Equating these two, one gets

$$\lambda p_n = \mu p_{n+1} \qquad (2-16)$$

It is left to the reader to show that this simple balance equation does in fact satisfy Eq. (2–15). The intuitive concept of balancing rates of departure from a state to rates of entering that state thus not only allows a balance equation to be set up (Eq. (2–15)), but a solution to be obtained as well! Repeating Eq. (2–16) n times, one finds very simply that

$$p_n = \rho^n p_0 \qquad \rho \equiv \lambda/\mu \qquad (2-17)$$

To find the remaining unknown probability p_0, one must now invoke the probability normalization condition

$$\sum_n p_n = 1$$

[KELL] F. P. Kelly, *Reversibility and Stochastic Networks,* John Wiley & Sons, Chichester, U.K., 1979.

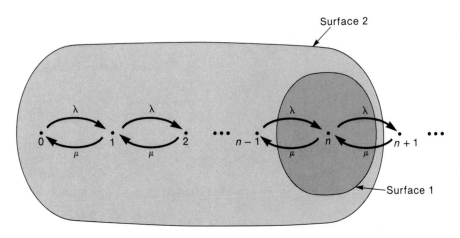

Figure 2-11 Flow balance, M/M/1 queue

For the case of an infinite M/M/1 queue, one finds very simply that $p_0 = (1 - \rho)$, *if* $\rho < 1$, and one gets, finally,

$$p_n = (1 - \rho)\rho^n \qquad \rho \equiv \lambda/\mu < 1 \qquad (2-18)$$

as the equilibrium state probability solution for the M/M/1 queue. Note the necessary condition $\rho = \lambda/\mu < 1$, which was alluded to earlier. Its significance, again, is that for equilibrium to exist the arrival rate or load on the queue must be less than the *capacity* μ. If this condition is violated for this infinite queue model, the queue continues to build up in time, and equilibrium is never reached. Mathematically, the fact that equilibrium exists only for $\rho < 1$ is noted by considering the limiting case $\rho = 1$. Since $p_0 = (1 - \rho)$, $p_0 = 0$. But from Eq. (2-16), $p_1 = \rho p_0 = 0$, $p_2 = 0$, ... All the stationary probabilities are thus zero, a contradiction, and equilibrium does not exist. More detailed and general discussions of conditions for equilibrium in the context of Markov processes appear in [COX] and [KELL].

The M/M/1 state probability distribution of Eq. (2-18) is called a *geometric distribution*. An example, for $\rho = 0.5$, appears in Fig. 2-12. Note that since the probability the queue is empty is $p_0 = 1 - \rho$, the probability that the queue is nonempty is just ρ, the utilization.

Now consider the extension of this analysis to the case of a *finite queue*, accommodating at most N packets. A little thought indicates that the governing balance equation, Eq. (2-15), is unchanged except for the two boundary points

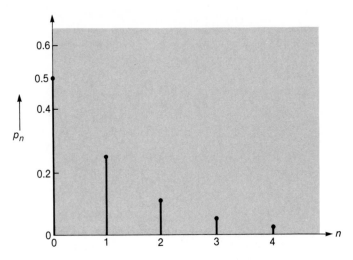

Figure 2–12 M/M/1 state probabilities, $\rho = 0.5$

$n = 0$ and $n = N$. Equation (2–17) is still the solution. The only difference now is that the probability p_0 that the queue is empty changes to accommodate the normalization condition summed over a finite number of states:

$$\sum_{n=0}^{N} p_n = 1$$

It is left for the reader to show that

$$p_0 = (1 - \rho)/(1 - \rho^{N+1}) \tag{2–19}$$

in this case, so that

$$p_n = (1 - \rho)\rho^n/(1 - \rho^{N+1}) \tag{2–20}$$

for the finite M/M/1 queue.

In particular, the probability that the queue is full is p_N, given by

$$p_N = (1 - \rho)\rho^N/(1 - \rho^{N+1}) \tag{2–21}$$

But this should be the same as the probability of blocking: the probability that customers (packets or calls) are turned away and not accepted by the queue. This may be demonstrated by the following simple argument, which will be found useful later in some of our data network performance analysis. Consider the picture of a queue shown in Fig. 2–13. This does not have to be a finite M/M/1 queue. It can be *any* queueing system that blocks customers on arrival. A load λ,

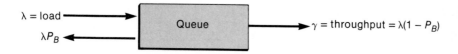

Figure 2-13 Relation between throughput and load

defined as the average number of arrivals/sec, is shown applied to a queue. With the probability of blocking given by P_B, the net arrival rate is then $\lambda(1 - P_B)$. But this must be the same as the throughput γ, or the number of customers served/sec for a conserving system. We thus have

$$\gamma = \lambda(1 - P_B) \tag{2-22}$$

as shown in Fig. 2-13. (A more detailed discussion, in the context of blocking in circuit-switched systems, appears in Chapter 10.)

One can calculate the throughput another way, however, by focusing on the output of the system. In particular, for a single-server queue, the type of queue under discussion, the average rate of service would be μ, in customers/sec served on the average, *if* the queue were always nonempty. Since the queue is sometimes empty, with probability p_0, the actual rate of service, or throughput γ, is less than μ. More precisely, $\gamma = \mu(1 - p_0)$, since $(1 - p_0)$ is the probability that the queue is nonempty. (So long as there is at least one customer in a queue, the average rate of service will be μ.) $(1 - p_0)$ is thus the (single-server) utilization. As a check, consider the infinite M/M/1 queue. There is no blocking in that case, and the throughput $\gamma = \lambda$, the average arrival rate. We must thus have $\lambda = \mu(1 - p_0)$, or $p_0 = 1 - \rho$, $\rho = \lambda/\mu$, just as found earlier! In the case of the finite M/M/1 queue, we equate the net arrival rate $\lambda(1 - P_B)$ to the average departure rate $\mu(1 - p_0)$ (see Fig. 2-14), to obtain

$$\gamma = \lambda(1 - P_B) = \mu(1 - p_0) \tag{2-23}$$

It is left to the reader to show, using Eq. (2-19) in Eq. (2-23), that the blocking

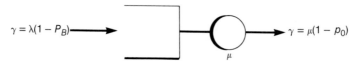

Figure 2-14 Throughput calculation, single-server queue

probability is in fact given by

$$P_B = p_N = (1 - \rho)\rho^N/(1 - \rho^{N+1}) \qquad (2-24)$$

for the finite M/M/1 queue.

This equation for the blocking probability can be used for a simple design calculation. Specifically, what should the queue size N be to provide a prescribed blocking performance P_B? From Eq. (2–24) this also depends on ρ. For a small blocking probability, Eq. (2–24) may be simplified. With $\rho < 1$ and $N \gg 1$ we have

$$P_B \doteq (1 - \rho)\rho^N \qquad \rho^{N+1} \ll 1 \qquad (2-25)$$

This is just the probability that the *infinite* queue is at state $n = N$ and indicates that for small P_B the probability that the finite queue is at state $n = N$ could just as well be calculated by assuming an infinite queue. Truncating the infinite queue at $n = N$ does not affect the queue statistics measurably *if $\rho^N \ll 1$.*

As examples of the use of Eq. (2–25), let $\rho = 0.5$. Then for $P_B = 10^{-3}$, $N \doteq 9$ customers that have to be accommodated. For a packet-switched network, the concentrator need only be capable of handling 9 packets. If $P_B = 10^{-6}$ is desired (one customer in 10^6 is rejected on the average), and $\rho = 0.5$, the number rises to $N \doteq 19$. Larger values of ρ (increased traffic) give rise to correspondingly larger values of N. As a simple example, say a concentrator in a packet-switched network handles packets that are 1200-bits long on the average. For a 2400-bps capacity line, the average transmission capacity is $\mu = 2$ packets/sec that can be delivered to the line. For $\rho = 0.5$, $\lambda = 1$ packet/sec is the allowable load on the concentrator. For a blocking probability of 10^{-3}, then, the line queue should be capable of accommodating 9 packets of average length 1200 bits. For $P_B = 10^{-6}$, this rises to 19 such packets.

Returning now to Eq. (2–20), the expression for the equilibrium state probability of the finite M/M/1 queue, we note that the condition $\rho < 1$ is no longer required for the equilibrium probabilities to exist. Since the queue is finite, the probabilities exist and are given by Eq. (2–20) for *all* values of ρ. Note in particular that as the load λ increases with respect to the capacity μ ($\rho = \lambda/\mu \to \infty$), the queue fills up more often and, in the limit of $\lambda \to \infty$, stays at state $n = N$ with a probability of 1. From Eq. (2–20), $p_n \to 0$, $n \neq N$, as $\rho \to \infty$, and $p_N \to 1$, $\rho \to \infty$. The region $\rho > 1$ is said to be the *congested* region; the higher queue states are more probable. The blocking probability $P_B = p_N$ approaches 1 as $\rho \to \infty$. This is shown in Fig. 2–15, which plots P_B as a function of the normalized load $\rho = \lambda/\mu$. It is readily shown, using L'Hôpital's rule, that $P_B = 1/(N + 1)$ at $\rho = 1$. For $N = 9$, as an example, $P_B \doteq 10^{-3}$ at $\rho = 0.5$, rises to $P_B \doteq 0.1$ at $\rho = 1$, is approximately 0.5 at $\rho = 2$, and continues to increase to 1 as $\rho \to \infty$. This indicates the queue is often blocked for ρ in the congested region.

The throughput of the queue, closely equal to the load λ for small ρ, eventu-

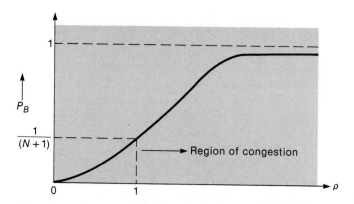

Figure 2–15 Blocking probability, finite M/M/1 queue, congested region

ally levels off and approaches the throughput capacity μ as ρ increases. The specific expression for the normalized throughput γ/μ as a function of the normalized load $\rho = \lambda/\mu$ is obtained by using Eq. (2–19) in Eq. (2–23). This gives

$$\gamma/\mu = (1 - p_0) = \rho(1 - \rho^N)/(1 - \rho^{N+1}) \qquad (2-26)$$

Equation (2–26) is sketched in Fig. 2–16. At $\rho = 1$, it is readily shown that $\gamma/\mu = N/(N + 1)$ as indicated in Fig. 2–16. For $N = 9$, then, $\gamma/\mu = 0.9$, at

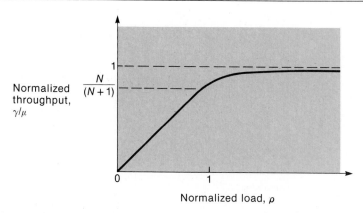

Figure 2–16 Throughput-load characteristic, finite M/M/1 queue

$\rho = 1$. Above this value of load, the normalized throughput levels off to values approaching 1 more and more closely.

We focus now on the uncongested region, $\rho < 1$, for which it is sufficient to use the infinite buffer analysis. From the state probabilities p_n, given by Eq. (2–18), one can calculate various statistics of interest. In particular, consider the average number of customers $E(n)$ (packets or calls queued) appearing in the queue, including the one in service. From the definition of the mean value of random variables, we have immediately

$$E(n) = \sum_{n=0}^{\infty} np_n = \rho/(1 - \rho) \qquad (2-27)$$

using Eq. (2–18) and carrying out the indicated summation. Equation (2–27) demonstrates the well-known queueing phenomenon that all of us have experienced in everyday life. When there is a relatively low load on the system ($\rho = \lambda/\mu \leq 0.5$, say), the average number of customers in the system is relatively small (<1 for $\rho < 0.5$). As ρ increases, approaching 1, the average number increases dramatically, rising because of the $(1 - \rho)$ term in the denominator. In a real, finite queue system, the number would, of course, not shoot up as fast in the vicinity of $\rho < 1$, but Eq. (2–27) for the infinite queue does provide a good model for the finite queue case. Equation (2–27) is plotted in Fig. 2–17.

From a comparison of Figs. 2–15 to 2–17, one can make some statements about queueing performance that will be reiterated elsewhere in the book in discussing the performance of networks. As the load on the system increases,

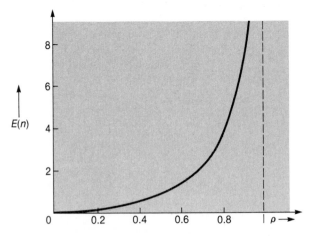

Figure 2–17 Average queue size, M/M/1 queue

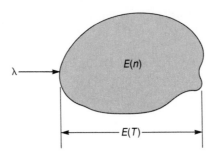

Figure 2–18 The environment of Little's formula

the throughput increases as well. More and more customers are blocked, however, and the average number of customers in the queue $E(n)$ also increases rapidly. This increase in $E(n)$ translates itself into increased time delay in the queue. There is thus a typical trade-off in performance: As the load increases the throughput goes up (a desirable characteristic) but blocking and time delay also increase (undesirable characteristics). In Chapter 5, in discussing congestion in more detail, we shall in fact see that queueing deadlocks can occur at high load. In this case, only apparent when two or more queues forming a network attempt to pass customers (packets) to one another, *nothing* will move, and the throughput drops to zero!

To find the time delay through the queue (this includes time spent waiting on the queue in addition to the service or transmission time), one invokes a simple formula that we shall be using throughout this book. The formula, called appropriately Little's formula after the individual who first proved it [LITT], says simply that a queueing system, with average arrival rate λ and mean time delay $E(T)$ through the system, has an average queue length $E(n)$ given by the expression

$$\lambda E(T) = E(n) \qquad (2-28)$$

The relations among these three quantities are diagrammed in Fig. 2–18. We shall provide a proof of this expression in the next section. Suffice it to say here that the relation is very general and is valid for all types of queueing systems, including priority disciplines. The parameter λ is interpreted to be the arrival

[LITT] J. D. C. Little, "A Proof of the Queueing Formula $L = \lambda W$," *Operations Res.*, vol. 9, no. 3, 1961, 383–387.

rate *into* the system. It thus corresponds to our throughput γ. This should be apparent since customers that are turned away cannot contribute to delays in the system.

Applying Eq. (2–28) to the M/M/1 queue under discussion as shown in Fig. 2–19, one has immediately, using Eq. (2–27) for $E(n)$,

$$E(T) = E(n)/\lambda = 1/\mu/(1 - \rho) \qquad (2-29)$$

This expression for the average time delay $E(T)$ through the M/M/1 queue has an interesting interpretation. For $\rho \ll 1$, $E(T) = 1/\mu$, exactly the average service time. This is the case, from Eq. (2–27), when there are few customers on the average in the queue. Hence very little time is spent waiting in the queueing system on the average, and the time delay is almost always due to service or transmission time. As the normalized load or traffic intensity increases, however, typical queueing behavior is experienced, with $E(T)$ beginning to rapidly increase. This is shown in Fig. 2–20 for $E(T)$ as a function of ρ. The normalized delay, $E(T)/1/\mu$—or $\mu E(T)$, the time delay relative to transmission time—is plotted in the figure. For $\rho = 0.5$, for example, the average delay doubles, to $2/\mu$. The average wait time in the queue at this point equals the average service time. For $\rho = 0.8$, the average delay is $5/\mu$, so that, on the average, there are $4/\mu$ units of wait time.

It is apparent that for the single-server queue the following simple relation between the average wait time $E(W)$ and the average delay $E(T)$ through the queue must hold:

$$E(T) = E(W) + 1/\mu \qquad (2-30)$$

The connection between $E(T)$ and $E(W)$ is diagrammed in Fig. 2–19. Little's theorem enables us to find an explicit relation for the average number of

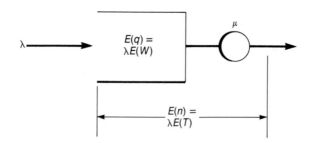

Figure 2–19 Little's formula applied to M/M/1 queue

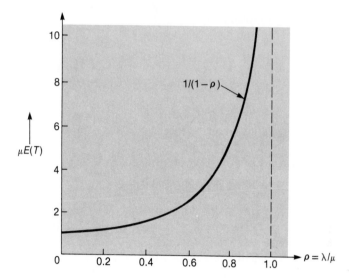

Figure 2-20 Normalized time delay, M/M/1 queue

customers $E(q)$ waiting in the queue (see Fig. 2-19). This must be given by

$$E(q) = \lambda E(W) = \lambda E(T) - \lambda/\mu$$
$$= E(n) - \rho \qquad (2-31)$$

(Recall that Little's formula is very general and that it can also be applied to a portion of a queueing system, as shown in Fig. 2-19.) As a check, ρ in Eq. (2-31) must represent the average number of customers in service. This is obviously less than one, since a customer is either in service or not. Focusing on the service station itself, there is a probability $p_0 = 1 - \rho$ that no one is in service (the queue is empty), hence a probability $(1 - p_0) = \rho$ that *one* customer is in service. The average is thus just ρ. The result for the M/M/1 queue, $p_0 = 1 - \rho$, may be generalized to *any* single-server queue.* From Eq. (2-31), ρ is always the average number in service. Hence $p_0 = (1 - \rho)$ must always be true.

Because of the great utility and generality of Little's formula, we devote the next section to a simple derivation.

---

* The queue must be *work conserving*, in the sense that if there is work to be done — i.e., a customer to be served — the customer will be served, and that a customer is not rejected once admitted into the system.

2–3 Little's Formula, $L = \lambda W$[†]

Consider a queueing system as shown in Fig. 2–21. For simplicity we consider the waiting section only. Let $A(t)$ represent the cumulative arrivals to the queue at time t, and let $D(t)$ represent the cumulative departures that go into service after waiting. Then $L(t) = A(t) - D(t)$ is the number waiting in the system at time t. No assumptions need be made about the arrival process or the departure process. We only stipulate that all customers entering the system are ultimately served (this is then a work-conserving system).

   In particular, let customers arrive at time $t_j, j = 1, 2, \ldots . A(t)$ then represents the number of such arrival times, up to the time t. A typical state of the function $A(t)$ appears in Fig. 2–22. At each arrival, $A(t)$ increases by 1. Let the customers depart, moving to the service station, at times $t_j' \geq t_j$. To simplify the discussion initially, say that the service discipline is FIFO (the result obtained will be shown to be general, independent of this initial assumption). For this case, the departure times must increase monotonically, $t_1' < t_2' < t_3' < \ldots . A$ typical set of departure times appears in Fig. 2–22 with the corresponding curve for the cumulative departures $D(t)$ indicated as well. Note that departures coincide with arrivals when the system is empty. The wait times, W_j, for customers are also indicated. These obviously represent the time each customer spends in the system between arrival and departure. From Fig. 2–22 one can write, by inspection, simple relations between the different quantities shown that lead directly to Little's formula. Consider a starting time 0 and a later time τ, at both of which $A(t) = D(t)$. Examples in Fig. 2–22 include t_2', t_5', and t_6'. Let $n(\tau) = A(\tau) - A(0)$ be the number of arrivals in the interval τ. Then the mean arrival rate in the interval $(0,\tau)$ is just

$$\lambda(\tau) = n(\tau)/\tau \tag{2–32}$$

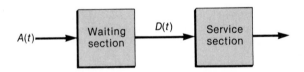

Figure 2–21 Derivation of Little's formula for a queueing system

† This section follows the approach of Kobayashi in [KOBA].

[KOBA] H. Kobayashi, *Modeling and Analysis: An Introduction to System Performance Evaluation Methodology*, Addison-Wesley, Reading, Mass., 1978.

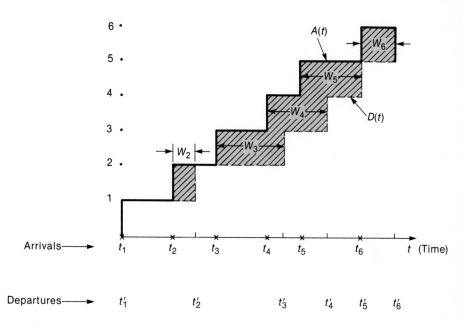

Figure 2-22 Arrivals and departures, FIFO (FCFS) queueing system

(An example in Fig. 2-22 would be six arrivals in the interval between t_1 and t_6.)

Focus now on the cross-hatched areas in Fig. 2-22. The cumulative area in the interval $(0,\tau)$ is just the function

$$\int_0^\tau L(t)dt$$

since $L(t) = A(t) - D(t)$, as noted earlier and as indicated in Fig. 2-22. Since this area is made up of a series of rectangles of unity height and width W_j as shown we obviously must have the equality

$$\sum_{j=1}^{n(\tau)} W_j = \int_0^\tau L(t)dt \tag{2-33}$$

But consider the quantity

$$\overline{W}(\tau) \equiv \sum_{j=1}^{n(\tau)} W_j / n(\tau) \tag{2-34}$$

This is the *average waiting time* in the interval $(0, \tau)$.

Consider also the expression

$$\bar{L}(\tau) \equiv \int_0^\tau L(t)dt/\tau \qquad (2-35)$$

This must represent the *average number of customers* in the system in the interval $(0,\tau)$, since $L(t)$ is the number at time t. From Eq. $(2-33)$, there is a close connection between $\overline{W}(\tau)$ and $\bar{L}(\tau)$. In particular, using Eqs. $(2-34)$ and $(2-35)$ in Eq. $(2-33)$ and recalling the defining relation Eq. $(2-32)$ for the arrival rate $\lambda(\tau)$, we have simply

$$n(\tau)\overline{W}(\tau)/\tau = \bar{L}(\tau) = \lambda(\tau)\overline{W}(\tau) \qquad (2-36)$$

This is just Little's formula derived for the special case of FIFO service and over a finite interval $(0,\tau)$. Now let $\tau \to \infty$ and assume that the quantities of interest all approach definite limits:

$$\overline{W}(\tau) \to \overline{W},\ \lambda(\tau) \to \lambda,\ \bar{L}(\tau) \to \bar{L}$$

We then get Little's formula

$$\bar{L} = \lambda\overline{W} \qquad (2-37)$$

with $\bar{L}$ the average number of customers in the queueing system, $\overline{W}$ the average waiting time, and λ the arrival rate. This result is easily extended to include the service station as well.

That Little's formula Eq. $(2-37)$ holds generally for *any* service discipline is shown as follows. Consider

$$\sum_{j=1}^{n(\tau)} W_j = \sum_{j=1}^{n(\tau)} (t'_j - t_j)$$

from Fig. $2-22$. This may be rewritten as

$$\sum_{j=1}^{n(\tau)} W_j = \sum_{j=1}^{n(\tau)} t'_j - \sum_{j=1}^{n(\tau)} t_j$$

Written in this form, it is apparent that $\sum_j W_j$ depends only on the sum of the departure times and not on the difference $t'_j - t_j$ used in the derivation assuming FIFO service. The individual departure times may depend on the service discipline, but Little's formula applied to any discipline holds nonetheless.

As an example, consider the last come–first served (LCFS) discipline. Typical plots for this case appear in Fig. $2-23$. The reader is asked to demonstrate, using this figure, that the equality Eq. $(2-33)$ holds here as well.

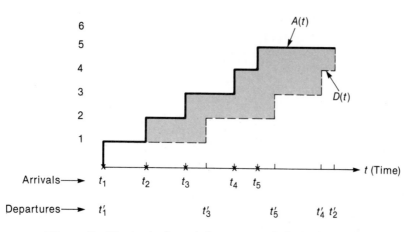

Figure 2-23 Arrivals and departures, LCFS discipline

2-4 State-dependent Queues: Birth-death Processes

The M/M/1 queue analysis carried out in detail earlier is readily extended to single-queue systems in which arrival and departure (service) rates are dependent on the state of the system. The processes are often called *birth-death processes* [COX]. We carry out a simple analysis of such a system in this section and provide some typical examples. A multiserver exponential queueing system, of the type M/M/m, will be found to fall in this class. A blocking system with no waiting room for customers, which often arises in the modeling of circuit-switched systems, will provide another example. Other examples appear in chapters that follow, particularly Chapter 5, on congestion control in packet-switched networks, and Chapter 10, on the traffic analysis of circuit-switched networks.

The state-dependent generalization of our previous M/M/1 model, with Poisson arrivals and exponential service distribution, is made for a system in state n simply by defining the probability of one arrival in the infinitesimal interval $(t, t + \Delta t)$ to be $\lambda_n \Delta t + 0(\Delta t)$, with the probability of no arrivals defined to be $(1 - \lambda_n \Delta t) + 0(\Delta t)$. The memoryless assumption is again invoked so that arrivals in the interval $(t, t + \Delta t)$ are independent of arrivals in other intervals. The arrival process is thus another example of a Markov process [COX], [PAPO]. A special case is our earlier Poisson process, with $\lambda_n = \lambda$, a constant, independent of state. This arrival process is often called a birth process, since $\lambda_n \Delta t$ can be visualized as representing the probability of "birth" of a customer, given n customers already in the system. Note, as previously, that with this model at most *one* customer at a time can arrive in the system in the interval $(t, t + \Delta t)$.

In a similar manner, the state-dependent departure or death process is generalized from our previous Poisson departure process to be one for which the probability of departure of *one* customer in the infinitesimal interval $(t, t + \Delta t)$ is defined to be $\mu_n \Delta t + 0(\Delta t)$, given the number of customers in the system is n. The probability of departure of no customers is $(1 - \mu_n \Delta t) + 0(\Delta t)$, and the memoryless assumption is again invoked. The state-dependent departure or death process is similarly an example of a Markov process.

Combining these two processes, as was done in the case of the M/M/1 queue, again letting $\Delta t \to 0$ and taking the system to be in statistical equilibrium, the balance equation governing the operation of the combined birth-death process or state-dependent queueing system, at equilibrium (see Fig. 2–24), may be written by inspection:

$$(\lambda_n + \mu_n)p_n = \lambda_{n-1}p_{n-1} + \mu_{n+1}p_{n+1} \tag{2-38}$$

Equation (2–15) for the M/M/1 queue is a special case. The corresponding state diagram appears in Fig. 2–25. The parameter λ_n represents the rate of arrival of a customer, given the system is in state n; p_n is the equilibrium probability that the system is in state n; μ_n is the rate of departure of a customer, given the system is in state n. The balance equation may again be obtained by equating the rate of departure *from* state n (the left-hand side of Eq. (2–38)) to the rate of arrival at state n (the right-hand side of Eq. (2–38)). Alternatively, from Fig. 2–25, one can obtain Eq. (2–38) by enclosing the state n with a closed surface, as was done in Fig. 2–11, and then equating the "flux" leaving the state to the "flux" entering it.

The same argument indicates that the solution to Eq. (2–38), extending the earlier M/M/1 analysis, is given by

$$\lambda_n p_n = \mu_{n+1}p_{n+1} \tag{2-39}$$

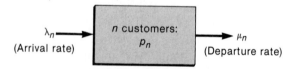

Figure 2–24 State-dependent queueing system

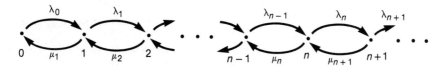

Figure 2-25 State diagram, state-dependent queue (birth-death process)

This is to be compared with the earlier equation, Eq. (2-16), representing the solution for the M/M/1 queue with state-independent Poisson arrival and departure processes. Using Eq. (2-39), it is left to the reader to show that the equilibrium probability p_n for the birth-death process, or state-dependent queueing system, is given by

$$p_n/p_0 = \prod_{i=0}^{n-1} \lambda_i / \prod_{i=1}^{n} \mu_i \qquad (2-40)$$

The unknown state probability p_0 for a finite queue holding at most N customers is again found by invoking the normalization condition $\sum_{n=0}^{N} p_n = 1$. Such a queueing system will *always* be stable. For the infinite queue ($N \to \infty$), stability is again ensured by having $p_0 > 0$.

Some examples of the application of these results are of interest. Consider first the case of two outgoing trunks (transmission links) connecting a statistical concentrator or packet switch to a neighboring packet switch, or node, in a packet-switched network. Data packets use either one of these two trunks randomly. What effect does adding a second trunk have on the operation of the system? Assume that packets arriving at the output queue driving the double transmission-link facility obey a Poisson process with average rate λ. Packets are again assumed to be exponentially distributed in length, with an average length $1/\mu$ in sec. The model for the resultant queueing system is then the one in Fig. 2-26. It is precisely that of an M/M/2 queue.

Consider the operation of this system now. If only one packet is available for transmission, it is immediately serviced by either trunk, at the service rate μ. If two or more packets are available, both trunks are occupied. Because of the exponential service length assumption made, the probability of *either* trunk completing service in the interval $(t, t + \Delta t)$ is $\mu \Delta t$, and the probability of *one* completion in the same interval is $2\mu \Delta t$. (Under the exponential assumption the probability of *both* trunks completing a transmission in the same infinitesimal interval is $0(\Delta t)$, and hence goes to zero as $\Delta t \to 0$). But this system is precisely that of a birth-death system, with $\lambda_n = \lambda$, independent of n, and $\mu_n = \mu$, $n = 1$,

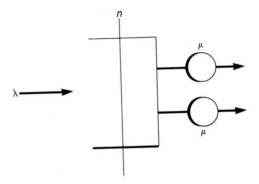

Figure 2-26 M/M/2 queue

$\mu_n = 2\mu$, $n \geq 2$. For the M/M/2 queue, then, one has, from Eq. (2-40),

$$p_n/p_0 = (\lambda/2\mu)^{n-1}\left(\frac{\lambda}{\mu}\right) = 2\rho^n, \qquad n \geq 1 \qquad \rho \equiv \lambda/2\mu \qquad (2-41)$$

The parameter ρ has been defined in terms of 2μ here, since the effective service rate is 2μ for the state $n \geq 2$. With at least two packets in the queue, the system serves at twice the rate of a single-trunk system, reducing the effective traffic intensity correspondingly, and hence acting to reduce queue congestion (as measured by the time delay) as well. These qualitative considerations will be borne out quantitatively. Allowing the queue to be an infinite one for simplicity, one finds, from the normalization condition, that

$$p_0 = (1-\rho)/(1+\rho) \qquad \rho \equiv \lambda/2\mu \qquad (2-42)$$

and

$$p_n = \frac{2(1-\rho)}{(1+\rho)}\rho^n \qquad n \geq 1 \qquad (2-43)$$

The average queue occupancy is readily shown to be given by

$$E(n) = \sum_{n=0}^{\infty} np_n = \frac{2\rho}{(1-\rho^2)} \qquad \rho \equiv \lambda/2\mu \qquad (2-44)$$

This is always less than the average queue occupancy for the M/M/1 case, $E(n)|_{M/M/1} = \rho/(1-\rho)$, $\rho \equiv \lambda/\mu$, as expected. The average time delay in the queue, wait time plus service time, is readily obtained using Little's formula:

$$E(T) = E(n)/\lambda = 1/\mu/(1-\rho^2) \qquad \rho \equiv \lambda/2\mu \qquad (2-45)$$

This is, of course, always less than the M/M/1 result. Note in addition that because of the 2μ service rate for the state $n > 1$, the M/M/2 queue can operate out to twice the arrival rate of the M/M/1 queue: $\lambda < 2\mu$. Adding the additional server thus improves both the time delay and the throughput performance. This is diagrammed in Fig. 2–27, which shows the normalized time delay $\mu E(T)$ plotted versus the normalized arrival rate $\lambda/2\mu$. Also shown is a time-delay load curve for an M/M/1 queue with twice the service capacity, 2μ. The performance of this system is always better than that of either of the other two systems. In terms of performance, it is always more effective to double the transmission capacity than to add a second trunk at the original capacity, *if justified by cost considerations* or required by reliability considerations. The reason is obvious: The packet transmission time is halved, so that at low utilization ($\rho \ll 1$) more packets are being served per unit time. For at least two packets in the system, the probability of completion is the same in the two systems (the M/M/2 queue with service rate μ and the M/M/1 queue with service rate 2μ). As the traffic utilization increases, the average time delays of the two strategies (the one adds a second server, the other doubles the service rate) approach one another.

Consider now, in the M/M/2 case, the significance of the traffic intensity parameter ρ, defined here as $\lambda/2\mu$. As in the case of the M/M/1 queue, one argues that the maximum possible throughput is 2μ. This is not attained here

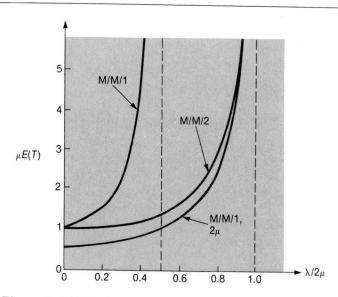

Figure 2–27 Performance characteristic, M/M/2 queue

because, with probability p_0, the queue may be empty; with probability p_1, only one server is utilized. The average throughput for the two-server case is therefore just

$$\gamma = \mu p_1 + 2\mu(1 - p_0 - p_1) \qquad (2-46)$$

For the infinite M/M/2 queue, with no blocking, this should be just the average arrival rate or load on the system λ. Introducing the values for p_0 and p_1 for the infinite queue case from Eq. (2–42) and Eq. (2–43), respectively, one finds this to be precisely the case. This is the reason for introducing the parameter $\rho = \lambda/2\mu$ to represent the ratio of load to maximum transmission capacity of the system.

The second and third examples we discuss can be considered together. The second example extends the M/M/2 case just treated to the case in which a server is made available to any customer entering the system. In both the packet-switched and the circuit-switched cases this implies that the number of transmission links or trunks is always equal to the number of packets or calls desiring transmission. Thus there is never any queueing up for service or any possibility of blocking. The model for this is simply $\mu_n = n\mu$, for all n, if exponential service is again assumed. We again take the arrival rate to be Poisson, with average rate λ. For the infinite queue case, the queue structure one gets is then called the M/M/∞ queue. For this case, from Eq. (2–40), one finds quite readily that

$$p_n/p_0 = (\lambda/\mu)^n/n! \qquad (2-47)$$

and that

$$p_0 = e^{-\rho} \qquad \rho \equiv \lambda/\mu \qquad (2-48)$$

The probability of state occupancy is, in this case, given by the Poisson distribution.

The third example is called a "queue with discouragement." It models a system with customer flow control at the input. Specifically, for this model we let the state-dependent arrival rate λ_n be given by $\lambda_n = \lambda/(n + 1)$, with λ a known constant. Only a single server is available in this case, so that $\mu_n = \mu$, for all n. This example could be used to model moviegoers or shoppers who arrive at the movie theatre or supermarket and find only a single line to serve them. When the line becomes too long (n is large), they are discouraged and turn away. In the context of packet transmission the arrival model represents one in which the maximum possible arrival rate is λ. As the queue length n increases, a system controller discourages packet arrivals (either by blocking or by shunting some arrivals elsewhere), so that the actual arrival rate decreases as n increases. (Flow control in packet networks will be described in Chapter 5.) For this example, a little thought will indicate that the probability p_n of queue occupancy is precisely that given by Eq. (2-47). For an infinite queue, p_0 is given by Eq. (2-48) as well. The two examples, the $M/M/\infty$ queue and the queue with discouragement, thus have the same solution for the probability of state. The average state occupancy is given in both cases by

$$E(n) = \sum_{n=0}^{\infty} np_n = \rho = \lambda/\mu \qquad (2-49)$$

either by using Eqs. (2-47) and (2-48) and carrying out the summation indicated or by invoking the property of the Poisson distribution noted earlier (Eq. (2-2)). This average number of customers is always less than $\rho/(1-\rho)$, the average number in the $M/M/1$ model, showing the benefit to be derived by either increasing the number of servers or controlling the input arrival rate.

The two examples do differ, however, in their time delay-throughput characteristics. This should be the case since, despite the identical solution for the probability of state in the two cases, they do represent different physical situations. Take the $M/M/\infty$ queue first. Its throughput is just λ, the applied load, since $\lambda_n = \lambda$ for all values of n. From Little's formula, then, the average time delay is

$$E(T) = E(n)/\lambda = 1/\mu \qquad (2-50)$$

using Eq. (2-49). But this is precisely the result expected since, with the number of servers always equal to the number of customers (packets or calls) in the system, there is no queueing, and the time delay is just the average service or transmission time $1/\mu$. As a check, consider the throughput γ, as calculated at the system *output*. For a state-dependent departure (death) process this must be the average departure rate, found by averaging over all the departure rates, μ_n, for all n. Specifically, in this case, with $\mu_n = n\mu$, we have

$$\gamma = \sum_{n=0}^{\infty} \mu_n p_n = \mu \sum_{n=0}^{\infty} np_n = \mu E(n) = \lambda \qquad (2-51)$$

invoking Eq. (2-49) to obtain the last result.

Consider the third example now, that of the queue with discouragement. Here, since the input arrival rate is state dependent, one must average over all the states to find the average arrival rate or throughput of the system. We thus have

$$\gamma = \sum_{n=0}^{\infty} \lambda_n p_n = \mu(1 - e^{-\rho}) \qquad \rho = \lambda/\mu \qquad (2-52)$$

using the Poisson distribution of Eqs. (2-47) and (2-48) for p_n, the definition of λ_n in this case, and carrying out the indicated summation. In this case the throughput could have been obtained much more simply, as indicated by the form of the result in Eq. (2–52), by recalling that for a single-server queue of capacity μ, the throughput is just $\mu(1 - p_0)$. From Eq. (2–48) one immediately obtains Eq. (2–52).

Using Eqs. (2–49) and (2–52), one now finds the normalized average time delay in the system to be given by

$$\mu E(T) = \mu E(n)/\gamma = \rho/(1 - e^{-\rho}) \qquad \rho = \lambda/\mu \qquad (2–53)$$

For small ρ ($\rho \ll 1$), the time delay is just $1/\mu$, the average transmission or service time, as expected. The throughput is just $\gamma = \lambda$ in this case, from Eq. (2–52). As the parameter $\rho = \lambda/\mu$ increases, the throughput approaches the maximum value μ, while the normalized time delay continues to increase linearly with ρ, reflecting the fact that the average queue occupancy $E(n) = \rho$, a linear increase with ρ. No limit is required on ρ in this case to ensure stability, since the flow (discouragement) control serves to keep the system stable even though ρ increases indefinitely. Alternatively stated, the Poisson distribution form for p_n, given by Eqs. (2–47) and (2–48), ensures that $p_0 > 0$ for all ρ, and the system remains stable as ρ increases. Equation (2–47) implies that higher states become more probable as ρ increases, with a corresponding increase in the average state occupancy $E(n)$.

The fourth example is a special case of the M/M/∞ queue, with a finite number of servers and no waiting room. Specifically, let $\mu_n = n\mu$, $1 \le n \le N$, $\lambda_n = \lambda$, and block all arrivals if $n = N$. This is often written as an M/M/N/N system. It models a system in which customers arrive according to a Poisson arrival process with average rate λ and always find a trunk (transmission link) available until a maximum number of trunks is occupied. At this point customers arriving are blocked. This model, with the addition of the exponential service time assumption, has been used for many years as a basic design model for telephone exchanges; it will be discussed and used in detail in Chapter 10. A conceptual diagram appears in Fig. 2–28. We focus on the circuit-switched (telephone) terminology. N servers (trunks or transmission links) are available to handle calls. When all trunks are occupied, further calls arriving are blocked. Since there is no queueing (no waiting room) allowed in this system, it is referred to as a pure loss system. The performance parameter of interest in the circuit-switched (telephone) application is the probability of blocking P_B. The solution here is readily obtained. Since $\mu_n = n\mu$ and $\lambda_n = \lambda$, one finds, using Eq. (2–40) and invoking the normalization condition

$$\sum_{n=0}^{N} p_n = 1$$

that

$$p_n = \frac{\rho^n/n!}{\displaystyle\sum_{\ell=0}^{N} \rho^\ell/\ell!} \qquad \rho = \lambda/\mu \qquad (2–54)$$

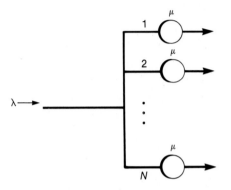

Figure 2–28 M/M/N/N system: no waiting room

In particular, blocking occurs with $n = N$, so that the blocking probability is given by

$$P_B = \frac{\rho^N/N!}{\displaystyle\sum_{\ell=0}^{N} \rho^\ell/\ell!} \qquad \rho = \lambda/\mu \tag{2–55}$$

This equation for the blocking probability is often called the Erlang-B distribution, Erlang distribution of the first kind, or the Erlang loss formula, after the great Swedish engineer A. K. Erlang, who first studied the traffic performance of telephone systems using a statistical approach in the early part of the twentieth century. We shall refer to this equation in detail in Chapter 10 in our discussion of circuit-switched traffic analysis. There we shall adopt the symbol A in place of ρ to represent the total normalized load or traffic intensity λ/μ on the system. The units of A are given in terms of Erlangs.

It is left for the reader to show that the average number of calls in the system is

$$E(n) = \rho(1 - P_B) \qquad \rho = \lambda/\mu \tag{2–56}$$

As the traffic intensity ρ increases, $P_B \rightarrow 1$ and $E(n) \rightarrow N$. The throughput γ, in calls per unit time accepted by the system and hence delivered at the output, is again

$$\gamma = \lambda(1 - P_B) \tag{2–57}$$

Averaging over the state-dependent service rate at the output of the system, one also finds

$$\gamma = \sum_{n=0}^{N} \mu_n p_n = \mu E(n) \tag{2–58}$$

invoking the definitions of μ_n and of $E(n)$, respectively. As the traffic builds up, the throughput approaches its maximum value of $N\mu$. This corresponds to the case $\rho = \lambda/\mu \gg N$, in which situation most of the calls arriving are being blocked as well; thus $P_B \to 1$. The average delay through the system, for those calls accepted, is just $1/\mu$, the service time (called the holding time in telephone practice). As a check, we have, invoking Little's formula,

$$E(T) = E(n)/\gamma = 1/\mu \qquad (2-59)$$

from Eq. (2-58).

Other examples of state-dependent queue analysis will be encountered throughout the book, in analyzing quantitatively the performance of packet-switched, circuit-switched, and integrated networks.

2-5 M/G/1 Queue: Mean Value Analysis

In the previous section we extended the M/M/1 queue analysis to the case of state-dependent arrival and service times (the birth-death process). The state-dependent model is extremely useful and is frequently used, as indicated by some of the examples. However, it still relies on the Markov memoryless property for both the arrival process and the service-time distribution. In this section we extend the analysis to one other case, that of a *general* service-time distribution. Packets or calls may thus have an arbitrary (but known) length or service distribution. However, the arrival process will be taken to be Poisson, a single server is assumed, and the queue buffer size (waiting room) is taken to be infinite. Such a queue is called an M/G/1 queue, using the Kendall notation, with G obviously standing for general service distribution.

For simplicity, in this section we shall focus on *average* (mean) occupancy and *average* time delay only. More general discussions appear in books on queueing theory. The book by L. Kleinrock is a particularly good example [KLEI 1975a]. A brief discussion appears in [SCHW 1977], Chapter 6. We shall show that the average queue occupancy $E(n)$ and the average time delay through the queue $E(T)$ are given, respectively, by the following rather simple-looking expressions:

$$E(n) = \left(\frac{\rho}{1-\rho}\right)[1 - \frac{\rho}{2}(1 - \mu^2\sigma^2)] \qquad (2-60)$$

[KLEI 1975a] L. Kleinrock, *Queueing Systems. Volume 1: Theory,* John Wiley & Sons, New York, 1975.

[SCHW 1977] M. Schwartz, *Computer-Communication Network Design and Analysis,* Prentice-Hall, Englewood Cliffs, N.J., 1977.

and

$$E(T) = \frac{E(n)}{\lambda} = \frac{1/\mu}{(1-\rho)} \left[1 - \frac{\rho}{2}(1 - \mu^2\sigma^2)\right] \qquad (2-61)$$

These are called the Pollaczek-Khinchine formulas after two Russian mathematicians. The parameter ρ is again given by $\lambda/\mu = \lambda E(\tau)$, with λ the average (Poisson) arrival rate and $E(\tau) = 1/\mu$ the average service time. The parameter σ^2 is the variance of the service-time distribution.

Note that both expressions appear closely related to the corresponding results in the M/M/1 case. (These are the leading terms, before the brackets, in both equations.) This is a remarkable result: The average queue occupancy (and corresponding time delay) for a queue with Poisson arrivals and *any* service-time distribution is given by the result obtained for an exponential service distribution with the same average service time, multiplied by a correction factor. This correction factor, the term in brackets in Eqs. (2–60) and (2–61), is seen to depend on the ratio of the variance σ^2 of the service distribution to the average value squared, $1/\mu^2$.

Recall that the variance of the exponential distribution is $\sigma^2 = 1/\mu^2$, i.e., the square of the average value. Setting $\sigma^2 = 1/\mu^2$ in Eqs. (2–60) and (2–61), then, one obtains the results derived earlier for the M/M/1 queue. As σ^2 increases, with $\sigma^2 > 1/\mu^2$, the corresponding average queue occupancy and time delay increase as well. For $\sigma^2 < 1/\mu^2$, on the other hand, the average queue occupancy and time delay decrease relative to the M/M/1 result. As a special case, let all customers (packets or calls) have the *same* service length $1/\mu$. Then for $\sigma^2 = 0$,

$$E(n) = \frac{\rho}{(1-\rho)} \left(1 - \frac{\rho}{2}\right) \qquad \sigma^2 = 0 \qquad (2-62)$$

and

$$E(T) = \frac{1/\mu}{(1-\rho)} \left(1 - \frac{\rho}{2}\right) \qquad \sigma^2 = 0 \qquad (2-63)$$

A queue of this type, with fixed customer service time, is called an M/D/1 queue. The letter D represents *deterministic* service time. This is then a special case of the M/G/1 queue, with the smallest possible queue occupancy and delay. Note that for ρ not too large, $E(n)$ and $E(T)$ may be obtained (conservatively) by using the M/M/1 results. For $\rho \to 1$, the M/D/1 results differ by 50 percent from the M/M/1 values.

The interesting thing is that for the general M/G/1 results of Eqs. (2–60) and (2–61) the dominant behavior of average queue occupancy and time delay is always the $1/(1-\rho)$ term of the denominators. *All* infinite buffer queues, no matter what the service distribution, thus tend to exhibit the same queue-congestion behavior as $\rho = \lambda/\mu \to 1$. Those with larger variance in their service distribution produce larger queue occupancy and time delay, on the average.

This is to be expected since the increased variance means a large spread in the service times, with a correspondingly higher probability that *longer* service times, leading to more congestion, will be encountered.

Equation (2−61) may be used to obtain a compact, general form for the average wait time $E(W)$ on the queue. Specifically, recall that $E(T)$ and $E(W)$ are related by the expression

$$E(W) = E(T) - 1/\mu \qquad (2-64)$$

Substituting Eq. (2−61) for $E(T)$ into Eq. (2−64) and simplifying, one obtains the following simple result for the average wait time $E(W)$ in an M/G/1 queue:

$$E(W) = \frac{\lambda E(\tau^2)}{2(1-\rho)} \qquad (2-65)$$

The term $E(\tau^2)$ is the second moment of the service-time distribution, given by

$$E(\tau^2) = \sigma^2 + 1/\mu^2$$

This is obviously a simpler expression to remember than the Pollaczek-Khinchine form of Eq. (2−61). It will be used in the next section in discussing the wait time in priority queues.

The derivation of Eq. (2−60) for the average queue occupancy $E(n)$ proceeds in a manner different from that used in the M/M/1 analysis. Because of the condition of general service-time statistics, with the corresponding lack of a memoryless property for service departures, one can no longer set up a simple balance equation for the states of the queue. Instead we use an approach that focuses on the change in buffer (queue) occupancy at the conclusion of service, i.e., the departure time. Specifically, label the time at which the jth customer departs the queue by the number j. The number of customers (packets or calls) remaining in the queue just after the departure is represented by the number n_j. A timing diagram indicating these quantities appears in Fig. 2−29.

A simple relation may be written connecting the number of customers in the queue at time j to the number at time $(j-1)$. Let v_j be the number of customers arriving during the service interval of the jth customer. Then n_j and n_{j-1} are obviously connected by the relation

$$\begin{aligned} n_j &= (n_{j-1} - 1) + v_j & n_{j-1} &\geq 1 \\ &= v_j & n_{j-1} &= 0 \end{aligned} \qquad (2-66)*$$

Equation (2−66), representing the queue dynamics for a general service-time distribution, enables us to find the steady-state statistics of the queue *at service*

---

* If the queue empties at $(j-1)$, we wait until the next customer arrives and in turn completes service. During this service v_j customers may arrive.

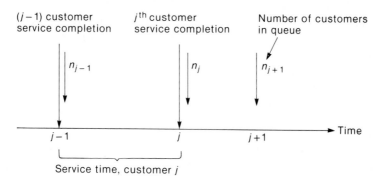

Figure 2–29 Timing diagram, general service-time distribution

completion times. For the special case of Poisson arrivals, on which we focus shortly, one can argue, because of the memoryless property of the arrivals, that the statistics found are the same at all points on the time axis.

Equation (2–66) may be rewritten in the following compact form:

$$n_j = n_{j-1} - u(n_{j-1}) + v_j \qquad (2-66a)$$

where $u(x)$ is the unit step function defined as

$$\begin{aligned} u(x) &= 1 \qquad x \geq 1 \\ &= 0 \qquad x \leq 0 \end{aligned}$$

Now let the time $j \to \infty$ and assume that equilibrium has set in. Taking expectations on both sides of Eq. (2–66a), and noting that $E(n_j) = E(n_{j-1}), j \to \infty$, we get

$$E(v) = E[u(n)] \qquad (2-67)$$

$E(v)$, the average number of arrivals in a service interval, will be expected to be less than one for an infinite queue in order to maintain a stable queue, as required for equilibrium. We shall show shortly that $E(v) < 1$, and in fact corresponds to the utilization ρ for a single-server queue, first encountered in the M/M/1 case. To demonstrate this, we note from the definition of the unit step function $u(n)$ that

$$E[u(n)] = \sum_{n=1}^{\infty} p_n = \text{prob. } (n > 0) \equiv \rho \qquad (2-68)$$

Here p_n is, as throughout this chapter, the probability that the queue is in state n. From Eq. (2–68) we thus have $p_0 = (1 - \rho) > 0$, as found first in the M/M/1 case.

Equations (2–66) and (2–66a) may be used to obtain the state probabilities of the M/G/1 queue, using transforms or moment-generating functions [KLEI 1975a], [SCHW 1977]. For our purposes, as noted earlier, it suffices to find average queue occupancy only. To do this, we employ a trick. Square the left- and right-hand sides of Eq. (2–66a), take expectations, and again let $j \to \infty$. Noting that $E[u^2(n)] = E[u(n)] = E(v) = \rho$, that $E[nu(n)] = E(n)$, and assuming v_j and n_{j-1} to be *independent* (as is the case for Poisson arrivals), one obtains the following interesting result after some simplification:

$$E(n) = \frac{\rho}{2} + \frac{\sigma_v^2}{2(1-\rho)} \qquad (2-69)$$

We have used $E(v) = \rho$ here; σ_v^2, yet to be found, is the variance of the number of customers arriving in a service interval.

To proceed, i.e., to find σ_v^2 specifically, we now invoke the Poisson arrival assumption. We also let the (general) service-time probability density function be given by $f_\tau(\tau)$, with the parameter τ representing the service time. To find σ_v^2 and $E(v)$ we must know the probability $P(v = k)$ that exactly k customers arrive in a service interval. For example, it is apparent from their definitions that

$$E(v) = \sum_{k=0}^{\infty} kP(v = k) \qquad (2-70)$$

and that

$$\sigma_v^2 = \sum_{k=0}^{\infty} [k - E(v)]^2 P(v = k) \qquad (2-71)$$

Defining the conditional probability $P(v = k|\tau)$ as the probability of k arrivals in τ sec, $P(v = k)$ is obviously given by

$$P(v = k) = \int_0^{\infty} P(v = k|\tau)f_\tau(\tau)d\tau \qquad (2-72)$$

Focusing now on the M/G/1 queue case, for Poisson arrivals we have

$$P(v = k|\tau) = (\lambda\tau)^k e^{-\lambda\tau}/k! \qquad (2-73)$$

Inserting Eq. (2–73) into Eq. (2–72), and then into Eq. (2–70), one finds after interchanging integration and summation that

$$E(v) = \lambda E(\tau) \equiv \rho \qquad (2-74)$$

$$E(\tau) = \int_0^{\infty} \tau f_\tau(\tau)d\tau \qquad (2-75)$$

The definition of $E(v)$, the average number of arrivals in a service interval, as the utilization ρ is thus validated another way: Eq. (2–74) indicates that $E(v)$ is, for Poisson arrivals, given by the average customer arrival rate λ times the average service time $E(\tau)$, extending the definition used in the M/M/1 case.

The calculation of σ_v^2 for the case of Poisson arrivals proceeds in a manner similar to that of the calculation of $E(\tau)$. Again inserting Eq. (2–73) into Eq. (2–72) and then inserting the result into Eq. (2–71) (the defining equation for σ_v^2), interchanging summation and integration, and then simplifying the result, one finds that

$$\sigma_v^2 = \rho + \lambda^2\sigma^2 \qquad (2-76)$$

with σ^2 the variance of the service-time distribution. In deriving this result, use is made of the fact that the variance of the Poisson distribution equals its mean value. Details are left to the reader.

Introducing Eq. (2–76) into Eq. (2–69), one obtains the desired Pollaczek-Khinchine form, Eq. (2–60). Equation (2–61) for the average time delay (wait time plus transmission time) through the queue follows directly from Little's theorem. Alternatively, a much simpler form of the Pollaczek-Khinchine formula results if one focuses on the average wait time $E(W)$ in the queue. This was noted earlier and the simple, compact form Eq. (2–65) for $E(W)$ was derived from Eq. (2–61) for the time delay $E(T)$ through the queue. Equation (2–65) will be found very useful in the next section in calculating the wait time in nonpreemptive priority queueing systems.

2–6 Nonpreemptive Priority Queueing Systems

The need to provide priority to certain classes of customers in a queueing system arises in many applications. Priority classes are used in many computer systems, in the computer control of digital switching exchanges, for deadlock prevention in packet-switching networks, and so forth. A simple example taken from packet switching serves to provide the necessary motivation at this point. Consider a packet-switching network that transmits, in addition to the normal data packets, control packets of much shorter length that carry out the vital operations of signaling, congestion notification, fault notification, routing change information, and so on. (Some of these control functions will be explored in the chapters following.) It is vital in many cases to ensure rapid distribution of these control messages. Yet without establishing a priority, they could easily queue up behind much longer data packets, delaying their arrival at the necessary destination points.

As a specific example, consider a network connected by 9600 bps transmission links. We designate data packets by the label 2, control packets by the label 1. Data packets are on the average 960 bits long, or $1/\mu_2 = 0.1$ sec, with a variance $\sigma_2^2 = 2(1/\mu_2)^2$, or $E(\tau^2) = 3(1/\mu_2)^2$. Control packets, on the other hand, are all 48 bits long, so that $1/\mu_1 = 5$ msec. In this case of fixed packet lengths $\sigma_1^2 = 0$. Focus on a single queue, served FIFO, that drives the outgoing transmission link. Let 20 percent of the total traffic be due to the short control packets, 80 percent to the much longer data packets. We thus have $\lambda_1 = 0.2\lambda$, $\lambda_2 = 0.8\lambda$, with λ the composite arrival rate at the queue in packets/sec. It is apparent that without priority the queueing of the combined traffic stream can be modeled as an M/G/1 queue. The combined traffic intensity is $\rho = \rho_1 + \rho_2$. Since packets of either kind arrive randomly, with rates λ_1 and λ_2, respectively, the second moment of the composite stream is given by the weighted sum of the second moments:

$$E(\tau^2) = \frac{\lambda_1}{\lambda} E(\tau_1^2) + \frac{\lambda_2}{\lambda} E(\tau_2^2)$$

Say that the effective ρ is 0.5, to be specific. The total arrival rate is then $\lambda = 6.17$ packets/sec, and using Eq. (2–65), the average waiting time for *either* type of packet is readily found to be 148 msec. The 48-bit control packets, requiring 5 msec for transmission, frequently may be trapped behind the much longer 100-msec data packets, and must wait, on the average, 148 msec for transmission! The obvious solution is to provide a higher priority to the control packets, enabling them to bypass the lower priority data packets on arrival and to move directly to the head of the queue.

Two types of priority are normally employed: nonpreemptive and preemptive. In the former case, higher-priority customers (packets in our example) move ahead of lower-priority ones in the queue but do not preempt lower-priority customers already in service. In the preemptive priority case, service on a lower-priority customer is interrupted and only continued after all arriving high-priority customers have been served. We focus here on the nonpreemptive priority case only.

Returning to the example just given, the case of short control packets and longer data packets, we shall show that providing nonpreemptive priority to the control packets essentially halves their wait time to 74.5 msec, on the average, while increasing the wait time of the data by an inconsequential amount. If this reduction is not sufficient in a real situation, preemptive priority may have to be employed. (There is a price paid, however: Customers whose service is interrupted must be so tagged. This entails added processing time and overhead, which may reduce some of the benefits theoretically obtained through preemption.)

To be more general, say there are r classes of customers to be served at a queue. Their respective arrival rates are $\lambda_1, \lambda_2, \ldots, \lambda_r$, each representing a Poisson stream. The average service time is $1/\mu_k$ for the kth class, $k = 1$, $2, \ldots, r$. The highest-priority class is taken to be 1, the lowest r, in descending order as labeled. We now show how one computes the average waiting time for any class, assuming nonpreemptive service. Consider class p $(1 \le p \le r)$ in particular. Let a typical customer of this class arrive at an arbitrary time t_0. Its random waiting time W_p (Fig. 2-30), measured from its arrival time until it enters service, is due to contributions from three sources. It must wait a random amount of time T_0 until the customer currently in service completes service. It must wait a random number T_k units of time until all customers of priority k lower than or equal to p, already enqueued at the arrival time t_0, complete service. Finally, it must wait a random time T'_k to service customers of each class k of priority lower than p arriving during the wait time W_p.

Putting these observations together, we write

$$W_p = T_0 + \sum_{k=1}^{p} T_k + \sum_{k=1}^{p-1} T'_k \tag{2-77}$$

Taking expectations term by term, the average waiting time $E(W_p)$ of priority p is obviously given by

$$E(W_p) = E(T_0) + \sum_{k=1}^{p} E(T_k) + \sum_{k=1}^{p-1} E(T'_k) \tag{2-78}$$

To find each of the three average times in Eq. (2-78), note first that $E(T_k)$ is due to an average number $E(m_k)$ customers of category k waiting in the system. Each requires $1/\mu_k$ units of service on the average, so that we immediately have

$$E(T_k) = E(m_k)/\mu_k \tag{2-79}$$

But by Little's formula, we also have $E(m_k)$ related to the average wait time $E(W_k)$. Specifically,

$$E(m_k) = \lambda_k E(W_k) \tag{2-80}$$

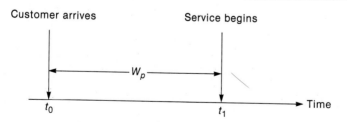

Customer arrives   Service begins

W_p

t_0 t_1 Time

Figure 2-30 Waiting time, queueing system

Combining Eqs. (2–80) and (2–79), we immediately have

$$E(T_k) = \rho_k E(W_k) \qquad \rho_k \equiv \lambda_k/\mu_k \tag{2-81}$$

Now consider the term $E(T'_k)$ in Eq. (2–78). This is due, on the average, to $E(m'_k)$ customers of class k arriving during the interval $E(W_p)$. Since the arrival rate is λ_k and each customer again requires, on the average, $1/\mu_k$ units of service, we immediately have

$$E(T'_k) = \lambda_k E(W_p)/\mu_k = \rho_k E(W_p) \tag{2-82}$$

Consider the remaining term $E(T_0)$ in Eq. (2–78). This is the residual service time of a customer in service. For the work-conserving nonpreemptive queueing system under discussion here (the server always serves a customer if one is waiting to be served), this is independent of queue discipline: It must be the same if customers of all k classes are served with the same priority, in their order of arrival. From Eq. (2–65), the average wait time of an M/G/1 queue, we find that

$$E(T_o) = \lambda E(\tau^2)/2 = \sum_{k=1}^{r} \lambda_k E(\tau_k^2)/2 \tag{2-83}$$

This generalizes the two-class example described earlier.

Using Eqs. (2–82) and (2–81) in Eq. (2–78) and solving for the wait time of each class recursively, starting with the highest priority class 1, one readily shows that

$$E(W_p) = E(T_0)/(1 - \sigma_p)(1 - \sigma_{p-1}) \tag{2-84}$$

with $\sigma_p \equiv \Sigma_{k=1}^{p} \rho_k$, $\rho_k = \lambda_k/\mu_k$, and $E(T_o)$ given by Eq. (2–83).

As an example, consider the two highest-priority classes. For these we have

$$E(W_1) = E(T_0)/(1 - \rho_1) \tag{2-85}$$

and

$$E(W_2) = E(T_0)/(1 - \rho_1)(1 - \rho)$$
$$= E(W_1)/(1 - \rho) \qquad \rho = \rho_1 + \rho_2 \tag{2-86}$$

The highest priority, class 1 customers, queue up as in an M/G/1 system of a single class, seeing themselves only, except for the added residual service time $E(T_0)$ that is due to customers of all classes that might have been in service.

In the special case of two priorities only, the case considered in the example described at the beginning of this section, Eqs. (2–85) and (2–86) provide the average waiting times for the two classes, respectively. Recall that with $\rho = 0.5$, and for the numbers given earlier, the average wait time $E(W)$ without priority is 148 msec. From Eq. (2–85), for the same example, we get $E(W_1) = 74.5$ msec. $E(W_2)$, using Eq. (2–86), turns out to be 149 msec. The average wait time of the higher-priority control packets has thus dropped to almost half of the original value with no priority used, while the lower-priority data packets have had their wait time increased by 1 msec in 148, an inconsequential change. This demonstrates the improvement possible through the use of priority queueing.

Although the effect on the lower priority packets in this example is negligible, in other situations it might be much more noticeable. The fact is that as some priority classes (the higher ones) improve their performance, others deteriorate. Interestingly, it is simple to show that a conservation law is at work here. In particular, from Eq. (2–84), it is simple to show that the weighted sum of wait times is always conserved. In particular, one finds that

$$\sum_{k=1}^{r} \rho_k E(W_k) = \rho E(W) \tag{2–87}$$

with $E(W)$ given by Eq. (2–65), the wait time of the FIFO M/G/1 queue. As some wait times decrease, then, others must increase in order to compensate. As a check, in the two-priority example discussed here we had, without priority, $\rho E(W) = 74$ msec. With priority, $\rho_1 E(W_1) + \rho_2 E(W_2) = 74$ msec as well. This conservation law is a special case of a more general conservation law for work-conserving queues first developed by Kleinrock [KLEI 1965]. (See also [KLEI 1976].)

Problems

2–1 Refer to Fig. 2–3 in the text. Calculate the probability of k independent events in the m intervals Δt units long, if the probability of one event in any interval is p, while the probability of no event is $q = 1 - p$. Show how one obtains the binomial distribution of Eq. (2–4).

2–2 In Problem 2–1 let $p = \lambda \Delta t$, λ a proportionality factor. This then relates the binomial distribution to the Poisson process. Let $\Delta t \to 0$, with $T = m\Delta t$ fixed. Show that in the limit one gets the Poisson distribution of Eq. (2–1). Show that the mean value $E(k)$ and the variance σ_k^2 are both equal to λT. What is the probability that no arrival occurs in the interval T? Sketch this as a function of T. Repeat for the probability that *at least* one arrival occurs in T.

[KLEI 1965] L. Kleinrock, "A Conservation Law for a Wide Class of Queueing Systems," *Naval Res. Logist. Quarterly*, vol. 12, 1965, 181–192.

[KLEI 1976] L. Kleinrock, *Queueing Systems, Volume II: Computer Applications*, John Wiley & Sons, New York, 1976.

2–3   Calculate and plot the Poisson distribution given by Eq. (2–1) for the three cases $\lambda T = 0.1, 1, 10$. In the third case try to carry the calculation and plot out to at least $k = 20$. (Stirling's approximation for the factorial may be useful here.) Does the distribution begin to crowd in and peak about $E(k)$ as predicted by the ratio $\sigma_k/E(k) = 1/\sqrt{\lambda T}$?

2–4   Carry out the details of the analysis leading to Eqs. (2–10) and (2–11), showing that sums of Poisson processes are Poisson as well.

2–5   Refer to the time-dependent equation (2–12a) governing the operation of the M/M/1 queue. Start at time $t = 0$ with the queue empty. (What are then the values $p_n(0)$?) Let $\lambda/\mu = 0.5$ for simplicity, take $\Delta t = 1$, and pick $\lambda \Delta t$ and $\mu \Delta t$ very small so that terms of $(\Delta t)^2$ and higher can be ignored. Write a program that calculates $p_n(t + \Delta t)$ recursively as t is incremented by Δt and show that $p_n(t)$ does settle down eventually to the steady-state set of probabilities $\{p_n\}$. Pick the maximum value of n to be 5. The set of steady-state probabilities obtained should then agree with Eq. (2–20). *Note:* Eq. (2–12a) must be modified slightly in calculating $p_0(t + \Delta t)$ and $p_5(t + \Delta t)$. You may want to set the problem up in matrix-vector form.

2–6   Derive Eq. (2–15), governing the steady-state (stationary) probabilities of state of the M/M/1 queue, in two ways:

1. from the initial generating equation (2–12)
2. from flow balance arguments involving transitions between states $n - 1$, n, and $n + 1$, as indicated in Fig. 2–11.

2–7   As a generalization of the M/M/1 queue analysis, consider a birth-death process with state-dependent arrivals λ_n and state-dependent departures μ_n. (See Figs. 2–24 and 2–25.) Show, by applying balance arguments, that the equation governing the stationary state probabilities is given by

$$(\lambda_n + \mu_n)p_n = \lambda_{n-1}p_{n-1} + \mu_{n+1}p_{n+1}$$

(See Eq. (2–38).) Show that the solution to this equation is given by Eq. (2–40).

2–8   Consider the M/M/1 queue analysis. Show that the stationary state probability p_n is given by

$$p_n = \rho^n p_0 \qquad \rho \equiv \lambda/\mu$$

in two ways:

1. Show that this solution for p_n satisfies Eq. (2–15) governing the queue operation.
2. Show that the balance equation $\lambda p_n = \mu p_{n+1}$ or $p_{n+1} = \rho p_n$ satisfies Eq. (2–15). Then iterate n times.

Calculate p_0 for the finite M/M/1 queue and show that p_n is given by Eq. (2–20).

2–9   Show that the blocking probability P_B of the finite M/M/1 queue is given by $P_B = p_N$ by equating the net arrival rate $\lambda(1 - P_B)$ to the average departure rate $\mu(1 - p_0)$ and solving for P_B.

2–10 Consider a finite M/M/1 queue capable of accommodating N packets (customers). Calculate the values of N required for the following situations:

1. $\rho = 0.5$, $P_B = 10^{-3}$, 10^{-6}
2. $\rho = 0.8$, $P_B = 10^{-3}$, 10^{-6}

Compare the results obtained.

2–11 The probability p_n that an infinite M/M/1 queue is in state n is given by $p_n = (1 - \rho)\rho^n$ $\rho = \lambda/\mu$.

a. Show that the average queue occupancy is given by

$$E(n) = \sum_n np_n = \rho/(1 - \rho)$$

b. Plot p_n as a function of n for $\rho = 0.8$.
c. Plot $E(n)$ versus ρ and compare with Fig. 2–17.

2–12 The average buffer occupancy of a statistical multiplexer (or data concentrator) is to be calculated for a number of cases. (In such a device the input packets from terminals connected to it are merged in order of arrival in a buffer and are then read out first come–first served over an outgoing transmission link.) An infinite buffer M/M/1 model is to be used to represent the concentrator.

1. Ten terminals are connected to the statistical multiplexer. Each generates, on the average, one 960-bit packet, assumed to be distributed exponentially, every 8 sec. A 2400-bits/sec outgoing line is used.
2. Repeat if each terminal now generates a packet every 5 sec, on the average.
3. Repeat 1. above if 16 terminals are connected.
4. Forty terminals are now connected and a 9600-bits/sec output line is used. Repeat 1. and 2. in this case. Now increase the average packet length to 1600 bits. What is the average buffer occupancy if a packet is generated every 8 sec at each terminal? What would happen if each terminal were allowed to increase its packet generation rate to 1 per 5 sec, on the average? (*Hint:* It might now be appropriate to use a *finite* M/M/1 model with your own choice of buffer size.)

2–13 Consider the finite M/M/1 queue holding at most N packets.

a. Show that the blocking probability is $P_B = 1/(N + 1)$ at $\rho = 1$.
b. Plot $P_B = p_N$ for all values of ρ, $0 \le \rho < \infty$ for $N = 4$ and $N = 19$.
c. The throughput is defined to be $\gamma = \lambda(1 - P_B)$. Sketch γ/μ (the normalized throughput) as a function of $\rho = \lambda/\mu$ (normalized load) for $0 \le \rho < \infty$, $N = 4$, and $N = 19$. Compare.

2–14 Refer to Problem 2–12. Find the mean delay $E(T)$ and the average wait time $E(W)$ in each case.

2–15 Use Fig. 2–23 to prove that Little's formula applies in the LCFS service discipline.

2–16 Draw your own arrival-departure diagrams for an arbitrary queueing system, as in Figs. 2–22 and 2–23, comparing the FIFO and LCFS service disciplines. Carry out your own proof of Little's formula.

2–17 Consider the M/M/2 queue discussed in the text. Derive Eq. (2–43), the expression for the probability of state occupancy and Eq. (2–44), the equation for the average queue occupancy. Plot $\mu E(T)$ (normalized time delay) versus λ, the average arrival rate (load on the system) for the M/M/2 queue, and compare with two single-server cases: an M/M/1 queue with service rate μ and an M/M/1 queue with service rate 2μ. Check Fig. 2–27.

2–18 Refer to the multiple (or *ample*) server and queue with discouragement examples discussed in the text. Show that the state probability distribution and the average queue occupancy are given, in both examples, by Eq. (2–47) with Eq. (2–48), and Eq. (2–49), respectively. However, the average time delay and throughput are different in the two cases. Calculate these two quantities in both cases and compare.

2–19 Consider a queue with a general state-dependent departure (service) process μ_n.

    a. Explain why the average throughput is given by $\gamma = \sum_{n=1}^{N} \mu_n p_n$.

    b. Take the special case of the M/M/2 queue, $\mu_n = \mu$, $n = 1$; $\mu_n = 2\mu$, $n \geq 2$. Show that $\gamma = \mu p_1 + 2\mu(1 - p_0 - p_1)$. Show that this is just $\gamma = \lambda$, if p_1 and p_0 are explicitly calculated using Eq. (2–41).

2–20 A queueing system holds N packets, including the one(s) in service. The service rate is state dependent, with $\mu_n = n\mu$, $1 \leq n \leq N$. Arrivals are Poisson, with average rate λ.

    a. Show that the probability the system is in state n is the Erlang distribution of Eq. (2–54).

    b. Show that the average number in the system is given by Eq. (2–56).

    c. Show that the average *throughput* is $\gamma = \mu E(n)$, in two ways:

        1. Use $\gamma = \sum_{n=0}^{N} \mu_n p_n$
        2. $\gamma = \lambda(1 - P_B)$, P_B the blocking probability.

    d. Little's theorem says that $E(T) = E(n)/\gamma = 1/\mu$ here. Explain this result (i.e., there is *no* waiting time).

2–21 A queueing system has two outgoing lines, used randomly by packets requiring service. Each transmits at a rate of μ packets/sec. When both lines are transmitting (serving) packets, packets are blocked from entering—i.e., there is no buffering in this system. Packets are exponentially distributed in length; arrivals are Poisson, with average rate λ. $\rho = \lambda/\mu = 1$.

    a. Find the blocking probability, P_B, of this system.

    b. Find the average number, $E(n)$, in the system.

    c. Find the normalized throughput γ/μ, with γ the average throughput, in packets/sec.

    d. Find the average delay $E(T)$ through the system, in units of $1/\mu$. (Alternatively, find $E(T)/1/\mu$.)

2–22 A data concentrator has 40 terminals connected to it. Each terminal inputs packets with an average length of 680 bits. Forty bits of control information are added to each packet before transmission over an outgoing link with capacity $C = 7200$ bps.

Twenty of the terminals input 1 packet/10 sec each, on the average.
Ten of the terminals input 1 packet/5 sec each, on the average.
Ten of the terminals input 1 packet/2.5 sec each, on the average.
The input statistics are Poisson.

a. The data units transmitted (called frames) are exponentially distributed in length. Find (1) the average wait time on queue, *not including* service time and (2) the average number of packets in the concentrator, *including* the one in service.

b. Repeat if the packets are all of constant length.

c. Repeat if the second moment of the frame length is $E(\tau^2) = 3(1/\mu)^2$; $1/\mu$ is the average frame length.

2-23 Refer to the derivation of the Pollaczek-Khinchine formulas in the text.

a. Derive Eq. (2-69), the general expression for the average number of customers in the queue.

b. For the case of Poisson arrivals, calculate $E(v)$ and σ_v^2, following the procedure in the text, and show that Eqs. (2-74) and (2-76) result.

2-24 Two types of packets are transmitted over a data network. Type 1, control packets, are all 48 bits long; type 2, data packets, are 960 bits long on the average. The transmission links all have a capacity of 9600 bps. The data packets have a variance $\sigma_2^2 = 2(1/\mu_2)^2$, with $1/\mu_2$ the average packet length in seconds. The type 1 control packets constitute 20 percent of the total traffic. The overall traffic utilization over a transmission link is $\rho = 0.5$.

a. FIFO (nonpriority service) is used. Show that the average waiting time for either type of packet is $E(W) = 148$ msec.

b. Nonpreemptive priority is given to the control packets (type 1). Show that the wait time of these packets is reduced to $E(W_1) = 74.5$ msec, whereas the wait time of the data packets (type 2) is increased slightly to $E(W_2) = 149$ msec.

2-25 Show that the average wait time of class p in a nonpreemptive priority system is given by Eq. (2-84).

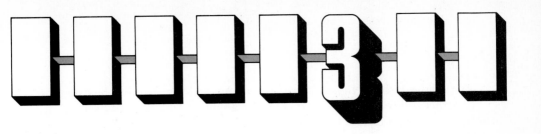

Layered Architectures in Data Networks

Chapter 1 introduced the study of data networks by first distinguishing briefly between packet switching and circuit switching, the two most common ways of transporting data through a network. In the case of packet switching, relatively small blocks of data called packets are individually transmitted across a network. They are stored and forwarded at each node along a path, sharing buffer and link transmission resources with all other packets being transmitted across a given link on the path.

Two modes of transmission are generally distinguished: the connection-oriented or virtual circuit mode, in which packets are constrained to follow one another in order, arriving in sequence at the destination; and the connectionless or datagram mode, in which packets may arrive out of order. In the virtual circuit case, a connection must first be established between source and destination before communication can begin. In either case, as just noted, the packets share buffer and transmission facilities with packets of other users while traversing the network. In the case of circuit switching, end users desiring to communicate do so over a path consisting of *dedicated* (private) channels or trunks. This path must first be set up before communication begins.

Packet switching and circuit switching refer to technologies used to transfer data from one end of a network to another. As noted in Chapter 1, however, there are two aspects to the data communication process. There is always a need to transmit data from one end of the network to the other in a timely and

71

cost-effective fashion. In addition, on reaching the ultimate user, the data must be correct and recognizable by that user in a syntactic and semantic sense. This problem is not a technological one in the case of voice communication, when human beings at either end of a network converse in real time. (It is up to the end users to establish the basis for understanding once the technology has assured relatively clear, low-noise communication.)

In the case of data transmission, however, the question of recognizable communication is paramount. A Fortran compiler interacting with a terminal *must* be addressed in a specified form. A printer *must* receive data in a prescribed form. An application involving the transfer of files between two possibly dissimilar computers requires that the commands and control character strings at either end be precisely specified. Specific protocols must be established and agreed to by all end users involved in a particular exchange of information before communication can begin.

To take care of these two separate problems of communication — one involving the timely and correct transport of data across a network, the other the delivery of the data to an end user in a recognizable form — layered communication architectures have been established. This concept, introduced in Chapter 1, is represented in Fig. 3 – 1 in a form similar to that used in Chapter 1. Two groups of layers, the lower set and the upper (higher) set of layers, are distinguished. Figure 3 – 1(a) indicates that the lower, network layers are used by the network nodes to direct the data from one end user (shown here as a host computer) to another. The higher layers shown are charged with the task of providing the data to the end users in a form they can recognize and use. These higher layers reside in the hosts. In the example of Fig. 3 – 1(a) the hosts outside the network proper are shown containing both higher and lower (network) layers. In some networks (ARPAnet is one example) the hosts contain the higher layers only; the network layers appear only in the network nodes. In some cases the communication nodes to which the hosts are shown connected are actually part of the hosts. All three cases appear in practice.

Figure 3 – 1(b) provides an example of a particular layered communications architecture, the Reference Model for Open Systems Interconnection (OSI), which was approved in 1983 as an international standard by the International Organization for Standardization (ISO) [ISO 1983] and by the CCITT [CCITT].* This seven-layer model was discussed briefly in Chapter 1. The bottom three layers of this architecture provide the networking capability,

---

[ISO 1983] *ISO International Standard 7498, Information Processing Systems — Open Systems Interconnection — Basic Reference Model,* Geneva, Oct. 1983.

[CCITT] *CCITT Draft Recommendation X.200, Reference Model of Open Systems Interconnection for CCITT Applications,* Geneva, June 1983.

* CCITT is the International Telephone and Telegraph Consultative Committee of the International Telecommunications Union (ITU).

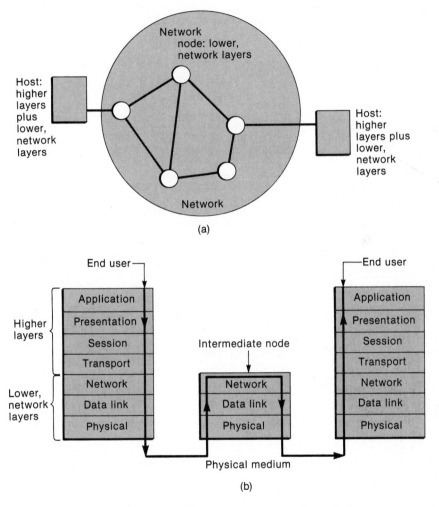

Figure 3-1 Layered architecture, OSI example
  a. Subdivision of layers
  b. OSI seven-layer architecture

whereas the upper four layers carry out the processing, at either end, required to present the data to the end user in an appropriate, recognizable form. (The work of establishing this Reference Model was begun under the auspices of ISO in 1978, as noted in Chapter 1. CCITT later agreed , in the interests of true international standardization, to support this effort as well. Both organizations have been cooperating in preparing international computer communication recommendations and standards.)

The abstract example of Fig. 3 – 1(b) shows information moving down from the uppermost application layer to the network layer at the entry point to the network. It is this layer that handles the routing through the network. To carry out this function, it requires the services of the data link layer below, which ensures correct, error-free transmission of a packet over the next link selected as part of the route through a network. In the example of Fig. 3 – 1(b), one intermediate node is shown connected between source and destination. In practice there might be a number of other intermediate nodes. The packet arriving over the physical medium connecting the source node with the intermediate node is directed up to the network layer at that node which determines the next portion of the path on the route through the network. (Recall from Chapter 1 that this "intermediate node" could itself be a network, or series of networks, in more complex situations involving the interconnection of networks.)

The separation into the two basic tasks of guiding data through the network and then processing it for final delivery to the end user, with further subdivision into the seven architectural layers shown, allows a whole variety of networking approaches to be used, while still maintaining the desired ability of two (or more) end users to "openly" communicate with one another. An example has already been provided and was noted as well in Chapter 1; an "intermediate node" could very well be another network. Local area networks can be accommodated by appropriately modifying the physical, data link, or network layers, or portions thereof. So long as modifications conform to the OSI concept and affect only one layer at a time, open systems interconnection can still proceed. The layers above need not be aware of a change, so long as they still receive the interface signals and services expected from the layers below. Although the stress here is on packet-switched networks because of the context in which these data communication protocols arose, there need not be a packet-switched network providing the actual network service. Data messages could be transmitted using circuit-switched technology, provided that it is either hidden from or made consistent with the transport-layer protocol.

The OSI Reference Model of Fig. 3 – 1(b) is the architecture we shall stress in much of the material following. Specifically, after this brief introduction to the concept of network architectures (following the presentation in Chapter 1), we shall describe the seven layers of the OSI model in more detail. We then give a brief description of X.25, a three-layer data network interface architecture. We follow with a description of the seven layers of the IBM Systems Network Architecture (SNA). In the chapters following we return to some of the layers in more detail, presenting quantitative analyses of performance where possible. We start with the data-link layer in Chapter 4, with emphasis placed on HDLC (high-level data link control), the international standard for data link control developed by ISO. In the next two chapters we discuss the role of the network layer, describing first, in Chapter 5, congestion control mechanisms carried out at this layer and then moving on, in Chapter 6, to the network routing function. Chapter 7 concludes the discussion of layered architectures, at least in the form of the OSI model, by focusing on the transport layer. Mentioned briefly in that chapter is the IP or internet protocol, lying within and at the top of the network layer, and developed specifically to handle communication between disparate networks. Another transport layer protocol, TCP (transmission control protocol), developed as part of ARPAnet activities, is also discussed.

Examples of other architectures are provided as well in the chapters following. In addition to more detailed descriptions and analyses of some features of X.25 and SNA, particularly at the network layer, we touch on Digital Equipment Corporation's DNA and Burrough's BNA. Both of these companies, like many others, have announced their full support of the OSI architecture, with the aim of adopting ISO standards at each layer as they appear.

3 – 1 OSI Standards Architecture and Protocols*

As noted in Chapter 1 and as indicated again with reference to Fig. 3 – 1(b), the OSI Reference Model provides a seven-layered skeletal architecture around which specific protocols can be designed that enable disparate users to "openly" communicate. The choice of seven layers was dictated by the usual engineering compromises required to develop a viable, cost-effective "product" [ZIMM]: (1) the need to have enough layers so that each one is not too complex in terms of developing a detailed protocol with correct and executable specifications; (2) the desire not to have so many layers that the integration and description of layers becomes too difficult; and finally (3) the desire to select natural bounda-

---

* Much of the material in this section is based on a special issue of the *Proceedings of the IEEE* devoted to the OSI standards architecture and protocols [IEEE 1983b]. Specific papers in that issue are referenced as the need arises.

[IEEE 1983b] "Open Systems Interconnection (OSI)—New International Standards Architecture and Protocols for Distributed Information Systems," H. C. Folts and R. desJardins, eds., *Proc. IEEE*, vol. 71, no. 12, Dec. 1983.

[ZIMM] H. Zimmermann, "OSI Reference Model—The ISO Model of Architecture for Open Systems Interconnection," *IEEE Trans. on Comm.*, vol. COM-28, no. 4, April 1980, 425–432; reprinted in [GREE].

[GREE] P. Green, ed., *Computer Network Architectures and Protocols*, Plenum Press, New York, 1982.

ries, with related functions collected into one layer and very different functions separated into different layers. Hopefully, this results in a layered architecture with minimal interactions across the boundaries and with layers that can be easily redesigned, or whose protocols can be changed without changing the interfaces to the other layers.

The Reference Model itself is just a construct, an abstract architecture that must then be fitted with detailed service requirements for each layer and with standard protocols that produce the desired services. This has been the task of many committees in both the ISO and the CCITT. Service requirements have been established for each layer, and protocols defined for a number of these requirements. We describe these next.

It has already been noted that the upper four layers, normally resident in the hosts or other intelligent systems desiring to communicate, handle functions involved with ensuring that the information (data) exchanged between the two end parties is delivered in a correct, understandable form. The application layer is concerned with the semantics of the information exchanged; the presentation layer handles the syntax. The session and transport layers together provide a means for setting up and taking down a connection between the two parties, for ensuring orderly delivery of data once the connection is set up, for distinguishing between normal or expedited data if desired, for detecting errors and repeating portions of messages if necessary, as well as other fuctions described in this section. The transport layer is the one that normally interfaces with the network or networks through which the user information is to be passed. The lower, network layers are to be designed to make the "connection" between transport layers at either end of the path appear to be transparent.

In the paragraphs following we provide some more detail on the purposes of the different layers. We begin with the *application layer,* the highest one in the seven-layer hierarchy. It is this layer that ensures that two application processes, cooperating to carry out the desired information processing on either side of a network, understand one another. As just noted, this layer is responsible for the semantics of the information being exchanged between the application processes. (This is in contrast to the presentation layer below, which is concerned with the representation or *syntax* of the information being exchanged.)

It is clear not all applications and the application processes associated with them can be or should be standardized. An application involving a company's private order-entry system may not be able to be standardized. Certain specific tasks carried out in the banking industry may not lend themselves to OSI standard development, although the banking industry might want to standardize on them. But certain procedures are *common* to all application protocols. OSI standardization activities focus on these procedures, called *common application service elements.* They include, for example, tasks involved in setting up and then terminating an association between application processes [BART]. The application layer can be visualized as consisting of two parts: these elements common to all processes, which interface with the presentation layer, and those specific to the particular application(s) involved.

In addition to these procedures common to most application protocols, three application-layer services and protocols based on them are being developed within the framework of the OSI Reference Model: *virtual terminal* [LOWE], *file* [LEWA], and *job transfer and manipulation* [LANG] *services and protocols.* Management protocols for the application layer are being developed as well [LANG]. The virtual terminal service, as the name implies, is used to provide terminal access to a user process in a remote system. The file service provides for remote access, management, and transfer of information stored in the form of files. With the OSI model, possibly incompatible file systems can thus work together. The job transfer and manipulation service allows distributed job processing to be carried out, involving the functions of job submission, job processing, and job monitoring.

Figure 3-2 demonstrates, as an example, how the OSI virtual terminal service fits within the OSI Reference Model construct. The communication aspects common to distributed terminal access of a remote user process are abstracted out to form a common virtual terminal service. Protocols providing this service form a part of the application layer at either end of the network. An application entity at either end actually executes the protocols. In the OSI approach the two entities at the same layer in the Reference Model communicate as peers, using the communication network services of the layers below. The virtual terminal service provides the appropriate communication services and protocols. Local mapping, not within the purview of the OSI model, is then required at either end to accommodate the real interfaces used by people and application programs [LOWE].

It was noted in Chapter 1 and restated earlier in this chapter that each layer in the OSI Reference Model provides a service to the layer above. Moving down from the application layer, the *presentation layer* just below is the one that isolates application processes in the application layer from differences in representation

---

[BART] Paul D. Bartoli, "The Application Layer of the Reference Model of Open Systems Interconnection," *Proc. IEEE,* vol. 71, no. 12, Dec. 1983, 1404–1407.

[LOWE] Henry Lowe, "OSI Virtual Terminal Service," *Proc. IEEE,* vol. 71, no. 12, Dec. 1983, 1408–1413.

[LEWA] Douglas Lewan and H. Garrett Long, "The OSI File Service," *Proc. IEEE,* vol. 71, no. 12, Dec. 1983, 1414–1419.

[LANG] A. Langford, K. Naemura, R. Speth, "OSI Management and Job Transfer Services," *Proc. IEEE,* vol. 71, no. 12, Dec. 1983, 1420–1424.

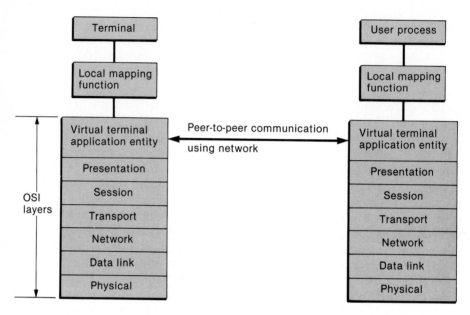

Figure 3-2 OSI virtual terminal application-layer services

and syntax of the data actually transmitted. The presentation layer provides a means for application entities desiring to communicate, to exchange information about the syntax of the data transmitted between those entities. This can either be in the form of names if both communicating systems know the syntax to be used or a description of the syntax to be used if one party does not have that knowledge. Where the syntax of the information sent differs from that used by the receiving system, the presentation layer must provide appropriate mapping. In addition to managing the syntax between systems, the presentation layer must initiate and terminate a connection, manage the layer states, and handle errors [HOLL].

The relationship of the two layers, application layer and presentation layer, is shown schematically in Fig. 3-3 [HOLL, Fig. 2]. Note that the figure indicates the division of the application layer into the two parts mentioned earlier: services common to all applications (setting up and terminating an association between application processes, for example) and those generic to a specific

[HOLL] Lloyd L. Hollis, "OSI Presentation Layer Activities," *Proc. IEEE*, vol. 71, no. 12, Dec. 1983, 1401-1403.

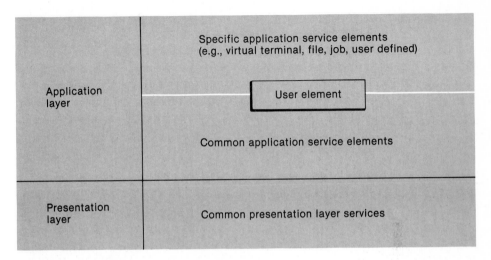

Figure 3 – 3 Structure, OSI application and presentation layers (from [HOLL, Fig. 2], © 1983 IEEE, with permission)

application. Among this group are the three OSI-defined services described earlier: virtual terminal, file, and job transfer. The user element shown indicates that additional capabilities are needed to interface the rest of the application process to the portions shown in the application layer.

The next layer down in the OSI model, as indicated in Fig. 3 – 1(b), is the *session layer* [EMMO]. This layer provides the appropriate service to the presentation layer above. In essence, it manages and controls the dialogue between the users of the service, the presentation entities (and, of course, the application layer above). A session connection between the users must first be set up, and parameters of the connection negotiated through one or more exchanges of control information. The dialogue between the session service users, the presentation entities (and the application entities above them), can consist of normal or expedited data exchange. It can be duplex; i.e., two-way simultaneous, with either user transmitting at will, or half-duplex, with only one user transmitting at a time. In this latter case a data token is used to transfer control of data transmission from one user to another.

The session layer provides a synchronization service to overcome any errors detected. In this service, synchronization marks can be placed in the data stream

[EMMO] Willard F. Emmons and A. S. Chandler, "OSI Session Layer: Services and Protocols," *Proc. IEEE,* vol. 71, no. 12, Dec. 1983, 1397 – 1400.

by the session service users. Should an error be detected the session connection can be reset to a defined state, and the users can move back to a designated point in the dialogue flow, discard part of the data transfer, and then restart at that point. The session layer also provides an activity management function, if desired. Using this function the dialogue can be broken into activity subsets, each separately identified if desired. A dialogue can then be interrupted and continued at any time, starting at the next activity or section of data transfer.

In summary, then, the session layer enables its users to conduct an orderly dialogue, repeating sections deemed to be in error and allowing users to interrupt the dialogue and continue it at any later time, passing control of the dialogue back and forth between the two entities if desired.

Moving down to the *transport layer* as indicated in Fig. 3 – 1(b), it is apparent that this layer provides services to the session layer. Recall, as pointed out both in Chapter 1 and in the early pages of this chapter, that communication networks come in many types and forms. Although we are by implication focusing on packet-switched networks in this chapter and Chapters 4 – 9 following, the network(s) underlying the communication facility could very well be circuit switched. The network could be of the local-area type, covering a relatively small geographic distance; it could be a metropolitan-area network, a wide-area, geographically dispersed network, or a combination of all these types. Relatively slow satellite links could be used, or noisy terrestrial links might be included. It is the purpose of the transport layer to shield the session layer from the vagaries of the underlying network mechanisms. The service it provides to the session-layer entities above is that of a reliable, transparent data-transfer mechanism. This is expressed in terms of quality-of-service requirements [KNIG].

Quality of service is defined in terms of throughput (number of transport-layer data unit octets transmitted per unit time), transit delay, residual error rate (ratio of improper or lost data units to total number transmitted), and transfer failure probability, all during data transmission. In addition, time delay during the establishment of a transport connection, the connection-release delay, and establishment/release failure probabilities can be specified as part of the quality-of-service requirements. The requirements specified are to be met at minimum cost.

Two types of data transmission at the transport layer can be distinguished: connection-oriented and connectionless transmission. These are similar to the concepts of virtual circuit and datagram models of operation, respectively, at the network layer (described in detail in Chapters 5 and 6). The current transport-layer protocol focuses on the connection mode of transmission. Work is

[KNIG] K. G. Knightson, "The Transport Layer Standardization," *Proc. IEEE,* vol. 71, no. 12, Dec. 1983, 1394–1396.

going on in the international standards community to develop an international standard for connectionless transmission. In the former case, as the name indicates, a transport connection must first be established before data transfer can begin. The establishment-phase quality-of-service parameters noted in the last paragraph implicitly assume the existence of this phase, as well as a disconnect phase following completion of data transmission. The connectionless mode allows the immediate transmission of transport-layer data units with no prior transport-layer setup or connection needed. The need for connectionless data transmission arose out of developments in the local area network field, with very high data rates and correspondingly rapid transfer of data possible between two users. The connection phase slows the transfer rate down unnecessarily in these cases [CHAP].

Five classes of protocols have been defined as part of the transport-layer standard [KNIG], [STUD]. These classes reflect different applications and different network connections. Network connections range from those with acceptable error rate and an acceptable rate of network errors, to those with unacceptable error rates. The various classes provide increasingly better quality of service, countering possible service degradation introduced by the underlying network(s). This involves means for carrying out error detection and recovery; segmenting and reassembly of data; possible multiplexing of several transport connections within a single network connection or, conversely, splitting of traffic over more than one network connection to increase throughput; flow control to reduce congestion and improve response times; and so forth. Details of some of these procedures appear in Chapter 7, which is devoted to the transport layer.

The four layers just described briefly, from the application layer down to the transport layer, constitute the higher or upper layers of the OSI hierarchy, as shown in Fig. 3 – 1. Protocols corresponding to these layers reside in the hosts, terminals, display systems, and other intelligent stations that might be involved in the end-to-end data-communication process. Protocols for the lower layers reside in the nodes of the network(s) and are actively involved in the relaying or routing of messages from a network origination (source) point to a destination. The network layer and the data link layer are described in detail in Chapters 4 – 6. Chapter 9, on local area networks, picks up again with descriptions of other data link and physical-layer standards. We thus devote relatively little space to the lower layers here. In addition, because a variety of networks have developed, each with its own characteristics, specific *network* standards have not

[CHAP] A. Lyman Chapin, "Connections and Connectionless Data Transmission," *Proc. IEEE*, vol. 71, no. 12, Dec. 1983, 1365–1371.

[STUD] P. von Studnitz, "Transport Protocols: Their Performance and Status in International Standardization," *Computer Networks*, vol. 7, 1983, 27–35.

been developed. Thus, as already noted, networks can be of the packet- or circuit-switched type. Within the packet-switched category, either connection-oriented (virtual circuit) service or connectionless (datagram) service is possible.

The X.25 network-interface recommendation of CCITT, which has fast become a de facto worldwide standard, plays the role of a lower-three-layer architecture, covering the lowest three layers of the OSI model. We shall discuss this architecture briefly later in this chapter and then in detail in Chapter 5. We must emphasize, however, that X.25 is not a *network* architecture, but a recommended *interface* to networks.

An *internet protocol* (IP) standard has been developed by ISO as the uppermost part of the network layer [CALL]. As its name indicates, this protocol allows users located on different networks to communicate. These networks could be local area networks of various types, X.25 packet-switched networks, circuit-switched networks, satellite networks, and so on.

In any case, whether the X.25 recommendation, the IP standard, or some other network layer protocol is used, the purpose is again the same: The network layer provides the actual communication service for the transport layer above it [WARE]. The transport layer is thus shielded from the details of the communication networks actually used for data transmission and can provide the necessary transparent pipe to the end users. (Although the transport layer is shielded by the network layer below it, the quality of service it can provide to the users obviously is critically dependent on the actual network(s) used. There is no way in which the transport layer can overcome delays introduced by a relatively high-delay transport network, for example.) The network layer will generally multiplex many users, will provide some form of error control, and will, under the connection-oriented type of service, guarantee sequential delivery of data packets.

Under connection-oriented network service, three phases of transmission again exist, just as in the layers above. First there is an establishment or call setup phase, during which the service users (the transport-layer entities) agree on the service characteristics to be provided. Data transfer then follows, followed in turn by the release or disconnect phase. Quality-of-service parameters, similar to those introduced during our discussion of the transport layer, may be introduced as well [WARE]. These include, as previously, such parameters as connection establishment/release delay, throughput, transit delay, and residual rate. A common way of controlling throughput and transit delay over a network or cascade of networks, is by invoking some form of flow or congestion control.

[CALL] Ross Callon, "Internetwork Protocol," *Proc. IEEE*, vol. 71, no. 12, Dec. 1983, 1388–1393.

[WARE] Christine Ware, "The OSI Network Layer: Standards to Cope with the Real World," *Proc. IEEE*, vol. 71, no. 12, Dec. 1983, 1384–1387.

A number of methods of providing this control are discussed in detail in Chapter 5, with emphasis on the impact of these methods on throughput and transit delay.

Below the network layer there appears, of course, the *data link layer.* As its name indicates, the purpose of this layer is to provide a link-level service for the network layer. The service guarantees error-free, sequential transmission of data units over a link in a network. To accomplish this, the data units carry synchronization, sequence number, and error-detection fields, in addition to other control fields, plus data. (The data carried is generally the packet delivered end to end in a network.) A detailed discussion appears in Chapter 4, with additional material appearing in Chapter 9.

Two principal applications and their corresponding protocols can be identified in the "real world" of networks. In the case of wide-area or geographically distributed networks, such as shown in Fig. 3–1(a), the link protocol is carried out on each link of a path from end to end (see also Chapter 1). The network layer above (Fig. 3–1b) can thus ignore the link and can assume that a packet introduced at one end of a link arrives correctly and in sequence at the other end. In the case of local area networks (LANs), however, there is generally only one link, the network connecting the end users. This is shown in Fig. 3–4, using a bus-type LAN as an example. For example, Stations A and D, if communicating, generally will have some provision for detecting and correcting errors. This is considered to be at the data link level. The network layer, if no relaying of messages were involved, would be a null one.

Standards have been developed for both cases. In the case of wide-area networks, the ISO-developed HDLC (high-level data link control), described in

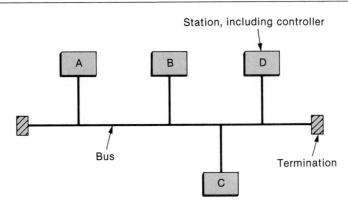

Figure 3–4 Bus-type local area network (LAN)

Chapter 4, is becoming the international data link layer standard. Connection-establishment, data transfer, and disconnect phases are defined in order to ensure that a link is up and running properly. In the data transfer phase, data unit sequence numbering and a defined error-control procedure are used to provide the desired service to the network layer above.

Three standards have been developed by an IEEE standards committee, IEEE Committee 802 for Local Area Network (LAN) Standards. These standards provide data link and physical-layer protocols for three different ways of communicating over short distances (several km or less). The bus topology of Fig. 3–4 is used for a contention-type protocol called Ethernet; both the bus topology and a ring topology are used with controlled protocols involving the passing of tokens (permits to transmit). Details appear in Chapter 9.

3–2 Unified View of OSI Protocols

A discerning reader will have noted that many of the terms used in our description of the various OSI layers appear to duplicate themselves layer by layer. Thus a typical layer provides a set of services to the entity of the layer above. The layer above is called the *service user;* the one below, the *service provider.* In the connection-oriented case, such services include provisions for connection establishment and connection release in addition to data transfer. These three sequential phases of operation of the protocols at a given layer are shown schematically in Fig. 3–5. Here two systems, A and B, have been identified as desiring to communicate. The peer entities at a given layer of the OSI architecture must first establish a connection, then transfer the data as required, and finally close down or release the connection.

In the connection-establishment phase the two peer entities at either end of the connection negotiate a set of parameters to be used during data transfer. During the data-transfer phase, errors must be detected and some form of error control exerted. This constitutes another set of services provided by the service provider. Other services may also be provided during data transfer. The data blocks or units generally have to be numbered in sequence. Blocks may be segmented or, conversely, concatenated to increase their size. Flow control may be exerted to keep a transmitting entity at one end of the connection from overwhelming the receiving entity at the other end. Multiplexing or splitting may be used to improve throughput-time delay performance characteristics (quality of service). Synchronization may be used. Full-duplex or half-duplex transmission may take place. In the latter case, a token (permission to transmit) must be passed from one entity to the other. Data may be transmitted normally or in expedited form.

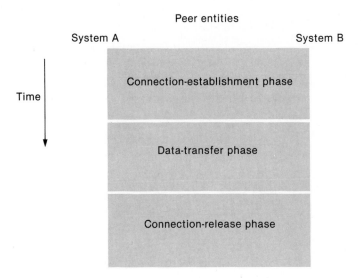

Figure 3 – 5 Three phases of communication at a layer, connection-oriented transmission

The *specific* requirements and services differ from layer to layer, but the concepts are similar. Thus errors at the data link layer generally refer to bit errors or errors in sequence numbering; at a higher level they may refer to a network failure or disconnect. This leads to a unifying concept, that of layer N in the architecture providing a service for layer $N + 1$ above it [DAY], [LINI]. This is shown schematically in Fig. 3 – 6. This figure shows a number of peer entities at each layer. They may exist in the same system, in which case they are not visible from the outside and do not come under the OSI concept. If they do correspond to entities in different systems, peer protocols govern their cooperation. The dashed lines with arrows connecting two entities at the same (N)-layer in Fig. 3 – 6 provide an example of entities in different systems communicating via peer protocols: the presentation-layer protocol at the presentation layer, the transport-layer protocol at the transport layer, and so on.

N-entities in the N-layer provide service to the $(N + 1)$-entities in the $(N + 1)$-layer above. The N-entity is thus the service provider to the $(N + 1)$-

[DAY] John D. Day and Hubert Zimmermann, "The OSI Reference Model," *Proc. IEEE,* vol. 71, no. 12, Dec. 1983, 1334 – 1341.

[LINI] Peter F. Linington, "Fundamentals of the Layer Service Definitions and Protocol Specifications," *Proc. IEEE,* vol. 71, no. 12, Dec. 1983, 1341 – 1345.

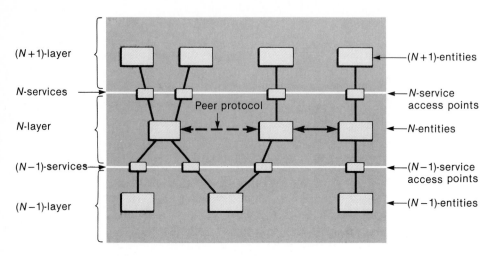

Figure 3–6 $(N + 1)$-entities and N-services

entity, the service user. Examples of such service already mentioned in describing the individual OSI layers include error control, flow control, multiplexing, and splitting. Figure 3–6 shows two entities in layer $(N + 1)$ sharing the services of a single entity in layer N. This would correspond to a multiplexed connection. The same N-entity in layer N is shown splitting its connection among two entities in layer $(N - 1)$. The services of layer N to layer $(N + 1)$ are provided at the interface between the layer, at the points called N-service access points.

In the case of connection-oriented communication, the service access points must set up connections between entities at the same layer. Thus as one of its services, the N-layer establishes connections between N-service access points. Connections can be one to one, one to many (broadcast communications), or many to one.

The data transferred between peer entities contains both user data, passed on from the layer above, and protocol-control information added at the layer in question (see Fig. 3–7). PDU stands for protocol data unit, the block of data containing both protocol-control information (PCI) added at the layer in question, as shown, and user data, originating at the layer above. The $(N + 1)$-PDU, in crossing the interface between the $(N + 1)$- and N-layers, is mapped onto the N-service data unit, N-SDU, as shown. This mapping may be one to one as shown, or N-SDUs, if too long, may be segmented to form multiple N-PDUs after the appropriate protocol-control information, N-PCI, is added. Concatenation of data units to form longer data units is also possible.

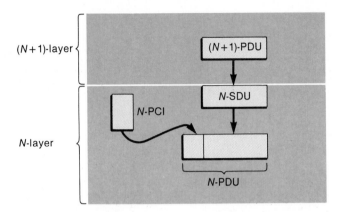

Figure 3–7 Data units in OSI architecture

The length of the N-PDU transmitted between peer N-entities would be negotiated during the connection phase. An N-PDU arriving at its destination, the peer N-entity in the N-layer at the destination, would have the control header, the N-PCI, stripped off before being sent up to the (N + 1)-layer at the destination. Segmentation or concatenation would be reversed as well.

In addition to the N-protocol data units exchanged between peer entities at the N-layer, N-interface-control information is exchanged between an (N + 1)-entity and an N-entity in the layer below in order to coordinate their operation. Flow control of data units can be exerted between peer entities in the same layer, as well as between (N + 1)- and N-entities across the interface at the service access point.

Following the concept of the N-entities providing a service to the users, the (N + 1)-entities, OSI has standardized on the use of four basic service primitives at each level in the architecture to provide the interaction between the service user at one level and the service provider at the level below. These service primitives provide the basic elements for defining an exchange between service users; the four types are *request, indication, response,* and *confirm* [LINI, p. 1343]. Not all four types may necessarily be used at the different layers; not all may be used in different phases of an exchange between peer entities at a given layer. The four primitives are represented schematically in Fig. 3–8. Some or all of these primitives may be used in each of the three phases of the communication process. Examples will be provided later.

Note from Fig. 3–8 that the *request* is issued by a service user at a given (N + 1)-layer in system A to invoke a procedure of the service provider protocol

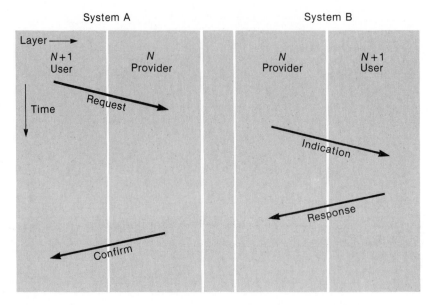

Figure 3–8 Four basic primitives, OSI architecture

at layer N below. This results in an N-layer message, sent as an N-PDU (or set of N-PDUs) to system B. The receipt of the N-PDU at layer N of system B sometime later causes an *indication* primitive to be issued by the service provider at that layer. This indication primitive is issued in either of two cases: It may indicate that the service user at B is to invoke a procedure of the protocol at its $(N + 1)$ level or that the service provider at layer N has invoked a procedure at the peer service access point. In summary, the indication primitive at system B is issued, in this example, in response to the request primitive issued earlier in system A.

The *response* primitive, as shown in Fig. 3–8, is issued by the service provider at layer $(N + 1)$ in system B in response to the indication appearing earlier at the service access point between layers N and $(N + 1)$ of system B. This response primitive is a directive to the protocol of layer N to complete the procedure previously invoked by the indication primitive. The protocol at layer N generates a protocol data unit that is transmitted across the network, appears at layer N in system A, and results in a *confirm* primitive being issued by the service provider at system A. This then results in the completion, at the service access point between N and $(N + 1)$ in system A, of the procedure previously invoked by the request at that point [LINI, p. 1343].

To summarize, a request by the service user at layer $(N + 1)$ of system A results sometime later in an indication primitive directed to the peer service user at system B. This user replies with a response primitive that eventually reaches the initiating service user at system A as a confirm primitive.

As a concrete example of the applicability of these primitives, consider the data link layer, the second layer of the OSI architecture [CONA]. Its purpose is to provide services to the network layer. These services are grouped into the three phases described earlier and portrayed in Fig. 3–5: the establishment, data transfer, and release phases. In the notation of the network and data link layers, establishment services are used to establish a logical connection between two network entities at either end of a link. Data transfer services ensure that network data units (the *packets* to which we have made reference previously) arrive correctly and in sequence on the other side of the link, once it is established. Release or termination services handle the release of the logical connection after data transfer has been completed. For simplicity we focus on the use of the primitives in these phases only for the HDLC protocol. This protocol, discussed in detail in the next chapter, has been established as an international standard by ISO for network point-to-point applications, among others. The X.25 protocol incorporates a subclass of this protocol as its data link protocol as well.

Three different categories of the basic four primitives are distinguished and are so labeled, corresponding to each of the three phases of communication at this layer. They appear in a typical exchange across the link in Fig. 3–9 [CONA, Fig. 7]. Note the use of the terms CONNECT. request, DATA. request, and DISCONNECT. request, with corresponding modifiers used on each of the three other primitives. Similar modifiers appear at each of the layers in the OSI architecture in which these primitives are used.

Figure 3–9 assumes that the data link in question is currently not in use by the network for the transmission of data. The logical connection must thus first be established before data can be transferred. The network layer entity at one end of the link, in this case system A, initiates the establishment of the connection by issuing a CONNECT. request primitive to its data link layer, the service provider in this case. As a result of this request the link layer protocol sets in motion the establishment of the logical connection.

HDLC incorporates a number of modes of operation and corresponding classes of procedure. (See Chapter 4; [CARLD] provides more detail.) The particular mode desired, in this example an asynchronous balanced mode appropriate to data link peers at either end of a point-to-point link, is conveyed in

---

[CONA] J. W. Conard, "Services and Protocols of the Data Link Layer," *Proc. IEEE*, vol. 71, no. 12, Dec. 1983, 1378–1383.

[CARLD] D. E. Carlson, "Bit-oriented Data Link Control Procedures," *IEEE Trans. on Comm.*, vol. COM-28, no. 4, April 1980, 455–467; reprinted in [GREE].

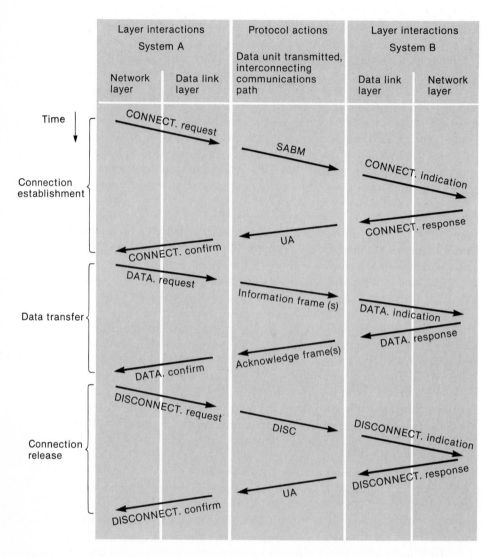

Figure 3 – 9   Use of primitives, HDLC protocol (from [CONA, Fig. 7], © 1983
IEEE, with permission)

the CONNECT. request primitive. Connect primitives generally carry mode-operation parameters (parameters indicating with whom connection is to be made, quality of service required, and so forth). In the case of HDLC, sequence numbers are used to number the data link data units during the data transfer phase. There is a normal modulo 8 sequence-number mode, and an extended sequence-number mode of module 128 (see Chapter 4). The numbering mode desired would appear as a parameter setting in the CONNECT. request primitive as well.

After receiving and then processing the CONNECT. request primitive, the data link protocol in system A causes a data link layer protocol data unit to be issued for transmission to station B at the other end of the link. HDLC incorporates a number of control-data units within its repertoire of operation. One of them, called the Set Asynchronous Balanced Mode (SABM) data unit, applies to the mode of operation here, and is shown being transmitted to station B. Other control-data units are used to establish the other modes of operation of HDLC [CARLD].

Data link layer data units for the HDLC protocol are given a special name: *frames*. The format of a typical HDLC frame appears in Fig. 3-10. The beginning and end of the frame are denoted by a special eight-bit synchronizing character *F*, 01111110. *A* is an address field, *C* a control field, and *FCS* a field used for error detection. (See Chapter 4 for details.) The *I*, or information, field carries data (the packet) transferred down from the network layer. Control frames such as SABM do not carry an *I*-field. Each is separately designated by appropriate setting of the bits in the *C*-field (Chapter 4).

The SABM frame, on reaching the system B side of the link, causes the data link entity there to initialize its variables and issue a CONNECT. indication primitive. The network-layer entity, on agreeing to the communication request, replies with a CONNECT. response primitive. This causes the data-link protocol at system B to transmit an unnumbered acknowledgement (UA) frame back to the originating data link entity at station A. This in turn issues a CONNECT. confirm primitive, indicating completion of the connection. The network layer, *at either end,* may now begin transmission of data. Figure 3-9 shows station A

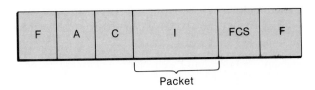

Figure 3-10   HDLC frame format

beginning the data-transfer phase with issuance of a DATA. request primitive. One or more information frames may now be transmitted to station B. These contain, as already noted, *I*-fields (Fig. 3 – 10), the data units or *packets* handed down from the network layer above. These in turn will generally carry the actual user data from the application layer plus control information passed down from the layers between (Fig. 3 – 7). The *I*-field is thus variable in length. The specific format of the HDLC-frame (Fig. 3 – 10) enables a receiving data-link entity to locate the *I*-field precisely and to strip off all but that field from a received I-frame on ascertaining that it has been received error free. The *I*-field is then sent up, untouched, to the receiving network-layer entity above.

HDLC requires a positive acknowledgement of I-frames received correctly to be sent back to the sending data link entity. (Provision is also made for a negative acknowledgement if a frame is received out of sequence or with bits detected to be in error — see Chapter 4.) The DATA. indication and DATA. response primitives shown in Fig. 3 – 9 are used to denote correct receipt of an I-frame. A positive acknowledgement arriving at the sending side causes the data link entity there to issue a DATA. confirm primitive. This process continues so long as the two systems desire to continue in the data-transfer phase. If either party wishes to release the connection, a DISCONNECT. request primitive must be issued. The resultant release phase, as portrayed in Fig. 3 – 9, is self-explanatory. The DISC frame shown being sent is another HDLC control frame used to convey the disconnect request. The DISCONNECT. response primitive shown being issued in Fig. 3 – 9 does not always have to be used in data link protocols. In some protocols the *local* (i.e., disconnect-initiating) data link entity issues the DISCONNECT. confirm in response to a DISCONNECT. request. As shown, it also initiates transmission of a control message to the other system, notifying that system that the connection has come to an end.

Although not explicitly stated, it is assumed in the previous discussion that successful physical-path connection has been made before initiating the logical link connection phase. This connection service is provided by the physical layer below. The four basic primitives at that layer are used in a similar manner to establish the physical-layer connection [McCL]. Details of physical-layer standard protocols appear in [BERT].

The transport and session layers of the OSI architecture provide two other examples of the use of the four basic primitives in the three phases of the information-transmission process. Consider the transport layer first. Recall from the brief discussion earlier that five classes are defined for the OSI proto-

[McCL] F. McClelland, "Services and Protocols of the Physical Layer," *Proc. IEEE,* vol. 71, no. 12, Dec. 1983, 1372 – 1377.

[BERT] H. V. Bertine, "Physical Level Protocols," *IEEE Trans. on Comm.,* vol. COM-28, no. 4, April 1980, 433 – 444; reprinted in [GREE].

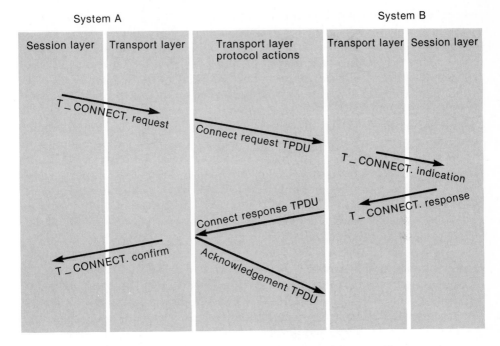

Figure 3-11 Connection establishment, transport layer

cols at this layer. All five use the four basic primitives in their connection-establishment phase. The sequence of primitives involved is shown in Fig. 3-11. Note how this sequence follows precisely the form of Fig. 3-8. The same use of the corresponding primitives at the data link layer was shown in Fig. 3-9. The connection-establishment phase is used to establish the class to be used, as well as to negotiate options within classes, and quality of service desired. The choice of class is made by the transport entities based on the user requirements incorporated in the T_CONNECT. request and response primitives, plus the quality of the underlying network service [KNIG]. The T_CONNECT. request primitive, from the session layer above (the service user), carries the called transport address, the calling transport address, option and quality-of-service parameters, plus, for some classes, transport-service data. (See Chapter 7 and [STUD] for more details.) The T_CONNECT. indication carries similar information. The T_CONNECT. response and T_CONNECT. confirm primitives carry the responding address, options, quality-of-service parameters, and, again for some classes, transport-service user data.

Communication between the transport-layer entities is carried out by transport protocol data units, labeled TPDU for short. Note that this is consistent with the notation introduced earlier, in Fig. 3 – 7, for the data units at a given layer in the OSI architecture. In the case of the connection-establishment phase of Fig. 3 – 11, the TPDUs are *control* TPDUs that carry the identities of the calling and called session entities as well as parameters necessary for negotiation of the class and options. During the data-transfer phase *data* TPDUs are transmitted. The three TPDUs shown being transmitted in Fig. 3 – 11 are the connection-request and connection-response TPDUs, each corresponding, of course, to the respective primitive, plus an acknowledgement TPDU. This *three-way exchange* ("hand shake") is prescribed for the class 4 transport protocol. The other classes require only a two-way exchange (no acknowledgement TPDU).

The release or termination phase of the transport layer only uses two of the four primitives: T_DISCONNECT. request and T_DISCONNECT. indication. The former, as always, is issued by the service user, the session layer; the latter by the service provider, the transport layer. The T_DISCONNECT. request primitive indicates to the local transport layer that it is to terminate the transport connection. The T_DISCONNECT. indication primitive notifies its service user that the transport connection has been terminated. No replies are expected. This is thus called an *abrupt* termination service. The disconnect service can be used to refuse to establish a transport connection or to abruptly terminate a connection already established. In this case any data in the process of being transferred on the connection is deemed lost, with the service user then responsible for taking appropriate corrective action.

Rejection of the request to establish a transport connection is portrayed in Fig. 3 – 12. In part (a) of the figure, the called transport user rejects the connection. In part (b), the local service provider (the local transport layer) rejects the connection. As part of its parameters, the T_DISCONNECT. indication primitive carries with it the reason for the rejection. Transport service-user data may also be carried by the two T_DISCONNECT. primitives, with no guarantee of delivery.

Figure 3 – 13 provides two examples of the disconnect phase once a transport connection has been completed and the data transfer phase has been initiated. Abbreviated notation has been used to simplify the figure. A number of possibilities exist: The release of the connection may be initiated by a transport service user, as shown in Fig. 3 – 13(a); it may be initiated by both transport service users; by both providers, as shown in Fig. 3 – 13(b); or independently, by a service user and provider at either end. In all cases the termination is abrupt, with data already being transferred over the connection lost.

Of the five transport-protocol classes, one, called class 0 or the simple class, has no provision for disconnection. The highest, class 4, has as an option a

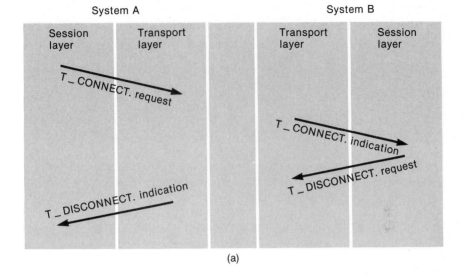

(a)

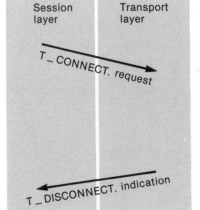

(b)

Figure 3–12 Rejection of transport connection-establishment request
a. Rejection by called transport user
b. Rejection by local transport layer (service provider)

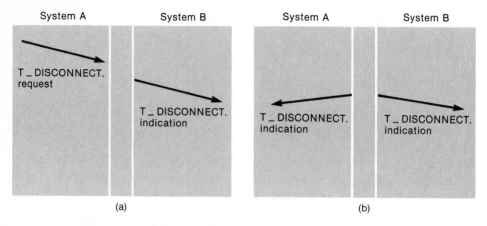

Figure 3–13 Disconnect mechanisms, transport layer
a. Release initiated by a service user
b. Release initiated by both service providers

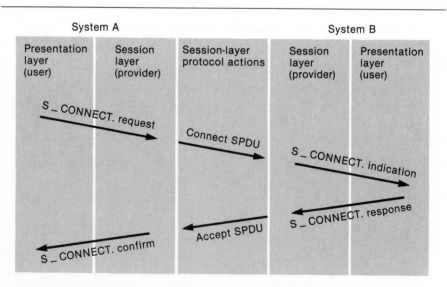

Figure 3–14 Connection establishment, OSI session layer (after [EMMO, p. 1399], © 1983 IEEE, with permission)

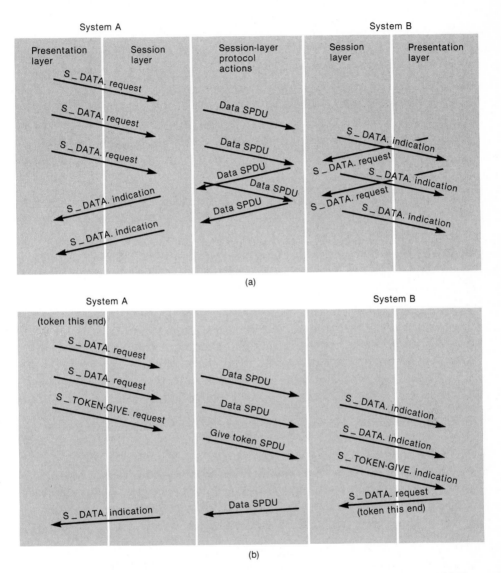

Figure 3–15 Data transfer phase, session layer (after [EMMO, p. 1399], © 1983 IEEE, with permission)
a. Duplex
b. Half-duplex

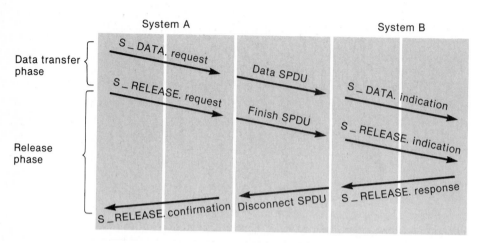

Figure 3 – 16 Orderly release of session connection (after [EMMO, p. 1399], © 1983 IEEE, with permission)

"graceful close" feature. This service, which requires the use of two new primitives, T _ CLOSE. request and T _ CLOSE. indication, allows a transport connection to be terminated without loss of data [NBS 1983c].

The data-transfer phase of the transport layer protocol uses the request and indication primitives only. The primitives come in two flavors: T _ DATA. for normal data service and T _ EXPEDITED _ DATA. for the delivery of expedited data. Some discussion appears in Chapter 7, with details provided in [NBS 1983c]. The expedited data feature allows a limited amount of data to bypass the end-to-end flow control exerted on normal data.

The OSI session layer uses the concept of four basic primitives in each of its three phases as well, in keeping with the general philosophy of a unified approach to the protocol layers. The primitives are also used to accommodate the token-management, session-synchronization, and activity-management services described earlier in this chapter, as well as other services provided by this layer [EMMO]. In the interest of simplicity we consider only the three phases of operation. Because these phases are very similar to those already described at the data-link and transport layers, we minimize the verbiage as well. Fig. 3 – 14 shows the actions involved in establishing a session connection. Fig. 3 – 15, on

[NBS 1983c] *Specification of a Transport Protocol for Computer Communications, vol. 3: Class 4 Protocol*, Inst. for Computer Sciences and Technology, National Bureau of Standards, Gaithersburg, Md., Feb. 1983.

the data-transfer phase, is somewhat different from the previous figures illustrating the data transfer phase, because of the two possibilities of half-duplex (transfer of token) and full-duplex data transfer. Note that the half-duplex case requires the use of the primitives S_TOKEN_GIVE. request and indication, with no confirmation made, to allow transfer of control from one system to another. In keeping with the notation introduced earlier, session-layer protocol data units passed between community entities at this layer are labeled SPDUs. These can again be of two types: data or control. Finally, Fig. 3–16 shows the sequence of primitives issued and SPDUs transmitted during an orderly—i.e., confirmed—release of a session connection. These figures are based on those appearing in [EMMO].

3–3 X.25 Protocol

The X.25 network-interface recommendation of the CCITT was approved by that body in 1976. A revised version appeared in 1980, and a second revision in 1984. The X.25 recommendation covers the connection of data terminals, computers, and other user systems or devices to packet-switched networks. The user systems are called generically *data-terminal equipment* (DTE). Their connection to the network is made via network equipment termed *data circuit-terminating equipment* (DCE). A DTE normally wishes to establish communication with another DTE (another user system) and uses the network for this purpose. That DTE will in turn be connected to a DCE that controls its access to the network. The network is responsible for managing communications between DCEs. X.25 regulates data flow between DTE and DCE at each end of the network only. Figure 3–17 demonstrates this concept pictorially.

X.25 is thus an *interface* specification only. It governs the interactions between a DTE and the DCE to which it is connected. Details of the communication between DCEs, using the network connecting them, are left to the network owner or operator. These details are hidden from the DTEs, the users of the packet-switched service.

X.25 is organized as a three-layer architecture, corresponding to the lowest three (network) layers of the OSI model. Figure 3–18 portrays the three layers of X.25. Figure 3–19 shows the relationship, layer by layer, to the OSI architecture. We introduce X.25 at this point because it provides a useful example of a layered architecture that has been in use for a number of years, predating the ISO announcement of the OSI model; for more detailed discussion see Chapter 5. Although X.25 predates the OSI architecture, cooperative efforts between the CCITT and the ISO have resulted in a conformance between X.25 and the lowest three layers of the OSI architecture, as shown schematically in Fig. 3–19.

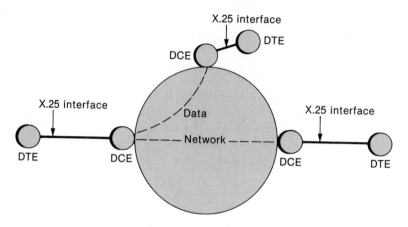

Figure 3–17 X.25 concept

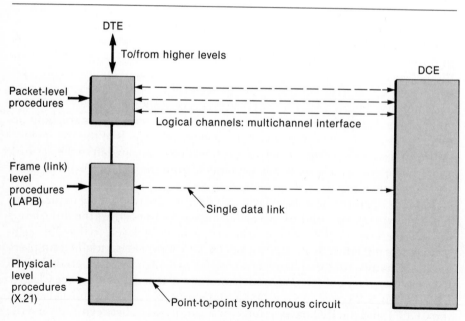

Figure 3–18 X.25 layers

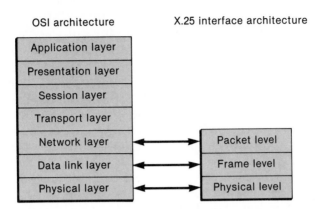

Figure 3-19 Relation between OSI and X.25 architectures

Returning to Fig. 3–17, note that in this example one DTE is shown communicating with two different geographically dispersed DTEs. The X.25 specifications originally included a datagram mode of operation in addition to virtual circuit service. In the 1984 revision, datagram service was dropped because of a lack of interest on the part of implementers. The stress here is thus on virtual circuit service. There are a number of datagram packet-switched networks that *internally* handle packets as datagrams, but *externally*, at the X.25 interfaces to the user DTEs, convert the operation to a virtual circuit one. (Examples appear in Chapter 5.)

Now focus on the X.25 layers shown outlined in Fig. 3–18. As in the OSI architecture of Fig. 3–1(b), the lowest physical layer ensures that a valid physical connection exists between the DTE and the DCE. CCITT protocol recommendation X.21 [BERT] is used for this purpose. It was noted earlier in this chapter that the link-level protocol at the data link layer is a subset of HDLC, labeled LAPB (balanced link access procedures). The data units or frames traversing the link in either direction, from DTE to DCE, have precisely the format of Fig. 3–10, described earlier. (Details appear in Chapter 4.)

It is at the third, network layer that X.25, as an interface architecture, is actually distinguished. This layer is called the *packet level* in X.25 terminology. As noted, X.25 focuses on virtual circuit (VC) connections. Logical channel numbers are assigned for this purpose on a particular X.25 connection (Fig. 3–18, packet level). Up to 4095 such connections are available between any DTE and the DCE to which it interfaces. A 12-bit address field is used for this purpose. This implies that up to the same number of VC calls may be going on simultaneously between a given DTE and all of the other DTEs in the network.

Data packets from a DTE each carry their own 12-bit logical channel number once a call is set up. All are multiplexed on to or share the same data link and use the same service provider, the data link layer below.

Each DTE-DCE interface assigns its own set of logical channel numbers. Hence a complete VC, end to end between two DTEs communicating with one another, may use different logical channel numbers at the two interfaces at each end of the virtual circuit. (Two such VCs are sketched in Fig. 3 – 17.)

As was the case with the various layers in the OSI architecture, three phases of communication are required for VC operation. These are, in order, the call-establishment or setup phase, the data-transfer phase, and the disconnect or call-clearing phase. (Permanent VCs may be used, in which case call setup is not needed.) The call-establishment phase is diagrammed in Fig. 3 – 20. The arrows indicate the various control packets (network-layer protocol data units) transmitted between the DTE and the DCE to which it is connected. These are all peer-level data units, transmitted between the packet-level protocols in each of the two systems (DTE and DCE). They thus correspond to the "protocol action" portion of the corresponding figures for the OSI data link (Fig. 3 – 9), the transport layer (Fig. 3 – 11), and the session layer (Fig. 3 – 14). The packets of Fig. 3 – 20 should *not* be confused with the primitives shown being issued in Figs. 3 – 9, 3 – 11, and 3 – 14. Recall that primitives cross the interface between adjacent layers in a given system and hence are not distinguishable to the outside. The packets of Fig. 3 – 20 are protocol data units *at the same network layer*, are transported between two distinct systems, and hence are distinguishable to the outside.

These call-setup control packets correspond to the SABM and UA PDUs of the data link layer of Fig. 3 – 9, to the connect-request TPDU and connect-response TPDU of the transport-layer actions of Fig. 3 – 11, and to the connect/accept SPDUs of the session-layer protocol actions of Fig. 3 – 14.

The X.25 call-request packet of Fig. 3 – 20 notifies the DCE to which it is directed that the DTE issuing it wishes to establish communications with another DTE connected to the network. The packet carries a logical channel number to be used for the DTE-DCE logical connections, calling and called DTE addresses, parameters specifying the characteristics of the call, plus up to 16 octets of data. This information is transmitted through the network using the network's own procedures and reaches the destination DCE. That system then transmits the same type of packet, carrying the same information, except for a possibly different logical channel number, to the destination (called) DTE. Since this packet is issued by a DCE, its name is changed to incoming call packet. The DTE replies with a call-accepted packet, which, on being forwarded through the network, appears as a call-connected packet issued by the calling DCE as a response to the calling DTEs original call request. Data transfer, using data packets (network-layer data PDUs) can now begin. The call-clearing procedure is similar, as shown in Fig. 3 – 20. The one difference is that a clear request

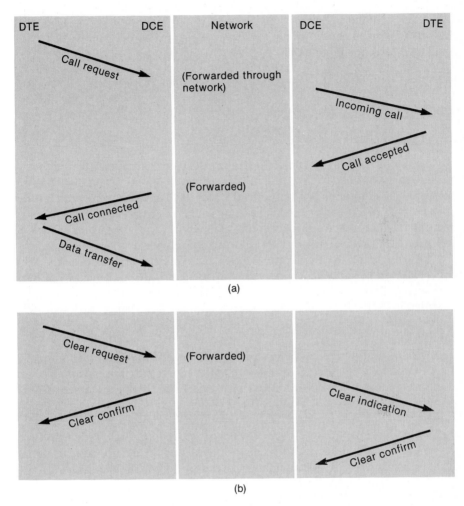

Figure 3–20 X.25 procedures and packets for setting up and clearing a virtual
circuit
a. Call-setup sequence
b. Call-clearing sequence

packet is *immediately* acknowledged by the DCE receiving it, without waiting for
a confirmation from the other DTE.

Like the OSI layers described previously, the X.25 protocol has provisions
for error control and flow control at the packet (network) level, for expedited
data transmission, and for other operations similar to those incorporated in the

OSI layer protocols. Further discussion of X.25 appears in Chapter 5, with emphasis given to the flow (congestion) control mechanism used. Packet formats used appear there as well.

3–4 Systems Network Architecture (SNA)

IBM's Systems Network Architecture, a seven-layered architecture designed to provide interconnection between IBM products, was announced in 1974 after extensive development beginning in the late 1960s. Enhancements and new releases have been announced regularly since. SNA preceded ISO developments and hence does not strictly conform, layer by layer, to features of the OSI architecture. There are similarities in function, however, as we shall see.

The discussion in this chapter is again quite brief, with details on the SNA network layer in particular appearing in Chapter 5. Good overviews of SNA appear in [SCHU] and [SNA 1982]. Detailed specifications, layer by layer, are provided in [SNA 1980] as well as in other IBM publications on various aspects of the architecture. A special issue of the *IBM Systems Journal* is devoted to SNA [IBM].

As is the case with other layered architectures, including the OSI standard, the purpose of SNA is to provide for reliable, timely communication between disparate end users, possibly located far apart. The architecture may also be visualized as being grouped into two categories: (1) a higher-layer grouping of four layers that is involved with setting up and maintaining a connection (called a *session* in SNA terminology) between end users, as well as being concerned with the syntax and semantics of the data exchanged, and (2) a lower-layer grouping of three layers that provide the network transport capability end to end.

End users in the SNA context are the same as those listed earlier for the OSI architecture; examples include terminal users, workstations, application programs, device media such as printers and graphics display devices, and memory storage devices. End users access an SNA network through access ports or

[SCHU] Gary D. Schultz, "Anatomy of SNA," *Computerworld,* vol. 15, no. 11a, March 18, 1981, 35–38.

[SNA 1982] *Systems Network Architecture, Technical Overview,* GC30-3073-0, IBM, Research Triangle Park, N.C., 1982.

[SNA 1980] *Systems Network Architecture Format and Protocol Reference Manual: Architectural Logic,* SC30-3112-2, IBM, Research Triangle Park, N.C., 1980.

[IBM] "Systems Network Architecture," Special issue, *IBM Systems J.,* vol. 22, no. 4, 1983, 295–466.

connection resource managers called *logical units* or *LUs.* The LUs at either end in turn establish the session or logical connection along which end-user data is transported. One LU can support several end users and can support sessions to multiple LUs.

To help in the management of the network, two other resource managers are defined: a *physical unit* or *PU,* which manages the communication resources at a given node (these comprise the data links and communication channels serving the node), and a *system services control point* or *SSCP,* which manages all resources within a subset of a network called a *domain.* All three "units" — the LUs, the PUs, and the SSCPs — comprise the group of *network-addressable units* (NAUs) in an SNA network. Each "unit" has a unique network address and is capable of being addressed from anywhere within the network, or from outside as well.

The PUs, together with an SSCP that oversees them, ensure that the communication links are available and ready to be used. The SSCP helps in setting up (and taking down) a session, provides control and maintenance support for its domain, maintains a directory and routing tables, communicates with the other SSCPs across the network, and so forth. It serves essentially as a centralized control for all the nodes in its domain.

An SNA network looks like any of the others mentioned in this chapter and in Chapter 1. Figure 3–21 provides a simple example. The SNA network is made up of interconnected nodes. Each node contains *one* PU, responsible for management of its links and channels. It may contain many LUs. Four kinds of nodes are defined, each designated as a different PU type (see Fig. 3–21). PU_T1, the lowest class, is generally made up of low-function terminals and controllers. The class PU_T2 consists generally of high-function terminals, distributed processors, and cluster controllers — devices controlling, in turn, terminals, display systems, and other lower-function devices. Both classes, PU_T1 and PU_T2, comprise the group of *peripheral* nodes, as indicated in Fig. 3–21. These do not participate directly in the operation of the transport (backbone) network. Instead they are always attached to the two other nodes defined next. The PU_T4 node-type is generally a communications controller. (An example is the 3705 running on NCP, the Network Control Program.) The PU_T5 is generally a host computer. Both PU_T4 and PU_T5 nodes are called *subarea nodes.* These nodes are interconnected to form the transport network. A PU_T4 or PU_T5, together with the peripheral nodes connected to it, comprise a subarea. (In practice, networks are often made up primarily of the PU_T4 communication controllers, with the PU_T5s providing network control but not being involved directly with the transport or backbone network function.) The network of Fig. 3–21 contains five subareas.

The SSCP resides in a PU_T5 or host node. One SSCP in turn controls a *domain,* made up of PU_T4, PU_T2, and PU_T1 nodes. (The PU_T2 and

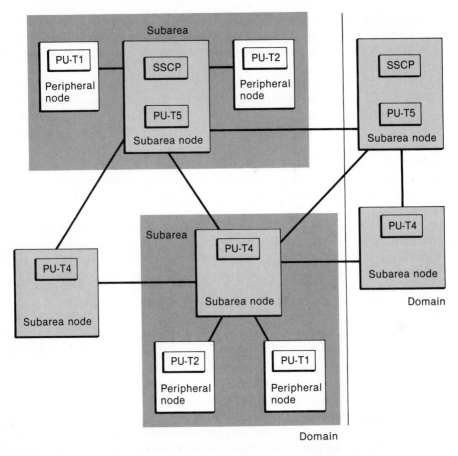

Figure 3–21 An SNA network

PU_T1 nodes are always connected to PU_T4 and PU_T5 nodes, as already noted. A PU_T4 node may reside in *two* domains.) A network is in turn made up of one or more domains; the network of Fig. 3–21 contains two. As just noted, the transport or backbone portion of this network consists of all five subarea nodes and the seven links interconnecting them. The SSCP in each domain is responsible for control of all resources in its domain.

Although a node may generally correspond to a system or device, it is possible to have more than one node (i.e., multiple PUs) in a given physical box. User implementations may provide variations on this. As an example, the 8100 distributed processing system, originally implemented as a PU_T2 node, can now contain *both* a PU_T2 and a PU_T5 in one box. This more recent implementation is used to combine two SNA networks. In one network, the 8100 serves as a peripheral node, homing on a PU＿T4 or a PU＿T5 and communicating with other systems or users in that network through the subarea node to which it is attached. In the other network, the 8100 acts as a host computer, containing the SSCP and able to participate fully as a subarea node. Communications between an end user in one network and an end user in the other network will then pass through the 8100.

Now consider a typical node. An end user desiring to communicate with another end user, at the same node or elsewhere, does so through an LU that sets up a session with an LU that the other end user accesses. One LU is always designated as a primary, the other as a secondary, with a half-session set up at each end. Communication between the two end users is carried out using the layered architecture (SNA) shown in Fig. 3–22.

Note the similarities, at least superficially, with the OSI architecture of Fig.

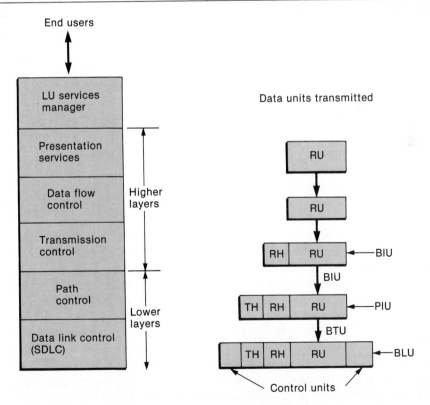

Figure 3–22 SNA and its data units

3–1(b). The LU services manager, to which an end user connects, participates in setting up and taking down sessions (the PU and SSCP are involved as well, as will become apparent shortly), and manages application services. A part of its function is thus comparable to that of the application layer of OSI. The LU service manager may in general manage multiple LUs. Each LU and the half-session established by it are associated with the next three layers down in the architecture: *presentation services, data-flow control,* and *transmission control.*

The presentation services layer provides data transformation, carries out encoding and compression of data, and does display formatting, all comparable to the presentation layer of the OSI model. Data-flow control assigns sequence numbers and handles the logical chaining of user messages. It provides for the correlation of requests and responses between users, controls the serial multi-plexing ("bracketing") of transactions during the period the session is active, and so forth. Transmission control handles end-to-end flow control (called *session pacing* in the IBM nomenclature), carries out data enciphering and deciphering, and does sequence-number checking. The data-flow control and transmission control layers of SNA together carry out some of the functions, and provide some of the services, of the session and transport protocol layers of the OSI architecture. Note, however, that there is no one-to-one mapping of functions between layers of the two architectures.

Path control in Fig. 3–22 is comparable to the network layer of OSI. It provides the routing and congestion (flow) control function associated with the network layer. There is only one path control at a node, with half-sessions of all the LUs at the node multiplexed onto it. (Figure 3–23, to be discussed shortly, shows the multiplexing function of path control very clearly.) A detailed discussion of SNA path control appears in Chapter 5. Path control in turn uses the services of data link control, which is responsible, as in the OSI architecture, for ensuring error-free transmission of data units over the connections between nodes. The IBM data link control is called SDLC, for *synchronous data link control.* SDLC is very similar to, but predates HDLC, whose development was based on it. Most of the control functions of SDLC, and the formats of the data units transmitted (Fig. 3–10), are identical to these of HDLC [SCHW 1977].

Data units transmitted across layers and between peer layers at the LUs at either end of the session are also shown in Fig. 3–22. The basic unit of data transmitted by an end user is called a request or response unit (RU). (As the name indicates, a response unit is a reply to a request unit.) This can be either a data unit carrying user data or a control-data unit introduced as part of the control of a session. The RU moves down through the data-flow control layer, where the sequence number is added, to transmission control, where the RH header is added. RH and RU combined form a basic information unit, or BIU for short, which is sent to path control. At this point the transmission header, TH, is added. (Figure 5–27, provides details of the format of the TH.) The

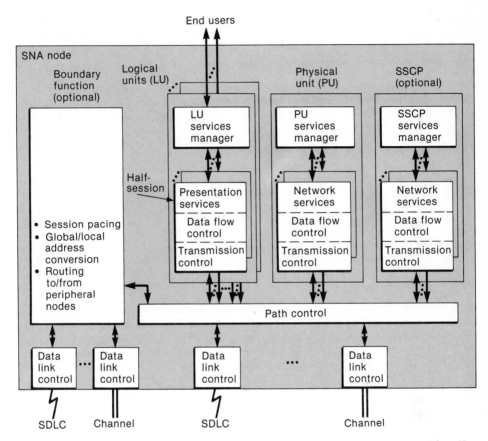

Figure 3–23 Structure of an SNA node as of 1981 (from [SCHU, Fig. 4], © 1981 CW Communications Inc., Framingham, MA 01701; from *Computerworld* with permission)

resultant path information unit, PIU, can either be passed down unchanged to the data link layer, segmented or blocked (Figs. 5–19 and 5–20). The data unit passed down to data link control, comparable to the packet of X.25 or the network data unit of OSI, is called a basic transmission unit, or BTU, in the SNA terminology. The final data unit transmitted over a link, the frame of X.25, is called the basic link unit, or BLU (Fig. 3–22).

In addition to communication between LUs, communication is required between SSCP and PU, between LU and SSCP, between PU and PU, and between SSCP and SSCP. Sessions must be set up for each of these interactions. The layered architecture applies here as well. As a result, an SSCP node, for example, will have three different kinds of session going at one time. This is demonstrated by Fig. 3–23, taken from Schultz [SCHU, Fig. 4]. (This figure shows the SNA node in 1981. Some modifications, particularly at the services manager level, have been made since.) The figure shows end users associated with LUs. It also indicates that a given end user, or LU, may have multiple half-sessions, as pointed out earlier, each comprising the three layers, presentation services, data flow control, and transmission control. Both the PU and the SSCP have services managers associated with them and may have multiple sessions going on. However, all sessions of all types are multiplexed onto the one path-control layer. It is at this point that routing decisions are made. Locally generated messages (PIUs) and those arriving from other nodes in the network on incoming data link control lines are distributed, using information carried in the transmission header. A PIU may be routed locally, up to one of the half-sessions terminating on path control, or it may be sent down to the data control driving a particular outgoing link or channel, for transmission to another node. Details are provided in Chapter 5.

Also shown in Fig. 3–23 is a boundary function that connects this subarea node with peripheral nodes attached to it. The full figure represents a PU_T5 node with PU_T1s and PU_T2s attached to it. A PU_T4 node would not have the SSCP section, labeled optional. If peripheral nodes are not attached, the boundary function section does not appear.

It was noted earlier (see Fig. 3–22) that the user data units, the RUs, are generated for the LU to LU session at the presentation services layer. Among the services provided by this layer is that of formatting data to be displayed or printed. This function is similar to that of the OSI presentation layer. To help in this and other functions, SNA has defined different types of LUs, each established to carry on a particular type of session [SNA 1982]. LU type 1, as an example, is designed to support communication between an application program and data-processing terminals. LU type 2 is used for an application program communicating with a single display terminal in an interactive mode. LU type 3 corresponds to the case of an application program communicating with a single printer. LU type 4 enables data-processing terminals, connected as peripheral nodes, to communicate. LU type 6 corresponds to program-to-program communication. Protocols and the accompanying data formats have been defined for all of these LU types [HOBE].

These LU types, with the accompanying data formats, provide the semantic relations and syntactic descriptions required to have end users communicate

[HOBE] V. L. Hoberecht, "SNA Function Management," *IEEE Trans. on Comm.*, vol. COM-28, no. 4, April 1980; reprinted in [GREE].

meaningfully and reliably. The SNA approach—grouping different kinds of application programs, systems, and devices into categories for which protocols and data formats are prescribed—thus corresponds to the OSI approach of establishing semantic relationships at the application layer and syntactic descriptions at the presentation layer. Note from Fig. 3-23 that the presentation services layer is replaced by network services in both the PU and the SSCP half-sessions. The protocols provided by these services handle network operation. They are used to control SSCP-LU, SSCP-SSCP, and SSCP-PU sessions required for the proper functioning of the physical network and the SNA constructs embedded in it. The LU has a network services component as well, not shown in Fig. 3-23. The grouping of services at this layer in SNA, whether of the presentation or network services type, of all three network-addressable units (LU, PU, SSCP), is generically called function management data (FMD) services. The layer is referred to as the FMD services layer, and RUs generated or received by this layer are thus function management data units.

The response/request units are thus of two types: RUs containing user data formatted by LU presentation services, and control RUs involved in the operation of the network services protocols. A large number of such control RUs are defined in SNA. We provide one small subset only in concluding this section. We describe briefly how an LU-LU session is set up in SNA. It is left to the reader to compare this with the setting up of an application–application connection in the OSI architecture, using the constructs described briefly earlier in this chapter. (Recall again that Chapter 7 describes the OSI transport layer services, protocols, and data units in more detail.)

There are two basic distinctions between SNA and the OSI architecture in setting up end-user connections. One is that the SSCP is actively involved in SNA; the other is that SNA requires one LU to be primary, the other secondary. There are no comparable constructs in OSI. SNA tends to be a centrally controlled architecture; OSI focuses on peer interactions. One reason for the SNA philosophy is that, from the beginning, it was geared to the interaction of subservient systems and devices with a large host system. Control was always exerted by the host, and the SSCP concept reflected this type of control. The LU types, most of which concern communication between application programs and terminals or devices, thus also reflect this orientation toward centralized control or a primary-secondary relationship.

Figure 3-24, taken from [SNA 1982, Fig. 5-1], indicates pictorially how an LU-LU session is set up. The sequence of messages (RUs) transmitted is sometimes called a *bind triangle*. Note that the three cases shown correspond to different ways of initiating a session: a primary LU can initiate the session, a secondary LU can initiate it, or a third LU can attempt to activate a session between two other LUs. (An example would be one in which a terminal user sets up a session between an application program and a printer.) In all three cases an

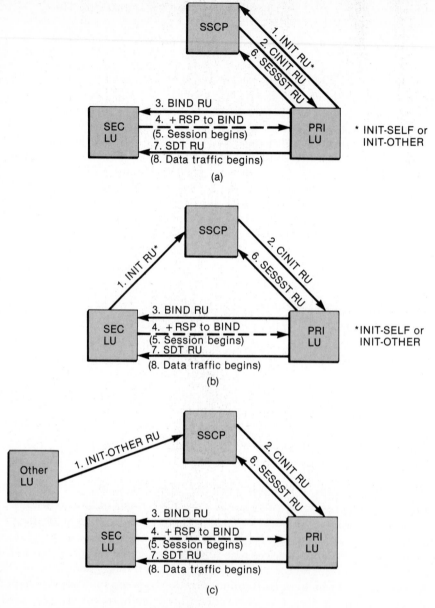

Figure 3–24 Starting an LU-LU session, same domain (from [SNA 1982, Fig. 5–1, p. 5–3] with permission)
  a. Primary LU initiates LU-LU session
  b. Secondary LU initiates LU-LU session
  c. A third LU initiates LU-LU session

initiate or INIT RU is sent by the initiating RU to the SSCP.* This control RU comes in two varieties as indicated, depending on whether the initiating LU wishes to begin a session itself or is doing this on behalf of another LU. The INIT RU, identified as such by a 3-byte header, carries the names of the originating and destination LUs, a mode name identifying the set of roles and protocols to be used in conducting the session, the class of service (COS) desired for the end-to-end route (this is described in Chapter 5), and an indication as to whether the destination LU is the primary or secondary LU. Other parameters are transmitted as well. The indication as to whether the destination LU is to be the primary or secondary LU means that the initiating LU selects the primary LU in the case of two peer LUs attempting to communicate. (In the case of communication between an application program residing in a host and a device or terminal, the host LU is normally the primary one.)

The SSCP, on receiving the INIT RU, determines whether it is a valid request (the two LUs must both be able to support the type of session, i.e., the LU-type required, passwords must be valid, and so on). If so, it resolves the name of the destination LU to a network address, determines whether a path exists and whether the destination LU is available, and then selects the parameters to be used by the LUs when their session has been activated. Network services in the SSCP then prepares and sends the CINIT RU to the primary LU. Note that this function is carried out in all three cases of Fig. 3-24. The primary LU responds by sending a BIND RU to the secondary LU. This is the session-activation request. It is similar to the connect-data units in the OSI architecture (Figs. 3-11 and 3-14). The BIND RU establishes the protocols for the session. The sender (the primary LU) proposes conditions for the session. These are based on the mode specified by the initiating LU and on parameters set by the SSCP. The LU type, for example, is mapped onto a set of profile fields in the BIND RU.

The secondary RU may accept the request, replying with a positive response (+ RSP in Fig. 3-24), or may reject it. Some of the different session types allow a negotiable bind. In that case the + RSP returned to the primary LU either can carry the same parameters as the BIND, indicating that the secondary LU accepts them, or can carry different parameters, which the primary LU can, in turn, either accept or reject. An example is the transmission of compressed data during the session. This can be proposed by the primary LU. If non-negotiable, the secondary LU must accept or reject this request. If negotia-

---

* The example of Fig. 3-24, as indicated by the title, is that of communication between LUs in one domain. Communication across domains requires the cooperation of the two SSCPs involved, one in each domain, with a special control RU sent between them for this purpose.

ble and the secondary LU does not handle compressed data, it will reply in the negative, giving the primary LU an opportunity to accept or reject.

The successful operation of this LU-LU session-initiation procedure requires the existence of LU-SSCP sessions. These must thus have been set up earlier. SSCP-PU sessions must also have been established earlier to set up the physical connections of the network [SNA 1982].

The remaining RUs in Fig. 3 – 24 are self-explanatory: the session started (SESSST) RU is a notification from the primary LU to the SSCP that the LU-LU session has been activated successfully. (If the process of session initiation is unsuccessful, a bind failure (BINDF) RU is sent instead.) The start data (SDT) RU, with a positive response expected, completes the LU-LU session activation.

Problems

3 – 1   Each layer in the OSI Reference Model carries out a variety of similar-sounding functions. An example is the multiplexing and splitting of traffic connections. Draw a picture of a network consisting of a set of nodes connected together by links. (Figure 3 – 1(a) shows one simple example.) Let some of the neighboring nodes be connected by multiple links. Superimpose a number of intelligent systems at various nodes in the networks as the users desiring to communicate with one another. Show by example, in your network, how both multiplexing and splitting might be carried out at the transport layer, the network layer, and the data link layer. *Hint:* Note that the transport layer, as an example, "sees" an end-to-end pipe, from source to destination. The multiplexing process thus implies that several "users" in the same system and above the transport layer use the same "pipe" end to end. Splitting means that one user would use several transport connections. The same concept applies at the network and data link layers. (However, "users" may not necessarily correspond to the same physical system.)

3 – 2   Error-control procedures are specified in each layer of the OSI Reference Model. Indicate the layers at which each of the following errors might occur:

   1. Noise on the transmission link converts a 0 bit to a 1.
   2. A packet (a network protocol data unit) is routed to the wrong destination.
   3. A frame (a data link protocol data unit) is received out of sequence.
   4. A packet-switching network delivers a data unit to a terminal attached to it out of sequence.
   5. A printer printing halfway through a line is suddenly commanded, by mistake, to return to the beginning of the line.
   6. During a half-duplex mode session the transmitting user starts receiving data from the user at the other end.

3-3   List one or more "errors" that might occur at each of the layers in the seven-layer OSI model.

3-4   One of the purposes of the data link layer is to ensure sequenced, correct delivery of data units (frames) transmitted across the link. A sequence number field thus appears in the frame and is incremented by 1 on each successive frame. In many networks the physical link between two neighboring nodes may actually consist of multiple physical channels in parallel to improve the throughput. Frames are sequenced (randomly or in ordered fashion) onto any one of these parallel channels.

   a. What provision must be made at the receiving side to ensure that the frames delivered to the data link layer there do arrive in correct sequential order? Draw a block diagram showing how to accomplish this.

   b. In most data link protocols frames received out of order are considered to be in error and are dropped at the receiver. (The transmitter then must carry out an error-recovery procedure, which is described in Chapter 4.) Is this data-link-layer requirement in conflict with the multiple physical channel implementation just described? Explain.

3-5   A number of users at two ends of a network are multiplexed onto the same transport layer connection (the "transport pipe" of Fig. 1-7). They thus all share the same transport-layer id. Transport protocol data units are sequentially numbered as transmitted. Draw a diagram showing how data units from a number of users, delivered to the transport layer at random, are multiplexed onto the transport pipe. Show how the transport id and transport-layer sequence number are appended to each.

3-6   Packets from a multiplicity of network sources share a particular link in a network. In traversing the link each packet has a header added that contains control information. The resultant data link data unit is called a frame. The control information consists of the link receiving address, a sequence number incremented by 1 with each new frame, and an error detection control field. Draw a diagram of a physical link connecting nodes A and B in a network. Show different packets that correspond to different source-destination pairs in the network using this link and arriving randomly at node A. Sketch the resultant sequence of frames moving across the link from A to B. Explicitly identify the header of each and show what happens when the frames arrive at point B. Compare with Problem 3-5.

3-7   Refer to Fig. 3-11. Assume that the network layer that provides network service for the transport layer under consideration there cannot guarantee correct, sequenced delivery of any data units given to it by the transport layer. (This is discussed in detail in Chapter 7.)

   a. Show why the so-called "three-way handshake" of Fig. 3-11 (the connect request, the connect response, and the acknowledgement transport protocol data units shown there) is necessary. What might happen if the acknowledgement TPDU were not used?

   b. Timers are used to "ensure correct operation" of the transport layer protocol: When the protocol sends out a TPDU, it sets a timer. If no reply is received within a specified time, the TPDU is repeated.

1. Why are these timers necessary?
2. How could the length of time be specified? (*Hint:* Draw a picture of a "typical" network or set of interconnected networks. Superimpose the "transport pipe" on the network.)

3–8   Consider two geographically separated systems, connected by a network, whose application layers wish to set up a session. The data links in the network are all operative, so there is no need to invoke the connection phase of the data link layers throughout the network. A virtual circuit at the network layer, however, must be established before communication through the network can take place. The virtual circuit consists of a path (route) through the network connecting the two network nodes to which the two systems are connected.

a. Start with one system's application transmitting an interface message (an A_CONNECT. request primitive) to its presentation layer, requesting that a connection be made to the remote application. Show with a diagram what primitives and peer-level protocol data units, at all layers down to and including the network layer, will have to be transmitted, until an A_CONNECT. confirm primitive is received back at the initiating system. (This concludes the application-layer establishment phase.) *Hint:* No data units can be transmitted through the network until the network-layer connection phase, setting up the virtual circuit, is concluded. The network layer then moves to its data-transfer phase, during which it can transfer control-data units for the layers above. Why must this process be repeated for each layer above?

How many network protocol data units (packets) will have been transmitted through the network by the time the A_CONNECT. confirm primitive has been received? (Assume that a three-way handshake is used at the transport layer.)

b. In the procedure just outlined, primitives at each level generate protocol data units that must be issued separately and transmitted to the other side of the network. Can you come up with another scheme that concatenates or bundles together the primitives issued by each layer, reducing the number of packets actually transmitted? Compare with part a. (See Fig. 3–24 for the approach used in IBM's SNA. However, that architecture uses a central supervisory system, the SSCP. Not shown in Fig. 3–24 is the exchange of messages required to set up the routes over which the various bind triangle messages are sent.)

3–9   Sequence numbers are used on each of the logical channels of X.25 to provide sequential transmission and reception of packets on each of the channels. The packets are in turn multiplexed onto the data link layer and appear as the information field of the HDLC frames sent across the X.25 link (Fig. 3–10). The HDLC frames carry separate sequence numbers in the C (control) field (Fig. 3–10). There is no connection between the sequence numbers at the X.25 packet level and those of the frame level. (This is in keeping with the concept of layered architectures to isolate layers from one another.) Demonstrate this to your own satisfaction by showing packets corresponding to three different logical channels, each separately sequenced in number, arriving in random order at the data link layer and then being transmitted, in FIFO order, as frames. Pick arbitrary sequence numbers at each of the two levels. Limit sequence numbers to the 3-bit range, 0 to 7.

3–10 Draw your own picture of an SNA network, incorporating at least three domains, with at least three subarea nodes in each. At least one of the nodes in each domain should be a PU _ T5 node, with one of them chosen as the SSCP.

    a. Consider a single domain first. A user at a terminal is connected to one of the subarea nodes; that user first sets up a session with an application program residing in a computer located at another PU _ T5 node. The same user then sets up a session between the application program and a printer at still another node. Show how this is done, using Fig. 3–24 as a model. The application program is to be taken as the primary LU in both cases. Assume that the communication paths or routes required to send messages over the network between the nodes containing the primary and secondary LUs, as well as the SSCPs, have already been set up through previous exchanges of messages.

    b. Repeat a. with the initiating user-terminal located in one domain, the other LUs in another domain. Communication is then with the SSCP in one's own domain; SSCPs in different domains in turn send special RUs to one another. Assume here as well that all routes between nodes required to carry the various messages have been set up previously.

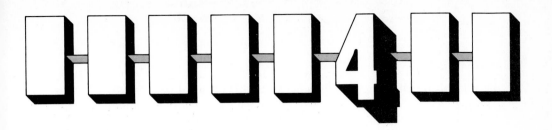

Data Link Layer: Examples and Performance Analysis

In the previous chapter we provided an overview of data networks, stressing the concept of layered architectures. In this chapter and the two following, we discuss the data link and network layers in more detail, focusing on performance questions where possible and describing simplified examples of practical protocols. The transport layer of the OSI and ARPA developed architectures are described in Chapter 7.

As already noted, the data link layer of any communication architecture must ensure orderly, correct delivery of packets between neighboring nodes in a network. A variety of protocols have been developed for this purpose. The HDLC (high-level data link control) protocol, developed by the International Organization for Standardization (ISO), essentially has become the international standard for this purpose. We shall discuss HDLC briefly later. Before doing that, however, we describe the main aspects of a data link control (DLC) protocol and various ways of designing such a protocol. We then carry out a simple throughput performance evaluation of idealized versions of two protocols: the stop-and-wait protocol and the go-back-N protocol. The performance evaluation of the HDLC protocol, which is essentially a go-back-N protocol, then builds on the simple procedure used to analyze the idealized versions of these two protocols. We conclude the chapter with a brief discussion of a third protocol, a selective repeat scheme.

As mentioned in Chapter 3, a link protocol must incorporate rules or procedures that enable communication between two ends of a link to be established (this is called the *establishment* or *connect* phase of the protocol). It must also provide for orderly transfer of data packets* during the subsequent *data-transfer* phase. Finally, the protocol must include procedures for *terminating* communication, either when the link is no longer needed or when a link becomes noisy or fails, aborting the connection. We consider only the data-transfer phase in this chapter. The reader is referred to [SCHW 1977, Chap. 14] for a detailed discussion of all three phases, as well as references to papers on the subject. The establishment and termination rules for the HDLC protocol are discussed in papers referenced later in this chapter.

Communication between two ends of a link is innately asynchronous. Thus individual packets must carry synchronization information, generally in the form of a synchronization field, to enable bit registration to be carried out. Packets received in error must be so recognized. An acknowledgement procedure must be established to indicate whether packets have been received correctly or incorrectly. (This is part of the general problem, at *all* levels of a layered architecture, of establishing valid, foolproof procedures for dealing with errors or other contingencies once they are detected.) Finally, packets must be numbered to ensure orderly delivery to the next higher layer (the network layer) at the receiving end of a link.

To carry out these functions, control fields are added to packets. Among these control fields are a synchronization field (or fields), an error-checking field, sequence-number fields, and other fields described later in the discussion of HDLC; examples appeared in Chapter 3. The combination of the data packet plus these added control fields is commonly called a *frame*. It is the frame, therefore, that is actually transmitted between neighboring nodes of a link. The packet represents the data portion of the frame. A hypothetical example appears in Fig. 4–1.

Various ways exist to carry out the error-detection and correction process. One way is to add sufficient check bits (so-called parity bits) in the error-checking field to *correct* a specified number of errors [SCHW 1980a, Chap. 6]. Although this procedure has been used very successfully in deep-space communications and other point-to-point communication systems, as well as in providing error checking for computer systems, digital magnetic tape systems, and other devices used with computer systems [LINS], it has not generally been adopted

---

* Data packets represent information transmitted down from the network layer above.
[SCHW 1977] M. Schwartz, *Computer-Communication Network Design and Analysis,* Prentice-Hall, Englewood Cliffs, N.J., 1977.
[SCHW 1980a] M. Schwartz, *Information Transmission, Modulation, and Noise,* 3d ed., McGraw-Hill, New York, 1980.
[LINS] Shu Lin and Daniel J. Costello, Jr., *Error Control Coding: Fundamentals and Applications,* Prentice-Hall, Englewood Cliffs, N.J., 1983.

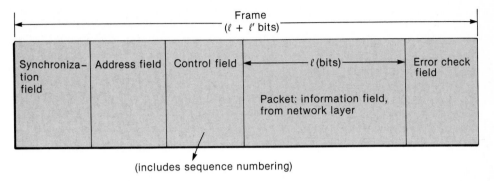

(includes sequence numbering)

Figure 4 – 1 Example of a data link frame

for data networking. The basic reason is that too many check (parity) bits are needed to ensure the level of error correction required. Instead *error detection with retransmission,* also commonly called ARQ (automatic repeat request) error detection, has been universally adopted.

A number of versions of ARQ procedures are possible: Each correctly received frame may be individually acknowledged by a special acknowledgement frame (ack), or the acknowledgement may be embedded as a control field in data frames flowing in the reverse direction. (Special ack frames must also be made available in this latter case since data frames in which to embed the ack may not always be available.) Negative acknowledgements (naks) may also be used to indicate an error condition. In either instance — ack alone or the ack/nak case — timeouts must be used to avoid deadlock situations: The transmitter, not receiving a reply (ack or nak) within a specified interval timeout after transmission, repeats the frame in question. (If this procedure were not built into the protocol, the transmitter might wait forever for an ack/nak that had itself been lost or delivered in error and discarded!) To accommodate the timeout procedure, therefore, packets must be held in a transmit buffer until correctly acknowledged.

There are also a number of ways of handling the response to ack/naks. The three most commonly identified procedures are the following:

1. *Stop-and-wait protocol.* In this procedure only one frame at a time can be transmitted. The transmitter then waits for an ack/nak. If either a nak arrives (if used) *or* a timeout expires, the frame is retransmitted. Only when an ack arrives is the packet in question dropped from the transmit buffer. Stop-and-wait communication between two stations A and B is portrayed in Fig. 4 – 2.

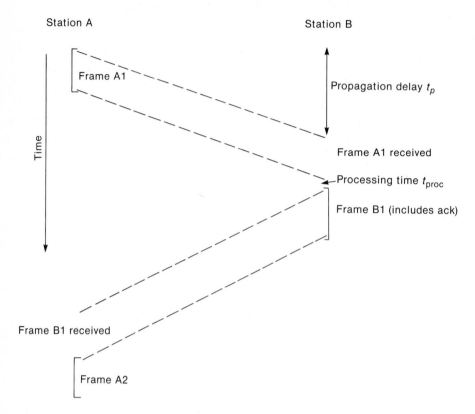

Figure 4-2 Stop-and-wait protocol

This is an appropriate protocol for half-duplex transmission, in which the two sides alternate transmission. However, this protocol obviously suffers from reduced throughput in the full-duplex case (independent transmission in either direction), particularly if the link propagation delay is significantly longer than the packet transmission time. (If propagation delays are negligible, due either to short link lengths or to relatively low transmission rates, the stop-and-wait protocol does not result in significantly reduced throughput. Quantitative justification of these intuitive statements will be presented shortly.)

2. *Go-back-N* or *continuous transmission.* Here data frames are transmitted continuously, if available, without waiting for an ack. On receipt of a nak, or expiration of the timeout without receipt of an ack/nak, the frame in question and all frames following are retransmitted. A simple example appears in Fig. 4–3. Frame transmission from A to B is shown with B acknowledging each frame as received. In part (a) frame 3 is assumed to have been received in error, with a nak then generated. Part (b) shows the expiration of a timeout at A *before* receipt of either a nak or an ack. For simplicity only the timeout of frame 3 is shown. Note that *each* frame transmitted has a similar clock, set just after the transmission is completed; frame 3 happens to be the only one for which the timeout expires before receipt of a nak or an ack. (If negative acks are *not* used in a protocol, two possibilities may have occurred: An error was detected in frame 3 at B, in which case B does nothing, or a positive ack from frame B may have been

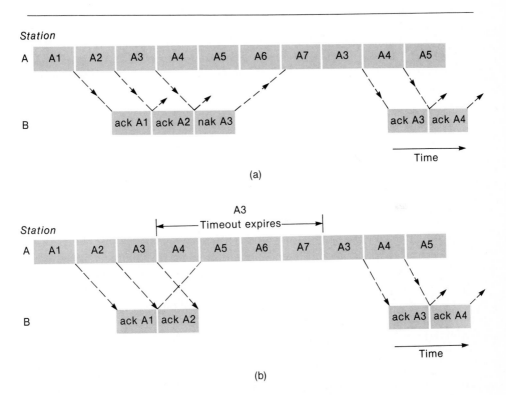

(a)

(b)

Figure 4–3 Go-back-N protocol: handling of ack and timeout expiration. Data traffic from A to B only. (Arrows represent direction of transmission.)
  a. Negative acknowledgement of frame 3
  b. Expiration of timeout

mangled during transmission, in which case A repeats 3 and all frames following.)

The pipelining of frames obviously improves the throughput performance, as we shall see. In practical versions of a go-back-N protocol (HDLC is an example) not all frames need be acked. An ack may positively acknowledge a given frame and all frames preceding it.

3. *Selective repeat procedure.* Here only the frame negatively acknowledged, or the frame for which the timeout has expired without receiving a positive ack, need be retransmitted. This obviously improves the throughput performance over the go-back-N case. A reordering buffer is required at the receiver, however, since frames may now be retransmitted and received out of order. The buffers required are particularly large for satellite transmission with long propagation delays, long pipelines, and consequently many packets en route between transmitter and receiver. For this reason and the added cost of implementing the protocol, selective repeat (or *selective reject*, as it is sometimes called), has not received commercial acceptance. With modern VLSI technology the picture may change. Cost-effective reordering buffers are in fact coming into the marketplace.

In all three of the cases just discussed, the assumption has been implicitly made that frame numbering is no problem. In practice the sequence-number field is finite. Once the maximum frame number has been transmitted, with no ack as yet received (this is called "sequence-number starvation"), the transmitter must cease to deliver frames. Only when an ack arrives can new frames be sent out. This phenomenon reduces the throughput rate and may, in particular, pose a problem over satellite links and other long propagation delay links. We shall discuss this problem and the resultant appropriate choice of sequence-number field size when carrying out the throughput analysis of the HDLC protocol. Because go-back-N schemes, including the version implemented in the HDLC protocol, are used extensively, we focus on the throughput analysis of these procedures.

We provide a simplified version of the analysis of both the stop-and-wait and the go-back-N strategies in the next two sections. The selective repeat analysis is deferred until the end of Section 4–3. An infinite sequence-number scheme is assumed in the simplified analyses of the idealized protocols. In the HDLC analysis (Section 4–3) a finite sequence-number field is taken into account. We also assume fixed-length data frames; a constant, known round-trip propagation delay between transmitter and receiver; and a fixed, known processing delay at the receiver. Some of these assumptions are relaxed in references cited later. The basic distinguishing characteristics and comparative throughput performance of the various protocols are nevertheless retained with these simplifying assumptions.

4–1 Stop-and-Wait Protocol

For the throughput analysis of this protocol under the assumptions just noted we refer to Fig. 4–4. Let t_I be the time required to transmit a frame (data packet plus control bits). Let t_{out} be the timeout interval, as shown. This is then at least equal to the round-trip propagation delay $2t_p$ plus receiver processing time t_{proc}, plus the time to transmit an acknowledgement. The propagation delay and processing time are indicated in Fig. 4–2. To simplify figures and analysis, we henceforth incorporate the processing time t_{proc} in the propagation time t_p. The time $t_T = t_I + t_{out}$ is then the minimum time between successive frames, as shown in Fig. 4–4.

To determine the maximum possible throughput we shall assume that A transmits to B only, with B replying with an ack or a nak. In this case the transmission time of the ack or nak is given by a fixed value t_s. Transmitter A is further assumed to be operating in a *saturated* state: It always has a frame waiting for transmission. The throughput expression to be found will thus represent an upper limit on the throughput performance. (The same saturated condition will be assumed to exist for both the idealized go-back-N and HDLC analyses following, allowing comparisons in performance to be made.) Figure 4–5 indicates the basic topology assumed: Station A is shown transmitting a frame of length t_I to station B. As noted, the one-way propagation delay (with processing time included) is labeled t_p.

At the expiration of t_{out}, by the assumptions made earlier an ack will have arrived indicating that a new frame may be transmitted if available; if not, the original frame outstanding will be retransmitted. (In this idealized version of the protocol either a nak arriving in place of an ack or the expiration of t_{out} without receipt of an ack will trigger the retransmission of the frame.) Hence the system may transmit at most one packet/t_T sec, which represents the *maximum possible*

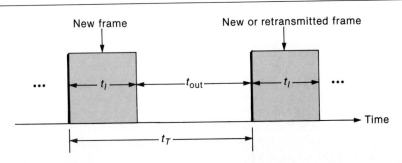

Figure 4–4 Frame transmission for stop-and-wait throughput analysis

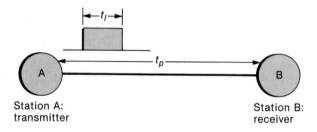

Figure 4–5 Link-level transmission between stations A and B

system throughput. The actual maximum throughput will be less because of possible retransmissions.

Say that the probability of a frame being received in error at station B is p. Then it is apparent that with no limit on the number of retransmissions (in practice, if the number of retries with no success were to exceed some threshold, the link would have to be deemed "down," and upper layers of the communication architecture would have to take some action), the average time for a *correct* transmission to be received is given by

$$t_V = t_T + (1 - p) \sum_{i=1}^{\infty} i p^i t_T = t_T/(1 - p) \tag{4-1}$$

This expression indicates that to get to the ith retry the frame must have been delivered in error i times. The probability of receiving it correctly on the ith retry is just $(1 - p)$. Possible errors in the reverse path (from B to A) that may destroy acks have been neglected here. (As noted earlier, the throughput calculation here represents an upper limit. One obtains maximum possible throughput by letting A do all the transmission. Separate ack frames from B to A are then generally much shorter than data frames, and the chance of their being received in error is correspondingly much smaller.) The increase in average time between transmission of packets due to retransmission is shown clearly by the factor $(1 - p)$ in the denominator of Eq. (4–1).

In the saturated case t_V represents the average time between transmission of correct frames. The *maximum throughput*, in packets/sec delivered, is then just the reciprocal of t_V, or

$$\lambda_{\max} = 1/t_V = (1 - p)/t_T = (1 - p)/a t_I \tag{4-2}$$

where the parameter $a \equiv t_T/t_I \geq 1$ has been introduced to relate the throughput to the data frame length t_I.

If we now let λ be the actual frame (or packet) arrival rate at the transmitter, we must have as the normalized throughput for the stop-and-wait protocol

$$\rho \equiv \lambda t_I \leq (1 - p)/a < 1 \qquad (4-3)$$

This expression shows explicitly the dependence of throughput on frame error probability p and the ratio a of time between frames to frame-transmission time. If the time to receive an ack/nak is negligible ($2t_p + t_s \ll t_I$), the timeout t_{out} is also negligible, $a = 1$, and

$$\rho < 1 - p \qquad a = 1 \qquad (4-4)$$

The same result appears in a much more thorough analysis by D. Towsley and J. K. Wolf [TOWS] of the stop-and-wait scheme using discrete-time analysis. Their work focuses on waiting times and queue lengths for ARQ schemes.

Examples involving the use of the simple expressions just obtained are deferred until the go-back-*N* analysis following, when both protocols are considered together.

4 – 2 Go-Back-*N* Protocol

In half-duplex transmission, with each side of a link using the link alternately, one might naturally use a stop-and-wait link protocol. Most modern systems use full-duplex transmission, however, in which case it appears more natural to keep transmitting frames continuously, if a frame is available, rather than to wait for a positive acknowledgement from the receiver. Continuous transmission will improve the throughput of the link, particularly if the propagation delay is not negligible compared with the frame transmission time. As noted earlier, the go-back-*N* scheme uses this continuous transmission strategy and serves as the basis of the HDLC and other modern ARQ link protocols. Recall that in this scheme, as contrasted to a selective-repeat scheme, all outstanding frames following one received in error must be retransmitted.

In this section we determine the throughput performance of an idealized go-back-*N* protocol using the same approach adopted for the stop-and-wait scheme. In this idealized protocol, sequence numbers can again be as large as possible, and no specific distinction is made between a nak and a timeout strat-

[TOWS] D. Towsley and J. K. Wolf, "On the Statistical Analysis of Queue Lengths and Waiting Times for Statistical Multiplexers with ARQ Retransmission Schemes," *IEEE Trans. on Comm.*, vol. COM-27, no. 4, April 1979, 693–702.

egy for determining an error occurrence. In the HDLC discussion and analysis to follow, both finite sequence numbers and a combination of nak/timeout error-control strategies are taken into account.

Since frames may be transmitted continuously one after the other in the go-back-N scheme without waiting for a positive acknowledgement, the minimum time between transmissions is t_I, the frame transmission time, rather than t_T as in the stop-and-wait case. The maximum throughput is increased correspondingly. Using the same assumptions as in the stop-and-wait scheme—saturated transmission from station A to station B, fixed frame length t_I, and fixed timeout interval t_{out} (at the end of which an ack arrives or a retransmission takes place)—we have, as the average transmission time of a frame (see Fig. 4–6),

$$
\begin{aligned}
t_V &= t_I + (1 - p) \sum_{i=1}^{\infty} i p^i t_T \\
&= t_I \left[\frac{1 + (a - 1)p}{1 - p} \right]
\end{aligned}
\tag{4-5}
$$

after carrying out the indicated sum and combining terms. Note that the only

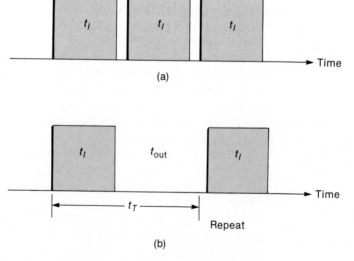

Figure 4–6 Go-back-N analysis
a. No errors
b. Error case

difference between this expression and Eq. (4–1) for the stop-and-wait protocol is that t_I appears here in the first term in place of t_T there. The parameter p again represents the probability a frame is received in error, and $a \equiv t_T/t_I = 1 + (t_{\text{out}}/t_I)$.

Because of the assumption of saturation conditions, the maximum possible throughput is given by

$$\lambda_{\max} = 1/t_V = (1 - p)/t_I[1 + (a - 1)p] \qquad (4-6)$$

and the normalized throughput $\rho \equiv \lambda t_I$ for any frame (packet) arrival rate λ is then bounded by

$$\rho = \lambda t_I < (1 - p)/[1 + (a - 1)p] \qquad (4-7)$$

Note, as required, that when $a = 1$ (propagation and processing delays are negligible compared with the frame transmission time t_I) these simple results are the same as those obtained for the stop-and-wait protocol.

Consider some simple examples now to clarify these throughput results:

Example 1. Take $a = 4$, $p = 0.01$. Then for the stop-and-wait protocol, the normalized throughput ρ is bounded by $0.99/4$, while the corresponding bound for the go-back-N scheme is about 0.96. In this case, even with frames retransmitted on the average of once in 100 frames, the go-back-N scheme outperforms the stop-and-wait scheme by a factor of four. For $a > 1$ it clearly pays to use a continuous transmission strategy.

Example 2. Terrestrial link. Say the frames are 1200 bits long and the line speed is $C = 9600$ bps. Then $t_I = 125$ msec. If the link is less than 100 miles long the round-trip propagation delay is less than 2 msec, using 1 msec/100 miles as a typical propagation-delay figure. If the receiver processing time t_{proc} is much less than 125 msec, and one uses $t_{\text{out}} = 2t_p + t_{\text{proc}} + t_s$, with $t_s \ll t_I$, $t_{\text{out}} \ll t_I$ and the two strategies perform about the same.

For safety's sake, however, a larger value of t_{out} would normally be chosen. The *minimum* time to receive an acknowledgement is $2t_p + t_s$, with t_{proc} again included in t_p. This assumes separate ack packets. If acks are embedded in frames traveling in the reverse direction (B to A), the time to receive an ack rises to at least $2t_p + t_I$. To be safe, let $t_{\text{out}} = 2t_p + 2t_I$ to include various possibilities. In this case $t_T = t_I + t_{\text{out}} = 2t_p + 3t_I$, and $a \equiv t_T/t_I = 3 + 2t_p/t_I$. The improvement in throughput performance of the go-back-N scheme over the stop-and-wait strategy is then apparent. In this case the stop-and-wait strategy reduces the throughput by a factor of $a > 3$ over the continuous go-back-N scheme.

If 56-kbps links are used instead of 9600-bps links, the improvement in performance becomes even greater. For 1200-bit frames t_I is now $1200/56{,}000 \doteq 21$ msec. For 1000-mile terrestrial links, $2t_p$ will be 20 msec, $t_{\text{out}} = 2t_p + 2t_I = 62$ msec, and $a = 83/21 \doteq 4$.

Example 3. Satellite link. With two earth stations on the ground communicating via a satellite that constitutes a link, the one-way propagation delay between the two stations is $t_p = 250$ msec [SCHW 1980a]. The two-way propagation delay (corresponding to *four* traverses between the ground and the satellite repeater!) is then 500 msec (Fig. 4–7). For 4800-bps transmission and 1200-bit frames, $t_l = 250$ msec. Again taking $t_{out} = 2t_p + 2t_l$ and neglecting processing delay for simplicity, we then have $a = t_T/t_l = 5$. With $C = 9600$ bps, $t_l = 125$ msec, and $a = 7$. With $C = 48$ kbps, $t_l = 25$ msec, and $a = 23$. The stop-and-wait protocol is clearly inadequate in this case, and continuous transmission schemes must be used. In discussing the HDLC protocol later as applied to a satellite link (Section 4–3), we shall find that relatively large sequence-number fields are also required to keep the throughput up. This is apparent from the large values of a noted in the satellite-link case. With a 9600-bps transmission rate and 1200-bit frames, $t_l = 125$ msec and two consecutive frames will "fill" the one-way propagation path. With $C = 48$ kbps, $t_l = 25$ msec and 10 consecutive frames are needed to fill the path. If a maximum sequence number of 8 is used (as is common over a terrestrial link), the throughput will clearly suffer. A maximum sequence number of 128 is thus suggested for this case. We shall explore this point in more detail later when evaluating the use of the HDLC protocol over a satellite link. In our idealized model of the go-back-N protocol, with no limit on sequence numbers, this problem of "sequence-number starvation" does not arise.

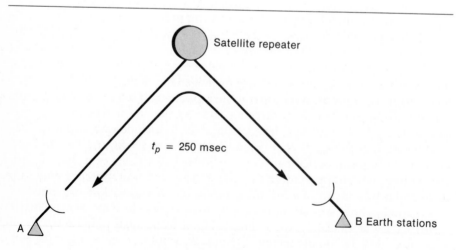

Figure 4–7 Satellite link

The go-back-*N* retransmission scheme has been analyzed in detail by A. G. Konheim [KONH 1980] using a discrete time analysis approach similar to that used in [TOWS]. Konheim's maximum throughput result agrees with Eqs. (4-6) and (4-7). He has analyzed the selective repeat strategy as well, obtaining wait-time distributions for both protocols. These analyses refer, of course, to the common two-station, single-link case of Figs. 4-5 and 4-7. With satellite transmission one may want to transmit identical frames simultaneously to multiple destinations (receivers) in broadcast mode. The throughput calculations here are more complex, but they indicate that in this case the go-back-*N* scheme becomes inadequate, its throughput dropping rapidly with even a moderate number of receiving stations. (This is due to the requirement that *all* stations must receive the data correctly.) The selective repeat strategy, however, continues to perform well [SABN].

4-2-1 Throughput Efficiency and Optimum Packet Length

The simple throughput analyses just carried out also provide an estimate of the best *packet* length (the data or information field of the frame) to be used in packet networking. In most networks this is chosen to be about 1000 bits.

Recall that the frame has necessary control fields appended to the packet data or information field (Fig. 4-1). It is apparent that if the packet (data) length is small, the system is operating inefficiently. The link is transmitting control bits rather than real data in this case. On the other hand, if the packet length is made too long, the frame is more likely to be received in error, necessitating more retransmissions and resulting in reduced throughput. There thus exists an optimum packet length in the sense of maximizing the *data* throughput. The analysis indicates that the specific length, however, is not too critical since a broad range of values provides about the same throughput. The optimum choice will be found to depend on the error characteristics of the link and on the number of control bits used.

To find the optimum packet length one needs a model of the error characteristics of the link. For the satellite link we make the assumption that bits are *independently* prone to error, with a bit error probability p_b. This assumption turns out to be valid in the satellite case, where errors are due primarily to random noise introduced over the satellite-earth transmission path [SCHW

[KONH 1980] A. G. Konheim, "A Queueing Analysis of Two ARQ Protocols," *IEEE Trans. on Comm.*, vol. COM-28, no. 7, July 1980, 1004-1014.

[SABN] K. Sabnani and M. Schwartz, "Performance Evaluation of Multidestination (Broadcast) Protocols for Satellite Transmission," National Telecommunications Conference, New Orleans, Nov. 1981.

1980a]. Letting the length of the packet (data) field in the frame be ℓ bits (Fig. 4-1), with a total of ℓ' bits used in the remaining (control) fields, it is apparent that the frame error probability (the probability that *at least one* bit is received in error) is given by

$$p = 1 - (1 - p_b)^{\ell+\ell'} = 1 - q_b^{\ell+\ell'} \qquad (4-8)$$

with $q_b = 1 - p_b$. For very small p_b such that $(\ell + \ell')p_b \ll 1$,

$$p \doteq (\ell + \ell')p_b \ll 1 \qquad (4-8a)$$

This model of bits independently prone to error turns out not to be as valid over terrestrial links, where *burst errors* may often occur. In this case an error burst, as the name indicates, will wipe out a number of bits in sequence. However, experiments with data transmission over terrestrial links indicate that the frame error probability in that case is proportional to the length of the frame [BURT]. This corresponds, of course, to the form of Eq. (4-8a), with the parameter p_b appearing there not a bit error probability but rather a proportionality constant. For a small error probability the form of Eq. (4-8a) is thus appropriate to either satellite or terrestrial links, while Eq. (4-8) must be used for satellite links in the case of larger error probabilities.

The variation of data throughput with packet (data) length ℓ, and the value of ℓ that maximizes this throughput, are now obtained readily using the error probability expressions of Eqs. (4-8) and (4-8a). Let the transmitting station A again be in a saturation state, transmitting λ_{max} frames/sec. The average *data* rate, in bits/sec of data delivered to receiving station B, is then, for the go-back-N scheme,

$$D = \lambda_{max}\, \ell = (1-p)\ell/t_I[1 + (a-1)p] \qquad (4-9)$$

from Eq. (4-6). (If $a = 1$ one gets the expression for the stop-and-wait scheme as well.)

Letting $t_I = (\ell + \ell')/C$, with C the link transmission rate capacity in bits/sec, we have, for the normalized data rate,

$$D/C = \left(\frac{\ell}{\ell + \ell'}\right)\left[\frac{(1-p)}{1 + (a-1)p}\right] \le 1 \qquad (4-10)$$

This expression shows clearly the effect on the data transmission rate of packet length ℓ, control field length ℓ', and bit-error probability. The timeout interval is represented by the normalized parameter a. Figure 4-8 is a plot of the normalized data rate, D/C, as a function of packet (data) length ℓ for two cases.

[BURT] H. O. Burton and D. D. Sullivan, "Errors and Error Control," *Proc. IEEE,* vol. 60, no. 11, Nov. 1972, 1293-1301.

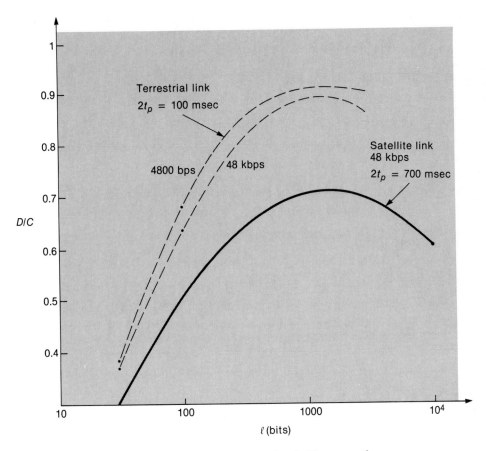

Figure 4-8 Throughput, go-back-N protocol

$$p_b = 10^{-5} \quad \ell' = 48 \text{ bits}$$

One case represents a terrestrial link with a round-trip propagation delay plus processing delay taken to be 100 msec. The timeout interval is then $t_{out} = 100 + 2t_I$, in msec. Two values of transmission capacity C have been used in the calculations: 4.8 kbps and 48 kbps. The second case is that of a satellite link, with the round-trip propagation delay plus processing delay chosen as 700 msec. The timeout interval is then $t_{out} = 700 + 2t_I$, in msec. In both cases the control-field length is $\ell' = 48$ bits (this is the HDLC value), and the bit-error probability is taken to be $p_b = 10^{-5}$. For the terrestrial case, Eq. (4-8a) has been used to calculate p in Eq. (4–10); for the satellite case, Eq. (4–8) has been used. As noted earlier, the throughput rate peaks at values of ℓ in the range of 1000–2000 bits for these (realistic) examples. For small values of ℓ, the inefficiency of the system, represented by the term $\ell/(\ell + \ell')$ in Eq. (4–10), keeps the data throughput rate low. (A frame carries relatively few data bits.) For large ℓ, however, the probability of retransmission p increases, driving the throughput down.

Under certain approximations an explicit expression for the optimum packet length ℓ, producing peak throughput, may be readily obtained. Specifically, let $a \doteq 1$ (or $(a - 1)p \ll 1$) in Eq. (4–10). Then it is readily shown, by differentiating Eq. (4–10) with respect to ℓ and setting the derivative to 0 (this assumes that ℓ is a continuous variable), that the optimum value of ℓ is given by

$$\ell_{opt} = \frac{\ell'}{2}\left[\sqrt{1 - \frac{4}{\ell'\log_e q_b}} - 1\right] \qquad a = 1 \qquad (4\text{–}11)$$

As an example, if $\ell' = 48$ bits and $q_b = 1 - 10^{-5}$, the numbers used in Fig. 4–8, $\ell_{opt} \doteq 2200$ bits. This agrees fairly well with the curves of Fig. 4–8 even though a is greater than 1 in all cases. It is apparent from Eq. (4–11) that if $p_b \ll 1$, as in the example here, and in particular if $\ell' p_b \ll 1$, a much simpler approximate expression for ℓ_{opt} is obtained:

$$\ell_{opt} \doteq \sqrt{\ell'/p_b} \qquad \ell'p_b \ll 1 \qquad (4\text{–}11a)$$

This also gives $\ell_{opt} \doteq 2200$ bits in the example presented here.

Consider now the special case for which the error probability is given by $p = (\ell + \ell')p_b$, with p_b a *constant*. This is the terrestrial-error model noted earlier. It is left to the reader to show that for this case the optimum packet length is given by

$$\ell_{opt} = \sqrt{\ell'/p_b} - \ell' \qquad p = (\ell + \ell')p_b \qquad (4\text{–}12)$$

independent of the parameter a. Note that this is similar to the approximation of Eq. (4–11a), as expected, for with p_b very small, Eq. (4–8) for p reduces to Eq. (4–8a), just the terrestrial model. A value of $p_b = 10^{-5}$ is typical for terrestrial data communication using normal telephone carrier facilities. Packet lengths of about 1000 bits thus provide maximum data throughput, and such lengths are in fact typical of packet lengths in practice. Lower bit-error rates are obtainable with satellite transmission. If the bit-error probability p_b in that case is reduced to 10^{-7}, the optimum packet length for the examples considered here increases by 10 to $\ell' \doteq 20{,}000$ bits.

4-3 High-level Data Link Control (HDLC)

We now focus more specifically on the HDLC protocol, which as already noted has fast become an international standard. This protocol followed, and in many respects is based on, the IBM SDLC (synchronous data link control). American standards activities, paralleling the ISO work and interacting with it, resulted in the American National Standards Institute (ANSI) data-link control procedure standard ADCCP (advanced data communication control procedure). HDLC and ADCCP are closely related and will not be distinguished specifically in this discussion [CARLD]. The CCITT X-25-recommended data-link procedures, LAPB (balanced link access procedures), a subset of HDLC, will be described in some detail later. All these protocols and others like them are examples of bit-oriented protocols, in which the frame structure used eliminates a specific dependence on byte or character formatting [SCHW 1977].

In this section we first describe the basic philosophy and operation of HDLC, with reference made to the tutorial paper by Carlson [CARLD]. We then outline a throughput performance analysis of one common mode of operation of HDLC, following the work of Bux, Kummerle, and Truong [BUX 1980]. The analysis is similar to that carried out in the last section for the idealized go-back-N protocol, but because it focuses on a model of a real protocol, it captures the effect of finite sequence numbering and a specific error-control procedure. This enables us to compare the idealized throughput analysis with the analysis for a real protocol.

The standard frame format for HDLC (ADCCP and SDLC have the same format) appears in Fig. 4-9. Note that the number of overhead (control) bits is $\ell' = 48$, just the number used earlier for calculations. The eight-bit flag sequence 01111110 that appears at the beginning and end of a frame is used to establish and maintain synchronization. Because the flag appears at the beginning and end of the frame there is no need to prescribe an information field structure. The information field (packet) delivered from the network layer above can be any desired number of bits. Extended versions of the frame structure of Fig. 4-9 are available as well: The address, control, and block-check fields can all be increased to allow additional addressing, improved error detection, and increased sequence numbers. Since the flags appearing at the

---

[CARLD] D. E. Carlson, "Bit-oriented Data Link Control Procedures," *IEEE Trans. on Comm.*, vol. COM-28, no. 4, April 1980, 455-467; reprinted in [GREE].

[BUX 1980] W. Bux, K. Kummerle, and H. L. Truong, "Balanced HDLC Procedures: A Performance Analysis," *IEEE Trans. on Comm.*, vol. COM-28, no. 11, Nov. 1980, 1889-1898.

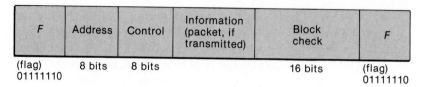

F	Address	Control	Information (packet, if transmitted)	Block check	F
(flag) 01111110	8 bits	8 bits		16 bits	(flag) 01111110

Figure 4–9 HDLC standard frame format

beginning and end of a frame contain six consecutive ones, that sequence may not appear anywhere else in the frame. Bit stuffing is used to eliminate this possibility: a zero is inserted at the transmitter any time that five ones appear outside the F fields. The zeros are removed at the receiver. If seven ones appear anywhere in the frame (six ones followed by an additional one), the frame is declared in error.

Three types of frames are defined to handle information flow, supervisory and control signals, and responses to all of these:

- I (information transfer) format
- S (supervisory) format
- U (unnumbered) format

S- and U-frames carry no information field. They are used strictly for supervisory and control purposes. The eight-bit control field in the frame determines which type of frame is being transmitted, and, for the S- and U- frames, which specific control signal is being transmitted. Figure 4–10 breaks the eight-bit control field down for the three types of frame. A zero in the first bit of the control field corresponds to an I-frame. The bit pairs 10 and 11 appearing as the first two bits indicate S-frame and U-frame, respectively, as shown. The two S bits in bit positions 3 and 4 of the S-frame allow four different S-frames to be transmitted. The five M bits in the U-frame allow 32 different U-frames to be transmitted. The three-bit number $N(S)$ in the I-frame represents the sequence number of the I-frame. Mod-8 sequence numbering is thus standard with normal HDLC. Each successive I-frame has its sequence number incremented by one. When the transmitter reaches its maximum sequence number it is forced to stop transmitting until a frame in the reverse direction is received, acknowledging an outstanding packet. The $N(R)$ bits in the I- and S-frames are used to acknowledge I-frames received. The number $N(R)$ acknowledges the receipt of $N(R)$-1 and any frames preceding that number not already acknowledged. $N(R)$ indicates that the receiver is *expecting* I-frame number $N(R)$. Thus $N(R) = 5$ (bit

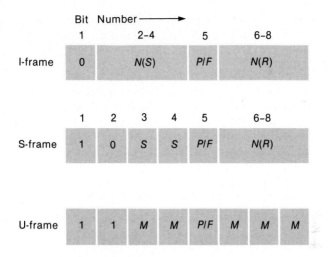

Figure 4–10 Control field for the three types of frames

pattern 101) acknowledges the receipt of I-frame number four (and any preceding frames not yet acknowledged) and indicates that I-frame number five is expected. The $N(S)$ and $N(R)$ fields can be extended to seven bits to allow mod-128 sequence numbering. The transmitter buffers all frames not yet acked positively. Once acked positively the frame can be purged and its sequence number used again.

Three modes of operation are defined for the HDLC protocol [CARLD]:

1. *Normal response mode* (NRM). This mode of operation is used in a centralized control environment. It is suited for polled multipoint operation, in which a single primary station communicates with one or more subservient secondary stations, the latter being allowed to initiate transmission only in response to the primary command. (Polling is described in more detail in Chapter 8.)

2. *Asynchronous response mode* (ARM). This mode is similar to NRM except that the secondary station does not need permission from the primary station to initiate transmission.

3. *Asynchronous balanced mode* (ABM). This mode is for point-to-point link transmission only, with both stations serving as equal partners. A class of procedures for this mode forms the basis for the link level of the X.25

protocol. It is identified there as LAPB [RYBC], which we describe briefly in the following paragraphs. Our discussion focuses exclusively on the asynchronous balanced mode of operation of HDLC (or ADCCP).

In HDLC data transmission and error detection and recovery are handled through the use of the I-frames and S-frames; we discuss only these here. U-frames are used in the connect/disconnect phases of the data link procedure, as well as to provide extended sequence numbering if desired [CARLD]. As noted earlier, there are four possible S-frames, of which only three are used in ABM. These three control frames, with their corresponding pair of S bits and the functions for which they are prescribed, are tabulated as follows:

NAME	S	S	FUNCTION
Ready to receive (RR)	0	0	$N(R)$ acks all frames received up to and including $N(R) - 1$.
Not ready to receive (RNR)	1	0	This provides a flow control for a temporarily busy condition. $N(R)$ also acks all frames up to and including $N(R) - 1$.
Reject (REJ)	0	1	$N(R)$ *rejects* all frames from $N(R)$ on. It positively acknowledges all frames up to and including $N(R) - 1$.

Recall that the I-frame also carries an $N(R)$ field that is used for acknowledging frames up to and including $N(R) - 1$. HDLC thus provides both a "piggyback" feature, with the ack function embedded in an I-frame transmitted in the reverse direction, and a separate acknowledgement, called the RR (ready to receive) frame, that can be used to signal a positive ack in the absence of an I-frame or to expedite the delivery of an ack if so desired. The REJ (reject) frame provides a negative ack (nak) feature if so desired. Note that the ABM subset of HDLC uses the go-back-N feature: The REJ-frame rejects *all* I-frames from $N(R)$ on. Hence they must *all* be retransmitted. (The unbalanced modes of HDLC also provide for selective reject if desired. The fourth S-frame, labeled SREJ for selective reject, calls for retransmission of a particular I-frame.)

Corresponding to each of the three modes of operation just noted are classes of procedure with defined functions and options. For the ABM mode these are called balanced asynchronous class (BAC) procedures. The defined functions recognize the use of I, R, and RNR (not ready to receive) frames, as

[RYBC] Antony Rybczynski, "X.25 Interface and End-to-End Virtual Circuit Service Characteristics," *IEEE Trans on Comm.*, vol. COM-28, no. 4, April 1980, 500–510; reprinted in [GREE].

well as certain prescribed U-frames used to set up and disconnect the link. Two options are added: the use of REJ (option 2) and the restriction that I-frames be *commands* only (option 8). The composite class of procedures is called the BAC 2,8 class and corresponds to the X.25 link-level LAPB class of procedures. [CARLD], [RYBC].

The concept of *command* appears in the use of the address field (Fig. 4–9) and the P/F bit (Fig. 4–10). In the unbalanced modes of HDLC, with their recognition specifically of one primary station and one or more secondary stations, the address is always that of the secondary station. In the balanced (ABM) mode the address is always that of the *responding* station. Since an I-frame is always a *command* in option 8, the address must be that of the receiving station. RR and RNR frames may be either commands or responses. If the former, the address is that of the receiving station. If the latter, the frames carry their own address. REJ frames are always considered *responses* and carry their own addresses. The P/F bit enables the command-response mechanism to be carried out. A 1 set in a command is defined to be a P. The response to a $P = 1$ must carry $F = 1$. In the normal response mode the P/F bit is used for polling. In the ABM mode, with the BAC 2,8 class of procedures, an I-frame sent with the P bit set equal to 1 requires an S-frame response (RR, REJ, RNR) with $F = 1$, since the I-frame cannot be a response. An RR with the P bit set will force an S-frame with $F = 1$ to be sent in reply. This P/F procedure is called a *checkpointing* procedure. One example of its use is to force an immediate ack. On receipt of an I-frame with $P = 1$, an RR with $F = 1$ will be sent by the recipient immediately, ahead of any I-frames waiting to be transmitted. (This thus induces a nonpreemptive priority.) Details appear in [CARLD].

What are some reasons for invoking the P/F checkpointing procedure, in addition to expediting error detection? The procedure can be used to check whether an operational data link is present; it can be used to force an early acknowledgement of a particular I-frame, rather than having it indirectly acked later by a higher-numbered $N(R)$; it can be used to force transmission of an REJ (i.e., a nak) in case of an error, rather than relying on a timeout mechanism. (This might in certain circumstances speed up error recovery and reduce the number of frames that might have to be retransmitted in the event of an error.) Finally, the procedure could be used for preparing to take a link down (disconnect). In this case it could be used to clean up outstanding acks or other control signals. It is important to note that the P/F checkpointing procedure, like others in the HDLC repertoire, is optional. Although various procedures are defined specifically, others are left undefined, allowing flexibility in the use of the data link control.

To provide a better understanding of the use of the various S-frames in the ABM mode of HDLC, we first follow the typical flow of frames back and forth between two stations A and B. An example appears in Fig. 4–11, which shows

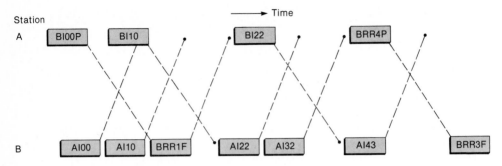

Figure 4–11 Example of HDLC (ABM) error-free transmission

only the data transfer mode. (See [CARLD, p. 461] for other examples, including the use of U-frames for setting up the link connection.) Error-free transmission is assumed. (The error case is treated later, in connection with the throughput analysis of HDLC.)

The notation used, adapted from [CARLD], is as follows:

1. *I-frames.* Address, I, $N(S)$, $N(R)$, P (optional; 1 if used, 0 otherwise)

    Thus AI10P refers to an I-frame from station B addressed to station A, with $N(S) = 1$, $N(R) = 0$ (i.e., B is *expecting* frame 0 from A), and the P/F bit set to 1. (A reply with $F = 1$ is thus expected.)

2. *S-frames.* Address, id, $N(R)$, P/F

    The address, as just noted, is that of the receiver if a command; itself, if a response. The id is RR, RNR, or REJ for each of the three S-frames used. $N(R)$ again acks all frames up to and including $N(R) - 1$. P/F is written as P if the frame is a command and as F if a response to a P. (In either case the P/F bit is set to 1.)

Station A initiates transmission to station B in the example of Fig. 4–11 by sending an I-frame numbered 0. Station A requests an immediate ack by setting $P = 1$. Since A has not as yet received any I-frames from B, $N(R)$ is set at 0, indicating that I-frame 0 is expected from B. In the meantime station B independently sends two I-frames in succession, AI00 and AI10. (Station B is also expecting I-frame number 0.) On receiving BI00P, station B immediately sends an ack BRR1F. This acknowledges receipt of I-frame 0 from A and indicates that frame 1 is expected. Station B later receives I-frame BI10 from A and then acknowledges this frame with its own I-frame AI22. B follows this frame with

AI32. Both frames are later acked by station A with an RR-frame BRR4*P*. This frame might be used, for example, if station A were now preparing to disconnect. Station B replies with BRR3*F*.

4–3–1 Throughput Analysis, Balanced HDLC Procedure

Following this introduction to some of the aspects of the HDLC protocol, we now proceed to a throughput analysis of the procedure. We follow Bux et al. [BUX 1980], but for a simplified case only, noted later. Bux et al. also provide a time-delay analysis. The reader is referred to their paper, and other papers referenced there, for details of both the throughput and the time-delay analyses.

The basic differences between this analysis and that of the idealized go-back-*N* analysis carried out earlier are that finite sequence numbering is included here and that specific error-control features are modeled as well. This analysis serves two purposes: It enables us to compare the idealized infinite sequence number throughput result with the finite number case, and it provides an exercise in modeling a real rather than an idealized protocol.

It was noted earlier, in discussing some aspects of the HDLC protocol, that certain features are not explicitly defined but are left to the implementor to allow flexibility in use. This is true specifically of the error-control mechanism. Although the S-frames REJ, RR, RNR, and the checkpointing (*P/F* bit) mechanism are all available for use in error control, specific roles for their use are left to the implementor. For example, although receipt of REJ rejects I-frame *N(R)* and all I-frames following, this nak feature does not necessarily have to be used. One could choose to implement only the positive ack (either RR or an I-frame) with the timeout feature. However, only an analysis of a specific implementation can be carried out. For this reason Bux et al. provide some specific rules for the use of error control. These involve the use of the REJ frame, the timeout mechanism, and the use of checkpointing (the *P/F* bit). Specifically, the following stipulations are made:

1. *REJ recovery.* This is assumed to be always used where possible to speed up system recovery. However, it can be used only once for a given frame. It cannot be invoked on repeats of that frame.

2. *Timeout recovery.* This must *always* be used, in addition to REJ recovery; without this feature an isolated I-frame or the last in a sequence of I-frames could not be recovered if garbled. (Why not?) In addition, since an REJ frame may be used only once per I-frame, multiple losses of a given frame must be handled through a timeout.

3. *Checkpoint (P/F) use.* At the end of a timeout the transmitter sends RR*P*.

With this assumption, a reply from the receiver is required after the expiration of each timeout interval.

Using these three rules, we are in a position to calculate the link throughput. Following the procedure adopted in our earlier analyses of the stop-and-wait and go-back-N protocols, we consider the case where one station only, station A, is transmitting. This station is assumed in addition to be in a saturated state: It *always* has frames to send. As noted earlier, this provides the maximum possible throughput. The receiving station B then acks with RR or acknowledges negatively with REJ.

Let the sequence number modulus be M. The condition on the sequence number $N(S)$ is then

$$0 \le N(S) \le M - 1 \qquad (4-13)$$

As noted earlier, for normal HDLC $M = 8$; the extended version has $M = 128$. At most $M - 1$ frames can then be outstanding. (Consider the case in which all M frames, 0 through $M - 1$, are outstanding and hence unacknowledged. Since $N(R)$ acks all frames up to and including $N(R) - 1$ and indicates that $N(R)$ is expected, a difficulty arises immediately.) As lower-number frames are acked, their numbers may be used again. This mod-M sequence-numbering mechanism with acknowledgement thus establishes a variable *window* of sequence numbers that can be used. An example appears in Fig. 4–12 for $M = 8$. Initially it is assumed that frames 1–5 are outstanding and unacknowledged. A window of two frames, 6, 7, remains (recall that $M - 1$ frames at most may be outstanding). In part (b) of Fig. 4–12 an ack bearing $N(R) = 3$ is assumed to have arrived. This acks frames 1 and 2, allowing a window of frames 6, 7, 0, 1 to be used.

In carrying out the analysis we again assume that all frames are of *fixed* length. I-frames are of length t_I sec, $t_I = (\ell + \ell')/C$, with ℓ the packet (information field) length, ℓ' the control bits in the frame, and C the link capacity in bps. All S-frames are of length $t_s = \ell'/C$, with $\ell' = 48$ bits for the HDLC protocol. The one-way propagation delay is again taken to be t_p sec, with processing time included. The round-trip acknowledgement time is then

$$t_{\text{ack}} = 2t_p + t_s \qquad (4-14)$$

and we again take

$$t_{\text{out}} = 2t_p + 2t_I > t_{\text{ack}} \qquad (4-15)$$

(The acknowledgement time is defined to be the time between transmission of a frame and receipt of an ack.) We take the case $t_{\text{ack}} \le (M - 2)t_I$. This simplifies the analysis considerably and is not unduly limiting. This case covers such examples as both terrestrial and satellite transmission with 1000-bit frames transmitted at 48-kbps link speed, using $M = 8$ for the terrestrial case and

$M = 128$ for the satellite case. For smaller frame lengths (or lower transmission speeds) or smaller sequence numbers, such that $t_{ack} > (M - 2)t_I$, one must, of course, use the analysis covering that case as presented in [BUX 1980].

Before proceeding with the analysis we provide a simple example of frame transmission in the presence of errors to show how the three error-control rules stipulated by Bux et al. are used (see Fig. 4–13). The maximum sequence number M is taken to be four for simplicity. At most three I-frames can then be outstanding at any time. Since station A *always* has I-frames waiting, they are transmitted consecutively, one after the other, until either $M - 1 = 3$ frames are outstanding, a timeout has occurred with no ack or nak, or a go-back-N repeat of frames is required due to an error. The notation used in Fig. 4–13 is the same as in Fig. 4–11 except that addresses are dropped. Addresses need not be included since *all* I-frames emanate from station A by hypothesis. Frames 0–2 are shown proceeding uneventfully from station A to station B, each one being individually acknowledged by an appropriately numbered RR-frame. I-frame 3 is shown being hit by an error during transmission. The receiver in this example ignores the faulty frame. When frame 0 arrives it is out of sequence, however (station B is expecting frame 3), and REJ3 is returned to A. In

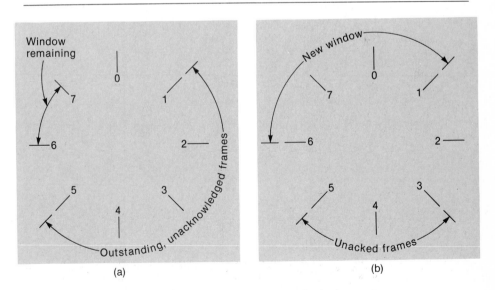

Figure 4–12 Sequence number and window concept
a. Prior to ack arrival
b. Ack with $N(R) = 3$ arrives

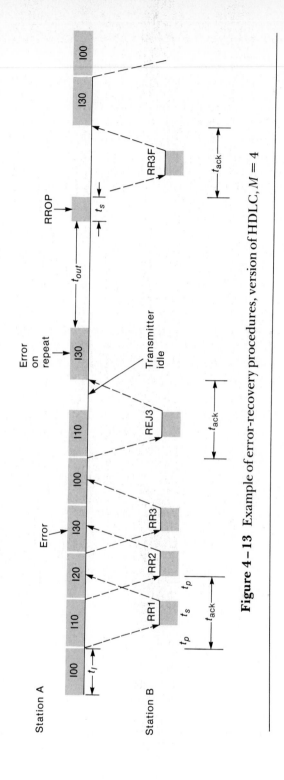

Figure 4-13 Example of error-recovery procedures, version of HDLC, $M = 4$

the meantime, station A transmits frame 1 and then stops transmitting. (Why?) When it receives REJ3 it repeats frame 3, and again follows this with frames 0 and 1 (not shown here to avoid cluttering the figure). In this example frame 3 is again hit by an error during transmission. Station B does nothing since it has already sent REJ3 once. The timeout at A then runs out. Using the Bux et al. rule, RR0P is sent from A to B to force an ack. (Why is there a 0 in this frame as well as in all other frames from A to B?) B replies with RR3F and, on receipt at A, I30, I00, . . . are again sent. (In other versions of HDLC one might simply repeat the faulty frame and all frames following on expiration of the timeout.)

The approach adopted to calculate the throughput for this model of the HDLC protocol is the same as the one used earlier to analyze the stop-and-wait and go-back-N protocols. One calculates t_V, the virtual transmission time of a typical frame, and then inverts this to find the maximum frame (packet) throughput. The essential difference here is that either REJ *or* timeout recovery may have to be evoked on an error occurrence, and one must distinguish between the two. Note again that with the implementation rules assumed here, *either* REJ *or* timeout recovery will take place after an initial frame error, but *only* timeout recovery can be used during subsequent errors on retransmissions of the same frame.

Again letting p be the probability that a frame is received in error, the virtual transmission time, or the average time required to transmit a frame of length t_I, may be written as

$$t_V = t_I + pE(T_1) + \sum_{n=2}^{\infty} (1 - p)p^n(n - 1)T_2$$
$$= t_I + pE(T_1) + \frac{p^2T_2}{(1 - p)} \tag{4-16}$$

Here T_1 represents the random time required to transmit the first repeat, $E(T_1)$ is its average value, and T_2 is the average time required for transmission on subsequent retransmissions. Since the recovery mechanism (timeout recovery) will always be the same on retransmissions beyond the first, because of the rule used here, T_2 will be the same on each retransmission. T_1 will differ, however, depending on which recovery mechanism (REJ or timeout) happens to be invoked.

Note that Eq. (4–16) is very similar to Eq. (4–5), the equation for virtual transmission time in the (idealized) go-back-N protocol. In fact, it is apparent that if $E(T_1) = T_2 = t_T$, one gets precisely Eq. (4–5).

As just stated, T_1, the time required to transmit the first repeat of a faulty frame, depends on the error-recovery procedure invoked. One must thus determine the condition for each to occur and average accordingly. In addition, in both cases T_1 is found to depend on a parameter L, defined to be the *maximum* number of I-frames following the one that is disturbed before recovery is started. This will be $M - 2$ if the sequence space (window) has expired (recall that $M - 1$ frames at most can be outstanding at any one time) *or* the number of frames transmitted during a timeout interval, whichever is less. In the latter case a frame undergoing transmission at the transmitter when the timeout interval runs out is allowed to complete transmission. (The timeout counter thus invokes a nonpreemptive interrupt on I-frames awaiting transmission.) L may thus be written in the form

$$L = \inf \{M - 2, \lfloor t_{out}/t_I \rfloor + 1\} \tag{4-17}$$

with inf (inferior) meaning the "lesser of" and $\lfloor a \rfloor$ representing the largest integer not exceeding a.

As an example, consider a satellite link with $2t_p = 700$ msec, 1000-bit frames, and $C = 48$-kbps transmission rate. Using $t_{out} = 2t_p + 2t_I$, $\lfloor t_{out}/t_I \rfloor + 1 = 36$. For $M = 8$, $L = 6$; for $M = 128$, $L = 36$. Clearly "sequence number starvation" can occur on a satellite link with a sequence number space of $M = 8$.

1. *Recovery via REJ.* Consider now the calculation of T_1 in the case where error recovery is invoked via the REJ mechanism. Clearly this will occur if the REJ mechanism comes into play *before* the effect of the timeout expiration is felt. As will be seen from studying some typical examples, the condition for this to happen is that *not all L I-frames following the one under consideration are disturbed.* (If *all* L frames following *are* disturbed timeout recovery will be invoked.)

   The detailed calculations depend in turn on two ranges of values for L.

   a. L small (for example, t_{out} is relatively small)

   An example appears in Fig. 4–14(a). An error occurs during transmission of frame 1. We then calculate the time T_1 required to complete the first retransmission of this frame. In this example frame 2 is also shown disturbed during transmission. The timeout for frame 1 expires before any frame from B arrives. Frame 3, in the process of being transmitted during the expiration of the timeout, is allowed to complete, and the S-frame RR0P is then transmitted to force an ack from station B. But B receives frame I30 out of order (it is expecting I10) and immediately generates REJ1. This causes I10 to be retransmitted by A. The S-frame response RR1F to station A's RR0P after expiration of the timeout arrives later and is ignored. The REJ mechanism thus is invoked before the timeout mechanism, due to receipt of the undisturbed I30 by B. Had I30 been disturbed, REJ1 would *not* have been sent by B, and the timeout mechanism would have taken over (RR1F in reply to RR0P from A). Hence the condition for REJ to take precedence is that noted earlier: Not all L frames following the one under consideration (frame 1

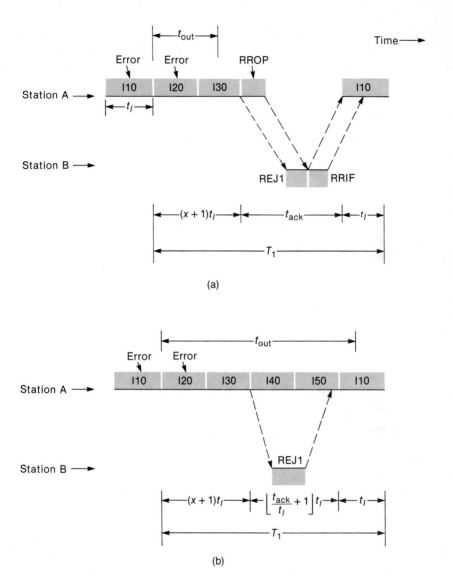

Figure 4–14 Example of REJ recovery
a. $Lt_I < (x + 1)t_I + t_{ack}$
b. $Lt_I \geq (x + 1)t_I + t_{ack}$

in this example) are disturbed. Note that in this case channel A to B is idle on receipt of the REJ frame. (This is due either to expiration of a time-out, as in this example, or because the sequence window is closed. $M - 1$ unacknowledged frames are then outstanding in this case.)

There are in general $x \leq L - 1$ frames disturbed *after* the one under consideration. In the example of Fig. 4–14(a), $x = 1$. As indicated in the figure, in general

$$T_1 \equiv \tau(x) = (x + 1)t_I + t_{ack} + t_I \qquad (4-18)$$

The probability of this event happening is just $p^x(1 - p)$ (x disturbed frames and one frame received correctly). This case corresponds to L small enough so that

$$Lt_I < (x + 1)t_I + t_{ack} \qquad (4-19)$$

b. L large (for example, say that t_{out} is now larger)

Specifically, let

$$Lt_I \geq (x + 1)t_I + t_{ack} \qquad (4-20)$$

Referring to Fig. 4–14(b), it is apparent that REJ recovery is now invoked simply because the timeout on frame 1 is so long as to expire after the REJ frame asking for a repeat of the frame has been received from station B. In this example the numbers have been chosen to have the sequence number space in I-frame units, $(M - 2)t_I$, longer than the timeout interval. Had we chosen $(M - 2) < \lfloor t_{out}/t_I \rfloor + 1$, we would have obtained the same result, with the condition that $Lt_I = (M - 2)t_I \geq (x + 1)t_I + t_{ack}$.

From Fig. 4–14(b) it is apparent that for this case we have

$$T_1 \equiv \tau(x) = (x + 1)t_I + [\lfloor t_{ack}/t_I \rfloor + 1]t_I + t_I \qquad (4-21)$$

again with probability $(1 - p)p^x$. Comparing with Eq. (4–18), we note that Eqs. (4–21) and (4–18) are very similar, the only difference being that the quantity in brackets in Eq. (4–21) is replaced by t_{ack} in Eq. (4–18). Here a frame is in the process of being transmitted on receipt of the REJ and it is allowed to go to completion, lengthening the effective time T_1, as contrasted with the case of Fig. 4–14(a). There the REJ arrived with the channel from A to B empty.

2. *Timeout recovery.* This recovery mechanism is invoked if *all* L I-frames following the one initiating the error-recovery procedure are disturbed. This happens with probability p^L. Then either the timeout runs out or the sequence window is exhausted ($M - 1$ frames are oustanding), whichever comes first. Thus there are again two cases to consider. An example of each appears in Fig. 4–15. Since all frames sent from A are disturbed, no REJ frame can be sent from B, and timeout alone controls the recovery.

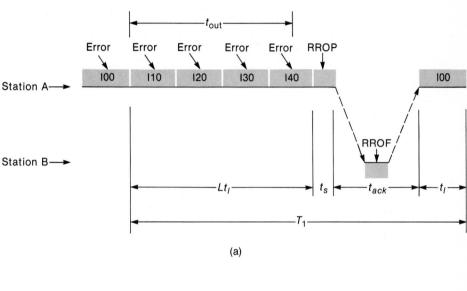

(a)

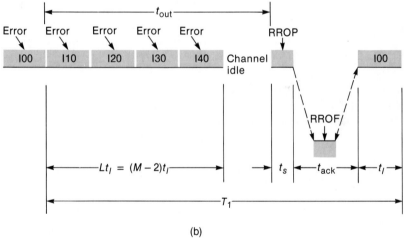

(b)

Figure 4–15 Timeout recovery
a. $t_{out} \leq Lt_I$
b. $t_{out} > Lt_I$

a. $t_{out} \le Lt_I$

Then $L = \lfloor t_{out}/t_I \rfloor + 1 < M - 2$ in this case. As is apparent from Fig. 4–15(a),

$$T_1 = Lt_I + t_s + t_{ack} + t_I \equiv \tau(L) \qquad (4-22)$$

b. $t_{out} > Lt_I$

Then clearly $L = M - 2$, the sequence window is exhausted before the timeout expires, and there is an interval between the two events during which the channel from A to B is idle. This case appears in Fig. 4–15(b). It is apparent that for this case

$$T_1 = t_{out} + t_s + t_{ack} + t_I \equiv \tau(L) \qquad (4-23)$$

almost the result of Eq. (4–22) except for the slight difference in the first term.

Combining the results of the REJ and timeout recovery cases, by weighting each with the respective probability of occurrence, we get

$$E(T_1) = \sum_{x=0}^{L-1} (1 - p)p^x \tau(x) + p^L \tau(L) \qquad (4-24)$$

The functions $\tau(x)$ and $\tau(L)$ depend on the two cases just considered.

Finally we must calculate T_2, the average transmission time for each repeat of a frame beyond the first retransmission. (See Eq. (4–16).) Clearly T_2 must be the function $\tau(L)$ defined in either Eq. (4–22) or Eq. (4–23), since timeout recovery only can be invoked for retransmissions beyond the first. (This was one of the rules assumed here in defining the error-recovery procedure.) Thus

$$T_2 = \tau(L) \qquad (4-25)$$

Summarizing the analysis carried out above in step-by-step fashion, we have as the average virtual transmission time for the balanced HDLC procedure

$$t_V = t_I + pE(T_1) + \frac{p^2}{(1 - p)} T_2 \qquad t_{ack} \le (M - 2)t_I \qquad (4-16)$$

Here

$$E(T_1) = \sum_{x=0}^{L-1} (1 - p)p^x \tau(x) + p^L \tau(L) \qquad (4-24)$$

and

$$T_2 = \tau(L) \qquad (4-25)$$

The functions $\tau(x)$ and $\tau(L)$ are in turn given by

$$\tau(x) = (x + 1)t_I + t_{ack} + t_I \qquad Lt_I < (x + 1)t_I + t_{ack} \qquad (4-18)$$

otherwise

$$= (x + 1)t_I + [\lfloor t_{\text{ack}}/t_I\rfloor + 1]t_I + t_I \qquad (4-21)$$

and

$$\tau(L) = Lt_I + t_s + t_{\text{ack}} + t_I \qquad t_{\text{out}} \le Lt_I \qquad (4-22)$$

otherwise

$$= t_{\text{out}} + t_s + t_{\text{ack}} + t_I \qquad (4-23)$$

The coefficient L is again defined as

$$L = \inf\{M - 2, \lfloor t_{\text{out}}/t_I\rfloor + 1\} \qquad (4-17)$$

The maximum frame (packet) throughput rate is in turn given by

$$\lambda_{\text{max}} = 1/t_V \qquad (4-26)$$

The relative data throughput D/C is finally found as in the case of the idealized go-back-N protocol, as

$$D/C = \lambda_{\text{max}}\ell/C = \ell/t_V C \qquad (4-27)$$

We reproduce in Figs. 4-16 and 4-17 the results of calculations using these equations as well as equations obtained in [BUX 1980] for the condition $t_{\text{ack}} > (M - 2)t_I$. (These curves appear as Figs. 4-8 and 4-9, respectively, in [BUX1980].) Figure 4-16 plots the relative throughput D/C versus packet length ℓ, in bits, for a terrestrial link using a sequence number space of $M = 8$, assuming $p_b = 10^{-5}$ and taking $t_p = 50$ msec. The control field size is $\ell' = 48$ bits, as appropriate for HDLC. Two curves appear, one for a link capacity $C = 4800$ bps, the other for $C = 48$ kbps. Figure 4-17 is for a satellite link, with a number of cases shown plotted. Curves for $M = 8$ and 128 appear, showing the effect of changing sequence number size on throughput. Two bit-error probabilities, $p_b = 10^{-5}$ and 10^{-7}, have been included as well. The link capacity used is $C = 48$ kbps, and t_p has been taken equal to 350 msec. Note that these numbers are precisely those used in Fig. 4-8 for the idealized go-back-N protocol, enabling comparisons to be made. Discontinuities appearing in the curves of Figs. 4-16 and 4-17 are due to the $\lfloor t_{\text{ack}}/t_I\rfloor$ terms appearing in the analysis and to the assumptions of the model.

Note first how closely the curve for 4800 bps in the terrestrial case (Fig. 4-16) agrees with the comparable curve in Fig. 4-8 for the idealized go-back-N strategy. It is apparent that the finite sequence number field of $M = 8$ is ample in this case. The curve for $C = 48$ kbps in Fig. 4-16, however, shows a sharp dropoff at the small packet end. This is obviously due to sequence-number starvation in this case. As long as packet lengths are kept above 500 bits in length, however, there appears to be no problem.

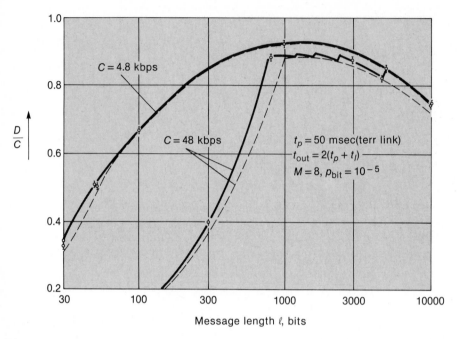

Figure 4–16 Throughput efficiency D/C versus message length ℓ (terrestrial links) (from [BUX 1980, Fig. 8, © 1980 IEEE, with permission])

The satellite case is drastically different. It is apparent from Fig. 4–17 that the sequence-number field *must* be extended to $M = 128$ to provide appropriate throughput over a broad range of packet sizes. The $M = 128, p_b = 10^{-5}$ curve is in substantial agreement with the comparable curve of Fig. 4–8 for the idealized go-back-N protocol with no limit on sequence-number size. The primary factor limiting throughput in this case is the noise characteristic assumed for this channel. Reducing the bit error probability to $p_b = 10^{-7}$ improves the performance considerably. However, this implies substantially higher costs: higher power transmitters, lower temperature receivers, larger antennas, and so on [SCHW 1980, pp. 432, 433]. Alternatively, as noted earlier in this chapter, selective repeat strategies could be used to enhance the performance of a satellite link for data transmission [LINS, Chapter 15 and references therein]. This requires reordering packets at the receiver since there is no longer any guarantee that frames transmitted or retransmitted will be received in order. A reordering buffer obviously is required at the receiver for this purpose.

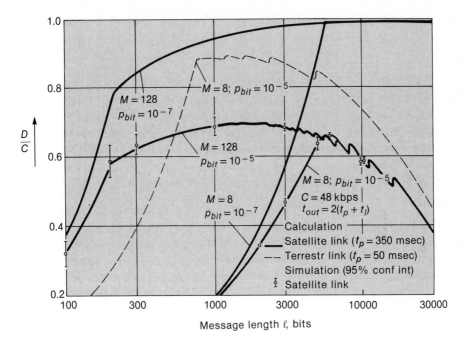

Figure 4–17 Throughput efficiency D/C versus message length ℓ (satellite links) (from [BUX 1980, Fig. 9, © 1980 IEEE, with permission])

A simple throughput analysis of an idealized selective-repeat scheme, with no limit on the number of retransmissions of a given frame, shows the maximum throughput improvement possible through the use of this strategy. This idealized scheme requires the use of an infinite reordering buffer at the receiver. Since individual frames only must be retransmitted, frames transmitted and received correctly after one is found to be in error need not be retransmitted. The only reduction in throughput is thus that due to the need to retransmit a *given* frame detected to be in error. Reduction in throughput is given precisely by the average number of retransmissions required. This differs specifically from the go-back-N case, where transmission of subsequent frames (packets) was held up until a particular frame had cleared the system. Again letting p be the probability of receiving a frame in error, the average number of transmissions required to get a frame through the link successfully is just

$$E(n) = (1 - p) \sum_{i=1}^{\infty} i p^{i-1} = 1/(1 - p) \tag{4-28}$$

The normalized data rate is then

$$\frac{D}{C} = \frac{1}{E(n)}\left(\frac{\ell}{\ell + \ell'}\right) = \frac{(1-p)\ell}{(\ell + \ell')} \tag{4-29}$$

Here as earlier $p = 1 - (1 - p_b)^{\ell+\ell'}$, with p_b the bit error rate. Note by comparison with Eq. (4–10) for the go-back-N case that the reduction in throughput due to retransmission of subsequent frames given by the $(a - 1)p$ term in the denominator there has been eliminated. The throughput should thus be considerably enhanced over the go-back-N ack when the probability p of a frame error becomes significant.

Equation (4–29) is precisely the special case of Eq. (4–10) when $a = 1$, i.e., when the round-trip propagation delay in the go-back-N case is negligible. This was the case that was assumed in deriving Eq. (4–11) for the optimum packet length. Hence the optimum choice of packet length, ℓ, is the same as that found earlier: in the vicinity of 1000 bits for $p_b = 10^{-5}$ and 10,000 bits for $p_b = 10^{-7}$. The data throughput for these probabilities of bit error will be higher than for the go-back-N scheme because the dependence on a has been eliminated; for small p_b the optimum throughput region will be quite broad as well. This is shown in Fig. 4–18, a plot of the relative data throughput for the selective repeat case versus packet length ℓ. The same frame and channel characteristics as in Fig. 4–8 have been assumed: $\ell' = 48$ bits, $p_b = 10^{-5}$. The 48-kbps, satellite, go-back-N curve of Fig. 4–8 has been repeated in Fig. 4–18 so that comparisons can be made. Note the improved throughput *theoretically* possible using the selective repeat strategy. A finite buffer for the selective repeat case will reduce this idealized throughput [LINS, pp. 464–474].

The relative dependence of the go-back-N and selective repeat strategies on the bit-error probability is shown in Fig. 4–19. Here a fixed packet length of $\ell = 1000$ bits has been chosen, and the relative data throughput (throughput efficiency) of the two schemes plotted as a function of p_b. Equations (4–10) and (4–29) have been used for this purpose. (Actually, $(\ell + \ell'/\ell)D/C$ has been plotted for simplicity.) The only difference between the two curves is then the factor $[1 + (a - 1)p]$ in the denominator of Eq. (4–10) for the go-back-N scheme, which is due to the forced retransmission of all frames following the one received in error. A satellite channel has again been assumed, with $2t_p + t_s = 700$ msec and $C = 48$ kbps to allow comparison with previous curves. For these values the parameter $a = 35.1$. It is apparent from these curves that the efficiency of the go-back-N scheme in the vicinity of $p_b = 10^{-5}$ is quite sensitive to error probability. If p_b drops to 10^{-4}, for example, the throughput decreases by a factor of 3.6. The throughput of the selective repeat scheme is constant in the vicinity of $p_b = 10^{-5}$ and begins to show sensitivity to increased error probability only in the vicinity of $p_b = 10^{-3}$. These results are, of course, illustrative. The specific values of p_b at which the throughput begins to drop rapidly for the

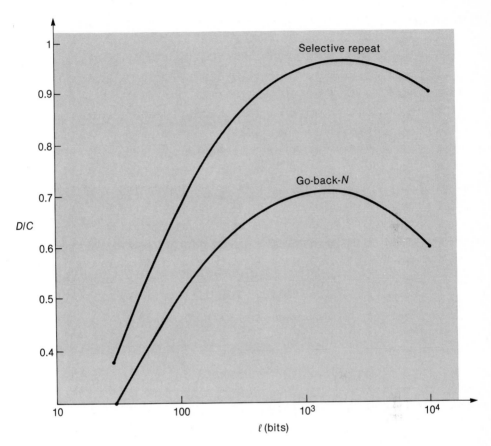

Figure 4-18 Selective repeat and go-back-N procedures

$$C = 48 \text{ kbps}$$
$$2t_p = 700 \text{ msec}$$
$$p_b = 10^{-5} \quad \ell' = 48 \text{ bits}$$

two schemes depend on the packet length ℓ and the parameter a, in the case of go-back-N. The performance of the selective repeat strategy deteriorates because of the finite buffers required. Still, similar results are obtained for other choices of packet length, and an order-of-magnitude increase in the bit-error probability at which the throughput begins to drop appears to be obtainable through the use of the selective-repeat scheme [LINS, pp. 464, 465].

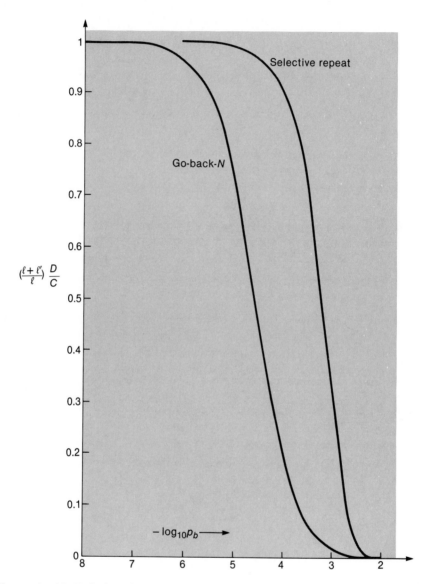

Figure 4–19 Relative data throughout versus bit error probability, satellite channel

$$\ell \; = 1000 \text{ bits}$$
$$\ell' = 48 \text{ bits}$$

Problems

4-1 Derive Eq. (4-5), the expression for the average transmission time for a frame in the go-back-N protocol.

4-2 Plot Eq. (4-10), the normalized data throughput rate D/C for the go-back-N protocol, as a function of packet length ℓ for the following two cases:

1. $\ell' = 48$ bits, the link length is 1500 km, the propagation speed over the link is 150,000 km/sec, processing delay is 30 msec, $p_b = 10^{-5}$, $C = 1200$ bps. Bit errors are independently distributed. Take $t_{out} = 2t_p + 2t_I$, as in the text, with processing time included in the propagation delay.
2. Same as case 1, except that $C = 9600$ bps.

Compare the two cases.

4-3 **a.** A *stop-and-wait* protocol uses positive acknowledgements (acks) and timeouts only. (No negative acknowledgements are used.) The ack arrives at the transmitter at time t_{ack} after transmission of a frame. Take $t_{out} > t_{ack}$. Frames are all of length t_I and are always available for transmission. A frame is received in error with probability p. Show that the maximum frame throughput is

$$\lambda_{max} = (1 - p)/t_I[1 + (b - 1)p]$$

where $b = (t_I + t_{out})/(t_I + t_{ack}) > 1$.
Compare with Eq. (4-2) in the text.

  b. A *go-back-N* protocol is used in place of the stop-and-wait protocol. Again take $t_{out} > t_{ack}$. Show that the maximum frame throughput is now given by Eq. (4-6) and is *independent* of t_{ack}. Why does t_{ack} *not* appear here? Equation (4-6) was derived for the case of $t_{out} = t_{ack}$. Why does it apply as well to the procedure used here? How would the maximum frame throughput expression change if negative acknowledgements were also used? Which procedure provides a potentially greater throughput?

4-4 Consider a data link protocol.

  a. The protocol uses positive acknowledgements with timeout. Will the protocol work correctly (i.e., with the data eventually getting through) if, because of lack of knowledge, t_{ack} is chosen $> t_{out}$? Explain.

  b. The protocol uses both negative and positive acks. Is a timeout function then needed? Explain.

  c. A data link has the property that occasionally a frame may be misdirected and eventually arrive much later than expected. In the meantime the transmitter will have retransmitted the frame in question, and possibly frames following. Can one guarantee that the receiver will detect the original (duplicate) frame and discard it on arrival? Explain.

4-5 Consider the following go-back-N data link protocol:

1. Packets are acknowledged individually by the receiver. The acknowledgement (ack) is received at the transmitter t_{ack} sec after a packet completes transmission.

2. For packets received in error:

    a) The *first* transmission of a packet received in error is negatively acked. The nak appears at the transmitter t_{ack} sec after transmission of the packet.

    b) Subsequent retransmissions of a packet received in error are ignored by the receiver. The transmitter times out t_{out} sec after the previous retransmission and then repeats the packet.

$$t_{out} > t_{ack}$$

Each packet is t_I sec long. The probability that a packet is received in error is p. Find the maximum packet transmision rate, λ_{max}, in packets/sec. Make any simplifying assumptions needed.

4-6 An idealized data link protocol incorporates both positive acknowledgements (acks) and negative acknowledgements (naks). The nak is always used the first time a frame is rejected (assume that the timeout is never required); all subsequent repeats of that frame received in error are dropped by the receiver, and recovery is effected by the timeout mechanism only. Show that the average time to transmit a frame is given by

$$t_V = t_I + pt_T + \frac{p^2 t'_T}{(1-p)}$$

Here p = probability that a frame is received in error, t_I = frame length, $t_T = t_I + t_{ack}$, $t'_T = t_I + t_{out}$ ($t_{nak} = t_{ack}$). Compare with Eq. (4–16), the expression obtained for the model of the HDLC protocol described in the text. Explain the differences. Show that one gets Eq. (4–5) if $t_{ack} = t_{out}$.

4-7 Plot the normalized data throughput rate, D/C, given by Eq. (4–10) as a function of packet length ℓ for the three examples of Fig. 4–8, and compare with the curves shown there.

4-8 Taking $(a-1)p \ll 1$ in Eq. (4–10), show that the optimum packet length is given by Eq. (4–11). Check the predictions of this equation with the curves of Fig. 4–8.

4-9 Find the optimum packet length ℓ and the corresponding data rate, D, in bps, for the following cases:

    a. $\ell' = 48$ bits, propagation speed on the link is 150,000 km/sec, link length is 300 km, there is negligible processing delay, $p_b = 10^{-5}$, $C = 1200$ bps.

    b. Same as the first case, but $C = 9600$ bps.

    c. Satellite link, $\ell' = 48$ bits, $t_p = 0.25$ sec, $p_b = 10^{-5}$, $C = 56$ kbps.

    d. Same as the third case, but $p_b = 10^{-7}$.

4-10 Following is a simplified infinite-sequence-number model of the HDLC protocol analyzed in this chapter. An RR control frame is used to acknowledge correct receipt of an I-frame of length t_I. An REJ frame is used to reject a frame received in error, on the first transmission of that frame only. Subsequent retransmissions of the frame, if received in error, are ignored; a timeout t_{out} is then used to control retransmission. Both the RR and the REJ frames are received at the transmitter at time t_{ack} after transmission of an I-frame. Take $t_{out} \geq t_{ack}$. Find an expression for the maximum frame transmission time λ_{max} and compare with the HDLC analysis results in the text. Make any assumptions required on the length of t_{out} and t_{ack} to simplify the analysis. (For example, t_{out} is a multiple of t_I.)

4-11 Refer to Problem 4-2. Calculate and plot D/C versus ℓ for the two examples given there, using the results of the HDLC analysis (see Eq. (4-16) and the equations following, as summarized after that analysis). Take $M = 8$. Compare with the results of Problem 4-2. It might be helpful to program the various equations and solve by computer. Assume that bit errors are independent, bit by bit.

4-12 Consider the terrestrial error-probability model discussed in the text. Show that the optimum packet length for this model is given by Eq. (4-12).

4-13 Two neighboring nodes carry out the HDLC asynchronous balanced-mode protocol. Depict a typical full-duplex sequence of frames flowing between the two nodes, including the use of the error-recovery procedures discussed in the text. (This diagram should be a composite of Figs. 4-11 and 4-13.) Take $M = 4$ to simplify the figure.

4-14 Refer to Fig. 4-13 in the text. Why is the transmitter idle after transmitting frame 1, as shown?

4-15 A suggestion is made that the RR frame in HDLC be used not only for positive acknowledgements, but for negative acknowledgements as well. After all, it is argued, both the RR and the REJ (nak) frames positively acknowledge receipt of frame $N(R) - 1$ and all frames preceding and indicate that frame $N(R)$ is expected. They are thus implicitly carrying out the same function. Hence why not use RR in all cases? Explain why this could not be possible.

4-16 The maximum data throughput efficiency D/C for the balanced HDLC protocol is to be calculated for the following example:

$$M = 8, \quad p_b = 10^{-5}, \quad \ell' = 48 \text{ control bits}, \quad C = 2400 \text{ bps}, \quad t_p = 50 \text{ msec}$$

Assume that bit errors are independently distributed.

a. Take $\ell = 100, 1000, 2000, 10^4, 30,000$ bits. Plot D/C versus ℓ. *Note:* You might want to program this problem and use the computer.

b. Compare your results with those of Fig. 4-16 (see [BUX 1980, Fig. 8]). Compare also with comparable results for the idealized go-back-N protocol with an infinite sequence number.

4-17 It is noted in the text that no more than $(M - 1)$ unacknowledged frames can be outstanding at any one time if M is the maximum sequence number. Explain why this is a necessary condition for the protocol always to work correctly. A counterexample, showing what might happen if M unacknowledged frames *were* outstanding, would be helpful. *Hint:* Consider the following possible scenario. Station A transmits I-frames $0, 1, 2, \ldots, M - 1$, and then stops. Station B, silent to this point, now transmits an I-frame. How does this differ from the case in which A is always silent?

4-18 Explain by an example, with one node serving as transmitter and the other as receiver, why the selective repeat scheme requires a reordering buffer at the receiver. Show a series of sequentially numbered frames being transmitted. A copy of each frame is stored at the transmitter until acknowledged positively. A number of the frames are "hit" by errors. Indicate with diagrams how error recovery is carried out. Sketch a block diagram of transmitter and receiver incorporating a "waiting for ack" buffer at the transmitter and a reordering buffer at the receiver.

4-19 Compare the go-back-N and selective repeat strategies using examples of your own choosing. Select at least one example in which the performance of the two differs substantially and one in which they perform about the same.

4-20 The asynchronous balanced mode of HDLC incorporating classes of procedure BAC 2,8 allows only one REJ frame to be outstanding at any one time. Once the I-frame requested is received, another REJ frame can be sent when necessary. Provide an explanation.

4-21 Explain the following statement in the X.25 specification concerning its data link layer (comparable to the asynchronous balanced mode of HDLC): "If a (receiver), due to a transmission error, does not receive (or receives and discards) . . . the *last I frame* in a sequence of I frames, it will not detect an out-of-sequence exception condition, and therefore will not transmit REJ." [X.25, Sec. 2.3.5.4].

[X.25] *Draft Revised Recommendation X.25, Interface between Data Terminals Operating in the Packet Mode in Public Data Networks,* Study Group VII, CCITT, Geneva, Feb. 1980; reprinted in *Computer Comm. Rev.,* vol. 10, nos. 1 and 2, Jan./April 1980, 56–129.

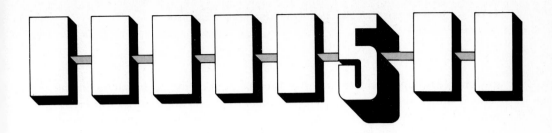

Network Layer: Flow Control and Congestion Control

In Chapter 3 we presented the concept of *layered network architectures,* describing in some detail the principles and philosophy of the seven-layer OSI Reference Model of the International Organization for Standardization (ISO). We also discussed some aspects of the X.25 protocol and the IBM Systems Network Architecture (SNA) briefly, indicating how they fit into the general picture of layered architectures.

In Chapter 4 we described the link layer in detail, focusing on the internationally recognized protocol HDLC (high-level data link control). The link layer represents the second layer of all architectures. (The bottom layer, as already noted, is the physical layer that ensures that a transmission path is available between two adjacent nodes in a network.) In this chapter and the next one we continue our discussion of the layered concept, moving up one layer to the network layer. This is the layer which, among its other functions, carries out routing in a network and provides flow control to ensure adequate, timely delivery of packets from one end of the network (the source) to the other (the destination). It uses the data link layer below to provide the necessary reliable transfer of blocks of data across a physical link.

In this chapter we discuss the flow control function. In Chapter 6 we discuss routing procedures in networks. We shall try to be as quantitative as possible in this chapter, but will introduce the necessary modeling and analysis in the context of practical examples. Thus we start by describing the X.25 interface protocol in more detail. This enables us to introduce the concept of a virtual circuit (VC) and a window-type mechanism for controlling flow. We model and compare two VC window flow controls, using queueing models for the nodes along the virtual circuit. We then describe the network or path-control layer of SNA, indicate how end-to-end flow control (called pacing) is maintained in that architecture, and proceed to model a simplified version of the flow-control mechanism. The mechanism is of the window type, allowing comparisons to be made with the two other window controls introduced earlier.

Using queueing models for the nodes in a packet-switched data network results in an overall model called a *queueing network*. A great deal of work in recent years has gone into studying the properties of these networks, since performance questions relating to such important service criteria as network throughput and time delay can be answered quantitatively using such network models. The major problem is computational: Even for moderate-sized networks the analysis gets out of hand. We thus describe some computational procedures developed to reduce the computational burden. The queueing network model of a real network enables us to formulate and, in some cases, answer questions of optimum network control. We also describe and analyze some input-node flow control mechanisms (as contrasted to end-to-end control mechanisms) used in datagram networks. DEC's Digital Network Architecture (DNA) uses this type of control. IBM's Network Control Program (NCP) uses a similar mechanism to control buffers in its network concentrator, the 3705; this mechanism is used in addition to the end-to-end pacing control noted in the last paragraph.

5–1 X.25 Protocol

The material in this section reviews and extends the discussion of X.25 in Chapter 3 (see Section 3–3). Recall that the X.25 protocol, although strictly speaking not a network architecture, is a layered protocol that exhibits many of the properties of network architectures. X.25 is an *interface* recommendation: It spells out the detailed interface protocols required to enable data terminal equipment (DTE) to communicate with data circuit-terminating equipment (DCE), which in turn is presumed to provide access to a packet-switched data network. This has already been noted in Chapter 3. The basic idea appears in Fig. 5–1, which repeats Fig. 3–17. DTEs that desire to communicate through a data network do so by establishing X.25 interfaces to the network. The network then handles delivery of the necessary data between the two DCEs. The X.25 architecture assumes that the data network is of the packet-switched type, al-

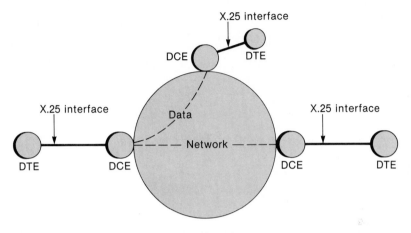

Figure 5 – 1 X.25 concept

though this is not necessary since only the interface protocols between the network and the users (DTEs) are prescribed. The DTEs can be any type of intelligent terminal or computer equipped to handle the X.25 protocol. So-called *PADs* (packet assemblers-disassemblers) to which simple terminals can connect are available to make these terminals appear as DTEs.

Three layers are prescribed for the X.25 protocol. These layers are shown schematically in Fig. 5 – 2 (which appeared previously as Fig. 3 – 18). The lowest layer is, of course, the physical layer. CCITT Recommendation X.21 [BERT], [FOLT 1980a] is the physical-layer procedure adopted for X.25. The link or frame-level protocol that uses the point-to-point synchronous circuit established by X.21 consists of a set of procedures called LAPB (balanced link access procedures). LAPB procedures form a subset of and are compatible with HDLC and ADCCP, which were described in some detail in Chapter 4. As noted there, the link layer uses a frame structure to carry the data (packets) reliably between the two ends of the link. It is the third layer, the packet layer, that is fully prescribed in X.25. This layer corresponds to the network layer in network architectures.

[BERT] H. V. Bertine, "Physical Level Protocols," *IEEE Trans. on Comm.*, vol. COM-28, no. 4, April 1980, 433 – 444; reprinted in revised form in [GREE].

[FOLT 1980a] H. C. Folts, "Procedures for Circuit-switched Service in Synchronous Public Data Networks," *IEEE Trans. on Comm.*, vol. COM-28, no. 4, April 1980, 489 – 495; reprinted in revised form in [GREE].

[GREE] P. Green, ed., *Computer Network Architectures and Protocols*, Plenum Press, New York, 1982.

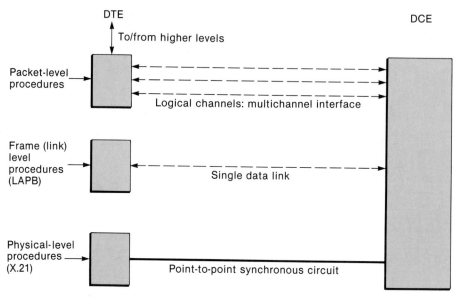

DTE

DCE

To/from higher levels

Packet-level procedures

Logical channels: multichannel interface

Frame (link) level procedures (LAPB)

Single data link

Physical-level procedures (X.21)

Point-to-point synchronous circuit

Figure 5–2 X.25 layers

Recommendation X.25 was developed by Study Group VII of the CCITT.* Study Group VII is responsible for data communications over *public* data networks† and issues X-series recommendations. (Study Group XVII, on the other hand, is responsible for data communications over *telephone* facilities and issues V-series recommendations.) X.25 was initially approved by CCITT in 1976 and revised in 1977, 1980, and 1984. It defines an international *recommendation* (*not* a standard!) for the exchange between DTE and DCE of data as well as control information for the use of packet-switched data transmission services. Four types of service are defined: virtual circuit, permanent virtual circuit, datagram,‡ and fast select.

Datagram service implies that packets—the data transmitted at the third (packet) layer—are transmitted into the network one at a time with no guaran-

---

* CCITT is an acronym for the International Telegraph and Telephone Consultative Committee of the International Telecommunications Union (ITU), a specialized agency of the United Nations.

† Its charter thus contrasts with that of ISO, which is made up of standards bodies from each nation represented and which is geared more to commercial enterprises.

‡ Datagram service has never been implemented and was dropped from the X.25 recommendation in the 1984 revision.

tee of final delivery to the destination and no guarantee of orderly (sequenced) delivery. Each datagram must carry complete address and control information to enable it to be delivered to the proper destination DTE. Virtual circuit service, on the other hand, ensures orderly (sequenced) delivery of packets in either direction between the two end users. Permanent virtual circuit service indicates that the connection between the end users is *always* available. Regular, or switched, virtual circuit service requires a virtual circuit (VC) to be established whenever data communication between two end users is desired. In fast-select service the control packet that sets up the VC may (optionally) carry data as well. Datagram and VC service are examples of connectionless and connection-oriented service, respectively (see Chapters 1 and 3).

As already noted, a virtual circuit looks like a dedicated data connection to the two DTEs communicating with one another. In actuality, links along the path between the end users may be shared by many virtual circuits. This is indicated in Fig. 5 – 2, where multiple logical channels are shown using the single data link. In the case of TYMNET, a U.S. public packet-switched network, a complete virtual circuit is established from one user end of the network to the other [SCHW 1977], [TYME]. IBM's SNA uses the same procedure. Canada's Datapac, however, although it provides an X.25 interface between DTEs and DCEs in the network, actually uses datagram techniques for transmitting packets within the network. This means that datagrams must be stored and resequenced at either end of the network before being shipped as packets to the user. Datagrams lost during transmission must also be retransmitted within the network to provide the VC guarantee to the external users [SPRO]. The Datapac network is described briefly in the next chapter, as is TYMNET.

The obvious purpose of the X.25 packet level is to provide procedures for handling each of these services, including the call setup (connect) and clearing (disconnect) procedures required for VC service. In addition, provision for flow control must be made both to ensure that one user does not overwhelm the other user with packets and to maintain timely, efficient delivery of packets. (In the next section we focus on this latter function, also called congestion control.) Provision must also be made to handle errors at the packet level, to abort or restart VCs if necessary, and so forth. Note that all these functions are required of any layer in the abstract multilayer network architecture described in Chapter 3. (The OSI Reference Model conceptually includes these features in every layer. The data link layer discussed in Chapter 4 provides some of these features

[SCHW 1977] M. Schwartz, *Computer-Communication Network Design and Analysis*, Prentice-Hall, Englewood Cliffs, N.J., 1977.

[TYME] L. Tymes, "Routing and Flow Control in TYMNET," *IEEE Trans. on Comm.*, vol. COM-29, no. 4, April 1981, 392 – 398.

[SPRO] D. E. Sproule and F. Mellor, "Routing, Flow, and Congestion Control in the Datapac Network," *IEEE Trans. on Comm.*, vol. COM-29, no. 4, April 1981, 386 – 391.

at that level.) The details of the packet-level procedures, as well as the various packet formats required to implement them, appear in the formal recommendation [X.25]. We shall summarize just a few of them here to give the flavor of the recommendation, as well as to introduce the window flow-control mechanism we shall model in the next section. Detailed discussions of the recommendation appear in [FOLT 1980b] and [RYBC].

Virtual circuit and datagram services are provided through logical channels made available for each DTE-DCE physical connection (Fig. 5–2). As will be seen later, these channels are given by numbers in a logical channel field in each packet. For each link connection, 4096 such logical channels are provided. All datagrams flow on the same datagram channel. Permanent VCs, if they exist, are each assigned a logical channel; each switched VC is then assigned a logical channel as it is created. A DTE may then carry 4095 simultaneous calls or "conversations" with users elsewhere through the DCEs to which they are connected. Both control and data packets flow over the logical channels. In the case of switched VCs these packets are used to set up calls, transfer data once a call is set up, and then clear the call once it is completed. Procedures at the packet level are specified to handle all of these. Procedures (and packets corresponding to these) are specified as well for confirming (acknowledging) call setup requests and clear requests. All packets are transmitted as the *I*-field in frames using the LAPB link protocol described in the last chapter. (I-frames at the link level may therefore not actually be carrying user data. *All* packets handed down from the packet level are treated alike. Recall from the discussion in Chapters 3 and 4 that lower levels in layered architectures do not normally "see" or change what is being passed down to them. Aside from segmentation or concatenation, a layer simply surrounds the data unit of the layer above with appropriate header fields and passes it on as its own transmission or data unit.)

As noted earlier, error-handling procedures are also specified by the X.25 recommendation. Three types are spelled out: reject procedures involving a single packet, reset procedures for handling error conditions on a single VC, and restart procedures involving the entire packet level. These procedures not only specify how error conditions are to be handled, but also how reinitialization is carried out. Control packets are designated for each procedure.

[X.25] *Draft Revised Recommendation X.25, Interface between Data Terminals Operating in the Packet Mode in Public Data Networks*, Study Group VII, CCITT, Geneva, Feb. 1980; reprinted in *Computer Comm. Rev.*, vol. 10, nos. 1 and 2, Jan./April 1980, 56–129.

[FOLT 1980b] H. C. Folts, "X.25 Transaction-Oriented Features—Datagram and Fast-Select," *IEEE Trans. on Comm.*, vol. COM-28, no. 4, April 1980, 496–499; reprinted in revised form in [GREE].

[RYBC] Antony Rybczynski, "X.25 Interface and End-to-End Virtual Circuit Service Characteristics," *IEEE Trans. on Comm.*, vol. COM-28, no. 4, April 1980, 500–510; reprinted in revised form in [GREE].

As examples of some of the packet-level procedures, consider the call setup and cleardown processes for a switched VC. Figure 5 – 3, presented previously as Fig. 3 – 20, indicates the principal features; control-packet types are indicated under the arrows. Thus a DTE that wishes to establish a call with another DTE sends a call-request packet to its DCE requesting that a VC be established. The DCE then forwards this request through the network. (How the forwarding is done is beyond the purview of X.25; this is left strictly to the discretion of the network.) After the destination DCE receives the request, that DCE sends an incoming call packet to the destination DTE. If the call can be accepted, a packet to that effect is returned to the destination DCE. After it moves through the network, the acceptance arrives at the source DCE; there it generates a call-connected packet. The source DTE then enters the data-transfer phase and can begin sending data. As shown in Fig. 5 – 3, the clearing phase is similar except that a clear request by a DTE does not have to wait for a confirmation from the other DTE. The request is confirmed by its own DCE, as shown. The VC and the logical channel associated with it are then released.

Procedures and protocols in network architectures must be stated very clearly to avoid ambiguity and vagueness. Special high-level languages, finite state machines, and combinations of these have been used to provide the detailed specifications. Procedures in X.25 are specified in terms of state diagrams. An example is the state diagram for VC call setup, which appears in Fig. 5 – 4 [X.25, Fig. 2.2]. The states of a DTE-DCE pair appear as ovals in the diagram. An arrow indicates the issuance of a packet type and the consequent shift to a new state. The device that issues the packet is indicated at the tail of the arrow. Thus, under normal conditions the system rests in the ready state, state 1. A DTE issuing a call request moves to state 2, DTE waiting state. Upon receiving the call-connect packet from its DCE, the DTE moves to state 4, the data-transfer state. Similarly, a DCE, upon receiving an incoming call request from elsewhere in the network, issues an incoming-call packet and moves to state 3, DCE waiting. The system moves to state 4, data transfer, when the DTE issues a call-accepted packet in reply.

State diagrams (and the procedures they specify) must be complete and must account for all possible contingencies. In the case of Fig. 5 – 4 an additional state, state 5, call collision, appears. This accounts for the joint events of incoming call and local call request, both of which want the same logical channel, being received by the DCE at the same time. The procedure adopted is to cancel the incoming call in favor of the local call request.

Packet formats have the general form shown in Fig. 5 – 5(a) [X.25]. Note that 12 bits are set aside for the logical channel identification. (Bits 1 – 4 of octet 1 are used to designate a logical channel group number; the second octet identifies a channel within a group. The 12 bits provide the 4096 possible logical channels described earlier.) The first bit in octet 3 is used to indicate whether the

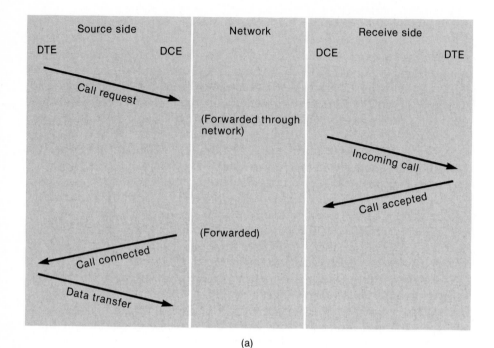

(a)

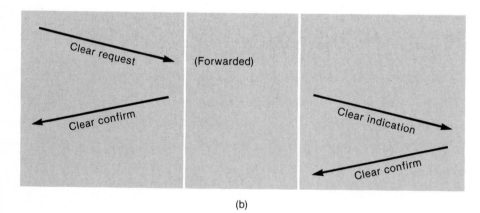

(b)

Figure 5–3 X.25 procedures and packets for setting up and clearing a virtual circuit
   a. Call setup sequence
   b. Call-clearing sequence

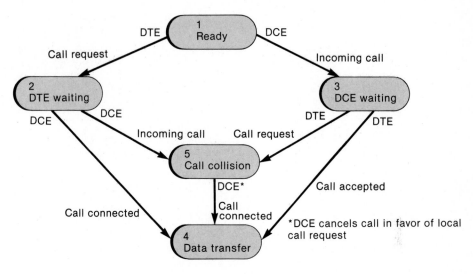

Figure 5–4 X.25 call setup (from [X.25, Fig. A2.2(a)], with permission from CCITT Recommendation X.25 [CCITT Red Book 1984])

packet is a control or data packet. Figure 5–5(b) shows, as an example, the format of the call-request (and incoming-call) packet. Note the 1 that appears in the C/D bit. The rest of octet 3 serves to identify the control packet type. Four bits each are used to indicate the lengths of the calling DTE and the called DTE address fields, respectively. Those fields then follow. Call-user data (up to 16 octets) may be appended at the packet end if the fast-select feature is used. The GFI field will be described later. As mentioned previously, all of these packets are transmitted as I-fields in the LAPB (HDLC) frame protocol.

Now consider the data-transfer phase. Successive data packets, on a given logical channel (VC), are numbered consecutively. Two types of numbering are used, just as in LAPB (or HDLC), the layer below. The two types are modulo 8 numbering and extended numbering, modulo 128. The corresponding data packet formats for these two cases appear in Fig. 5–6. Both conform to the general picture of Fig. 5–5(a). The C/D bit is set to 0 to indicate data packet. Bits 5 and 6 in the GFI field of the first octet are used to differentiate between the two types of data packet: 01 in bits 6 and 5, respectively, as shown, indicate mod 8; the sequence 10 indicates mod 128. For the mod 8 packet there is a three-bit $P(S)$-field in octet 3. The $P(S)$-field for the mod 128 data packet is given by seven bits in octet 3 as well. Note that this sequence numbering, with the accompanying $P(R)$-field shown in Fig. 5–6, is very similar to that employed in the data link LAPB (HDLC) layer carrying these packets. As in the case of the control packets, the entire packet appears as the I-field of the LAPB frames

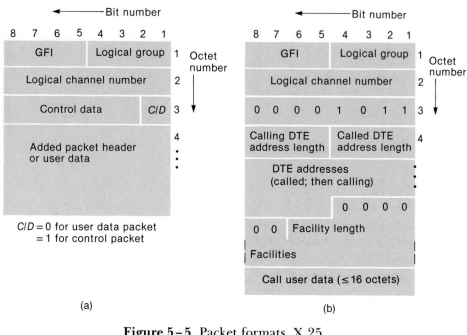

Figure 5–5 Packet formats, X.25
a. General format
b. Call request packet

transmitted between DTE and DCE. The LAPB numbering, on the lower link level, encompasses *all* VCs. The link layer does not distinguish between logical channels. (It does not know they exist!) It multiplexes all VCs together and provides an error-free transmission medium for *all* packets flowing at the packet layer (control packets as well as data packets).

The Q bit (qualifier bit), which appears as bit 8 in octet 1 of the data packet formats of Fig. 5–6, allows two-level data to be transmitted if desired. (One level might be of higher priority, for example.) Normally the bit is set to 0. The M bit in octet 3 of Fig. 5–6(a) and octet 4 of Fig. 5–6(b) is called the "more data" bit. A 1 indicates that more data is coming; it is used when DTEs at either end of the VC have different, locally selected packet sizes. The D bit in octet 1 will be discussed later.

5–1–1 X.25 Flow-control Mechanism

The packet sequence-number field $P(S)$ and the $P(R)$-field are used for *flow control* at this (packet) level of X.25. In addition to these numbers, a window of

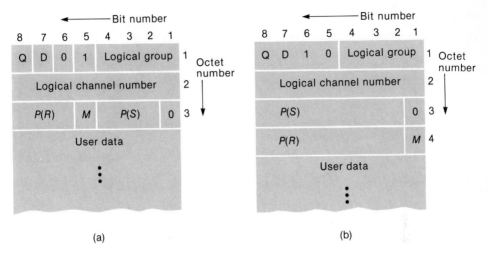

Figure 5-6 Data-packet formats, X.25
a. Modulo 8
b. Extended numbering, mod 128

size w is maintained on each logical channel. The specific value of w is negotiable on setting up a logical channel, with a default value of 2 used. Possible choices of w and how the value chosen relates to performance objectives will be discussed in the next section, which covers the modeling and analysis of window flow-control mechanisms. The window represents the maximum number of data packets that may be outstanding in any one direction (DTE to DCE; DCE to DTE).

The use of the $P(S)$- and $P(R)$-fields is similar to that of the $N(S)$- and $N(R)$-fields in HDLC, except that a window w also appears. Thus $P(S)$ is incremented for each data packet transmitted on a given logical channel (VC). On reception, if $P(S)$ agrees with the number expected (in sequence and within the window size chosen), the receiver (DTE or DCE as the case may be) accepts it. As with the $N(R)$-field in HDLC, $P(R)$ acknowledges delivery of all data packets up to and including $P(R)$-1, in the reverse direction.

Three control packets, with the same names as those adopted for S-frames in HDLC, are used in the data-transfer phase as well. These packets are called receive ready (RR), receive not ready (RNR), and reject (REJ). The REJ packet, used in single-packet error control, is transmitted from DTE to DCE only. The RR and RNR packets carry the $P(R)$-field and participate in the flow-control process. The interplay of the two number fields, $P(S)$ and $P(R)$, and the window w is diagrammed for $w = 4$ and mod 8 sequence numbering in Fig. 5-7. Assume first that two packets have been transmitted, starting with $P(S) = 0$. The next sequence number used will be $P(S) = 2$, as shown in Fig. 5-7(a). With $w = 4$ only

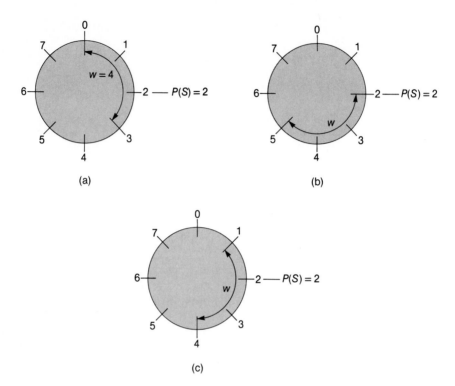

Figure 5–7 X.25 window flow-control mechanism, $w = 4$, mod 8 sequence
numbers
   a. Initial state
   b. Case 1: $P(R) = 2$ received
   c. Case 2: $P(R) = 1$ received

two more packets may be transmitted, as shown. Two cases are then diagrammed: In case 1 (Fig. 5–7b), $P(R) = 2$ is received on a return packet. This
acknowledges delivery of packets 0 and 1 for this logical channel. Four more
packets may now be transmitted, and the lower end of the window shifts to
$P(R) = 2$. In case 2 (Fig. 5–7c), $P(R) = 1$ is assumed received on a return packet.
This acknowledges delivery of packet 0, and the lower end of the window shifts
to $P(R) = 1$. Thus there are only three packets $(P(S) = 2, 3, 4)$ within the window
that may be transmitted.

    Control of the flow of packets for a VC in either direction (DTE to DCE;
DCE to DTE) may be exerted by the receiver by holding back or providing
prompt sending of $P(R)$. Thus if the transmission window is not advanced by the

receiver sending $P(R)$ (see Fig. 5 – 7), the number of packets outstanding within a window will be exhausted, and transmission will cease until a higher value of $P(R)$ is received. Quicker stopping of flow is obtained by having the receiver transmit an RNR packet. Transmission begins again (if numbers within a window are available) upon receipt of an RR packet from the receiver. Flow control on each VC is thus maintained by the receiver; it is used to prevent a receiver from being overwhelmed with data it cannot handle. (A typical example is "speed control": a high-speed computer or data link that "talks to" a lower-speed terminal. Unless flow control is exerted the terminal buffers may overflow.) This window flow-control mechanism provides *congestion control* as well. Here the objective is not necessarily to provide "speed matching" as in flow control, but to prevent resources (buffers and transmission links) along a VC path from becoming congested, which could lead to deadlock situations. Congestion-control mechanisms are used in all data networks and will be studied in detail in the following sections.

Implicit in the preceding description of the X.25 packet-level window flow control is the fact that the DCE and the DTE are exerting control on each other. This is one possible mechanism in the X.25 protocol. Recall that X.25 is an *interface* protocol only. However, virtual circuits are used to connect DTEs through a network. It may be desirable to have the DTEs that communicate with one another exert control on the flow over the VC. In this case $P(R)$ indications are sent not by a DCE to its DTE, but rather by the receiving DTE at the other end of the network. The window flow-control mechanism thus covers the *entire* VC from one end to the other. The window w then refers to the number of packets allowed on the VC *end to end.* Either of these two flow-control procedures (end-to-end or DTE-DCE) may be invoked under the X.25 protocol. The D bit (delivery confirmation bit) in a data packet (bit 7, octet 1 — Fig.5 – 6) carries this information. If D is set to 1, the $P(R)$ number in the packet is transmitted to the far end of the VC, end-to-end flow control is invoked, and the window w refers to the complete end-to-end virtual circuit.

In the next section we model end-to-end window flow-control mechanisms (the DTE-DCE control is then a special case). We compare two extreme cases; normal X.25 operation usually falls between the two. In one case, each data packet is acknowledged as soon as it is received. The lower end of the window thus keeps sliding regularly from number to number. This mechanism has been called a sliding-window mechanism. In the other case, acknowledgement (the $P(R)$ number) is withheld until an entire window has been received at the receiver. This reduces the number of acknowledgements transmitted back to the transmitter but reduces the packet throughput correspondingly. The two mechanisms are compared quantitatively.

Table 5 – 1 [X.25, Table 3.1] summarizes the various packet types used in X.25. All conform to the general format of Fig. 5 – 5(a). The call setup and

TABLE 5 – 1 X.25 Packet Types and Their Use in Various Services				
Packet Type		**Service**		
From DCE to DTE	From DTE to DCE	VC	PVC	DG*
Call Setup and Clearing				
Incoming call	Call request	X		
Call connected	Call accepted	X		
Clear indication	Clear request	X		
DCE clear confirmation	DTE clear confirmation	X		
Data and Interrupt				
DCE data	DTE data	X	X	
DCE interrupt	DTE interrupt	X	X	
DCE interrupt confirmation	DTE interrupt confirmation	X	X	
Datagram				
DCE datagram	DTE datagram			X
Datagram service signal				X
Flow Control and Reset				
DCE RR	DTE RR	X	X	X
DCE RNR	DTE RNR	X	X	X
	DTE REJ*	X	X	X
Reset indication	Reset request	X	X	X
DCE reset confirmation	DTE reset confirmation	X	X	X
Restart				
Restart indication	Restart request	X	X	X
DCE restart confirmation	DTE restart confirmation	X	X	X
Diagnostic				
Diagnostic*		X	X	X

(from [X.25, Table 3.1], with permission from CCITT Recommendation X.25[CCITT Red Book 1984])

* Not necessarily available on all networks. (Datagram service has been eliminated in the 1984 X.25 Recommendation.)

 VC = Virtual call
PVC = Permanent virtual circuit
 DG = Datagram

clearing packets for VC service have already been described, as have the data packets and some of the flow-control packets. As noted earlier, a reset packet, initiated by the DTE only, reinitializes a single VC. Reset causes are listed in the X.25 document [X.25, Table 6.5]. The restart packet initializes the *entire* packet-level DTE-DCE interface. Lists of uses (error conditions giving rise to the restart procedure) also appear in the X.25 document [X.25, Tables 6.6, A3.2, A3.3].

5–2 Analysis of Window Flow-control Mechanisms

In the brief description of aspects of the X.25 data interface recommendation in the previous section, we introduced the concept of window flow control. In this section we model two window flow-control mechanisms and analyze them quantitatively. This approach serves several purposes. Obviously, it enables us to compare various control mechanisms. It focuses attention on performance measures used in network design and how to incorporate them in the analysis. It also introduces the concept of *queueing networks* to represent the model of a data network (or portion thereof). This concept will be exploited later in Section 5–4 in more generality.

As already noted, flow- and congestion-control procedures are always incorporated in data network operations at the network layer to prevent congestion from developing. Window-type mechanisms have been widely adopted for end-to-end control, and this is one of the reasons we analyze models of these mechanisms in this section and in the following one, which covers SNA control mechanisms. Other types of congestion control have also been suggested or used, and some of these will be described later in this chapter. Control mechanisms are always needed to guard against stochastic fluctuations in traffic that may arise, which may temporarily deplete network resources (nodal buffers and transmission links) and cause congestion to develop.

When it occurs, congestion manifests itself in two ways: Time delays will increase markedly in the network, and the throughput, measured in packets per unit time delivered to their destinations, may begin to *decrease* as offered load increases. Both effects are diagrammed in Fig. 5–8. As shown in the figure, the decrease in throughput with an increase in offered load beyond a specified operating point is due to the blocking of finite resources in the network (buffers and finite line capacity). If an offered load is high enough, a *deadlock* situation may even prevail: All buffers are filled, traffic ceases to flow, and the throughput drops to zero. Strategies have been proposed to prevent the occurrence of

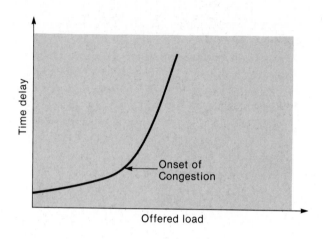

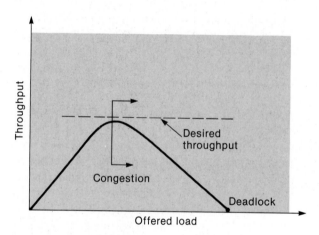

Figure 5–8 Onset of congestion in a network

deadlock [GIES], [KLEI 1976]. Deadlock is not considered explicitly in the virtual circuit window flow-control analysis described in this section. Rather, it is assumed that the control mechanism eliminates the congestion region of the throughput curve of Fig. 5–8; infinite buffer models used for the analysis do not exhibit the deadlock behavior of that curve. In Section 5–5, which covers

[GIES] A. Giessler et al., "Flow Control Based on Buffer Classes," *IEEE Trans. on Comm.*, vol. COM-29, no. 4, April 1981, 436–443.

[KLEI 1976] L. Kleinrock, *Queueing Systems. Volume 2, Computer Applications,* John Wiley & Sons, New York, 1976.

congestion control by input-buffer limiting, the phenomenon of deadlock does appear in the model used, since it incorporates a finite buffer. The use of (more realistic) finite buffers in the VC model of this section would complicate the analysis immeasurably with little gain in insight.

The purpose here is to compare window control mechanisms on the basis of their time delay-throughput characteristics. We assume that these mechanisms, if properly designed, do not let deadlock develop. To carry out the analysis we first develop a model for the virtual circuit, then show how the window control mechanism may be easily modeled on top of this.

5 – 2 – 1 Virtual Circuit Model

Consider a virtual circuit covering M store-and-forward nodes from source to destination in a packet-switched network. A typical example appears in Fig. 5 – 9(a). Three virtual circuits are shown in this small network example, all emanating from the same source node 1. VC1 traverses $M = 3$ store-and-forward nodes (including the source node), while VC2 and VC3 each cover two store-and-forward nodes. The virtual circuits overlap in three of the network links. If we focus on only one VC, a little thought will indicate that it may be modeled by the M queues in series of Fig. 5 – 9(b). This is an extension of the single-queue model for the store-and-forward process at a node in a network that was treated in Chapter 2. The parameter λ represents, as usual, the average packet-arrival rate (load) to the VC, and Poisson arrival statistics are assumed. As was the case previously, this is the simplest model possible: A single server (single transmission link) services each queue.

Propagation delay has been neglected. The two sources of delay in each queue are thus queueing delay (wait time) and transmission time. The latter depends, of course, on packet length and transmission-link capacity. For the ith queue in the VC ($1 \leq i \leq M$), the transmission rate or capacity is denoted by μ_i packets/sec. (The data link control on each link along a VC is ignored here. Recall that by virtue of the network layering concept used, the data link layer provides the VC with an error-free, transparent transmission medium on each link along the path. Occasional bit errors will cause link frames, and hence packets within them, to be retransmitted. This phenomenon occurs relatively infrequently and is assumed to have negligible effect on the VC time delay; thus it is ignored. The transparency of the data link layer enables us to decouple the characteristics of the two layers.)

A more general (and more valid) model would show the complete network represented by interconnected queues with virtual circuits superimposed and interacting. The analysis of such a queueing network is much more complex and is deferred until Section 5 – 4. It is apparent from Fig. 5 – 9 that even if one focuses on a single VC, as is the case here, two or more VCs may share nodal queues and transmission links, giving rise to queues with multiple customers.

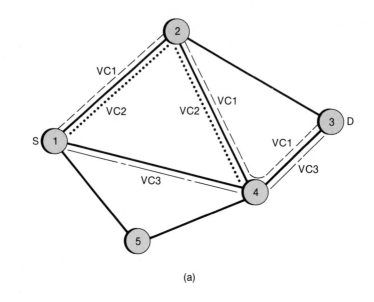

(a)

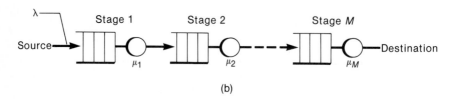

(b)

Figure 5–9 Single virtual circuit model
a. Virtual circuits in a network
b. Virtual circuit queueing model

Figure 5–9(b) thus should really show multiple customer streams arriving at various queues along the VC and exiting as they move along their own VC. The simplest model that takes these added customer types into account assumes that they use up a portion of the transmission capacity on a given link, reducing the capacity available to the VC user. This concept is made more rigorous and is justified mathematically in [PENN] and [SCHW 1977]. The μ_i's shown in Fig. 5–9(b) are assumed to have been reduced accordingly.

[PENN] M. C. Pennotti and M. Schwartz, "Congestion Control in Store and Forward Tandem Links," *IEEE Trans. on Comm.*, vol. COM-23, no. 12, Dec. 1975, 1434–1443.

Even with the relatively simple form of Fig. 5–9(b), it turns out to be extremely difficult to analyze the model, except for certain special cases, unless one additional assumption is made. The quantity $1/\mu_i$ at the ith node represents the average time required to transmit a packet over its outgoing link. This was emphasized in Chapter 2. The transmission time depends in turn on average packet length in bits and the line speed (capacity) in bits/sec. The assumption that must be invoked is that a packet traversing a cascade of queues in series, as in Fig. 5–9(b), has its length selected randomly and independently, from a given probability distribution, as it reaches each new queue. As an example, if the packets are distributed exponentially in length (the case to be studied in detail here), with average value $1/\mu$, one must assume that a typical packet changes its length randomly from node to node as it traverses the VC! This independence assumption was first used by L. Kleinrock [KLEI 1972] in his pioneering work on packet-switched networks; it has since been validated by simulation a number of times. One reason given for the validity of the assumption is that different packet streams are often statistically multiplexed at a given outgoing link, so that the contribution of any one stream is small compared with the others. If they all have the same length distribution, a typical packet may in fact act at each queue along its path as if it were of randomly varying length. Simulation studies comparing the performance of different single-VC window control mechanisms have shown that the assumption is quite good if M (the number of hops along the VC) is not too big (typical of most networks). For large M ($M \geq 6$) the absolute results of analysis differ appreciably from the simulation results, but the results for the comparative performance of different mechanisms are found to be valid [SCHW 1982a].

If one adopts the independence assumption (is there really a choice?) and takes the packet length to be distributed exponentially, the analysis becomes tractable. The single-VC model of Fig. 5–9(b) becomes in fact a cascade of independent M/M/1 queues for which various parameters of interest may be written almost by inspection. (The exponential packet-length distribution in turn implies Poisson departures from any queue. The corresponding Poisson arrivals at the next queue ensure the M/M/1 characteristic.) The queueing model of Fig. 5–9(b) under these assumptions turns out to be a special case of an *open queueing network* of *product form*, which is discussed later in this chapter (Subsection 5–4–2). The term "product form," which we will define explicitly later, means that the probability of state of the composite queueing network may be written as products of state probabilities for each queue. The implication then is that they appear to act independently.

[KLEI 1972] L. Kleinrock, *Communication Nets: Stochastic Message Flow and Delay*, McGraw-Hill, New York, 1964; reprinted, Dover Publications, New York, 1972.
[SCHW 1982a] M. Schwartz, "Performance Analysis of the SNA Virtual Route Pacing Control," *IEEE Trans. on Comm.*, vol. COM-30, no. 1, Jan. 1982, 172–184.

5−2−2 Sliding-window Model

The discussion up to this point has focused on validating the queueing network model of Fig. 5−9(b) for a single virtual circuit. How does one now superimpose the flow-control mechanism on this model? We describe the sliding-window control first. Recall from Section 5−1 that in this control mechanism each packet is individually acknowledged as it arrives at its destination. The acknowledgement, on arriving back at the source, shifts the window forward by one, allowing another packet to enter the network if all packets in a window have been previously transmitted. An example appears in Fig. 5−10. We now use the letter N to represent the window size; the letter w will be used to describe the next window control mechanism that we model and analyze.

By virtue of the sliding-window control, packets can be transmitted onto the VC so long as there are fewer than N along the VC. If N have already entered the VC no more are allowed in until an acknowledgement packet arrives, sliding the window forward by one. How arriving packets are handled when the window is depleted (N unacknowledged packets are in transit along the VC) gives rise to a variety of control models. We choose the simplest case: Packets are assumed blocked and lost to the system if, on arrival, N packets are outstanding along the VC. This is often a realistic case. Consider two examples. In the first example, that of interactive terminals accessing a network, terminal keyboards might be locked under conditions of congestion. The user must wait until a later time to transmit his or her data. The packets he or she might have transmitted are thus

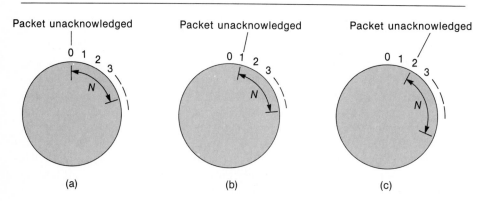

Figure 5−10 Sliding-window control
   a. Initial state
   b. Packet acked
   c. Next packet acked

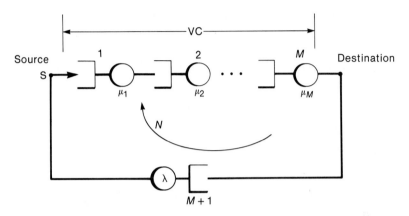

Figure 5-11 Sliding-window control model

assumed lost. In the second example, the packets might be the result of a multiprogrammed computer working on a particular task. If the computer system finds input to this particular VC (i.e., connection to a particular destination) blocked, it can simply suspend work on this task and go on to another job. The blocked-packet model appears appropriate in this example as well.

We now assume that VC acknowledgements are transmitted back through the network at highest priority. Although not the case for X.25 DTE-DCE interface acknowledgements, it might be for the packet-switched network that carries the packets. IBM's SNA (described in Section 5-3), for example, handles acknowledgements in this manner. If this is not the case, an analysis may be carried out by representing the VC ack path as a cascaded series of queues as well. The assumption made here of highest priority return of acknowledgement packets provides a "best-case" performance evaluation.

With this assumption we essentially neglect the time delay incurred by acknowledgements returning from the destination end of the VC back to the source end. This enables us to model the sliding-window control, superimposed on the VC, as the closed-system model of Fig. 5-11. Here source and destination are shown linked through an added artificial queue labeled $M + 1$, whose service rate is λ, precisely the input rate to the VC of Fig. 5-9(b). A fixed number N of packets are shown circulating through the closed system. Note now how this closed-system model captures the sliding-window mechanism. If N packets are in transit along the VC (the upper M queues), the bottom queue $M + 1$ is empty and cannot serve. This models the blocking condition that packets encounter on arrival at a time of depleted window. The instant that one of the N packets along the VC arrives at the destination, it appears at queue $M + 1$ (this is the assumption of zero delay time for acknowledgements back to the source), and the source may now deliver packets at its Poisson rate λ. This is always the case if fewer than N packets are en route along the VC, with the remainder then stored in the artificial queue $M + 1$.

What can we now do with this model? What we would like to find are the end-to-end statistics of packets moving along the virtual circuit: What is the time delay in getting from the source to the destination? What is the throughput of this control mechanism (clearly less than λ because of blocking or, in the model, due to depletion of the $M + 1$ buffer)?

Even without any analysis one can get a qualitative feel for the operation of the sliding-window control. Let the packet arrival rate λ increase. Without the control the queues at the nodes along the VC (Fig. 5 – 9b) would begin to build up, the end-to-end time delay would increase rapidly, and congestion would ensue. With the control, congestion is limited since no more than N packets can be in transit along the VC at any one time. The smaller the value of N, the lower the end-to-end time delay. The price paid, however, is that the throughput is reduced correspondingly. As N increases the throughput increases, but the time delay goes up as well. There is thus a tradeoff between time delay and throughput that, as we shall see, recurs in all of our congestion-control mechanisms. The best control scheme, then, is the one that provides the shortest time delay for a given throughput or, equivalently, the highest throughput for a given time delay. The dependence on M, the length in hops traversed of the virtual circuit, is also easily understood. As M increases, the minimum end-to-end time delay due to transmission alone increases as well, since, in a store-and-forward network as assumed here, the packet must be stored and retransmitted at each node. This minimum delay thus increases linearly with M and is the value obtained for $N = 1$. But the throughput decreases as M increases precisely because the minimum time to get through the network increases.

Now let the number of packets N allowed on the VC increase. This increases both the throughput and the time delay. For N not-too-big the effect on throughput is more dramatic. As N continues to increase, however, the throughput saturates at its maximum possible value (the smallest capacity in any of the links of Fig. 5 – 11), while the time delay continues to increase. There thus exists an optimum value of N in terms of providing increased throughput with not too large an increase in end-to-end time delay. We shall find that the best $N \doteq M$. This value is in fact often recommended for network operation!

To go any further, however, one must resort to quantitative analysis. This requires the introduction of methods of analyzing the closed system of Fig. 5 – 11. This closed system is a special case of what is known as a *closed queueing network* [KOBA], [SAUE], [KELL]. Such networks are amenable to solution under conditions similar to those noted earlier for the open network of the VC without control (Fig. 5 – 9b). The solutions again become those of product form; we consider such networks later in this chapter. Suffice it to say at this point that exponential packet length statistics and Poisson arrivals provide one example of statistics that ensure the product-form solution for this network. Tools thus exist to solve for the desired statistics of this network.

We now invoke an even simpler method of analysis that is particularly appropriate to this problem. As just noted, we are principally interested in evaluating the end-to-end performance of the sliding-window mechanism as represented by its time-delay – throughput characteristic. These are properties of the *overall* virtual circuit. The details of wait times or queue lengths at individual nodes along the virtual circuit are not of specific interest in this case. As such, we can utilize an interesting theorem first proven by K. M. Chandy, U. Herzog, and L. S. Woo. This theorem states that for product-form networks (the M-stage VC model with exponential servers is a special case, as noted earlier), any subnetwork may be replaced by one composite queue with state-dependent service rate. The network remaining retains exactly the same statistical behavior [CHAN], [SAUE, pp. 105, 106].

Consider as an example the closed queueing network of Fig. 5 – 11. The theorem states that the M queues in cascade between points S and D could be removed and replaced by a single state-dependent queue with no change in the statistical behavior between points S and D. Since we are interested in the end-to-end properties (time delay and throughput) of the VC sliding-window control only, this provides an extremely valuable simplification in our case.

The theorem is variously called the Norton, aggregation, or decomposition theorem of queueing networks. It is described quite readily in the general network context by reference to Fig. 5 – 12. A closed queueing network (Fig. 5 – 11 provides one example; more general examples appear later in this chapter) is shown with a single queue Q pulled away from it. Say that one desires to find the probabilities of state of that queue only. If the network between points A and B can be replaced by a single state-dependent queue, the methods of

[KOBA] H. Kobayashi, *Modeling and Analysis: An Introduction to System Performance Evaluation Methodology*, Addison-Wesley, Reading, Mass., 1978.

[SAUE] C. H. Sauer and K. M. Chandy, *Computer Systems Performance Modeling*, Prentice-Hall, Englewood Cliffs, N.J., 1981.

[KELL] F. P. Kelly, *Reversibility and Stochastic Networks*, John Wiley & Sons, Chichester, U.K., 1979.

[CHAN] K. M. Chandy, U. Herzog, and L. S. Woo, "Parametric Analysis of Queueing Networks," *IBM J. Research and Development*, vol. 19, no. 1, Jan. 1975, 43 – 49.

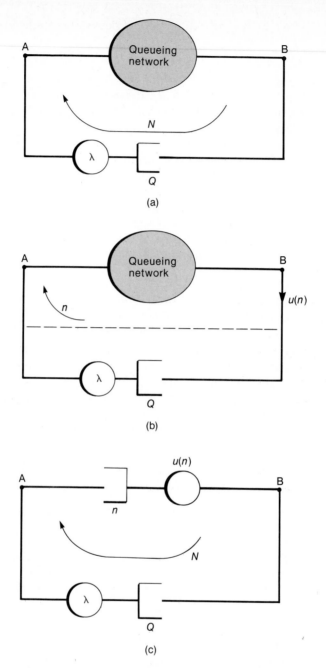

Figure 5–12 Use of Norton's theorem in queueing networks
a. Original network
b. Norton's theorem applied
c. Equivalent model

Chapter 2 enable us to solve for the probabilities of state of *either* queue quite simply, providing the desired result. The Norton theorem states that if the queueing network is of product form (the probabilities of state of the entire network are given by products of state probabilities in each queue), the single state-dependent queue representing that network may be found by shorting out points A and B as shown in Fig. 5-12(b), allowing n packets (customers) to circulate throughput the new closed network, and calculating the resultant "short-circuit" throughput $u(n)$ (packets/sec), as indicated. The throughput $u(n)$ may be found by the recursive methods described later in this chapter. The much-simplified equivalent model to be solved is the one shown in Fig. 5-12(c): A Norton's equivalent queue with service rate $u(n)$, n the state of the queue, replaces the entire network between points A and B of Fig. 5-12(a). The probabilities of state of the lower queue (and hence statistical parameters of interest between points A and B) are identical in the two cases.

Now let N be the total number of packets in the closed network of Fig. 5-12(c). If the upper queue is in state n, the lower queue is in state $N - n$. It thus suffices to find the probabilities of state of the upper queue. But note that the equations of state for this queue are just those of a generalized birth-death process with birth (arrival) rate λ and state-dependent death (service) rate $u(n)$. It is apparent that p_n, the probability that the upper queue is in state n, is just

$$p_n = p_0 \, \lambda^n \Big/ \prod_{i=1}^{n} u(i) \qquad (5-1)$$

with p_0 found by invoking the usual probability normalization condition

$$\sum_{n=0}^{N} p_n = 1 \qquad (5-2)$$

We now apply the Norton theorem to the study of the sliding-window control mechanism embodied in the closed queueing network of Fig. 5-11. We first simplify the analysis considerably and still get useful and significant results by taking the service rates $\mu_1, \ldots, \mu_M$ all to be of the same value μ. This implies that all links along the VC are of the same transmission capacity and that each carries, on the average, the same total traffic. In general, these are valid characterizations of network operation. However, a little thought indicates that other conditions are satisfied by this simplification as well. Say, for example, that the links along the VC fall into either of two classes: Each has a *net* service rate (capacity less the load of the non-VC traffic) μ_1 or $\mu_2 \gg \mu_1$. Then the bottleneck links are those of net capacity μ_1, little queueing is expected at the links with service rate μ_2, and those nodes may be safely removed from the model. In effect, the M-hop VC behaves very nearly like one with a smaller number of hops.

This special case of the homogeneous VC with sliding-window control appears in Fig. 5–13(a). The Norton equivalent for this simple, symmetric network turns out to have the state-dependent service characteristic

$$u(n) = n\mu/[n + (M - 1)] \tag{5-3}$$

and is shown as such in Fig. 5–13(b). The simplicity of this form for $u(n)$ indicates that it should be found readily from Fig. 5–13(a), and this is precisely

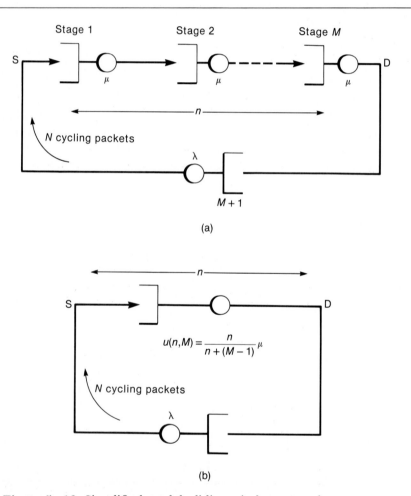

(a)

(b)

Figure 5–13 Simplified model, sliding-window control
a. Sliding-window control, symmetric virtual circuit
b. Norton equivalent, cyclic queue network

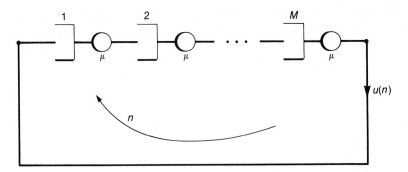

Figure 5-14 Norton equivalent calculation, Fig. 5-13(a)

the case. Consider points S and D shorted out, as required, in Fig. 5-13(a). The resultant network whose throughput is to be found appears in Fig. 5-14.

The throughput $u(n)$, with n packets distributed among the M queues shown, is always less than or equal to μ. In fact it must be given by

$$u(n) = \mu \cdot \text{prob(a queue is nonempty)} \qquad (5-4)$$

Since all queues are identical, by symmetry, the probability that a queue is nonempty is the same for all queues. Focus on the last one, M, as an example. The probability this queue is nonempty is just the probability that at least one of the n packets shown circulating through the network resides at this queue. But given n packets distributed randomly among M identical queues, the calculation of the desired probability (or the equivalent p_0, the probability that a queue is empty) is just the combinatorial one of calculating the probability that at least one of n identical balls will fall into one of M boxes. This produces the desired expression $n/[n + (M - 1)]$. Details are left to the reader as an exercise.

$u(n)$ in Eq. (5-3) applied to Eq. (5-1) provides the desired state probabilities of the upper queue in Fig. 5-13(b). One finds quite readily that

$$p_n/p_0 = \rho^n \binom{M - 1 + n}{n} \qquad (5-5)$$

with $\rho \equiv \lambda/u$, the normalized applied load, and p_0 given by

$$\frac{1}{p_0} = \sum_{n=0}^{N} \rho^n \binom{M - 1 + n}{n} \qquad (5-6)$$

The expression $\binom{M - 1 + n}{n} \equiv \dfrac{(M - 1 + n)!}{(M - 1)!n!}$ represents the usual notation for the binomial coefficient (the number of combinations of $M - 1 + n$ quantities taken n at a time). All end-to-end (source-to-destination) statistics of interest can now be found from Eq. (5–5). The throughput γ of the window-controlled VC, for example, is given by averaging over all the N possible service rates:

$$\gamma = \sum_{n=1}^{N} u(n)p_n \qquad (5-7)$$

By Little's formula the end-to-end time delay $E(T)$ through the VC is just the ratio of the average number of packets $E(n)$ along the VC to γ:

$$E(T) = E(n)/\gamma = \sum_{n=1}^{N} np_n/\gamma \qquad (5-8)$$

Equations (5–7) and (5–8) can be plotted readily as a function of the load with N fixed, or as a function of N with λ fixed, to determine the operating performance of the sliding-window control [SCHW 1980c].

Rather than do this, however, we carry out a much simplified analysis, obtaining the desired results almost by inspection! We focus on the region of very heavy load, the congested region, with $\lambda \geq \mu$, to see how well the window control functions and to determine the value of N suggested for "appropriate" control performance. First take the worst possible case, that of $\lambda \gg \mu$; i.e., the load is much heavier than the capacity. In particular, let $\lambda \rightarrow \infty$ in Fig. 5–13. It is then apparent that as soon as a packet arrives at the destination and is acknowledged at the source, another packet is immediately outputted onto the VC. The VC in this limiting case is thus *always* held at the state N, $E(n) = N$, and the throughput γ is just

$$\gamma = u(N) = N\mu/[N + (M - 1)] \qquad \lambda \rightarrow \infty \qquad (5-9)$$

From Little's formula we also have immediately

$$E(T) = N/\gamma = [M - 1 + N]/\mu \qquad \lambda \rightarrow \infty \qquad (5-10)$$

A plot of normalized end-to-end time delay $\mu E(T)$ versus normalized throughput γ/μ under this condition of very heavy load appears in Fig. 5–15 for the typical case of a three-hop VC, $M = 3$. Note from Eq. (5–10) that the minimum time delay $E(T) = M/\mu$ occurs at $N = 1$. This is precisely as expected. With a single packet allowed on the VC at any one time, there is no queueing delay and the end-to-end delay is just that of packet transmission through M store-and-forward queues in succession. This is shown in Fig. 5–15 with the minimum time delay given by $\mu E(T) = 3$ for $M = 3$. As the window size N

[SCHW 1980c] M. Schwartz, "Routing and Flow Control in Data Networks," NATO Advanced Study Inst.; New Concepts in Multi-user Communications, Norwich, U.K. Aug. 4–16, 1980; Sijthoff and Nordhoff, Neth.

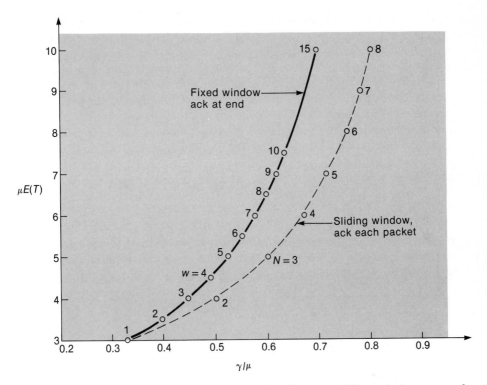

Figure 5-15 Time delay-throughput tradeoff curve, sliding-window control, $M = 3$, $\lambda \to \infty$

increases the time delay goes up proportionally, as shown by Eq. (5-10). From Eq. (5-9) the throughput γ initially goes up linearly with N as well, but then levels off, approaching a maximum value of μ more and more slowly as N increases. Equations (5-9) and (5-10) may be combined to produce a simple queueing-type expression for the time-delay–throughput tradeoff characteristic:

$$\mu E(T) = (M - 1)/[1 - (\gamma/\mu)] \qquad \lambda \to \infty \qquad (5-11)$$

This equation is valid at integer values of N only. It demonstrates the fact that as γ approaches μ more and more closely (by increasing N correspondingly), small changes in γ give rise to large changes in $E(T)$. This behavior is manifested by Fig. 5-15.

What value of N should be used? Various criteria lead to the same value. It is obvious from the form of Eq. (5-11) that the slope of the tradeoff curve increases as γ (actually N) increases. In particular there exists a value of γ at which the tangent to the curve of Eq. (5–11), if extrapolated back to $\gamma = 0$, intersects at $E(T) = 0$. It is left to the reader to show that this corresponds to the value $\gamma = \mu/2$, at which point $\mu E(T) = 2(M - 1)$, somewhat less than twice its smallest value. From Eq. (5–9) or Eq. (5–10) it is apparent that this corresponds to the window size $N = M - 1$. Increasing N beyond this point results in steadily increasing time delay, while the throughput levels off more and more. Another criterion sometimes used to find a "good" value for N is to find the value that maximizes the ratio $\gamma/E(T)$ or its normalized equivalent $\gamma/\mu/\mu E(T)$. This has been called the "power" of the system for obvious reasons. Although a rather arbitrary measure, it does convey the idea that one would like to maximize both γ and $1/E(T)$. Since both are not possible simultaneously, one chooses a single measure to optimize the product of the two. It is again left to the reader to show that this produces exactly the same value of N, $N = M - 1$.

The discussion thus far has focused on the worst case, $\lambda \rightarrow \infty$. As noted previously, one would more realistically like to control the VC at a load in the vicinity of the capacity μ. Here too the analysis turns out to be almost trivial. If we refer to the original window model of Fig. 5–13(a) and set $\lambda = \mu$, we note that the model looks just like that for $\lambda \rightarrow \infty$, except that there are now $M + 1$ queues to consider rather than M. From Eq. (5–9) the throughput must now be given by

$$\gamma = N\mu/(N + M) \qquad \lambda = \mu \qquad (5-12)$$

The new time-delay expression is found similarly except that one must reduce the time delay by $M/(M + 1)$ to account for the fact that only the delay between S and D in Fig. 5–13(a) (the M-node VC delay) has physical significance. This allows us to immediately write, using Eq. (5–10),

$$\mu E(T) = \left[\frac{M}{M + 1} \right] (M + N) \qquad \lambda = \mu \qquad (5-13)$$

Equations (5–12) and (5–13), as well as Eqs. (5–9) and (5–10), can be found from Eqs. (5–7) and (5–8) as well, but obviously not as simply. A proper operating point in this case of $\lambda = \mu$, using the same reasoning applied earlier, is to choose $N = M$.

The value of window size has an interesting intuitive basis as well and has been suggested as a rule of thumb for network operations. Consider the case of *fixed-length* packets of length $1/\mu$ sec. Say that we want to control their insertion into the M-node VC so that both time delay and throughput are *simultaneously* optimized. The obvious "Maxwell demon" solution is to put precisely one packet into each node and let each move synchronously through the VC. This window of $N = M$ packets produces a maximum throughput of μ since a packet arrives at the next node just as one has completed transmission. (We neglect propagation delay here.) All transmission links are thus always operating at maximum capacity. The end-to-end time delay is just M/μ, the time to transmit a packet across M store-and-forward nodes, since packets are served as soon as they enter a node. This procedure is clearly infeasible; it requires an infinite store of packets that move deterministically and synchronously across the VC. But it does provide some physical justification for the choice $N = M$, found through our statistical analysis.

The sliding-window control itself has an interesting interpretation. Say that one is able to control the arrival rate of packets as a function of the number of packets, n, already in the VC. Thus if n gets to be large one might want to reduce the rate; if n is small, one wants to increase the rate. Let the resultant rate be $\lambda(n)$. The control mechanism, although no longer necessarily a sliding-window one, has the same model as that of Fig. 5–11 but with $\lambda(n)$ replacing λ. It turns out that the product-form solution still holds in this case [KOBA], the same Norton equivalent may be used, and the state probabilities p_n may be found just as in the sliding-window case. Calculations have been carried out for a number of possible control functions $\lambda(n)$ to see how their performances compare [SCHW 1980c]. Rather than compare suggested strategies, however, one can find an *optimum* control. This is defined to be the function $\lambda(n)$, which maximizes the throughput γ with the end-to-end time delay $E(T)$ bounded by the two values $L \leq \mu E(T) \leq L + 1$. The parameter L is some arbitrary design value chosen so that $L \geq M$. In addition one puts a constraint on the control function $\lambda(n)$ such that $\lambda(n) \leq c$. (The parameter c represents the maximum arrival rate at the VC.) This constrained optimization problem turns out to have the sliding-window control as its solution! [LAZA 1983]. Thus the *best* control under those conditions is one for which $\lambda(n) = c$, $n < N$, with N a window that depends on the allowable time delay L. For $n > N$, $\lambda(n) = 0$. This is also an example of a so-called bang-bang control (a control with two possible states).

5–2–3 Acknowledge-at-End-of-Window Control

In this discussion of the X.25 flow-control mechanism we noted that not every packet is necessarily acknowledged, as is the case with the sliding-window control. Acknowledgements may be withheld by a receiver until it is ready to receive packets. The acknowledgement at that time then acks all prior unacked packets. As noted earlier, this flow-control mechanism can be used to "speed match" a transmitter-receiver pair if the transmitter is transmitting packets at

[LAZA 1983] A. A. Lazar, "Optimal Flow Control of a Class of Queueing Networks in Equilibrium," *IEEE Trans. on Automatic Control*, vol. AC-28, no. 8, Aug. 1983, pp. 1001–1007.

too high a rate for the receiver to accept them. This method of withholding acknowledgements (sometimes called "credits") also reduces the number of acknowledgements and hence overhead introduced into the system. The price paid, however, is a reduction in the VC throughput.

Our purpose in this subsection is not to discuss the flow-control "speed matching" mechanism, but rather to quantitatively study the method of withholding transmission credit as a *congestion-control* mechanism. How, for example, does withholding transmission credit compare with the sliding-window control? In particular, we take the extreme case in which permission is withheld until all packets in a window have been received. This reduces the overhead from one ack/packet to one ack/window. The throughput, in packets/time delivered to the destinations, is reduced as well, as noted in the last paragraph. The sliding-window mechanism and this one, with acknowledgement provided at the end of the window, bound the performance of a class of mechanisms, with permission to transmit withheld until some portion of a window has been received. As noted earlier, the sliding mechanism provides optimal control. The analysis in this subsection will enable us to bound the deterioration in performance produced by mechanisms that withhold transmission credit.

Since the preferred choice of window sizes may differ in different mechanisms, we replace the value N used in the sliding-window case with the letter w. Consider the mechanism now whereby the last packet in a window of w only is acknowledged. A counter C is assumed kept at the source. This is decremented by one whenever a packet is transmitted out onto the virtual circuit. The counter is initially set at w, and when it reaches zero, indicating that w packets are in the network, packets are blocked from entering the network. As previously, we assume that the acknowledgement, triggered by the last packet in the window, is received at the source at the instant the last packet in the window of w reaches the destination. A model depicting this mechanism appears in Fig. 5–16, which shows w packets circulating in the closed network. The w-box shown stores up to $w - 1$ packets and then fires on arrival of the wth packet, reinitializing the count C, which must be zero at that time, back to w. Note that C is represented by a queue served at the VC Poisson arrival rate λ.

The only difference between the model shown in Fig. 5–16 and that of the sliding-window control of Fig. 5–13(a) is the w-box of Fig. 5–16. This one difference, however, makes the analysis potentially much more complex, for the emptying of the box into the C-count queue at the end of a window represents a *bulk arrival process* [KLEI 1975a]. The closed network of Fig. 5–16 no longer retains the product-form property, and the Norton theorem equivalent for the VC, found so useful in the sliding-window case, no longer provides an exact

---

[KLEI 1975a] L. Kleinrock, *Queueing Systems, Vol. 1: Theory,* John Wiley & Sons, New York, 1975.

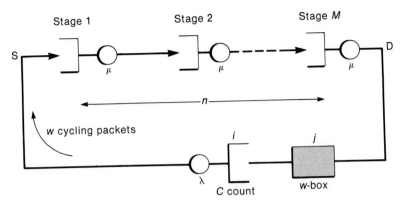

Figure 5−16 Window control with ack at end of window

solution. We nonetheless go ahead and use the Norton equivalent anyway, as an *approximation.* (We really have no choice, since non-product-form networks can only be solved approximately, or by exhaustive enumeration, setting up appropriate balance equations for the entire network.) Simulation has verified that this approximation is valid for $M \le 3$ [SCHW 1982a]. For longer VCs ($M > 3$), the results of the analysis using the Norton approximation are no longer valid in an absolute sense, but *relative* performance comparisons with other mechanisms can still be inferred quite accurately [SCHW 1982a].

Using the Norton equivalent for the VC, the closed queueing network to be solved appears in Fig. 5−17. Because of the introduction of the added w-box the state of this network must be represented by *two* numbers i, j to represent the combined state of the C-count queue and the w-box. Alternatively, one of these may be eliminated and replaced by the state of the upper (Norton equivalent) queue, given by $n = w - (i + j)$. Two-dimensional balance equations for the probabilities of state p_{ij} can readily be written; from these equations all parameters of interest can be found by numerical computation. (There are the order of $w^2/2$ equations to be solved simultaneously.)

Rather than solve this general set of equations, however, we again focus on the two special values of offered load, $\lambda \to \infty$ and $\lambda = \mu$. As in the sliding-window case, computations are simplified considerably when $\lambda \to \infty$. The $\lambda = \mu$ solution is then found using the same trick used earlier: The solution is similar to that of the $\lambda \to \infty$ case, but with M replacing $M - 1$ and the time-delay expression adjusted to account for M rather than $M + 1$ queues.

In the very heavy traffic case, $\lambda \to \infty$, the two-dimensional state equations involving i, j reduce to equations in j only. This is apparent from Fig. 5−17. For

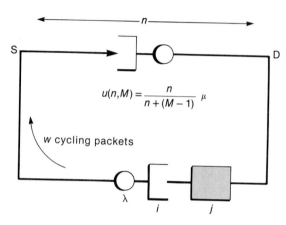

Figure 5–17 Queueing model, ack at end of window

with $\lambda \to \infty$, the C-count queue will instantaneously empty its contents of w packets into the VC as soon as they are received from the w-box. We thus need only solve for the state probabilities p_j of the w-box (or, equivalently, $p_n = p_{w-j}$ of the upper Norton queue). From Fig. 5–17, with the C-count queue eliminated, it is apparent that the appropriate balance equations for the w-box are given by

$$u(w)p_0 = u(1)p_{w-1}$$

$$u(w-1)p_1 = u(w)p_0$$

$$\vdots$$

$$u(1)p_{w-1} = u(2)p_{w-2}$$

Here $u(n) = n\mu/[n + (M-1)]$ as in Eq. (5–3) and as indicated in Fig. 5–17. Solving these equations for p_j, we find

$$p_j/p_0 = u(w)/u(w-j) \qquad (5-15)$$

Summing over all the states, with the sum set equal to 1, we get

$$p_j = \frac{(w-j) + (M-1)}{(w-j)[w + (M-1)T_w]} \qquad (5-16)$$

with the parameter T_w defined to be the finite sum given by

$$T_w = \sum_{j=1}^{w} \frac{1}{j} \qquad (5-17)$$

The throughput γ and the average end-to-end time delay $E(T)$ are easily evaluated using Eq. (5–16). The throughput is obtained either by averaging $u(n)$ in the upper (Norton) queue of Fig. 5–17 over all values of $n = w - j$, and noting from the balance equations (5–14) that $u(j)p_{w-j}$ is a constant, or by noting that whenever the w-box of Fig. 5–17 fires, w packets are inserted into the VC. The throughput is given by

$$\gamma = \mu w / [w + (M - 1)T_w] \tag{5-18}$$

This is to be compared with the corresponding throughput equation (Eq. 5–9) found in the sliding-window case. They are of exactly the same form, with $(M - 1)T_w/w$ replacing $(M - 1)/N$ in the sliding-window case.

To find the average time delay along the virtual circuit we again use Little's formula. Letting $p(n)$ be the probability of state of the virtual circuit Norton equivalent, we have, for the average packet occupancy of the VC,

$$E(n) = \sum_{n=1}^{w} np(n) = \frac{\gamma}{\mu} \left[\frac{1+w}{2} + (M - 1) \right] \tag{5-19}$$

after noting that $n = w - j$, $p(n) = p_{w-n}$, using Eq. (5–16), and evaluating the resultant expression. From Little's formula, then, the normalized time delay is just

$$\mu E(T) = \frac{E(n)}{\gamma/\mu} = \left(M - 1 + \frac{1+w}{2} \right) \tag{5-20}$$

This is precisely in the same form as Eq. (5–10) for the sliding-window end-to-end delay, with the window size N there replaced by $(1 + w)/2$ here. The explanation for this form is very simple. In this heavy-traffic case of the acknowledge-at-end-of-window control, the number of packets along the VC ranges from 1 to w. The *average* number is $(1 + w)/2$, which is comparable to the *fixed* number N moving along the VC in the sliding-window case under heavy traffic. For $N = (1 + w)/2$, the average time delays will be the same. Because the number of packets along the VC in the ack-at-the-end case fluctuates between 1 and w, rather than being kept fixed as in the sliding-window case, the throughput will be reduced. It is readily shown that the throughput found from Eq. (5–18) is always below that of the sliding-window case given by Eq. (5–9); i.e., $T_w/w > 1/N$, with $N = (1 + w)/2$.

The time-delay–throughput tradeoff curve for a three-hop ($M = 3$) VC is shown in Fig. 5–18. Note the reduction in throughput incurred by acknowledging at the end of a window. This deterioration in performance increases as the window size increases. Similar results are obtained for the case of $\lambda = \mu$. As noted earlier, one simply replaces $(M - 1)$ by M and reduces the time-delay expression to correspond to M rather than $(M + 1)$ queues. Details are left to the reader.

The optimum operating point for the sliding-window control in the case of $\lambda = \mu$ was previously shown to be $N = M$. The corresponding window size for the ack-at-the-end control would be $w = 2M - 1$. As an example, for $M = 3$ hops, this corresponds to a window of five packets, contrasted to the window of three packets in the sliding-window case.

Acknowledging receipt of a packet before the end of a window increases the throughput for the same time delay, approaching that of the sliding-window case. The performance of such a scheme would be expected to lie between the two curves of Fig. 5–18. As noted previously, earlier acknowledgement requires more acks to be sent. In the next section, where we discuss and analyze the IBM SNA control mechanism, we shall see that it is possible to improve the performance to almost that of the optimum sliding-window control and yet retain the feature of one acknowledgement per window.

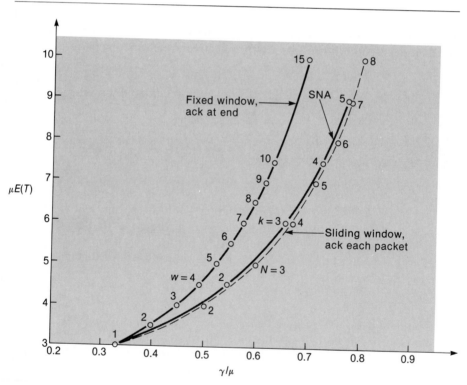

Figure 5–18 Comparison of ack at end and sliding-window controls, $\lambda \to \infty$, $M = 3$

5-3 SNA Path Control

Recall from Chapter 3 that the IBM Systems Network Architecture (SNA) layer that corresponds to the OSI Reference Model network layer is called *path control*. This is the layer that controls routing and the flow of messages throughout the network. As shown in Fig. 5-19, path control receives a basic information unit (BIU) from transmission control, the layer above, and appends a transmission header (TH) to form a path information unit (PIU). The PIU may be passed down to the data link control with no change, or segmenting or blocking (concatenation) may first be carried out. BIUs will be segmented into multiple PIUs if too long (Fig. 5-20a), or PIUs may be blocked or combined into a longer unit if too short (Fig. 5-20b). (This is only done in special circumstances, to be noted later.) In all cases, the final unit passed down to data link control is called the basic transmission unit (BTU). The BTU in turn forms the information field of the synchronous data link control (SDLC) basic link units (BLUs), comparable to HDLC frames, that are actually transmitted between adjacent nodes, at the data link (SDLC) level in SNA networks. The specific fields comprising the transmission header added at the path control level will be described later, after path control functions are discussed and analyzed in detail.

Consider the five-node SNA routing or communication network shown as an example in Fig. 5-21. Subarea or routing nodes only are shown in this figure. Recall from Chapter 3 that these may only be PU_T4 (concentrator) or PU_T5 (host) nodes, forming the communication backbone network. Other

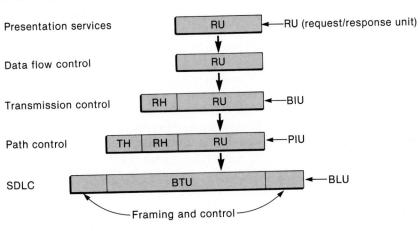

Figure 5-19 SNA data units

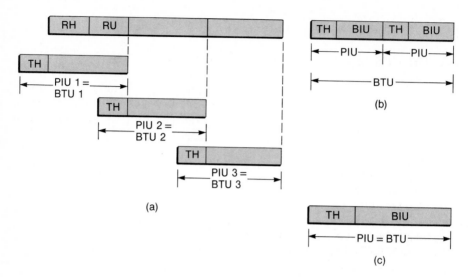

Figure 5–20 Types of BTUs
a. Segmenting (BIU too long)
b. Blocking (PIU too short)
c. No change

nodes (PU_T1, PU_T2, or sometimes PU_T5 hosts) may in turn be connected to these subarea nodes, forming a more complete network. We show only the communication network in Fig. 5–21. This network has nodes interconnected by groups of transmission links, called transmission groups (TGs). Multiple groups between nodes (such as TG2 and TG7 in Fig. 5–21) may be defined as well. Single-link TGs are the only ones over which blocking of PIUs is allowed.

As noted in Chapter 3, in SNA terminology two users or logical units (LUs) communicate via a session (a half-session at either end) established between the two LUs. Messages for a given session are routed by path control over a physical path called an *explicit route* (ER). Such a route consists of a set of nodes with TGs connected between them. An ER is given a number called an *explicit route number* (ERN). Two ERs are shown between nodes N1 and N5 in the network of Fig. 5–21. ER1 consists of N1, TG2, N2, TG5, and N5. ER2 consists of N1, TG1, N3, TG3, N4, TG6, and N5. Messages assigned to a specific multilink TG, as part of an explicit route using that TG, may use different links and thus may arrive out of order at the node on the other end of the TG. For this reason they *must* be reordered before being sent further along the explicit route. Messages between source and destination nodes currently are constrained by the SNA

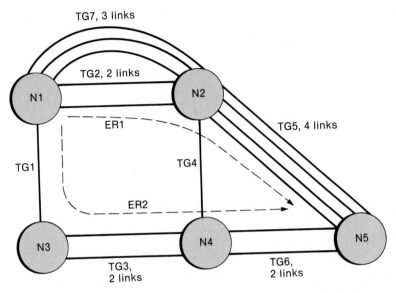

Figure 5–21 Example of SNA routing network, subarea nodes only

architecture to follow the same physical path. The paths in the two directions, however, may have different numbers (i.e., different ERNs).

In addition to having explicit routes comprising physical paths between specified source-destination nodes in a network, SNA defines *logical duplex paths* between source-destination nodes. These are called *virtual routes* (VRs) and correspond to the virtual circuits we have been discussing up to now. A maximum of 48 VRs may be defined between subarea pairs, and multiple sessions may use the same VR. Path control in SNA *multiplexes* sessions from individual transmission controls, one for each half-session, in the layer above it at the two ends of a virtual route. Multiple VRs may share one explicit route. The relation between half-session, virtual route, and explicit route is shown in Fig. 5–22 [SCHU]. The virtual route is known only at the two ends, the subarea nodes, of a given path. Routing along the specific path, the explicit route, is determined by the ERN and the destination (subarea) address. An intermediate node along the explicit route uses these two fields, as carried in the transmission header, to direct messages to the appropriate outgoing transmission group.

[SCHU] Gary D. Schultz, "Anatomy of SNA," *Computerworld,* vol. XV, no. 11a, March 18, 1981, 35–38.

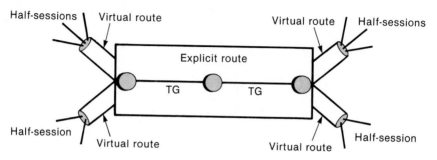

Figure 5–22 Relation between sessions, virtual routes, and explicit route in SNA

The 48 possible virtual routes between subarea pairs comprising source and destination nodes are obtained by defining 16 VR numbers and three levels of priority on each. The VR number and priority combine to form the VRID. Each entry in a VRID table at a subarea source node is in turn associated with a user *class of service* (COS) specification. The class of service depends on the type of data to be transmitted (for example, batch or interactive) and the service desired (such as high throughput, low delay, or cryptographic, among others). A mapping is specified at a source node between class of service and VRID. The VRID list appears in order of performance, and a session is assigned to the first VR on the list that is available. The overall mapping, used in setting up a path between subarea nodes in the process of establishing a session between them, then looks as follows:

$$COS \rightarrow VRID \rightarrow VR \rightarrow ER$$

The appropriate explicit route or path to use between any two subarea nodes is not currently specified by SNA, but rather is user dependent. Ways of selecting "appropriate" paths will be described in the next chapter, in which routing in networks is discussed.

A simplified diagram of message (data unit) flow within path control, showing segmenting or blocking, as well as routing, is shown in Fig. 5–23 [SNA 1980, Fig. 3–3]. A VR control element (VRC) is shown connected to an ER control (ERC) which is in turn connected to a transmission group control (TGC). All three functions appear within path control and hence have PC in their labels. At the higher end, VRC communicates with transmission control TC. Segmenting, if required, is done by VRC. The transmission header (TH) is added here as

[SNA 1980] *Systems Network Architecture Format and Protocol Reference Manual: Architectural Logic*, SC30-3112-2, IBM, Research Triangle Park, N.C., 1980.

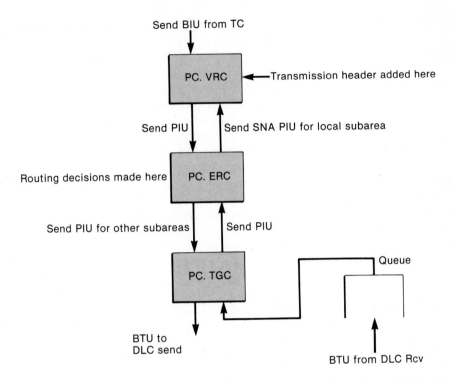

Figure 5–23 Simplified diagram, SNA path control (from [SNA 1980, Fig. 3.3] with permission, © IBM)

well. The resultant PIU (Fig. 5–20) is passed on to explicit route control (ERC). This control carries out routing for the node; it decides, based on the transmission header field, which TG to use for transmission. In a corresponding way, routing decisions are made by ERC on PIUs passed up from the transmission group control (TGC): Messages may be sent down to an appropriate transmission group, if destined for another subarea, or, if destined for this subarea, passed up to VRC. Transmission group control will block PIUs, if necessary, to form the basic transmission unit, but only on single-link TGs.

5–3–1 Virtual Route Pacing Control

With this brief introduction to SNA path control, we now focus on congestion control at this level. SNA has flow-control mechanisms, called *pacing control,* built in at multiple levels in the architecture [SCHU]. Session pacing, at the transmission control (TC) level, is used to control the rate of receipt of messages between the two ends of a session. SDLC, at the link level, has link-level pacing built in with the use of the RR- and RNR-frames discussed in the last chapter. End-to-end *congestion control*, which is carried out on each virtual route at the path control level, is termed *virtual route pacing control*. The actual mechanism used is quite complex and is described in detail in the SNA manual [SNA 1980, Chap. 3], as well as in a basic paper by J. D. Atkins [ATKI]. We shall summarize the mechanism here and then analyze a simplified version. This will enable us to compare this control procedure and the window control mechanisms discussed earlier. We shall use the term packet to represent the basic transmission unit as previously, even though SNA does not use this term.

The SNA virtual route pacing control works as follows. A fixed window k is initially established for a given virtual route, and a pacing count PC is set at this value. The pacing count is decremented every time a packet enters the virtual route:

$$PC = PC - 1 \qquad (5-21)$$

Packets are withheld from the network if the pacing count is zero. PC is thus comparable to the count C defined for the acknowledge-at-end-of-window control (Fig. 5–16). We use the letter k to represent the window size here, again to distinguish it from the window size parameters N and w defined previously.

In the SNA mechanism the *first* packet in a given window generates a special response packet (the acknowledgement) when it arrives at the destination. This virtual route pacing response (VRPRS) is returned to the source at expedited priority. On arriving at the source it causes the current pacing count to be incremented by the window size:

$$PC = PC + k \qquad (5-22)$$

This is the basic mechanism that we model and analyze in this section. Additional mechanisms, not considered here, are invoked when any nodal link buffer along the path exceeds a given threshold (the current window is then decremented by one on arrival of the VRPRS at the source), when a second larger threshold is exceeded (the window is then reduced to a specified minimum), and when the pacing count is zero on receipt of the VRPRS (the window is incremented by one). A detailed description of the complete control procedure, with accompanying flow chart, appears in the paper by Atkins [ATKI].

[ATKI] J. D. Atkins, "Path Control: The Transport Network of SNA," *IEEE Trans. on Comm.*, vol. COM-28, no. 4, April 1980, 527–538; reprinted in [GREE].

Focusing on the *fixed* window mechanism (i.e., ignoring the window variations just described, which we assume occur slowly compared with the operation of the basic mechanism), it becomes apparent after a little thought that the pacing count, under stationary operation, varies between 0 and $2k - 1$:

$$0 \leq PC \leq 2k - 1 \qquad (5-23)$$

The number of packets allowed over the virtual route varies between these limits as well. Recall that in the acknowledge-at-end-of-window control, considered in the previous section, the count C varied between 0 and w. As soon as w packets were en route along the VC no further packets could be transmitted until the *last* packet in the window arrived, reducing the throughput as a result. In the SNA case, the *first* packet in the window of k is acknowledged, allowing the PC count to be increased to $2k - 1$ on arrival of the VRPRS at the source. This acknowledgement of the first packet in a window, with the concurrent increase of the count, allows the throughput deterioration of the previous scheme to be overcome. We shall see, in fact, that the performance of this SNA mechanism approaches that of the optimum sliding-window control quite closely, but with a reduction of acknowledgement overhead to one ack (VRPRS) per window. (For the same performance value, however, window sizes will be different.)

The operation of the fixed-window virtual route pacing count control, as just outlined, is readily shown to be captured by the closed-network model of Fig. 5-24. As previously, the M-hop virtual route (or virtual circuit) is represented by M queues in tandem. VRPRS packets are assumed to arrive at the source as soon as they are generated at the destination. Corresponding to the w-box of Fig. 5-16 we now have a k-box, storing up to $k - 1$ packets. As the kth packet arrives it triggers the discharge of the k-box, and the PC-count queue is incremented by k instantaneously. Thus this model provides the same type of bulk arrival process described earlier in the case of the acknowledge-at-end-of-window control. The only difference between this model and the one shown in Fig. 5-16 is the use of $2k - 1$ packets, which circulate around the closed network. It is left to the reader to show that the use of $2k - 1$ packets is required to capture the effect of acknowledging the first packet in the window.

Note from the model of Fig. 5-24 that when the PC-count queue is empty, $j \leq k - 1$ packets have been delivered to the destination and are waiting in the fictitious k-box, while $(2k - 1) - j$ packets are in transit somewhere along the virtual route. The λ server shown cannot serve, and packets are blocked from entering the virtual route. If $PC > 0$, packets leave the PC queue at an average rate of λ packets per unit time, obeying the same Poisson arrival statistics as invoked in the previous control models.

The analysis of the model of Fig. 5-24 is very similar to that of the acknowledge-at-end-of-window analysis. As there, one assumes that the M-hop virtual

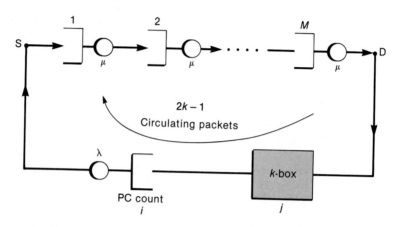

Figure 5–24 Model of SNA virtual route pacing control

route can be represented by the Norton equivalent with state-dependent service, again an approximation only, giving rise to the model of Fig. 5–25. The two integers, i and j, shown in Fig. 5–25 represent, respectively, the PC count and the state of the k-box. Based on the discussion just concluded, we have

$$0 \le i \le 2k - 1, \quad 0 \le j \le k - 1, \quad 0 \le i + j \le 2k - 1$$

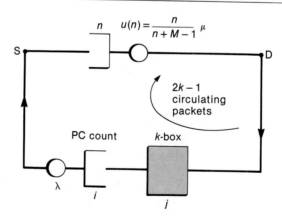

Figure 5–25 Queueing model, virtual route pacing control

The state n, representing the number of packets moving along the M-node virtual route, is just $n = (2k - 1) - (i + j)$. As previously, given the joint statistics of (i, j), one can calculate the statistics of n, and from this the desired performance measures of the controlled virtual route.

The joint state probabilities p_{ij} of the integer pair (i, j) may again be found by setting up two-dimensional balance equations, using the rate parameters λ and $u(n)$ to relate the transitions between states. A simple count indicates that there are $(3k + 1)k/2$ state probabilities to be found. It is left to the reader to show that a typical balance equation for finding p_{ij} in the interior of the two-dimensional (i, j) region is given by

$$\{\lambda + u[2k - 1 - (i + j)]\}p_{ij} = \lambda p_{i+1,j} + u[2k - (i + j)]p_{i,j-1} \quad (5-24)$$

Similar although reduced equations may be written along the four boundaries $i = 0$, $j = 0$, $(i + j) = 2k - 1$, and $j = k - 1$. These two-dimensional equations represent an extension of the finite boundary, one-dimensional birth-death state diagram, introduced briefly in Chapter 2, to two dimensions. These equations are examples of Markov chains.

As in the previous section, the solution of these two-dimensional balance equations, although readily carried out for k not-too-large, is not really necessary for our purpose. As was the case there, analysis for the special values of the load parameter λ, $\lambda \to \infty$ $(\lambda \gg \mu)$ and $\lambda = \mu$ may be carried out almost trivially. As already noted, these two values essentially delimit the high-traffic region, the region in which the congestion control is expected to operate. The performance of the control mechanism and its comparison with other mechanisms may again be inferred from its operation at these two load points.

Following the procedure of the previous two sections, consider first the heavy traffic case $\lambda \to \infty$. From Figs. 5-24 and 5-25 this implies that the load (the offered traffic) is so high that most arriving packets are turned away $(PC \to 0)$. Yet whenever the k-box fires (the first of k packets in a window has arrived at the destination and been acknowledged), there are enough new packets present at the source to deposit k such packets into the virtual route. The number of packets along the virtual route thus fluctuates between k and $2k - 1$. (Compare this with fixed N in the sliding-window case and fluctuation between 1 and w in the acknowledge-at-end-of-window case.) The two-dimensional problem reduces to a one-dimensional one (the PC count $i = 0$) as represented by the state diagram of Fig. 5-26. Note that because the k-box of Fig. 5-25 cannot fire in the interval $0 \leq j \leq k - 1$, only upward transitions, or births, appear, with the rates shown. Since the k-box fires and empties on the arrival of a packet when in state $j = k - 1$, the state diagram shows itself wrapping back around from $j = k - 1$ to $j = 0$. Letting p_j be the probability that the k-box is in state j, $0 \leq j \leq$

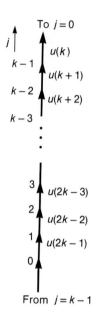

Figure 5–26 State diagram, virtual route pacing control, $\lambda \to \infty$

$k-1$, as the equations that relate the state probabilities in this special case of $\lambda \to \infty$,

$$u(2k-1)p_0 = u(k)p_{k-1}$$
$$u(2k-2)p_1 = u(2k-1)p_0$$

$$\vdots$$

$$(5-25)$$

$$u(k)p_{k-1} = u(k+1)p_{k-2}$$

Note how similar this set of equations is to those of Eq. (5–14), which describe the comparable balance equations for the acknowledge-at-end-of-window control. Again, the one difference is that here the state of the virtual route fluctuates between k and $2k-1$; there the state of the equivalent virtual circuit fluctuated between 1 and w.

From Eq. (5–25) the desired probability of state p_j is readily shown to be given by

$$p_j/p_0 = u(2k-1)/u(2k-1-j) \qquad (5-26)$$

Note that this is precisely the form of Eq. (5–15), with $(2k-1)$ replacing w there. Solving for p_0 by invoking the usual probability measure relation

$$\sum_{j=0}^{k-1} p_j = 1$$

and substituting in for the Norton throughput $u(n) = n\mu/[n + (M-1)]$, one finds

$$p_j = \frac{(2k-1-j) + (M-1)}{(2k-1-j)[k + (M-1)S_k]} \qquad (5-27)$$

Here the parameter S_k represents the finite sum

$$S_k = \sum_{j=k}^{2k-1} \frac{1}{j} \qquad (5-28)$$

This is again similar to expression (5–16), found previously for the state probabilities in the acknowledge-at-end-of-window control model. $2k-1$ replaces w there, and the sum S_k corresponds to the sum T_w there. One would thus expect the throughput and end-to-end delay expressions here to be comparable to these found earlier. This turns out to be the case. It is left to the reader to show, following the same procedure as used in the previous analysis, that the normalized throughput of the virtual route with M hops is given by

$$\gamma/\mu = k/[k + (M-1)S_k] \qquad \lambda \to \infty \qquad (5-29)$$

while the normalized end-to-end delay is given by

$$\mu E(T) = M - 1 + \left(\frac{2k - 1 + k}{2}\right) \qquad \lambda \to \infty \qquad (5-30)$$

Equations (5–29) and (5–30) are to be compared with Equations (5–18) and (5–20), respectively, for the comparable expressions for the acknowledge-at-end-of-window control.

Comparing with Eqs. (5–9) and (5–10) for the sliding-window control, it is apparent that all three mechanisms obey similar equations in the heavy-traffic case. Since the analysis for the case $\lambda = \mu$ (the load value at which the congestion region is entered) is identical, with M replacing $(M-1)$, one gets comparable equations for this case as well.

The equations describing the performance of all three control strategies may be written in the following unified form:

1. $\lambda \to \infty$ (heavy traffic) case

$$\mu E(T) = M - 1 + N' \qquad (5-31)$$

$$\gamma/\mu = 1/[1 + (M-1)/K] \qquad (5-32)$$

2. $\lambda = \mu$ case

$$\mu E(T) = \left(\frac{M}{M+1}\right)(M+N') \qquad (5-33)$$

$$\gamma/\mu = 1/(1 + M/K) \qquad (5-34)$$

The two parameters, N' and K, depend on the control strategy, as described in the following table:

CONTROL PROCEDURE	N'	K
Sliding window	N	N
Ack at end of window	$(1+w)/2$	w/T_w
VR pacing control	$(3k-1)/2$	k/S_k

It is apparent from Eq. (5–31) or Eq. (5–33) that all three schemes produce the same normalized time delay if N' is chosen the same in all three cases. Thus, as an example, if $N = 4$ is the window size in the sliding-window case, $w = 7$ and $k = 3$ in the two other strategies provide the same end-to-end time delay. It was shown in the previous section that the throughput for the acknowledge-at-end-of-window control procedure is correspondingly less than that for the sliding-window control, this reduction in throughput increasing with N'. The parameter k/S_k, which governs the throughput performance of the VR pacing control scheme, is found to be at most 4 percent less than N, if N in the sliding-window case is chosen to equal $(3k-1)/2$. This indicates that the SNA control mechanism tracks the sliding-window mechanism very closely for all lengths of the virtual circuit and for all window sizes. This is demonstrated by the example of Fig. 5–18, which compares the time-delay–throughput performance of all three control schemes for the special case of $\lambda \to \infty$ and a three-hop virtual circuit.

Simulation bears out these results [SCHW 1982]. In particular, as noted earlier, the approximations made turn out to be valid if the VC is not too long. The Norton approximation for the two bulk-arrival models is found to be quite accurate so long as the bulk arrival is "not too large." This means that $N' < 2M$ in the cases simulated. Since we have indicated earlier that an appropriate operating point is $N' \doteq M$, the Norton approximation provides useful results. The independence assumption used in the analysis—which requires each packet to have its lengths randomly chosen from the same distribution as it arrives at each node along the VC—turns out to provide increasingly poorer approximations to the performance curves as the VC length M increases. But the curves for all three control mechanisms continue to track one another, so that relative comparisons can still be made [SCHW 1982a].

The one serious drawback to the VC models of this chapter is that they focus on a *single* isolated VC only. They are applicable if the effect of other VCs using the same transmission link can be modeled as simply reducing the transmission capacity available. This is obviously not true in many practical situations, where VCs may share multiple links in tandem, may carry different amounts of traffic with different length packets, and may vary in length. The most general approach is to analyze multiple VCs in the context of closed queueing networks. This will be touched on briefly in Subsection 5–4–3, where closed-network analysis is described. The problem there, aside from the computational one of analyzing complex queueing networks, is that results can only be obtained for specific examples of networks, and only for the sliding-window control. No general results are available for the "best" choice of window size for each of a number of interacting VCs. No insight into the operation of the control in general has been developed by studying a variety of examples. The best that can be done is to pick the window size and characterize the performance of *isolated* VCs, as done in this section.

5–3–2 SNA Transmission Header

It was noted earlier that the SNA path control or network level appends a transmission header (TH) to the data unit handed down from transmission control (Figs. 5–19 and 5–23). The resultant data unit is called the path information unit (PIU). The transmission header is used to carry out the routing and flow-control functions at each node along a virtual route. It is of interest to briefly describe the format of this header.

SNA actually defines six types of transmission header [ATKI], [SNA 1980]. Here we discuss only the one involved in communication between adjacent subarea nodes that comprise an SNA communication network. The format of this header, identified by the label FID4 and consisting of 26 bytes, is presented in Fig. 5–27 [SNA 1980, pp. 2-16–2-22, D-2, D-3]. We shall discuss only portions of this format.

It is apparent from a perusal of Fig. 5–27 that the virtual route is identified by the virtual route number and 2-bit transmission priority field of byte 3. The priority bit pattern 00 is used to represent low priority, 01 means medium priority, and 10 refers to high priority. The explicit route number (ERN) onto which the VC is mapped (in this direction only), is given by the 4 bits of byte 2. Destination and origin subarea addresses are provided by bytes 8–11 and 12–15, respectively.

The mechanism for implementing virtual route pacing control, as described earlier, is assigned to a number of bits. Recall that the first data unit (PIU) in a window of k is the one to trigger an acknowledgement (the VRPRS) on arriving at the destination end of the virtual route. This PIU is identified by

Byte

0	Format identification, FID4: 0100 TG sweep indicator (0 or 1) ER and VR support indicator (0 or 1) Network priority (0 or 1)	Reserved byte

2	Initial explicit route number (4 bits) Explicit route number (ERN) (4 bits)	Virtual route number (VRN) (4 bits) 2 reserved bits Transmission priority field (2 bits)

4	VR change window indicator (0 or 1) TG non-FIFO indicator (0 or 1) VR sequencing and type indicator (2 bits) TG sequence number field (bits 4–15)

6	VR pacing request (0 or 1) VR pacing response (0 or 1) VR change window reply indicator (0 or 1) VR reset window indicator (0 or 1) VR send sequence-number field (bits 4–15)

8	Destination subarea address field (4 bytes)

12	Origin subarea address field (4 bytes)

16	Bits 0-2, reserved SNA indicator (0 or 1) Mapping field (2 bits) Bit 6 reserved Expedited flow indicator (0 or 1)	Reserved byte

18	Destination element field

20	Origin element field

22	Sequence-number field

24	Data-count field

Figure 5–27 Transmission header, SNA FID4 format (from [SNA 1980, p. 2–16] with permission, © IBM)

setting bit 1 (the VR pacing-request bit) in byte 6 of its transmission header to 1. The VRPRS data unit is in turn identified by setting the VR pacing-response bit number 2 in byte 6 to 1. The network-priority bit in byte 0 is set to 1 at the same time, which provides rapid return of the VRPRS to the origin node.

Other bits are used to carry out the window-variation parts of the pacing-control mechanism noted earlier. For example, minor congestion on a transmission group along the explicit route on which this PIU is traveling is indicated by setting the VR change-window indicator of byte 4 to 1. Subsequent transmission groups along the ER leave the indicator on. The indicator is in turn carried back by the next VRPRS, by setting the change-window reply-indicator bit in its byte 6 to 1. When the VRPRS arrives at the source (origin) node, the window will be decremented by one. If heavy traffic is experienced anywhere along an explicit route, the node that experiences the congestion will set the VR reset-window indicator bit of byte 6 to 1 on the first PIU passing through in the direction *toward* the source. On arrival there, this PIU will cause the window to be set to the minimum specified. Note that these actions affect *all* VRs passing through a node that experiences congestion. Intermediate nodes do not distinguish among the VRs that use them.

These actions taken at an intermediate node that is experiencing congestion are shown schematically in Fig. 5–28. How the congestion is measured is not spelled out by SNA; this is left to the network implementer to specify. A simple measure might be to indicate minor congestion by a buffer threshold on an outgoing transmission group that is being exceeded. Heavy (major) congestion might manifest itself by a higher buffer threshold being exceeded. Another

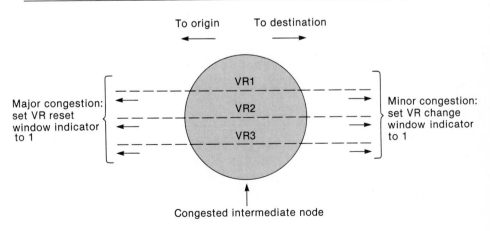

Figure 5–28 Actions taken at a congested SNA node

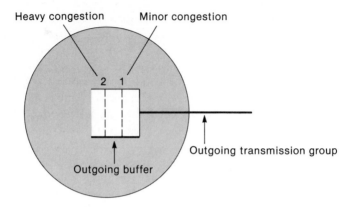

Figure 5–29 Possible measure of nodal congestion

measure might include the outgoing packet rate on a transmission group exceeding either of two thresholds. Figure 5–29 portrays the buffer-threshold concept schematically. Threshold 1 is used to measure minor congestion; threshold 2 measures heavy congestion.

Other bytes in the transmission header of Fig. 5–27 not discussed here are described in detail in the SNA literature [SNA 1980].

5–4 Queueing Networks

In discussing the virtual circuit model earlier in this chapter we noted that this model was a special case of a class of models that involve more generally the interconnection of a multiplicity of queues. For obvious reasons, these are called queueing networks. It is apparent that a store-and-forward packet-switched network of the kind considered in Chapters 1 and 3, in which packets queue up for transmission over appropriate outgoing links, can be modeled by a queueing network. A simple five-node example appears in Fig. 5–30. (The links are shown half-duplex only, directed to the two destinations, to keep the figure relatively uncluttered. In practice more links, generally full-duplex transmission, would be used and additional queues would appear.) Queueing networks appear as well in the modeling, for performance analysis, of large computer systems with multiple I/O devices attached [KOBA], [SAUE]. The analysis of queueing networks has in fact received an enormous impetus since the mid-1970s as a result of the use of such network models for evaluating the

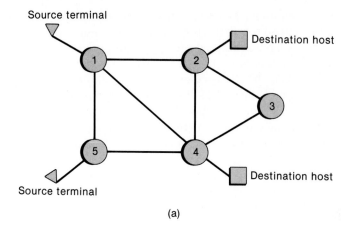

(a)

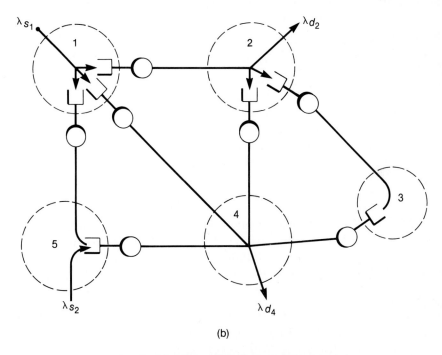

(b)

Figure 5-30 Example of queueing network
a. Five-node network
b. Queueing network model

performance of computer systems and of computer communication networks. Other applications arise regularly in the operations research field [KELL].

These models are particularly useful when they give rise to so-called product-form solutions. Analysis can then be readily carried out, subject to computational difficulties when the network gets too large. Much of the literature of queueing networks since the mid-1970s has been devoted to these two problem areas: conditions under which product-form solutions arise and improved algorithms for reducing the computational complexity.

Two generic classes of queueing networks can be distinguished; these are portrayed schematically in Fig. 5–31. In the first case, as in the model of Fig. 5–30(b), traffic is shown entering and leaving the network. The corresponding network is said to be of the *open-network* type. If packets are not dropped anywhere in the network, it is apparent, from flow-conservation arguments, that the net packet-arrival rate into the network, from external sources, must equal the departure rate summed over the various destinations; i.e., in Fig. 5–31(a), $\lambda_s = \lambda_d$. In Fig. 5–30(b), $\lambda_{s_1} + \lambda_{s_2} = \lambda_{d_2} + \lambda_{d_4}$. The second case, in which packets (more generally *customers*) are confined to the network, circulating indefinitely with no arrivals or departures, corresponds to the *closed-network* class. Note that the open networks of Figs. 5–30(b) and 5–31(a) are readily converted to the closed-network class by setting source and destination rates to zero. An example appears in Fig. 5–31(b). Note that both Fig. 5–31(a) and Fig. 5–31(b) *generalize* the queueing-network model of Fig. 5–30, which was developed specifically for packet-switched networks where queueing at processing points is neglected. In this case queueing is encountered at outgoing links only, just the case we have been discussing throughout this chapter. Therefore traffic cannot be split at the output of a queueing service station, as shown in Fig. 5–31. Of necessity, packets that leave an output-transmission queue and that are transmitted out on its associated transmission link arrive at the next node to which they are connected. More generally, a node would have nodal processing associated with it (see the models described in Chapter 2), and the packets leaving the nodal processor could move to one of a number of output transmission queues. Figure 5–31 would then be used to represent this more general case.

Closed-queueing models arise in communication networks, as we have seen, by the introduction of window-type controls. Figure 5–13(a) provides an example. These models also arise in computer system models when the use of multiprogramming is studied [KOBA], [SAUE]. The models shown in Fig. 5–31 are more general than might appear in a packet-switched network; they include the possibility of feedback and other features that might arise in incorporating computer processing at a node.

If, in all these models, arrival statistics are Poisson, the service times at each node are taken to be exponential and distributed independently from node to

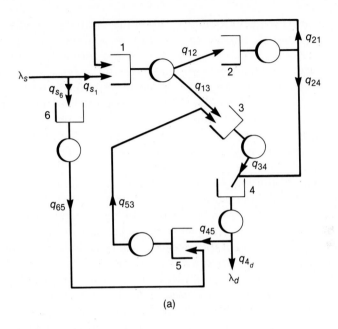

(a)

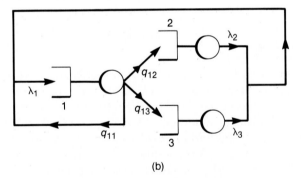

(b)

Figure 5–31 Open and closed queueing networks
   a. Open queueing model
   b. Closed queueing model

node, and the route from one queue to another is selected randomly, with fixed routing probabilities designated by the q symbols shown in Fig. 5–31, the networks are readily shown to be of the product form. We shall prove this statement shortly. Such networks are said to be of the *Jackson* type, after J. R. Jackson, who first showed they give rise to product-form solutions [JACK]. More generally, he showed that the product form is retained if external arrivals depend on the *total* number of packets (customers) in the network and if service times depend on the state of their respective queues (i.e., if they are of the generalized death-process type). Pioneering work on closed queueing networks of the exponential type was done by Gordon and Newell [GORD].

More general conditions for the existence of product-form solutions have been found by a number of investigators since the mid-1970s, including, among others, Chandy, Baskett, Muntz, Kobayashi, Reiser, and Kelly [BASK], [KOBA], [SAUE], [KELL]. The proof that a Jacksonian network, with first-come, first-served service, is of the product form is readily carried out by showing that a global balance equation is satisfied by a product-form solution. More generally, local balance conditions, with balance invoked at each queue and for each class of customers, have been used to prove the product-form condition [BASK], [SAUE]. Lazar and Robertazzi have developed a geometric approach to studying the existence of product-form solutions [LAZA 1984].

5–4–1 Product-Form Solution, Exponential Network

As just noted, we shall focus here on the Jacksonian network with exponential servers only. Consider an open network of this type, with M queues (or stations) and with only one source and destination pair, for simplicity. A portion of such a network appears in Fig. 5–32. The Poisson arrival rate at the source is labeled λ. A typical queueing station showing various probabilities of moving on to the next station (the routing probabilities), including returning for service to itself, appears in Fig. 5–33. (Note again that this is a generalization of modeling communications in a packet-switched network.) The symbol q_{ij} represents the

---

[JACK] J. R. Jackson, "Job Shop-like Queueing Systems," *Management Science,* vol. 10, no. 1, 1963, 131–142.

[GORD] W. L. Gordon and G. F. Newell, "Closed Queueing Systems with Exponential Servers," *Operations Research,* vol. 15, no. 2, 1967, 254–265.

[BASK] F. Baskett et al., "Open, Closed, and Mixed Networks of Queues with Different Classes of Customers," *J. ACM,* vol. 22, no. 2, 1975, 248–260.

[LAZA 1984] A. A. Lazar and T. G. Robertazzi, "The Geometry of Lattices for Markovian Queueing Networks," *Columbia Research Report,* July 1984.

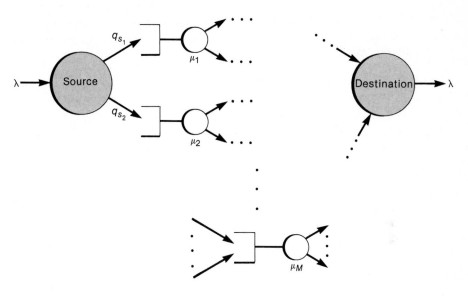

Figure 5-32 Open queueing network model, exponential servers

probability that a packet (customer) completing service at queue i is routed to queue j. The queue service rate is, as usual, labeled μ_i.

We must obviously have the condition

$$q_{id} + \sum_{j=1}^{M} q_{ij} = 1 \qquad 1 \le i \le M \qquad (5-35)$$

with d representing the destination. In practice, a number of the q's might be zero, indicating that no packets move on to the queues in question.

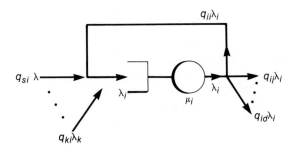

Figure 5-33 Typical queue in network

Continuity of flow provides another necessary condition at the input to, and output of each queue: Letting λ_i be the packet-arrival rate (or departure rate for an infinite queue with no blocking), we have

$$\lambda_i = q_{si}\lambda + \sum_{k=1}^{M} q_{ki}\lambda_k \qquad 1 \le i \le M \tag{5-36}$$

The objective is, as always, to determine the probability $p_i(n_i)$ that queue i is in state n_i. Because of the interconnection of queues, however, the various probabilities are not necessarily independent. We first calculate the joint probability

$$p(n_1, n_2, \ldots, n_i, \ldots, n_M) \equiv p(\mathbf{n}) \tag{5-37}$$

We shall prove that in the open network case

$$p(\mathbf{n}) = \prod_{i=1}^{M} p_i(n_i) \tag{5-38}$$

$$p_i(n_i) = (1 - \rho_i)\rho_i^{n_i}$$
$$\rho_i \equiv \lambda_i/\mu_i < 1 \tag{5-39}$$

Equation (5–38) is an example of a product-form solution: The various queues, even though interconnected and coupled through the continuity expressions (5–35) and (5–36), behave as if they are independent! More remarkably, they each appear as M/M/1 queues, with the familiar state probability form of Eq. (5–39)! (The interconnection and dependence of queues then arises through the ρ_i factors, which must obey the continuity equations (5–35) and (5–36).) In the case of state-dependent service times, not considered here, similar results, involving familiar solutions of state-dependent queues, also arise [KOBA]. We shall use the result of Eqs. (5–38) and (5–39) in the next chapter in our discussion of routing through networks.

In the closed-network case with N circulating packets (customers), we shall prove a somewhat different and more general version of the product-form result:

$$p(\mathbf{n}) = \prod_{j=1}^{M} \rho_j^{n_j}/g(N, M) \tag{5-40}$$

The normalization constant $g(N, M)$ must be chosen to have the sum of $p(\mathbf{n})$ over all possible states equal 1. The evaluation of this constant for large networks ($M \gg 1$) and large numbers of packets ($N \gg 1$) can lead to computational problems. We shall discuss a simple recursive scheme, first developed by Buzen, that is quite useful if the networks are not too large [BUZE]. Interestingly, all

[BUZE] J. P. Buzen, "Computational Algorithms for Closed Queueing Networks with Exponential Servers," *Comm. ACM*, vol. 16, No. 9, Sept. 1973, 527–531.

statistical quantities of interest can be derived from this constant. (The constant is analogous to the partition function of statistical mechanics, from which equations of state and physical parameters can be found.)

5–4–2 Open Queueing Networks

Consider the case of open queueing networks first. By extending the balance equation concept for one and two dimensions (discussed in Chapter 2 and earlier in this chapter), we can set up a multidimensional *global balance equation* for the vector probability $p(\mathbf{n})$. We then show that this global equation is satisfied by the product-form solution of Eqs. (5–38) and (5–39). Specifically, we equate all the possible rates of leaving state $\mathbf{n}$ to the rates of entering the state. This results in the global balance equation (5–41), shown next. To simplify the notation we use the expression $p(\mathbf{n} - \mathbf{1}_i)$ to represent $p(n_1, n_2, \ldots, n_i - 1, n_{i+1}, \ldots, n_M)$. The unit vector $\mathbf{1}_i$ thus represents a change of one unit from state n_i. Similarly, we use $p(\mathbf{n} - \mathbf{1}_i + \mathbf{1}_j)$ to represent $p(n_1, n_2, \ldots, n_i - 1, \ldots, n_j + 1, \ldots, n_M)$. With this notation the global balance equation, equating the total rate of leaving state $\mathbf{n}$ to the rate of entering $\mathbf{n}$, is readily written as

$$\left[\lambda + \sum_{i=1}^{M} \mu_i \right] p(\mathbf{n}) = \lambda \sum_{i=1}^{M} q_{si} p(\mathbf{n} - \mathbf{1}_i)$$

$$+ \sum_{i=1}^{M} q_{id} \mu_i p(\mathbf{n} + \mathbf{1}_i) \qquad (5-41)$$

$$+ \sum_{i=1}^{M} \sum_{j=1}^{M} q_{ji} \mu_j p(\mathbf{n} + \mathbf{1}_j - \mathbf{1}_i)$$

As just stated, the left-hand side of Eq. (5–41) represents the summed rate of leaving state $\mathbf{n}$: There can be an arrival with rate λ, or a departure from any one of the M queues, with corresponding rate μ_i. The right-hand side of Eq. (5–41) represents the sum of the rates over the various ways of entering state $\mathbf{n}$. As an example, $\mathbf{n}$ can be entered from state $\mathbf{n} - \mathbf{1}_i$, with queue i at state $n_i - 1$, by an arrival at queue i. This can happen in two ways: A packet can arrive from the external source with rate λq_{si}, or it can arrive from queue $j(1 \leq j \leq M)$ with rate $q_{ji}\mu_j$. In the latter case queue j must have been at state $n_j + 1$. A packet can also exit queue i directly to destination d, giving rise to an additional term in Eq. (5–41).

To solve this global equation we first eliminate q_{si} by substituting in the flow conservation equation (5–36). It is then left to the reader to show that the resultant equation is satisfied by the following equality:

$$\lambda_i p(\mathbf{n} - \mathbf{1}_i) = \mu_i p(\mathbf{n}) \qquad (5-42)$$

(Note that this then implies that

$$\lambda_j p(n - 1_i) = \mu_j p(n - 1_i + 1_j)$$

as well.) Specifically, on substituting Eq. (5–42) into the reduced form of Eq. (5–41), one finds all terms canceling except for the final set:

$$\lambda p(n) = \sum_{i=1}^{M} q_{id} \mu_i p(n + 1_i)$$

But with $\lambda_i p(n) = \mu_i p(n + 1_i)$, this is readily reduced to the equality

$$\lambda = \sum_{i=1}^{M} q_{id} \lambda_i$$

just the condition for flow conservation from the source to the destination. This proves that the balance equation (5–41) is in fact satisfied by Eq. (5–42).

Expanding the short-hand notation of Eq. (5–42), we have

$$p(n) = \left(\frac{\lambda_i}{\mu_i}\right) p(n_1, n_2, \ldots, n_i - 1, \ldots, n_M)$$

Repeating n_i times, we get

$$p(n) = \left(\frac{\lambda_i}{\mu_i}\right)^{n_i} p(n_1, n_2, \ldots, 0, \ldots n_M)$$

$$\underset{\text{state of } i\text{th queue}}{}$$

Now doing the same for each queue, we finally obtain

$$p(n) = \prod_{i=1}^{M} \left(\frac{\lambda_i}{\mu_i}\right)^{n_i} p(0) \tag{5–43}$$

To find $p(0)$—i.e., the probability that all M queues are empty—one sums the global state probability $p(n)$ over all possible states, setting the resultant sum, as usual, equal to 1. We thus have

$$\sum_{n} p(n) = p(0)S = 1 \tag{5–44}$$

with the sum S given by

$$S \equiv \sum_{n} \left[\prod_{i=1}^{M} (\lambda_i/\mu_i)^{n_i}\right] \tag{5–45}$$

For a feasible solution S must be finite. If so, theory indicates that the solution is unique. With $S < \infty$ the sums and products may be interchanged. One then gets

$$S = \prod_{i=1}^{M} \sum_{n_i=0}^{\infty} \rho_i^{n_i} = \prod_{i=1}^{M} (1 - \rho_i)^{-1} \tag{5–45a}$$

carrying out the indicated sum and using ρ_i as usual to represent λ_i/μ_i. From Eqs. (5-44), (5-45a), and (5-43) we get the final result:

$$p(\mathbf{n}) = \prod_{i=1}^{M} (1 - \rho_i)\rho_i^{n_i} \qquad (5-46)$$

Note that $\rho_i < 1$ to keep $S < \infty$. Equation (5-46), the solution for the global probability of state for an open queueing network, is of the product form and in fact indicates each of the M queues can be treated as an M/M/1 queue. Arrival rates λ_i, $i = 1$ to M, are uniquely determined by reference to the flow-conservation equation (5-36), with the routing probabilities q_{ki} given by Eq. (5-35).

Although proven here for the special case of one source-destination pair, Eq. (5-46) applies generally to *any* open network with Poisson arrival rates and exponential service rates. Implicit in the derivation is the assumption that service rates are *independent*. This is precisely the assumption made in our analysis earlier of end-to-end window controls: We assumed that packets moving along a path from source to destination have their lengths selected *independently* at each outgoing link. This independence requirement will be invoked again in the next chapter when we apply Eq. (5-46) to the study of routing in networks.

As a simple example of the applicability of Eq. (5-46), consider the average end-to-end time delay incurred by exponentially distributed packets flowing through a network as they move from source to destination. Congestion control is not invoked in this simple case. Say that the packet arrival rate at the source node is Poisson, with average rate λ. The packets follow a virtual circuit such as one of those shown in Fig. 5-9(a) as modeled by Fig. 5-9(b). If the average service time at link i in this M-hop virtual circuit is $1/\mu_i$ (due possibly to different transmission capacity on each link or to interference from traffic that reduces the capacity available), the end-to-end time delay is that due to M cascaded M/M/1 queues; it is given by

$$E(T) = \sum_{i=1}^{M} \frac{1}{\mu_i - \lambda_i} = \sum_{i=1}^{M} \frac{1/\mu_i}{1 - \rho_i} \qquad \rho_i = \lambda_i/\mu_i \qquad (5-47)$$

More specifically, consider the five-node example network of Fig. 5-34. Link transmission in one direction only is considered for simplicity. The γ_i's represent the (Poisson) arrival rates at each of the three nodes shown. The branching (routing) probabilities on each of the links leaving a node are indicated by the numbers next to the arrows. The numbers shown in parentheses alongside each link represent the net packet load, in packets/time, accessing that link. This is found by simply summing the arrival rates at a node and multiplying by the probability of accessing a given link. As a check, note that the sum of the outgoing traffic rates at nodes 3 and 4 is 6 packets/sec, just the sum of the traffic rates at nodes 1, 2, and 5, as expected from flow-conservation considerations.

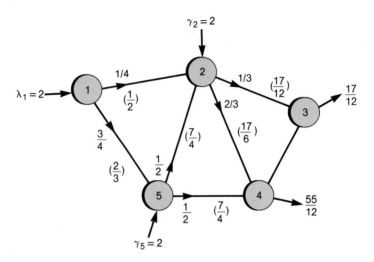

Figure 5–34 Time-delay calculation, example five-node network

It is apparent that to ensure a stable system the packet-service rate, or transmission capacity, of each link must be greater than the load on each link. As an example, say that the transmission capacity of all links is identical and equal to 3 packets/sec. The average time delay from link 1 to link 3, routing via node 2, is

$$E(T_{13}) = \frac{1}{3 - 1/2} + \frac{1}{3 - 17/12}$$

The average end-to-end delay from node 1 to node 4, routing directly from node 5 to node 4, is

$$E(T_{14})_d = \frac{1}{3 - 3/2} + \frac{1}{3 - 7/4}$$

The average end-to-end delay from node 1 to node 4, going via node 2, is

$$E(T_{14})_2 = \frac{1}{3 - 3/2} + \frac{1}{3 - 7/4} + \frac{1}{3 - 17/6}$$

Other time delays are left to the reader to calculate.

Consider in general an M-link network with service capactiy of μ_i packets/sec on the ith link and λ_i packets/sec accessing that link on the average. The time delay along any path (a cascaded set of links, each one modeled as an M/M/1 queue) is then the sum of the M/M/1 time delays constituting that path, as apparent from the two examples just provided. But now say that we wish to know the average *networkwide* time delay, averaged over all M links. This is readily found by invoking both Little's formula and the product-form results of this section. Let the network be enclosed by the surface shown in Fig. 5–35. From Little's formula the network time delay $E(T)$, averaged over the entire network, is just

$$\gamma E(T) = E(n) \tag{5-48}$$

with γ the net arrival rate into the network and $E(n)$ the average number of packets (customers) contained within it. But $E(n) = \sum_{i=1}^{M} E(n_i)$, with $E(n_i)$ the average number of packets either in service or enqueued at link i. $E(n_i)$ is in turn given by $\lambda_i T_i$, with $T_i = 1/(\mu_i - \lambda_i)$ the M/M/1 average time delay of link i. The average time delay is thus just

$$E(T) = \frac{1}{\gamma} \sum_{i=1}^{M} \lambda_i T_i = \frac{1}{\gamma} \sum_{i=1}^{M} \frac{\lambda_i}{\mu_i - \lambda_i} \tag{5-49}$$

(This expression neglects propagation delay over the links.) Equation (5–49) will be used in the next chapter in discussing routing at the network level. For, as indicated in Eq. (5–36), the λ_i's depend on the probabilistic routing parameters q_{ki} (Figs. 5–31 to 5–33). The routing problem for packet-switched networks consists of assigning the flows $\lambda_i, i = 1, \ldots, M$ (alternatively, the set of probabilities q_{ki}) to minimize the average packet time delay $E(T)$. Algorithms for carrying out the minimization will be discussed in the next chapter.

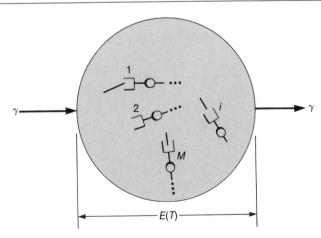

Figure 5–35 Little's formula applied to general open queueing network

5–4–3 Closed Queueing Networks

The product-form solution for closed queueing networks with exponential service times is proven in a manner similar to that carried out in the previous subsection, on open queueing networks [KOBA]. One obtains, as the product-form solution for the case of M queues, the same form as that found in Eq. (5–43):

$$p(\boldsymbol{n}) = p(n_1, n_2, \ldots, n_M) = \prod_{i=1}^{M} \left(\frac{\lambda_i}{\mu_i}\right)^{n_i} p(0) \qquad (5-43)$$

As previously, n_i is the number of packets at queue i, and $1/\mu_i$ is again the average service time (taken to be exponential) at queue i. However, there are two basic distinctions. Since there are no net arrivals or departures in a *closed* network, the link loads λ_i, $i = 1, \ldots, M$ cannot be related to a net external arrival rate λ as shown in Eq. (5–36). An example appears in Fig. 5–36. The flow-conservation expression (5–36), equating flow into and out of queue i to the net flow arrival rate from all queues directed at queue i, still holds, but only *relative* values of the λ_i's can be found. All values may be scaled up or down by the same fixed constant without changing the results. The ratio λ_i/μ_i is thus no longer the actual utilization of queue i. It is an arbitrary number that can be varied at will by adjusting all λ_i's so that they still satisfy Eq. (5–36). The actual utilization of each queue will be one of the parameters calculated later.

The other distinction between closed and open queueing networks consists of the fact that a fixed number N of packets must be taken to be circulating throughout the closed network. This limits the number of states in each queue to at most $N + 1$ (including zero). In fact the sum of *all* queue states in the

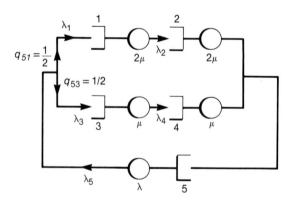

Figure 5–36 Example of a closed queueing network: window control of two virtual circuits; N = window size

network must at all times equal N. This constraint, to be invoked among all M queues, must be used in calculating the product-form normalization constant $p(\mathbf{0})$. As a result the reduction of Eq. (5–43) to Eq. (5–46), the multiple M/M/1 form, is not possible in the closed-network case. Instead, we write the product-form solution for closed queueing networks in the form

$$p(\mathbf{n}) = \frac{1}{g(N, M)} \prod_{i=1}^{M} (\lambda_i/\mu_i)^{n_i} \qquad (5\text{–}50)$$

with the normalization constant $g(N, M)$ written in place of $1/p(\mathbf{0})$ in Eq. (5–43).

Interestingly, we shall find that all the statistical parameters of interest for a closed network can be obtained in terms of $g(N, M)$. There is no need to use $p(\mathbf{n})$ explicitly. The analysis of closed queueing networks thus depends on the evaluation of the normalization constant. For large networks this becomes a computational problem, as already noted, and a variety of recursive techniques have been proposed for reducing the computational burden [KOBA], [SAUE]. In the interest of simplicity, we describe only one such technique due to Buzen [BUZE]. Buzen's algorithm, for queues with exponential service times, states simply that

$$g(n, m) = g(n, m - 1) + \rho_m\, g(n - 1, m) \qquad \rho_m \equiv \lambda_m/\mu_m \qquad (5\text{–}51)$$

with initial starting conditions given by

$$g(n, 1) = \rho_1^n \qquad n = 0, 1, 2, \ldots, N \qquad (5\text{–}52)$$

and

$$g(0, m) = 1 \qquad m = 1, 2, \ldots, M \qquad (5\text{–}53)$$

As noted in Eq. (5–51), we use the symbol ρ_i to represent λ_i/μ_i in carrying out the recursion.

Starting with $g(n, 1)$ and $g(0, m) = 1$, one increases n and m in steps of one and uses Eq. (5–51) to build up $g(n, m)$ recursively until $g(N, M)$ is found. Some examples will be provided later. To prove the recursive relation Eq. (5–51) we first start with the definition of $g(N, M)$. From Eq. (5–50) it is apparent that $g(N, M)$ must be chosen to have the sum of $p(\mathbf{n})$ over all possible states equal to one. The states correspond to all values of $n_i, i = 1, 2, \ldots, M$ such that $n_i \geq 0$ and $\sum_{i=1}^{M} n_i = N$. More precisely, we define a function $F(N)$ given by

$$F(N) = \left\{ \mathbf{n} \,|\, n_i \geq 0, \text{ all } i = 1, 2, \ldots, M, \text{ and } \sum_{i=1}^{M} n_i = N \right\} \qquad (5\text{–}54)$$

Then we must have, from Eq. (5–50), with $\rho_i \equiv \lambda_i/\mu_i$,

$$\sum_{\mathbf{n} \in F(N)} p(\mathbf{n}) = 1 = \frac{1}{g(N, M)} \sum_{\mathbf{n} \in F(N)} \prod_{i=1}^{M} \rho_i^{n_i} \qquad (5\text{–}55)$$

From this we get

$$g(N, M) = \sum_{n\epsilon F(N)} \prod_{i=1}^{M} \rho_i^{n_i} \qquad (5-56)$$

(Note that if *all* possible positive values of n_i are allowed, with the constraint of Eq. (5–54) removed one obtains in place of $g(N, M)$ the sum S of Eq. (5–45), from which the M/M/1 result of Eq. (5–45a) is readily obtained.)

To prove the recursive relation Eq. (5–51), we define the function $g(n, m)$, in agreement with Eq. (5–56), as

$$g(n, m) = \sum_{n\epsilon F(n)} \prod_{i=1}^{m} \rho_i^{n_i} \qquad (5-57)$$

where

$$F(n) = \left\{ \boldsymbol{n} \middle| n_i \geq 0 \quad \text{and} \quad \sum_{i=1}^{m} n_i = n \right\} \qquad (5-58)$$

With $m > 1$ and $n > 0$, we now decompose the sum of Eq. (5–57) into two disjoint sets, one for $n_m = 0$, the other for $n_m > 0$. (Recall that n_m is the state of queue m.) Since $\rho_m^{n_m} = 1$ for $n_m = 0$, we get

$$g(n, m) = \sum_{n\epsilon F(n)} \prod_{i=1}^{m-1} \rho_i^{n_i} + \sum_{\substack{n\epsilon F(n) \\ n_m > 0}} \prod_{i=1}^{m} \rho_i^{n_i} \qquad (5-59)$$

The first term in Eq. (5–59) is just $g(n, m - 1)$. To evaluate the second term we factor ρ_m out of $\rho_m^{n_m}$, leaving $\rho_m^{n_m-1}$. Since $n_m > 0$, $n_m - 1 \geq 0$. Calling this n'_m, it is apparent that $n'_m + \sum_{i=1}^{m-1} n_i = n - 1$. It is left to the reader to show that the second term is then just $\rho_m g(n - 1, m)$, completing the proof of Eq. (5–51).

As an example of the use of the recursive relation (5–51), consider the model of the sliding-window control given by Fig. 5–13(a). Note that here there are $M + 1$ queues to be considered. For a three-hop virtual circuit one must determine $g(N, 4)$. By continuity we have

$$\lambda_1 = \lambda_2 = \cdots = \lambda_{M+1}$$

(Recall that these flows are arbitrary, defined to within an arbitrary constant. See Fig. 5–36 for an example of two two-hop virtual circuits.) Then we have

$$\rho_1 = \rho_2 = \cdots = \rho_M = \lambda_1/\mu$$

$$\rho_{M+1} = \lambda_1/\lambda = \frac{\lambda_1}{\mu} \frac{\mu}{\lambda} = \rho_1/\rho$$

with $\rho \equiv \lambda/\mu$. Since the flows are arbitrary, the calculation may be simplified

TABLE 5–2 $g(n, m)$, 3-hop Sliding-window Control

$n \downarrow$	$m \rightarrow$ 1	2	3	4
0	1	1	1	1
1	1	2	3	$3 + \dfrac{1}{\rho}$
2	1	3	6	$6 + \dfrac{1}{\rho}\left(3 + \dfrac{1}{\rho}\right)$
3	1	4	10	$10 + \dfrac{1}{\rho}\left[6 + \dfrac{1}{\rho}\left(3 + \dfrac{1}{\rho}\right)\right]$
4	1	5	15	$15 + \dfrac{1}{\rho}\left\{10 + \dfrac{1}{\rho}\left[6 + \dfrac{1}{\rho}\left(3 + \dfrac{1}{\rho}\right)\right]\right\}$

considerably by setting $\rho_1 = \rho_2 = \cdots = \rho_m = 1$. Then $\rho_{M+1} = 1/\rho$. Applying the recursive algorithm (5–51) to this example, the window control of a three-hop VC, we get the entries for $g(n, m)$ shown in Table 5–2. In Table 5–3 we take the same example with $\lambda = \mu$, or $\rho = 1$. These tables will be used later in evaluating the performance of the sliding-window control for a three-hop VC. (The results will, of course, agree precisely with those obtained using the Norton theorem approach earlier in this chapter.)

As another example of a closed queueing network, one with random routing included, consider again the network shown in Fig. 5–36. This might model a sliding-window control superimposed on two parallel virtual circuits, one with transmission capacity μ at each link, the other with higher-speed links of capacity 2μ. There is a probability of $1/2$ of routing to either VC. The ρ_i's in this

TABLE 5–3 $g(n, m)$, 3-hop Sliding-window Control, $\rho = 1$

$n \downarrow$	$m \rightarrow$ 1	2	3	4
0	1	1	1	1
1	1	2	3	4
2	1	3	6	10
3	1	4	10	20
4	1	5	15	35
5	1	6	21	56
6	1	7	28	84

TABLE 5-4 $g(n, m)$ for Network of Fig. 5-36

		1	2	3	4	$m \rightarrow$ 5
						$\rho_m \rightarrow$
		1	1	2	2	$4/\rho$
	0	1	1	1	1	1
	1	1	2	4	6	$6 + 4/\rho$
$n \downarrow$	2	1	3	11	23	$23 + \dfrac{4}{\rho}(6 + 4/\rho)$
	3	1	4	26	72	$72 + \dfrac{4}{\rho}\left[23 + \dfrac{4}{\rho}(6 + 4/\rho)\right]$
	4	1	5	57	201	$201 + \dfrac{4}{\rho}\left\{72 + \dfrac{4}{\rho}\left[\left(23 + \dfrac{4}{\rho}(6 + 4/\rho)\right)\right]\right\}$

example, to be used in finding $g(N, M)$ by application of the recursive relation (5-51), are given by

$$\rho_1 = \lambda_1/2\mu, \quad \rho_2 = \rho_1, \quad \rho_3 = \lambda_3/\mu = 2\rho_1, \quad \rho_4 = \rho_3$$
$$\rho_5 = \lambda_5/\lambda = 2\lambda_3/\lambda = 4\rho_1/\rho, \quad \rho = \lambda/\mu$$

Since the λ_i's are arbitrary, we let $\lambda_1 = 2\mu$ in this example. This gives us

$$\rho_1 = \rho_2 = 1, \quad \rho_3 = \rho_4 = 2, \quad \rho_5 = \frac{4}{\rho}$$

The resultant values of $g(n,m)$ for this example, obtained using Eq. (5-51), appear in Table 5-4. Values for the special case of $\lambda = \mu$ are tabulated in Table 5-5.

TABLE 5-5 $g(n, m)$, Same as Table 5-4, $\rho = \lambda/\mu = 1$

		1	2	$m \rightarrow$ 3	4	5
	0	1	1	1	1	1
	1	1	2	4	6	10
$n \downarrow$	2	1	3	11	23	63
	3	1	4	26	72	324
	4	1	5	57	201	1497

We now demonstrate how all desired statistics of the closed queueing network can be obtained from the normalization constant $g(N, M)$. As noted previously, there is then no need to find and use $p(\mathbf{n})$ explicitly in the analysis. (It is shown in [PENN, appendix] and [SCHW 1977, pp. 249–253] that the normalization constant and the moment-generating function defined as the M-dimensional transform of $p(\mathbf{n})$ are related. All statistical parameters of interest can be found from the moment-generating function. This thus accounts for the same property of $g(N, M)$.

Consider first $p(n_i \geq k)$, the probability that the number of packets in queueing station i equals or exceeds k. This is shown readily to be given by the following equation, which involves $g(N - k, M)$ and $g(N, M)$:

$$p(n_i \geq k) = \rho_i^k g(N - k, M)/g(N, M) \tag{5-60}$$

These two parameters can be obtained from the appropriate entries in the table for $g(n, m)$, found using the Buzen recursive technique.

The proof of relation (5–60) is relatively straightforward and is similar to that used in deriving the recursive relation (5–51). Specifically, from the definition of $p(n_i \geq k)$, we must have

$$p(n_i \geq k) = \sum_{\substack{\mathbf{n} \in F(N) \\ \text{and } n_i \geq k}} p(\mathbf{n}) = \sum_{\substack{\mathbf{n} \in F(N) \\ n_i \geq k}} \frac{1}{g(N, M)} \prod_{j=1}^{M} \rho_j^{n_j} \tag{5-60a}$$

using Eq. (5–50) and using the symbol ρ_j again to represent the arbitrary ratio λ_j/μ_j. We now use the same trick as previously: We introduce a new state parameter $n_i' = n_i - k$ in place of n_i in Eq. (5–60a) and factor out ρ_i^k. Using $n_i' \geq 0$ in place of n_i in $\mathbf{n}$ and noting that N must be smaller by k as a result, we then have

$$p(n_i \geq k) = \frac{\rho_i^k}{g(N, M)} \sum_{\mathbf{n} \in F(N-k)} \prod_{j=1}^{M} \rho_j^{n_j} = \rho_i^k \frac{g(N - k, M)}{g(N, M)} \tag{5-60}$$

applying definition (5–56) to $g(N - k, M)$.

As a special case, let $k = 1$. The probability $p(n_i \geq 1) = 1 - p(n_i = 0)$ is just the utilization of the server in a single-server queue. For a single-server queue of service rate μ_i, the queue throughput γ_i, or the average number of packets serviced per unit time, is just $\gamma_i = \mu_i p(n_i \geq 1)$. We thus have

$$\gamma_i = \mu_i p(n_i \geq 1) = \mu_i \rho_i g(N - 1, M)/g(N, M) = \lambda_i g(N - 1, M)/g(N, M) \tag{5-61}$$

invoking Eq. (5–60) and the definition of ρ_i. (The reader concerned with the prior statement that the ρ_i's and λ_i's are arbitrary numbers, bound only by flow-conservation equations, should note that as the λ_i's are scaled up and down, the $g(n, m)$ entries change as well. The throughputs γ_i, $i = 1, 2, \ldots, M$ are *absolute* and remain unchanged. They depend on the queueing network configuration, the branching probabilities q_{ik}, and the total number of packets N in the closed network.)

Consider now the probability $p(n_i = k)$. This is the (marginal) probability that queue i has k packets (including the one in service). Using Eq. (5–60), this is written readily as

$$p(n_i = k) = p(n_i \geq k) - p(n_i \geq k + 1)$$
$$= \frac{\rho_i^k}{g(N, M)} [g(N - k, M) - \rho_i g(N - k - 1), M] \qquad (5–62)$$

Finally, we show how the expectations of queue lengths may be obtained from tables of $g(n, m)$. Specifically, the average number of packets at queue i is given by

$$E(n_i) = \sum_{k=0}^{N} k p(n_i = k) = \sum_{k=0}^{N} k[p(n_i \geq k) - p(n_i \geq k + 1)]$$

Writing the indicated sum out term by term and canceling identical terms that appear, one obtains the well-known identity

$$E(n_i) = \sum_{k=1}^{N} p(n_i \geq k)$$

Using Eq. (5–60), we then have as the desired identity

$$E(n_i) = \sum_{k=1}^{N} \rho_i^k g(N - k, M)/g(N, M) \qquad (5–63)$$

We now provide some examples, using Tables 5–2 through 5–5, of the utility of these identities. Consider first the cyclic closed queueing network of Fig. 5–37. This diagram can represent a virtual circuit with a sliding-window control applied in the case $\lambda \to \infty$ (see Fig. 5–11), or it can represent the model needed to obtain the Norton equivalent of the M-hop virtual circuit. (Recall from Fig. 5–12 that the Norton equivalent of a product-form queueing net-

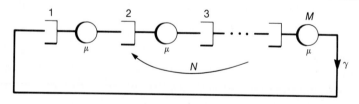

Figure 5–37 Cyclic queueing network

TABLE 5–6	Throughput, γ/μ, Fig. 5–37	
N	$M = 3$	$M = 4$
1	1/3	1/4
2	2/4	2/5
3	3/5	3/6
4	4/6	4/7
5	5/7	5/8
6	6/8	6/9

work is obtained by shorting the two ends of the network and calculating the throughput $u(n)$ as a function of the number of packets, n, circulating through the closed queueing network.) We calculate the throughput first, using Eq. (5–61), for two cases, $M = 3$ and $M = 4$. Table 5–3 applies in this example. Since $\rho_i = 1$, we have simply

$$\gamma/\mu = g(N - 1, M)/g(N, M)$$

The appropriate values of $g(n,m)$ for $M = 3$ and $M = 4$ come from columns 3 and 4, respectively, of Table 5–3. The resultant normalized throughputs are tabulated in Table 5–6. They have been purposely written in fractional form to show that they obey the simple relationship $N/(N + M - 1)$, noted previously in Eq. (5–3) (the throughput of the Norton equivalent of the M-hop VC) and in Eq. (5–9) (the normalized throughput of the sliding-window control with $\lambda \rightarrow \infty$).

Consider now the average number of packets in each of the queues of Fig. 5–37. By symmetry they should all be the same and equal to N/M, with N the total number of packets circulating in the network. It is left for the reader to check that this is in fact the case, by applying Eq. (5–63) and the appropriate entries of Table 5–3.

We now take the network of Fig. 5–36 as the second example. The throughput γ_1 of both queue 1 and queue 2, for the special case of $\lambda = \mu$ and $N = 3$ packets in the network, is, from Table 5–5 and Eq. (5–61), just $2\mu(63/324) = 0.389\,\mu$ packet/sec. The throughput γ_3 of queues 3 and 4 is the same since packets are equally likely to be routed to either of the two paths shown. As a check, the throughput through queue 5 is $\gamma_5 = \lambda \rho_5 g(N - 1, M)/g(N, M) = 0.778\,\mu$ for $N = 3$ and $\lambda = \mu$. If we now allow $N = 4$ packets into the network of Fig. 5–36 the throughput of either VC increases to $0.43\,\mu$ packet/sec. A plot of throughput versus window size N can be obtained, if desired, by enlarging Tables 5–4 and 5–5 by using the recursive algorithm for $g(n,m)$. Other values of the load can be considered as well by adjusting $\rho = \lambda/\mu$

in Table 5 – 4 appropriately. Examples of other networks to which the recursion of $g(n, m)$ and the calculation of statistical parameters may be applied are left to the reader as exercises.

5 – 4 – 4 Mean Value Analysis

In many analyses of closed queueing networks, only mean values of various performance measures are required. For example, we have been stressing the calculation of the average throughput and average time delay of packets in a packet-switched network. The computational burden required to calculate the normalization constant $g(N, M)$ can be reduced by the use of *mean value analysis.* This very simple recursive procedure, introduced by Reiser and Lavenberg [REIS 1980], [REIS 1982], calculates average queue lengths, average time delay at a queue, and queue throughput directly without involving product-form notation and the necessity of calculating the normalization constant $g(N, M)$.

We demonstrate the procedure in the context of the cyclic closed queueing network of Fig. 5 – 38. This is of course just the model we encountered in studying the sliding-window control of a VC (Fig. 5 – 11) and is a generalized version of the cyclic queue model of Fig. 5 – 37.

Critical to mean value analysis is the calculation of the average time delay $\bar{t}_i$ at queue i in a closed queueing network. We focus on the instant that a packet arrives at this queue. The average delay the packet will experience is its own service time $1/\mu_i$ plus the time to serve the packets waiting in the queue in front of it, including the one in service. Thus

$$\bar{t}_i = \frac{1}{\mu_i} + \frac{1}{\mu_i} \times \text{(average number of packets on arrival)} \qquad (5-64)$$

The *arrival theorem* for closed queueing networks of the exponential type states that the state of the system at the time of arrival has a probability distribution equal to that of the steady-state distribution of the network with one packet (customer) less [LAVE]. This theorem, which is intuitively satisfying (an arriving packet or customer sees the system in a state not including itself), provides us with a measure of the average number of packets seen on arrival, as needed in Eq. (5 – 64). Specifically, the time delay at the ith queue, with N packets in the

[REIS 1980] M. Reiser and S. S. Lavenberg, "Mean-value Analysis of Closed Multichain Queueing Networks," *J. ACM,* vol. 27, no. 2, April 1980, 313 – 322.

[REIS 1982] M. Reiser, "Performance Evaluation of Data Communication Systems," *Proc. IEEE,* vol. 70, no. 2, Feb. 1982, 171 – 195.

[LAVE] S. S. Lavenberg and M. Reiser, "Stationary State Probabilities of Arrival Instants for Closed Queueing Networks with Multiple Types of Customers," *J. Appl. Prob.,* vol. 17, 1980, 1048 – 1061.

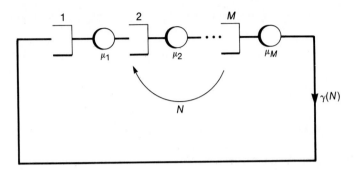

Figure 5-38 More general cyclic queueing network

network, is related to the average number waiting with $(N-1)$ packets in the network. We thus have, from Eq. (5-64),

$$\mu_i \bar{t}_i(N) = 1 + \bar{n}_i(N-1) \tag{5-65}$$

with $\bar{n}_i(N)$ defined to be the average number of packets enqueued at queue i with N in the overall network.

Equation (5-65) is precisely of the recursive form desired. The connection between $\bar{t}_i$ and $\bar{n}_i$ required to complete the algorithm is made by invoking Little's formula. Thus, from Fig. 5-38, it is apparent that one must have

$$\gamma(N) \sum_{i=1}^{M} \bar{t}_i(N) = N$$

In addition, at an individual queue, we have

$$\gamma(N)\bar{t}_i(N) = \bar{n}_i(N)$$

The complete recursion for the cyclic network of Fig. 5-38 is now given by

$$\bar{n}_i(0) = 0 \qquad i = 1, \dots, M \tag{5-66}$$

$$\mu_i \bar{t}_i(N) = 1 + \bar{n}_i(N-1) \qquad i = 1, \dots, M \tag{5-67}$$

$$\gamma(N) = N \bigg/ \sum_{i=1}^{M} \bar{t}_i(N) \tag{5-68}$$

$$\bar{n}_i(N) = \gamma(N)\bar{t}_i(N) \qquad i = 1, \dots, M \tag{5-69}$$

As an example, take the special case $\mu_1 = \mu_2 = \cdots = \mu_M$, which gives rise to the network of Fig. 5-37, whose properties we studied in the previous section and in earlier ones. Applying Eqs. $(5-66)-(5-69)$, we get, for various values of N,

$$\mu \bar{t}_i(1) = 1, \qquad\qquad \gamma(1)/\mu = 1/M, \qquad \bar{n}_i(1) = 1/M,$$

$$\mu \bar{t}_i(2) = 1 + \frac{1}{M} = \frac{M+1}{M}, \quad \gamma(2)/\mu = 2/(M+1), \quad \bar{n}_i(2) = 2/M,$$

$$\mu \bar{t}_i(3) = 1 + \frac{2}{M} = \frac{M+2}{M}, \quad \gamma(3)/\mu = 3/(M+2), \quad \bar{n}_i(3) = 3/M$$

and so forth. Note that we have $\gamma(N)/\mu = N/(M+N-1)$, just the value of throughput for this network found in previous subsections. (See Eq. $(5-3)$ as an example.)

Mean value analysis can be extended easily to more general closed queueing networks of the exponential type [REIS 1980], as well as to the case of queues with state-dependent service rates [REIS 1981].

5–5 Input-buffer Limiting for Congestion Control

In the earlier parts of this chapter we focused on end-to-end window-type methods for controlling congestion in a network. These methods are used commonly at the network layer in architectures that incorporate the virtual circuit concept. The two ends of a VC to be controlled are well defined in this case. Networks designed on the basis of the datagram approach cannot incorporate an end-to-end control at the network level readily. Instead they commonly incorporate congestion control at the *input* to the network, blocking newly arriving packets if the number already present at an input queue exceed some threshold. This approach has been adopted, for example, by Digital Equipment Corporation in its Digital Network Architecture (DNA) [WECK], [DEC, Sec. 5]. The control appears as part of the DNA routing layer, comparable to the OSI network layer. The Canadian Datapac network uses a similar approach [SPRO]. In both of these examples *input-buffer limiting* is applied: A distinction is made

---

[REIS 1981] M. Reiser, "Mean-value Analysis and Convolution Method for Queue-Dependent Servers in Closed Queueing Networks," *Performance Evaluation*, vol. 1, 1981, 7–18.

[WECK] S. Wecker, "DNA: The Digital Network Architecture," *IEEE Trans. on Comm.*, vol. COM-28, no. 4, April 1980, 510–526; reprinted in [GREE].

[DEC] *DECnet, Digital Network Architecture, Routing Layer Functional Specification*, Version 2.00, Digital Equipment Corp., Maynard, Mass. May 1, 1983.

between input packets—packets newly arriving at a node—and transit packets—those already in the network and arriving from another node. Transit packets are favored over input packets in the allocation of nodal buffers. In the simplest allocation strategy, a maximum number of input packets is allowed into the node. If that number is reached, further input packets are blocked. Transit packets may continue to be accepted, however, until the total nodal buffer pool is depleted.

In this section we provide a simple analysis of this input-buffer-limiting strategy [LAM 1979]. Other allocation strategies, however, can also be defined. One simple extension of the input-buffer-limiting scheme, in which input packets are blocked if the *sum* of the input and transit packets exceeds a specified level, has been found by analysis to provide improved time-delay–throughput performance over the original scheme [SAADS].

The analysis of a number of input-buffer-limiting strategies is carried out quite readily if one assumes a homogeneous network with each node identical to every other node. A simple example is the ring network of Fig. 5–39. The parameter λ_I represents the average load, in packets/time, arriving at each node in the network. Since all nodes look alike, a typical node may be pulled out and examined in isolation. Conditions at both the input and the output must be adjusted, however, so that the node "sees" precisely what it does in the original network. For the case of the ring network of Fig. 5–39 the isolated node with total buffer capacity N appears as shown in Fig. 5–40. λ_T represents the (attempted) arrival rate of transit packets from the previous node. These are blocked with probability P_{B_T}, so that the actual net arrival rate at the node is

$$\gamma_T = \lambda_T(1 - P_{B_T})$$

as shown. Some of the transit packets may be destined for this (typical) node. This is accounted for by assigning a fixed probability p_T of exiting at this node. The rate of departure at the node is thus $p_T\gamma_T$. Input packets are shown being blocked with a probability P_{B_I}, so that their net arrival rate at the node is $\lambda_I(1 - P_{B_I}) \equiv \gamma_I$. By flow continuity, we must have as many packets entering the node, on the average, as leaving, which gives rise to the following relation among these different parameters:

$$\gamma_T = \lambda_T(1 - P_{B_T}) = (1 - p_T)\gamma_T + \gamma_I \tag{5-70}$$

[LAM 1979] S. S. Lam and M. Reiser, "Congestion Control of Store-and-Forward Networks by Input Buffer Limits: An Analysis," *IEEE Trans. on Comm.*, vol. COM-27, no. 1, Jan. 1979, 127–134.

[SAADS] S. Saad and M. Schwartz, "Input Buffer Limiting Mechanisms for Congestion Control," ICC, June 1980, Seattle.

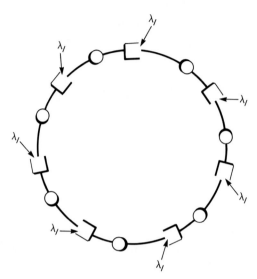

Figure 5–39 Homogeneous ring network

From Eq. (5–70) it is also apparent that

$$\gamma_T = \gamma_I / p_T = \bar{n}\gamma_I \qquad (5-71)$$

with $\bar{n}$, the average number of links (hops) traversed by a packet in the network, just equal to $1/p_T$. As an example, if $p_T = 0.3$, the average number of hops traversed is 3.3. If $p_T = 0.5$, so that transit packets have a 0.5 chance of exiting at

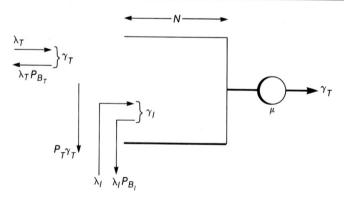

Figure 5–40 Model of node in network of Fig. 5–39

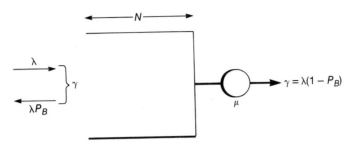

Figure 5-41 Simplified model, no congestion control

each node, the average distance between source and destination is 2 links. By adjusting the parameter p_T we thus adjust the average source-destination distance in the network. We shall pursue this ring network example in the analysis that follows. Lam and Reiser have treated a much more complex network example, with multiple outgoing links at a node, the acknowledgement retransmission buffers explicitly modeled, and so forth [LAM 1979]. It suffices for our purposes to focus on this much simpler example, which retains the most important features of the model.

We first demonstrate via straightforward analysis that the model of Fig. 5-40 predicts that congestion will arise if no controls are used and the packet-arrival rate (load) is increased steadily [SCHW 1979]. In this case, because of the finite buffer model, we shall actually demonstrate a downturn in the throughput and an approach to deadlock. (This was not possible in the VC model of the earlier part of this chapter because of the infinite queue models used.) With no control exerted no distinction is made between transit packets and newly arriving packets. Both are blocked with the same probability P_B. (This is the probability that the finite queue is full.) Since there is no distinction made between transit packets and newly arriving packets, we use the one parameter λ to represent the (attempted) rate of input to the queue. This leads to the simplified picture of Fig. 5-41. (Note, by comparing with Fig. 5-40, that λ actually represents λ_T, the transit-packet-arrival rate. The newly arriving net packet-arrival rate γ_I of Fig. 5-40 is just balanced by the departing packets $p_T\gamma_T$. Since $\lambda_T = \bar{n}\lambda_I$, an increase in the load λ_I is immediately reflected in the transit-packet-arrival rate λ_T, so that either parameter may be used in this noncontrolled case.)

[SCHW 1979] M. Schwartz and S. Saad, "Analysis of Congestion Control Techniques in Computer Communication Networks," *Proc. Symp. on Flow Control in Computer Networks*, Versailles, Fr., Feb. 1979; North-Holland Publishing Co., 113-130.

To calculate the blocking probability P_B, we note that packets getting through to the next node may in turn be blocked with the same probability P_B. (Recall that this model assumes a homogeneous network.) This corresponds to saying that the effective link capacity has been reduced to $\mu' = \mu(1 - P_B)$ [PENN]. Alternatively, each packet requires an average of $1/(1 - P_B)$ retries before finally being transmitted successfully. The effect is to again reduce the maximum throughput (effective link capacity) to $\mu' = \mu(1 - P_B)$. Assuming that one may use finite M/M/1 analysis in this case (this is an unverified assumption but we proceed undaunted anyway!), we write as the blocking probability of the model of Fig. 5–41

$$P_B = \rho'^N (1 - \rho')/(1 - \rho'^{N+1}),$$

$$\rho' \equiv \lambda/\mu' = \rho/(1 - P_B), \quad \rho \equiv \lambda/\mu \tag{5-72}$$

Note that this gives rise to a nonlinear equation in P_B to solve. This is readily done by trial and error or by iteration.

Using Eq. (5–72) to calculate P_B, one can find the throughput $\gamma = \lambda(1 - P_B)$ as a function of the load λ. In Fig. 5–42 we sketch three curves of normalized throughput γ/μ as a function of normalized load, $\rho = \lambda/\mu$. Note that all three curves, for $N = 1$-, 5-, and 8-packet-capacity queues, demonstrate the congestion phenomenon: They all start with $\gamma/\mu = \rho$ for small ρ, reach a peak value, and then plunge rapidly with increasing load, approaching the "deadlock" situation (zero throughput) for $\rho = 1$.

Specific points on these curves can be found for certain special cases. It is left to the reader to show, for example, that for the special case of $N = 1$, $\gamma/\mu = \rho(1 - \rho)$ is the specific equation obtained. For $\rho' = 1$, one finds the solution, using L'Hôpital's rule, to be given by $\rho = N/(N + 1)$, $\gamma/\mu = \rho^2 = [N/(N + 1)]^2$. These values, for $\rho' = 1$, appear to correspond to the maxima of the curves.

The model of Fig. 5–41 — the special case of Fig. 5–40 in which no distinction is made between input and transit packets — predicts deadlock in the high-traffic (load) case. We now introduce control of the input packets and show that by proper design of the control function the deadlock situation can be eliminated. As noted earlier, we focus on the input-buffer-limiting scheme analyzed by Lam and Reiser [LAM 1979]. Specifically, let n_I and n_T be the number of input packets and transit packets in the queue, respectively. The control function, very simply, is to block further input packets if $n_I = N_I < N$. The threshold N_I is the control parameter to be found. Transit packets, on the other hand, are allowed to enter until all buffers are filled. We thus have these two conditions in the controlled situation:

$$0 \leq n_I \leq N_I \tag{5-73}$$

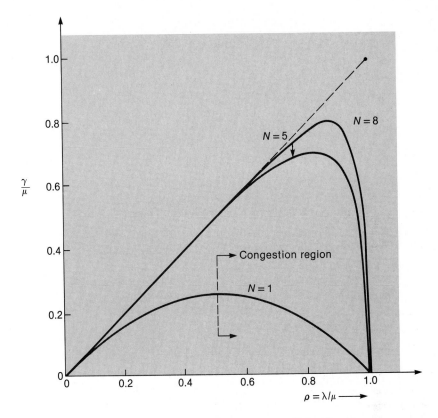

Figure 5-42 Throughput curve of Fig. 5-41

and

$$0 \le n_I + n_T \le N \tag{5-74}$$

The state space of the queue is represented by the two numbers (n_I, n_T). The region of operation of the queue, corresponding to Eqs. (5-73) and (5-74), is given by the two-dimensional diagram of Fig. 5-43.

As in our previous queueing analyses, the performance of this input-buffer-control mechanism, given by both the throughput γ_I of Fig. 5-40 and the average queueing time delay $E(T)$ as a function of the load λ_I, can be determined once the state probabilities $p(n_I, n_T)$ are found. These may be calculated in the

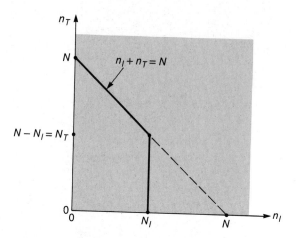

Figure 5–43 State space, input-buffer-limiting procedure

following manner: Let the service rate of the queue be a fixed, known quantity μ'. (The determination of this service rate is discussed later.) Say that the queue is at state (n_I, n_T), with n_I input packets and n_T transit packets sitting in the queue, including the one in service. For a FIFO discipline, with either packet type served (whichever is at the head of the queue), the chance that an input packet is in service is $n_I/(n_I + n_T)$, while the chance that a transit packet is in service is $n_T/(n_I + n_T)$. The rate of service of input packets is thus $\mu'n_I/(n_I + n_T)$, while the rate of service of transit packets is $\mu'n_T/(n_I + n_T)$. This gives rise to a *state-dependent* service rate for each of the two types of packets (customers), the input packets and the transit packets. A similar approach has been used in analyzing the effect of *two* types of users (customers) accessing a link along a virtual circuit [PENN], [SCHW 1977, pp. 249, 250].

With this justification for the use of state-dependent service rates, the queueing model corresponding to Figs. 5–40 and 5–43 becomes that of a *two*-dimensional birth-death process, with upward transitions along the n_I-axis driven at a rate λ_I, upward transitions along the n_T-axis driven at a rate $\lambda_T(1 - p_T)$ (the net transit-packet-arrival rate once transit packets destined for this node are accounted for), and downward transitions along either axis driven at the appropriate state-dependent rate calculated previously. These two-dimensional rates in the interior of the state space of Fig. 5–43 are portrayed schematically in Fig. 5–44.

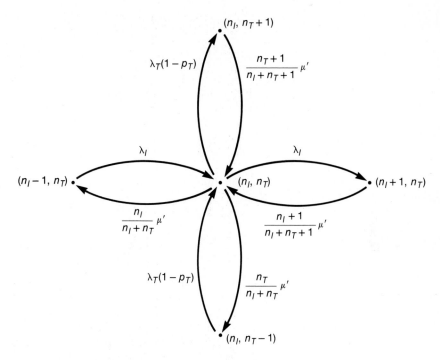

Figure 5–44 Transition rates, interior region, Fig. 5–43

The two-dimensional balance equation corresponding to the state transitions of Fig. 5–44 is given by

$$[\mu' + \lambda_I + \lambda_T(1 - p_T)]p(n_I, n_T) = \lambda_I p(n_I - 1, n_T)$$

$$+ \lambda_T(1 - p_T)p(n_I, n_T - 1)$$

$$+ \frac{\mu'(n_I + 1)}{(n_I + 1 + n_T)} p(n_I + 1, n_T) \quad (5-75)$$

$$+ \frac{\mu'(n_T + 1)}{(n_I + 1 + n_T)} p(n_I, n_T + 1)$$

It is left to the reader to show that Eq. (5–75) is satisfied by the following product-form solution [LAM 1977], [PENN], [SCHW 1977, p. 250]:

$$p(n_I, n_T) = \frac{1}{S} \frac{(n_I + n_T)!}{n_I! n_T!} \rho_I^{n_I} \rho_T^{n_T} \tag{5-76}$$

Here S is the usual normalization constant found by summing $p(n_I, n_T)$ over the two-dimensional state space of Fig. 5-43. Thus

$$S = \sum_{\substack{0 \le n_I \le N_I \\ 0 \le n_I + n_T \le N}} \frac{(n_I + n_T)!}{n_I! n_T!} \rho_I^{n_I} \rho_T^{n_T} \tag{5-77}$$

(The notation means, sum over all values of n_I and n_T with the constraints shown.)

The parameters ρ_I and ρ_T are given, respectively, by $\rho_I = \lambda_I / \mu'$; $\rho_T = \lambda_T (1 - p_T) / \mu'$; $\lambda_T = \lambda_I (1 - P_{B_I}) / p_T (1 - P_{B_T})$, from Eq. (5-70), Eq. (5-71), and the definition of λ_I.

The only remaining question is that of calculating the effective service rate μ'. There is a problem here. This parameter depends on the transit-packet blocking probability P_{B_T}, which in turn depends on the state probability $p(n_I, n_T)$, which is itself to be determined from μ'! This is the same dilemma noted earlier in analyzing the congestion-prone model of Fig. 5-41. As was the case there, the deadlock phenomenon and its possible alleviation by the control mechanism under study here can only be modeled by assuming that the actual queue service rate μ is reduced by blocking at the next node. We are thus forced to write, as was done earlier,

$$\mu' = \mu(1 - P_{B_T}) \tag{5-78}$$

Here P_{B_T} is the transit blocking probability shown in Fig. 5-40. (Note again that it is the *transit* packets, those exiting the node, that may be blocked at the next node. In the discussion of the model of Fig. 5-41 no distinction was made between input and transit packets, and the symbol P_B was used to designate blocking probability.)

We resolve the dilemma, as has been done a number of times already (including the congestion analysis of the model of Fig. 5-41), by ignoring it: We assume that the birth-death model of Fig. 5-44 holds in this case, even though μ' itself is a function of probabilities to be found, and use the product-form solution (Eq. 5-76), now an approximation, to calculate quantities of interest. (Simulation studies have been used to verify the input-buffer analysis [LAM 1979].)

Specifically, to find μ' in Eq. (5-78) we must calculate the transit-packet blocking probability P_{B_T}. The throughput, $\gamma_I = \lambda_I (1 - P_{B_I})$, is determined by

[LAM 1977] S. S. Lam, "Queueing Networks with Population Constraints," *IBM J. Research and Development*, vol. 21, July 1977, 370-378.

calculating the input-packet blocking probability P_{B_I}. The transit packet blocking probability P_{B_T} is simply the probability that the queue is filled. This requires summing over all the states that correspond to $n_I + n_T = N$—those appearing along the line $n_I + n_T = N$ shown in Fig. 5-43. We thus have

$$P_{B_T} = \sum_{n_I + n_T = N} p(n_I, n_T) = \sum_{n_I=0}^{N_I} p(n_I, N - n_I) \qquad (5-79)$$

where the notation of the first sum means, sum over all possible values of n_I and n_T with the constraint that $n_I + n_T = N$.

The input packets are blocked whenever $n_I = N$ or $n_I + n_T = N$. Hence P_{B_I} is found by summing $p(n_I, n_T)$ over all the states that correspond to these two constraints, i.e., by summing the states that appear along the two corresponding lines of Fig. 5-43. Thus

$$P_{B_I} = P_{B_T} + \sum_{n_T=0}^{N-N_I-1} p(N_I, n_T) \qquad (5-80)$$

As the simplest possible example take $N = 2$ and $N_I = 1$. The buffer can thus accommodate a maximum of two packets. Input packets are blocked if there is already one such packet in the queue. There are a total of five states in this system, as shown in Fig. 5-45. From Eq. (5-76), the five state probabilities are given, in order, by

$$\mathrm{Sp}(0,0) = 1, \quad \mathrm{Sp}(1,0) = \rho_I, \quad \mathrm{Sp}(0,1) = \rho_T, \quad \mathrm{Sp}(1,1) = 2\rho_T\rho_I, \quad \mathrm{Sp}(0,2) = \rho_T^2$$

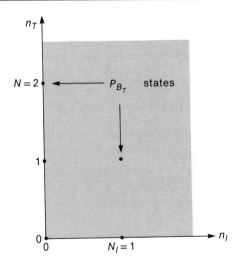

Figure 5-45 Example: $N = 2$, $N_I = 1$

The normalization constant S is then just

$$S = 1 + \rho_I + \rho_T + 2\rho_I\rho_T + \rho_T^2$$

The transit-packet blocking probability is the probability that two packets of either kind are in the queue. It is thus given by

$$P_{B_T} = p(1,1) + p(0,2) = (2\rho_I\rho_T + \rho_T^2)/(1 + \rho_I + \rho_T + 2\rho_I\rho_T + \rho_T^2)$$

The input-packet blocking probability is in turn given by

$$P_{B_I} = P_{B_T} + p(1,0)$$

Since ρ_I and ρ_T are both functions of μ', and hence P_{B_T} from Eq. (5–78), iteration must be used to find P_{B_I} and P_{B_T}. Specifically, for a given load λ_I, one would first choose initial values for P_{B_I} and P_{B_T}. This provides an initial value of ρ_I and hence ρ_T. S is then found, as are the initial values of the $p(n_I,n_T)$'s. One then finds P_{B_T} and P_{B_I} again, and repeats the process, iterating until the changes in P_{B_T} and P_{B_I} are below some specified threshold. Details are left to the reader in exercises at the end of this chapter.

Curves for a more realistic example appear in Fig. 5–46. Plotted here is the throughput γ_I as a function of the load λ_I. Both are shown normalized to the link capacity μ, in packets/time. In this example, the maximum nodal buffer size has been taken as $N = 8$ packets. The average packet path length has been chosen as $\bar{n} = 2.5$ hops, with $p_T = 1/\bar{n} = 0.4$ the probability that transit packets exit at any node. The control parameter N_I is shown varying from 1 to 6 input packets. Note that the downturn in throughput, leading to deadlock, is reduced as N_I is reduced, as predicted. At $N_I = 2$, in this example, the curve straightens out, with the downturn in throughput eliminated, and the throughput reaching a maximum of about 35 percent of capacity. A tighter control, with at most $N_I = 1$ input packet allowed in at any one time, also eliminates congestion, but at reduced throughput.

More extensive calculations by Lam and Reiser, for other examples and with more complex models, as already noted, lead to similar conclusions [LAM 1979]. They in fact provide a simple design value for the control parameter N_I. They find, from observation of their results, that congestion is eliminated if

$$N_I/N < 1/(1 + \bar{n}) \qquad (5-81)$$

with $\bar{n} = 1/p_T$ the average number of hops in a routing path (as already noted). This agrees with the results of Fig. 5–46: With $\bar{n} = 2.5$, $N_I/N < 0.29$ should eliminate congestion. Figure 5–46 shows that this occurs at $N_I/N = 0.25$.

A simpler and intuitively satisfying form for this design parameter is found by defining $N_T = N - N_I$ as the buffer space assigned *solely* to transit packets. Using Eq. (5–81), another form for the appropriate choice of the control

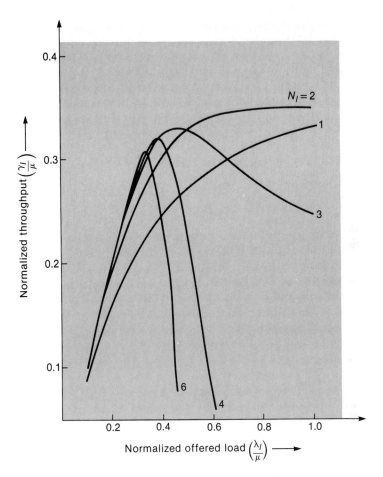

Figure 5-46 Throughput characteristic versus offered load, input-limiting scheme; $N = 8, \bar{n} = 2.5$

parameter N_I may then be written

$$N_I/N_T < 1/\bar{n} = p_T = \gamma_I/\gamma_T \qquad (5-82)$$

from Eq. (5-71). In the example of Fig. 5-46, $p_T = 0.4$, and $N_I/N_T = 2/6 = 0.33$ eliminates congestion.

As noted earlier in this section, congestion-control strategies other than the simple input-buffer-limiting scheme described in detail here may also be designed. One such scheme, which provides tighter control over the input packets and which appears, through analysis, to provide better performance than the input-buffer-limiting control, has been called the *additional buffer allocation scheme* [SAADS]. In this control mechanism, input packets are blocked if both input *and* transit packets equal the number N_I. Transit packets continue to be allowed in until the nodal buffer, with N-packet capacity, is filled. In addition, transit packets receive nonpreemptive priority over the input packets. Mathematically then, with this control, input packets are allowed in only if

$$0 \le n_I + n_T < N_I \tag{5-83}$$

If the total number of packets equals N_I, $N_T = N - N_I$ additional buffers are allocated to the transit packets. This scheme is diagrammed in Fig. 5–47, which may be compared with Fig. 5–43, for the input-buffer-limiting procedure.

Figure 5–48 is a plot of the throughput characteristic for this scheme for the same example used earlier (Fig. 5–46) for the input-buffer-limiting scheme. Note that by adjusting N_I to 4 in this case one again eliminates congestion. Reducing N_I further (applying tighter control) decreases the throughput. A comparison with Fig. 5–46 indicates that the throughput performance of the additional buffer allocation scheme is somewhat better than that of input-buffer limiting, but not by much.

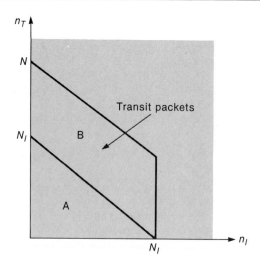

Figure 5–47 Additional buffer-allocation scheme

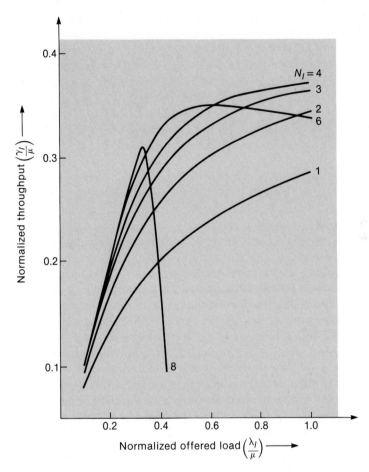

Figure 5–48 Throughput characteristic versus offered load, additional buffer allocation; $N = 8$, $\bar{n} = 2.5$

However, the performance comparison thus far is incomplete. Average time delay must be considered as well, as was done in the case of the end-to-end control mechanisms discussed earlier in this chapter. Using Little's formula the average time delay T is given simply by

$$T = [E(n_I) + E(n_T)]/\gamma_I \qquad (5-84)$$

with $E(n_I)$ and $E(n_T)$ the average number of input and transit packets, respectively, queued at a node, and γ_I the average throughput, as noted earlier. (This may be readily checked by multiplying numerator and denominator by n, the total number of nodes in the network. In the numerator one then gets the total number of packets residing in the network on the average; the denominator corresponds to the networkwide net arrival rate.) Both $E(n_I)$ and $E(n_T)$ may be obtained from the nodal state probabilities $p(n_I, n_T)$.

The time-delay characteristic for the same example used thus far, that of an eight-packet buffer with $\bar{n} = 2.5$ hops, on the average, appears in Fig. 5–49. Curves for both the input-buffer-limiting scheme (labeled ILS) and the additional-buffer-allocation scheme (ABA) have been superimposed for values of the control parameter N_I that were previously found to provide congestion-free (nondroop) performance for the control schemes. Note, by comparison with Figs. 5–46 and 5–48, that the time-delay performance is much more sensitive to the choice of N_I than was the throughput performance. Thus reducing N_I from 2 to 1 in the ILS case reduces the time delay by a factor of 2, while the throughput of Fig. 5–46 is affected much less. The same is true for the ABA scheme. But note also that the ABA scheme, which provides a tighter control on input packets from the beginning, results in a much lower time delay than does the ILS strategy. This is demonstrated particularly clearly in Fig. 5–50, which plots the power γ_I/T versus offered load for the two schemes and for various values of the control parameter N_I. It is apparent that because of the large reduction in time delay as the control parameter N_I is reduced, the maximum value of power is *not* obtained for the values of N_I that produce maximum throughput. In particular, it appears that the tightest control possible ($N_I = 1$) should be used if time-delayed performance is considered as well. The ABA also outperforms the ILS scheme on the basis of time delay. Similar results are obtained for other examples.

The schemes described thus far have focused on control of the total buffer pool at a node. Some additional improvement in control performance can be obtained by adjusting the buffers allocated to individual links as well [SAADS], [IRLA]. Digital Equipment Corporation has incorporated this added level of control in its input-limiting congestion-control mechanism [DEC].

Control of congestion by providing priority to transit packets over newly arriving input packets is a special case of control mechanisms in which transit packets receive increasingly higher priority as they approach their destination [RAUB]. It has been proved that schemes of this type can be used to prevent

[IRLA] M. I. Irland, "Buffer Management in a Packet Switch," *IEEE Trans. on Comm.*, vol. COM-26, no. 3, March 1978, 328–337.

[RAUB] E. Raubold and J. Haenle, "A Method of Deadlock-free Resource Allocation and Flow Control in Packet Networks," ICCC, Toronto, Aug. 1976.

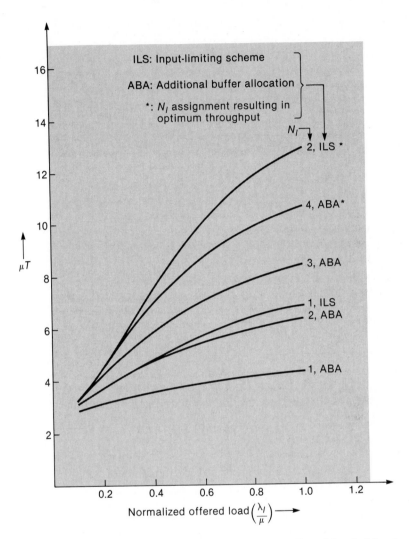

Figure 5-49 Time-delay characteristic versus offered load, $N = 8$

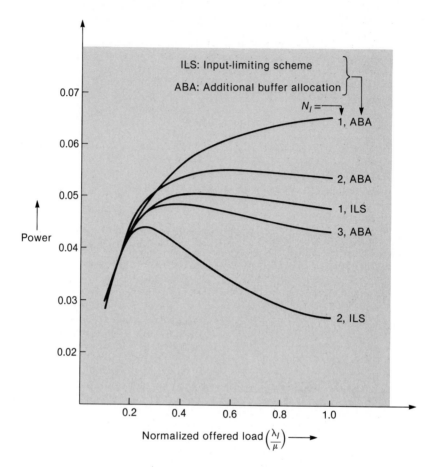

Figure 5–50 Power versus offered load, $N = 8$

deadlocks in a network [GUNT]. The German GMD network has implemented a dynamic buffer-allocation strategy in which added buffers are allocated to transit packets as they move closer to their destination. Simulations have been used to verify both improvement in performance and prevention of deadlock

[GUNT] K. D. Günther, "Prevention of Deadlocks in Packet-switched Data Transport Systems," *IEEE Trans. on Comm.*, vol. COM-29, no. 4, April 1981, 512–524.

[GIES]. Simulation studies at the British National Physical Laboratory have also shown that input packet-transit packet priority schemes provide increased improvement in performance [PRICE].

Problems

5–1 Using virtual circuit service, a DTE has two calls connected to two distant DTEs; an intermediate packet-switched network is used.

    a. A third call is to be set up with a third DTE through the packet-switched network. Show the steps that have to be taken, both at the local DTE-DCE interface and the remote DCE-DTE interface, to set up the call. Indicate the packet types that have to be transmitted at both ends. Include logical channel number assignments at both ends. Show how the data link layer at the local side multiplexes this new call assignment with the two ongoing ones. (Assume that data packets from the two ongoing calls arrive randomly from either end. Differentiate these packets by their own logical channel numbers. Use appropriate packet-level sequence numbers for each.) Include appropriate data link sequence numbers for the multiplexed frames in the two directions.

    b. One of the three calls connected is now cleared. Show the steps required, at both ends of the connection, to carry out this procedure. Provide a diagram of a typical sequence of frames transmitted in both directions, at the local side, involving all three logical channels as the clearing procedure is being carried out.

5–2 A DCE receives a call setup request from a remote DTE through the network to which the DCE is connected. It assigns an unused logical channel number to the call and transmits an incoming call packet, with that number, to the appropriate DTE connected to it. In the meantime, this latter local DTE has independently selected the same logical channel number for a new call it wants to set up and so indicates in a call request packet directed to the DCE. Indicate, with the help of a diagram, what happens.

5–3 Consider the data-transfer phase of the packet level of the X.25 protocol. A three-bit sequence number field is used. Sketch a typical exchange of packets between the DTE and its DCE in two cases:

    1. The D-bit is not set ($D = 0$).
    2. The D-bit is set to 1.

The window size is $w = 2$ in the first case and $w = 4$ in the second case.

[PRICE] W. L. Price, "Data Network Simulation Experiments at the National Physical Laboratory, 1968–1976," *Computer Networks*, vol. 1, 1977, 199–208.

5-4 Indicate, with the aid of a diagram, how the RR and RNR packets might be used during the X.25 data transfer phase to maintain flow control between a DTE and its DCE. Use a three-bit sequence number field and set the window $w = 2$.

5-5 Refer to the sliding-window control model of Fig. 5-13(a). The Norton equivalent of the M-node virtual circuit between source S and destination D is to be found. The object is to calculate the Norton equivalent throughput $u(n)$ of Fig. 5-14. One approach uses combinatorics.

  a. Show that the total number of ways n identical packets (balls) can be distributed randomly among M queues (boxes) is given by

$$\binom{n + M - 1}{n} \equiv C(M)$$

  b. Check this result for $n = 3$, $M = 4$.
  c. Show that

$$C(M - 1) = \binom{n + M - 2}{n}$$

  represents the number of combinations that result in *zero* packets (balls) in any queue (box). Hence show that the probability that a queue is empty is $p_0 = C(M - 1)/C(M) = (M - 1)/(n + M - 1)$. Use this to calculate the Norton equivalent throughput given by Eq. (5-3).

5-6 Problem 5-5 provides one way of finding the Norton equivalent of an M-hop virtual circuit. The form of the Norton equivalent throughput, $u(n)$, given by Eq. (5-3) seems so simple that it would appear that one should be able to derive it *directly* from combinatorics without resorting to the intermediate use and calculation of the function $C(M)$ defined in that problem. The author has not been able to do this. It is thus left as an open problem (and challenge!) to interested readers.

5-7 Refer to Fig. 5-12(c).

  a. Show that the probability of state of the upper queue for the case of the Norton equivalent model of the M-node virtual circuit with sliding-window control of Fig. 5-13 is given by Eqs. (5-5) and (5-6). Start with the state-dependent service characteristic $u(n)$ given by Eq. (5-3).
  b. Using Eqs. (5-5) and (5-6), show that the normalized average throughput of the sliding-window control with a window size of N is given by

$$\gamma/\mu = \sum_{n=0}^{N} u(n) p_n / \mu$$
$$= B(N - 1)/B(N)$$
$$B(N) = \left[1 + M\rho + M\left(\frac{M + 1}{2}\right)\rho^2 + \cdots + \frac{M(M + 1) \cdots (M + N - 1)\rho^N}{N!} \right]$$

c. Plot γ/μ as a function of ρ for $M = 3$ and $N = 1, 2, 3, 4$. Compare and explain the results.

d. Let $\rho = \gamma/\mu \to \infty$. Show that $\gamma/\mu \to N/(M + N - 1)$. Check with the results of part c. Show that this simple result for the limiting throughput of the sliding-window control can be contained directly (Eq. 5–9) from Fig. 5–13(b) without any calculation. Show that the average time delay across the virtual circuit in this case is given by Eq. (5–10).

e. Check the expression in (b). by calculating $\lambda(1 - P_B)$.

5–8 Consider the throughput-time delay performance equations (5–9) and (5–10) for the sliding-window control in the case of $\lambda \to \infty$. As N increases, *both* throughput and time delay increase. Show that the value of N that maximizes the ratio of throughput to time delay $(\gamma/\mu)/\mu E(T)$ is $N = M - 1$. This measure of performance of the congestion control mechanism has been called "power." Why would you want to maximize this ratio? Explain. Locate this point on the sliding-window performance curve of Fig. 5–15. Does it seem an appropriate operating point? Explain.

5–9 Refer to Fig. 5–13 for the sliding-window control. Take the case $\lambda = \mu$.

a. Show that the throughput of the controlled virtual circuit and the time delay across it are given by Eq. (5–12) and Eq. (5–13), respectively.

b. Show that the window size N that maximizes the "power" defined as $(\gamma/\mu)/\mu E(T)$ is given by $N = M$. Compare with the result of Problem 5–8.

5–10 Plot $\mu E(T)$ versus γ/μ for the sliding-window control for $M = 4$ hops, for the two cases $\lambda \to \infty$ and $\lambda = \mu$. Compare with Fig. 5–15 for $M = 3$ and $\lambda \to \infty$. Locate the point $N = M$ on both your curve and Fig. 5–15. This is the point that maximizes the "power" defined in Problem 5–9. Does this appear to be an appropriate operating point? Explain.

5–11 Demonstrate that the model of Fig. 5–16 captures the acknowledge-at-end-of-window control mechanism.

5–12 Consider the balance equations (5–14) that arise in the acknowledge-at-end-of-window control mechanism in the heavy-traffic case $(\lambda/\mu \to \infty)$.

a. Redraw Fig. 5–17 for this case and focus on the w-box. Draw its state diagram and show that it has upward transitions only, except for state $w - 1$, which wraps back around to state 0. Label the transitions and show how the balance equations (5–14) arise from this state diagram.

b. Solve Eq. (5–14) to obtain Eq. (5–15).

c. Let $u(n) = n\mu/[n + M - 1)]$, the Norton-equivalent throughput function for the M-hop virtual circuit. Show that the probability of state occupancy of the w-box is given by Eqs. (5–16) and (5–17).

d. Derive the throughput-time delay performance expressions given by Eqs. (5–18) through (5–20). Show that for a given time delay the throughput in this case is always less than that for the sliding-window control.

5–13 Obtain the throughput-time delay performance equations for the acknowledge-at-end-of-window control mechanism for the case $\lambda = \mu$. Show that the window size that maximizes the "power," $(\gamma/\mu)/\mu E(T)$, is given by $w = 2M - 1$. Plot $\mu E(T)$ versus γ/μ for $M = 4$, $\lambda \to \infty$, and $\lambda = \mu$. Superimpose these curves on those obtained for a sliding-window control for the same cases (Problem 5–10) and compare. Compare with the curves of Fig. 5–18 ($M = 3$ case) as well. Judging from your curves, what would appropriate operating points (window size w) be? Compare with that obtained using the "power" criterion.

5–14 Refer to the discussion of the IBM SNA virtual route pacing-control mechanism in the text.

    a. Show that Fig. 5–24 models the (fixed-window) SNA control.

    b. Approximating the M-hop virtual route (circuit) by the Norton equivalent in this case, one obtains the queueing model of Fig. 5–25. Draw the two-dimensional state diagram that represents transitions among the various states (i, j). Show that the balance equation in the interior of the two-dimensional region is given by Eq. (5–24). Write appropriate balance equations along the four boundaries as well.

    c. Take the cases $k = 2$ and $k = 3$ for an $M = 3$-hop virtual route. Program the balance equations for these two cases and solve explicitly for the state probabilities p_{ij} for a number of values of $\rho = \lambda/\mu$. (You might take $\rho = 0.5$, 1, 2, and 5 as examples.) Calculate the normalized throughput γ/μ and the normalized time delay $\mu E(T)$ using these values of p_{ij} and compare with the corresponding values obtained using the heavy traffic and $\lambda = \mu$ analysis in the text. (See Eqs. (5–31) through (5–34).) The heavy-traffic values appear in Fig. 5–18 as well.

5–15 Refer to Fig. 5–25 for the IBM SNA virtual route pacing-control model. Take $\lambda/\mu \to \infty$. Show that the balance equations (5–25) represent the resulting system. Solve these to obtain Eq. (5–27). From this, show that the throughput performance equation is given by Eq. (5–29), while the normalized end-to-end time delay is given by Eq. (5–30). Verify the corresponding equations for the $\lambda = \mu$ case as given by Eqs. (5–33) and (5–34), using the proper entries in the table that accompanies these equations.

5–16 The sliding-window and SNA VR pacing-control mechanisms are to be compared for the two cases $\lambda = \mu$ and $\lambda \to \infty$. Set the window size in each case such that the end-to-end time delays are the same. Show that the throughput obtained for the VR pacing mechanism is only a few percent less than that for the sliding-window mechanism.

5–17 Consider the SNA VR pacing control mechanism. Take $M = 4$ hops, $\lambda = \mu$. Plot $\mu E(T)$ versus γ/μ. Compare with Fig. 5–18 as well as the results of Problems 5–10 and 5–13.

5–18 Figure P5–18(a) represents the model of a window-type congestion control mechanism. N is the window size. n represents the state of the (upper) system under control, including packets in service.

    a. Show that the upper (controlled) system behaves just like the finite queueing system of Fig. P5–18(b).

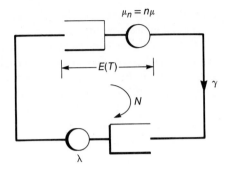

Figure P5–18a

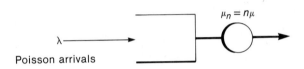

Holds at most N packets

Figure P5–18b

---

b. Show that the probability p_n of state n of the controlled system is given by

$$p_n = \rho^n/n! \left/ \sum_{j=0}^{N} \rho^j/j! \right. \qquad \rho \equiv \lambda/\mu$$

c. Find the throughput γ and time delay $E(T)$ indicated in Fig. P5–18(a). Can you explain the time delay result?

d. Take the case $\lambda \to \infty (\lambda/\mu \gg 1)$. Find γ and $E(T)$ by inspection and compare with the results of part c.

5–19 Starting with Eq. (5–41), the global balance equation for a FIFO, exponential, open queueing network, and eliminating q_{si} as indicated, show that the resultant equation is satisfied by Eq. (5–42). Hence show that the solution to Eq. (5–41) is given by Eq. (5–46).

5–20 Refer to Fig. 5–34. The link capacities are all $\mu = 3$ packets/sec. Find the average end-to-end delays from 1 to 3 via (1) node 2 and (2) nodes 5 and 2.

5–21 Refer to Fig. 5–34. The link capacities are all $\mu = 3$ packets/sec.

a. Find the *networkwide* average time delay $E(T)$, as given by Eq. (5–49).
b. The routing probabilities are now changed to $q_{15} = 1/4$, $q_{12} = 3/4$, $q_{24} = 1/2$. Find $E(T)$ and compare with the value found in part a.

5–22 Consider the packet-switched network shown in Fig. P5–22. Terminals 1, 2, and 3 generate Poisson traffic at the rate of $\gamma_1 = 1/2$, $\gamma_2 = 1$, and $\gamma_3 = 1\ 1/2$ packets/sec, respectively, as shown. All packets are $1/\mu = 1/6$ sec long, on the average, exponentially distributed. All are destined for node 4, as shown. All line capacities are $\mu = 6$ packets/sec, as shown.

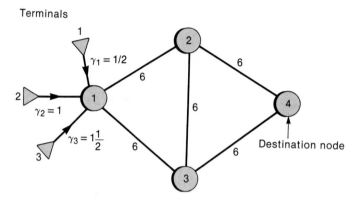

Figure P5–22

a. Find the average time delay $\overline{T}$ from node 1 to node 4 if the packets go directly through node 2.
b. Repeat a. if packets follow the path $1-2-3-4$.
c. Packets are now routed randomly through the network, from node 1 to node 4, with the following routing probabilities:

$$q_{13} = 1/3, \quad q_{23} = 3/4, \quad q_{34} = 1$$

Find the average network-wide delay given by Eq. (5–49).

5–23 Prove the recursive relation (5–51) (Buzen's algorithm), filling in all the details left out in the derivation in the text.

5–24 Refer to the network of Fig. 5–36. Take $\lambda = \mu$.

a. Check the entries of Table 5–5.
b. Use this table and the appropriate equations in the text to find the time delay-throughput characteristic, $E(T)$ versus γ, for *each* of the VCs for $N = 1$, 2, 3, 4. (*Hint:* How *does* one find the time delay across each VC?)

5-25 Repeat Problem 5–24 for the two cases (1) $q_{51} = 1/3$, (2) $q_{51} = 2/3$. Compare and discuss these results with those of Problem 5–24. *Note:* A separate $g(n, m)$ table has to be calculated for each of these two cases.

5-26 Consider the VC shown in Fig. P5–26. A sliding-window mechanism is used to control this VC. Use Buzen's algorithm to find the time delay-throughput characteristic for $\lambda \gg 2\mu$. Take $N = 1, 2, 3, 4, 5, 6$. Compare with Fig. 5–15. Do the results agree with what you would expect?

Figure P5–26

5-27 Use Buzen's algorithm to find the Norton equivalent (Figs. 5–12 and 5–14) of an M-hop virtual circuit. Find $u(n)$ for various values of n and M and show that it is given by the expression $u(n) = n\mu/(n + M - 1)$. Show, as a check, that the average number of packets in each queue is n/M.

5-28 Refer to Problem 5–26. Solve the problem using mean value analysis and compare with the results of Problem 5–26.

5-29 Refer to Figs. 5–41 and 5–42.

    a. Let $N = 1$. Show that $P_B = \rho$ and $\gamma/\mu = \rho(1 - \rho)$. (See Eq. 5–72.) Sketch γ/μ as a function of ρ. Find $\mu E(T)$ and sketch as a function of ρ.

    b. Let $N = 5$. Sketch γ/μ versus ρ and compare with the curve in Fig. 5–42.

    c. For *any* N, show that for $\rho' = 1, \rho = \dfrac{N}{N+1}, \gamma/\mu = \left(\dfrac{N}{N+1}\right)^2$. Show that for

$$\rho' \to \infty, \ \rho \to 1, \ 1 - P_B \to \frac{1}{\rho'} \to 0, \text{ and } \frac{\gamma}{\mu} \to (1 - P_B) \to 0. \text{ Show that for}$$

$\rho \ll 1, \gamma/\mu \doteq \rho$.

Can you use these values to sketch γ/μ versus ρ for any N? Compare with parts a and b. above, as well with the curve for $N = 8$ in Fig. 5–42.

5-30 Refer to Section 5–5, on input-buffer limiting. Take $N = 2, N_I = 1$. Write and use a simple computer program to solve for P_{B_T} and P_{B_I} iteratively. Use this to find the average throughput $\gamma_I = \lambda_I(1 - P_{B_I})$ as a function of load λ_I for several values of λ_I. Take $\bar{n} = 2$. Compare with the results of Problem 5–29 (Fig. 5–42). One possible procedure: Initialize with $P_{B_T} = P_{B_I} = 0$. Keep repeating until, at iteration $K + 1$,

$$\sqrt{(P_{B_T}^{K+1} - P_{B_T}^{K})^2 + (P_{B_T}^{K+1} - P_{B_I}^{K})^2} < \epsilon \qquad \text{(a stopping constant)}$$

You can try any other initial values and stopping threshold.

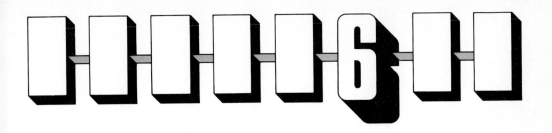

Network Layer: Routing Function

In the previous chapter we discussed flow control and congestion control, which are carried out at the network layer of network architectures. The other major function of the network layer, as noted in Chapters 1 and 3, is to handle the routing of packets as they move through the network. The packets may be constrained to follow a virtual circuit or they may be moving in a datagram mode. In either case a path or set of paths must have been set up to connect a given source with a given destination. Most commonly the appropriate routing path is mapped into entries in a routing table at each node along the path, indicating to which outgoing link a given packet is to be directed. A packet with a specified identification (this could be source and destination addresses, the virtual circuit number, the destination address only for some procedures, and so on) is directed by an entry in the routing table to the appropriate outgoing link.

An example appears in Fig. 6–1: Associated with each packet id is a next-node assignment or, equivalently, the outgoing link over which the packet will be transmitted. Although the id shown in Fig. 6–1 is a pair of numbers, in some cases, described later, the id might just be the destination address.

Routing algorithms used to establish the appropriate routing paths or the equivalent routing table entries in each node along a path constitute the subject of this chapter. These algorithms can be variously classified. They can be classified as to whether the paths are centrally established or set up in a decentralized manner, with each node carrying out a specified routing algorithm. They can be

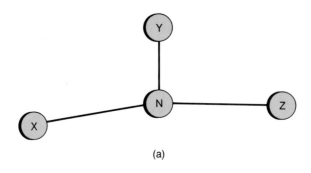

(a)

Packet indentification	Next node assignment
(1,4)	X
(1,5)	Y
(2,4)	X
(2,5)	Z
(2,6)	Y
.	.
.	.
.	.

(b)

Figure 6–1 Routing at a node in a network
       a. Current node and neighbors
       b. Routing table

classified as to response of adaptation: Is an algorithm static, the routing path changing only in response to topological changes in the network (a node/link coming up or going down)? Does it adapt, even if slowly, to traffic changes in the network? Or does it provide dynamic response to traffic conditions measured throughout the network?

Most algorithms are based on assigning a "cost" measure to each link (or possibly even each node) in the network. The "cost" could be a fixed quantity related to such parameters as link length, speed, or bandwidth of link (transmission capacity), whether secure or not, estimated propagation delay, or some combination of these. The cost could include the average traffic expected at a given hour on a given day, it could include measured estimates of link traffic, buffer occupancy, measured error conditions on the link, and so forth. The classification might then be made according to the amount of information used in running the algorithm and how often this information changes.

Finally, algorithms can be classified on the basis of performance objective. Here two classes of algorithms have been discussed most often in the literature. The first class, generally called the class of *shortest-path* algorithms, provides a least-cost path between a given source-destination pair, the "cost" of the path defined to be the linear sum of the costs of each hop in a given path. Included are dynamic algorithms in which the "cost"/hop may actually be an estimated time delay along that hop, as well as quasi-adaptive algorithms in which the cost changes as error conditions or average traffic along a link change. Most networks in operation, as well as network architectures specified in detail at the network level, incorporate shortest-path algorithms of one form or another. [SCHW 1980b]. We shall consider this class primarily in this chapter.

The second class provides routing paths that minimize the *network-wide* average time delay, using estimates of average traffic entering the network, on a source-destination basis. The use of this performance measure leads generally to *multiple routing paths* for packets flowing between a given source-destination pair. The routing technique based on this performance measure has thus been called a *bifurcated* routing procedure. (The prefix bi- is actually a misnomer since, with k outgoing links at a given node, packets may be assigned probabilistically to each link.) We shall describe this procedure briefly. It has not been adopted as a routing procedure by any network or network architecture, although from time to time suggestions do arise as to its possible use. It has been used primarily in the topological design of networks [SCHW 1977].

It is apparent that different classes of users will require different routing paths. For example, bursty interactive traffic flowing between a low-speed terminal and a computer will require a relatively low response-time path. Such traffic should thus avoid long propagation-delay links (satellite links, for example). This is readily accomplished in a shortest-path algorithm by increasing the link cost of long propagation-delay links for this class of users. File transfers are most readily effected over high-bandwidth (high-capacity) links, if available. Assigning a relatively low cost to such links for this class of users will accomplish this goal in a shortest-path algorithm. Traffic that requires a measure of security must be routed over secure links only, and this can again be accomplished by assigning link costs appropriately.

In the next section we describe bifurcated routing briefly. We then go on to discuss shortest-path routing in detail. Since a variety of algorithms have been developed to provide least-cost routing paths, we focus first on two generic

[SCHW 1980b] M. Schwartz and T. E. Stern, "Routing Techniques Used in Computer Communication Networks," *IEEE Trans. on Comm.*, vol. COM-28, no. 4, April 1980, 539–552; reprinted in [GREE].

[SCHW 1977] M. Schwartz, *Computer-Communication Network Design and Analysis*, Prentice-Hall, Englewood Cliffs, N.J., 1977.

algorithms, one often used for centralized routing, the other adopted in some networks for decentralized, or distributed, routing. Variants of these two algorithms can then be distinguished readily. We outline the routing procedures used in some real networks and network architectures and relate them, where possible, to the two algorithms described in detail. We conclude the chapter with a discussion of the comparative performance of various shortest-path algorithms of the distributed type: Although they all provide the same least-cost routing paths, they differ in their convergence time when changes in routing tables must be made, in the number of control (nondata) packets required to be transmitted during convergence, in the complexity of computation required, in storage (memory) required, and so forth. To make the discussion concrete, a comparative performance evaluation of two shortest-path algorithms applied to simple networks is carried out.

6–1 Bifurcated Routing

As just noted, bifurcated or multiple-path routing arises when the network-wide average time delay is to be minimized. With link propagation delays neglected, this time delay was shown in the previous chapter (Eq. 5–49) to be given by the expression

$$E(T) = \frac{1}{\gamma} \sum_{i=1}^{M} \lambda_i T_i = \frac{1}{\gamma} \sum_{i=1}^{M} \frac{\lambda_i}{\mu_i - \lambda_i} \qquad (6-1)$$

Here γ represents the total external load on the network (the sum of all arrival rates to the network), λ_i is the flow, in packets/time, over link i, and μ_i is the transmission capacity, in packets/time, of link i. Poisson arrivals, exponential packet lengths, and the independence assumption have all been used in arriving at this expression for the average time delay, averaged over all M links in an open queueing network.

An example of a five-node, seven-link network is shown in Fig. 6–2. The average traffic flows, or external loads, between a number of the nodes are shown designated by the notation γ_{ij}. Here i is the source node, j the destination node. Full-duplex links are assumed in this example. The flow γ_{ij} is, in general, different from γ_{ji}. The total external load on the network is then

$$\gamma = \sum_{i=1}^{N} \sum_{j=1}^{N} \gamma_{ij} \qquad (6-2)$$

for an N-node network. For an N-node network there are in general $N(N-1)$ possible source-destination pairs, and the γ_{ij}'s can be visualized as entries in an $N \times N$ traffic matrix consisting of the average traffic arrival rates of traffic flowing between the various nodes.

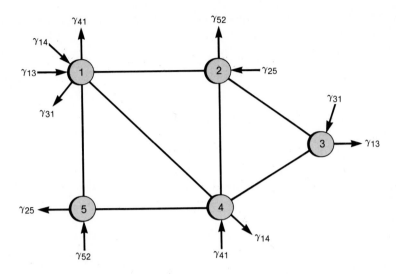

Figure 6–2 Example network, typical external loads shown

Equation (6–1) may be modified to include propagation delay, processing delay, and other fixed delays along a link by adding a composite fixed delay term T_{fi} to the M/M/1 queueing delay, $T_i = 1/(\mu_i - \lambda_i)$ of the equation. It is also useful to rewrite the time delay $E(T)$ in terms of bits/sec traffic flow parameters f_i rather than the packets/sec flows λ_i used in Eq. (6–1). This is readily done by recalling that, with link-transmission capacity C_i bits/sec and an average packet length, in bits, $1/\mu'$, we must have $\mu_i = \mu' C_i$. (The average packet length, $1/\mu'$, is assumed to be the same for all links.) The traffic flow over link i, in bits/sec, is then $f_i = \lambda_i/\mu' < C_i$. It is then easily shown that Eq. (6–1), as modified to include the fixed-link delays T_{fi}, may be rewritten in the equivalent form

$$E(T) = \frac{1}{\gamma} \sum_{i=1}^{M} \left[\frac{f_i}{C_i - f_i} + f_i T_i' \right] \tag{6–3}$$

with T_i' representing the fixed link delay term $\mu'T_{fi}$. The routing strategy used determines the link flows f_i, $i = 1, 2, \ldots, M$. The "best" routing strategy is one that assigns the f_i's that minimize $E(T)$. As shown next, this may be formulated as a multicommodity flow problem.

Specifically, say that there are $F \le N(N-1)$ source-destination pairs in the network. (Not all nodes may be communicating with one another.) We then say that there are F commodities to be distributed in the network. The objective of the routing strategy is to assign paths for each of these commodities to minimize the overall average time delay. To simplify the notation, assign a number k

$(k = 1, 2, \ldots, F)$ to each commodity (source-destination pair (i, j)). Say that the average flow over link i, in bits/sec, due to commodity k is f_i^k. Then we must have the total flow in link i given by

$$f_i = \sum_{k=1}^{F} f_i^k \qquad (6-4)$$

The assignment of f_i^k for all commodities and for all links in the network determines the paths and hence the routes for all commodities.

   Consider now a typical node ℓ. At this node (and for every other node in the network) the packet flows must be conserved, commodity by commodity. The average incoming-message flow due to commodity k, whether generated at node ℓ, or whether coming from elsewhere in the network, must equal the average outgoing flow of commodity k. This is precisely the extension of the continuity of flow equation (5–36) of Chapter 5. (Note again, as stated there, that that equation is somewhat more general, allowing packets leaving a queue to be split probabilistically among several possible paths. Here, with nodal processing neglected, packets queue up for a particular transmission link. Each queue is associated with its own link, and there is no routing of packets at the output of the queue. With this model, routing involves selecting the appropriate outgoing link and neighboring node to which a packet is directed. Instead of M possible paths from a given output, there are N possible links, at most, to be included in the flow-conservation equation here.) We then have, as the continuity of flow equation at node ℓ, for commodities entering the network at source node i and exiting at destination node j,

$$\sum_{m=1}^{N} f_{m\ell}^{ij} - \sum_{n=1}^{N} f_{\ell n}^{ij} = \begin{cases} -r_{ij}, & \text{if } \ell = i \\ r_{ij}, & \text{if } \ell = j \\ 0, & \text{otherwise} \end{cases} \qquad (6-5)$$

Here $r_{ij} \equiv \gamma_{ij}/\mu'$ represents the average traffic in bps generated at node i and destined for node j. We have labeled links by the two nodes to which they are connected. The sums of incoming flows $f_{m\ell}^{ij}$ and outgoing flows $f_{\ell n}^{ij}$ are then taken over links connected to node ℓ.

   An example, for commodity (1,4) at node 1 in Fig. 6–2, appears in Fig. 6–3(a). More generally, flows f_{21}^{14}, f_{41}^{14}, and f_{51}^{14} could have been included as well. From simple reasoning in this case, it is apparent that they would all be zero. (Looping can only *increase* time delay!) A more general full-duplex example does appear in Fig. 6–3(b). Node 2 here is assumed to have as neighboring nodes, nodes 1, 3, and 4. The flows corresponding to commodity (5,6) (nodes 5 and 6 are *not* connected to node 2 in this example) are indicated in the figure.

   A conservation equation of the type of Eq. (6–5) must be written for each of the F commodities at each of the N nodes. There are thus FN such equations. In

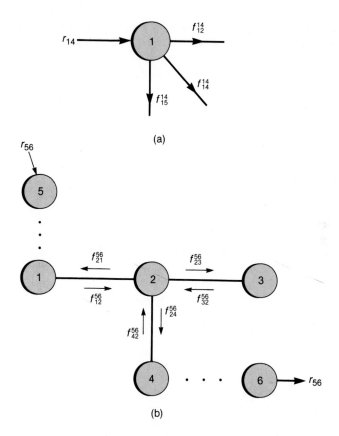

(a)

(b)

Figure 6 – 3 Examples of flow conservation
a. From Fig. 6 – 2
b. More general example

the most general case, with each node sending packets to every other node in the network, $F = N(N - 1)$, and there are $N^2(N - 1)$ such equations.

The object of the routing strategy is to find each flow $f_{m\ell}^{ij}$ in the network, $f_{m\ell}^{ij} \geq 0$, satisfying the conservation equations, such that the average network time delay $E(T)$ is minimized. This gives rise to a constrained optimization problem. (One additional obvious constraint is the capacity constraint $f_i < C_i$. This constraint is normally ignored since $E(T) \to \infty$ as $f_i \to C_i$, and its effect is automatically brought into play.) Various algorithms have been utilized to solve this constrained optimization problem. They all use iteration to converge to the desired constrained optimum. They all rely on the fact that $E(T)$, the object function to be minimized, is convex in the flows, giving rise to just one global minimum to be found. One iterative technique that has been found very useful is the flow-deviation method first described by M. Gerla in his doctoral dissertation [GERL], [FRAT].

The comparative efficiency or performance of iterative algorithms can generally be assessed in terms of the complexity of computations required, memory requirements, and rate of convergence (number of iterations required). The flow deviation technique, although relatively efficient in terms of computational requirements, is found to converge rather slowly [BEST]. Another algorithm, the extremal flows method [CANT], uses a steepest-descent or gradient-type iteration to speed up the rate of convergence. A third algorithm, the gradient projection technique, which uses a gradient search modified to account for the constraints of the problem, has been found to converge faster than the flow deviation technique, but its computational complexity is greater [SCHW 1976]. It appears primarily suited to bifurcated routing determination for small networks or for those in which the number of commodities F is relatively small. (Only a subset of the nodes would be expected to communicate with one another.)

The reader is referred to the references just cited for details of the algorithms involved. An example of bifurcated routing in a small network, representing a simplified model of one incorporating a satellite as one of the nodes, appears in Fig. 6–4. The optimum flow distribution is in this case shown in kbps or packets/sec (a packet = 1000 bits here), as are the various link capacities. The normalized ground-satellite propagation delay has been chosen as 0.5 unit. Optimum flows only are shown in this figure; all other flows are zero. Note that multiple routing arises quite naturally in this example.

[GERL] M. Gerla, *The Design of Store-and-Forward Networks for Computer Communications*, Ph.D. dissertation, Dept. of Computer Science, UCLA, 1973.

[FRAT] L. Fratta, M. Gerla, and L. Kleinrock, "The Flow Deviation Method: An Approach to Store-and-Forward Communication Network Design," *Networks*, vol. 3, 1973, 97–133.

[CANT] D. G. Cantor and M. Gerla, "Optimal Routing in a Packet-Switched Computer Network", *IEEE Trans. on Computers*, vol. C-23, no. 10, Oct. 1974, 1062–1069.

[BEST] M. Best, "Optimization of Nonlinear Performance Criteria Subject to Flow Constraints," 18th Midwest Symposium on Circuits and Systems, Concordia University, Quebec, Aug. 1975, 438–443.

[SCHW 1976] M. Schwartz and C. K. Cheung, "The Gradient Projection Algorithm for Multiple Routing in Message-switched Networks," *IEEE Trans. on Comm.*, vol. COM-24, no. 4, April 1976, 449–456.

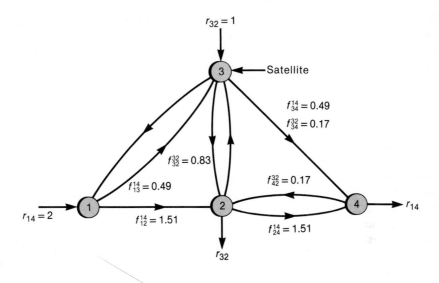

Capacities and propagation delays assumed:

Link index $i,j \longrightarrow$	1,2	1,3	2,3	2,4	3,1	3,2	3,4	4,2
C_{ij}	3	2.5	2.5	3	3	2	2	3
T_{ij}	0	0.5	0.5	0	0.5	0.5	0.5	0

Figure 6-4 Example of network with bifurcated routing (all units in kbps or packets/sec; 1 packet = 1000 bits)

6-2 Shortest-path Routing

It was noted at the beginning of this chapter that most packet-switched networks in current operation use some form of shortest-path routing in carrying out the routing function at the network layer [SCHW 1980b]. Some networks do the routing in a centralized fashion, establishing the paths between source and destination nodes at a centralized network management center and then distributing the resultant routing information to all nodes in the network. Other networks carry out decentralized routing, each node exchanging cost and routing information with its neighbors on an iterative basis, until routing tables at the nodes converge to the appropriate shortest-path entries.

In this section we first describe two generic algorithms for carrying out shortest path computations. We label them algorithms A and B, respectively, for simplicity. Both, or variants of both, are used in various operating networks to carry out the routing function. Algorithm B, in particular, lends itself quite readily to decentralized routing. We describe a decentralized version of the algorithm that, in various forms, has served as the basis for decentralized routing in a number of networks and network architectures.

The use and implementation of shortest-path algorithms has a long history. The reader is referred to the literature for a comparative discussion of some of the algorithms [FORD], [FRAN], [HITC]. In some networks, particularly those that involve the static assignment of routes, alternative routes are made available, ranked in order of cost, to provide alternative paths in the event of failure along an existing route. These routes can be calculated using algorithms that find the *k* shortest paths. The reader is referred to the literature on this problem [SHIE].

Algorithm A, which we now describe , is due to Dijkstra [DIJK], [AHO]. Algorithm B is a form of Ford and Fulkerson's algorithm, one of a class of algorithms originally due to Floyd [FORD], [FRAN], [HITC]. It is described as well, with added examples, in [SCHW 1977, pp. 238–241]. We follow the approach of [SCHW 1980b] in describing these two algorithms.

Consider the network of Fig. 6–5 [SCHW 1980b]. The numbers associated with each link represent link costs. (For the sake of simplicity the cost in either direction along a link is taken to be the same. More generally, the cost could be different in the two directions.) Algorithm A finds the shortest paths from a source to all other nodes. To do this, it requires global topological knowledge, i.e., a list of all nodes in the network and their interconnections, as well as costs for each link. It thus lends itself to centralized computation, with complete topological information available at a central database. It is also the basis of the ARPA network *decentralized* routing algorithm, discussed in Subsection 6–3–2, in which each node in the network maintains its own global database. In the example of Fig. 6–5, the objective is to find the shortest path from

[FORD] L. R. Ford, Jr., and D. R. Fulkerson, *Flows in Networks,* Princeton University Press, Princeton, N.J., 1962.

[FRAN] H. Frank and I. T. Frisch, *Communication, Transmission, and Transportation Networks,* Addison-Wesley, Reading, Mass., 1971.

[HITC] L. E. Hitchner, *A Comparative Study of the Computational Efficiency of Shortest Path Algorithms,* Univ. of California, Berkeley, Operations Research Center Report ORC 68-25, Nov. 1968.

[SHIE] D. R. Shier, "On Algorithms for Finding the *k* Shortest Paths in a Network," *Networks,* vol. 9, no. 3, 1979, 195–214.

[DIJK] E. W. Dijkstra, "A Note on Two Problems in Connection with Graphs," *Numer. Math.,* vol. 1, 1959, 269–271.

[AHO] A. V. Aho, J. E. Hopcroft, and J. D. Ullman, *The Design and Analysis of Computer Algorithms,* Addison-Wesley, Reading, Mass., 1974.

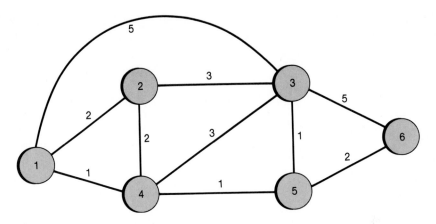

Figure 6-5 Example network

node 1, as the source, to all other nodes in the network. The algorithm does this in a step-by-step fashion, building a shortest-path tree, rooted in the source node (node 1 in this example), until the furthermost node in the network has been included. By the kth step the shortest paths to the k nodes closest to the source have been calculated. These are defined to be within a set N. We present the algorithm informally, leaving it to the reader in accompanying problems at the end of this chapter to program it and work out other, much larger examples.

Let $D(v)$ be the distance (sum of link weights along a given path) from source 1 to node v. Let $\ell(i,j)$ be the (given) cost between node i and node j. There are then two parts to the algorithm: an initialization step, and a step to be repeated until the algorithm terminates:

1. *Initialization.* Set $N = \{1\}$. For each node v not in N, set $D(v) = \ell(1,v)$. (We use ∞ for nodes not connected to 1; any number larger than the maximum cost or distance in the network would suffice.)

2. *At each subsequent step.* Find a node w not in N for which $D(w)$ is a minimum, and add w to N. Then update $D(v)$ for all nodes remaining that are not in N by computing

$$D(v) \leftarrow \text{Min}[D(v), D(w) + \ell(w,v)]$$

Step 2 is repeated until all nodes are in N.

The application of algorithm A to the network of Fig. 6-5 is demonstrated by the successive steps shown in Table 6-1. Circled entries in the distance

TABLE 6-1   Algorithm A Applied to Network, Fig. 6-5

Step	N	D(2)	D(3)	D(4)	D(5)	D(6)
Initial	{1}	2	5	1	∞	∞
1	{1,4}	2	4	①	2	∞
2	{1,4,5}	2	3	1	②	4
3	{1,2,4,5}	②	3	1	2	4
4	{1,2,3,4,5}	2	③	1	2	4
5	{1,2,3,4,5,6}	2	3	1	2	④

columns represent the minimum value of $D(w)$ at that step. (Ties are resolved randomly.) The corresponding node w is then shown added to N. The values of $D(v)$ are then updated as required. Thus, at step 1, after initialization, node 4, with minimum $D(4) = 1$, is added to set N. At step 2, node 5, with $D(5) = 2$, is added to N, and so on. After step 5 all nodes are in N, and the algorithm terminates.

As the algorithm is run, resulting in the entries shown step by step in Table 6-1, the shortest-path tree rooted in source node 1 is built at the same time: As a node is added to set N it is connected to the appropriate node already in N. The resultant tree for the network of Fig. 6-5 appears in Fig. 6-6(a). Entries in parentheses below each node represent the step at which that node is added to the tree. From the shortest-path tree for node 1 one obtains the routing table at node 1, indicating, by destination, the outgoing link to which to direct packets. A similar routing table would be generated for each node as the source node. In the case of centralized computation each routing table would then be transmitted to its corresponding node. In the case of decentralized, or distributed routing, as in the ARPA algorithm discussed later, each node would carry out its own computation, using the same global information and generating its own tree and corresponding routing table [McQU 1980].

Now consider algorithm B. It also has two parts: an initialization step and a shortest-distance calculation part that is repeated until the algorithm has been completed. Here the shortest distance represents the distance a given node is away from a node 1, say, considered as the *destination* node. It ends with all nodes labeled with their distance away from node 1 (the destination node), and a label as to the next node into the destination node, along the shortest path. Construction of a routing table using algorithm B requires repeated or parallel applica-

[McQU 1980] J. M. McQuillan et al., "The New Routing Algorithm for the ARPAnet," *IEEE Trans. on Comm.*, vol. COM-28, no. 5, May 1980, 711-719.

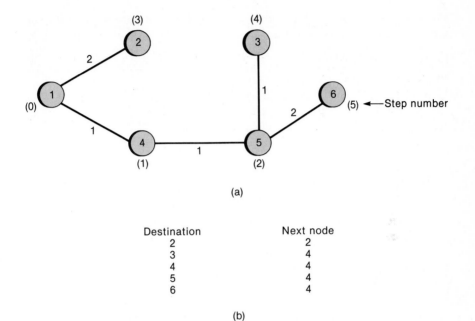

(a)

Destination	Next node
2	2
3	4
4	4
5	4
6	4

(b)

Figure 6-6 Algorithm A applied to Fig. 6-5, node 1 as source node
a. Building of tree
b. Routing table, node 1

tion of the algorithm for each destination node, resulting in a *set* of labels for each node, each label giving routing information (next node) and distance to a particular destination.

We again provide only an informal presentation of the algorithm. Each node v has the label $(n,D(v))$, where $D(v)$ represents the *current* value of the shortest distance from the node to the destination and n is the number of the next node along the currently computed shortest path.

1. *Initialization.* With node 1 the destination node, set $D(1) = 0$ and label all other nodes $(\cdot, \infty)$.

2. *Shortest-distance labeling of all nodes.* For each node $v \neq 1$, do the following: Update $D(v)$ by using the current value $D(w)$ for each neighboring node w to calculate $D(w) + \ell(w,v)$ and performing the operation

$$D(v) \leftarrow \operatorname*{Min}_{w}[D(w) + \ell(w,v)]$$

TABLE 6-2   Algorithm B, Fig. 6-5

Cycle	Labels, Node →	2	3	4	5	6
Initial		$(\cdot,\infty)$	$(\cdot,\infty)$	$(\cdot,\infty)$	$(\cdot,\infty)$	$(\cdot,\infty)$
1		(1,2)	(1,5)	(1,1)	(4,2)	(5,4)
2		(1,2)	(5,3)	(1,1)	(4,2)	(5,4)

Update the label of v by replacing n with the adjacent node that minimizes the expression just given and by replacing $D(v)$ with the new value found. Repeat for each node until no further changes occur. (The algorithm then terminates, with all nodes labeled both with their shortest distance from node 1 and with the neighboring node next in, along the shortest path.)

The network of Fig. 6–5 again provides a simple example (other examples appear, as noted, in [SCHW 1977, pp. 239–241]). Invoking part 2 of the algorithm in the cyclic order, nodes 2–6, one finds the algorithm terminating after two cycles. The cycles, and resultant labels at the end of each cycle, appear in Table 6–2. (Any other order of scanning the nodes would produce the same final result.)

Thus, note in cycle 1 that node 2 is "closest" to its neighbor 1. Its resultant new cost is $D(v) = D(1) + \ell(1,2) = 2$. Its label then becomes (1,2), as shown. Moving to node 3, its choice of costs is either $D(2) + \ell(2,3) = 5$ or $D(1) + \ell(1,3) = 5$. Selecting $D(1) + \ell(1,3)$ arbitrarily (the other choice would have done just as well), one gets (1,5), as shown in the table. The other entries follow in the same manner. On cycle 2, node 3 now has five neighboring nodes with non-infinite $D(w)$ among which to choose. The minimum value of $D(w) + \ell(w,3)$ is given by $D(5) + \ell(5,3) = 3$, and the label of 3 is changed to (5,3), as shown. No other nodes change their labels, and the algorithm terminates. Again, the shortest-path tree rooted at node 1, as shown in Fig. 6–6(a), can be obtained by scanning the labels of each node: Node 2 is connected to 1, node 3 to 5, node 4 to 1, node 5 to 4, and node 6 to 5. Alternatively, the routing table or next-node entry at each node *for destination 1* is precisely the first entry in each two-part label, as noted earlier. To obtain the complete routing table at each node, algorithm B would have to be repeated in turn for each node taken as a destination node.

An algorithm similar to that of algorithm B is used in the centralized routing computations for Tymnet [RAJA]. Routing in TYMNET will be discussed in detail in Subsection 6–3–1.

[RAJA] A. Rajaraman, "Routing in TYMNET," European Computing Conference, London, May 1978.

6-2-1 Decentralized Version of Algorithm B

As noted earlier, algorithm B lends itself easily to implementaton in a distributed or decentralized sense. Baran's "hot potato" routing strategy, developed in pioneering work by Rand Corporation on computer-communication networks in the early 1960s, is related to this algorithm and probably represents its first suggested use for decentralized routing in data networks [BARA]. The ARPAnet originally used a version of this algorithm, with costs given by estimated time delays in the network [McQU 1974], [SCHW 1977, p. 220]. The Michigan MERIT network, interconnecting computers at three Michigan state universities, used the decentralized minimum-hop version of this algorithm [TAJI]. The Swedish TIDAS network, used for controlling an extensive Swedish power grid, used a modified version [CEGR], as does the Canadian public packet-switched network Datapac [SPRO]. Digital Equipment Corporation and Burroughs Corporation use versions of this algorithm in their communication architectures. (A summary of some of these routing procedures appears in Section 6-3.) Because of the extensive use of this algorithm and its innate simplicity, it is of interest to describe it in some detail. We do that briefly in this subsection. In Section 6-4 of this chapter we devote some time to a performance evaluation of the algorithm and a comparison with related algorithms.

Note that because of the nature of this decentralized routing algorithm, in which nodes pass control messages back and forth until the algorithm is completed at each node, with convergence to shortest-path routing assured, the algorithm can be used for datagram routing only. This is in fact a general characteristic of decentralized algorithms of the type we shall consider: There is no guarantee that during transmission of packets the algorithm may not be invoked somewhere in the network, changing the path after convergence. The implication then is that packets sent on one path may arrive on another, *possibly arriving out of order*. This poses no problem with datagram transmission. It does pose a problem with virtual circuit routing, where paths are set up *before* packets

[BARA] P. Baran, "On Distributed Communication Networks," *IEEE Trans. on Comm. Syst.*, vol. CS-12, March 1964.

[McQU 1974] J. M. McQuillan, *Adaptive Routing Algorithms for Distributed Computer Networks*, BBN Report 2831, May 1974 (also available as Ph.D. dissertation, Division of Engineering and Applied Physics, Harvard University, 1974).

[TAJI] W. D. Tajibnapis, "A Correctness Proof of a Topology Information Maintenance Protocol for Distributed Computer Networks," *Communications of the ACM*, vol. 20, July 1977, 477-485.

[CEGR] T. Cegrell, "A Routing Procedure for the TIDAS Message-switching Network," *IEEE Trans. on Comm.*, vol. COM-23, no. 6, June 1975, 575-585.

[SPRO] D. E. Sproule and F. Mellor, "Routing, Flow, and Congestion Control in the Datapac Network," *IEEE Trans. on Comm.*, vol. COM-29, no. 4, April 1981, 386-391.

flow, and where the same path must be maintained throughout the existence of a session. The decentralized version of algorithm B is in fact prone to looping during the convergence interval: Data packets sent out from a given node may, during the operation of the algorithm, find themselves back at the same node some time later! This may again not represent a significant problem in datagram routing — the packet will eventually make its way to the destination. It is intolerable, however, in virtual circuit routing. We shall have more to say about looping in the last section of this chapter.

To carry out the algorithm in a distributed fashion, each node in the network must maintain *two* tables. One, called a *distance table* here, contains the cost to each destination via each outgoing link (or the equivalent via each neighboring node). Such a table, extending the routing table concept of Fig. 6 – 1, appears in Fig. 6 – 7(b). As in Fig. 6 – 1(a), N is the node at which the table is kept, and X, Y, Z are its neighbors. The table shown in Fig. 6 – 7(c) is the *routing table* referred to previously. The routing-table entry for each destination node is just the minimum-cost, next-node assignment of the distance table. Note that here the packet identification shown in Fig. 6 – 1(b) is the destination node address. All packets headed toward the same destination follow the same path with this algorithm. (In the next section we shall describe examples from virtual circuit routing in which the destination address alone does not suffice to provide appropriate routing.)

For the algorithm to work a node must be able to detect a cost change (including a node/link failure or a node/link coming up) on any one of its links. If a change occurs, all nodes detecting this change update their distance tables and check for changes in the minimum distance entries to all destination nodes. A node w detecting such a change (or changes) in its distance table changes the corresponding entry in the routing table and sends a control message of the form $[d, D_d(w)]$ ($D_d(w)$ is the minimum distance from w to the destination d), for each such change, to all its neighbors.* These control messages, with the new routing information for each destination affected, may be bundled together or sent separately, one for each destination involved. (In the one case of a new link coming up, the new neighbor at the other end of the link receives a copy of the complete routing table.) These neighbors in turn update their distance and routing tables, adding the link cost to each distance measured on the link received. If the routing tables change (i.e., minimum updated distances change), this new information is sent in turn to all neighbors, including the one from which the original change was received, using the same form of control message. This exchange of messages and update of tables continues until all activity

---

* We follow, in somewhat modified form, the description and algorithms presented by Tajibnapis [TAJI]. His specific algorithms refer to minimum-hop, rather than least-cost, routing, as is the case here.

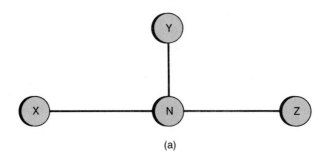

(a)

Distance table			
Destination	Cost via neighboring node		
node→ X	X	Y	Z
A	3	②	5
B	5	6	③
C	4	4	②
D	③	6	5
⋮			

Routing table		
Destination	Next neighbor	Cost
A	Y	2
B	Z	3
C	Z	2
D	X	3
⋮		

(b)   (c)

Figure 6-7 Decentralized routing
a. Current node and neighbors
b. Distance table
c. Routing table

ceases. At this point convergence has taken place and data messages will again be forwarded along shortest-distance (least-cost) paths. Tajibnapis has shown, for the special case of minimum-hop routing (the weight or cost of each link is 1), that the algorithm does converge to the minimum-hop path in finite time if a finite number of message changes occur during the time in question [TAJI]. J. Hagouel has proven similar convergence for the more general case of least-cost, rather than minimum-hop, routing [HAGO].

[HAGO] J. Hagouel, *Issues in Routing for Large and Dynamic Networks,* Ph.D. dissertation, Columbia University, New York, 1983.

We now present the algorithm in more detail, again using a somewhat informal presentation for ease of understanding. The algorithm has three parts, one for each of three possible events: a link (or node) coming up, the cost of a link changing, receiving a control message. We use a notation similar to that of algorithms A and B. Thus we use the letter v to represent the node carrying out the algorithm. $D_d(v)$ is the *minimum* distance from node v to destination d. w represents one of node v's neighbors. $\ell(w,v)$ is the link cost from w to v. Entry (d,w) in v's distance table contains the cost $C_d(v,w)$ from v to destination d via neighboring node w. The minimum of $C_d(v,w)$ over all neighboring nodes w is just $D_d(v)$. Let the neighboring node for which the minimum cost is obtained be n. The row d entry in v's routing table then consists of the pair of numbers $(n,D_d(v))$. (See Figs. 6–7(b) and (c).)

Part 1. Link m,v comes up   On recognizing this event, node v does the following:

1. Sets entry (m,m) of the distance table to $\ell(m,v)$. (This is the distance from v to destination node m *via* m.)

2. Computes $\min_w C_m(v,w)$. Say that the neighboring node for which the minimum occurs is node p.

3. If $\min C_m(v,w) = D_m(v)$ (i.e., the minimum distance to node m is unchanged), set $n = p$. (The routing table next-node entry is set to p.) Else

$$D_m(v) \leftarrow \min_w C_m(v,w)$$

the routing table entry for row m is changed to $(p,D_m(v))$, and the control message $[m,D_m(v)]$ is sent to all neighbors of v.

4. Let the set of destinations (rows in v's routing table) be numbered a, b, c, . . . , r. Then v sends the control messages $[a,D_a(v)]$, $[b,D_b(v)]$, . . . , $[r,D_r(v)]$ to node m. (These would probably all be bundled together in one message, as already noted.)

Note: This procedure, which is used when a link comes up, is applicable as well when a new node m comes up. In that case a new (destination) row and a new (neighbor) column of the distance table, as well as a new row of the routing table, might have to be created for node m.

Part 2. Cost of link m,v changes   This part is very similar to part 1. On noting that the cost of link m,v has changed by $\Delta(mv)$, node v does the following:

1. All entries in column m of v's distance table are changed by $\Delta(mv)$. (In the special case when the link goes down, the entries are set to ∞, or a very high number, larger than all possible costs).

2. For all destination nodes d, do the following:
Compute $\min_w C_d(v,w)$. This occurs for node p, say.

3. If $\min_w C_d(v,w) = D_d(v)$, set $n = p$.
Else

$$D_d(v) \leftarrow \min_w C_d(v,w)$$

the routing table entry for row d is changed to $(p, D_d(v))$, and the control message $[d, D_d(v)]$ is sent to all neighbors. (Note again that control messages may be bundled together as one message. Messages are *not* sent for those destinations for which no cost change has occurred.)

Part 3. A control message $[d, D_d(w)]$ is received by v If $d = v$, so that the destination is v itself, v ignores it. If $d \neq v$, v does the following:

1. $$C_d(v,w) \leftarrow D_d(w) + \ell(w,v)$$

(If $D_d(w) = \infty$ or a very large number, $C_d(w)$ is set equal to the same number.)

2. Compute

$$\min_w C_d(v,w)$$

This occurs for node p, say.

3. If $\min_w C_d(v,w) = D_d(v)$, set $n = p$. Else

$$D_d(v) \leftarrow \min_w C_d(v,w)$$

the routing table entry for row d is changed to $(p, D_d(v))$, and the control message $[d, D_d(v)]$ is sent to all neighbors. (As noted earlier, a control message is sent to *all* neighbors, including the one from which the original message was received, if and only if a change occurs in the distance table minimum cost.)

Note that the distance table row minimization, for a given destination, after updating the appropriate entry, is the same as the operation

$$D(v) \leftarrow \min_w [D(w) + \ell(w,v)]$$

carried out in algorithm B. The algorithm thus corresponds to a distributed version of algorithm B, with all nodes now cooperating in the convergence process.

Although convergence to the shortest path is guaranteed, routing table entries may change during the convergence period, giving rise to possible loops during that interval. This possibility has already been noted. Algorithm B may also have entries changing during convergence, as noted in the example described earlier (Table 6–2).

TABLE 6-3 Distributed Algorithm Applied to Network, Fig. 6-5, Destination 1

Initialize. Node 1 comes up. Nodes 2, 3, 4 carry out part 1 of algorithm.

Tables, Node 2

Distance				Routing	
Via Node →	1	3	4	*Via*	*Cost*
	②	∞	∞	1	2

Node 2 sends [1,2] to all neighbors

Tables, Node 3

Distance						Routing	
	1	2	4	5	6	*Via*	*Cost*
	⑤	∞	∞	∞	∞	1	5

Node 3 sends [1,5] to neighbors

Tables, Node 4

Distance					Routing	
	1	2	3	5	*Via*	*Cost*
	①	∞	∞	∞	1	1

Node 4 sends [1,1] to neighbors

Tables, Node 5

Distance				Routing	
	3	4	6	*Via*	*Cost*
	∞	∞	∞	—	∞

Tables, Node 6

Distance			Routing	
	3	5	*Via*	*Cost*
	∞	∞	—	∞

Iteration 1. All nodes receiving messages carry out part 3 of algorithm, make table changes.

Tables, Node 2

Distance				Routing	
Via Node →	1	3	4	*Via*	*Cost*
	2	8	3	1	2

Send [1,4]

Tables, Node 3

Distance						Routing	
	1	2	4	5	6	*Via*	*Cost*
	5	5	④	∞	∞	4	4

Send [1,4]

Tables, Node 4

Distance					Routing	
	1	2	3	5	*Via*	*Cost*
	1	4	8	∞	1	1

Send [1,2]

Tables, Node 5

Distance				Routing	
	3	4	6	*Via*	*Cost*
	6	②	∞	4	2

Send [1,2]

Tables, Node 6

Distance			Routing	
	3	5	*Via*	*Cost*
	⑩	∞	3	10

Send [1,10]

Iteration 2. All nodes receiving messages carry out part 3 of algorithm, make table changes.

Tables, Node 2

Distance			Routing	
Via				
Node → 1 3 4			*Via Cost*	
2 7 3			1	2

Tables, Node 3

Distance					Routing	
1 2 4 5 6					*Via Cost*	
5 5 4 ③ 5					5	3

Send [1,3]

Tables, Node 4

Distance				Routing	
1 2 3 5				*Via Cost*	
1 4 7 3				1	1

Tables, Node 5

Distance			Routing	
3 4 6			*Via Cost*	
5 2 12			4	2

Tables, Node 6

Distance		Routing	
3 5		*Via Cost*	
9 ④		5	4

Send [1,4]

Iteration 3. All nodes receiving messages carry out part 3 of algorithm, make table changes.

Tables, Node 2

Distance			Routing	
Via				
Node → 1 3 4			*Via Cost*	
2 6 3			1	2

Tables, Node 3

Distance					Routing	
1 2 4 5 6					*Via Cost*	
5 5 4 3 9					5	3

Tables, Node 4

Distance				Routing	
1 2 3 5				*Via Cost*	
1 4 6 3				1	1

Tables, Node 5

Distance			Routing	
3 4 6			*Via Cost*	
4 2 6			4	2

Tables, Node 6

Distance		Routing	
3 5		*Via Cost*	
8 4		5	4

No messages sent, algorithm has converged.

We now provide an example of the operation of the algorithm, again using Fig. 6–5. Other examples, for the equal-link cost, minimum-hop case only, appear in [TAJI]. To compare with the operation of algorithms A and B earlier, we describe the operation of this decentralized algorithm as node 1 in Fig. 6–5 comes up, with node 1 as the destination node. Table 6–3 shows row 1 (corresponding to destination 1) of the distance and routing tables at each of the five nodes other than node 1, at the end of each iteration of the algorithm. An iteration corresponds to the receipt of a control message by at least one node. (For simplicity, we assume that all nodes operate in synchronism. In real life, this would of course not be the case. The lack of synchronism and unequal propagation delays, ignored here, could under certain conditions give rise to race conditions. A simple example is provided in Subsection 6–4–1.

To initialize the algorithm, nodes 2, 3, 4, the neighbors of node 1, recognize that node 1 is coming up and, carrying out part 1 of the algorithm, change their distance table entries in column 1 ("via node 1") from ∞ to their respective link costs $\ell(1,v)$, $v = 2, 3, 4$. (All entries in all tables are set initially to ∞ before the algorithm begins to operate.) Changes in the minimum distance, resulting in a change in the routing table entries and in control messages being sent to all neighbors, are shown encircled. Thus, in this example, nodes 2, 3, 4 all change their minimum distance to 1 and send control messages to that effect to all their neighbors. This is indicated as well in Table 6–3; messages to node 1 are not indicated.

Nodes that receive control messages carry out part 3 of the algorithm (i.e., add the link cost to the incoming cost, enter in the distance table, compute the minimum distance, enter in the routing table if a change has occurred, and send a control message to that effect to all neighbors). This is labeled iteration 1 in Table 6–3; the resultant table entries at each node for destination 1, at the end of this iteration, also appear in the table. Note that although distance table entries at all nodes have changed as a result of this iteration, the routing table entries in nodes 3, 5, and 6 only have changed. These three nodes in turn send control messages with the new cost (distance) values to destination 1 to all their neighbors. Iteration 2 is carried out in response to these control messages. At the end of this iteration routing-table entries in nodes 3 and 6 change, and messages with the new cost values are sent to their respective neighbors. The algorithm repeats one more time, for a third iteration, with no change in routing table entries. Thus no control messages are sent, and the algorithm stops.

Now note the resultant routing-table entries at the end of iteration 3. These correspond precisely to the labels indicated for each node at cycle 2 in Table 6–2, the completion of the operation of algorithm B earlier! The distributed algorithm has converged to exactly the same routing-table entries as those of algorithm B. Note also, as pointed out in the discussion of that algorithm, that the next-neighbor entries of the routing table at each node (the column labeled "via") give rise to the shortest-path tree, rooted in destination 1, that was obtained by the application of algorithm A, as shown earlier in Fig. 6–6(a). Thus node 2 in Table 6–3 for the distributed algorithm points to node 1, as in the tree of Fig. 6–6(a). Node 3 in Table 6–3 points to node 5, again as in the tree of Fig. 6–6(a), and so forth. In Fig. 6–6 node 1 is considered the source node, and the tree portrays the shortest paths to all other nodes as destinations. For this example, with link costs in either direction implicitly assumed the same ($\ell(w,v) = \ell(v,w)$ here), the shortest path tree *to* node 1 is the same as the tree *from* node 1.

We now point out some of the properties of this distributed algorithm, as apparent from this one example. First note that because nodes may change their routing-table entries from iteration to iteration, as is the case here, data packets transmitted during the time of the operation of this algorithm (i.e., after a cost change has occurred and before convergence has taken place) may find themselves at a node visited earlier. This phenomenon, noted earlier, is called *looping*. In datagram routing, where cost changes do not occur too often, this may not pose too great a problem, particularly since the algorithm is so simple to apply. The algorithm has thus been adopted in a number of networks, some of which were mentioned earlier. More details are provided in Section 6–3. In the case of the ARPA network, the original routing algorithm of this type had estimated time delays as link costs [McQU 1974]. Estimates were updated every 2/3 sec. Because of the relatively brief time available to estimate delay, the estimates turned out to be relatively unstable, adaptivity was quite rapid, different nodes could have different views of network conditions, and looping became a serious problem [McQU 1978]. As a result the algorithm was abandoned, and a new one developed [McQU 1980]. A summary of the new algorithm, related to algorithm A as noted earlier, appears in Subsection 6–3–2.

The possible presence of loops that use this distributed algorithm appears in the example just described. Consider the distance table entries after convergence, iteration 3 in Table 6–3. Note, for example, that although the minimum-cost entry in node 4's table is a cost of 1 via node 1 directly, the second-best entry is a cost of 3 via node 5. Node 4 thus thinks that there is an alternate path to 1 via node 5. But that path leads right back on itself! (Note from the routing table for node 5 that that node routes to node 4.) If link (1,4) were to fail (see Fig. 6–5), node 4 would at first choose its neighboring node 5, with an apparent cost of 3, as the appropriate node to route to for destination 1. It would send messages to that effect (part 2 of the algorithm) to all its neighbors. These in turn would update their distance tables, as prescribed in part 3 of the algorithm,

[McQU 1978] J. M. McQuillan, G. Falk, and I. Richer, "A Review of the Development and Performance of the ARPAnet Routing Algorithm," *IEEE Trans. on Comm.*, vol. COM-26, no. 12, Dec. 1978, 1802–1811.

and send messages to *their* neighbors in turn. Data packets proceeding from node 4 to destination 1 would for a time find themselves routed to node 5 and then back to node 4. After a number of iterations, however, the algorithm would converge to the new shortest-path (least-cost) entries for destination 1, and any looping would cease. Details are left to the reader.

Looping can be reduced somewhat, although not eliminated completely for this algorithm, by some simple changes. As an example, in "split-horizon" routing, first implemented on the TIDAS network [CEGR] and then on the Canadian Datapac network [SPRO], the minimum-cost entry for a given destination is sent to all neighbors *except* the one to which the route points. This is done by excluding the cost via a particular neighboring node from the calculation of the minimum-cost entry to be sent to that neighbor. In a related scheme, called the predecessor algorithm, a neighboring node determined to be the routing or successor node to a given destination is so notified. That node then sets the cost entry in its distance table for the predecessor node from which it received this message to a very large number. In the event of a cost change (or link failure) it initially avoids looping back to its predecessor on the shortest-path tree [SCHW 1980c]. A comparison of this last algorithm with the generic distributed one described in this section appears in Subsection 6 – 4 – 4.

Loop-free distributed routing algorithms have been presented in the literature [MERL], [JAFF]. These are also discussed briefly in Subsection 6 – 4 – 3.

In addition to convergence and looping considerations for distributed routing algorithms, performance issues include convergence time and the number of control messages transmitted during convergence, noted at the beginning of this chapter. In the example just described (Table 6 – 3), convergence of the generic algorithm is fairly rapid, requiring three iterations for the simple network of Fig. 6 – 5. The total number of control messages exchanged during convergence, obtained from Table 6 – 3 by counting the number sent at the end of each iteration, is 24. This turns out to be a rather high number and is due to the innate "blindness" (based on its simplicity) of the algorithm: Messages are sent to all neighbors. This is another reason why attempts have been made to improve on the generic algorithm, in addition to a desire to reduce looping. Again, a more detailed discussion of these performance issues (convergence time and number of control messages transmitted) is deferred until Subsection 6 – 4 – 4.

[SCHW 1980c] M. Schwartz, "Routing and Flow Control in Data Networks," NATO Advanced Study Inst., New Concepts in Multi-user Communications, Norwich, U.K., Aug. 4 – 16, 1980; Sijthoof and Nordhoof, Neth.

[MERL] P. M. Merlin and A. Segall, "A Fail Safe Distributed Routing Protocol," *IEEE Trans. on Comm.*, vol. COM-27, no. 9, Sept. 1979, 1280 – 1287.

[JAFF] J. M. Jaffe and F. H. Moss, "A Responsive Distributed Routing Algorithm for Computer Networks," *IEEE Trans. on Comm.*, vol. COM-30, no. 7, July 1982, 1758 – 1762.

6–3 Examples of Routing in Networks and Network Architectures

It was pointed out earlier that most networks in operation use some variant of least-cost, or shortest-distance, routing. Networks based on virtual circuit routing tend to use centralized computations to determine the routing paths. Control packets are then used to set up the path chosen. Datagram networks, on the other hand, use distributed routing techniques. In this section we describe a representative group of networks (and network architectures) in both categories, focusing on the routing procedures used in each.

6–3–1 Virtual Circuit-oriented Networks

As discussed in Chapter 5, a virtual circuit represents a specific path between source and destination nodes along which data packets are constrained to flow. Normally a call-setup or connection-establishment phase and a clear or call-disconnect phase are associated with data transfer. Since multiple virtual circuits may exist between the same source-destination pair, routing cannot be done on the basis of source-destination addresses only. Data packets must carry an indication of virtual circuit identification as well. Examples of such ideas mentioned in the last chapter include the X.25 logical channel identification and the IBM SNA virtual route/explicit route numbers.

Although, conceptually, virtual circuits can be numbered globally in a network, this leads to inordinately large VC fields in the data packets. (Typically thousands of VCs may be active in a given network.) Instead, a number of techniques are commonly used to reduce the VC id field size. It was noted in Chapter 5 that at most 48 virtual routes with 16 distinct explicit routes (paths) can be assigned between a source-destination pair in SNA. Routing is then done on the basis of explicit route number and destination address. An explicit routing table at each node associates an appropriate outgoing transmission group with the destination address and explicit route number. Such a table at a particular node appears in Fig. 6–8. The letters represent the transmission groups to which packets with the corresponding address, route number pair are directed. By changing the explicit route number for a given destination, a new path will be followed. This introduces alternate route capability. If a link or node along the path becomes inoperative, any sessions using that path can be reestablished on an explicit route that bypasses the failed element. Explicit routes can also be assigned on the basis of type of traffic, types of physical media along the path (satellite or terrestrial, for example), or other criteria, as already noted. Routes could also be listed on the basis of cost, the smallest-cost route being assigned first, then the next-smallest-cost route, and so forth.

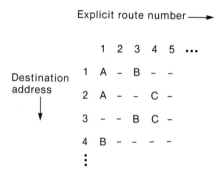

Figure 6–8 Explicit routing table, SNA

In TYMNET, which is described in detail next, each link in the network has a set of logical record numbers. The combination of (link, logical record number) provides a unique VC identification and serves to reduce the size of the VC field. The logical record number on a given incoming link must then be swapped to a new logical record number on the appropriate outgoing link as part of the routing process.

TYMNET   TYMNET is a computer-communication network developed in 1970 by Tymshare, Inc., of Cupertino, California. It has been in commercial operation since 1971. Originally developed for time-shared purposes, TYMNET later took on a network function as well and currently serves as one of the two public packet-switched networks in the United States (GTE Telenet, discussed later, is the other such network). It has over 1000 nodes, almost all of which are connected to at least two other nodes in the network, giving rise to a distributed topology with alternate path capability. A detailed description of an earlier, much smaller version of the network, including a map of the network at that time, appears in [SCHW 1977]. Routing and flow control in the network are still handled essentially as they were in the earlier version, although some significant changes have had to be made because of the much larger network size and increased traffic usage [TYME]. The network covers the United States and Europe, with connections also made to the Canadian Datapac network. Trans-Atlantic lines are cable with satellite backup. Satellites are avoided for interactive users where possible because of the substantial delay involved.

[TYME] L. Tymes, "Routing and Flow Control in TYMNET," *IEEE Trans. on Comm.*, vol. COM-29, no. 4, April 1981, 392–398.

Individual user data, in the form of logical records, each preceded by a 16-bit header incorporating an 8-bit logical record number (discussed later) and an 8-bit packet character count, are concatenated to form a packet of at most 66 8-bit characters, including 16 bits of header and 32 bits of checksum for error detection [SCHW 1977]. These logical records can range in length from a few characters to a maximum of 58 characters. (Packets are transmitted as soon as available, without waiting for a logical record of specified size to be assembled.)

TYMNET routing is set up centrally on a virtual circuit, fixed-path basis by a supervisory program running on one of four possible supervisory computers in the network.

A least-cost algorithm is used to determine the appropriate path from source to destination node over which to route a given user's packets. The path is newly selected each time a user comes on the network and is maintained unchanged during the period of the user connection or session. The algorithm used by the supervisor is a modification of Floyd's algorithm, a variation of our algorithm B. Details of the algorithm used appear in [TYME] and [RAJA].

The routing path construction carried out by the central supervisor incorporates the concept of a class of service, similar to that used in SNA. Link weights vary depending on the type of traffic transmitted. For example, low-speed interactive users are steered away from satellite links by increasing the link weight in that case. Computers that are passing files between them, however, will be assigned a wider-band satellite link as compared with a lower-bandwidth terrestrial link. Link weights also depend on link utilization and error conditions detected on the link. Specifically, the number 16 is assigned to a 2400 bps link, 12 to a 4800 bps link, and 10 to a 9600 bps link. A penalty of 16 is added to a satellite link for low-speed interactive users. This shifts such users to cable links, as just noted.

A penalty of 16 is also added to a link if a node at one end complains of "overloading." The penalty is 32 if the nodes at both ends complain. Overload is experienced if the data for a specific virtual circuit has to wait more than 0.5 sec before being serviced. This condition is then reported by the node to the supervisor. An overload condition may occur because of too many circuits requesting service over the same link, or it may be due to a noisy link with a high error rate, in which case the successive retransmissions that are necessary also slow down the effective service rate. The penalty used in this case serves to steer additional circuits away from the link until the condition clears up.

In the absence of overloading, the algorithm tends to select the shortest path (least number of links) with the highest transmission speed. As more users come on the network, the lower-speed links begin to be used as well. In lightly loaded situations, users tend to have relatively shorter time delays through the network. The minimum-hop paths, favored in the lightly loaded case, also tend to be more reliable than ones with more links. Users coming on in a busy period may experience higher time delays due both to congestion and to the use of lower-speed lines. The use of the overload penalties tends to spread traffic around the network, deviating from the shortest-path case but attempting to reduce the time delay. In practice, the average response time for interactive users is 0.75 sec [RAJA].

In the event that a node or a trunk along the VC path fails, the supervisor is notified, automatically determines a new routing path, and again notifies the nodes involved. The notification in either case (original or rerouting) consists of sending a route-building control packet that contains a list of all nodes along the route to the originating node. The originating node in turn sends this control packet out over the link to the next node on the list, using an unassigned logical record number for this purpose. This 8-bit logical record number will be the same number carried by the following data logical records using the VC. The 8-bit logical record field allows up to 256 channels to share any one link. One channel or number is reserved for a node to communicate with the supervisor (it thus corresponds to a "permanent VC"), while one channel is reserved for communications with the neighboring node.

The next node to which the routing packet is addressed in turn swaps the incoming logical record number to a new, unassigned number on the appropriate outgoing link to the node following on the list. Path setup proceeds in this way until the final (destination) node on the list is reached. (The average path in TYMNET tends to be about three links long. In the older version of the network, the supervisor did the necessary swapping of link and logical record numbers.) Each node on the path acknowledges receipt of the routing information.

Once the path is set up, data logical records for the VC in question, following in the wake of the routing-control packet, are routed in the same manner: At each node along the path, the incoming (link, logical record number) pair is swapped to the appropriate outgoing pair, and the logical record is then queued up for transmission over the outgoing link designated. The swapping or routing function of a node is established at the time the control packet comes through by setting entries in routing tables called *permuter tables* in TYMNET terminology. A permuter table is associated with each link at a node.

Each logical record number in either direction on the link is associated with an entry in the table. That entry corresponds in turn to the address of a pair of buffers at the node, one for each direction of data flow (inbound and outbound). For L links at a node, L permuter tables are needed, each receiving up to 256 buffer addresses. An error-free packet arriving at a node is disassembled into its component logical records. Each logical record is steered by the permuter-table entry to its appropriate buffer. Data in buffers destined for terminals and/or computers associated with this node is then transferred to the appropriate device. This node thus represents the destination node for these logical records.

Logical records waiting in transit buffers are handled differently. A packet for a given outgoing link is created, under program control, by scanning sequentially the entries in the permuter table for that link. As each buffer address is read, a determination is made as to whether its pair has had data entered. If so, the data is then formed into a logical record with its corresponding new logical record number. This logical record is incorporated in the packet and is transmitted out over the link.

A specific example appears in Fig. 6–9 [RIND]. Fig. 6–9(a) shows a typical two-link virtual circuit connecting nodes numbered 5, 7, and 10. In this example, terminal data enters the network via a terminal port at node 5, destined for a host computer connected to node 10. The link connecting nodes 5 and 7 is labeled 1 as seen at the node-5 side, and 2 as seen at the node-7 side. Similarly, the link connecting nodes 7 and 10 is labeled 3 at the node-7 side and 1 at the node 10-side.

Figure 6–9(b) portrays the logical record number assignments and permuter table entries in detail, node by node. (Only eight possible logical record numbers have been assumed for simplicity.) Logical record numbers 4 and 6 have been assigned to this virtual circuit over the two links shown, respectively. At node 5, the entry node, the number 3 in entry 4 in the permuter table for link 1 indicates that data with logical record number 4 is to be found in buffer 2, the pair of buffer 3.

At node 7, data coming from link 2 is stored in buffers designated by the contents of the permuter table for link 2 at that node. Continuing with this example, data arriving at that link with logical record 4 is to be further transmitted over outgoing link 3 to node 10. Its outgoing logical record number is to be changed to 6. Note that to accomplish this the contents of entry 4 of permuter table 2 and entry 6 of permuter table 3 are paired. Data arriving over incoming link 2 is stored in buffer 8. This data is read out over link 3 when the entries for the permuter table for that link are scanned, with entry 6 pointing to buffer 9, the pair of buffer 8. At node 10, the destination node for this virtual circuit, data arriving with logical record 6 is stored in buffer 100 of that node and is then transferred to the appropriate host.

It is of interest to consider flow control in the TYMNET network at this point. Flow control in TYMNET is carried out on a per node, per VC basis [TYME], [SCHW 1977]. For each logical channel on each link a window size or maximum number of bytes to be sent is established, depending on the channel-throughput class. Permission to transmit beyond this point is required from the receiving node. The receiving node can ask for additional data or can withhold permission. Note that this is similar to the SNA pacing concept and to X.25 VC

[RIND] J. Rinde, "Routing and Control in a Centrally Directed Network," 1977 Nat. Comput. Conf., *AFIPS Conf. Proc.*, vol. 38, 1977, 211–216.

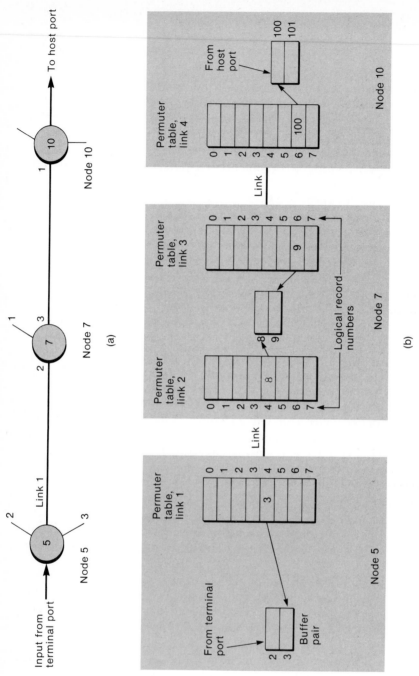

Figure 6–9 Virtual circuit routing, TYMNET
a. Typical virtual circuit: two links, terminal to host port
b. Permuter tables and logical record numbers

flow control, discussed in Chapter 5, except that it is carried out on each link along the path rather than end to end. If a window at a node is exhausted, indicating that the VC buffers are full, this node will in turn withhold permission from the preceding node along the path. The resultant "back pressure" thus moves backward along the VC, to the originating node, turning off the transmission at that point. Analysis of a simplified model of this type of "local control" indicates that its performance is similar to that of the sliding-window control [PENN], [SCHW 1977, Chap. 11].

If we compare the SNA routing philosophy with that of TYMNET, we note that they are similar. Both use a virtual circuit routing approach, both require central computation of a shortest-path route (this would normally be carried out by an SSCP for the nodes under its jurisdiction in the case of SNA), and both use control packets to set up a routing path. (In the SNA case an explicit route must first be activated before a virtual route may be mapped onto it. This is done by transmitting a specific activate-command packet from node to node along the path [AHUJ].) The basic difference is one of implementation: In TYMNET routing a path is newly set up every time a user signs on for a session. In the SNA case, precalculated fixed paths may be loaded into source node memory. TYMNET routing is thus more dynamic, but SNA routing allows alternate route capability. In the event of a failure, for example, the source node in the SNA case could select a new path by choosing a new explicit route number. In TYMNET routing the supervisor must recalculate a new path in the event of failure.

GTE Telenet   As already noted, GTE Telenet and TYMNET constitute the two public packet-switched networks in the United States. Telenet was organized as a network in 1974. As of early 1983 GTE Telenet consisted of over 200 switches and 40 56-kbps trunks, plus concentrators and access lines connected in hierarchical fashion [KRON]. As in the case of TYMNET, the network covers the continental United States with connections to the Canadian Datapac and overseas carriers as well.

Routing in this network is carried out in a two-level hierarchy. A backbone network covers the continental United States. Connected to nodes on the backbone are access lines and lower-level switches. Data traffic will thus be routed first to the appropriate backbone switch, then along the backbone network to

[PENN] M. C. Pennotti and M. Schwartz, "Congestion Control in Store-and-Forward Tandem Links," *IEEE Trans. on Comm.*, vol. COM-23, no. 12, Dec. 1975, 1434–1443.

[AHUJ] V. Ahuja, "Routing and Flow Control in Systems Network Architecture," *IBM Syst. J.*, vol. 18, no. 2, 1979, 298–314.

[KRON] R. Kronz, S. Lee, and M. Sun, "Practical Design Tools for Large Packet-switched Networks," Infocom '83, San Diego, Calif., April 1983.

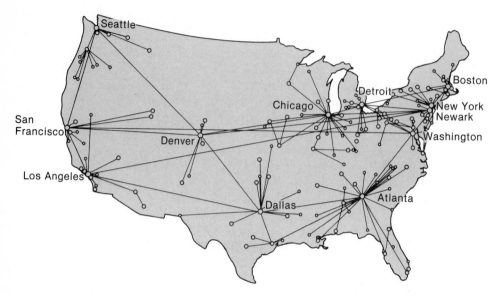

Figure 6–10 GTE Telenet common-carrier network (1982) (from [GTET, Fig. 1–1] with permission)

---

the appropriate destination backbone node. Backbone nodes are arranged to serve geographic areas that correspond to telephone area codes. Network addresses are assigned using the telephone area code plus additional digits that identify switches and access lines within an area code. A map of the network as of mid-1982, showing both the backbone network and switches homing in on the backbone, appears in Fig. 6–10 [GTET, Fig. 1–1], [KRON]. Note that at that time there were 12 backbone nodes interconnected by some 21 56-kbps trunks. Each node, or *hub* in GTE Telenet terminology, consists of a cluster of backbone switches, plus backbone trunks, access lines, and lower-level (level II) switches connected to them. A typical hub cluster appears in Fig. 6–11 [KRON, Fig. 6]. The level-II switches are in turn also arranged in a cluster, which serves traffic in one or more telephone area codes [KRON]. Redundancy is used in establishing network access paths.

GTE Telenet uses virtual circuit routing, based on the X.75 protocol [WEIR]. X.75 was developed by the CCITT as an interface protocol between

---

[GTET] *Functional Description of GTE Telenet Packet Switching Networks*, NFD-005.014, GTE Telenet Communications Corp., Vienna, Va., May 1982.

[WEIR] D. F. Weir, J. B. Holmblad, A. C. Rothberg, "An X.75-based Network Architecture," ICCC, Dec. 1980, Atlanta, Ga., 741–750.

packet-switched networks. (This contrasts with X.25, which allows DTEs to connect to a network.) It is essentially an enhanced version of X.25 and is completely compatible with it. It uses the same call setup and packet-handling procedures as X.25. The enhanced features include higher bit rates (56 kbps) at the physical level, extended sequence numbering at the frame level (making the system more efficient when transmitting via satellite), and an additional utility field for the packet-level call-setup packets. A multilink (trunk-group) protocol, similar to SNA's transmission group concept, is used for handling a given VC, providing load sharing and graceful recovery from link failure. As in the case of X.25 (see Chapter 5) a call is established by transmitting a call-request packet. In the Telenet use of the protocol for routing, the call-request packet threads its way through the entire backbone network, setting up an appropriate route, rather than just being used at the interface between DTEs and DCEs, as in X.25, or between networks, as in X.75 (see Chapter 5). Other X.25 control-packet types, discussed in Chapter 5, are used to acknowledge acceptance of a call, to clear calls, and so forth.

GTE Telenet was originally designed on a datagram routing basis. A major reason cited for switching to the VC concept is the cost improvement possible. The current X.75 packet-level header is 3 bytes long, contrasted with the 16 bytes required with the datagram approach. Higher line utilization is then possible with lower costs per packet resulting. A reconnect procedure must be used, however, to reestablish a path if a trunk group or transmit node fails [WEIR].

Routing, established, as noted, at VC setup time, follows a combined centralized-distributed approach [WEIR]. Possible routing paths are determined

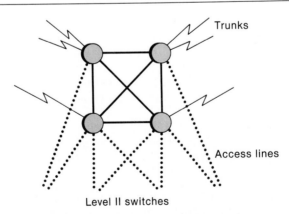

Figure 6-11 Hub cluster of backbone switches, GTE Telenet (from [KRON, Fig. 6] with permission. © 1983 IEEE)

by the Network Control Center (NCC) and are then transmitted in the form of routing tables to the various backbone switches. These consist of a set of choices of outgoing links for each nodal entry divided into two classes: a primary set of preferred choices, corresponding to shortest-path routes, and a secondary set of less optimal, alternate routes. At call setup time a secondary choice is invoked only if all primary links are out of order. A link-selection algorithm at each node is invoked at call setup time by the arriving VC call-request packet to determine the specific outgoing link to be used for a given destination node. The outgoing link chosen, of the set that may be used, is the one that has maximum excess capacity. This quantity is determined in turn by link capacity, current line utilization (estimated regularly by the frame level at a node), and an estimate of the total number of VCs passing through in each direction of traffic on a link. The idea of using both centralized and locally determined information for routing is similar to the concept of delta routing as suggested by Rudin [RUDI 1976]. Although possible routing paths are essentially established by the centralized NCC, the final routing decision is made locally, as just noted. The French network, TRANSPAC, uses a similar approach.

If no outgoing link is available at a given node during call setup, the VC for this call is cleared back to the previous node. That node then tries an alternative link, using the same link-selection algorithm. This process continues until a path is found if one exists and is represented in the routing tables. If none exists the call is blocked. Temporary routing loops or the existence of a nondeliverable (infinite) path are detected by maintaining a loop count and a "node visited" list in the utility field of the X.75 call-request packet. Both are incremented at each node along the VC path. If either one indicates a problem, the call is cleared back to the previous node, and the route-selection process is repeated.

The routing tables transmitted to the nodes by the NCC are based on topology and flows in the network. (Traffic flows have been found to be very stable and predictable from day to day and at specific times during the day [WEIR].) Link or node failures produce alarm messages that are sent to the NCC, which in turn distributes this information throughout the network.

The Telenet switches consist of members of the multiprocessor TP4000 packet-switch class [NEWP], [OPDE]. The generic switch consists of line-processing units (LPUs) for handling lines to terminals and hosts (DTEs) and net-

[RUDI 1976] H. Rudin, "On Routing and 'Delta-Routing': A Taxonomy and Performance Comparison of Techniques for Packet-switched Networks," *IEEE Trans. on Comm.*, vol. COM-24, no. 1, Jan. 1976, 43–59.

[NEWP] C. B. Newport and P. Kaul, "Communication Processors for Telenet's Third Generation Packet-switching Network," Eascon 77, Arlington, Va., Sept. 1977.

[OPDE] A. Opderbeck, J. H. Hoffmeier, R. L. Spitzer, "Software Architecture for a Microprocessor-based Packet Network," National Computer Conference, Anaheim, Calif., June 1978.

work trunks (at the X.25 frame level), and a master CPU involved with the X.25 packet level, which carries out the routing. The VC identification at a given node consists of a line-processing unit (LPU) id, the line number for that unit, and the logical channel (12 bits in the X.25 format) on that line. Routing at the packet level thus consists of mapping the incoming LPU id, line number, and logical channel number into the corresponding outgoing numbers. This is done using a trunk-to-trunk VC state table. Only active VCs are maintained in the table. As in X.25, then, up to 4095 calls may be in operation over any given line.

A specific VC reconnect procedure has been implemented by GTE Telenet to handle recovery from failures along the VC [WEIR]. The detailed features of the reconnect process, which is invoked during a potential reconnect, are established at call-setup time through the exchange of information by the two call endpoints. During the call itself the VC endpoints are responsible for determining the need to reconnect, for taking actions to set up a new path, and for recovering lost packets. Specifically, failure of a trunk line or a transit node triggers a clear-indication packet that is sent to both VC endpoints. This then clears the VC affected. The source node then creates a reconnect-request packet that is sent to the destination via the normal call-establishment routing process. The control block for the original call is mapped to a new VC with a new logical number. At the destination the new VC is linked to the original one as well. A reconnect-accept packet is then returned to the source.

TRANSPAC*   TRANSPAC, the French public packet-switching service, began operation in December 1978. Like GTE Telenet, it was designed to follow the virtual circuit packet-handling procedures of the X.25 interface protocol, and the routing procedures discussed here reflect this orientation [DANE]. Using facilities of the international record carriers, TRANSPAC is accessible to TYMNET, GTE Telenet, the Canadian Datapac, and other worldwide data networks.

For purposes of reliability, at least two 72 kbps lines, following different physical paths, connect each node to the remainder of the network. Each node consists of a control unit (CU) (a CII Mitra 125 minicomputer) to which are attached a number of switching units (SUs). Each incident link is controlled by an SU, which executes all data link procedures. The SUs also execute the access protocols for customers connected to the node. Routing is handled by the CU,

---

* Most of this section was written by Prof. T. E. Stern of Columbia University and is based, in part, on information provided by J. M. Simon of TRANSPAC. It originally appeared in [SCHW 1980b]. The author is indebted to Prof. Stern for permission to reproduce this material.
[DANE] A. Danet et al., "The French Public Packet-switching Service: The Transpac Network," ICCC, Toronto, Aug. 1976, 251–260.

using information from the Network Management Center (see the next paragraph).

Network control is partially decentralized through six local control points that handle a certain amount of statistics gathering and that perform test and reinitialization procedures in case of node or line failures. However, general network supervision, including the bulk of the routing computation, is exercised through a single Network Management Center (NMC).

Routes in TRANSPAC are assigned on a single path per VC basis. The algorithm of interest to us here is the one that governs the assignment of a route to a switched virtual circuit. The call request takes the form of a call packet, which is emitted by equipment connected to the originating network node and requests connection to a specified destination. The path that eventually will be retained by the switched VC is identical to that taken by the call packet as it is forwarded through the network. Routing of the call packet is effected through routing tables stored at each node; as previously, the tables associate a unique outbound link with each destination node. The network as configured currently has two classes of nodes. One class is connected in a distributed fashion, with alternative route capability. The second class consists of nodes homing in via a single link to a node of the first class. Node 5 in Fig. 6–12 is an example of a node of the first type; node 6 is a node of the second type. Messages destined to nodes of the second type are routed to the "target" node to which they are connected. In Fig. 6–12, messages destined for node 6 have node 5 as a target node. This is somewhat similar to Telenet hierarchical routing.

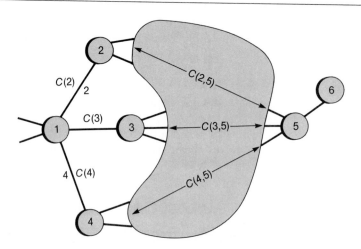

Figure 6–12 Routing example, TRANSPAC

The routing tables for the network are constructed in an essentially central-ized fashion, using a minimum-cost, i.e., shortest-path, criterion. Link costs are defined in terms of link resource utilization. Thus the cost assigned to a link varies dynamically with the network load. We shall describe the method of evaluation of link cost first and then the routing algorithm [SIMO]. Consider a full-duplex link k connected to nodes m and n. Let $C_m(k)$, $C_n(k)$ be the cost assigned to link k as perceived by nodes m and n, respectively, and let $C(k) =$ Max $[C_m(k), C_n(k)]$ be the "combined" estimate of link cost. The quantities $C_i(k)$ are the basic data on which routing computation is based; they are determined locally by each node's CU, which gathers estimated and measured data from its associated SUs. Link cost is defined as a function of the level of utilization of two types of resources: line capacity and link buffers. The utilization of these quanti-ties is evaluated by both estimation (based on the parameters of the active VCs using the link) and measurement. The cost $C_i(k)$ is set to infinity if either the link is carrying its maximum permissible number of VCs or it has exceeded a preset threshold of buffer occupancy. Otherwise, $C_i(k)$ is defined as a piecewise-con-stant increasing function of average link flow, quantized to a small number of levels and including a "hysteresis" effect. A typical function is shown in Fig. 6–13, with the arrows indicating the way that link cost changes as a function of changing utilization. The nodes send updated values of their $C_i(k)$'s to the NMC whenever a change occurs; these events are infrequent due to the combined effect of coarse quantization and hysteresis. At the NMC, the costs perceived by the nodes at both ends of each link are compared to form $C(k)$ as just defined.

The major part of the routing computation takes place at the NMC, but some local information is used at each node. The procedure is illustrated by an example in Fig. 6–12, in which a call packet arriving at node 1 (which may be either the originating node or an intermediate one) is to be forwarded through one of the adjacent nodes 2, 3, 4 to the target node 5, and finally to the destination node 6. Let $C(k,n)$ (computed by the NMC) be the total cost asso-ciated with the minimum cost path between nodes k and n. Node 1 determines the "shortest" route to node 5 by choosing the value of k that minimizes $C(k,5) + \text{Max } [C(k), C_1(k)]$, $k = 2, 3, 4$. In this way node 1 chooses the interme-diate node that would have been chosen by the NMC unless the value of $C_1(k)$ has changed recently. Ties are resolved by giving priority to the shortest-hop path. Because of the way in which link costs are defined, the routing procedure becomes a minimum-hop method upon which is superimposed a bias derived from the level of link-resource utilization.

[SIMO] J. M. Simon and A. Danet, "Controle des ressources et principes du routage dans le reseau Transpac," *Proc. Symp. on Flow Control in Computer Networks*, Versailles, Fr., Feb. 1979; J. L. Grange and M. Gien, eds., North-Holland, Amsterdam, 33–44.

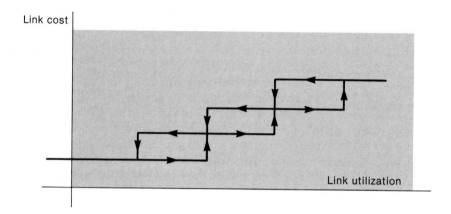

Link cost

Link utilization

Figure 6–13 A typical link cost relation, TRANSPAC

The TRANSPAC routing algorithm has many of the features of a typical centralized-routing procedure, but just as in GTE Telenet its operation departs from being purely centralized by allowing the final routing decision to be made locally, based on a combination of centrally and locally determined information. This too is similar to the concept of "delta routing" suggested by Rudin [RUDI 1976].

By examining the topology of the network, one can deduce the order of magnitude of the computational load at the NMC. The $C(k,n)$'s must be determined for all k and n belonging to the subset of all possible target nodes. For a 12-node network in 1980, only 6 nodes were in this category. Furthermore, rather than doing a complete shortest-path computation to determine these quantities, the designers chose to limit the shortest-path computation to a minimization over a prescribed subset of four or five paths joining each pair of nodes. Thus the computation of all pertinent $C(k,n)$'s involved at most 75 path-length evaluations.

6–3–2 Datagram-oriented Networks

As noted in the introduction to this section, most current networks and network architectures tend to fall in the category of either virtual circuit-oriented or datagram-oriented networks. Routing in a representative group of VC-type networks has just been described. We now move on to a similar discussion of some datagram-oriented networks and network architectures. We start with the ARPAnet, the best known of these networks and a forerunner of all of them. It was noted earlier in this chapter that the original ARPA routing algorithm had to be changed because of problems with stability. That algorithm, the distributed version of algorithm B, serves as the basis for routing in the Canadian Datapac network and in the two network architectures, BNA (Burroughs) and DNA (DEC), that we describe after the discussion of ARPAnet. Questions of stability do not arise in these cases because no attempt is made to estimate time delays over short intervals of time.

ARPAnet   ARPAnet, a pioneering experiment in computer networking, began operation in 1969 with four nodes interconnected. By 1983 the network contained more than 100 communication nodes interconnecting considerably more than that number of host computers. The network, conceived and directed by the U.S. Defense Advanced Research Projects Agency (DARPA) as a research project into all aspects of computer and data networking, has played a pivotal role in the development of packet-switched technology worldwide. ARPAnet is no longer considered a research network but rather handles a great deal of operational traffic for the Defense Department and other U.S. government agencies as well as university computers connected to it. However, a great deal of research activity still goes on using the network, particularly in the area of higher-level protocols.

Details of earlier aspects of the ARPAnet design philosophy, topology, architecture, and network protocols appear in [SCHW 1977] and in references cited therein. Here we focus primarily on routing within the network.

As already noted a number of times, a new routing algorithm was introduced in 1980 after a series of "patches" to the original algorithm could not solve problems encountered over the years of its use. The new algorithm uses a modified version of Dijsktra's shortest-path-first algorithm, our algorithm A, running independently on all nodes. In essence, each node keeps a complete (global) topological database that is updated regularly as topological changes occur. The ARPA routing philosophy has consistently been that of routing packets (datagrams) along paths of (currently) estimated minimum time delay, so that time-delay estimates must also be disseminated when significant changes occur. Both new and old algorithms have been well documented; we provide a summary here. Details of the new algorithm appear in [McQU 1980] and [ROSE 1980]. A discussion of the old algorithm appears in [McQU 1978].

In the old ARPA routing scheme, nodes *cooperated* in carrying out the shortest-path computation. In the new scheme each node does the calculation for itself. The basic problem is then to maintain (reliably) the same database at

---

[ROSE 1980] E. C. Rosen, "The Updating Protocol of ARPAnet's New Routing Algorithm," *Computer Networks*, vol. 4, Feb. 1980, 11–19.

each node and to disseminate changes rapidly and reliably to the databases at all nodes.

There are three parts to the algorithm:

1. The *measurement process,* that of estimating time delay. In the old algorithm this was done about two times a second, leading to estimates that were sometimes statistically not significant and hence to instabilities in the routing. In the new algorithm estimates are made over an interval of 10 seconds.

2. The *update protocol* for disseminating information measured. A flooding technique is used here to reach all nodes in the network as rapidly as possible.

3. The *shortest-path calculation,* by each node, to set up the routing paths. A modified version of the Dijskstra algorithm (algorithm A) is used here, with some pruning used to eliminate unnecessary calculations (i.e., no calculations are made for portions of a shortest-path tree rooted in a given node that are unaffected by topological or cost changes in the network).

Measurement of the time delay at a given node is carried out on a packet-by-packet basis by time-stamping individual packets with the arrival time and the sent time (the time of the final successful transmission if retransmissions are needed). This provides the time delay at the node. To this time delay is added the transmission time (done by table lookup indexed by packet length and line speed) and the propagation delay, a known constant for each outgoing link. Averaging these packet delays over a period of 10 seconds for a given outgoing link provides the statistical estimate of time delay over that link. A new estimate is transmitted to all neighbors, for subsequent flooding throughout the network, if the new estimate differs from the previous one by more than 64 msec. If this threshold is not exceeded, the process is repeated at intervals of 10 sec with the threshold reduced by 12.8 msec each time. This ensures at least one update every 60 sec. However, when a topological change occurs (e.g., a line goes down or comes up), an update is sent out immediately.

The update policy, as just noted, uses a flooding procedure: Update packets, averaging 176 bits in length, are sent out, as received, over all lines, including the ones from which they are received. The "echo" thus serves as an acknowledgement. Each packet carries a 6-bit sequence-number field. Since this number (up to 64 outstanding update packets from a given node) may not be sufficient to prevent problems with cyclic wraparound, particularly if part of the network goes down for some time, an age field, initially set to 64 sec, also is contained in each packet. This field is decremented at each node once per 8 sec. When the age field goes to zero, the update packet is considered too old and is flushed out of the network.

In an interesting paper analyzing possible synchronization problems that might arise with a routing algorithm such as that of ARPAnet, Radia Perlman of DEC notes that the 8-sec decrement interval may be too coarse [PERL]. If a packet stays at a given node less than 8 sec it is not decremented. In this case an update packet could conceivably loop "forever" without being detected as such. However, no such problems have been noted in the ARPAnet thus far [McQU 1980], [ROSE 1981]. The one major problem encountered, which shut the network down for several hours, involved a hardware malfunction in reading the sequence number from tabular information [ROSE 1981]. (In two successive retransmissions of an update packet, bits were apparently dropped, causing the packet to appear like three update packets. These packets flooded the network, filling buffers with cyclic repetitions of the three packets.) Duplicate update packets are not flooded further. Measurements under normal operation at the time the network was 60 nodes in size indicate that an update takes about 100 msec to propagate through the network.

As an additional feature, to ensure appropriate resynchronization after a failure, a line that has been down and has been restored is put into a wait state for one minute, with no data transmitted over it. Routing packets are transmitted, however. This is done to ensure that an update packet from each node in the network (generated at least once every minute) will have been sent over the network.

The routing algorithm, running independently at each node and using the global database generated by the update packets, generates a shortest-path tree rooted in the node in question. As noted earlier, the algorithm is a modified form of our algorithm A [McQU 1980]. It makes selective and incremental changes in the tree only. For example, if a delay on a given link that is *not* in the tree increases, no computation is made. Other conditions where no recalculation needs be done are singled out as well [McQU 1980].

The running time of the algorithm at a node corresponds to the size of the subtree in nodes, which turns out to be the same as the average path length in hops [McQU 1980]. This is in turn proportional to $\log n$, with n the number of nodes in the network. For the 60-node ARPAnet, the average running time turned out to be 2 msec (with an actual range of $1-40$ msec). The total update processing time (including maintenance of the topological data base) averaged 4 msec. The storage requirement for the algorithm — the global database and the algorithm itself — is 2000 16-bit words. To put this in perspective, the absolute minimum storage per destination for a very simple algorithm such as our distrib-

[PERL] R. Perlman, "Fault-tolerant Broadcast of Routing Information," *Computer Networks,* vol. 7, no. 6, Dec. 1983, 395–405.

[ROSE 1981] E. C. Rosen, "Vulnerabilities of Network Control Protocols: An Example," *Computer Comm. Rev.,* July 1981, 11–16.

uted version of algorithm B is $C \log n$ bits for the address and cost parameters, indexed by outgoing link. n is the network size, in nodes, and C is some constant. The total storage is thus $Cn \log n$. For $n = 60$, this is $360C$. For three outgoing lines per node, and five to eight levels of cost quantization, $C \doteq 9$. Thus 3600 bits (one-tenth the size of the ARPA tables) are required. The old ARPA algorithm required about one-third the storage of the new one; this indicates the added cost in memory required to maintain a *global* database at each node.

Measurements indicated that the line overhead for carrying out the routing averaged less than 1 percent while CPU overhead was less than 2 percent. The algorithm tended to route traffic on minimum-hop paths. (These numbers obviously depend on network traffic. The minimum-hop routing obtained in practice probably indicates low utilization.) The algorithm is not loop free because of inconsistent distributed databases during the update cycle, and a smaller number of loops have been detected during the transient intervals while the databases and routing tables are being updated.

Datapac   The Datapac network, the TransCanada Telephone System's packet-switched network, provides end-to-end virtual circuit (VC) service to users accessing the network but differs from GTE Telenet and TRANSPAC in using datagram routing to accomplish this. Specifically, a user accessing the network (at either end) sees an X.25 virtual circuit interface. A call to a destination point must be set up before data transfer can take place. The end-to-end VC service prescribed means, of course, that packets must be delivered to the destination in the order in which they are sent, without loss and with no duplicates. The underlying mechanism for accomplishing this, however, is that of datagrams that are routed independently throughout the network. The end nodes must therefore be responsible for restoring any datagrams lost during transmission, reordering datagrams that arrive out of sequence, and discarding duplicates of datagrams already received. We now summarize methods for converting the basic datagram network service to a logical end-to-end VC service. Details appear in [SPRO].

A user who accesses the Datapac network appears to see the same end-to-end VC service as in GTE Telenet and TRANSPAC. The difference, of course, is that in those networks a prescribed route is actually set up for each VC, with data packets then constrained to follow the path selected. Here the VC exists logically end to end but datagrams are sent through the network independently.

The Datapac network began service in mid-1977 with five nodes and by mid-1980 had grown to 14 nodes interconnected by 24 56-kbps trunks [SPRO]. It provides connections to TYMNET and GTE Telenet in the United States, as well as to the Canadian international carrier Teleglobe for transcontinental traffic.

We now summarize the actual handling of a call. As just noted, standard X.25 procedures are used for the VC call-establishment, data-transfer, and disconnect phases. As an example, an X.25 call-request packet received from a DTE on a customer line on a free logical channel creates a source VC process. A datagram delivered to the network is then used to convey the call request to the destination and the appropriate (VC) destination link. If a free logical channel is available at that link, a process is created there to handle the VC, and this process sends the call-request packet on to the destination DTE. If the latter responds with a call-accepted packet, an acknowledgement is returned through the network (again in datagram form) to the source VC process, which notifies the source DTE of the establishment of the VC. Note that these procedures are precisely those described in Chapter 5 for the X.25 protocol.

In the data-transfer phase, VC flow-control parameters determine whether a data packet from the customer line is to be accepted and buffered into local-line processor memory. Packets, under end-to-end VC flow control, are then individually injected as datagrams into the network. Each datagram is individually acknowledged on arrival at the destination. This is required to provide the necessary end-to-end VC service. If the acknowledgement is not received at the source before a 3 sec timeout, the datagram is retransmitted. The Northern Telecom SL-10 multiprocessor switch serves as a node in this network. In this configuration local processors control the customer access (X.25) lines. Trunk processors control the network (datagram) trunks. Common memory modules provide the necessary communication between the two classes of processors for X.25/datagram conversion. Figure 6–14, taken from [SPRO, Fig. 4], shows the conversion between X.25 and datagram service.

Datapac routing is similar to the old ARPA routing algorithm, as already noted. Link (trunk) speed is used in assigning a cost to the link. Two tables are again kept. A routing table provides the "best" outgoing trunk group for each destination. A second table provides the end-to-end cost (called a "delay" estimate in the Datapac terminology) to reach each destination node via each outgoing trunk. Routing updates are made, as in the ARPA case, when the absolute change in the "delay" estimate exceeds a threshold. New information is then passed to neighboring nodes, as in the distributed version of algorithm B. The one difference is that the update is not sent back over the transmission group over which it came, to reduce the incidence of loops. As noted earlier, this is the "split-horizon" routing procedure used for the TIDAS network [CEGR].

Congestion control for the Datapac network is carried out both through the X.25 end-to-end VC sliding-window method and through the use of input-buffer limiting. In this latter case, as described in Chapter 5, input (local) datagrams are dropped if the number of common memory blocks available is less than a specified threshold. Both local and transit datagrams may be dropped

VC = Virtual circuit
LP = Line processor
TP = Trunk processor

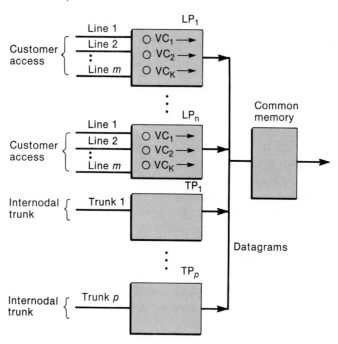

Figure 6–14 Datagram flows into SL-10 common memory (from [SPRO, Fig. 4] with permission. © 1981 IEEE)

if the common buffers begin to be depleted. This causes the datagram 3-sec timeout retransmission strategy to be invoked, except that no more than four attempts (after 12 seconds) are allowed. The VC call is then cleared.

In the case of VC calls that require the use of other networks to reach the destination (internetwork calls, as in the case of international calls), three levels of routing are used. The specific procedure required is established at call-setup time, in the routing of the call-request packet. At level-1 routing, the next (adjacent) network through which to route the call is established. This is done at call-setup time using a static table that maps each destination (remote) network into a specific adjacent network. Once the adjacent network is established, level-2 routing finds the "best" gateway to get to that network. An adjacent network-routing table is used and is maintained by the same (dynamic) process as in normal Datapac routing. Such a table is kept at each node and establishes the "nearest" gateway on the basis of an overall minimum "delay" estimate, based on the number and speed of internodal trunks to get to the gateway as well as the number and speed of links at the gateway. Finally, level-3 routing is the normal Datapac (datagram) routing; it is used to get to the "best" gateway.

BNA (Burroughs Network Architecture)   BNA, described in [BURR], is a completely distributed architecture: All nodes (hosts) are peers of one another. There is no notion of a central manager to help establish communications between users, as in the SNA SSCP. BNA is thus related to and based on the ARPA network philosophy of complete autonomy of nodes. Since all communication must be between neighboring nodes, rather than with a manager as well, as in IBM's SNA, GTE Telenet, and TYMNET, routing can only be carried out on a distributed, decentralized basis.

The architecture sits below the application programs and provides network services for them. There are two functional levels to BNA: host services and network services. Host services are intended to extend normal operating-system functions to the network environment. Six functions, with appropriate protocols for each, are defined: remote file access, file transfer, "operator inquiry," control messages between hosts, logical transfer of terminals between hosts (i.e., the logical attachment of a terminal to a remote host), and remote tasking. Network services, consisting of three levels plus a network services manager, in turn provides the necessary transport functions for host services to carry out the desired interactions between remote application programs and terminals.

The three levels of network services are, in descending order:

- *Port level:* similar to transmission control in SNA or the OSI transport level; designed to provide a reliable transmission path from source to destination.

- *Router level:* similar to SNA path control or the OSI network level; routes messages over the networks.

- *Station level:* provides reliable transmission between neighboring nodes. Two types of stations are recognized: a BDLC station (BDLC is similar to HDLC) and an X.25 station. (Provision is made to have a "link" actually be a public data network.)

User-to-user communication is carried out using a subport at each end. The subports, embedded in a port at one end, are similar to the SNA LU and half-session concepts described in Chapter 3. Subports are created by the port level, using a three-way handshake (similar to the SNA bind triangle) at the time

[BURR] *Burroughs Network Architecture (BNA), Architectural Description, Reference Manual,* vol. 1, Burroughs Corp., Detroit, Mich., April 1981.

of dialogue initialization. Subports carry out message segmentation, sequence numbering, flow control, and data compression. A window mechanism similar to X.25 VC flow control is used to carry out flow control at the subport level.

The prime function of the router level, as the name indicates, is to carry out the routing of messages. The procedure and algorithm used are very much like those used by Datapac. The algorithm is again a version of the simple distributed algorithm (our distributed algorithm B) used originally by ARPAnet, except that time delays are *not* estimated. Here a link cost (called a *link resistance factor*) is determined for each link using a simple formula. This link cost is simply the maximum message-segment size divided by the link-transmission speed. It is further divided by the number of parallel links that accompany the actual physical connection (the SNA transmission group concept described in Chapter 5) and by a link efficiency factor, which varies depending on whether full-duplex or half-duplex transmission is being used. (The factor provides an approximation to the transit time, in msec, over the physical connection.) A nodal resistance factor is added in as well; used as this measure is the approximate message processing delay (in msec) through a given system.

Two routing tables are again kept at each node: One table carries, for each destination and for each outgoing link (to a neighboring node) from the node, the hop count and cost to that destination. An indication of destination reachability (whether the hop count and cost are below allowable thresholds or not) is kept as well. The second table, used for actual routing, indicates, for each destination, whether it is reachable; if it is, the minimum-cost neighbor, cost ("resistance factor"), and hop count are listed. As an interesting addition, the id of the next link is also indicated if this link is actually a public data network. Routing between nodes connected by a public data network is thus allowed as well (X.25 connections must then be made). Provision is made for transmitting two types of routing-control messages: One is transmitted between neighbors if the link characteristics between them change, and the other is used in the routing algorithm to propagate minimum-cost changes throughout the network.

To summarize routing in BNA: The algorithm is of the distributed (decentralized) type, service is of the datagram type, a class-of-service concept does not appear to be used (all messages are routed the same way), and routing tables are updated on an event-driven basis. A network of up to 255 nodes can be accommodated. In the following discussion of DNA (DEC) routing, we shall find the same algorithm used, with the one distinction that DEC uses two-level or hierarchical routing to increase the number of nodes that can be accommodated.

The widespread adoption of this basic distributed algorithm, our distributed algorithm B, in most datagram networks is one of the reasons we have described it in detail. Recall that it is simple, robust, and stable (*if* fixed link costs rather than adaptively varying time estimates per link are used). As noted earlier, this algorithm does require that a relatively large number of control packets be disseminated throughout the network. In addition, although its convergence time property is generally quite good, the convergence time can be shown to increase markedly if links differ significantly in their cost factors. This point will be noted again, more quantitatively, in the last section of this chapter, when we study the performance of the distributed version of algorithm B in more detail.

DNA (Digital Network Architecture)   DNA, the network architecture of Digital Equipment Corporation (DEC), has been evolving over a number of years, with plans to have it conform to the various layers of the ISO OSI Reference Model as they become standardized. In its phase IV embodiment, announced in 1983, DNA encompasses the layers shown in Fig. 6–15. Note that the routing layer, on which we focus here, is comparable to the network layer of the ISO model. The end communications layer corresponds to the ISO transport layer, and the session control layer corresponds, of course, to its counterpart in the ISO model.

Although DNA provides a datagram transport service only through its routing layer, a virtual circuit concept called a *logical link* connection does exist at the end communications and session control layers above. This is comparable to SNA's session and to BNA's port. The session control layer is involved in the establishment (connection phase) and the aborting (disconnection) of logical links. It carries out system-dependent aspects of logical link communications: It does local-name-to-global-address translation, requests logical links for end users, and issues connect requests to the layer below, the end communications layer. The end communications layer in turn is responsible for system-independent aspects of logical link communications. Since the DNA architecture is predicated on datagram routing, the end communications layer must assume that datagrams passed up to it from the routing layer below may arrive out of sequence and that some datagrams may not arrive at all. The end communications layer is thus responsible for reordering out-of-sequence datagrams and requesting retransmission of datagrams if they are missing. To carry out these functions and others required for logical link communications, this layer does sequence-number checking and error-control segmenting, manages the logical links (allocates buffers required by the routing layer, for example), and carries out logical link-level flow control (comparable to SNA session-pacing control).

The Datapac approach of providing the end user with a virtual service end to end, but using a datagram network to provide the underlying transport service, is similar to the DNA approach. The logical link in the DNA case corresponds to the X.25 VC in the Datapac case. ARPAnet uses a similar approach.

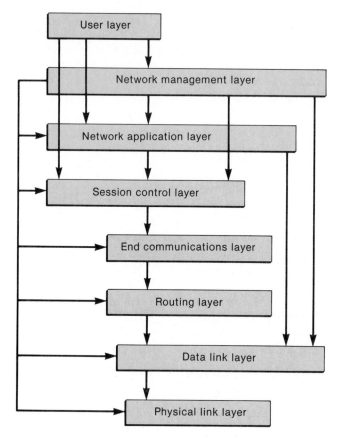

Figure 6–15 Layering in DNA, phase IV, 1983

As in all layered architectures, the end communications or transport layer communicates with the routing layer below via interface commands and responses. Datagrams, the basic units of data passed down to the routing layer by the end communications layer, have a routing header added to them. The resultant packet is the unit of data routed through the network by the routing layer at each node visited.

Routing in DNA is carried out in a two-level hierarchical fashion, using the same distributed routing algorithm at each level [DEC]. The algorithm is pre-

---

[DEC] *DECnet, Digital Network Architecture, Routing Layer Functional Specification,* Version 2.0.0., Digital Equipment Corp., Maynard, Mass., May 1, 1983.

cisely our distributed version of algorithm B, the old ARPAnet algorithm. As is true with all commercial implementations and as already noted, link costs (*circuit costs* in DNA terminology) are traffic independent. This is done precisely to ensure a stable routing procedure.

A typical DNA network, showing level-1 areas connected to a level-2 backbone network, appears in Fig. 6–16. Note that the two-level hierachy concept is similar to that used in GTE Telenet, except that there VC routing rather than datagram routing is used. Each node in a DNA network belongs to a specified area. The example of Fig. 6–16 shows three such areas. Least-cost distributed routing is carried out independently in each area. Certain nodes are designated level-2 nodes (see Fig. 6–16). They participate in level-1 routing in their respective areas and also carry out least-cost routing among themselves. The level-2 backbone network in Fig. 6–16 is shown emphasized with double lines.

A packet with a destination address in an area other than its own is routed first to the nearest level-2 node in its own area. It then follows the least-cost path to the required foreign area, after which it is again routed via a least-cost path to the destination node. Sixteen-bit global destination addresses, consisting of a 6-bit area address and a 10-bit address within an area, are used throughout. There are thus at most 63 areas and 1022 nodes within an area that may be addressed. (The destination address 0 within an area is always interpreted to mean the nearest level-2 router.) The function of partitioning a large network into areas is left to the network manager. See [HAGO] for a thorough discussion of routing in hierarchies, including methods of aggregating or clustering nodes to form areas. Earlier work appears in papers by Kleinrock and Kamoun [KLEI 1977], [KLEI 1980]).

Although link costs are unspecified in DNA (they are left up to the implementor of a network), a suggested set of costs is provided. As in the BNA case, proposed costs are made inversely proportional to link capacity. Thus a maximum value of 25 is suggested for links with capacity less than 4 kbps, one for links with capacity greater than or equal to 100 kbps, and rounded integer numbers inversely proportional to line speed for speeds in between. The algorithm thus tends to steer packets to the highest-speed lines from source to destination. A similar though coarser assignment strategy was described earlier for TYMNET. Unlike TYMNET and SNA, however, DNA makes no class-of-service distinction between types of messages to be delivered. All packets are treated alike, no matter what data they carry. The implementor of a DNA

[KLEI 1977] L. Kleinrock and F. Kamoun, "Hierarchical Routing for Large Networks: Performance Evaluation and Optimization," *Computer Networks*, vol. 1, 1977, 155–174.

[KLEI 1980] L. Kleinrock and F. Kamoun, "Optimal Clustering Structures for Hierarchical Topological Design of Large Computer Networks," *Networks*, vol. 10, 1980, 221–248.

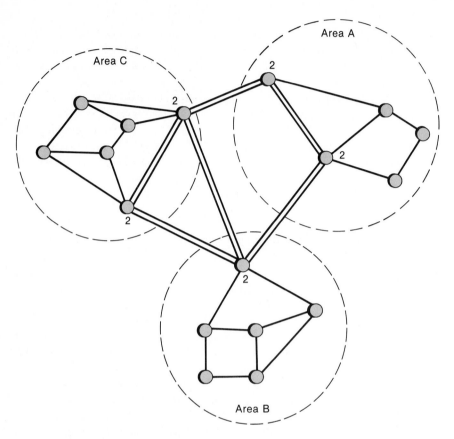

Figure 6 – 16 Hierarchical routing in DNA: an example

network specifies the cost assignment per link, as just noted. In addition, a maximum path cost (maxcost, sum of the link costs) must be specified at each of the two routing levels. This is again left to the implementor's discretion, with the one architectural stipulation that the maximum path cost be no more than 1022 (10 bits) at each of the two routing levels. Cost fields in the routing central messages are, correspondingly, 10 bits long.

As noted in our discussion of the distributed version of algorithm B, two tables must be kept at each node. Level-2 routing nodes must keep two such sets, one for each level of routing. Corresponding to the distance table in our earlier discussion is the *routing database* in DNA terminology. The subset that contains the least-cost outgoing link for each destination, the routing table in our earlier discussion, is called the *forwarding database* in DNA terminology.

The routing database at a DNA node contains information in addition to path cost to each destination via each outgoing link. The hop count (path length) to all destinations via each link is kept as well. It is used to ensure reachability of the different destinations. The minimum-cost entry for each destination and the path lengths corresponding to each are stored. The routing data base also keeps a record of each neighboring node's id and type (whether level 1, level 2, or an end node—a third category not involved in routing decisions).

DNA is designed to operate in a network environment that involves not only the point-to-point connection of peer-type packet-switched nodes, but X.25 networks and multiaccess local area networks as well. (Local area networks will be discussed in detail in Chapter 9.) Recall that the BNA architecture accommodates X.25 "nodes." In the DNA case the links themselves (*circuits*, in DNA terminology) are classified as to whether they are connected to X.25 networks or Ethernet networks (a type of local area network described in Chapter 9), or to normal nodes. This information is kept in the routing database as well. (It is worthwhile to point out here that the DNA link-level protocol is a proprietary protocol called the Digital Data Communications Message Protocol (DDCMP). Details appear in [WECK].)

It was just noted that the routing database contains hop-count information in addition to cost information for each destination. In the operation of the distributed routing algorithm both cost and hop-count numbers are exchanged with neighbors. If a specified hop-count threshold maxh (for a given destination) is exceeded at either routing level, that destination is declared to be unreachable, and packets marked for that destination are either dropped or returned to the sender. Although the specific value of maxh is implementation dependent, a suggested value is twice the largest path in the network. However, it cannot exceed the value 30. The hop-count field in routing control messages is thus kept to 5 bits. (This architectural specification is prescribed at both levels of routing.) The hop-count specification provides a safety feature in the convergence of the routing algorithm. Without it, as we shall see by example in the next section, convergence of the algorithm could take an inordinate amount of time in certain cases and for certain network configurations.

The forwarding database, corresponding to the routing table of our distributed algorithm B as already noted, is constructed from the routing database and consists, as previously explained, of the least-cost outgoing link for each destination, *if reachable.*

To carry out the necessary construction of these databases, to propagate routing-control messages to neighbors, and finally to route data packets appro-

---

[WECK] S. Wecker, "DNA: The Digital Network Architecture," *IEEE Trans. on Comm.*, vol. COM-28, no. 4, April 1980, 510–526; reprinted in [GREE].

priately using the information in the forwarding database, a number of processes are specified as part of the DNA routing layer. First is the *decision process,* whose function is to select the appropriate paths and maintain both the routing and the forwarding databases. It receives routing messages, uses them to carry out the decision algorithm (the distributed version of algorithm B), and then forwards any modifications to the routing and forwarding databases, if necessary. In carrying out the algorithm it first computes the minimum-cost path, then computes the path length (hop count), and finally determines whether *both* cost *and* length are within the specified limits set. In addition to forwarding modifications to the two databases, the decision process also notifies an *update process* to send routing messages to all neighbors when necessary.

The basic routing algorithm is event driven, responding to a change in link or neighboring-node cost and/or configuration, as noted in our earlier discussion of the distributed routing algorithm. However, provision must be made for periodic exchange of routing messages. This point was also mentioned in our discussion of the new ARPA algorithm. This change is necessary to prevent problems due to possible corruption of the routing database. (This could happen because of memory failure, a corrupted routing message, or a software error.) Transmitting routing messages periodically is one way to ensure the data bases are valid and up to date. The DNA architectural specification sets a *maximum* time of 10 min between such routing updates [DEC]. The *minimum* time between routing messages of any type is set at 1 sec.

A *forwarding process* in the routing layer uses the entries in the forwarding database to forward data packets to the appropriate neighbor. A *receive process* receives packets from the data-link layer and makes the decision as to where to send them. If the packet is a routing message, it is sent to the decision process. If it is a data packet, it can be destined for another node, in which case it is sent to the forwarding process, or if the current node is the destination, the packet is sent up to the end communications layer.

A conceptual diagram of these various processes and their interaction with the databases appears in Fig. 6–17.

In discussing the properties of the distributed version of algorithm B in the previous section, it was noted that this version is prone to looping. Data packets propagated during a convergence interval of the algorithm may find themselves returning to nodes already visited. If kept within bounds this may not pose too much of a problem. If packets loop too long, however, they become out of date and should be discarded. To accomplish this function, a node-visit field in the packet-route header is incremented by 1 at each node visited. If this number exceeds some maximum count maxv, the packet is deemed too old and is discarded. A suggested range for maxv is any number between maxh + 2 and 2 maxh. (Recall that maxh is the maximum hop count specified.) The absolute maximum value is 63, however, which keeps the number within a 6-bit packet-visit field [DEC].

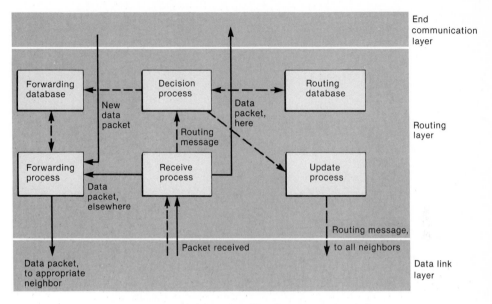

Figure 6-17 Routing layer processes, DNA
——— Data packets
------- Control paths

Two types of messages are handled by the routing layer, as already noted and as shown in Fig. 6-17. Data packets consist of datagrams handed down by the end communications layer above, to which a packet-route header is added. Routing messages are the control messages exchanged between neighbors that are used to update the routing algorithm. There are two types of data packets used in DNA — one with a 6-byte route header used for point-to-point or multi-point transmission (X.25 or normal data link layer protocol) and one with a 21-byte header used for Ethernet (multiaccess or broadcast) transmission. The former is referred to as *short-format* packet; the latter, a *long-format* packet.

The 6-byte route header for the short-format packet appears in Fig. 6-18(a). The format of the level-1 routing message appears in Fig. 6-18(b) [DEC]. The flag byte is used to distinguish between types of packet and, in the case of a data packet, whether it is to be returned to the sender if rejected. A 0 in bit 0 of the flag field indicates a data packet, a 1 indicates a control packet (a level-1 or level-2 routing message). Thus a short-format packet is denoted by 2 (010); a long-format packet, by 6 (110). The 2-byte address fields in the short-

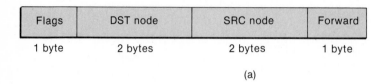

Flags	DST node	SRC node	Forward
1 byte	2 bytes	2 bytes	1 byte

(a)

Flags	SRC node	Reserved	Segment ...	Check sum
1 byte	2 bytes	1 byte		2 bytes

(b)

Figure 6–18 Formats, DNA routing layer (from [DEC] with permission. © 1983 Digital Equipment Corporation, all rights reserved)
a. Route header, short format packet
b. Level 1 routing message

format packet header of Fig. 6–18(a) contain the destination and source addresses, respectively. Six bits of the 1-byte forward field represent the nodes-visited field mentioned earlier. (This field is incremented by 1 at each node visited.)

The level-1 routing message of Fig. 6–18(b) consists of the source node address (the node sending the message), a 1-byte reserved field, and then a segment section containing a list of destination ids, with a 5-bit hop count and a 10-bit cost field for each. (Recall that destination 0 is used to designate the nearest level-2 router.) The 2-byte check sum shown is carried out over the segment section only and provides added security for the information carried therein. The level-2 routing message format is identical, with area ids used for the destinations. (Recall that this message is used to find the least-cost path in the backbone of a level-2 network.)

We have considered the routing function of the DNA routing layer only in this subsection. The routing layer also carries out congestion control, as is the case with all network layer embodiments. As noted in Chapter 5, DNA uses an input-buffer-limiting congestion-control mechanism in accordance with the datagram routing philosophy. However, input-buffer limiting is done on a per link (circuit) basis. Thus input (or originating) packets from the end communications layer are blocked if the number on a given output link queue exceeds an originating packet limit (OPL). Each OPL is initialized to the number of buffers necessary to present the circuit from idling [DEC, Sec. 5.2]. In addition, all packets, transit plus input, are subject to a so-called square-root buffer control:

A packet is rejected if the link (circuit) output queue exceeds a threshold given by the ratio of all nodal buffers to the square root of the number of active output links (circuits) [DEC]. An analysis of this square-root control strategy appears in [IRLA]. A comparison of input-buffer limiting by output link with limiting for the node as a whole appears in [SAADS]. Both references were also cited at the end of Chapter 5 in the discussion of input-buffer-limiting congestion-control mechanisms.

This discussion of the DNA routing layer, with particular emphasis on the datagram routing function carried out by this layer, concludes our discussion of routing procedures in some typical networks and network architectures. Note that the datagram-oriented networks and architectures, with the exception of the current ARPAnet, use as the routing algorithm the distributed version of algorithm B (the old ARPAnet algorithm) or a variant. As noted previously, this algorithm is simple, robust, and converges rather quickly in most network situations. It is prone to looping, however, and requires a relatively large number of routing-control messages to be exchanged during its convergence interval. In the next and final section of this chapter we provide a more formal performance evaluation of the algorithm and algorithms related to it. We conclude that the algorithm is quite suitable for relatively small networks, although problems may arise if link weights (costs) differ by large values. Modifications to the algorithm can reduce the number of control messages and incidence of loops somewhat. By going to hierarchical routing, as DEC has done, one can use the algorithm for routing over considerably larger networks. The repeated use of the algorithm, first at level 1, then at level 2, then again at level 3, no longer provides a guarantee that the complete end-to-end path from source to destination is of least cost. J. M. Jaffe of IBM Research has shown that the maximum increase in path *length* (hops, not cost) is three times if hierarchical rather than shortest-path routing is used [HAGO].

6-4 Performance Analysis of Distributed Routing Algorithms

In Section 6-2 we described two basic routing algorithms, our algorithms A and B, that serve in one form or another as prototypes for most shortest-path algorithms currently in use. Algorithm B, in particular, in its distributed form is

[IRLA] M. I. Irland, "Buffer Management in a Packet Switch," *IEEE Trans. on Comm.*, vol. COM-26, no. 3, March 1978, 328–337.

[SAADS] S. Saad and M. Schwartz, "Input Buffer Limiting Mechanisms for Congestion Control," ICC, Seattle, June 1980.

used in a number of networks and network architectures, as noted in the previous section. It is quite simple and easy to implement but does suffer from a number of potential problems, as noted earlier: It is prone to looping and can generate a sizable number of control messages during its convergence phase. In this section we attempt to quantify some of the earlier discussion. We also describe briefly some additional algorithms suggested in the literature that are either loop free or that reduce the number of control messages transmitted during the routing-table convergence interval.

A formal quantitative comparison of shortest-path distributed-routing algorithms is rather difficult to carry out, for two major reasons. First, there are a number of performance criteria that one might want to use to assess performance. The performance of various algorithms may very well differ; some may have better performance on certain objective functions; others, on other objective functions. Second, the relative performance of algorithms depends on specific network configurations. One can obtain general analytic results for certain regular graphs only. For "real" network configurations one must rely on computer analysis and/or simulations. In this section we provide general results only for a full-duplex loop of n nodes. We also refer to some computer analysis of a more complex network to provide some added insight into the performance of some algorithms.

What are some criteria one might want to use to evaluate the performance of a given shortest-path algorithm? Note that so long as the network is quiescent, routing tables remain unchanged. All shortest-path algorithms provide the same results for the routing tables. The differences in the various algorithms arise when they adapt to changes in network topology. We must thus compare algorithms as to how they perform in response to a change. Various criteria can be listed:

1. *Speed of response.* This is the time required for a change in topology to propagate through the network and for the routing tables to settle down to a new quiescent, shortest-path state. In this section we use as a measure of speed the average number of iterations required for the algorithm to converge to a new state.

   This performance measure is obviously extremely important in a dynamic environment, for the speed of response must be faster than the rate of change of network topology. Otherwise convergence will not occur and the routing algorithm will be useless.

2. *Number of control packets transmitted.* This is the information that must be transmitted networkwide to propagate the desired topological change and routing-table updates. The larger the number of control packets, the larger the overhead and the greater the possible congestion introduced by transmitting topological and routing change information.

3. *Computational complexity.* Algorithms require varying amounts of computation at a node to process the control packets, carry out the shortest-path computations, and update the routing tables. This also is reflected in processing time at a node, and hence in processing speed.

4. *Size of control packets.* Algorithms differ as to the amount of information transmitted per control packet.

5. *Buffer space required.* Algorithms differ as to memory space required to update and hold the routing tables.

6. *Looping and loop freedom.* "Loop freedom" refers to the transient, converging state of a routing algorithm. A loop-free algorithm ensures that data packets will not return to nodes already visited while the routing tables are in the process of changing.

In our quantitative comparison in this section we focus primarily on criteria 1 and 2: the speed of response, in number of iterations before the algorithm converges, following a topological change, and the total number of control packets networkwide that are transmitted up to the time of convergence. We compare four algorithms on the basis of these criteria. One is the distributed version of algorithm B; another, a decentralized loop-free algorithm proposed by Merlin and Segall [MERL]. As we shall see, this latter algorithm is more complex than the distributed algorithm B and converges more slowly, but requires fewer control packets to be transmitted and is, as noted previously, loop free during the convergence period. A third algorithm, called the predecessor algorithm, is a simple modification of the distributed algorithm B that reduces the number of control packets substantially. It is similar to the "split-horizon" algorithm used in the TIDAS and Datapac networks, noted earlier [CEGR], [SPRO]. In this algorithm not only is information kept as to the shortest-path outgoing link of a node, but the upstream or predecessor nodes in the shortest-path route to each destination are labeled as well. The fourth algorithm, which we mention for only the sake of reference, could be any one of a number of shortest-path algorithms. In a sense it is an idealized version of the ARPAnet algorithm [McQU 1980]. It assumes that each node has global topological information and can thus calculate shortest paths to all other nodes in the network. It serves as a basis for comparison of the three other algorithms just described. This algorithm is readily seen to be the "best" possible for criteria 1 and 2, and that is why we choose it as a basis for comparison. Since every node is assumed to have complete topological information, control packets can be sent along least-time-delay paths. (In the special case of fixed-cost links, these would correspond to minimum-hop paths.) The time for convergence is then just the time to cover the diameter of the network to the furthermost node. *No other algorithm can do better.* Again because of complete topological information only one control packet per node in the network need be sent. For an n-node network this corresponds to $(n - 1)$ packets. This is the smallest number possible required to notify all other nodes in the network of a change in link cost or network topology.

We first review some of the problems already noted with distributed algorithm B, then describe the two other algorithms briefly, focusing on the way in which they attempt to alleviate one or more of the problems. We also describe another loop-free algorithm, proposed by Jaffe and Moss [JAFF]. We then compare performance for the full-duplex loop noted previously.

6–4–1 Distributed Algorithm B

Among the problems associated with this algorithm, as noted earlier, are a tendency to loop during convergence and a large number of control packets issued during the convergence period. For a network of n nodes and the special case of minimum-hop routing (equal cost on each link), it is easy to see that the number of control packets is $O(n^3)$. ($O(\ )$ means "order of.") The convergence time for this algorithm in the special case of minimum-hop routing is the time for control messages to propagate to the outermost part of a network (the diameter of the network) and then back again, signaling the initiating node that all nodes have updated their tables. This is about double the time required for the "best" algorithm noted earlier. Since the diameter of the network is bounded by n, the number of nodes in a network (the pathological case in which all n nodes are strung out in a link), the convergence time is $O(n)$ iterations. At each iteration during convergence, each node in the network may be sending a control packet for each destination to each of its neighbors. This implies $O(n^2)$ packets per iteration (n nodes sending n packets each) or, finally, $O(n^3)$ packets during convergence, as just noted. Improvements in the algorithm will attempt to reduce this relatively large number of packets.

The looping problem and other potential problems associated with the algorithm can be pinpointed with some simple examples. Some of these may appear to represent pathological conditions, but they could be encountered as portions of much larger networks. Consider first the equal link cost, linear chain of Fig. 6–19(a). Take node 1 as the destination node. Then node 2 will point directly to node 1, with a cost of 1; node 3 will point to node 2, with a cost of 2; and node 4 will point to node 3, with a cost of 3. Node 2's distance table will show another entry, however: It will point to node 3, with a cost of 3. Node 3 will show an entry pointing to node 4, with a cost of 4. Now let link 1–2 fail. Communication to 1 should cease, since there is no alternate path. Yet node 2, thinking that there is another path to 1 via 3, with cost 3, will send a control message to that effect to 3. Node 3 will in turn update its tables and send messages to 2 and 4. The details are left to the reader, but a little thought will indicate that unless

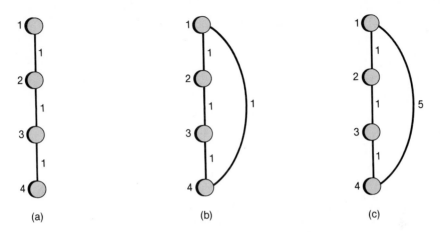

Figure 6-19 Simple network examples
    a. Linear chain, equal costs
    b. Full-duplex loop, equal costs
    c. Full-duplex loop, unequal costs

the maximum cost in any table is bounded at some finite number maxcost, control messages will keep propagating back and forth forever! The predecessor algorithm described in the next section eliminates this specific problem.

The steady increase of costs in Fig. 6-19(a) to a maximum value maxcost can be eliminated by connecting node 1 to node 4, as shown in Fig. 6-19(b), which allows alternate path capability and converts the network to a full-duplex loop. Consider node 1 as the destination and let a failure of link 1-2 occur again. It is left to the reader to show that the table entries stabilize to the new least-cost entries, 2 pointing to 3, 3 to 4, and 4 to 1, in one iteration. Only one message, sent from 2 to 3, is required in this case. But now consider the same network with one of the links, 1-4 in this example, increased to a cost of 5, as in Fig. 6-19(c). It is left to the reader to show that the algorithm now requires five iterations to converge, with eight control messages transmitted. Convergence time thus depends not only on topology but on link weights as well. The more unequal the weights, the larger the convergence time. Tajibnapis has noted this point in suggesting that a link of higher cost may be considered to act like an equivalent set of unity cost links in casade [TAJI]. In the limit, as the cost of link 1-4 in Fig. 6-19(c) gets larger and larger, the network approaches that of Fig. 6-19(a), with convergence time after a failure (or increase in link cost) limited by setting a maximum cost maxcost.

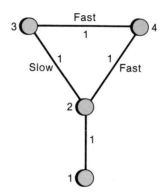

Figure 6-20 Effect of unequal delays

Implicit in all of these examples, including those discussed earlier in the chapter, is the assumption that transmission and/or propagation delays are the same for all links. Control packets thus move precisely one hop every iteration. In reality this is not the case, of course, and variations in link speed or propagation delay may lead to unexpected results as well. Consider as a simple example the network of Fig. 6-20. Let link 2-3 be a slow link, as indicated, while all the other links are fast. With node 1 as the destination node nodes 3 and 4 point to node 2, and node 2 in turn points directly to node 1. Now let link 1-2 fail. Node 2, thinking that it has alternate paths through nodes 3 and 4, with a cost of 3 each (why?), will send a message with cost 3 to both those nodes. The message to node 4 arrives and node 4 then chooses node 3 as the lower-cost (cost = 3) node to destination 1. It sends the cost message to both 2 and 3. Node 3 still has not received the original message from 2 and so still routes to node 2. Data packets originally sent from 4 to 2 may return to 4 to be forwarded to 3, in turn to be sent to 2, and so on. Eventually, when the message from 2 reaches 3, the system will begin changing its routing pattern and, again when costs reach maxcost, the algorithm will terminate. But it is apparent that difficulties may arise not only because of unequal costs on links but because of unequal time delays as well.

6-4-2 Predecessor Algorithm

Some of the problems of the distributed B algorithm can be alleviated by changing the algorithm slightly or by adding some additional information to be transmitted. The split-horizon technique mentioned earlier provides one such approach [CEGR]. Hagouel has suggested a number of others [HAGO]. We focus here on a related algorithm, called the predecessor algorithm [SCHW 1980c]. This algorithm requires $O(n^2)$ control packets to be transmitted.

The basic idea is to eliminate the unnecessary passing back and forth of control messages between neighboring nodes in the distributed B algorithm while convergence is taking place. These unnecessary control packets are generated because of the "blindness" of that algorithm. When a previous least-cost path is declared inoperable, a node chooses the second-best least-cost outgoing link. This could very well lead to a proposed path back through the node itself, as noted in a number of examples in this chapter. A node attempts to set up a route through its neighbor upstream from, or in the direction away from, the destination. (Any fault must lie between a node and its destination to require updating of the algorithm.) It does not know that the cost through this predecessor neighbor is in fact based on routing messages through itself. With the predecessor algorithm the predecessor (upstream) node is labeled as such and cannot be used in updating the costs of the routing table.

Specifically, the predecessor algorithm is described informally as follows. (We focus on one destination for simplicity. The algorithm is repeated for each destination.) Each node does the following on receipt of a routing (update) message from a neighboring node:

1. If a P (predecessor) message, label that node a predecessor node by setting its cost to P. (*Note:* The predecessor message may simply consist of setting one bit in an appropriate field. The value of P can be any number larger than the largest expected cost.)

2. If a cost-update message, increment the cost by the cost to this node, enter the new cost in the table, and determine the minimum cost. If there is a change in minimum costs, label the minimum-cost neighbor the successor node. Else do nothing. Send new cost to all nodes other than successor node. If the successor node is the same as before, stop. Else send a P message to new successor node.

In essence, then, a node on recalculating its cost to a destination will pass over its predecessor node, eliminating an immediate loop back to itself. More important, the node will not pass incrementally increasing cost messages back and forth between itself and its predecessor node.

The application of the predecessor algorithm to the network of Fig. 6–5 is left to the reader as an exercise. Setting up the routing tables at each node with node 1 as the destination is found to take the same number of iterations and about the same number of messages passed as the distributed B algorithm. (This was the example worked out in Table 6–3.) Now let link 1–4 in the network of Fig. 6–5 fail. Invoking the distributed B algorithm is found to require five iterations, with 20 control messages passed during this time. The predecessor

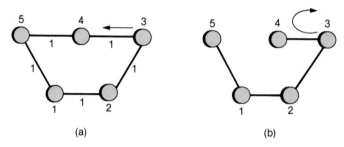

Figure 6–21 Example of data looping
a. Before link failure
b. After link 4–5 fails

algorithm does this in three iterations, requiring nine cost messages and one predecessor message. This demonstrates the improvement possible by eliminating the transmission of spurious messages.

Looping is not eliminated, however, by the use of the predecessor algorithm. Loops through nodes further away than the nearest neighbors may still be formed [HAGO, p. 255]. One would thus expect that the larger and more dense the network, the less the improvement provided by the predecessor algorithm. This has been observed in some computer experimentation with large networks [SCHW 1982b].

In Subsection 6–4–3 we describe, rather briefly, two loop-free algorithms. Note that loop freedom means no looping during the algorithm convergence interval. Looping of data can never be avoided completely, as demonstrated by Fig. 6–21 [HAGO, p. 222]. In that figure a packet is shown on its way to destination 5 via node 4. Just as it reaches 4, link 4–5 fails. Obviously the packet *must* either be returned to 3 or dropped at node 4.

6–4–3 Loop-free Distributed Routing Algorithms

We describe two loop-free algorithms in this section. One algorithm is due to Merlin and Segall [MERL]; the other, to Jaffee and Moss [JAFF]. Consider the Merlin-Segall algorithm first. This algorithm is quite different from the

[SCHW 1982b] M. Schwartz and T-K. Yum, "Distributed Routing in Computer Communication Networks," 21st IEEE Conference on Decision and Control, Orlando, Fla., Dec. 1982.

distributed B and the predecessor algorithms just described. As noted, it has been designed specifically to ensure loop freedom at all times during convergence. (Note again that all algorithms ultimately converge to the same routing-table structure.) Data packets always follow a path inbound to a destination node. After convergence of the algorithm these are shortest paths, as in all least-cost algorithms. During convergence they need not be shortest paths. For this algorithm each destination node is visualized as serving as the *sink* node in a shortest-path tree containing all other nodes in the network from which it may be reached. Each node other than the sink node knows its inbound (minimum distance) neighbor along the tree. Say that a link goes down. The neighbor on the inbound side of the tree starts a control message going toward the sink that there has been a failure (a change). The neighbor on the outbound side of the link broadcasts another message to all its neighbors telling them to set their distance to the sink to infinity. The inbound control message, on arriving at the sink, triggers an update cycle by that node, which broadcasts a distance measure to all its neighbors. These in turn update their distance to sink and broadcast this information to all their neighbors, except for the one from which the information was received. This process, constituting phase 1 of the update cycle, continues until the furthermost node currently reachable along the tree from sink has been reached and has heard from all its neighbors. (Update cycles are numbered to ensure that events within *one* cycle are counted.) At this point phase 2 of the update cycle begins. A node hearing from all its neighbors during this update cycle updates its tables and chooses the minimum-distance neighbor to sink. It then transmits this information to the minimum-distance neighbor. This phase-2 process repeats until sink has heard from all its neighbors, and the update ends. Sink then initiates a second update cycle, and the process repeats until the neighbors of sink report no change in their tables.

Note that the speed of response here is quite slow. The time to the conclusion of the first update cycle alone — requiring a control message to move down to sink, then the outbound propagation of phase 1 and the inbound propagation of phase 2 — is larger than that of the previous two algorithms in all equal-link-cost (minimum-hop) cases examined. One update cycle may be all that is required, but in general all one can prove for the Merlin-Segall algorithm is that the neighbors nearest to sink will have converged by the end of update cycle 1, their neighbors upstream in turn at the end of update cycle 2, and so forth [MERL]. Thus the process may be lengthy. As noted earlier, the algorithm does provide loop freedom during the convergence process since nodes on the downstream side of a change keep sending packets inbound at all times. The nodes on the upstream side hold their data packets until phase 2 is completed, in which case they then send the data packet along the new tree. (This may not yet be the shortest path.) But it is possible that a packet that has been held may have to retrace its path through a node already visited. This is similar to the case of Fig. 6–21.

The Jaffe-Moss loop-free algorithm is based on recognition of the fact that looping can occur only as a result of link-cost *increases* [STER] [HAGO]. The focus of the algorithm is then on that case. The algorithm relies on the concept of a shortest-path tree rooted in a destination node for each such node, as introduced in the Merlin-Segall algorithm. When a link increase is detected the entire tree is frozen for each destination affected — i.e., the successor entries are not changed until all nodes learn about the increase. This is shown to be a sufficient condition for the elimination of looping.

Freeze information is sent uptree from the node that detects a cost increase in the form of marked messages. A node that receives such a message from its successor node downstream on the tree enters the freeze state for that destination, increases its cost using the information carried in the message, and then in turn sends the new cost to its other neighbors. The freeze thus percolates uptree to the furthermost node on the tree.

Since the new message is sent to *all* neighbors other than the one from which it was received, a message may come to a node from one other than its successor node. The receiving node processes that message and sends an acknowledgment back to the sender. When a given node has received acknowledgments from all its neighbors except for its successor, it unfreezes, sends an acknowledgment to the successor, and calculates the new successor. Note that this unfreezing process starts at the furthermost node and works it way back to the originating node. This up-down cycle is similar to the one invoked in the Merlin-Segall algorithm, but convergence is guaranteed in *one* cycle rather than many. After unfreezing takes place, a node can use the normal distributed B algorithm with no looping guaranteed. Convergence is much more rapid than in the Merlin-Segall case.

The basic idea is to purge the nodes of old information before allowing them to proceed with the shortest-path calculation. Looping occurs because of the existence of out-dated information.

6–4–4 Comparative Performance

As noted earlier, some general comparison can be carried out for special classes of networks. We focus on a full-duplex loop of n nodes in this subsection. Two cases are considered. The first case is that of a minimum-hop network with all links of equal cost, taken to be one for simplicity. The network is shown in Fig. 6–22(a). In the second case, one link is assumed to have a weight ML different from the others. The network is shown in Fig. 6–22(b). (The examples of Figs.

[STER] T. E. Stern, "An Improved Routing Algorithm for Distributed Computer Networks," IEEE International Symposium on Circuits and Systems, Workshop on Large-scale Networks and Systems, Houston, Tex., Apr. 1980.

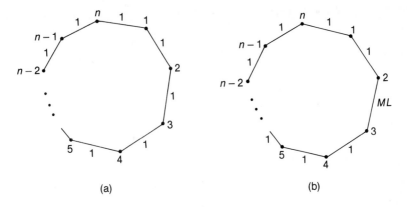

(a)   (b)

Figure 6-22 Full-duplex loop
a. Equal-cost case
b. One link different

6-19(b) and 6-19(c) are special cases.) In both cases an arbitrary link has been assumed to fail, and the speed of convergence and total number of control messages transmitted has been calculated, averaged over all possible link failures and all possible destinations.

In the first case, comparisons are made among the distributed B, predecessor, Merlin-Segall, and "best-possible" algorithms. The Merlin-Segall algorithm is found to require only one update cycle for this case. In the second case, the predecessor and distributed B algorithms are compared.

The results for case 1 (minimum-hop routing, Fig. 6-22(a)) appear in Table 6-4. Note, as expected, that the distributed B and the predecessor speeds of response are about one-half that of the "best possible," while the Merlin-Segall algorithm is slower by another factor of two. The distributed B algorithm has the number of control packets increasing as $O(n^3)$, as expected, while the predecessor and Merlin-Segall algorithms have the number going as $O(n^2)$.

Other smaller networks studied (these include 3-connected Moore graphs and simple networks given in [TAJI]) show that the distributed B and predecessor algorithms are superior in terms of speed of convergence in the minimum-hop case. For those users where looping during convergence is not a problem and the link costs do not differ too much in size, the simple distributed B algorithm, or our predecessor modification, are clearly good choices as decentralized algorithms. For large networks where the n^3 factor becomes significant, and essentially minimum-hop paths are used, the predecessor or a related algo-

TABLE 6–4   Full-duplex Loop, Equal Costs

Algorithm	No. of Hops to Converge	No. of Control Packets
"Best"	$\dfrac{n}{2} - 1$	$n - 2$
Distributed-B	$n - 3$, n even; $n - 2$, odd	$n(n - 1)(n - 2)/12 \sim n^3/12$
Predecessor	$n - 3$, n odd; $n - 2$, n even	$(n - 3)(n - 1) + 2$ n odd;
		$n(n - 4) + 10$ n even
Merlin-Segall	$2(n - 1)$	$n\left(\dfrac{5n}{2} - 1\right) \sim 5n^2/2$

rithm may be preferred. Finally, where loop freedom during convergence is of the essence, algorithms such as the Merlin-Segall one or the Jaffe-Moss one will prove useful. It is significant to note that the networks or architectures described in the previous section that use the distributed B algorithm tend to be relatively small in size. Datapac and TIDAS are small networks. The DEC architecture uses a two-level hierarchy to obtain the effect of a smaller routing network at each level. To avoid excessive looping, maxcost and maxhop fields are used in the control and data packets, as discussed earlier.

The results of the second case are tabulated in Table 6–5. They generalize the results of Table 6–4, of course, and again demonstrate the improved performance of the predecessor algorithm over the distributed B algorithm. As in case one, we assume that the network routing tables have converged to their shortest-path entries. One link is assumed to fail, and the routing algorithm, beginning with the nodes on either side of the fault, is invoked. The B entries in Table 6–5 represent results for the distributed B algorithm; the P entries represent results for the predecessor algorithm. As in case 1, the number of control packets transmitted represents transmissions summed over all destinations.

As was noted earlier, these results are rather artificial since they only demonstrate comparative performance for a special class of networks. Results for more complex networks of the type found in practice, with mesh-type topologies, are very dependent on the topology chosen. Nonetheless it is useful to study the performance for such networks; one obtains a feeling for performance under more realistic conditions. In particular, studies were made [SCHW 1982b] of the comparative performance of the distributed B algorithm and the predecessor algorithm applied to a prototype mesh network containing 57 nodes and 69 links arbitrarily chosen to have the same topology as the ARPA network of 1976 [SCHW 1977, p. 44]. The average connectivity of this network

TABLE 6-5 Full-duplex Loops, Unequal Costs

1. Distributed B Algorithm

(B1) Iterations Required for Convergence

$n + ML - 4 \qquad n - ML$ odd
$n + ML - 3 \qquad n - ML$ even

(B2) Total Number of Control Messages Generated during Convergence

(a) $ML \leq n$

$$\frac{k(k + 1)(2k + 1)}{3} + (k + 1)^2 + (k + ML)^2 + ML - 1$$

with

$$k = \begin{cases} (n - ML - 3)/2 & n - ML \text{ odd} \\ (n - ML - 2)/2 & n - ML \text{ even} \end{cases}$$

(b) $ML \geq n$

$\frac{1}{2}(ML - n - 1)(2n - 3) + n(n - 1) \qquad (n - ML)$ odd
$\frac{1}{2}(ML - n)(2n - 3) + n(n - 1) \qquad (n - ML)$ even

2. Predecessor Algorithm

(P1) Iterations Required for Convergence

(a) $ML \leq n$

$n + ML - 3 \qquad n - ML$ even
$n + ML - 2 \qquad n - ML$ odd

(b) $ML \geq n$

$2n - 3$

(P2) Total Number of Control Messages Generated

(a) $ML < n$

$(n - ML - 2)(n - ML) + 5\,ML - 3 \qquad n - ML$ even
$(n - ML - 3)(n - ML + 1) + 5(ML + 1) \qquad n - ML$ odd

(b) $ML \geq n$

$5n - 9$

(number of links incident on a node) is 2.4; its average diameter (average number of hops between source-destination nodes) is about 4.

In one set of 11 experiments, the 69 links were chosen randomly to have weights of 1, 5, 10, in the proportions 40 percent, 52 percent, and 8 percent, respectively. For each experiment a different link, again chosen randomly, was made to fail, and the two routing algorithms invoked. The average number of iterations required for convergence in the case of the distributed B algorithm was 26, with a standard deviation of 7.7 about this. (The actual range varied from 17 to 35.) For the case of the predecessor algorithm the average convergence time was 20 iterations, with a standard deviation about this of 8.6. (The range was 10–39.) These numbers appear reasonable, since the convergence of the algorithms should be loosely bounded by twice the average diameter (8) at the lower end and twice the network size ($2n = 114$) at the upper end.

The average number of control messages transmitted during convergence summed over all destinations was 4223 for the distributed B algorithm and 2398 for the predecessor algorithm. This represents an average reduction in message overhead of 46 percent of the predecessor algorithm compared with the distributed B algorithm. The standard deviation about this value was 14 percent, with the range of improvement varying from 17–66 percent (10 of the 11 values were in the range 30–66 percent). This result indicates that the use of the predecessor algorithm is probably advisable in place of the distributed B algorithm for mesh-type networks that use the latter procedure for routing. The improvement through the use of the predecessor algorithm is not as dramatic as in the case of the (rather artificial) full-duplex loop network. This is as to be expected: The predecessor algorithm is expected to reduce immediate-neighbor routing loops only. These are the only type of loop possible for the full-duplex loop network; for the mesh network many other types of loops, at various hops removed from a given node, are possible. These would presumably not be affected by the predecessor algorithm. Put another way, by its very nature the predecessor algorithm provides *local* rather than *global* network routing improvement.

The absolute number of control messages required varied quite wildly in the experiments. In the case of the distributed B algorithm the number ranged from 500 to 17,400, with a standard deviation of 4500 about the average of 4223. Similar results were obtained for the predecessor algorithm. It is thus difficult to make absolute judgments about the use of either algorithm in a large mesh-type network. Generally speaking, the expression $O(n^3)$ is expected to hold for the number of messages transmitted during convergence in the case of the distributed B algorithm, but specific values depend critically on the topology and, except for regular graphs such as the full-duplex loop, not much more analysis can be done.

A second, smaller set of experiments was carried out on the same topological network, but with link weights changed and proportions of links having the three weights of 1, 5, and 10 now changed to 65, 29, and 6 percent, respectively. Results similar to the first set of experiments were obtained. The average reduction of control messages with the use of the predecessor algorithm was now 36 percent; the predecessor algorithm used 1818 messages on the average to converge; and the distributed B algorithm required 2849 messages on the average. The number of iterations to converge was about the same for both algorithms, ranging from 17 to 20 for the predecessor scheme and 18 to 24 for the distributed B scheme.

Problems

6–1 Consider the network shown in Fig. 6–4. Show that any other choice of flows along the various links leads to longer time delay.

6–2 Consider the network shown in Fig. P6–2. Use both algorithms A and B to find

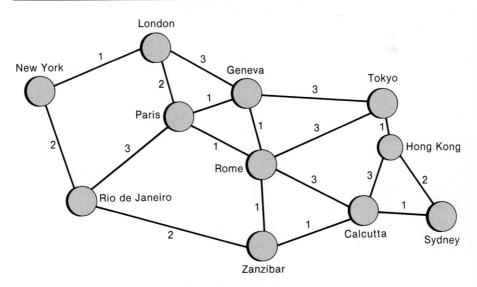

Figure P6–2

the shortest paths between Rome and each of the other cities. Link weights are as indicated.

Repeat for minimum-hop routing (link weights = 1). Compare.

6-3 Design a 10-node network of your own choice. There should be at least 20 links in the network. Link capacities are selected randomly and uniformly from 4800 bps, 9600 bps, and 14.4 kbps. Link weights are inversely proportional to the capacities. Use algorithms A and B to find the shortest path between one of the nodes and all other nodes.

6-4 Refer to the backbone portion of the GTE Telenet network shown in Fig. 6-10. (This consists of the 12 cities shown and the links interconnecting them.) Links (trunks) are all of 56-kbps capacity. Use algorithms A and B to find the shortest paths between Denver and all other cities.

6-5 Consider the backbone portion of the SITA network diagrammed in Fig. 1-2. Assign 9600-bps and 14.4-kbps link capacities randomly, in the proportion 2 to 1, to the links shown. Transcontinental links are all assumed to use satellite transmission. (*Note:* These capacity and satellite assignments have been arbitrarily chosen, for illustrative purposes only, and do not reflect the actual SITA configuration.) Use algorithms A and B to find the shortest-path routes between New York and all other cities. Link weights are inversely proportional to the link capacity. Satellite links are, in addition, weighted three times the weight of a terrestrial link of the same transmission capacity.

6-6 The distributed B algorithm is to be applied to the two full-duplex loop networks shown in Fig. P6-6. In both cases let 1 be the destination node.

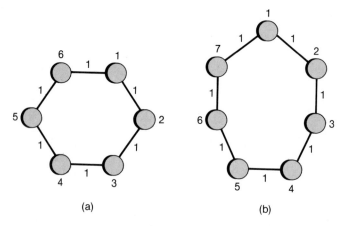

(a)   (b)

Figure P6-6

      a. Initialize by having node 1 "come up." Its neighbors start the algorithm going. (All entries, in all nodes, are initially set at ∞ to destination 1, prior to 1 "coming up.") Run the algorithm to completion. Show two tables—the distance and the routing tables—at each node after convergence. Count the *total number* of control messages passed until completion. How many iterations are required for convergence in this case?

      b. Repeat the algorithm for both networks if link 1–2 fails. Again count the number of messages passed and the number of iterations required until completion.

6–7 Repeat Problem 6–6 for the six-node network of Fig. P6–6(a), if link 2–3 has a weight of 4. Compare the number of iterations (essentially the convergence time) and the number of messages passed with the results of Problem 6–6.

6–8 Choose a mesh network with at least six nodes. Let the link costs be 1, 5, 10 arbitrarily selected. (*Note:* Do *not* pick a loop network or a fully connected one.) Use one node as the destination node. Repeat Problem 6–6. (First initialize and run the algorithm to completion. Then let a link fail and run it again. Keep count of the number of iterations and the total number of messages passed in both cases.)

6–9 Repeat Problem 6–6 but use the predecessor algorithm this time. Compare the number of iterations required and the control messages transmitted with those found in Problem 6–6.

6–10 Refer to Problem 6–8. Compare the predecessor and split-horizon algorithms with the distributed B algorithm for the same network used there.

6–11 *Note:* This problem is actually a project designed to compare routing strategies for a number of networks. The minimum goal is to compare the distributed B algorithm, the predecessor algorithm, and the split-horizon algorithm, all in distributed form, for at least three networks of 10, 20, and 30 nodes or more. One stipulation: The average degree of the networks chosen must be *at least* 2.5. (The average degree d is given by the average number of links connected to a node. For n nodes and ℓ links, $nd/2 = \ell$. Show this.) Let the weights of the links be 1, 5, or 10, chosen randomly so that they are equally distributed throughout the network.

      1. Initialize each node as a destination by having it come up, with each neighbor then beginning to carry out the algorithm. Determine the number of iterations required and the *total* number of control messages passed until convergence. Repeat for all three algorithms and compare.

      2. Let a link, randomly chosen, fail. Use each algorithm separately to determine new routing tables for *each* destination. Again determine the number of iterations and the total number of messages required, summed over all destinations. Again compare the three algorithms using the number of iterations and the total number of messages passed, summed over all destinations, as measures of performance.

      Correlate a given algorithm's performance, for both the initialization and recovery from link failure modes, with network size. Is there a discernible trend?

(The size is the maximum span of the network measured in links or hops spanned.) Do the predecessor and split-horizon algorithms provide any improvement over the distributed B algorithm? Which seems better? Can you explain the results obtained?

6-12 Refer to Fig. 6-19. Take node 1 as the destination node. Link 1-2 then fails.

    a. Use the distributed B algorithm to determine the distance and routing tables at each node for all three cases as the algorithm carries out its functions. Show that the example of Fig. 6-19(a) is unstable, requiring that the algorithm be terminated at each node at some (arbitrarily chosen) maximum value of cost. Compare convergence time (number of iterations) and the number of control messages transmitted for cases (b) and (c) of Fig. 6-19.

    b. Repeat using the predecessor algorithm and compare with the distributed B algorithm results.

6-13 Design a simple example, similar to that of Fig. 6-20, that demonstrates problems that might arise with the distributed B algorithm because of unequal delays on links. Do the predecessor and split-horizon algorithms exhibit this synchronization problem? Explain.

6-14 Refer to the Jaffe-Moss loop-free algorithm. Work out some examples to satisfy yourself that it is in fact loop free in the sense described in the text. (Refer to [JAFF] for details of the algorithm if necessary.)

6-15 Apply the predecessor algorithm to the example network of Fig. 6-5 with node 1 as the destination. Initialize and let all tables converge to their final settings. Then let link 1-4 fail and repeat the convergence process. Compare the number of iterations required to converge and the total number of control messages passed with those obtained for the distributed B algorithm.

6-16 Compare the performance of the distributed B and predecessor algorithms for the minimum-hop full-duplex loop of Fig. 6-22(a). Check your results with those of Table 6-4.

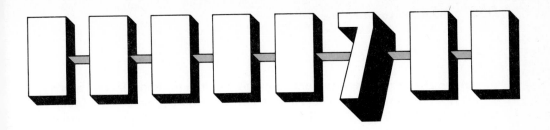

Transport Layer

Chapter 3 was devoted to an overview of the seven-layer open systems interconnection (OSI) architecture and protocols, which were developed by the International Organization for Standardization (ISO) to provide a standard way of allowing intelligent data systems (users) throughout the world to communicate. The IBM layered architecture, SNA, was introduced there briefly as well to provide another example of a layered communication architecture designed to enable users to communicate.

With Chapter 4, we began a more detailed and systematic study of communication architectures, working our way up from the data link layer. In that chapter, after an introductory discussion of a number of ways of implementing a data link control, we went on to describe HDLC in more detail. HDLC is the bit-transparent data link control that has been adopted by ISO as a standard for point-to-point and point-to-multipoint connections.

Chapters 5 and 6 were devoted to detailed discussions of the network layer of communication architectures, the purpose of which is to route packets from one end of a network to the other while providing a means for keeping possible congestion within tolerable limits. The network layer does this by relying on the services of the data link layer to provide correct, sequenced transmission of packets from one node on a path to its neighbor at the other end of a link.

In this chapter we move up one more layer to the transport layer. Figure 7–1 reviews the concept of the seven-layer OSI architecture, discussed in Chapters 1 and 3. As indicated there, the function of each layer is to provide a service to the layer(s) above it. The transport layer is the layer just above the network layer that provides the necessary reliable end-to-end data transmission service for the session layer and other layers above it. (These layers are called the *users* of the transport service.) In performing this service, the transport layer must in turn call on the network layer to actually provide an acceptable path through the network or interconnected networks that lie between the two end users that desire to communicate. The transport layer is the first one, moving up from the bottom, that always resides in the end-user system and that is not involved with aspects of the communication process relating to the network or networks through which data transmitted must actually flow.

In discussing issues at the transport layer in this chapter, we first describe the OSI Standard Transport Protocol (TP), focusing in detail primarily on one of the five classes defined for this protocol. Error-detection and error-recovery mechanisms play important roles in the operation of this transport protocol. They are vital to the correct operation of such protocols and play a key role in the correct operation of each of the various layers in all layered architectures. We thus single out, in Section 7–2, some of the problems that might arise in the operation of a transport protocol, with some of the suggested solutions that have appeared in the literature. This is followed by a detailed discussion of the error-recovery procedures adopted in the OSI transport protocol itself (Section 7–3). Some of this material is applicable to other layers of the architecture, starting with data link control. We then describe, in less detail, an older and pioneering transport-layer protocol, the Transmission Control Protocol (TCP)

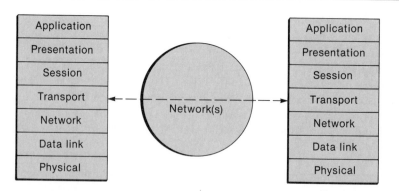

Figure 7–1 Review of seven-layer OSI architecture

developed for the ARPAnet communication activities. Some of the pioneering aspects of TCP have been adopted in the ISO-defined TP, and much of the work on error recovery was motivated by the development of TCP. We provide a brief comparison of these two protocols after describing both of them. Layered architectures such as IBM's SNA and DEC's DNA, described in Chapters 3, 5, and 6, were also developed before the detailed design of TP and contributed to its development, as did earlier architectures and protocols, such as those developed for the European Information Network and other networks and organizations.

7 – 1 OSI Transport Protocol

7 – 1 – 1 Overview

The OSI transport-layer protocol became an ISO international standard in 1984 after a number of years of intense activity and development. CCITT committees have cooperated in the standardization process. The European Computer Manufacturers Association (ECMA) has participated actively in defining the standard, as have representatives from the American National Standards Institute (ANSI) and the U.S. National Bureau of Standards.

The protocol was described briefly in Chapter 3 in connection with the overall discussion of OSI. As noted there and as just mentioned briefly, the purpose of the transport layer is to provide a transparent, reliable, end-to-end data transfer mechanism for the session layer and other layers (the users) above. It uses the services of the network layer below to help in this function but shields the upper layers from the details of the network connections and types of networks used.

To handle both the different types of data transfer that might be expected over a transport connection and the wide variety of networks that might be available to provide network services, five classes have been defined for the ISO transport protocol. These are labeled classes 0, 1, 2, 3, and 4. Class 0, for example, is compatible with an earlier CCITT recommendation for providing transport service for Teletex terminals. It is the simplest type of transport connection, with a minimum of functions defined. Its network connection is assumed to provide an acceptable error rate, as well as an acceptable rate of network connection failures. In this class the transport protocol is required only to set up a simple end-to-end transport connection and, in the data-transfer phase, to have the capability of segmenting data messages if necessary. It has no provision for recovering from errors and cannot multiplex several transport connections onto a single network connection. (Error conditions must be signaled by the network layer; these are then passed on to higher layers for any action required.)

Class 1 is a simple class as well but incorporates a basic error-recovery capability. Errors could be due to a network disconnect or failure, to receipt of a data unit for an unrecognized transport connction, and so forth. Classes 2, 3, and 4 are successively more complex, providing more functions to meet specified service requirements or to overcome problems and errors that might be expected to arise as the network connections become less reliable. Examples of such functions are the ability to multiplex multiple transport connections onto one network connection, to split one transport connection among several network connections, and to recover from a failure signaled by the network.

The choice of class is dictated by the quality of service requested by the transport service user (this will be discussed in detail later) as well as by the underlying network connection(s) available to provide the necessary services. To make this more explicit, the TP standard defines three types of network connections to be used in conjunction with the various classes. Type A is a network connection with both an acceptable error rate and an acceptable rate of failures that are signaled to the transport protocol. The basic connection assumed is thus quite good, packets are assumed not to be lost or misordered, and there is thus no need for the transport layer to provide failure recovery services, loss of data message services, and resequencing services, and so forth. This is the type of network connection over which class 0 is designed to operate. An example might be a network with virtual circuit services at the network layer.

Type-B network connections are defined to be those with an acceptable error rate but an unacceptable rate of signaled failures. A transport protocol in this environment must have provision to recover from errors when they occur. Class 1, as noted, falls into this category.

Finally, type-C network connections are those for which the error rate is not acceptable to the transport service user(s). This means that the transport protocol for this class must be capable of detecting and recovering from network failures, of detecting and correcting out-of-sequence, duplicate, or misdirected data messages, and so on. The network service in this situation is expected to be of relatively low quality, and the protocol must be designed to shield the transport-service users from possible problems. Some types of local area networks, datagram networks with packets (datagrams) possibly arriving out of order, urban networks with mobile nodes, and packet radio networks with fading radio transmission are all examples of networks in which some of these problems might arise. Table 7–1 summarizes this ISO classification of three types of network services.

Class 4, which will be discussed in detail in this section, is designed to work with type-C connections. Class 2, to which we will refer in passing, is designed to work with type-A network connections. As we shall see, the class-2 protocol,

TABLE 7 – 1 Network Services for Use with Transport Protocols

Type A
Network connection with both acceptable error rate and acceptable rate of signaled failures.

Type B
Network connections with acceptable error rate but unacceptable rate of signaled failures.

Type C
Network connections with error rate not acceptable to transport service user.

(based on [ISO 1984b, Sec. 5.3.1.2.4])

because it assumes a reliable network connection, does not require an acknowledgement data unit to be transmitted during the connection phase; the class-4 protocol does. Class 2 incorporates a multiplexing capability that class 0 does not; class 3, designed to work over a type-B connection, incorporates error-recovery procedures in addition to the multiplexing capability of class 2. Table 7 – 2 summarizes the functions of these five classes.

The basic references on the ISO Transport Protocol are two standards documents: *Information Processing Systems—Open Systems Interconnection—Transport Service Definition* [ISO 1984a] and *Information Processing Systems—Open Systems Interconnection—Transport Protocol Specification* [ISO 1984b]. Overviews of the protocol appear in papers by K. G. Knightson [KNIG] and P. von Studnitz [STUD]. More detailed, more formal specifications of the class-2 and class-4 protocols appear in a series of six volumes issued by the Institute for Computer Sciences and Technology of the U.S. National Bureau of Standards

[ISO 1984a] *ISO International Standard 8072, Information Processing Systems—Open Systems Interconnection—Transport Service Definition,* Geneva, 1984.

[ISO 1984b] *ISO International Standard 8073, Information Processing Systems—Open Systems Interconnection—Transport Protocol Specification,* Geneva, 1984; appears in *Computer Comm. Rev.,* vol. 12, nos. 3 and 4, July/Oct. 1982.

[KNIG] K. G. Knightson, "The Transport Layer Standardization," *Proc. IEEE,* vol. 71, no. 12, Dec. 1983, 1394–1396.

[STUD] P. von Studnitz, "Transport Protocols: Their Performance and Status in International Standardization," *Computer Networks,* vol. 7, 1983, 27–35.

TABLE 7–2   OSI Transport Protocol Classes

Class	Network Connection Type	Basic Functions Required of Class
0	A	Set up connection
1	B	Error recovery
2	A	Multiplexing
3	B	Error recovery and multiplexing
4	C	Error detection and recovery

Note: Classes 2, 3, and 4 allow multiple transport connections (of possibly varying classes) to be multiplexed onto the same network connection.

[NBS 1983a]–[NBS 1983f]. The NBS volumes specify the U.S. Federal Information Processing Standard (FIPS) for the transport protocol. The FIPS transport service definitions contain some optional features that are not part of the ISO transport protocol. Some of these features will be indicated in the discussion following. Aside from them, the protocols are identical.

In this section we focus principally on the class-4 protocol, with reference made occasionally to the class-2 version. The Transmission Control Protocol (TCP), which will be discussed in Section 7–5, is analogous to class 4 of the ISO TP, allowing some comparisons to be made.

As has been noted a number of times, the purpose of the transport layer is to provide a reliable, end-to-end transparent transport connection through a network or series of networks for two end users. As stated previously, the class-4 transport protocol does not rely on network services to provide a reliable con-

[NBS 1983a] *Specification of a Transport Protocol for Computer Communications, Vol. 1: Overview and Services,* Inst. for Computer Sciences and Technology, National Bureau of Standards, Gaithersburg, Md., Jan. 1983.

[NBS 1983b] *Specification of a Transport Protocol for Computer Communications, Vol. 2: Class 2 Protocol,* Inst. for Computer Sciences and Technology, National Bureau of Standards, Gaithersburg, Md., Feb. 1983.

[NBS 1983c] *Specification of a Transport Protocol for Computer Communications, Vol. 3: Class 4 Protocol,* Inst. for Computer Sciences and Technology, National Bureau of Standards, Gaithersburg, Md., Feb. 1983.

[NBS 1983d] *Specification of a Transport Protocol for Computer Communications, Vol. 4: Service Specifications,* Inst. for Computer Sciences and Technology, National Bureau of Standards, Gaithersburg, Md., Jan. 1983.

[NBS 1983e] *Specification of a Transport Protocol for Computer Communications, Vol. 5: Guidance for the Implementor,* Inst. for Computer Sciences and Technology, National Bureau of Standards, Gaithersburg, Md., Jan. 1983.

[NBS 1983f] *Specification of a Transport Protocol for Computer Communications, Vol. 6: Guidance for Implementation Selection,* Inst. for Computer Sciences and Technology, National Bureau of Standards, Gaithersburg, Md., June 1983.

nection. It thus incorporates means for detecting and handling a variety of error or fault conditions. It attempts to provide a specified quality of service to the transport-service users (the layers above) by choosing appropriate network connections, possibly multiplexing or splitting transport connections, choosing appropriate data unit sizes, allowing expedited data flow, maintaining appropriate flow control, and so forth.

Note how these services at the transport layer, in keeping with the abstract layer concept of the OSI Reference Model introduced in Chapter 3, fall into the same categories as those provided generally at all layers of the model: Each layer may multiplex or split connections (Fig. 3 – 6), segment or concatenate data units, provide error-detection and error-recovery procedures, and invoke flow control, among other features. Recall how these features were handled at the data link and network layers discussed earlier. We now find them appearing, albeit in a different form and at a higher level, at the transport layer as well.

In order for the transport protocol to provide the appropriate service requested by the transport service users, a number of quality-of-service parameters, as well as other features of the data-transfer phase following, are negotiated during its connection or establishment phase. The desired quality-of-service parameters are passed down from the transport service users to the transport protocol entity within the transport layer, which in turn will pass parameters down to the network layer below.

Although the primary purpose of the transport protocol is to provide a reliable end-to-end connection for the transfer of data between users, to do this properly management services must be made available as well. The services provided by the transport layer are thus grouped into two types:

1. Transport-connection management, responsible for establishing and then terminating a connection.

2. Data transfer.

The ISO version of the transport protocol provides an *abrupt termination* or disconnect service only. Data in transit between the two transport entities may thus be lost. The NBS FIPS version incorporates as an option a *graceful close,* with no loss of data in transit. In the data-transfer phase, provision is made for the transfer of higher-priority or *expedited* data outside the normal data flow.

Implicit in the preceding comments is the fact that there are the usual three phases to transport layer communications: a connection phase, a data transfer phase, and a disconnect (termination) phase. The bulk of the discussion in this chapter relates to *connection-oriented* communications. The ISO transport-layer standards documents 8072 and 8073 — [ISO 1984a] and [ISO 1984b], respectively — focus exclusively on this environment, in which a transport connection must be set up and then terminated. *Connectionless-mode transmission,* also available as an option, is described in detail in two ISO standards docu-

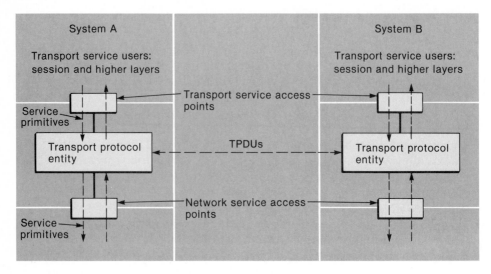

Figure 7–2 Transport layer interactions

---

ments: ISO 8072/DAD1, an addendum to the transport service definition that covers connectionless-mode transmission, and ISO 8602, the transport protocol that covers connectionless-mode transport service.* The NBS FIPS protocol incorporates as an option a connectionless data-transmission service. With connectionless-mode transmission a single unit of data is transmitted reliably from one transport user to another, without the need to set up and then terminate a connection. This feature could be used in an application such as a query/response system, in which the users did not want to incur the overhead of establishing and terminating a connection. Connectionless service is described briefly in this chapter.

Recall from Chapter 3 that the transport service user and a transport entity in the same system interact, across the interface between them, by exchanging transport service primitives. The transport entity and the network service provider in turn interact by exchanging network service primitives. The primitives are abstract representations of the interactions across the service access points, indicating that information is passed between the service user and the service provider. The OSI protocols do not specify how the primitives are to be implemented.

---

* These two documents are reprinted in draft proposal form in *Computer Commun. Rev.*, vol. 14, no. 4, Oct. 1984, 18–45.

Four types of primitives are prescribed, as indicated in Chapter 3: request, indication, response, and confirm (or confirmation). Peer transport entities in turn interact, or communicate, by exchanging transport protocol data units (TPDUs). Figure 7 – 2, which is a special case of Fig. 3 – 6 devoted to the transport layer (rather than the general *N*-layer described there), indicates the interactions just described. As noted in Chapter 3, the request and response primitives are generated by the users of the service; the indication and confirmation primitives are generated by the providers of the service. Figure 7 – 3 portrays the four primitives being passed between user and provider.

In the subsection following we describe the services specified for the transport layer, as well as how the primitives are used to convey the information required across the interface between the session layer (the transport service user) and the transport layer (the service provider). We then follow with a subsection that summarizes the features of the transport protocol, showing how the protocol provides the services requested.

7 – 1 – 2 Transport Services

It was noted earlier that transport services fall into two categories: those required to manage a connection and those needed to transfer data. The transport connection-management services in turn consist of those required to provide an end-to-end connection and those required to terminate or release a connection. The data-transfer services provide for the transmission of both

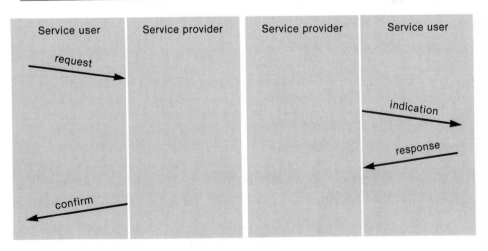

Figure 7 – 3 Abstract use of primitives

normal and expedited data. As noted, a connectionless option is also provided that allows a single unit of data to be transmitted.

We focus on connection-management services first, then follow with a discussion of data-transfer services.* The notation used to represent the primitives involved is the same as that used in Chapter 3. As an example, consider the request primitive passed down by the transport user (the session layer above) during the connection-establishment phase. This is labeled

T_CONNECT. request.

The word CONNECT in capital letters indicates the service provided, in this case that of establishing a connection; the "T" preceding identifies the service *provider*, in this case the transport entity. (Primitives that cross the interface between the transport layer and the network layer below use the letter N_, indicating that the *network* layer is the service provider in that case. See Figs. 3–14 through 3–16 for the use of S_ primitives, which are passed between the presentation layer and the session layer, the latter serving as the service provider for the presentation layer.)

The connection-establishment service at the transport layer is depicted in Fig. 7–4. The parameters carried in each of the four primitives are listed in Table 7–3. Consider the T_CONNECT. request primitive first. This obviously carries the request from the transport service user (the session layer and layers above) to set up a transport connection over which to transmit user data. The request includes the source and destination addresses (these may be local addresses or names, in which case the transport layer must map them into network service addresses), quality-of-service parameters, and up to 32 octets of user data. An indication as to whether expedited data service will be made available is included as well. (This feature is always specified in the NBS FIPS version of the transport protocols, classes 2 and 4.) The quality-of-service parameters represent the user's requirements for throughput (in octets/sec), transit delay, reliability, and priority of this connection, relative to other connections. As will be seen shortly, the transport protocol will in turn forward these parameters, in a connection-request transport protocol data unit, to the peer transport entity at the destination as part of the process of negotiation. The transport protocol uses these parameters to determine the network services required (what type of network connections, for example), which of the five transport classes to choose, the size of the sequence number space required, whether a transport layer checksum is needed, and so forth. Quality-of-service parameters are in turn passed down to network services.

The user throughput requirements are given in terms of average and minimum values for each direction of transmission, while the transit delay includes

---

* The material in this subsection is based on [ISO 1984a] and [NBS 1983a].

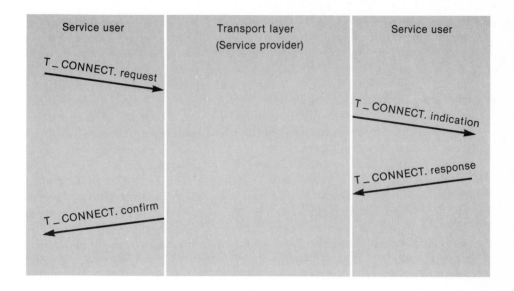

Figure 7 – 4 Connection-establishment primitives, transport layer

average and maximum allowable values. The reliability parameter refers to an acceptable residual error rate (including errors due to corrupted, duplicated, or lost data). This error rate will in turn determine whether or not a checksum at the transport layer is to be used. The choice of priority impacts on allocation of buffer resources, the type of transmission strategy to be used, and the allocation of connection resources. The NBS FIPS specifications include buffer manage-

TABLE 7 – 3 Transport Service Primitives, Connection Establishment

Primitive	Parameters
T_CONNECT. request	(Called Address, Calling Address, Expedited Data Option, Quality of Service, TS User Data)
T_CONNECT. indication	(Called Address, Calling Address, Expedited Data Option, Quality of Service, TS User Data)
T_CONNECT. response	(Quality of Service, Responding Address, Expedited Data Option, TS User Data)
T_CONNECT. confirm	(Quality of Service, Responding Address, Expedited Data Option, TS User Data)

(based on [ISO 1984a, Table 3, Sec. 2])

ment and security parameters as additional options. Buffer management is a strictly local concept, involving local buffer resources to be devoted to this connection. The buffer-management parameter is thus not passed on to the destination provider and user. Security parameters must be agreed on by the transport users at either end.

The quality of service may be reduced (i.e., lower throughput, longer delay, higher error rate, lower priority) by the transport service provider— in which case new parameters appear in the T_CONNECT. indication—or by the called transport user. In either case the parameters proposed by the destination system appear in the T_CONNECT. response. The T_CONNECT. confirm primitive in turn forwards to the calling transport user the parameters that appear in the indication primitive. Thus this process of negotiation can result either in the service requested by the calling user (the quality of service parameters are unchanged) or in reduced quality of service. Improvement of quality is not allowed.

The transport connection release service defined in the OSI transport service specification can be used to refuse a connection or to unilaterally terminate a connection in progress. Two primitives are defined for this purpose: T_DISCONNECT. request and T_DISCONNECT. indication. These primitives and their parameters appear in Table 7–4. The use of these primitives in refusing a connection is shown in Fig. 7–5. User and provider labels have been left out to simplify the figures.

In Fig. 7–5(a) the called transport service user refuses to accept a connection. In Fig. 7–5(b) the provider of the service (at the transport layer) rejects the proposed connection. In either case up to 64 octets of data may be transmitted along with the disconnect primitives.

Once a connection has been successfully established, as shown in Fig. 7–4, the transport protocol enters the data transfer phase. This phase may be terminated at any time by invoking the same T_DISCONNECT primitives. Four

TABLE 7–4 Transport Service Primitives, Connection Release

Primitive	Parameters
T_DISCONNECT. request	(TS User Data)
T_DISCONNECT. indication	(Disconnect Reason, TS User Data)
T_CLOSE. request*	—
T_CLOSE. indication*	—

* The T_CLOSE service is not provided in the ISO standard. It appears only in the NBS FIPS standard [NBS 1983a].

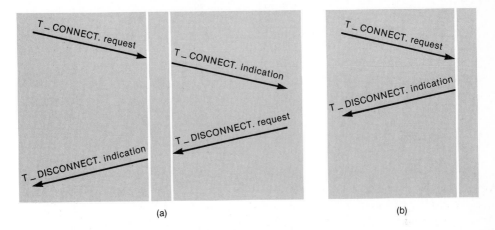

Figure 7 – 5 Connection-establishment rejection, transport service
a. User rejects establishment request
b. Service provider rejects establishment request

possible disconnect scenarios can be diagrammed [ISO 1984a, Fig. 4]; these appear in Fig. 7 – 6. In part (a) of the figure, one of the transport service users initiates release of the connection. In part (b), both users initiate release. The release requests at either end in this case are not synchronized in time in any way; either one can occur at any time without reference to the other. This absence of a time relationship is denoted by the symbol ~. The same lack of connection in time appears in parts (c) and (d) of Fig. 7 – 6. Part (c) shows the service provider releasing the connection. (This could be one or both of the peer providers.) T_DISCONNECT. indications are then sent to the users at either end. Finally, part (d) shows a user and provider initiating a release independently.

Release of a connection may occur for any of several reasons: data transmission is terminated, data cannot be delivered, resources are depleted, and so on. The request primitive in Fig. 7 – 6, from a user, normally indicates completion of a connection. No reason has to be given for invocation of this primitive. A connection terminated by the provider could be due to lack of resources (locally or at the other end), some unexpected problem at the provider level, a drop in quality of service, or even some unknown reason. The reason parameter in the T_DISCONNECT. indication primitive (Table 7 – 4) is used to convey to the transport service user an indication as to whether it is the provider or the user at the other end that invoked the disconnect procedure. The T_DISCONNECT. indication reason parameter in the case of connection rejection (Fig. 7 – 5) is

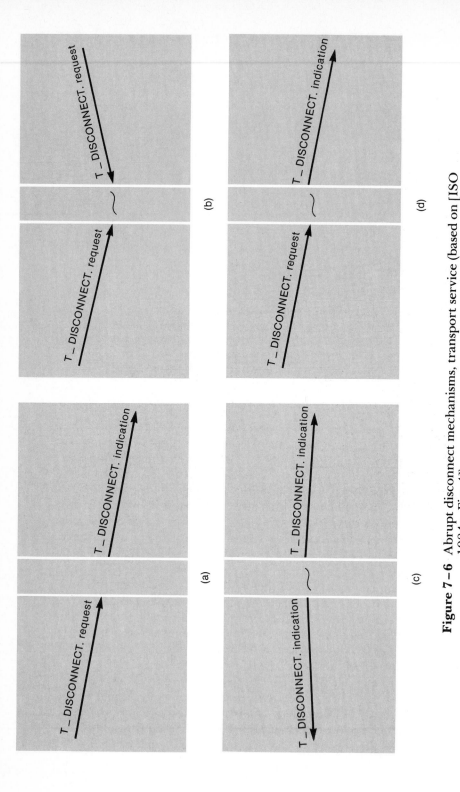

Figure 7-6 Abrupt disconnect mechanisms, transport service (based on [ISO 1984a, Fig. 4])

a. Transport service user initiates connection release
b. Both users initiate release
c. Transport service provider initiates release
d. A user and a provider both initiate release

used to convey the same information. In this case rejection might be due to too poor a quality of service, inability to complete the connection, called user not available or unknown, or another reason.

The transport connection-release procedures of Figs. 7 – 5 and 7 – 6 can be invoked at any time. Because of the abrupt nature of the disconnect process, there is no guarantee that user data will be delivered once the process has started. An optional connection-termination service, called T _ CLOSE service, has been defined in the NBS FIPS [NBS 1984a]; it terminates the connection only upon transmission of all user data. Two primitives shown in Table 7 – 4, T _ CLOSE. request and T _ CLOSE. indication, which carry no parameters, are used for this purpose. The T _ CLOSE service using these two primitives is diagrammed in Fig. 7 – 7 [NBS 1983a, Fig. 3.4]. This process of termination is also called a "graceful close" [NBS 1983a]. Note from Fig. 7 – 7 that complete closure is attained, with the transport connection released, only when two invocations each of the two primitives have been made. The T _ CLOSE. request primitive is issued by a transport service user when it wishes to terminate a connection, and then only when it has no more data to send. The transport-layer protocol delivers this request to its peer at the other end some time later, resulting in a T _ CLOSE. indication being issued to the transport user at the other end. This indicates to that user that it will be receiving no more data from the other end. When it in turn has completed delivery of its data to the transport layer, it issues its own T _ CLOSE. request and closes down the connection. The second T _ CLOSE. indication informs the termination-initiating user that no more data is coming, and it in turn closes its end of the connection.

The final service provided by transport services is, of course, *data transfer.* It has already been noted that the purpose of the transport layer is to provide a reliable end-to-end delivery service for transport user data. Two types of data service are available through the ISO transport service definition, as already noted: a normal data service and an expedited data service. Once the transport connection is successfully established, using the connection primitives described earlier and diagrammed in Fig. 7 – 4, data transfer can begin. Two primitives are defined for this service: request and indication. These primitives are listed in Table 7 – 5 for the two types of data service. Figure 7 – 8 diagrams the use of the two primitives in the case of normal data transfer. A similar diagram applies in the case of expedited data transfer. *Note that there is no confirmation of delivery by the receiving transport user.* The transport-layer protocol is responsible for correct delivery, as will be seen in the next subsection.

The data carried in each T _ DATA. request primitive is called a *transport service data unit* (TSDU) (see Fig. 3 – 7) and is delivered intact to the receiving transport user by the T _ DATA. indication primitive. TSDUs are not restricted in length. However, the transport protocol may segment a TSDU into multiple transport protocol data units (TPDUs), transmitted by a transport entity to its

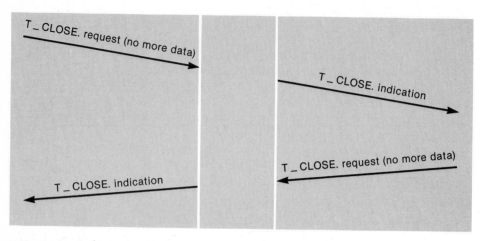

Figure 7 – 7 Use of graceful close, transport services (from [NBS 1983a, Fig. 2 – 4] with permission)

peer entity at the other end. TSDUs are always delivered in sequence to the receiving transport user. The T_EXPEDITED_DATA. request primitive (Table 7 – 5) passes a maximum of 16 octets of transport user data to its transport entity. Expedited data at the transport layer bypasses normal data end-to-end flow control (described in the next subsection). The transport layer must

TABLE 7 – 5 Transport Service Primitives, Data Transfer

Primitive	Parameters
T_DATA. request	(Transport Service User Data)
T_DATA. indication	(Transport Service User Data)
T_EXPEDITED_DATA. request	(Transport Service User Data)
T_EXPEDITED_DATA. indication	(Transport Service User Data)
T_UNIT_DATA. request*	(Called Address, Calling Address, Quality of Service, Security Parameters, TS User Data)
T_UNIT_DATA. indication*	(Called Address, Calling Address, Quality of Service, Security Parameters, TS User Data)
T_UNIT_DATA. confirm*	(Result, Reason)

* The T_UNIT_DATA. confirm primitive does not appear in the ISO connectionless transport service. It appears only in the NBS FIPS connectionless service option. Security parameters appear only in the NBS version of the primitives.

Figure 7–8 Data transfer service primitives, transport services

ensure that expedited data will not be delivered to the receiving transport user later than any normal data sent after it.

It was noted earlier that connectionless data service is available with the ISO transport protocol. The service, if used, sends data directly without requiring a transport connection to be set up. A single TSDU is sent by a transport user whenever it wishes.

Two primitives have been defined by ISO for this service. The primitives, together with the parameters they carry, are listed in Table 7–5. The T_ UNIT_DATA. request primitive passes a single TSDU to the transport layer, signaling that layer that it is to transfer the TSDU to the transport user designated by the called address. Note from Table 7–5 that each such request primitive independently carries source and destination transport user addresses, as well as quality-of-service parameters and, in the NBS version, security parameters. The quality-of-service parameters, as in the connection phase of the connection-oriented transport service described earlier, indicate the user's requirement for throughput, transit delay, reliability (error rate), and relative priority of a particular transaction. These parameters are used by the transport layer to select appropriate protocol options and are passed on to the network layer as well. The indication primitive passes the TSDU and the quality-of-service parameters on to the destination transport user. The ISO connectionless data service, unlike the connection-mode service, provides no guarantee of delivery of data units. It is thus up to the transmission service user, at a higher level in the architecture, to handle the question of retransmission of data units.

The NBS version of connectionless data service, called T_UNIT_DATA service [NBS 1983a], does provide for an indication of the success or failure of delivery of data units. It introduces a third primitive, the T_UNIT_DATA.

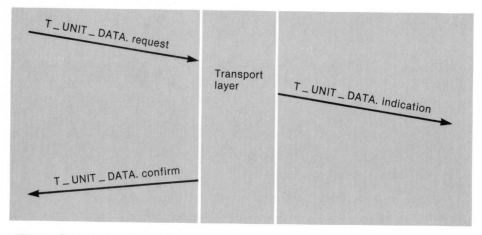

Figure 7–9 Primitives used for unit data transfer (from [NBS 1983a, Fig. 3–11] with permission)
The confirm primitive does not appear in the ISO connectionless transport service

confirm primitive, for this purpose (Table 7–5). The use of the three primitives in the NBS connectionless mode of service is diagrammed in Fig. 7–9 [NBS 1983a, Fig. 3.11]. (The confirm primitive shown does not appear in the ISO connectionless mode of service, as already noted.) Note that no response primitive appears. It is thus up to the *transport layer,* the service provider, to notify the sending transport user, using a T_UNIT_DATA. confirm primitive, as to whether the TSDU arrived successfully or not. The T_UNIT_DATA. confirm primitive carries a result parameter (Table 7–5) that indicates the success or failure of delivery. The reason parameter is used to indicate any reason(s) for a failure of delivery.

The use of the transport service primitives for the connection-oriented transport service is summarized in the state transition diagram of Fig. 7–10. This diagram, which focuses on only one end of a transport connection, indicates the possible allowed sequences of occurrence of the various primitives defined earlier [ISO 1984a, Fig. 5]. Arrows indicate the invocation of the primitives. The directions shown indicate the resultant transitions from one state to another. The idle state, 1, is the one from which connection initiation starts and is the one reentered to release a connection. Note that connection release, involving the use of either the T_DISCONNECT. request or the T_DISCONNECT. indication primitives, can be initiated at any time in either

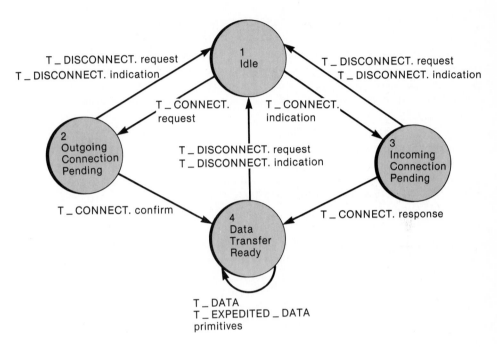

Figure 7–10 State transition diagram for possible allowed sequence of TS primitives at a transport connection endpoint (based on [ISO 1984a, Fig. 5])

the connection-establishment or data-transfer phases. States 2 and 3 correspond to the connection-establishment phase; state 4 represents the data-transfer phase.

7–1–3 Protocol Mechanisms

The previous subsection described the services required of the transport layer. We now summarize the protocol mechanisms used to provide these services. For the sake of simplicity we focus on class 4 of the transport protocol, with some reference made to class 2 as well. Protocols for all five classes are presented verbally in the standards document [ISO 1984b]. A brief summary of the protocol mechanisms for classes 2 and 4 appears in [NBS 1983a]. Detailed and *formal* specifications for the class 2 and class 4 protocols appear in [NBS 1983b] and [NBS 1983c], respectively.

Recall from our discussion at the beginning of this section that class 2 assumes that the underlying network is reliable and that it can be counted on to deliver the data as sent — unchanged, in order, and with no duplication. (As noted earlier, a virtual circuit connection at the network layer might be one example of such a network service.) Class 4 makes no such assumption. The transport protocol for this class is designed to work with a possibly unreliable network that may garble, lose or misorder, or duplicate data. The network layer cannot be counted on to notify the transport protocol as to a possible problem. (In the class-2 protocol, the transport layer is notified if the network layer fails to carry out its required functions.) Acknowledgements and various timeouts, described later, are used to cope with any problems that might arise.

As noted a number of times, first in Chapter 3 and then earlier in this chapter, the transport protocol entities at either end communicate by exchanging *transport protocol data units,* or TPDUs. (Recall that the standard OSI notation for data unit at layer $N + 1$ is $(N + 1)$-PDU; see Fig. 3 – 7. Figure 3 – 11, which will be repeated shortly, shows the specific use of TPDUs.) A TPDU is issued at either end as the result of a protocol action, just as is the case in HDLC (Chapter 4), at the X.25 packet level (Chapter 5), and in the other protocols and architectures mentioned briefly in Chapters 3, 5, and 6.

Ten TPDUs have been defined for use with the transport protocol class 4; They are listed in Table 7 – 6. Also shown is an eleventh TPDU, the graceful close request (GR) TPDU, defined by NBS for use with its graceful close option.

TABLE 7 – 6   Transport Protocol Data Units

TPDU Type	Code	Amount of Data Carried
CR, connection request	1110	≤ 32 octets
CC, connection confirm	1101	≤ 32 octets
DR, disconnect request	1000	≤ 64 octets
DC, disconnect confirm	1100	none
GR, graceful close request*	0011	none
ERR, TPDU error	0111	none
DT, data	1111	up to negotiated length
ED, expedited data**	0001	≤ 16 octets
AK, acknowledgement	0110	none
EA, expedited data acknowledgement**	0010	none
UD, unit data	0100	up to maximum value of network service data unit (NSDU)

* This is defined only in the NBS FIPS. It does not appear in the ISO standard specifications.
** These are the ISO symbols. NBS uses XPD and XAK, respectively.
(based on [NBS 1983c, pp. 219 – 236])

Most of these will be discussed in the material following. Primary references for this material are [NBS 1983a], [NBS 1983c], and [ISO 1984b]. An additional TPDU, the reject (RJ) TPDU, is defined by ISO for use with classes 1 and 3 [ISO 1984b]. Note that most of these data unit types conform to the primitives described in the previous subsection. It is apparent that receipt or issuance of a primitive is related to protocol actions, resulting in transmission of a TPDU at the transport layer.

Consider the use of these data units as the result of protocol actions. We first summarize the connection-establishment phase and then describe data transfer. The disconnect phase is treated in a similar manner. Figure 7 – 11 shows the exchange of connection TPDUs required for successful connection establishment in class 2 of the transport protocol. The diagram is very similar to that of Fig. 7 – 4, except for the protocol actions (the TPDUs exchanged) that are included. Should the receiving transport user decide not to accept the connection request, it would pass a T_DISCONNECT. request to its transport provider, as shown in Fig. 7 – 5(a). That transport entity would in turn issue a DR (disconnect request) TPDU (Table 7 – 6) in place of the CC (connect confirm) TPDU of Fig. 7 – 11. The DR, on arriving at the peer transport entity, would

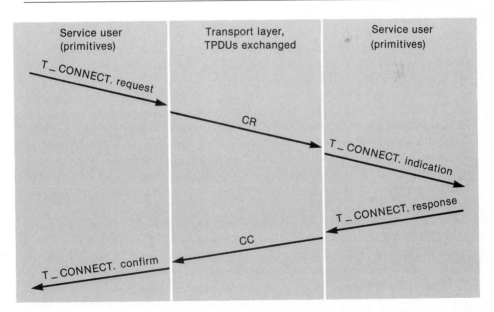

Figure 7 – 11 Exchange of TPDUs, successful connection, class-2 transport protocol

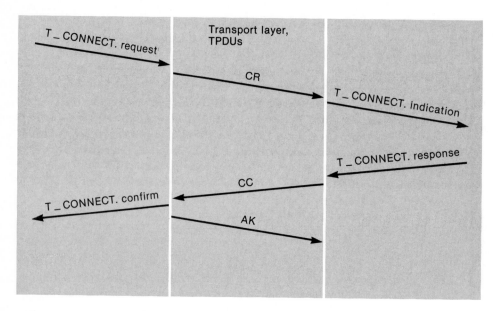

Figure 7–12 Exchange of TPDUs, successful connection, class-4 transport
protocol

cause the T_DISCONNECT. indication of Fig. 7–5(a) to be issued instead of
the T_CONNECT. confirm of Fig. 7–11.

Since network service in class 2 is reliable, there is no need to confirm the
CC TPDU. In the case of the class 4 transport protocol, however, with network
service inherently unreliable, the CC TPDU, acknowledging receipt of the
earlier CR (connect response) TPDU, must itself in turn be acknowledged. The
resultant "three-way handshake" is portrayed in Fig. 7–12. A data or expedited
data TPDU could be used in place of the AK (acknowledgement) TPDU (Table
7–6) shown there. This "three-way handshake" is similar to the "bind triangle"
used by SNA (see Fig. 3–24).

At least two reasons can be advanced for requiring an acknowledgement of
the CC TPDU. In one scenario the transmitter, having sent the CR TPDU,
crashes. The receiver sends the CC and then waits forever! This, of course,
results in a deadlock. An acknowledgement mechanism is required to avoid this
situation. This acknowledgement mechanism includes the AK TPDU of Fig.
7–12, with appropriate timeouts needed at either end of the connections.
(These timeouts are discussed later.) If the transmitter were really down, as

envisioned in this scenario, the receiver would timeout and retransmit the CC TPDU. After a number of such (unsuccessful) retransmissions, it would assume the connection could not be completed.

The second reason for requiring the three-way handshake for the class-4 protocol concerns the possibility of duplicate CR TPDUs. The receiver can normally be designed to detect and discard a duplicate CR for the same connection. However, it is possible for a duplicate CR to arrive some time after a connection has been made and *terminated*. The receiver would then send a CC in reply, thinking that a *new* connection was to be made. It expects acknowledgement of this connection. The transmitter, on receiving the erroneous CC, replies with a disconnect request (DR) TPDU, canceling the connection.

It was noted earlier that connection-establishment negotiation is carried out by the transport protocol. Negotiation includes such items as class number, flow-control option, maximum data TPDU size, whether to incorporate a checksum at the transport level, and quality-of-service parameters. This process of negotiation is made clearer by presenting the relevant TPDU formats. Figure 7 – 13(a) portrays the format, by octet, of the CR TPDU. All TPDUs have a

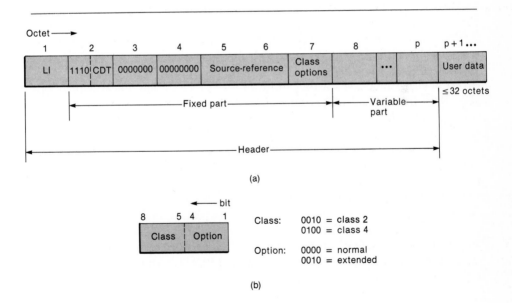

(a)

(b)

Figure 7 – 13 Format, CR TPDU (LI = Header length, in octets)
a. Overall format
b. Octet 7

similar format, as will be seen from the discussion following. (It might be of interest to the reader to compare transport-layer data unit formats with those at the network layer; specifically, the packet formats of X.25 presented in Figs. 5–5 and 5–6. Recall again that the network PDUs or data packets incorporate the TPDUs *within* them in their data fields.) The first octet, labeled LI, indicates the length, in octets, of the rest of the header (the entire CR TPDU, excluding any data carried). The maximum size of the header is 254 octets. This is prescribed for *all* TPDUs, not just the CR. The header is shown grouped into two parts, a fixed part and a variable part. The variable part optionally may carry a number of parameters, each preceded by its own identifying code and of specified or variable length. Note that the fixed part of the CR TPDU begins with the CR identifying code number, 1110 (Table 7–6). The 4-bit CDT *(credit allocation)* is used for flow control and will be described later. The source reference in octets 5 and 6 provides an abbreviated CR sender's id for the proposed transport connection, which will be used during the data-transfer phase. This abbreviated id serves a purpose similar to that of the X.25 logical channel number or of the virtual circuit number at the network layer below. (There is, of course, no connection between these numbers. Recall that multiple transport connections may be multiplexed onto one network connection or VC.) The complete sending (source) and destination addresses of the transport users at either end are carried in the variable part of the CR header. (The connect confirm or CC TPDU will in turn send its abbreviated reference number. During the data-transfer phase, only destination numbers are sent.)

Octet 7, shown broken out in expanded form in Fig. 7–13(b), carries the protocol class suggested for the connection — 2 or 4 — plus a flow control option. Data TPDUs and AK TPDUs, described later, contain sequence-number fields for use with error detection and flow control. Normal flow entails a 7-bit sequence number and a 4-bit credit field. (The concept of credit, similar to that of window at the network layer, is used for flow control, and will be described later in a discussion of the transport protocol flow-control procedure.) Extended numbering, the default option for class 4, is set at 31 bits, with a credit field of 16 bits. (Flow control is always used in class 4; it is an option in class 2. A one in bit 1 of octet 7 disables the flow-control feature in class 2.)*

Both the class to be used and the option selected are negotiable. The one strict rule, noted earlier, is that a service option may be reduced but never may be increased. The values returned by the CC TPDU can thus either be equal to

---

* Note again the reappearance, at each layer of the OSI architecture, of the same general concepts: flow control, sequence numbering, and error detection (described previously at the data link and network layers) plus multiplexing, segmenting, concatenation, and so forth. This idea of having layers $(N + 1)$, N, $(N − 1)$, . . . in a general layered architecture carry out conceptually similar functions was first noted in Chapter 3.

or less than those sent by the CR TPDU. The CC sender may reduce the sequence-number field suggested by the CR sender or leave it the same value but may not increase it. Similarly, if class 4 is selected as the protocol class, the CC sender may choose to reduce it to class 2. But if class 2 is initially selected, the CC sender *must* agree with this choice and so state in the CC.

Some of the parameters carried in the variable part of the CR header, their identifying binary code, and their length in octets appear in Table 7 – 7. Details appear in [ISO 1984b, Sec. 8.3] and [NBS 1983c, pp. 219 – 224].

The maximum TPDU size parameter represents the maximum value proposed for the data TPDUs transmitted during the data-transfer phase. This number varies in powers of two from 128 (since this is 2^7, the parameter length = 00000111) to 8192 (2^{13}, or 00001101). The CC sender can in turn *reduce* the maximum value proposed, in powers of two, but to no less than 128. The *transport service access point identifiers* are the source and destination addresses that are mapped to the two octet reference identifiers during data transmission. Security parameters are not specified and are left to the implementors.

The 2-octet checksum option available in the class 4 protocol is used, if desired, to provide further bit-error protection over and above that available at the data link level. Checksumming is mandatory for CR TPDUs but is an option for the data-transfer phase. (It is not available in any of the other transport protocol classes.) The additional option parameter shown in Table 7 – 7 is used to specify whether checksumming will be used. If *either* the CR *or* CC TPDUs request its use, it must be used. (The same parameter indicates whether expedited data service will be used or not; the default value is to use it. The NBS FIPS mandates its availability during data transfer.)

The checksumming procedure used is called the "Fletcher checksum" after the individual who first investigated its error-detection properties [FLET 1982]. This procedure is specifically designed to be carried out in software rather than with a hardware implementation. It uses integer arithmetic instead of the cyclic-redundancy checking (CRC) or polynomial checking done in HDLC and other data link protocols [SCHW 1977], [SCHW 1980a], [LINS]. The error-detection properties are not quite as good as those of a CRC proce-

[FLET 1982] John G. Fletcher, "An Arithmetic Checksum for Serial Transmission," *IEEE Trans. on Comm.,* vol. COM-30, no. 1, Jan. 1982, 247 – 252.

[SCHW 1977] M. Schwartz, *Computer-Communication Network Design and Analysis,* Prentice-Hall, Englewood Cliffs, N.J., 1977.

[SCHW 1980a] M. Schwartz, *Information Transmission, Modulation, and Noise,* 3d ed., McGraw-Hill, N.Y., 1980.

[LINS] Shu Lin and Daniel J. Costello, Jr., *Error Control Coding: Fundamentals and Applications,* Prentice-Hall, Englewood Cliffs, N.J., 1983.

TABLE 7-7   Selected Parameters, Variable Part of CR TPDU, Class-4 Protocol

Parameter	Code	Parameter Length (octets)
Maximum TPDU size	11000000	1
Calling transport service access point identifier (calling transport address)	11000001	variable
Called TSAP-ID	11000010	variable
Security parameters	11000101	variable
Checksum	11000011	2
Additional options (checksum/expedited service)	11000110	1
Quality-of-service parameters		
Acknowledge time	10000101	2
Throughput	10001001	12
Residual error rate	10000110	3
Priority of connection	10000111	2
Transit delay	10001000	8

dure, but computation is more efficient [FLET 1982]. In its application to the transport protocol, all bits in a TPDU (header plus data) are checksummed. An incoming TPDU is discarded if the checksum is found to be invalid.

The quality-of-service parameters listed in Table 7-7 are precisely those previously discussed in connection with the T_CONNECT. request primitive (Table 7-3). The throughput parameter, for example, would normally be the set of numbers passed down in the primitive that represents the user's requirements for throughput performance. (They may be reduced by the transport layer, based presumably on the availability of networks and network connections.) Four throughput numbers, each three-octets long, are specified. These are, in octets/sec, and in the order in which they appear in the variable part of the TPDU: average or target throughput in the source-destination (or calling-called) direction; minimum throughput in the same direction; average throughput in the destination-source direction; and minimum throughput in the same direction. These numbers are negotiable and can be reduced (but not increased) by the destination (called) transport user.

The residual error rate is defined to be the unreported TSDU loss, for a specified TSDU length. Three numbers appear. The first is the target value and the second is the minimum value, both in powers of 10^{-1}. The third number is the specified TSDU length, in powers of 2. Priority of connection was mentioned earlier. The transit-delay parameter consists of four numbers: the target (average) and maximum delays, in msec, in each of the two directions of transmission. All of these would normally be passed down in the T_CONNECT. request primitive, as discussed in the previous subsection, although they could be relaxed by the transport-layer protocol. They too are negotiable and can be accepted or reduced by the called transport user. The acknowledge-time parameter is not negotiable and is sent as an indication to the receiving transport entity of the maximum data TPDU acknowledgement time. This can then be used for setting acknowledgement timers.

The connect confirm (CC) TPDU format appears in Fig. 7 – 14. As shown by the data unit exchange of Figs. 7 – 11 and 7 – 12, the CC TPDU acknowledges receipt of the connect response (CR) TPDU; it represents a positive reply to the opening of the transport connection. The CC TPDU too chooses a source reference number for use during data transfer in place of its address, and forwards this number, plus the calling reference number — its destination (DST) reference number — back to the calling system. Negotiable options and parameters, as sent down from the called transport user through the T_CONNECT. response, are included as well. As indicated previously, these can have the same values carried in the CR TPDU or can be relaxed. In the class-2 case this completes the connection phase, and the data-transfer phase can begin. In the class-4 case (Fig. 7 – 12), an additional acknowledgement must still be sent.

The data-transfer phase includes the transmission of normal or expedited data units, the DT (data) TPDU or the ED (expedited) TPDU of Table 7 – 6, respectively. Sequence numbering of 7 or 31 bits is used, depending on the outcome of the connection-establishment negotiation process. Consider normal data transfer first. We indicated in the last subsection that a T_DATA. request primitive is used to transfer a transport service data unit (TSDU) from a transport user to its transport entity (Fig. 7 – 8). This TSDU may be segmented into one or more DT TPDUs. The two formats of a DT TPDU are shown in Fig. 7 – 15. Figure 7 – 15(a) shows the format for a 7-bit sequence number TPDU; Fig. 7 – 15(b) displays the format for the extended 31-bit case.

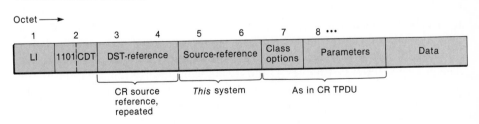

Figure 7 – 14 Format, CC TPDU

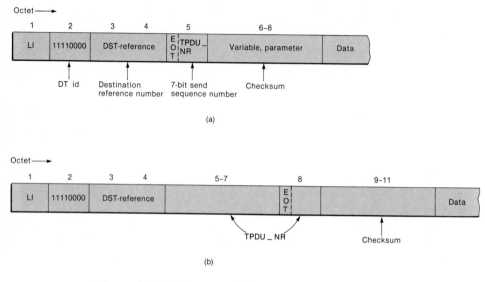

Figure 7–15 Formats, DT TPDU
 a. Normal, 7-bit sequence number
 b. Extended, 31-bit sequence number

The EOT bit shown in both cases is used to identify the last TPDU in the group representing a TSDU. The 4-bit sequence 1111 identifies the DT TPDU (Table 7–6). The 2-octet destination reference number is the number into which the destination transport service access point (TSAP) address is mapped during the connection negotiation process (see Figs. 7–13 and 7–14). A 2-octet Fletcher checksum, preceded by the checksum parameter, precedes the data if that option has been chosen during the connection phase.

We now focus on the send sequence number, labeled TPDU_NR in Fig. 7–15. This number is either 7 or 31 bits long. In the class 2 protocol it is used for flow control only, since DT TPDUs are deemed to arrive in order. (The network provides the reordering service.) In class 4 this number is used not only for flow control, but to reorder DT TPDUs that arrive out of order and to detect duplicate or lost TPDUs. As was the case with sequence numbering at both the data link level (HDLC, Chapter 4) and the network level (X.25, Chapter 5), the send sequence number at the transport level is incremented sequentially with each transmission of a data TPDU. Note that the sequence number space at this layer is generally much larger than at the layers below (7 or 31 bits, compared with 3 or 7 bits, respectively, at the lower layers).

Normal class-4 data transfer, with a single TSDU shown split into $n + 1$ DT TPDUs, is diagrammed in Fig. 7 – 16. The number that accompanies each DT TPDU is its send sequence number. Note that the last DT TPDU in the sequence that represents a complete TSDU is denoted by the symbol EOT. This data unit would have its EOT bit (Fig. 7 – 15) set to 1. Figure 7 – 16 shows the transport layer at the receiver issuing a T_DATA. indication primitive after receiving the complete TSDU. Data also could be passed to the receiving transport user piecemeal, before receipt of the complete TSDU. This is left as an implementation choice. Multiple T_DATA. indication primitives would then be invoked to carry the TSDU segments.

The AK TPDUs that appear in Fig. 7 – 16 are obviously used to acknowledge correct receipt of the DT TPDUs. The acknowledgement procedure is the same as that already described at lower layers of the architectures. An individual DT TPDU or a sequence of TPDUs may be acknowledged. The AK TPDU does this by carrying the number of the next expected DT TPDU, implicitly acknowledging receipt of all those sent before. The AK TPDU shown

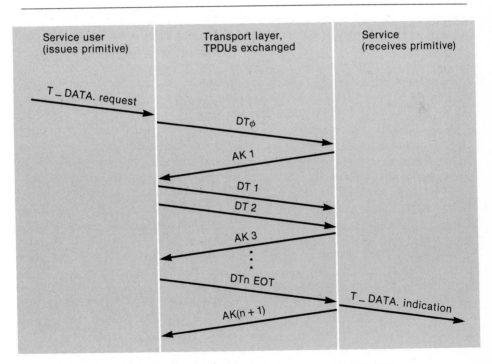

Figure 7 – 16 Normal data transfer, class 4, multi-TDPU message

here is the same as that used in the class 4-connection establishment procedure that requires a three-way handshake, which was diagrammed in Fig. 7 – 12. The format of the AK TPDU (Table 7 – 6) for the case of normal (7-bit) sequence numbering appears in Fig. 7 – 17. The fifth octet carries the 7-bit number (labeled "YR_TU_NR") that indicates the number of the next expected DT TPDU. The variable part of the format, beginning with octet 6, consists of a checksum field, if used during data transmission, plus additional number fields that will be described later. The format of the AK TPDU for the extended numbering case is similar, except for the use of a 31-bit YR_TU_NR field and a 7-bit CDT (credit) field.

The 4-bit CDT field of the AK TPDU of Fig. 7 – 17 is used to carry out *flow control*. As in the case of flow control at the data link layer and end-to-end window control at the network layer, flow control here is managed by the receiver. Its function here, as distinguished from congestion control at the network layer, is to prevent overutilization of the *receiver* resources. As is apparent from the wording, TP uses a credit scheme: The receiver sends an indication through the AK TPDU of how many DT TPDUs it is ready to receive. Unlike the window schemes at the network layer, described in Chapter 5, the receiver may reduce the credit if it senses that its resources (receiver buffer, for example) are becoming depleted. (In the window schemes described in Chapter 5, a window size is negotiated or established before data transmission begins. The window size may be reduced, as in the case of the SNA VR pacing control, if congestion is encountered in the network, but the receiver normally does not change the window size. The receiver in these schemes, as in the case of the data link layer, may withhold acknowledgements or may send a receive-not-ready (RNR) packet to indicate a temporary problem.)

Specifically, the transport protocol class-4 flow-control scheme works as follows: An AK TPDU carries the next expected DT TPDU (this is called the *lower window edge*) and the credit, which is the size of the window of DT TPDUs that may be sent by the TP protocol at the other side of the transport connec-

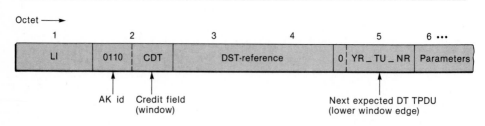

Figure 7–17 Format, AK TPDU, 7-bit sequence numbering

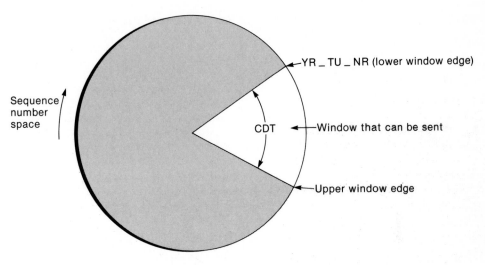

Figure 7 – 18 Flow control via credit allocation

tion. The number of the first DT TPDU that *cannot* be sent, called the uppe
window edge, is then the lower window edge plus the credit (modulo the sequence field size). This window scheme is diagrammed in Fig. 7 – 18. As an
example, say that the current lower window edge sent by an AK TPDU (the next
expected DT TPDU number) is 45. Say that the credit sent by the AK is 10. The
transmitter (the system receiving the AK) is now free to transmit 10 DT TPDUs
numbered 45 to 54.

Initial credit is provided in the CR TPDU for data flowing to the CR sender,
and in the CC TPDU for data flowing to the CC sender. (These numbers are
carried in the CDT fields in octet 2 of these TPDUs. See Figs. 7 – 13 and 7 – 14.)
From then on during data transfer, credit allocation is transmitted via the AK
TPDUs.

The window may be reduced on successive AK transmissions, or it may go
to zero. (This would, of course, be indicated by a zero in the CDT field of the AK
TPDU.) In the class-4 protocol the lower window edge (YR_TU_NR) *cannot*
be reduced, however, since it represents the next expected DT number and
implicitly acknowledges correct receipt of all DT TPDUs of lower number. In
the class-4 protocol the receiver sends an AK TPDU *only* when a DT arrives in
order or completes an appropriate sequence of numbers, since the AK is used
for positive acknowledgement. In class 2 the AK may be sent at *any* time, since
DT TPDUs are expected to always arrive in order. In the class-4 case the upper

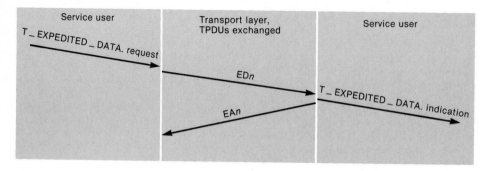

Figure 7–19 Expedited data flow

window edge can be reduced, or may remain the same even if the credit is being reduced, since the lower window edge can increase if the credit is nonzero.

Relatively little quantitative analysis has been carried out of flow control at the transport layer. Sunshine [SUNS 1977] has studied the impact of retransmission interval, window size, and other parameters on the throughput and end-to-end delay performance of a transport-type protocol, using a closed queueing network model similar to that introduced for window-control mechanisms at the network layer in Chapter 5. Hsieh and Kraimeche have extended the window control models of Chapter 5 to include a constraint introduced at the destination receiver on the maximum rate of delivery of data units it can accommodate [HSIE]. This is used to model the requirement that source and destination speeds be matched.

To conclude this discussion of the data-transfer phase of the transport protocol we describe the transmission of expedited data briefly. As noted earlier, a maximum of 16 octets of transport user data may be delivered in expedited fashion, bypassing the normal data flow control. The TPDUs exchanged in this mode of transmission are diagrammed in Fig. 7–19. As should be clear from our discussion in the last subsection, the expedited data mode of TP is invoked by a T_EXPEDITED_DATA. request primitive from the service user

[SUNS 1977] Carl A. Sunshine, "Efficiency of Interprocess Communication Protocols for Computer Networks," *IEEE Trans. on Comm.,* vol. COM-25, no. 2, Feb. 1977, 287–293.

[HSIE] W. Hsieh and B. Kraimeche, "Performance Analysis of an End-to-End Flow Control Mechanism in a Packet-Switched Network," *J. Telecomm. Networks,* vol. 2, 1983, 103–116.

that includes up to 16 octets of data to be delivered to the service user at the other end of the connection. As a result of this primitive the transport protocol at the transport layer sends out an expedited (ED) TPDU. (Recall that the NBS FIPS notation is XPD TPDU.) On receipt of the ED at the other end, the receiving transport protocol immediately acknowledges it with an expedited acknowledgement (EA) TPDU (labeled XAK TPDU in the NBS notation), and delivers a T_EXPEDITED_DATA. indication primitive, which carries the data to the receiving transport service user.

The ED TPDU is transmitted ahead of any pending normal data. Only *one* ED TPDU can be outstanding (not acknowledged) at any time. The receiver *must* accept the ED TPDU even if its normal data credit window is closed. The ED TPDUs are numbered and hence ordered in sequence as shown in Fig. 7 – 19. (Number *n* is there acked by EA number *n*.) A duplicate ED is acknowledged as well but is then discarded.

An expedited data TPDU is *guaranteed* not to be delivered to the transport user at the other end after normal data (DT) TPDUs sent after it. In the class-4 protocol this is accomplished by having the transmitting transport entity sus-

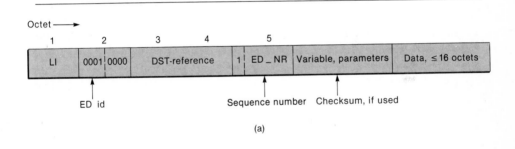

(a)

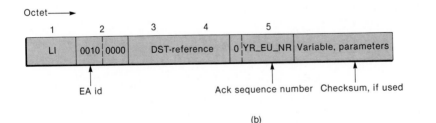

(b)

Figure 7 – 20 Expedited data formats
    a. Expedited data (ED) TPDU
    b. Expedited acknowledgement (EA) TPDU

pend transmission of normal DT TPDUs received from its transport user after the expedited data unit until the acknowledging EA TPDU has been received from the other end. The ED TPDU is kept at the head of the transmit queue until its EA arrives. The formats for both the ED and the EA TPDUs appear in Fig. 7–20. Note that a 7-bit sequence number is used for these TPDUs.

We conclude this section by summarizing the TP mechanisms and the transport-protocol data units sent as a result of the protocol actions. We do this by showing the exchange of TPDUs when a connection is released. Figure 7–21 portrays the abrupt-release phase of the transport protocol [NBS 1983a, Fig. 4.8]; Fig. 7–22 does the same for the graceful close [NBS 1983a, Fig. 4.9]. Note that Fig. 7–21 duplicates part of Fig. 7–6(a), with TPDUs included. Figure 7–22, without the TPDUs, appeared previously as Fig. 7–7. The TPDUs exchanged—the disconnect request (DR), the disconnect confirm (DC), and the graceful close request (GR)—were previously identified in Table 7–6. As noted earlier, the disconnect procedure of Fig. 7–21, specified in the ISO standard, provides abrupt termination of a connection. Any data already sent to either transport entity and not yet delivered is discarded. The graceful close procedure of Fig. 7–22, which requires the additional exchange of AK TPDUs, results in delivery of all pending data. Recall that this procedure is an option in the NBS FIPS class-4 specifications.

As indicated in Table 7–6, the DR TPDU contains up to 64 octets of data. It carries the destination reference for this connection and also a reason code that indicates the disconnect was initiated by the transport service user. The GR TPDU also carries the destination reference but contains no data. It does,

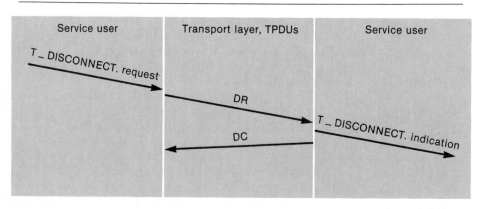

Figure 7–21 Connection termination, disconnect (from [NBS 1983a, Fig. 4–8] with permission)

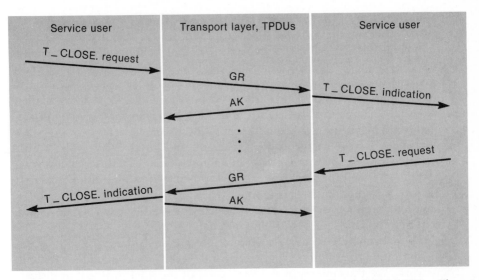

Figure 7-22 Connection termination, graceful close (from [NBS 1983a, Fig. 4-9] with permission)

however, carry the next normal data TPDU sequence number and is therefore subject to flow control. This is necessary in the class-4 protocol to ensure that the GR TPDU *is* the last data TPDU sent on this connection.

7-2 Error-detection and Error-recovery Mechanisms in Transport Protocols

The previous section described both the services provided by the OSI class-4 transport protocol (TP4) and the protocol mechanisms specified by the standard. We indicated in passing that the protocol must be designed to handle a variety of contingencies and error conditions. These include the requirement of detecting duplicate packets or packets arriving out of order when, as in the case of the class-4 protocol, the underlying network (or series of networks through which data might flow) cannot be relied on to provide correct, sequenced delivery of data. A flow-control procedure must be incorporated as well to prevent the receiver resources from being inundated by data that they are not prepared to accept.

This problem of the detection of error conditions and mechanisms to recover from them is a universal one for transport protocols designed to operate in the face of unreliable networks. A great deal of thought and study has gone into the development of procedures for handling these problems should they occur. Much of this work was motivated by the development of the ARPAnet transmission control protocol (TCP) and earlier versions of the transport protocol that preceded it. A number of the solutions to these problems first studied in the context of the ARPAnet transport protocol development, as well as other earlier transport protocols and problems encountered in other layers, have also been incorporated in the current OSI TP4 standard. Some of these solutions — for example, the three-way handshake for connection establishment (Fig. 7–12) — were noted in the last section. It is apparent as well that the use of sequence numbers and acknowledgement data units in TP class 4 is required to detect lost, duplicated, or out-of-sequence data TPDUs.

It is useful, however, to pose some of the problems that involve possible error conditions and solutions suggested for them in a somewhat more general framework. We thus interrupt our discussion of the TP4 protocol to discuss some of these transport protocol design issues in a broader sense. (The discerning reader will note that a number of the issues raised here — such as detection of out-of-sequence or duplicate data units, avoidance of deadlock situations, size of sequence-number space required, correct establishment of a connection, number of acknowledgements needed — occur at other layers as well and were mentioned briefly in our discussion of the data link layer.) After discussing these issues in general we return to the OSI TP4 in Section 7–3, where we describe how provision has been made for handling them. In the final section, on the ARPAnet-developed TCP, we note some of the solutions adopted there to these possible conditions.

The basic design issue that must be answered in the case of a transport protocol that uses an unreliable network or series of networks to transmit transport data units is this: The protocol is required to function correctly despite the fact that packets may remain undelivered for relatively long periods of time, during which connections may have been opened and closed a number of times [GARL]. It is apparent that data units must be numbered sequentially in order to detect duplicate data units or data units that arrive out of order. A positive acknowledgement with a timeout procedure must be incorporated as well to ensure that lost data units or those received in error are properly retransmitted. Sequence numbering and acknowledgements for the OSI TP4 were described in the last section.

[GARL] L. L. Garlick, R. Rom, J. B. Postel, "Reliable Host-to-Host Protocols: Problems and Techniques," Fifth Data Communications Symposium, Snowbird, Utah, Sept. 1977, 4–58—4–65.

Consider now a problem that arises in a transport protocol due to an old data unit having arrived after a connection has been closed and later reopened [SUNS 1978]. Figure 7-23 portrays an example of such a scenario between two users A and B. (Time lines are shown horizontal to simplify the figure.) A opens the connection and sends the first data unit, numbered 0. This arrives at B and is acknowledged with an indication that number 1 is the next data unit expected. A sends number 1, but this data unit is unaccountably delayed during transmission. A times out and retransmits the data unit. This is received and accepted by B, which then closes the connection. Some time later, A again opens a connection with B. It proceeds to initiate data transmission again by sending a new data unit numbered 0. It sends a second data unit after number 0 has been positively acknowledged. This data unit number 1 is also delayed during transmission. By this time the *old* data unit 1, sent during the previous connection, arrives, is mistaken for the *new* number 1, and is so acknowledged by B!

There are some obvious ways to avoid this problem: One might use a very large sequence number space, large enough to cover transmission of data units during the maximum possible delay of a data unit over the network (or series of networks) between the end-to-end uses. (See Fig. 7-24 for a hypothetical example of a "meandering path" through a network.) On the opening of a new connection within this interval of time, then, the sequence number would continue to be incremented, and a new number used with the initial data unit. This would allow the duplicate data unit to be recognized. A problem with this approach, however, aside from the possibly extra-large sequence field required, is that both sides of the connection must remember sequence numbers from connection to connection. A more serious problem arises if one or both of the user systems crashes. How does either side then remember the sequence number to use?

One solution is to delay starting a new connection until at least the maximum possible delay of a data unit through the entire network or set of networks that encompasses the transport path has been exceeded. One could then safely use the same initial sequence number again. Alternatively, one might separately label each connection. Such a label has been termed the *incarnation number:* a unique name or number that is sent within each data unit in a given connection [GARL]. As will be noted in the next section, TP4 uses the source and destination reference numbers (Figs. 7-13 through 7-15) for this purpose. TCP incorporates the concept of a socket address in the same manner. To guard against any problem with duplicate data units, the same number cannot be assigned to a new connection until the maximum data unit delay has been

[SUNS 1978] C. A. Sunshine and Y. K. Dalal, "Connection Management in Transport Protocols," *Computer Networks,* vol. 2, 1978, 454-473.

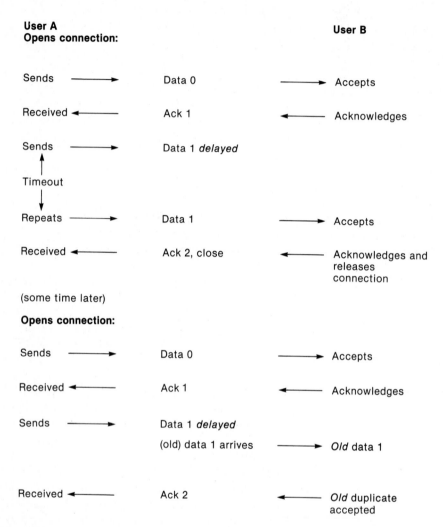

Figure 7 – 23 Duplicate data unit acceptance (after [SUNS 1978, Fig. 2] with permission)

exceeded. This time is called the *reference time* in the TP4 specifications [ISO 1984b]. It is referred to as the *maximum packet lifetime* in the literature on transport-protocol problems and solutions [GARL], [SUNS 1978].

The relation between the incarnation number (or its equivalent reference numbers or address) and the startup delay in opening a new connection is easily expressed as follows [GARL]: Let the reference time or maximum end-to-end delay in receiving a data unit be L (Fig. 7 – 24). Say that an r-bit field is used to represent the incarnation number. Then 2^r different numbers may be uniquely chosen. Since they must cover the entire reference time L—i.e., the same number should not be used again in the interval L—the delay in setting up a new connection should be at least equal to $L/2^r$. As an example, say that this delay is to be about 2 sec. Let r be 8 bits. Then the maximum data unit delay or reference time L that can be accommodated is 512 sec, or 8.9 minutes. This should appear to be sufficient for most purposes. If not, and if the delay between opening a connection is allowed to increase to 10 sec, for example, then $L = 45$ minutes can be accommodated. In the case of TP4, 16-bit fields are used for the reference numbers (Figs. 7 – 13 through 7 – 15). This reduces the effect of the problem even more. (One critical question, of course, is the estimate of the reference time L. This will be spelled out more specifically in the next section. It is clear, however, that one can only guess at this time in a very approximate way. Making it large enough to encompass almost any contingency in a sense eliminates the problem.)

The combination of the incarnation number and the sequence number uniquely identify a data unit. There is always the slight chance that *both* of these

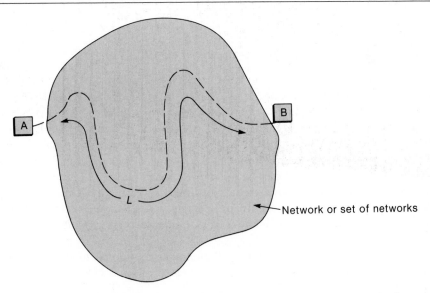

Figure 7 – 24 Maximum data unit delay L, end to end, in connectionless network delivery

numbers might wrap around just as the system crashes. This could lead to an ambiguity on recovery [GARL]. It is apparent, however, that this chance is slight indeed if the sequence number field is 7 or 31 bits (2^7 or 2^{31} possible data units) and the reference number is 16 bits long.

Another possible way to guard against old duplicate data units being mistaken for new ones is to *age* data units as they move through the network(s). They can then be dropped or destroyed once they exceed the age L, and they therefore never arrive at the receiver after time L [SUNS 1978]. This also requires waiting time L, however, before restarting a new connection if a system has failed. It also requires a clock with high-enough resolution (hence a large-enough "age" field in the data unit) to carry out this approach. For example, say that 1-sec resolution is required, and $L = 512$ sec. Then a 9- or 10-bit aging field is needed. But if a data unit takes less than 1 sec to be processed at an intermediate node in a network it is not aged. If this happens too often during a traversal of an end-to-end path, the aging procedure fails. Increasing the resolution to avoid this problem requires a correspondingly larger age field. This problem and ones like it have been raised, in a different context involving datagram routing at the network layer, by R. Perlman [PERL].

Consider now the case of connection establishment. We again demonstrate the need for a three-way handshake or some other way to detect out-of-sequence data units [SUNS 1978]. The basic problem here is again that of synchronizing on the starting sequence number in the first data unit transmitted. The transport protocol at each end of the connection could do this by keeping enough state information about the various closed connections to know what sequence number to expect and thus to recognize old, previously undelivered data units. This would require remembering the sequence numbers and incarnation numbers (if used) for each connection for the reference time L [SUNS 1978]. If a system were to fail, a connection could not be restarted for a period of time L.

An alternative procedure would have the user who initiates the connection send a specified starting sequence number with the connection-request message. (A default value could be zero, in which case the connect message would not have to carry this number. This is precisely what is done in TP4. Recall from Fig. 7–13 that the CR TPDU carries no sequence number. It does serve to resynchronize the data TPDU stream that follows to the defined initial value.) The procedure serves to synchronize the users of the two ends of the connection. The receiving system acknowledges with its own specified sequence number.

Two examples that indicate the problem arising appear in Fig. 7–25. In Figure 7–25(a), the two-way TP class-2 call-establishment procedure (see Fig.

[PERL] R. Perlman, "Fault-tolerant Broadcast of Routing Information," *Computer Networks,* vol. 7, no. 6, Dec. 1983, 395–405.

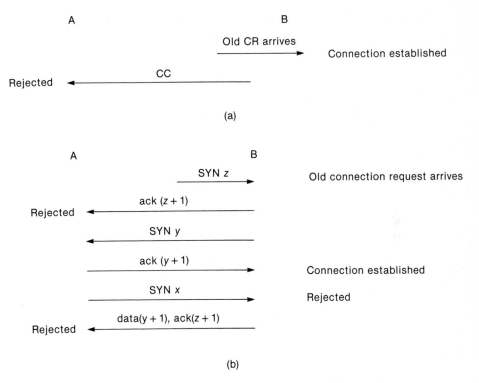

Figure 7 – 25 Possible deadlocks, call establishment, unreliable network
  a. TP class-2 procedure, unreliable network(s)
  b. Deadlock on connection establishment (after [SUNS 1978, Fig. 4] with permission)

7 – 11) is assumed to be used with an underlying unreliable network. The connection between A and B has been closed. An *old* CR TPDU, from A to B, that has been wandering around the network (see Fig. 7 – 24) now arrives. B thinks it is a new connection request, agrees to the opening of the connection, and returns a CC TPDU. A, on receiving the CC, rejects it, knowing that it is a reply to an old connection-establishment request. In the meantime B waits, and waits, and. . . . A second, somewhat more complex example, due to [SUNS 1978, Fig. 4] and sketched in Fig. 7 – 25(b) in modified form, shows a deadlock clearly developing. Here the connection-request data units are labeled SYN and carry a sequence number that will be used to resynchronize the data stream in the same direction.

An old connection-request data unit, SYN z, carrying the sequence number z, arrives at B. As in Fig. 7–25(a), B interprets this to mean that a new connection is desired. It returns ack $(z + 1)$ to A, which is promptly dropped there. In the meantime, B proceeds to send its own SYN y, indicating that *its* data transfer will begin with the number $(y + 1)$. A agrees to this (apparent) connection request from B, and replies with ack $(y + 1)$ (it is now expecting data unit $(y + 1)$). B now considers the connection established. A now follows with *its* start sequence number x, transmitting SYN x. (This is in reply to SYN y.) But this is rejected by B, since it thinks a connection has already been established. B now transmits its first data unit, which is rejected by A. A deadlock has been established. It is left to the reader to show that a three-way handshake (see Fig. 7–12) will resolve this problem.

A detailed discussion of the three-way handshake, references to its earlier use, and a treatment of other problems involving connection management in transport protocols appear in [SUNS 1978]. Connection termination, including a description of both the abrupt and the graceful close, is discussed as well. A timer-based protocol that can be used in lieu of the three-way handshake for both opening and gracefully closing a connection reliably has been proposed by Fletcher and Watson [FLET 1978]. The method requires rather specific knowledge of the worst-case data unit delay L in the network. It might thus be applied to smaller networks that cover a geographically confined area in which L can be controlled or bounded tightly. In such a situation the method would appear to save on the transmission of control messages and might be particularly useful in single-message, connectionless transmission [FLET 1978].

We now return to our discussion of the OSI class-4 transport protocol, describing some of the error-detection and recovery mechanisms incorporated specifically in that protocol. Since we focus on a particular protocol the discussion is much more detailed than the discussion of this section; some of the material overlaps the more general treatment here. However, the detailed description of the timer mechanisms used in TP4, as well as synchronization of the flow control credit fields, extends this discussion considerably.

7–3 Class-4 Transport Protocol Error-detection and Error-recovery Mechanisms

In this section, as just noted, we describe mechanisms specifically incorporated in the OSI class-4 transport protocol for detecting and recovering from error

[FLET 1978] J. G. Fletcher and R. W. Watson, "Mechanisms for a Reliable Timer-based Protocol," *Computer Networks*, vol. 2, 1978, 271–290.

conditions that might arise.* The three-way handshake for ensuring that a connection is established correctly has already been described and will thus not be discussed further. The stress will be on timer mechanisms incorporated with the various acknowledgements used in TP4 to ensure correct operation of the protocol.

Recall that for the class-4 TP, it is assumed that the underlying network does not guarantee reliable, sequenced delivery of packets. Hence TPDUs may be lost, duplicated, or arrive out of sequence. The transport protocol must detect these error conditions and recover from them. This was precisely the situation stressed in the previous section. In addition, TPDUs may arrive garbled. The checksum described earlier (Fig. 7-15 and Table 7-7) is used to detect this situation. If the checksum detects an error, the TPDU is dropped and hence is considered lost. Error recovery, as in the lost TPDU case, will then be invoked.

The primary vehicle for detection and recovery from these types of errors is the familiar use of acknowledgement and timer combinations. The NBS version of TP class 4 has 12 timers defined to handle various contingencies. In the material following we discuss most of them, as well as other mechanisms used to handle problems that may arise. The timers are required not only to recover from various error situations but to prevent deadlock in some cases as well.

As the first example of an error-recovery procedure, consider the one just noted above of detecting and retransmitting a lost TPDU. Recapitulating our discussion of the class-4 TP in Section 7-1, recall that five types of TPDUs are used in the connection-oriented mode of transmission in setting up a connection, transferring data, and then terminating the connection. These are, in the order of use in each of these phases, the CR, CC, DT, ED, and DR TPDUs. (See Table 7-6 and Figs. 7-12, 7-16, 7-19, and 7-21, which describe their use.) The GR (graceful close) request, as noted earlier, is treated like a DT TPDU, although it carries no data. Each of these TP data units is acknowledged in the class-4 procedure. A CR is acknowledged by a CC, for example; the CC in the three-way handshake is in turn acknowledged by an AK (or by a DT, as the case may be). (See Fig. 7-12.) DTs are in turn acknowledged by an AK (Fig. 7-16); the ED is acknowledged by an EA (Fig. 7-19); and the DR is acknowledged by a DC (Fig. 7-21). Because of the use of a positive acknowledgement for each of these TPDUs, a timer can be set on the transmission of each one. On expiration of the timeout specified, with no appropriate reply received from the other side of the transport connection, the TPDU in question is repeated. An example appears in Fig. 7-26. The data TPDU numbered 2 receives no acknowledgement after T1 units of time and so is repeated.

---

* We draw particularly on [NBS 1983a, pp. 45-51], [NBS 1983c, pp. 44, 45], and [NBS 1983e, pp. 2-13].

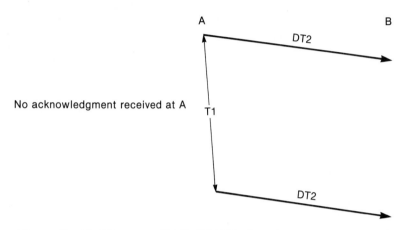

Figure 7–26 Timeout of DT TPDU, class-4 transport protocol

The five retransmit timers noted here appear in the list of 12 shown in Table 7–8, which are specified in the NBS FIPS document for the class-4 protocol [NBS 1983c, pp. 44–48]. The notation used is a mixed one, conforming to ISO for the values of timers and to NBS for their names.

These timers are all shown being set for the same interval T1. This time must be at least the time required for the TPDU and its acknowledgement each to transit the end-to-end transport connection, plus processing time at both ends. With the value E representing the average one-way end-to-end transit delay (Fig. 7–27), X the time to process a TPDU, and A_R the time for the remote transport entity to respond to a TPDU (Table 7–8), the timer value should obviously be given by at least

$$T1 = 2E + X + A_R \qquad (7-1)$$

This is in form very similar to the timeout t_{out} described in Chapter 4. The timeout there, however, at the *link* layer, involved the point-to-point propagation time from one node of a network to its neighboring node.

The end-to-end delay E in turn depends on the network or series of networks that form the transport connection. As an example, say that the network is a terrestrial one 5000-km long, with an average of three transmission hops required for end-to-end communication (Fig. 7–28). Say that 56-kbps transmission links are used, with data link frames 1120 bits long. [Packets, network data units, are 1070 bits long.) Traffic intensity throughout the network is $\rho = 0.5$. Then the propagation delay, using $c = 150,000$ km/sec as an average propagation velocity, is 33 msec. The frame-transmission time is 20 msec, and

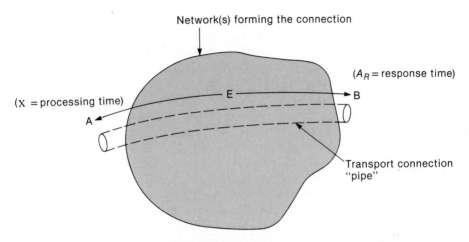

Figure 7–27 Transit delay over a transport connection

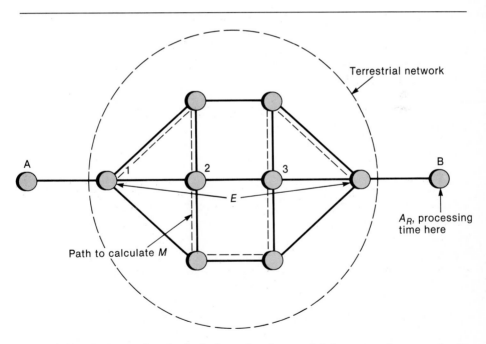

Figure 7–28 Example of calculation of end-to-end delay; queueing at nodes 1, 2, 3; end-to-end span = 5000 km

the queueing delay at one node, including transmission time, is about 40 msec, assuming that the M/M/1 approximation is valid (Chapter 2). For three hops this becomes 120 msec, for a total end-to-end delay of $E = 153$ msec. A satellite link would obviously increase this number (Chapter 4). Going through two or more interconnected networks would increase the transit delay as well.

For local area networks, which are discussed in Chapter 9, the propagation delay would be much smaller, on the order of 10 μsec for a 2-km network. Queueing, transmission, and processing delays would then account for the bulk of the delay that is used to set the timers.

In TP4 an unacknowledged TPDU may be retransmitted a maximum number of times, N, before some action must be taken. In the NBS FIPS version of TP class 4, a *give-up timer* is set any time the value N is reached, in the case of repeatedly retransmitted and unacknowledged CR, CC, DT, GR, and DR TPDUs (Fig. 7–29). The transport connection is closed if, at the expiration of this timer, the appropriate acknowledgement still has not been received. (It is then implicitly assumed that the network connection has been broken or that the transport entity at the other end has crashed.) An upper-bound $r$wait on this timer value, specified by the ISO TP4 standard, is given by Table 7–8:

$$r\text{wait} = 2M + A_R \qquad (7-2)$$

Here M is the maximum lifetime or transit delay of network service data units (NSDUs) in the underlying network or networks that provide the transport connection. An example of the calculation of M appears in Fig. 7–28. Note that M is normally much greater than the end-to-end transit delay E. A_R is again the time for the remote transport entity to acknowledge the receipt of a TPDU (Fig. 7–28). The factor of $2M$ obviously accounts for the maximum time to transit the network in both directions.

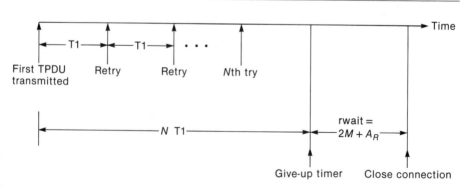

Figure 7–29 Use of give-up timer, TP4 protocol

TABLE 7 – 8   Timers, Transport Protocol, Class 4

Name[1]	Function	NBS FIPS Value[2]
initiate	CR TPDU retransmit	$T1 = 2E + X + A_R$
retransmit-CC	CC TPDU retransmit	T1
retransmit	DT TPDU retransmit	T1
x. retransmit	ED TPDU retransmit	T1
terminate	DR TPDU retransmit	T1
give-up[3]	Final check on unacked TPDUs	$\leq rwait = 2M + A_R$
give-up-xpd[3]	Final check on unacked ED TPDUs	
inactivity	Maximum time transport connection can remain inactive	I
window	Maximum time between transmission of AK TPDUs.	W
flow-control (optional)[3]	Synchronizes flow control information; AK may be transmitted if necessary.	
reference	Time during which source reference or sequence number cannot be reassigned	$L = 2M + A_R + R$
incoming-nc[3]	Time network connection is held open	

Notes
[1] These are names assigned in [NBS 1983c].
[2] Notation follows that of [ISO 1984b].
[3] Not defined [ISO 1984b].

Definitions
E = End-to-end transmit delay
X = Time for local TPDU processing
A_R = Time for remote transport entity to respond to receipt of TPDU
M = Bound on maximum lifetime of network service data units in the underlying network(s)
R = Persistence time; maximum time the local transport entity continues to retransmit TPDUs
N = Maximum number of retransmissions of a TPDU
L = Reference time = $2M + A_R + R$

As is apparent from Eq. (7 – 2) and Fig. 7 – 28, the give-up timer is used to account for any network packets that may have been routed over a particularly long path through the network(s). It is required for datagram networks, which have no tight bound on the end-to-end transit delay E. A give-up-xpd timer, in the NBS FIPS transport protocol class 4, is used to provide the same function for expedited transport data units (Table 7–8). Its value may be less than that of the give-up timer because of expedited delivery of these data units.

Both the give-up and the give-up-xpd timers provide a way to close transport connections after the TPDUs are retransmitted repeatedly with no acknowledgement. These timers are required in the case of underlying datagram networks. If the underlying network is of the connection-oriented (virtual circuit) type, they may not be needed, since the transit time E through the network or networks comprising the connection can be more readily bounded. In that case the transport connection might be closed if N retransmissions of a TPDU with no acknowledgement were reached.

In addition to providing a way to close a connection if transport data units are not getting through, a way must be provided to close a connection if no TPDUs have been received for a period of time. This is done by the use of the *inactivity timer* (Table 7–8). This timer is reset any time a TPDU from the peer entity is received. If the timer expires with no further TPDUs received, this is again used as an indication that the network connection has been broken or that the peer transport entity has crashed, and the transport connection is closed.

The value I of this timer is set at several times the value of a *window timer W*, which in turn is prescribed to be the maximum time that can elapse with no transmission of an AK TPDU. The distinction between I and W is diagrammed in Fig. 7–30. The window timer is reset each time an AK TPDU is sent. If the time W elapses with no further AK TPDUs sent, the previous AK TPDU is transmitted, carrying the same sequence number but with possibly different credit. The use of the window timer ensures some activity between transport peers in the absence of data flow. It prevents the following deadlock situation from developing: Say that one of the transport entities has exhausted its credit. It stops transmitting. If the other entity, the receiver, now sends no AK TPDU or sends one and it is lost, the transmitter remains silent. The use of the window timer ensures that successive AK TPDUs will be sent at intervals of no more than W units of time apart. If a number of these in succession do not get through, indicating a loss of the network connection or a problem with a transport entity, the inactivity timer will expire, and the transport connection will be closed.

The *reference time L* (Table 7–8) represents the length of time during which a sequence number or source reference (the source's id, used during a connection) cannot be reused to avoid confusion or ambiguity. This parameter was discussed at some length in the previous section. If a transport entity crashes, after restart all active references (source ids) must be timed out with this time interval. This point was made previously using somewhat different terminology. Recall that the reference time L represents the maximum length of time a TPDU would be expected to stay in the network(s). TPDUs thus would be flushed out in at most this time, so that spurious duplicate TPDUs could not interfere with a new transport connection. An upper bound on this time pro-

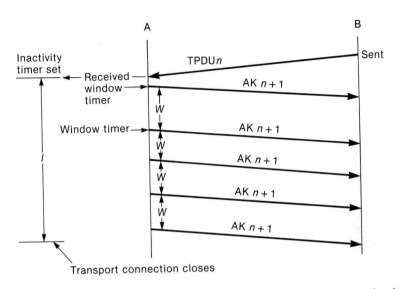

Figure 7-30 Use of inactivity and window timers: no TPDUs received for I units of time

vided in the TP4 protocol specification is the time R required to retransmit the maximum of N TPDUs (the *persistence* time) plus the bound on the give-up timer value $r$wait:

$$L = r\text{wait} + R = 2M + A_R + R \qquad (7-3)$$

(See also Table 7-8.)

The persistence time R is in turn at least equal to N T1 plus the local TPDU processing time X.

$$R \geq N\ T1 + X \qquad (7-4)$$

The relation between these different parameters is diagrammed in Fig. 7-31.

As an example of the choice and use of these numbers, consider the 5000-km wide-area network noted earlier (Fig. 7-28). For that example, with 56-kbps links, 1120-bit packets, three hops from one end of the connection to the other, and an average network utilization of $\rho = 0.5$, we found the round-trip transit delay to be $2E = 306$ msec. Say that the remote-entity processing time is $A_R = 34$ msec, while the time to do local TPDU processing is $X = 10$ msec. (These numbers are chosen arbitrarily here to obtain simple numbers with which to work.) Then the retransmit-timer value is calculated to be T1 = 350 msec. Let the maximum network lifetime of packets arbitrarily be $M =$

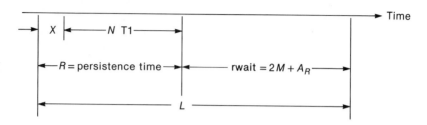

Figure 7–31   Reference time setting, TP class-4 standard

$3E = 459$ msec. The maximum give-up time is then, from Eq. $(7-2)$, rwait $= 952$ msec. Let the maximum number of retransmissions of a TPDU be $N = 5$. Then the persistence time is $R \geq 1760$ msec. Finally, the reference time is

$$L = 2.7 \text{ sec}$$

for these numbers. A reference number would not be reassigned for at least this interval of time to ensure that all TPDUs are flushed out of the system. Note that this is much less than the values for L mentioned in the previous section. Since this is the time, as noted previously, during which a sequence number should not be reused [ISO 1984b, sec. 7.4.3.6], this might also put a limit on the throughput, or number of DT TPDUs that can be transmitted on a transport connection in this interval of time. To check this, let the sequence number field be 7-bits long. This implies no more than $2^7 = 128$ data TPDUs transmitted in the interval 2.7 sec long. Since we started with 56-kbps lines and 1120-bit frames in this example, this seems quite adequate. (Say that the TPDUs are each 128-bytes or 1024-bits long. With 1120-bit, or 140-byte, HDLC frames, the network data units, the packets, are 1072-bits, or 134-bytes long. The numbers are thus quite consistent. The maximum link throughput is 50 frames/sec, or 135 frames in 2.7 sec.)

Consider another, more conservative calculation. Say that larger networks are used, each again covering 5000 km, with a typical path in one network running five hops end to end. Links are again 56 kbps, with frames also 1120 bits long. Let the transport path cover three such networks, each pair connected through a satellite link. An end-to-end path thus covers 15 hops for a total queueing plus transmission delay of 600 msec. Propagation delay is 599 msec (two satellite hops plus three terrestrial networks). The total end-to-end delay is then $E = 1.2$ sec. Letting $X = 10$ msec and $A_R = 34$ msec, as before, the retransmit timer value is then

$$T1 = 2E + X + A_R = 2.44 \text{ sec}$$

Let the maximum network lifetime of packets now be $M = 10E = 12$ sec because of the complexity of the networks involved. The maximum give-up time is now $rwait = 2M + A_R = 24$ sec. Let $N = 10$ be the maximum number of retransmissions of a TPDU. Then the persistence time is $R \geq N\ T1 + X = 24.4$ sec. Finally, the reference time is

$$L = 48.8 \text{ sec}$$

Note the difference in value for this use and that of the simpler network example worked out first. This indicates the wide variation in lifetime and reference time expected because of the many types of connections and networks encountered on a global scale. For this example extended sequence numbering appears to be required. The maximum possible throughput, limited by the 56-kbps links, is 50 HDLC frames per second. The same limit applies to TPDUs. In 50 sec one could thus transmit at most 2500 TPDUs. A sequence number field of 7 bits, covering 128 TPDUs, is thus clearly inadequate. A range of 2^{31} possible TPDUs would, on the other hand, be more than adequate. This indicates the need to negotiate the size of the sequence-number space during the connection phase.

It is apparent that the various timer values just discussed sometimes are only approximately known, at best. End-to-end transit delays depend on a variety of factors, as noted previously, and some of these delays may be difficult to estimate. One could always *overestimate* the various values used, which of course would reduce the throughput if carried to an extreme. Alternatively, what happens if the numbers are too small? In the case of the retransmit timers for the various TPDUs, this would result in increased retransmission of TPDUs and consequent overuse of resources. In the case of the reference timer, it is apparent that its value L could easily be chosen too small since the maximum lifetime of packets in a network or series of networks is not readily estimated.

With the reference time L too small and a source reference number reused too soon for another connection, duplicate TPDUs from the previous connection may arrive during the new connection. How are they recognized? In the case of duplicate data TPDUs, they would hopefully be detected because of their out-of-order sequence numbers. Duplicate CR (connect request) TPDUs, on the other hand, are handled by the connection-establishment three-way handshake (Fig. 7-12). Specifically, say that a duplicate CR TPDU arrives. This generates a CC TPDU. When the CC TDPU arrives at the CR sender, that entity, recognizing the duplication, sends a DR (disconnect request) TPDU instead of an AK TPDU.

One final problem that must be handled by the TP class-4 protocol relates to the credit mechanism used for flow control, as described earlier. The problem is that of ensuring that the transmitter and receiver at each end of the transport connection have the same information concerning the credit field. Recall that the credit is delivered to the transmitting side by the AK TPDU from the receiver. (Note that with full-duplex transmission each side plays the role of both a transmitter and receiver.) Say that the credit field is set to zero by an AK TPDU. Sometime later the receiver sends another AK that carries the same sequence number and increases the credit. Because of the unreliable properties of the network or networks underlying the transport connection, the second AK may arrive first. This would *increase* the credit, rather than *reduce* it!

To overcome this problem, AK TPDUs that reduce credit, or set the credit field to 0, carry a *subsequence number* in the variable parameter field (Fig. 7–17) that is increased for every credit reduction. This number is reset to 0 only when further DT TPDUs are acknowledged. The subsequence number is then used as part of the following algorithm to determine whether a transmitter should accept an AK TPDU as "in order" ([NBS 1983a, pp. 49, 50], [ISO 1984b, sec. 7.4.5.6]).

Accept an AK TPDU only if

1. It acks DT TPDUs with higher sequence numbers than previously; or,

2. It acks the same DT TPDU as the previous AK TPDU, but has a higher subsequence number; or,

3. It acks the same DT TPDU as the last AK TPDU and has the same subsequence number but carries increased credit.

The acknowledgement window timer W discussed earlier (Fig. 7–30 and Table 7–8) is required for this algorithm to function properly. This is left as an exercise to the reader to explain. The window timer is also used more generally to prevent deadlock, as already noted. The problem with the use of this acknowledgement window-timer mechanism is that it results in the periodic transmission of AK TPDUs when no data is flowing, possibly wasting resources. (The transmission of AK TPDUs must be frequent enough to ensure reasonable response, thus providing appropriate throughput and time delay.) There is a way to reduce the number of AK TPDUs sent: Provide acknowledgement of the AK TPDUs themselves. Such a mechanism is included as an option in the NBS FIPS TP class 4 protocol [NBS 1983a], [NBS 1983c]. With this option the AK TPDU contains a "flow-control confirmation" parameter that acknowledges receipt of an AK TPDU in the other direction. The receiver then knows, on receipt of this AK TPDU, that the sender has the correct values both for acknowledged DT TPDUs *and* the flow-control windows. A *flow-control timer* (Table 7–8) is required with this option. When the timer expires an AK TPDU will be transmitted, but only if the credit field is zero. This is necessary to prevent deadlock. (Why?) The number of AK TPDUs transmitted can thus be reduced. The window timer is still necessary, but its value W may be increased substantially, to just less than the inactivity interval I (Fig. 7–30). This ensures some activity between peer transport entities in the absence of data. The inactivity timer is still

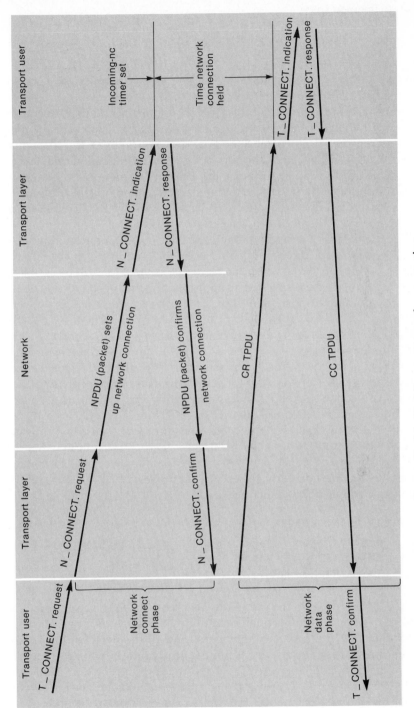

Figure 7-32 Holding a network connection

always needed to ensure that if one side of the connection crashes, the other side will eventually close the connection.

One final timer appears in Table 7 – 8. This is the incoming-nc timer, which represents the maximum time an incoming network connection is held open while the transport protocol machine awaits the arrival of a CR TPDU on that connection. Specifically, once the network layer at a destination side, using the N _ CONNECT. indication primitive, has indicated to the transport layer that a source transport entity would like to set up a connection, the destination entity awaits arrival of a connect response (CR) TPDU. If the TPDU does not arrive in the time specified, the network connection is closed.

The use of this timer is diagrammed in Fig. 7 – 32. Connection-mode network service is assumed here, with a network connection (i.e., a virtual circuit) required to be set up over which the transport connection will be established. Figure 7 – 32 further assumes that an appropriate network connection does not exist to carry the transport connection, in which case one must first be set up. The figure then shows the events that occur from the time a T _ CONNECT. request primitive is issued at the source transport user to the time the T _ CONNECT. confirm primitive is received in reply. The network connection is set up first, with the source transport entity issuing an N _ CONNECT. request primitive to its network service provider. The network-layer protocol in turn causes a network-connection packet to be transmitted through the network. (This is the VC setup packet; see Chapter 5 for an example from X.25.) On arriving at the destination side this causes the network-layer protocol at that end to issue an N _ CONNECT. indication to its (transport) user. This indicates that a CR TPDU will be coming shortly to set up the transport connection. The incoming-*nc* timer is set at this time. If the timer runs out before the CR TPDU is received, the network connection is closed down.

7 – 4 Summary, TP Class 4 — Finite State Machine

Figure 7 – 33, taken from the NBS FIPS TP class-4 protocol specification [NBS 1983c, Fig. 3.1], shows a portion of the finite-state machine that describes the operation of that protocol. It summarizes some of the protocol functions described in previous sections. The full finite state machine involves two additional figures similar to, and connected to, portions of Fig. 7 – 33 [NBS 1983c, Figs. 3.2 and 3.3]. Some state transitions that involve expedited data and receipt of out-of-sequence and erroneous TPDUs have been omitted to prevent overcrowding of the figure. Transitions numbered 1 – 23 relate to the connection-establishment phase; 24 – 34 involve the connection-termination phase; 37, 42,

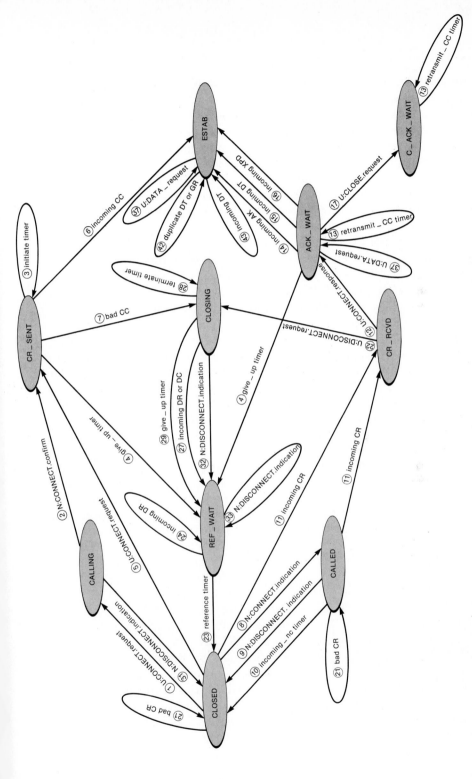

Figure 7–33 Class-4 finite state machine: connection-oriented transitions (from [NBS 1983c, Fig. 3–1] with permission)

385

and 43 are in the data-transfer phase. We describe a few transitions only to indicate the significance of this diagram.

Three types of transitions are shown: those involving primitives, TPDUs, and expiration of timers. The symbol U stands for the interface with the transport user; the symbol N, for the interface with the network entity.

Consider the transition labeled 1 as an example. This involves a transition from the CLOSED state, on receipt of a T_CONNECT. request at the user interface, to the CALLING state. The CLOSED state represents the one for which the transport connection is closed. The T_CONNECT. request indicates that the user wishes to establish a new transport connection. A network connection does not exist in this case and must first be set up. The transport protocol machine issues an N_CONNECT. request for this purpose, and then waits in the CALLING state for an N_CONNECT. confirm, confirming the establishment of the network connection (see Fig. 7–32). Transition 2 indicates that on receipt of the N_CONNECT. confirm, the transport machine moves to the CR_SENT state, after sending a connection request (CR) TPDU to the peer transport entity with which it wishes to establish the transport connection.

Now consider the case in which a network connection to the peer transport entity already exists, and the new transport connection can be multiplexed onto it. There is thus no need to set up a new network connection. The transport machine then moves directly to the CR_SENT state on receipt of the T_CONNECT. request from the transport user, after sending a CR TPDU. This is indicated by transition 5. While in the CR_SENT state the transport machine awaits the arrival of the connect confirm (CC) TPDU. On sending the CR TPDU the initiate timer is set, as discussed earlier (Table 7–8). If the timer expires (transition 3), the CR TPDU is retransmitted. Retransmission of the CR TPDU may take place a maximum of N times, in which case the give-up timer is set (Fig. 7–29). On expiration of this timer, transition 4, the transport entity notifies the transport user, by initiating a T_DISCONNECT. indication, that the transport connection cannot be established. It sets the reference timer (Fig. 7–31 and Table 7–8) and waits for its expiration. At the expiration of this time, the transport connection is considered closed (transition 23), and the source reference number is again available for use on another connection.

Returning to state CR_SENT, we now focus on transition 6. This transition indicates that the CC TPDU with successfully negotiated parameters has been received. The machine moves to the ESTAB state, indicating successful establishment of the transport connection. It now enters the data-transfer phase.

In the event that on arrival the CC TPDU carries negotiated parameters that are unacceptable, the connection cannot be established. This is indicated by transition 7. The transport protocol machine sends a disconnect request (DR) TPDU to the peer transport entity (recall the three-way handshake) and moves to the CLOSING state. Here it waits for a disconnect confirm (DC) TPDU. The terminate timer is also set at the time the DR TPDU is sent (Table 7 – 8). On receipt of the DC TPDU, the reference timer is set, and the machine moves to the REF _ WAIT state, already described.

Transitions between other states can be traced through in a similar manner. Details appear in [NBS 1983c], in which all transitions are described in detail. Figure 3.2 of that reference includes states involved in the termination of a successful connection using the graceful close procedure.

This concludes the discussion of the ISO transport protocol, class 4. In the following section we describe another transport protocol quite briefly. This is the transmission control protocol (TCP), developed originally for the ARPAnet and since adopted as a transport protocol standard by the U.S. Department of Defense. Its significant features are compared with those of ISO TP4.

7 – 5 Transmission Control Protocol (TCP)—Comparison with Transport Protocol, Class 4

Transmission Control Protocol (TCP), developed originally for use with ARPAnet [SCHW 1977], has been adopted as the transport protocol standard by the U.S. Department of Defense (DOD). It has also over the years become a de facto standard in much of the U.S. university community involved in computer networks and has been incorporated as well by a number of manufacturers into intelligent systems designed to communicate over local area networks. It is an older transport standard than the ISO TP standard described previously in this chapter. As we shall see, a number of its features have been incorporated into ISO transport protocol, class 4.

It is expected that TCP and ISO TP class 4 will coexist for some time, but that, as more manufacturers and users begin to adopt the ISO suite of transport protocols, use of TCP will begin to decrease. It is of interest nonetheless to describe briefly some of the features of TCP here, comparing them with TP class 4; details appear in the DOD TCP standards document referenced here as [TCP].

The purposes and functions of TCP are very similar to those of TP class 4: It is designed for use with packet-switched networks or interconnected sets of such networks, in an environment where the networks themselves cannot be counted

---

[TCP] *Transmission Control Protocol, Military Standard,* MIL-STD-1778, U.S. Department of Defense, May 20, 1983.

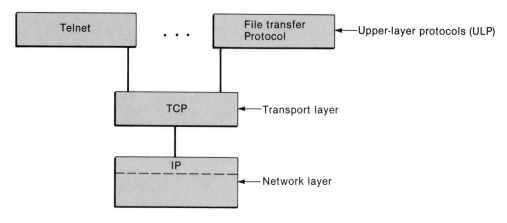

Figure 7 – 34 TCP in a layered architecture

on to deliver data in a reliable and orderly fashion. The purpose of TCP is to provide a connection-oriented data-transfer service between two users that delivers data reliably in sequenced order despite the possibility of damage, loss, duplication, and misordering of packets by the network service below. Data transmission is full-duplex and flow controlled, just as is the case with TP class 4. As shown in Fig. 7 – 34, TCP is meant to fit between a number of upper-layer protocols (ULPs) such as a file transfer protocol, a virtual terminal protocol, and others, and a DOD-developed subset of the network layer called the Internet Protocol (IP). This latter protocol was designed to support the interconnection of networks, using an internal datagram service for this purpose [POST]. It also grew out of the ARPAnet research activities, and in fact coexisted for a time with TCP as one comprehensive protocol. IP has been designated a DOD standard as well [IP]. The upper-layer protocol labeled Telnet that is shown in Fig. 7 – 34 was developed for ARPAnet as a terminal-remote host protocol.

As is the case with TP class 4, TCP has associated with it three phases of operation: the connection establishment, data transfer, and connection termination. A three-way handshake is used in the connection phase because of the innate unreliability of network services. TCP differs from TP4, however, in having a connection opened by a ULP in one of two modes, passive or active.

[POST] J. B. Postel, "Internetwork Protocol Approaches," *IEEE Trans. on Comm.*, vol. COM-28, no. 4, April 1980, 604–611; reprinted in [GREE].

[IP] *Internet Protocol, Military Standard,* MIL-STD-1777, U.S. Department of Defense, May 20, 1983.

With the passive open mode the ULP informs the transport protocol (TCP) that it is to wait (passively) the arrival of a connect request from some other system. With the active open mode the ULP designates another system with which a connection is specifically desired. The transport protocol then initiates the three-way handshake. Termination of a connection can be made, as in TP class 4, through a graceful close or an abrupt disconnect, termed an *abort* here. The passive open mode is designed to accommodate application processes such as a database manager that may be accessed by many remote users.

Other differences between TCP and TP class 4 occur during data transfer. TCP provides sequence numbering of octets within a block or unit of data, rather than numbering each unit of data sequentially, as is done in TP4. Each successive unit of data, called a *segment* in TCP terminology (and analogous to the DT TPDU in TP), may thus carry widely different though continuously increasing sequence numbers, depending on the number of octets in a segment. A credit-type sliding-window control, similar to that of TP class 4, is used in TCP as well.

TCP is termed a *stream-oriented* transport protocol: Data is delivered to TCP from the upper-layer protocol in continuous fashion; it is blocked arbitrarily by TCP into successive segments. There is no construct such as the ISO transport service data unit (TSDU), which may be blocked into smaller TPDUs for delivery to the peer user at the other end of the transport connection. A user (the sending ULP) may insist, however, on immediate transfer of its data by enabling a *push* function: On receiving this command, the TCP must immediately transfer the ULP data, plus any data waiting in TCP buffers. This push mechanism prevents possible deadlock situations from developing in which a remote ULP (the destination) waits for data that is held by the source TCP. An example might be an interactive terminal (the source) that communicates with an application process at a destination host. The end-of-the-line (carriage-return) key at the terminal might be designated to invoke the push function, causing each line to be delivered as soon as completed to the destination host via the two peer TCPs and the network or networks between.

TCP does not have an expedited data feature. It does provide an *urgent data* service: A specified number of octets within the system may be designated as urgent by the sending ULP and thus receive special attention by the receiving ULP.

A number of ULP/TCP interface primitives are defined in TCP as in TP class 4. Two types are identified: *service requests*, which cross the interface *from* the ULP to TCP, and *service responses*, which cross the interface in the reverse direction, from TCP *to* the ULP. These interface primitives are listed in Table 7–9, with an indication for each of the information carried [TCP, pp. 12–16]. We describe here a number of these primitives, in order of their use in connection establishment, data transfer, and connection termination. We follow with

TABLE 7-9   Primitives, TCP

Service-Request Primitives from ULP to TCP	Parameters
1. Unspecified passive open	Source port, ULP timeout (optional), precedence (optional), security (optional)
2. Fully specified passive open	Source and destination ports, destination address, ULP timeout (optional), precedence (optional), security (optional)
3. Active open	Source and destination ports, destination address, ULP timeout (optional), precedence (optional), security (optional)
4. Active open with data	Source and destination ports, destination address, ULP timeout (optional), precedence (optional), security (optional), data, data length, PUSH flag, URGENT flag (optional)
5. Send	Local connection name, data, data length, PUSH flag, URGENT flag, ULP timeout (optional)
6. Allocate	Local connection name, data length
7. Close	Local connection name
8. Abort	Local connection name
9. Status	Local connection name

Service Response Primitives from TCP to ULP	Parameters
1. Open id	Local connection name, source port, destination port (if known), destination address (if known)
2. Open failure	Local connection name
3. Open success	Local connection name
4. Deliver	Local connection name, data, data length, URGENT flag
5. Closing	Local connection name
6. Terminate	Local connection name, description
7. Status response	Local connection name, source port and address, destination port and address, connection state, amount of data (in octets) willing to be accepted by local TCP, amount of data (octets) allowed to send to remote TCP, amount of data (octets) awaiting acknowledgement, amount of data (octets) pending receipt by local ULP, urgent state, precedence, security, ULP timeout
8. Error	Local connection name, error description

(from [TCP, pp. 12-16])

some diagrams that indicate how the primitives might be used in a typical dialogue between two ULPs.

The two passive open-service requests correspond to the passive open mode mentioned earlier. A ULP passing either one of these primitives to the TCP indicates it is receptive to communication with another (remote) ULP. The fully specified passive open-service request specifies the particular remote ULP with which communication may take place. The unspecified passive open-service request indicates that the local ULP is ready to communicate with *any* remote ULP that might attempt to initiate communication. A full ULP address, called a *socket* in TCP terminology, is the concatenation of a global internetwork address and a port of that address. Note that the unspecified passive open primitive carries the local (source) ULP port address only. The fully specified passive open also carries the full address of the peer ULP with which communication can take place. This is given by the internetwork address and the ULP port at that address. The TCP replies in both cases with an open id service response primitive that specifies a shorter *local connection name* that will henceforth be used in place of the local ULP address in primitives passed across the interface.

The two active open-service requests, one with and one without accompanying data, are indications to the TCP that the ULP wishes to set up a connection with the destination ULP addressed. The TCP again replies with an open id service response primitive and then issues a special connection request message, called a SYN segment, comparable to the TP4 CR TPDU, which is delivered to the peer TCP, indicating that the source ULP would like to establish a connection.

On successful conclusion of a three-way handshake by the two TCPs (described more specifically later), each TCP issues an open success response primitive to its ULP, identified by the local connection name selected earlier. Data transfer may now begin.

During the data-transfer phase, data is transferred across the ULP/TCP interface using the send service request primitive. Note that the local connection name is again used to identify the ULP (see Table 7 – 9). The TCP will in turn map this name to the source and destination port addresses when it sends the data on to its peer TCP. If the PUSH flag in the send primitive is set, the data, and any data enqueued before it, must be transmitted immediately by the TCP. The URGENT flag, as noted earlier, indicates that the receiving ULP is to provide special attention to the data carried in this primitive. Data arriving at the destination TCP is delivered by that TCP to the destination ULP in the deliver service response primitive, using that ULP's local connection name as its address. Note that this primitive carries the URGENT flag. (There is no PUSH flag since the push mechanism is invoked only at the source side.)

Connection termination, as noted earlier, can be invoked using either a graceful close or an abrupt close procedure. The former uses the close service request and the closing service response primitives; the latter is invoked by the abort service request primitive. The terminate service response primitive is used in either case to finally close the connection. It is also used in other cases that require closing of the connection — for example, a service failure and the expiration of a connection timeout. The particular reason for the closing of the connection is carried in the description parameter of the primitive (Table 7 – 9).

The security and procedure parameters listed as optional in a number of the primitives in Table 7 – 9 are those defined in the Internet Protocol [IP]. Recall that security parameters were available as an option in TP4 (Table 7 – 7). Eight precedence levels are designated as possible options in IP. The underlying network(s) will attempt to deliver datagrams, with this option designated, in preferential order. The priority option of TP4 (Table 7 – 7) is comparable to the precedence option of TCP.

The ULP timeout option of TCP (Table 7 – 9) refers to the delivery of data once a connection is established. If no data is delivered to the destination in an interval of time called the ULP timeout, the connection is closed. Note that this option is listed in both the open and the send request primitives. If invoked in the former case, the timer is set as soon as the first segment of data is transmitted after connection establishment is completed. If an acknowledgement arrives within the time specified, the timer is canceled and then set again on transmission of the next data segment. Since each send request may (optionally) carry a ULP timeout, the parameter can be adjusted dynamically during the data transfer phase.

We shall indicate the use of these ULP/TCP primitives shortly in showing how a dialogue is carried out between two peer ULPs. First, however, we discuss the blocks of data or *segments* transmitted between peer TCPs. The basic format of a segment appears in Fig. 7 – 35. Unlike TP4, in which a number of different types of TPDU are specified (Table 7 – 6), all transport segments in TCP are of the same form. They differ only in whether flags in the fourth line of the header of Fig. 7 – 35 are set. Four segment types are specified: a data segment, an acknowledgement segment with the ACK flag so indicating, a SYN segment used to establish a connection (the SYN bit is then set), and a FIN segment used to terminate a connection (the FIN bit is then set). The FIN segment is the only one that may not carry data. Acknowledgements may thus be piggybacked on data segments.

Consider now a typical dialogue between two ULP/TCPs using the primitives and segments just described. We do this in three parts in Figs. 7 – 36 to 7 – 38, which correspond to the connection-establishment, data-transfer, and connection-termination phases, respectively. The figures are based on, although modified from, those appearing in [TCP, Figs. 2 – 4]. They are to be compared with Figs. 7 – 12, 7 – 16, and 7 – 22, described earlier for TP4.

Primitives are shown passing between the ULP and its TCP; segments, using the format of Fig. 7 – 35, are passed between the peer TCPs. Successive

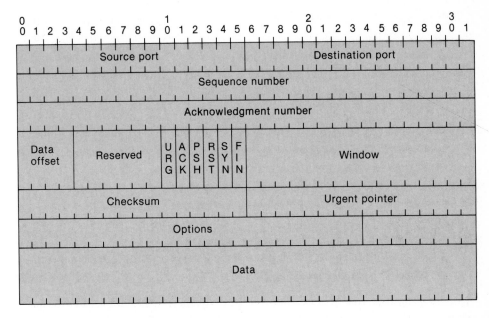

Figure 7–35 Format, TCP segment (from [TCP, Fig. 8])

steps in the dialogue are numbered consecutively. (Not shown, of course, are the IP and the other network layers below that carry out the actual delivery of the data in the form of packets.) ULP B first issues a passive open to its TCP, indicating that it is ready to establish a connection with another ULP. (Remember again that if this is a fully specified passive open, the remote ULP is specifically named; otherwise, any ULP may participate in a dialogue.) TCP B replies (step 2) with an open id, specifying a local connection name that will be used for the duration of any ensuing connection. ULP B now awaits a connection-establishment request. Sometime later, ULP A issues an active open to its TCP, indicating that it wishes to carry on a conversation with ULP B. TCP A replies with an open id, specifying the local connection name to be used between it and ULP A; it then sends a SYN segment, numbered 55, to TCP B, indicating that connection establishment is desired. This starts the three-way handshake; compare it with that of Fig. 7 – 12. (The segment includes the source and destination port addresses, as shown in Fig. 7 – 35. The primitive that contains this segment, which is passed down to the layer below, includes full address information.) Note that the SYN segment used to initiate connection establishment plays the role of the CR TPDU in TP4 (Fig. 7 – 12), although there is no negotiation of parameters at the TCP level. The SYN segment carries an initial

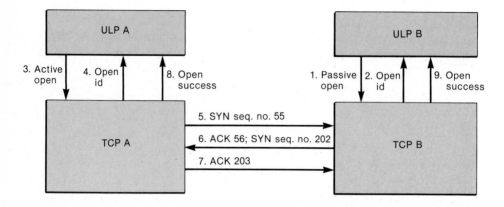

Figure 7–36 Connection establishment, TCP (based on [TCP, Fig. 2])

sequence number that will be followed in order, later, by data segments that begin with that number. (See step 11, Fig. 7–37. The SYN segment could carry data as well; in that case each octet in the segment would be numbered consecutively beginning with the number specified. This data would not be delivered to the destination ULP B until connection establishment was completed, following the three-way handshake.)

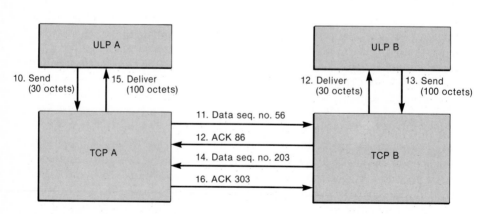

Figure 7–37 Data transfer, TCP (based on [TCP, Fig. 3])

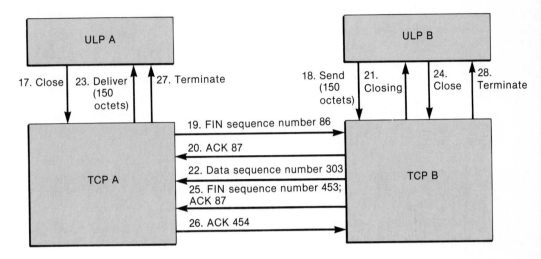

Figure 7-38 Graceful close, TCP

The initial sequence number used (number 55 in this example) is established by a clock-based initial sequence-number generator located at the TCP in question. The initial sequence number is arranged to cycle once every 4.6 hours; note from Fig. 7-35 that the sequence number is 32 bits long. The initial sequence-number generator increments its low-order bit about every 4 msec [TCP, sec. 6.1.11.4]. The same number is thus guaranteed not to be used again for another $2^{32} \times 4$ msec, or about 4.6 hours. In this way, the same connection, identified by the pair of sockets (port and address) at the two ends, can be reused with no danger of confusion of duplicate segments from different incarnations of the connection, so long as segments stay in the network no more than 4.6 hours. Recall from Section 7-2, on error-detection and recovery mechanisms, that this problem of duplicate packet detection is a critical one to address in transport protocols. The solution adopted in TCP is one of those listed in the earlier section [SUNS 1978].

Returning to Fig. 7-36, note that TCP B acknowledges receipt of the connection-establishment request SYN 55 from TCP A (step 6) by itself transmitting a SYN segment with its own initial number 202, with the ACK bit set and the acknowledgement field set to number 56. To complete the three-way handshake, TCP A in turn replies with the segment ACK 203 and notifies its ULP, with an open success response primitive, that the connection has been successfully established. The ULP timer may optionally be set now to await the delivery of data. TCP B, on receiving ACK 203, also notifies its ULP, by issuing an open success response primitive, that the connection is now established. That ULP can also optionally set the ULP timer at this time.

Figure 7–37 diagrams the (by now familiar) sequence of exchanges that take place during the ensuing data-transfer place. Only one scenario of many possible is shown here; it is to be compared with normal data transfer in TP4 as shown in Fig. 7–16. Since the sequence numbers in the two directions are not tied together, full-duplex transmission is possible, with either side sending data segments at any time. Acknowledgements need not be sent for every data segment received. As is already quite familiar to the reader, an ACK segment carries the number of the next expected data octet and implicitly acknowledges all octets up to that one. As indicated earlier, separate ACK segments need in fact not be sent. The acknowledgement can be piggybacked onto a data segment by setting the ACK bit and writing the next expected octet number into the 32-bit acknowledgement field (Fig. 7–35). Not diagrammed in Fig. 7–37 is the possibility of retransmission of data segments if an ACK segment is not received within a specified timeout. Details of the exchange in that case are left to the reader as an exercise. The one basic difference between the exchange shown in Fig. 7–37 and those diagrammed earlier in this book, at various architectural layers, is the use of octet sequence numbering. This is indicated in Fig. 7–37 starting with event 10, in which ULP A delivers 30 octets to its TCP by issuing a send request that carries the 30 octets of data. TCP A could choose to hold this data or segment it in any way it wishes. A PUSH indication in the send request, however, would force TCP A to deliver the data, plus any still not transmitted, immediately. In the example of Fig. 7–37 all 30 octets are delivered in the data segment beginning with number 56. (In TCP, the number of the first data octet transmitted is the initial sequence number plus one.) Since the last octet must be numbered 85, TCP B acknowledges receipt of the 30 octets by transmitting ACK 86. Note that the first data octet transmitted in the reverse direction, from TCP B to TCP A, is numbered 203, the initial sequence number 202 in that direction (Fig. 7–36) plus one.

Finally, Fig. 7–38 provides an example of the use of the graceful close mechanism in TCP. (The abrupt close is left to the reader as an exercise.) It is to be compared with the equivalent graceful close used in TP class 4 (Fig. 7–22). In this example ULP A decides to terminate the connection. It issues a close request primitive to TCP A to this effect (step 17), indicating that it has completed transmission of all data from its side. The close request implies the PUSH function [TCP], so that any data enqueued at the TCP would then be transmitted immediately. In the simple example described here, no data is waiting, so that TCP A can proceed immediately to transmit a FIN segment. This is indicated by step 19 in Fig. 7–38. The FIN segment is numbered 86 (the next in sequence after all data octets have been transmitted) and carries no data, as required for the graceful close. This segment is acknowledged by TCP B (step 20), and that TCP in turn issues a closing response primitive, notifying ULP B of the impending termination of the connection. Before this, however, ULP B had issued a send request that carries 150 octets of data (event 18 in Fig. 7 – 38). TCP B now transmits this information as data segment 303. The 150 octets are in turn delivered to ULP A after receipt at TCP A. ULP B issues a close request, indicating that it has no further data to send. TCP B follows with its own FIN number 453. (Why this number?) Note that it could have held back on the first ACK 87, acknowledging FIN 86 with its own FIN. TCP A in turn acknowledges receipt of FIN 453 (and implicitly the 150 octets of data received earlier) by sending ACK 454. Sometime later it issues a terminate response, and the connection is closed. TCP B follows with its own terminate to ULP B after receipt of ACK 454 and closes its side of the connection.

Since SYN and FIN segments all require positive acknowledgements, timers must be set for each, and the segments repeated if an ACK segment is not received within a specified timeout. This is the same procedure described in the previous section on TP4 timers. The issuance of a terminate response is delayed two *maximum segment lifetimes* [TCP] (equivalent to the reference time L in TP4), allowing segments that might have been delayed in the network(s) to finally arrive and be delivered.

We conclude this discussion of TCP by describing the fields in the TCP segment header format (Fig. 7 – 35) that have not as yet been discussed. The 16-bit window represents the number of data octets, beginning with the one specified in the acknowledgement field that the sender of the segment is willing to accept. This provides the same flow-control mechanism discussed earlier under TP4 (see Fig. 7 – 18). The same comments made earlier, particularly with respect to a zero window, apply here as well. Thus a sending TCP, with a zero window, must still transmit segments periodically to force ACK segments to be transmitted that indicate whether or not the window may have opened. This prevents deadlock, guaranteeing that knowledge of a reopened window is received, so that sending of data can be resumed. The periodic retransmission time suggested in TCP is two minutes.

To avoid a problem with out-of-order segments carrying old window information, it is suggested that both the sequence number and the acknowledgement number be checked when a send window is updated [TCP, Sec. 6.1.2.3]. Window information is considered valid either if the sequence number is higher than that of previous segments received, or, if the sequence number is the same, the highest acknowledgement number is chosen. Recall that the solution to this problem in TP4 involved the use of subsequence numbers.

Consider now the 16-bit urgent pointer field in the TCP segment header of Fig. 7 – 35. This field, in units of octets, is interpreted only if the URG flag bit is set. It then indicates the number of octets in the data stream following, beginning with the first octet of this segment, that are considered urgent. Specifically, the number of the urgent field, when added to the sequence number, points to the first octet of the data stream that is *not* urgent.

The 16-bit checksum that appears in the TCP segment header (Fig. 7–35) is calculated using a simple one's complement algorithm that covers the entire segment (header plus data) as well as a "pseudoheader" made up of source and destination addresses, among other parameters [TCP, Sec. 6.1.5].

It is apparent from the discussion here that TCP and TP class 4 share many of the same features, yet display some basic differences in their implementation philosophy. As noted a number of times, TCP is based on an octet-data stream philosophy, while TP4 focuses on blocks of data, data units, that are subdivisions of the TSDU passed down from the transport user. A segment in TCP may incorporate any number of octets, independent of the data passed down from the ULP, unless the PUSH function is invoked. The data stream concept, based on the smallest subdivision of data into octets and incorporating the PUSH function, was adopted for TCP as a help for interactive users. The choice of TSDU and the EOT indication following completion of transmission of the appropriate TPDUs in TP4 could presumably be designed to serve the same purpose in TP4. Thus the two protocols could be made to appear very much the same to interactive users.

TCP uses the active/passive open feature; TP4 allows any user to attempt communication with any other user. As a result of the active/passive open feature, TCP will establish only one call between a given source socket and a destination socket. In the case of TP4, if both ends initiate a connection establishment request simultaneously, two calls will be set up, each with different reference numbers, provided that multiple calls are allowed.

TCP normally uses the three-way graceful close to terminate a connection, although an abrupt close, with possible loss of data, can be initiated. TP4 allows only an abrupt disconnect, although, as noted earlier, the NBS FIPS version of TP4 does incorporate the graceful close. Both protocols use a three-way handshake to establish a connection; TP4 uses specific TPDUs (the CR, the CC, and the AK) for this purpose. TCP uses two SYN segments, each with a different sequence number, plus an ACK segment to complete the three-way handshake. The window flow-control procedure used is similar in the two protocols, although one is based on controlling data octets; the other, on controlling data TDPUs. The urgent feature of TCP has already been contrasted with the expedited data service of TP4. Recall that only one expedited data unit at a time can be outstanding without acknowledgement in TP4. There is no such limit on the number of urgent sections of the data stream in TCP.

Additional comparison between TCP and TP4 is left for the reader. Further comparison appears in an unpublished NATO technical memorandum [GROE].

[GROE] I. Groenback, *The TCP and ISO Transport Service—A Brief Description and Comparison,* NATO Technical Memorandum STC TM-726, SHAPE Technical Center, The Hague, Netherlands, Feb. 1984.

Problems

7-1 Explain, with the help of a diagram of a multinode mesh network, why a network that incorporates virtual circuit (connection-oriented) services provides a type-A connection to the transport service user, while a datagram (connectionless) network provides a type-C connection.

7-2 Explain, with the help of a diagram, what is meant by *multiplexing* at the transport layer. Could a given transport connection in turn be *split* among a number of virtual circuits? Explain. Provide a pictorial example of a number of transport users multiplexed onto one transport connection, with that connection in turn using the services of a number of network connections (virtual circuits).

7-3 Refer to Fig. 5-22. It portrays multiple SNA sessions multiplexed onto a single virtual route. Convert this figure to OSI terminology, indicating specifically into which OSI layer the virtual routes and half-sessions map, respectively. (As a hint, refer to Fig. 5-23 as well or reread the section on SNA path control in Chapter 5.)

7-4 Provide *specific* examples, corresponding to each of the four disconnect mechanisms of Fig. 7-6, as to why a transport connection might be terminated. Do this in the context of a specific example of a network, with two communicating users connected at two nodes in the network.

7-5 Refer to Figs. 7-6 and 7-7. Explain, with the help of diagrams if necessary, why the abrupt-disconnect mechanism may result in the loss of user data while the graceful close guarantees no loss of data. *Note:* A simple picture of a transport connection end-to-end "pipe" may be useful in explaining the difference. Alternatively, an example network with an end-to-end transport layer "pipe" superimposed could be used.

7-6 Use the state transition diagram of Fig. 7-10 to describe several scenarios that involve transport connection establishment, data transfer, and connection release. One example might involve a normal sequence of transitions that incorporate connection establishment, data transfer once the connection is established, and termination of the connection. Another example might involve expedited data. Other examples could involve abnormal situations with the connection establishment rejected (Fig. 7-5) or abruptly disconnected (Fig. 7-6). Note that Fig. 7-10 refers to *one* side of a transport connection only.

7-7 Sketch out, with the aid of a series of figures, a complete scenario that involves transport connection establishment between two transport service users, data transfer, and then connection termination. Let the maximum TPDU size be 128 octets, while the maximum TSDU length is 1024 octets. Choose an initial credit size of 8. Let the data transfer phase consist of the exchange of at least two TSDUs between the two sides of the connection. Use a 7-bit send sequence number for the DT TPDUs and show how successive DT TPDUs are numbered. Indicate the appropriate value of YR_TU_NR for each AK TPDU sent. Select some arbitrary credit control policy.

7-8   Explain, using a specific example of a problem that might occur without its use, why the three-way handshake of Fig. 7–12 is required with the class-4 transport protocol.

7-9   A transport connection is to be set up between a DTE at one end of an X.25 interface and a DTE remotely located at the end of another X.25 interface. Focus on one DTE-DCE interface. (Refer to Chapter 5 for details of the X.25 protocol.) Show how the CR TPDU of Fig. 7–13 fits into the X.25 data packet format of Fig. 5–6 and in turn into the HDLC information frame format that actually carries the data across the DTE-DCE interface.

7-10 *Note:* This problem is of project length and may require considerable time for completion.

Implementation details of the window credit flow-control scheme in the ISO class-4 transport protocol are not specified. It is thus of interest to compare a number of credit control strategies. Simulation is to be used to carry out the comparison. Specifically, develop a simulation model for the ISO transport protocol flow-control mechanism: Consider a transmitting side, a receiving side, and some network connection between. The network may be modeled most simply as an M/M/1 queue. (The service time chosen reflects the minimum end-to-end delay of the network.) Alternatively, a more sophisticated state-dependent model may used. (See the virtual circuit Norton equivalent model of Chapter 5.) Sunshine has suggested other models as well [SUNS 1977]. Let TPDUs be generated at the transmitter according to a Poisson process. (Some other process also could be modeled.) The TPDUs are then queued, pending transmission at some specified rate onto the network. TPDUs generated when the window is zero may be dropped (this is then a blocking model), or queued in a separate admission queue pending opening of the window. (An admission delay provides an additional cost in this case.) Let AK TPDUs carrying the credit indication and generated by the receiver arrive at the transmitter in zero time for simplicity. (A random-delay return path could be simulated as well.) Compare the throughput-delay characteristics obtained for a number of receiver credit control algorithms. One such algorithm has the credit allocated slam shut when the receiver queue reaches a certain threshold. It is opened fully when the queue drops below a second, lower threshold. More graceful reduction of credit size can be modeled as well. Another more graceful control has the credit allocation given as a function of both receiver buffer availability and transmitter queue occupancy.

7-11 Refer to Fig. 7–23.

   a. How would user B respond if the delayed data unit 1 arrived during the *same* connection? Would the system work properly?

   b. Provide another example of your own demonstrating the same problem posed in Fig. 7–23.

   c. Show how one avoids the problem posed in Fig. 7–23 by adding an incarnation number such as the one described in the text.

7-12 Refer to Fig. 7–25(b). Show how the deadlock shown is resolved using a three-way handshake.

7–13 Calculate the one-way end-to-end transit delay E for a transport connection that covers five links. Two of these are satellite. (Note that each satellite link consists in turn of two segments, one up to the satellite, the other down.) The terrestrial links each span 1500 km. Propagation delay on the satellite links is $c = 300,000$ km/sec; on the terrestrial links the delay is 150,000 km/sec. Transmission speed is 48 kbps over each link; data link frames are 960-bits long.

7–14 Repeat Problem 7–13 but assume that one of the terrestrial "links" is itself an X.25-based packet-switched network with an average transit delay of 150 msec.

7–15 Calculate the reference time L (Table 7–8 and Fig. 7–31) for the examples of Problems 7–13 and 7–14. Let the local TPDU processing time X be 10 msec, while the remote entity processing time is $A_R = 30$ msec. The maximum network lifetime of packets is $M = 5E$. The maximum number of retransmissions of a TPDU is $N = 5$.
How sensitive are the results to the choice of X, A_R, N, and M? What if X and A_R are doubled? What if N and M are doubled? What sequence number space would be required for these examples?

7–16 A subsequence number is used with the AK TPDU to handle a potential problem with the credit mechanism. This is described in Sec. 7–3, on TP4 error-detection and recovery mechanisms. Show that the algorithm used to accept an AK TPDU as "in order" is valid and accomplishes its desired function. Explain why the acknowledgement window time W (Fig. 7–30) is required for this algorithm to function properly.

7–17 The text describes an optional mechanism providing for acknowledgement of the AK TPDUs. This mechanism is used to reduce the number of AK TPDUs sent. A flow-control timer (Table 7–8) is required with this option. Explain this statement in the text: "When the timer expires an AK will be transmitted, but only if the credit field is zero. This is necessary to prevent deadlock."

7–18 Trace through a number of the transitions shown in the finite state machine representation of TP4 of Fig. 7–33. Connect these transitions with the equivalent explanations, in words, provided in earlier sections of the text.

7–19 Refer to Fig. 7–37. This diagram shows a typical exchange of data for the TCP protocol under error-free conditions. Modify the diagram to include the case in which one or more data segments are received in error, with an ACK segment then not sent. Data segments are then retransmitted after an appropriate timeout.

7–20 Show how Figs. 7–36 through 7–38 would appear if the ISO class-4 transport protocol were used in place of TCP. Use a similar numbering scheme to number the various events. Label the various TP4 primitives and TPDUs sent and received, and use appropriate sequence numbering for the various TPDUs. Use one TPDU per TSDU.

7–21 Repeat Fig. 7–38 using the abrupt close mechanism of TCP. (Note that this mechanism uses the abort and terminate primitives of Table 7–9.)

7–22 Sketch out an exchange similar to that of Figs. 7–36 through 7–38, this time using the active open and incorporating the PUSH function as well. Include segmentation of some of the data by one or both of the TCPs. Repeat the same exchange of data with the ISO class-4 transport protocol used instead, and compare with the TCP case.

7-23 TCP avoids a possible problem with flow control involving out-of-order segments that carry old window information by providing a check of both the sequence number and the acknowledgement number when updating a send window. In particular, as noted in the text, "window information is considered valid if either the sequence number is higher than that of previous segments received, or, if the sequence number is the same, the highest acknowledgement number is chosen." Explain this solution to the window problem using some simple examples. Compare with the TP4 solution using subsequence numbers. (See Problem 7-16.)

7-24 *Note:* This problem is more like a project and will require some time for completion. It involves detailed study of the transport-layer documents (either the ISO or the NBS versions referred to in the text) as well as comparable documents for one other alternative transport layer protocol. Examples include TCP, the data flow control and transmission control layers of SNA, and DEC's end communications layer (see Chapter 6).

a. Study and compare the transport-layer flow-control mechanism of the ISO class-4 transport protocol with that of one of the other transport protocols just noted. Comparison may range from a verbal comparison of the two mechanisms to a limited simulation of a subset of the mechanisms to analysis of a simple model of the schemes.

b. Code up a simplified subset of the transport-layer protocol mechanism for the data transfer phase. (The connection and disconnect phases could be included as well, time permitting.) Do this for the ISO class-4 transport protocol and one of the other transport protocols just noted. If two computers interconnected by a local area network or some other network are available, implement your programs on these machines and try transmitting some test messages. Compare the throughputs for the two protocols. Find the maximum possible throughput in each case and determine in what way it is limited by the protocol design, as contrasted with the processing speeds of the machines used and the throughput capability of the underlying network drivers and network used.

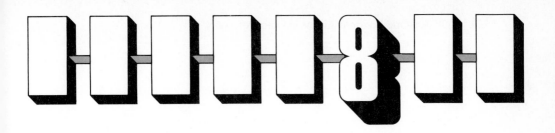

Polling and Random Access in Data Networks

Previous chapters have focused on the layered-architecture approach to packet-switched data networks. Starting with an overview in Chapter 3, we worked our way up layer by layer, culminating in the transport layer in Chapter 7. In this chapter we digress somewhat, concentrating on two major mechanisms for gaining access to a network. These two access mechanisms involve polling and random access. Descriptions and associated performance analyses of these two types of network access, together with a comparison of performance, are of interest not only for themselves, but because they also lead directly to a discussion of the two most common methods of accessing *local area networks* (LANs). LANs, a special type of network covering small geographic areas, are discussed in some detail in the next chapter. There we connect the operation of local area networks to the OSI Reference Model and show how the access controls for these special networks fit into the lowest two layers of that model, the physical and data link layers.

Polling and random access are of further interest since they have been used or suggested for use in interactive CATV (community antenna or cable television) networks and in packet radio networks. Polling techniques have been used for many years in controlling access, by terminals, to computers in such applications as airline reservation, banking, and retail and supermarket systems as well as many others. References to some of these applications appear later in this chapter.

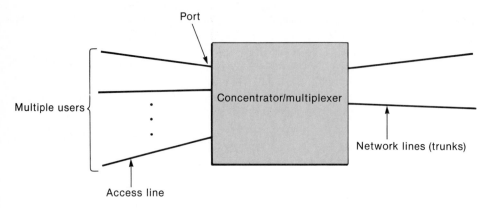

Figure 8-1   Access through dedicated port

Two distinct types of access to a network can be distinguished. In one, each user (terminal or computer) that accesses a network (through a packet switch, multiplexer, or concentrator) does so through a dedicated port. This is shown in Fig. 8-1; Figs. 6-11 and 6-14 provided specific examples. Ports may be scanned sequentially, or an interrupt mechanism may be used to accept user messages and either deposit them in appropriate buffers or send them out onto outgoing lines. There is no access contention in this case. Most commonly, in packet-switched networks user messages are stored and then read out first come-first served over an appropriate outgoing line, either as received, with header (control) information added, or split into smaller packets. Packet switches or concentrators carry out these operations. The access method used is then referred to as statistical or asynchronous time multiplexing. M/M/1 and M/G/1 queueing models have often been adopted to analyze this method of access. Some simple examples appeared in Chapter 2. More realistic models of this type of multiplexing have been studied as well, and analyses appear in [SCHW 1977, Chap. 7], and [HAYE, Chap. 5]. An older access method, called simply time-division multiplexing (TDM), allocates each user a prescribed portion (a slot) of an outgoing time frame and immediately delivers any incoming data onto the appropriate outgoing slot. Details of time-division multiplexing

[SCHW 1977] M. Schwartz, *Computer-Communication Network Design and Analysis,* Prentice-Hall, Englewood Cliffs, N.J., 1977.

[HAYE] J. F. Hayes, *Modeling and Analysis of Computer Communications Networks,* Plenum Press, New York, 1984.

appear in [SCHW 1980a]. An analysis appears in [HAYE]. A brief discussion appears in Chapters 10 and 11 of this book, in connection with circuit-switched systems.

The second access method, the one we stress in this chapter, involves the use of a *common* channel or medium that all users must access. An example might be a radio channel on which all users transmit at the same frequency. Packet radio, with users vying for the use of this common medium, is one example [KAHN 1978]. Interactive CATV, with users transmitting over one or more of the CATV bands, provides another example [McGA]. Satellite radio schemes also fall into this category [KAHN 1979], [JACO]. A coaxial cable or wire medium (discussed more specifically in the next chapter, which covers local area networks) is another example of a common channel. Users connected to the cable or wires at various points may want to communicate with one another or with a common control system. A number of possible configurations appear in Fig. 8–2.

In all these cases, whether radio or wire channel, there is contention for the common medium. Thus provision must be made for providing fair access to the channel to all users. There are basically two ways of accomplishing this. One way is by *controlling* access, either through a central controller or by passing control from one user to another in a decentralized fashion. *Polling* is the generic term used to describe this class of access strategies. In its decentralized version, used in local area networks, the term *token passing* is used instead.

The other method is *random access,* in which users transmit at will. Since two or more users may decide to transmit at the same time, packet "collisions" will result; there must thus be a way to resolve these collisions. A number of contention-resolution algorithms have been proposed, and comparative performance analyses made. We refer only to the simplest algorithm in this chapter; a detailed discussion appears in [HAYE, Chaps. 8 and 9].

Historically, polling techniques were the first to be used and are still commonly used to connect a multiplicity of terminals to a central controller. The term *multipoint* or *multidrop connection* is frequently used to refer to this type of

[SCHW 1980a] M. Schwartz, *Information Transmission, Modulation, and Noise,* 3d ed., McGraw-Hill, New York, 1980.

[KAHN 1978] R. E. Kahn et al., "Advances in Packet Radio Technology," *Proc. IEEE,* vol. 66, no. 11, Nov. 1978, 1468–1496.

[McGA] T. P. McGarty and G. J. Clancy, Jr., "Cable-Based Metro Area Networks," *IEEE J. on Selected Areas in Comm.,* vol. SAC-1, no. 5, Nov. 1983, 816–831.

[KAHN 1979] R. E. Kahn, "The Introduction of Packet Satellite Communications," National Telecommunications Conference, Nov. 1979, 45.1.1–45.1.8.

[JACO] I. M. Jacobs et al., "Packet Satellite Network Design Issues," National Telecommunications Conference, Nov. 1979, 45.2.1–45.2.12.

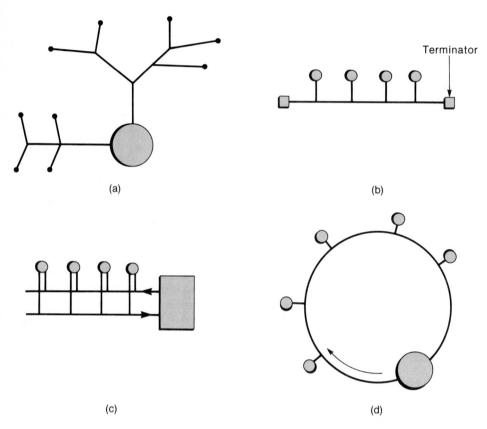

(a)

(b)

(c)

(d)

Figure 8–2 Cable/wire access topologies
a. Tree-type topology
b. Bus topology: decentralized control
c. Centralized control
d. Ring topology: centralized control

connection. Figures 8–2(a) and (c) provide examples. In this case, a number of terminals share the same communication medium (typically, leased telephone lines or private wires) to reduce communication costs. They are then individually polled to eliminate contention. Examples of the use of polling techniques appear in [SCHW 1977]. Two modes of polling are distinguished: *roll-call polling*, in which a central supervisor interrogates the terminals or stations connected, one after the other in some predetermined or adaptive order, and *hub polling*, in which control is passed sequentially from one user station to another. This latter method has been adopted—with the term token passing used instead—for local area networks. In the examples of Fig. 8–2, roll-call polling could be employed for the tree topology of part (a), for the two-wire topology of part (c), or for the ring with central control of part (d). CATV networks are encompassed by the tree topology of part (a), as are many rudimentary (and older) packet-switched networks, in which a number of terminals and/or computers desire access to a common host computer. Terminals and their associated controllers for a variety of applications have commonly been connected using topologies (c) and (d) of Fig. 8–2.

Hub polling is either of the decentralized or the token-passing variety (the latter used in local area networks), with permission to transmit passed sequentially from one designated user to another, or of the centralized type, with a central system reinitiating the poll one time per cycle. This kind of polling could be used with topologies (b), (c), and (d) of Fig. 8–2. In the bus topology of Fig. 8–2(b), packet energy propagates in either direction. Terminators must thus be provided at either end to absorb the energy. The other topologies allow energy propagation in one direction only, as shown.

Random-access techniques were first employed by investigators at the University of Hawaii in the early 1970s in pioneering experiments designed to connect a large number of geographically distributed users to a central computer by radio. This early system was called the Aloha system, and the basic random-access techniques discussed later in this chapter are referred to as Aloha access strategies. More recent random-access techniques used in bus-topology local area networks are variations of the Aloha strategies that are designed to improve performance. A discussion of the early work on the Aloha system appears in [ABRA].

In this chapter we discuss first the polling access strategy, with emphasis on time-delay–throughput performance evaluation. We then describe two types of Aloha access, pure Aloha and slotted Aloha, comparing them with polling in a selected set of examples in Section 8–3. Included are examples that apply to metropolitan-area networks using CATV technology. We conclude the chapter with a brief discussion of an improved random-access technique, CSMA/CD, which is used in random-access local area networks. The equations derived or presented in this chapter are used, with necessary modifications, in our comparison of token-passing LANs with random-access LANs in the next chapter.

[ABRA] N. Abramson, "The Aloha System," in *Computer Networks*, N. Abramson and F. Kuo, eds., Prentice-Hall, Englewood Cliffs, N.J., 1973.

8–1 Controlled Access: Polling

In discussing polling in this section we focus on centralized control of access for simplicity, although it should be apparent to the reader that the results obtained, particularly those for hub polling, are directly applicable to the decentralized environment found in local area networks. In the next chapter, as already noted, we shall see how the time-delay–throughput performance equation written here is readily modified to provide the performance characteristic of the token-passing LAN.

The model we shall use to analyze both roll-call and hub polling is that of Fig. 8–3, the topology shown earlier in Fig. 8–2(c). N stations are shown connected to a central controller with two common lines, one for receiving messages from the controller and the other for transmitting messages to it.

8–1–1 Roll-call Polling

Consider roll-call polling first. In this type of access strategy, as the name implies, stations are interrogated sequentially, one by one, by the central system, which asks if they have any messages to transmit. A station given permission to transmit does so on the inbound line shown, concluding with an indication to the central controller that its transmission is completed. The controller then sends a polling message using the outbound line to the next station on its list, repeating the same process. A station with no messages to transmit so indicates in a reply to the central controller on the inbound line. Once all stations have been given permission to transmit, a cycle is completed, and a new cycle begins. Stations may be polled more than once during a cycle, and polling may be made adaptive as well, responding to variations in traffic or based on priority considerations. We focus here on the simplest polling strategy, that of polling each of the N stations sequentially.

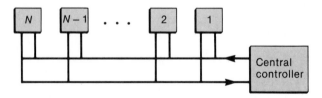

Figure 8–3 Model of a polling system

Figure 8–4 HDLC frame format

Polling in HDLC A number of data link protocols have been developed specifically for use with polling. An example of a protocol used for airline reservation systems appears in [SCHW 1977, Chap. 12]. Another more general example that we discuss briefly here is the polling procedure developed as part of HDLC and related bit-transparent data link control procedures (including ANSI's ADCCP and IBM's SDLC) [CARLD]. In Chapter 4 we discussed only the asynchronous balanced mode (ABM) of operation of HDLC. Recall that this mode is suited to the full-duplex operation of two connected point-to-point stations that would be used at the data link level of a packet-switched network. A special case of ABM is the LAPB mode used as the data link control in X.25. HDLC defines two other modes of operation for data transfer. One of these, the normal response mode (NRM), is defined specifically for use in a multipoint configuration such as that of Fig. 8–3 or Figs. 8–2(a), (c), and (d). In this mode, a single station designated the primary station (the central controller of Fig. 8–3) is responsible for control of the link, issuing commands to the other, secondary stations and receiving responses from them.

The frame structure of HDLC and related bit-oriented data control procedures first described in Chapter 4 is repeated here in Fig. 8–4 for easy reference. Recall that the 8-bit flag sequence 01111110 is used for the two *F*-fields, one at the beginning and the other at the end of a frame. The address of the *secondary* station appears in the *A*-field. The 16-bit frame check sequence (FCS) is used to detect errors in the frame as received. The *C*-field (control field) is used to differentiate the three types of frames used in HDLC: I, or information frames, those carrying data in the *I*-field, the S, or supervision frames, which contain no *I*-field; and the U, or unnumbered frames. The control field is further broken down in Fig. 8–5, as was done in Chapter 4, to show how the three types of frames are identified. Recall from Chapter 4 that $N(S)$ represents the sequence number of an I-frame; $N(R)$, which appears in both I and S-frames, acknowledges correct receipt of all frames up to and including $N(R)$-1 and

[CARLD] D. E. Carlson, "Bit-oriented Data Link Control Procedures," *IEEE Trans. on Comm.,* vol. COM-28, no. 4, April 1980, 455–467; reprinted in [GREE].

Bit number⟶	1	2	3	4	5	6	7	8
I-frame	0		N(S)		P/F		N(R)	
S-frame	1	0	S	S	P/F		N(R)	
U-frame	1	1	M	M	P/F	M	M	M

Figure 8–5 Control field format, HDLC

indicates that frame $N(R)$ is the next one expected. For a modulus of eight, the sequence number field indicated in Fig. 8–5, no more than seven unacknowledged I-frames may be outstanding.

The class of procedures defined for use with NRM, the unbalanced normal class (UNC), provides for the transmission of I-frames, as well as ready to receive (RR) and not ready to receive (RNR) S-frames, both as commands, from the primary station, and responses, from the secondary stations. Two options include the use of a selective reject (SREJ) S-frame and nonsequenced unnumbered information (UI) frames. This latter option allows nonurgent information, not requiring an acknowledgement, to be sent out from the primary station. In addition, a number of other U-frames are designated for use with setting up and disconnecting the data link connection [CARLD]. Note that the reject (REJ) frame used in the asynchronous balanced mode (ABM; see Chapter 4) does not appear in NRM.

During data transmission, the P/F bit (Fig. 8–5) is used to carry out the polling function in NRM. Specifically, the P bit set on a command from the primary station serves as a polling signal to the secondary station addressed. That station replies with as many I-frames as it desires to send, up to the maximum number possible without an acknowledgement, $M - 1$, with M the maximum sequence number. (In the case $M = 8$, the example used here, this represents a maximum of seven frames.) The *last* frame in the group has its F (response) bit set to 1, indicating that it has completed transmission in response to the poll. If the station polled has no information to send, it replies with an RR with $F = 1$.

Figure 8–6 diagrams an error-free data-transmission scenario between a primary station and three secondary stations, B, C, and D. Another example, using similar notation, appears in [CARLD, Fig. 3]. The case where errors occur, including the use of the SREJ frame, is described in [CARLD, Fig. 4]. Carlson also describes the use of U-frames in setting up and disconnecting a link.

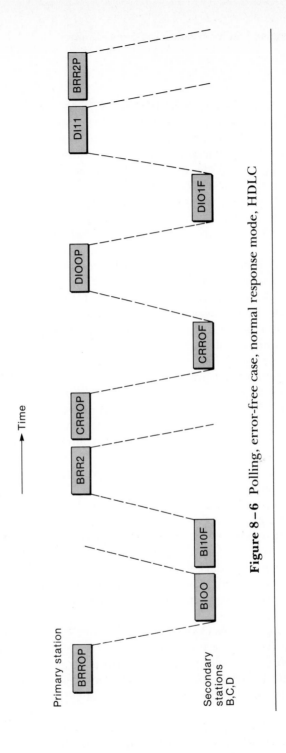

Primary station

Secondary stations B,C,D

Time

Figure 8–6 Polling, error-free case, normal response mode, HDLC

The notation of Fig. 8–6 is similar to that of Fig. 4–11. The first letter in each frame represents the address of the secondary station either being addressed or responding to a poll. The type of frame (I or RR) is then indicated, followed by $N(S)$ and $N(R)$ in that order for I-frames, or $N(R)$ for the RR-frames. The final P and F indication, if used, designate a poll and the end of a response to a poll, respectively. Additional examples of the use of the various commands and responses, particularly those involving error-recovery procedures, appear in [ISO 1979, Annex C].

In Fig. 8–6 the primary station initiates a polling cycle by transmitting an RR-frame with $P = 1$, addressed to station B. B replies with two I-frames, the second one including $F = 1$ to indicate completion of reply to this poll. The primary station acknowledges the two frames from B and then goes on to poll station C. C replies with an RR-frame, with $F = 1$, indicating that it has nothing to transmit at this time. The primary station then continues by polling station D. It does this by piggybacking permission to poll onto an information frame addressed to station D. D replies with one information frame. The primary station now finds that it has an additional I-frame to transmit to station D. It interrupts the polling cycle by sending an I-frame *without* the P bit set to station D. (What would happen if the P bit *were* set?) It then resumes the polling cycle by sending a polling message to station B. It could also have interrupted the polling cycle at any time to send a UI frame addressed to any one of the secondary stations or to all of them. For this latter case a group address, the all-1 setting of the address field, would be used. In the case of the UI frame no acknowledgement is required. The data so sent is thus not urgent, and unacknowledged loss could be tolerated.

The I-frame labeled DI11 in Fig. 8–6, shown being sent by interrupting the polling cycle, requires an acknowledgement and will be acked the next time station D is polled. The key idea in the NRM roll-call scheme of Fig. 8–6, which makes it differ from the full-duplex asynchronous balanced mode discussed in Chapter 4, is that secondary stations can only transmit when they are polled. This provides the necessary contention control over a common channel shared by a number of secondary stations, such as those in Fig. 8–3.

Roll-call polling analysis   How does one now determine the performance of a polling system such as the one just described for HDLC? We stress here the calculation of the *access delay* as experienced by users at each of the stations. This delay is measured from the time a data packet arrives at one of the stations that share the common channel to the time the packet begins transmission. It is then

[ISO 1979] *Data Communication—High Level Data Link Control Procedures—Elements of Procedures, International Standard ISO 4335,* International Organization for Standardization, Geneva, 1979.

comparable to the waiting time in queueing calculations (Chapter 2). To this time must be added the average frame-transmission time to obtain the inbound delay time. Comparable times will be calculated later for random-access schemes, enabling a delay comparison to be made.

The average access delay for Poisson arrival statistics will be found to be made up of two terms. One is precisely the M/G/1 wait time expression given by Eq. (2 – 65). The other, unique to polling systems, is given very nearly by one half the scan or cycle time. The scan is the time required to complete a poll of all stations. One half the scan time is thus intuitively the correct value for access delay at low-traffic values: It is the average time an arriving packet must wait in a polling system before transmission can begin. As traffic increases, additional delay is incurred because of the time required to serve packets that have arrived earlier. This is given by the M/G/1 waiting time expression Eq. (2 – 65).

The complete time-delay analysis of a polling system is quite complex, since the delays at all stations are interdependent. Thus the delay during a given polling cycle depends on the data waiting to be transmitted at each station. We shall therefore not attempt the analysis in this book; instead we shall quote results from the literature and refer the reader to the appropriate references for details of the analysis. However, we shall carry out the analysis of the average scan (cycle) time since not only is this very simple to do, but it does play an important part in the access delay calculation, as just noted.

Referring to Fig. 8 – 3, note that the time required to cycle once among the N stations has two components. There is a time required to transfer permission to transmit from one station to another, plus the time to actually transmit data once a station has been given permission to do so. The first time component is in turn given by the time required to transmit a polling message to each station in the system, the synchronization time for a station listening on the outbound line to recognize its address and take action to start transmitting, and the propagation time required for the polling signal to physically propagate out to the station in question and return with an indication that transmission is complete. This time required to transfer permission to poll from one station to another, which is unique to polling systems, is called the *walk time*. (The term comes from operations research literature on the analysis of periodic inspection and repair schemes of machines in offices and factories.) Walk time represents the minimum time required to scan the complete system and thus puts an irreducible minimum on the access delay experienced by users. This nonzero wait time did not appear in queueing systems studied earlier in this book. It provides a penalty at low traffic rates that the use of random-access techniques can overcome. They in turn will be shown to penalize users when the system operates at higher traffic intensities.

Figure 8 – 7 shows how the total cycle time t_c is made up of alternating walk times and station transmission times. The ith station, in particular,

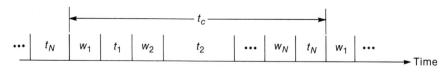

Figure 8–7 Scan or cycle time, polling system

$i = 1, \ldots, N$, is assumed to require a walk time w_i and a time t_i to transmit waiting frames when polled. (Note, as in Chapter 4, that the actual message transmitted out over the line is a frame that consists of the data packet plus required overhead. In the HDLC frame example of Fig. 8–4, the packet is the I-field passed down from the higher layers; the overhead consists of the six added 8-bit characters.) It is apparent from Fig. 8–7 and from this brief discussion that the scan or cycle time t_c is given by

$$t_c = \sum_{i=1}^{N} w_i + \sum_{i=1}^{N} t_i \qquad (8-1)$$

This scan time is, in general, a random variable, since it depends on traffic waiting at and then transmitted from each station. The walk times could also be random variables, since stations might, for example, require a random time to synchronize to the polling signal when it is received. The polling signal itself might vary randomly in length if the polling frames were themselves to carry data as shown in the HDLC example of Fig. 8–6. The following examples, however, will focus exclusively on the fixed walk-time case. Averaging Eq. (8–1) over the random station transmission and walk times, the average scan time $\bar{t}_c$ becomes

$$\bar{t}_c = \sum_{i=1}^{N} \bar{w}_i + \sum_{i=1}^{N} \bar{t}_i$$
$$= L + \sum_{i=1}^{N} \bar{t}_i \qquad (8-2)$$

Here $\bar{w}_i$ and $\bar{t}_i$ are, respectively, the average walk time and the average time to transmit waiting messages at station i; L is the total walk time of the complete polling system.

Now consider a model for analysis in which buffers at all stations are infinite and all data waiting is read out when the polling signal is received. (The analysis for a single buffer of fixed message length is provided in [HAYE, Chap. 7]. References to the literature appear both there, and in [SCHW 1977, Chap. 12].) Let the average arrival rate of packets at station i be λ_i packets/sec, and the average frame length in units of time be $\overline{m}_i$. (This is, of course, just the average frame length in bits, divided by the capacity of the channel, in bps. In the notation of Chapter 4, $\overline{m}_i = (\overline{\ell} + \ell')/C$, with $\overline{\ell}$ the average length of a packet, in bits, ℓ' the number of overhead bits added, and C, the channel capacity in bits/sec.) Then it is apparent that the average number of packets waiting to be transmitted when station i is polled is $\lambda_i \, \overline{t}_c$, and the time required to transmit these out on the line is

$$\overline{t}_i = \lambda_i \overline{t}_c \overline{m}_i = \rho_i \overline{t}_c \tag{8–3}$$

with $\rho_i \equiv \lambda_i \overline{m}_i$ the traffic intensity due to station i. Inserting Eq. (8–3) in Eq. (8–2) and simplifying the equation resulting, one finds that the average scan time $\overline{t}_c$ is given by

$$\overline{t}_c = L \left/ \left(1 - \sum_{i=1}^{N} \rho_i \right) \right.$$
$$= L/(1 - \rho) \tag{8–4}$$

with $\rho \equiv \sum_{i=1} \rho_i = \sum_{i=1}^{N} \lambda_i \overline{m}_i$ representing the total traffic intensity on the common channel.

Equation (8–4) is in the familiar queueing-delay form, first encountered in Chapter 2. With the assumption of infinite buffers at the stations it is apparent that the system will be statistically stable if and only if $\rho < 1$. The primary difference between Eq. (8–4) and the queueing formulas encountered previously is that the numerator here is the total system walk time

$$L = \sum_{i=1}^{N} \overline{w}_i$$

rather than the message transmission times found previously. For $\rho = 0$ — i.e., no traffic in the system — the minimum value of the scan time is just the walk time L, as noted earlier.

The scan time plays a critical role in the determination of access delay, as already noted. In particular, it was noted that for small ρ (i.e., very little traffic on the common channel), the average access delay should be just one half the scan time, $\overline{t}_c/2$. This is borne out by analysis. As already noted, that analysis is quite complex because of the interdependence of queues and so will not be attempted here. Instead we provide the results of an analysis of polling systems carried out by Konheim and Meister for the special case of a homogeneous polling system [KONH 1974]. They assume that each station has the same average packet-arrival rate λ, the same frame-length statistics, and the same

[KONH 1974] A. G. Konheim and B. Meister, "Waiting Lines and Times in a System with Polling," *J. ACM*, vol. 21, no. 3, July 1974, 470–490.

average walk time $\bar{w}$. Their analysis uses a discrete time approach, in which frame lengths are assumed to be multiples of a fixed slot time Δ, and frames at a station, when polled, are read out synchronously at the rate of one unit of the frame every Δ sec. (A unit could be a bit, a byte, or some fixed multiple of bits or bytes.) The assumptions made in the analysis are those of independent Poisson arrival processes at each station and geometric frame lengths, in multiples of the slot time, as already noted. The resultant expression for the inbound delay is quite simple in form and is reproduced in [SCHW 1977, Eq. (12–9)]. If one now lets $\Delta \rightarrow 0$ so that the discreteness of the time intervals disappears and takes limits appropriately, one finds the average access delay to be given by

$$E(D) = \frac{\bar{t}_c}{2}\left(1 - \frac{\rho}{N}\right) + \frac{N\lambda\overline{m^2}}{2(1 - \rho)}$$
$$= \frac{L}{2}\frac{(1 - \rho/N)}{(1 - \rho)} + \frac{N\lambda\overline{m^2}}{2(1 - \rho)} \qquad (8-5)$$

Here $\overline{m^2}$ is the second moment of the frame length, in (sec)^2, $\rho = N\lambda\bar{m}$, and all other terms are as already defined.

Recall that the access delay is the average time a packet must wait at a station from the time it first arrives until the time transmission begins. Access delay is thus comparable to the average wait time in an $M/G/1$ queue. But note in fact that the second term in Eq. (8–5) is precisely that of the average wait time $E(W)$ in an $M/G/1$ queueing system! (Compare with Eq. (2–65). $E(\tau^2)$ there is the second moment of the message-length statistics, precisely the meaning of $\overline{m^2}$ here; λ there is the total message arrival rate, just the meaning of $N\lambda$ here.) The entire polling system thus behaves like one equivalent $M/G/1$ queue, with all arrivals at all stations bundled together and transmitted in first come–first served, first in–first out (FIFO) fashion. The one difference is that there is an added wait time of $\bar{t}_c/2$ waiting for a given station to be polled.

Other analyses, using somewhat different models, provide similar although slightly different results. These are described in detail in [HAYE, Chap. 7]. Hayes includes a reference to an exact analysis of a model similar to that of Konheim and Meister, except that the discrete-time approach is not used. The result quoted there [HAYE, Eq. (7.43)] is identical to that of Eq. (8–5) here except that the first term has the factor $(1 + \rho/N)$ instead of the $(1 - \rho/N)$ appearing here. For $\rho < 1$, and the number of stations $N > 10$, the differences are clearly inconsequential. Equation (8–5) has been used in analyzing token-passing LANs, as will be noted in the next chapter.

To use Eqs. (8–4) and (8–5) in particular cases, one must obviously calculate the scan time. This requires, in turn, determining the walk time L quantitatively. Recall from our discussion that in the roll-call polling case the walk time is due to the polling-message transmission time, the necessary station synchronization time (in the LAN literature this is referred to as the latency time), and propagation delay. Specifically, let the polling message be a fixed value, t_p sec in length. (For the HDLC example of Fig. 8–6, this is the time required to transmit an RR-frame 48 bits long.) Let the time required per station to synchronize to a polling message be t_s sec. Let the *total* propagation delay for the entire N-station system be τ' sec. Then the total walk time for roll-call polling is given by

$$L = Nt_p + Nt_s + \tau' \qquad (8-6)$$

The propagation delay τ' depends both on the speed of propagation of energy in the medium and on the topology of connections of the N stations. It is the sum of the times required for signals to propagate back and forth between each station and the controller. Consider the topology of Fig. 8–3 as an example. Let the stations all be equally spaced, and let the round-trip (two-way) propagation delay between the controller and station N, as shown, be τ sec. It is then left to the reader to show that the overall propagation delay τ' is just

$$\tau' = \frac{\tau}{2}(1 + N) \qquad (8-7)$$

This result also applies to the ring topology of Fig. 8–2(d), in which all N stations are equally spaced. (Why can *any* topological configuration used for polling be mapped into a "virtual" ring of the form of Fig. 8–2(d)?) Note that Eq. (8–7) implies that the N stations being polled act as if they are all located at a point halfway along the extent of the access network. Another topology has the controller placed at the center of the string of N stations. Alternatively, for the ring of Fig. 8–2(d), let the ring consist of two wires, and let the controller first poll one half the stations clockwise starting at the left, then poll the remaining stations counterclockwise starting at the right. It is then left for the reader to show that the resulting overall propagation delay is reduced to about one half that of Eq. (8–7) [SCHW 1977, pp. 278, 279].

Reducing the propagation delay or any of the other terms in the walk time expression (8–6) reduces the scan time, of course, and from Eq. (8–5) results in a corresponding reduction in access delay. It is thus important to keep the walk time as low as possible in polling systems. This is borne out in examples both in this chapter and the next. Consider first the case of a polling network with 10 stations spaced 200 miles apart. (This might be an airline reservation system with each station serving in turn a number of agents' terminals. See [SCHW 1977, Chap. 12] and references cited there for a detailed description of such a system.) Say that the propagation delay is 2 msec/100 miles for this terrestrial network. Let the synchronization time t_s per station be 10 msec. The round-trip propagation delay is, of course, $\tau = 80$ msec. If the 10 stations are deployed as in Fig. 8–3, with the controller at the end of the polling network, $\tau' = 440$ msec, using Eq. (8–7). Total synchronization time is $Nt_s = 100$ msec. Say that the polling message is 48-bits long, as in HDLC, and let the line speed be 2400 bps. Then the polling message takes $t_p = 20$ msec per station to be transmitted. The total walk time is then $L = 740$ msec.

For very little traffic, this is effectively the scan time of the polling network, and the minimum access delay is then 0.37 sec. If the traffic increases to an intensity of $\rho = 0.5$, the average scan time $\bar{t_c}$ doubles to 1.48 sec. The access delay then increases to 0.70 sec, plus the average wait time, which depends, from Eq. (8–5), on message (frame) length statistics. If the line speed is increased from 2400 bps to 4800 bps, the walk time is reduced to $L = 640$ msec. This is again the scan time at $\rho = 0$, and the access delay for low traffic is one-half this value, or 0.32 sec. At $\rho = 0.5$, the scan time doubles to 1.28 sec. The reduction in walk time and scan time is not too great after we double the line speed here because of the large contribution to walk time made by the propagation delay; the network covers 2000 miles in this example.

To calculate the average access delay, we need the frame-length statistics. Say that the frames are on the average 1200-bits long, including 48 bits of overhead, and exponentially distributed in length. Then $\overline{m^2} = 2\,\overline{m}^2$. (Why is this so?) For this special case, the access delay expression of Eq. (8–5) simplifies to the form

$$E(D) = \frac{\bar{t_c}}{2}(1 - \rho/N) + \frac{\rho\overline{m}}{(1 - \rho)} \qquad (8\text{–}5a)$$

with $\rho \equiv N\lambda\overline{m}$, and the scan time $\bar{t_c}$ given by Eq. (8–4). (The second term in Eq. (8–5a) is the wait time for an M/M/1 queue. Why is this so?) For $\rho = 0$ the access delay is one half the scan time, as noted previously. For $\rho = 0.5$ the wait time adds a factor of $\overline{m}$ to the delay. In particular, for the same example under discussion, with 10 stations spaced out equally over a 2000-mile expanse, the average access delay for a 2400-bps line increases from 0.37 sec at $\rho = 0$ to $0.70 + 0.5 = 1.20$ sec at $\rho = 0.5$. For the 4800-bps line, the corresponding figures are 0.32 sec at $\rho = 0$, and 0.86 sec at $\rho = 0.5$. These results, plus additional calculations at $\rho = 0.8$, are summarized in Table 8–1. Details are left to the reader.

Now consider 10 stations spaced 10 miles apart. This might correspond to the case of data users on a single multipoint line that inputs data into a local switch (the controller) of a large geographically distributed data network. The 10 stations cover a distance out to 100 miles from the switch, which does the polling. It is obvious that the propagation delay component of the walk time has been reduced by a factor of 20 in this case to $\tau' = 22$ msec. Propagation time is now a small part of the total walk time. For this case, it is apparent that $L = 322$ msec for the 2400-bps line, and 222 msec for the 4800-bps line. The corresponding values of the access delay at $\rho = 0$ become $E(D) = 0.16$ sec for the 2400-bps line and 0.11 sec for the 4800-bps line. Table 8–1 tabulates these

TABLE 8 – 1   Roll-call Polling Example*

a. 200-mile Spacing/Station; All Times in Seconds

	2400-bps Line Speed			4800-bps Line Speed		
ρ	L	$\bar{t_c}$	$E(D)$	L	$\bar{t_c}$	$E(D)$
0	0.74	0.74	0.37	0.64	0.64	0.32
0.5	0.74	1.48	1.20	0.64	1.28	0.86
0.8	0.74	3.70	3.7	0.64	3.20	2.47

b. 10 mile Spacing/Station; All Times in Seconds

	2400-bps Line Speed			4800-bps Line Speed		
ρ	L	$\bar{t_c}$	$E(D)$	L	$\bar{t_c}$	$E(D)$
0	0.32	0.32	0.16	0.22	0.22	0.11
0.5	0.32	0.64	0.8	0.22	0.44	0.46
0.8	0.32	1.60	2.74	0.22	1.10	1.51

* $N = 10$ stations; 10-msec synch. time/station; 2-msec/100 miles propagation delay; 1200-bit exponentially distributed frames; 48-bit polling messages

values and values for $\rho = 0.5$ and 0.8 as well. These are to be compared with the values discussed previously for the much longer, 200-mile network. The decrease in access delay due to reduced propagation delay and to a higher line speed is apparent from the entries of Table 8 – 1. They show the impact of walk time and the corresponding scan time on the polling system's performance.

8 – 1 – 2 Hub Polling

As noted earlier, in hub polling control is transferred from station to station sequentially. In the topology of Fig. 8 – 3, for example, the controller might first poll station N, the one furthest away. This is shown schematically in Fig. 8 – 8, a repeat of Fig. 8 – 3 for the hub polling case. Station N, on completion of its transmission, appends the address of its neighbor, station $N − 1$, to its inbound message. That station, recognizing its address, in turn transmits any waiting messages, or, if there are none, a control message, with its neighbor's address appended. This process continues until station 1, completing its transmission, appends the address of the controller. That system then starts the cycle all over again. The resultant polling signal flowing around the system is indicated as the dashed line in Fig. 8 – 8. To carry out this process in a normal polling environment such as that of Fig. 8 – 8 requires that each station listen not only on the outbound line, as in roll-call polling, but on the inbound line as well. The

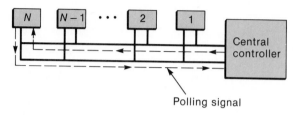

Polling signal

Figure 8–8 Hub polling with central control

resultant increased complexity and hence cost of the system has kept hub polling from being used as widely as roll-call polling in typical centralized polling networks. This is despite the fact that hub polling typically offers a substantial improvement in access delay, as we shall see. The application of hub polling to airline reservation systems is described in [SCHW 1977, Chap. 12].

It is apparent from this brief description, however, that hub polling lends itself to decentralized control. In the example of Fig. 8–3, the central controller could be a station just like the others, and polling would then proceed cyclically, as described, with each station addressing its neighbor or some other designated station. This is in fact what is done in token-passing schemes developed for local area networks, which we noted earlier in passing. The token in these schemes is simply a permit to transmit and is represented by the next station address; it was mentioned in the case of hub polling previously. Two types of token-passing strategies for LANs have been defined: a token-passing ring, exemplified by the ring of Fig. 8–2(d), with the centralized control replaced by a station just like the others, or a token-passing bus, using the bus topology of Fig. 8–2(b). In the latter case, each station maintains a next station address to which to pass the token. Note that the resultant configuration in the bus case might be called a "virtual" ring, since the token moves cyclically along the bus and back again. The primary difference is that energy in the bus topology travels in both directions — to both the left and right in Fig. 8–2(b) — while in the true ring energy travels in one direction only, as shown.

The use of decentralized control for hub polling in the case of the token-passing ring requires that one station be designated to absorb energy on the line once it has gone around once. This could be done by actually designating one station for this purpose. That station then functions like the controller in the centralized case. Alternatively, the station to which a data message is addressed could arrange to take the signal off the line. (In decentralized hub polling or token-passing schemes, messages are addressed to any station or set of stations on the ring or bus.) A third possibility, the one adopted for the token-passing ring described in the next chapter, is to have the transmitting station take the signal off the ring after one complete circuit. The lack of a centralized control poses other questions and problems that have to be addressed as well. Who starts the cycle going? What happens if a token is "lost" or if "multiple tokens" appear on the line? Answers to these questions will be provided in the next chapter.

The analysis of the hub-polling strategy is identical to that of roll-call polling; Eqs. (8 – 4) and (8 – 5) also apply in this case. The only difference is that the total walk time L is reduced through the use of hub polling. There are two reasons for this: First, propagation delay is reduced since there is no back-and-forth polling of stations. In addition, the polling message no longer enters into the picture since the next station address or a flag that grants permission to transmit appears in a field in the regular data message (frame) transmitted on the line. Since the address normally contributes little increase to the effective message length and hence has negligible effect on the access delay, we neglect it. For hub polling, then, the propagation time is just the round-trip-delay (or delay through a complete cycle) τ. The other component of the total walk time is the synchronization time Nt_s. For hub polling, we thus have

$$L_{\text{hub}} = \tau + Nt_S \qquad (8-8)$$

As already noted, all other polling equations still apply.

Consider the two examples discussed under the roll-call polling analysis. For the 2000-mile network, with stations strung out in a line as in Fig. 8 – 3, τ is the time to cover 4000 miles. This is 80 msec, instead of the 440-msec figure used for roll-call polling. (The propagation speed of energy on the line has again been taken as 2 msec/100 miles.) This shows the considerable reduction possible for long-distance polling networks. Using $t_S = 10$ msec rather arbitrarily as the station synchronization time, the total walk time becomes $L = 180$ msec in this case. This is to be contrasted with 740 msec and 640 msec, for 2400 bps and 4800 bps, respectively, in the case of roll-call polling (Table 8 – 1).

For the 100-mile network (10 miles spacing/station), the savings are obviously not as dramatic. Round-trip propagation delay is now $\tau = 4$ msec instead of $\tau' = 22$ msec, again using a figure of 2 msec/100 miles as the propagation speed in the medium. Both delays are short compared with the total synchronization time, so that term dominates. Total walk time is now $L = 104$ msec, contrasted with 320 msec and 220 msec for roll-call polling with 2400-bps and 4800-bps lines, respectively. Equations (8 – 4) and (8 – 5) can now again be used to calculate the scan and access delays for various traffic intensities and specified message statistics. Table 8 – 2 summarizes the results, again for the exponential frame-length distribution and an average frame length of 1200 bits. This is done for the 4800-bps line case only. Hub and roll-call polling are compared to show the significance of the results. Note, as pointed out earlier, that the differences are substantial for the long-distance system but considerably less for the smaller system.

TABLE 8–2   Hub and Roll-call Polling Compared*

a. 200-mile Spacing/Station; All Times in Seconds

		Hub Polling			Roll-call Polling	
ρ	L	$\bar{t}_c$	$E(D)$	L	$\bar{t}_c$	$E(D)$
0	0.18	0.18	0.09	0.64	0.64	0.32
0.5	0.18	0.36	0.42	0.64	1.28	0.86
0.8	0.18	0.90	1.41	0.64	3.20	2.47

b. 10-mile Spacing/Station; All Times in Seconds

		Hub Polling			Roll-call Polling	
ρ	L	$\bar{t}_c$	$E(D)$	L	$\bar{t}_c$	$E(D)$
0	0.10	0.10	0.05	0.22	0.22	0.11
0.5	0.10	0.21	0.35	0.22	0.44	0.46
0.8	0.10	0.52	1.23	0.22	1.10	1.51

* $N = 10$ stations; 10-msec synch. time/station; 2-msec/100 miles propagation delay; 1200-bit exponentially distributed frames; 48-bit polling messages; 4800-bps line speed

It is of interest to compare the effect of the message-length distribution on some of these results as well. Note that this distribution has an effect through the second moment term, $\overline{m^2}$, appearing in the wait-time portion of the access delay equation (8–5). Table 8–3 summarizes the results obtained, for the roll-call polling case only, for the same two examples utilized thus far. The average access delay $E(D)$ is shown for three different cases: $\sigma_m^2/(\overline{m})^2 = 0$, 1, and 2. σ_m^2 is the variance of the message (frame) lengths. The first case is the deterministic one: All frames are of the same length, 1200 bits. The second represents the

TABLE 8–3   Effect of Message-Length Distribution on Access Delay*

	$E(D)$ (sec)		
$\sigma_m^2/(\overline{m})^2 \rightarrow$	0	1	2
200-mile spacing/station	0.73	0.86	0.98
10-mile spacing/station	0.33	0.46	0.58

* Roll-call polling, examples of Table 8–2, $\rho = 0.5$

exponential distribution used to obtain the entries of Tables 8 – 1 and 8 – 2. The third represents an example of frames with wider variation about the mean than the exponential case. The access delay increases as expected, with increasing variance and hence $\overline{m^2}$. The absolute effect on access delay depends, however, on the one-half scan-time component of the access delay. Where this is larger, as in the 2000-mile system (stations spaced 200 miles apart), the walk time and hence the scan time are larger, and the effect of variations in the frame-length distribution is smaller.

Further discussion of polling access mechanisms is deferred until after we discuss random-access techniques. We then compare polling with random access, using the same two examples discussed thus far. In the next chapter, on LANs, we again compare the various schemes but for networks of much smaller extent and much higher bandwidths, features characteristic of local area networks.

8 – 2 Random-access Techniques

In the introduction to this chapter, we noted that the two basic ways of providing access to a common medium are controlled-access using polling and random access. (We leave out deterministic access, such as time division multiplexing, as noted in the introduction.) There are in turn a number of different types of random-access strategies. We focus in this section on the two simplest types, pure Aloha and slotted Aloha. We provide some references to the literature for a number of variations on these simple schemes, techniques designed to improve the performance obtainable through their use. We do provide a brief introduction in this chapter to the CSMA/CD (carrier-sense multiaccess with collision detection) technique, adopted for use in random-access local area networks. The discussion continues in the next chapter, in which we describe CSMA/CD-type LANs.

Random-access techniques, as the name implies, are completely decentralized. A user will essentially transmit at will, with possibly a few constraints depending on the particular access method adopted. These techniques range from the pure Aloha technique, in which a user (a station) transmits whenever it has a message (packet) to deliver to some destination, to techniques where the user is constrained to transmit in certain time intervals only, to more sophisticated techniques where a user "listens" before it transmits and then does so only if it senses an idle medium. Other variations include reservation techniques where a user may, through random access, request permission to transmit a full message at some reserved time [SCHW 1977]. Many other schemes have been suggested as well.

Because the essential idea of random access is a very simple one — transmit at will, with possibly a number of constraints — the access algorithm is generally easily implemented and is relatively inexpensive in practice. Random access has thus received widespread interest and, in a more sophisticated form such as CSMA/CD, has been implemented widely. It is particularly practical and useful at low levels of traffic.

No matter what the technique used, because of the random times at which users may decide to transmit, there is always the chance that two or more users will decide to transmit at overlapping times. This results in a "collision," which must first be recognized as such and then resolved. As the traffic intensity increases, the probability of collisions increases as well, leading to possible instabilities in the operation of these mechanisms. Throughput is limited as a result to some maximum value less than the channel capacity, the particular value depending on the original access mechanism and collision-resolution algorithm proposed. For these reasons, random-access techniques have been primarily suggested for applications that involve many bursty, interactive users, with each user attempting to transmit only infrequently or, at the other extreme, for applications that involve relatively few host computers communicating with one another. In this latter case, because there are very few users, the chance of a collision is reduced as well. In applications such as manufacturing, assembly plants, and other factory operations that require tight control of access delay, the use of controlled access has been preferred.

8-2-1 Pure Aloha

We begin a more quantitative discussion of random-access techniques by starting with the simplest scheme of all, pure Aloha. As noted earlier, this scheme was first adopted as a common channel-access strategy by workers at the University of Hawaii in the early 1970s. It is the forerunner of many random-access strategies that have been proposed and/or adopted since. In this scheme a user wishing to transmit does so at will. As a result two or more messages may overlap in time, causing a collision. There must be some way of recognizing the collision and so signaling to the users involved. This could be done by a central station designated for this purpose (in the original Aloha system all messages were beamed via radio to a central receiving site) or by the use of a positive acknowledgement with timeout method. In the case of a bus-type local area network this could be accomplished by having each station listen as it transmits and itself detect a collision. In any case, on detecting a collision, colliding stations attempt to retransmit the message in question, but they must stagger their attempts randomly, following some collision resolution algorithm, to avoid colliding again.

It turns out that this pure-Aloha access strategy, although very simple, is quite wasteful of bandwidth, attaining at most $1/2e \doteq 0.18$ of the capacity of the channel. To demonstrate this limit on throughput, we first require some definitions. Let there again be N stations contending for use of the channel. Each station transmits, on the average, λ packets/sec. Now take the specific case, for simplicity, where all messages transmitted are of the same fixed length, m, in units of time. (These messages again normally contain data plus necessary overhead.) In keeping with the notation commonly used in discussing Aloha-type systems—see [ABRA] and [SCHW 1977]—we now let the traffic intensity ρ, the fractional utilization of the channel by newly arriving packets, be written as a parameter S.

$$S \equiv \rho = N\lambda m \qquad (8-9)$$

(Recall that $1/m$, or μ, as used in earlier chapters, represents the channel capacity in units of packets/sec transmitted. $N\lambda/\mu = N\lambda m$ is thus the relative utilization of the channel, or throughput normalized to $\mu = 1/m$.) It is this parameter that we shall show is limited at most to $1/2e \doteq 0.18$.

We now assume that the arrival of packets at each station obeys a Poisson process. The total arrival rate is then Poisson, with parameter λN. As indicated earlier, the total traffic on the channel will consist of newly transmitted messages plus those that have been retransmitted. We now make the additional assumption that the retransmitted messages are Poisson distributed as well. This is obviously not true, since they do depend on collisions having taken place. Simulation studies indicate that the assumption is valid if the random retransmission delay time is relatively long [LAM 1974]. The total rate of packets attempting transmission over the channel, newly generated plus retransmitted ones, is then some number $\lambda' > \lambda$. The actual traffic intensity or utilization of the channel is then a parameter G given by

$$G = N\lambda'm \qquad (8-10)$$

Consider a typical message m-sec long, as shown in Fig. 8-9. It will suffer a collision with another message if the two overlap at any point. It is easy to see, by shifting the dashed (colliding) message in time, that collisions could come from an interval $2m$-sec long. The probability of *no* collision in an interval $2m$-sec long is just the probability that no Poisson messages are generated during that time. From Poisson statistics (Eq. 2-1), this probability is just $e^{-2N\lambda'm} = e^{-2G}$.

The ratio S/G represents the fraction of messages transmitted over the channel that get through successfully. This must equal the probability of suc-

[LAM 1974] S. S. Lam, *Packet Switching in a Multi-Access Broadcast Channel*, Ph.D. dissertation, Dept. of Computer Science, UCLA, April 1974.

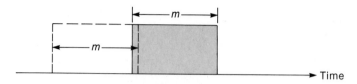

Figure 8-9 Collision between two messages

cess, i.e., the probability of no collision. We thus have, very simply, the pure Aloha throughput equation:

$$S = Ge^{-2G} \tag{8-11}$$

S is the normalized throughput (average packet arrival rate divided by the maximum throughput $1/m$) and G is the normalized carried load. S is thus the independent variable and G the dependent one. G plotted as a function of S gives rise to the two-valued curve of Fig. 8-10. Note that S is maximized at a value of $S = 0.5e^{-1} \doteq 0.18$ at $G = 0.5$. This is also shown by finding dS/dG from Eq. (8-11) and setting it equal to zero.

Note the interpretation from either Eq. (8-11) or Fig. 8-10. For small offered load S there are few collisions and $G \doteq S$. As S begins to increase, however, approaching its maximum value of 0.18, the number of collisions increases rapidly, increasing the number of retransmissions, which in turn increases the chance of a collision. The system becomes unstable, S drops, and G increases to a large value.

The fact that the maximum throughput of the Aloha system is limited to at most 18 percent of the line capacity may seem disconcerting at first. Yet in many applications this maximum transmission capability may be quite sufficient for practical use. Consider an example where interactive terminals are connected via a multipoint line to a central controller, as suggested by the various topologies drawn in Fig. 8-2. Say that the line capacity is 4800 bps. Messages inbound to the controller contend for the common medium; messages sent in reply by the controller are obviously controlled and can occupy almost the full bandwidth of a separate return channel. Say that a human user at a terminal inputs a 60-character message and receives a 400-character message in reply. Since the person at the terminal must take time to compose and type in his or her message and then wait to receive and read the reply, it is clear that there are human limits on the rate at which messages can be input. A typical figure might be to input one such message every 2 minutes. The average input rate per terminal is then 60 char/120 sec or 1 char/2 sec. If these are asynchronous terminals, with 10-bit characters [SCHW 1980a], the average input rate per terminal is 5 bits/sec! If

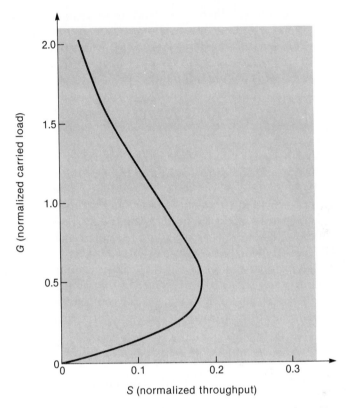

Figure 8-10 Throughput characteristic, pure Aloha system

10 percent of the line capacity is made available for random access with the Aloha technique, to ensure lying well below the 18 percent limit, 96 interactive users can be accommodated on this channel. If the line speed used were 2400 bps instead, this would mean that 48 such users could be accommodated. If user statistics indicate that a message is inputted every 3 minutes on the average, about 140 users could be accommodated on the 4800-bps line and 70 users on the 2400-bps channel. If input message lengths were reduced to 30 characters, the number of users could be doubled.

The point to be made here is that in many applications involving highly bursty interactive traffic, a simple scheme like pure Aloha could be used quite successfully and very simply. The original experiment at the University of Hawaii was designed precisely for this type of environment. Note also that we have purposely picked relatively low bandwidths, those associated with voice-grade telephone lines. Much higher bandwidths—available for example on a CATV system or packet radio—would allow correspondingly more interactive users to access the common medium [SAADT]. An example appropriate to metropolitan-area networks using CATV will be presented later when we compare random access with polling techniques.

What are the time delays involved in the pure Aloha scheme? To determine these we need first to define a retransmission strategy to use when collisions occur. As already noted, retransmission should be randomized to stagger the retransmissions and reduce the chance of a second collision or even more. One suggestion is to choose an arbitrary time interval and select a uniformly distributed random retransmission time within that interval. More precisely, let the time interval cover K message-unit times, of m units of time each. Retransmission then takes place within 1 to K such m-sec intervals after it is learned that a collision has taken place. Let the round-trip delay plus processing time required to obtain this knowledge be R m-sec intervals. (As noted earlier, a simple procedure would be one in which the destination system positively acknowledges each transmission. The source station would then timeout after R message-length intervals if a positive ack had not been received by that time. R must obviously be chosen, as discussed in Chapter 4, to account for round-trip delay plus processing time.) The average time required to successfully transmit a message is then

$$D = m\left[1 + R + E\left(R + \frac{K+1}{2}\right)\right] \qquad (8-12)$$

Here E represents the average number of retransmission attempts per message transmitted.

The average number of retransmission attempts per message transmitted should depend on the retransmission interval K. In fact, one would expect that for small K there are more collisions and hence more retransmissions; for large K there are fewer collisions. For large enough K, however, the dependence on K disappears and E is given by a quite simple expression. Specifically, we defined the parameter G as the normalized sum of the original attempts S and the retransmissions. It is then apparent that we must have

$$\frac{G}{S} = 1 + E \qquad (8-13)$$

and, from Eq. (8-11),

$$E = e^{2G} - 1 \qquad (8-14)$$

[SAADT] T. N. Saadawi and M. Schwartz, "Distributed Switching for Data Transmission over Two-Way CATV," *IEEE J. on Selected Areas in Comm.*, vol. SAC-3, no. 2, March 1985, 323–329.

As an example, take the N-station string of Fig. 8-3, which we discussed in connection with polling. We chose the message length there, quite arbitrarily, to be 1200 bits. For a 4800-bps line, $m = 0.25$ sec. Consider the case first of the 2000-mile system. The worst-case round-trip propagation delay, for the station furthest away from the controller, is 80 msec using 2 msec/100 miles again as the speed of propagation of energy. This is one-third of m and so will be considered negligible here. Let $K = 5$, and say that the system is operated at a normalized throughput of $S = 0.08$. Then $G = 0.1$, and $E = G/S - 1 = 0.25$. The total delay, *including* time to transmit a message, is then $D = 1.75m = 440$ msec. This appears quite tolerable and is comparable to the results obtained in the previous section for hub and roll-call polling. (See Table 8-2. Recall that the entries there were for *access* delay. The average message length and propagation delay must be added to make those calculations comparable to the one here.) As the traffic load increases, approaching a maximum value of 0.18, the behavior of the pure Aloha scheme becomes worse. A detailed comparison of Aloha access with polling will be presented after we discuss an improved version of pure Aloha, the slotted Aloha scheme.

8-2-2 Slotted Aloha

The limitation on maximum possible throughput of the pure Aloha scheme can be doubled by the simple expedient of slotting the time scale into units of time m (a message width) wide and allowing users to attempt transmission at the beginning of each slot time only. This scheme requires, of course, that all users in a system be synchronized in time. A simple example of the operation of this technique appears in Fig. 8-11. One message is shown being transmitted successfully; another suffers a collision.

Since messages can only be transmitted in the slot intervals as shown, collisions only occur when two or more users attempt transmission in the same time slot. The probability of a successful transmission, again assuming that retransmitted messages obey Poisson arrival statistics, is then given by e^{-G} (compare

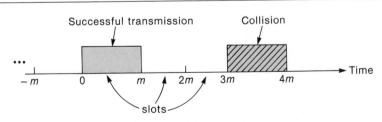

Figure 8-11 Transmission in slotted Aloha

with the same quantity in the pure Aloha case), and the throughput characteristic for slotted Aloha becomes

$$S = Ge^{-G} \qquad (8-15)$$

The normalized throughput S is easily shown to reach its peak value of $1/e \doteq 0.368$ at $G = 1$. The carried-load–throughput characteristic for slotted Aloha is shown plotted in Fig. 8–12 and is compared there with the corresponding characteristics for pure Aloha. It is apparent from this characteristic, with two values for G again possible for a given throughput S, that this access technique is subject to instability as well.

Studies of the instability problem in Aloha systems have appeared in [LAM 1974], [KLEI 1975b], [LAM 1975], [CARLA], and [FAYO]. A summary of the analysis appears in [HAYE, Chap. 8]. Control mechanisms designed to avoid this problem are suggested and analyzed in these papers as well. One simple control procedure is to increase the retransmission interval after each detected collision. This spreads user retransmission out and reduces the chance of a collision. It also increases the delay in the system and is the reason why very large retransmission intervals (very large K) are not chosen after the first collision. Interestingly, an algorithm similar to this is used with the CSMA/CD protocol designed for use on local area networks. The CSMA/CD collision-resolution algorithm is described in the last section of this chapter and in the next chapter.

The time-delay analysis of slotted Aloha is very similar to that of pure Aloha. The retransmission strategy is the same as that suggested for pure Aloha: On learning that a collision has taken place R slots after the attempted transmission, a station retransmits with uniform probability anywhere in an interval K slots long. This procedure is diagrammed in Fig. 8–13. Based on this retransmission procedure, it is apparent that the time required to successfully complete a transmission in slotted Aloha is given by

$$D/m = 1.5 + R + E\left[R + 0.5 + \frac{K+1}{2}\right] \qquad (8-16)$$

This equation is written in normalized form, in units of slots or message length

[KLEI 1975b] L. Kleinrock and S. S. Lam, "Packet Switching in a Multiaccess Broadcast Channel: Performance Evaluation," *IEEE Trans. on Comm.*, vol. COM-23, no. 4, April 1975, 410–423.

[LAM 1975] S. S. Lam and L. Kleinrock, "Packet Switching in a Multiaccess Broadcast Channel: Dynamic Control Procedures," *IEEE Trans. on Comm.*, vol. COM-23, no. 9, Sept. 1975, 891–904.

[CARLA] A. B. Carleial and M. E. Helman, "Bistable Behavior of Aloha-type Systems," *IEEE Trans. on Comm.*, vol. COM-23, no. 4, April 1975, 401–410.

[FAYO] G. Fayolle et al., "Stability and Optimal Control of the Packet Switching Broadcast Channels," *J. of the ACM*, vol. 24, July 1977, 375–380.

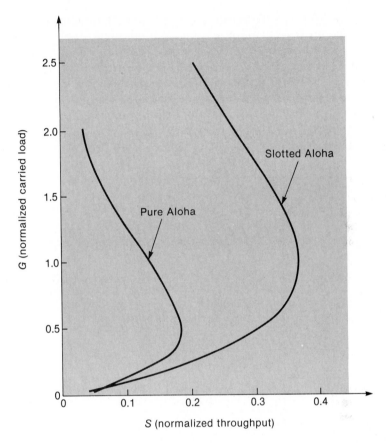

Figure 8–12 Throughput characteristics, slotted and pure Aloha

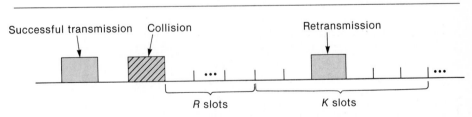

Figure 8–13 Collision resolution, slotted Aloha

m, to enable it to be plotted more readily. The only difference between this expression and Eq. (8–12) for pure Aloha is the extra factor of 0.5 units of time that appears in two places. This factor is required to account for messages that arrive after a slot interval has begun. Queueing (wait) time at a station has been neglected both here and in the prior expression (8–12) for pure Aloha, since with maximum throughput just a fraction of the line capacity in either case, the probability that a message has to wait for transmission would be expected to be small.

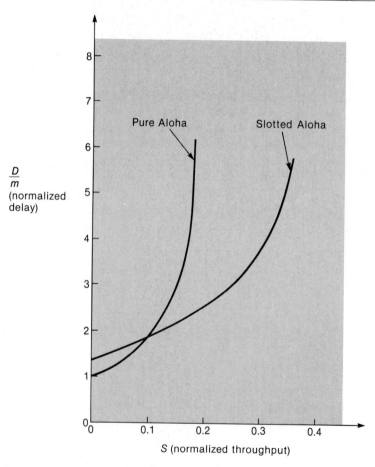

Figure 8–14 Delay-throughput characteristic, Aloha techniques, zero propagation delay, $K = 5$

The calculation of E, the average number of retransmissions once a collision has been detected, is again rather complex, since E depends on the retransmission interval K, as does S/G. A detailed analysis of the interrelations among K, E, S/G, and D appears in [SCHW 1977, Chap. 13], and in [LAM 1974] and [KLEI 1975b], from which this analysis is taken. It turns out, as noted earlier, that for very large K, one obtains the throughput characteristic Eq. (8 – 15), independent of K. Since $E = G/S - 1$, E is then written simply as

$$E = \frac{G}{S} - 1 = e^G - 1 \qquad (8-17)$$

for slotted Aloha. Both Eq. (8 – 15) and Eq. (8 – 17) turn out to be good approximations even for K as low as 5 ([LAM 1974, Figs. 3 – 3 and 3 – 4] or [SCHW 1977, Figs. 13 – 5 and 13 – 6]). We can thus safely use Eq. (8 – 17) in Eq. (8 – 16) to calculate the slotted-Aloha delay even for relatively small values of K.

An optimum value for the normalized retransmission interval K is found to exist just as in the case of pure Aloha. For, as noted earlier, with small K one reduces the time to schedule a retransmission but at the cost of more collisions, while with large K collisions are resolved more readily, but the scheduling time is increased [LAM 1974], [SCHW 1977, p. 296]. The optimum K, minimizing the time delay D, depends on the normalized throughput S as well as on the normalized round-trip propagation delay R. The variation of delay with K is quite insensitive to K, however, so long as one stays away from values of S close to its maximum value of $1/e \doteq 0.368$. A value of $K = 5$ appears to be a good compromise choice and will be used in the discussion to follow.

Figure 8 – 14 plots the normalized delay D for both pure Aloha and slotted Aloha, for the special case $R = 0$, using Eq. (8 – 12) and Eq. (8 – 16), respectively. The retransmission interval has been selected as $K = 5$.

8 – 3 Polling and Random Access Compared

The delay equations written for both the polling and the Aloha access techniques enable us to compare the two classes of schemes for examples such as those used earlier in studying polling. For the first example, we compare hub polling with the two types of Aloha access for a 2000-mile system such as that of Fig. 8 – 3. Two cases are considered: one in which traffic is generated at the rate of 1 message/minute at each station, the other in which the rate doubles to 2 messages/minute. We then allow the number of stations to vary, starting from a minimum value of $N = 10$. The overall walk time L will then increase as well in the case of hub polling. For this comparison we use the same numbers used in the polling examples: We take the line capacity to be 4800 bps; messages are 1200-

bits long and are of fixed length. In comparing the two types of access, we must use the *inbound delay* for the polling scheme. This is the access delay $E(D)$ described earlier plus the message transmission time m. We could incorporate an average inbound propagation delay as well but this is clearly negligible compared with m for these two examples. Using Eq. (8–5) for the access delay, the inbound-delay expression for the polling case, with fixed-length messages, is given by

$$D = \frac{\bar{t_c}}{2}(1 - p/N) + \frac{m}{2}\frac{p}{(1-p)} + m \qquad (8-18)$$

The scan time $\bar{t_c}$ is again given by Eq. (8–4), with the walk time L found using Eq. (8–8). The synchronization time t_S will again be taken, quite arbitrarily, as 10 msec/station. Note that the round-trip propagation delay τ must be retained in the expression for L so long as it is comparable to Nt_S.

Using Eq. (8–18) for the polling delay, and Eqs. (8–12) and (8–16) with $R = 0$ and $K = 5$ for the pure and slotted Aloha cases, respectively, one gets the curves of Figs. 8–15 and 8–16. Note that for relatively few stations (the low-traffic case) the pure Aloha scheme, the simplest of the three (just transmit at will, with provision made to retransmit randomly if a collision is detected), provides somewhat better time-delay performance, although not by much. For 20 stations and an average arrival rate/station of 1 packet/minute, all three schemes perform about the same. This drops to about 10 stations as the average arrival rate is increased to 2 packets/minute at each station.

These results tend to be typical of comparisons between random-access and controlled-access strategies: For low traffic (small numbers of stations in this example), random-access schemes generally outperform the controlled-access strategies. This is because the random-access delay, in the case of relatively few collisions, approaches the minimum possible—just the transmission time of the message (or frame, if we use the HDLC terminology). This is apparent from Figs. 8–15 and 8–16 for the pure Aloha case. (Recall that the slotted Aloha scheme has been penalized an additional one-half slot or message-transmission time because of random arrivals.) The hub-polling scheme has an irreducible total walk-time delay that provides a time-delay penalty at low-traffic rates. It is apparent that had we chosen to plot results for roll-call polling as well in Figs. 8–15 and 8–16, the time delays incurred for this scheme would have been substantially higher than those for hub polling. This conclusion is based on our prior discussion, as exemplified by the entries of Table 8–2.

The time-delay–throughput characteristic for the controlled access strategies—hub and roll-call polling—depends critically on the walk-time parameters, as has already been noted a number of times. Our choice of 2 msec/100 miles as the propagation delay in the example of Figs. 8–15 and 8–16, as well as the two earlier 10-station examples based on Fig. 8–3, is probably typical, but still an arbitrary choice. Note that this corresponds to an

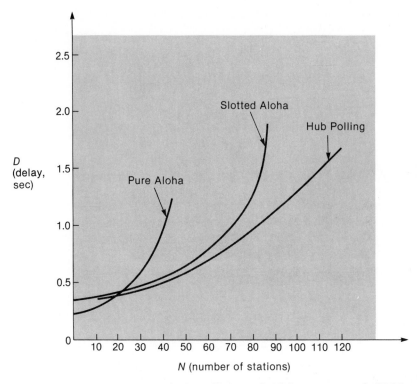

Figure 8–15 Inbound delays: hub polling and Aloha compared, 2000-mile system, $\lambda = 1$ packet/min/station, 4800-bps line capacity

average speed of light of 50,000 miles/sec for the system modeled. Measurements of propagation delay for typical voice-telephone networks, as would normally be used to provide the 2000-mile span here, provide values ranging from 4 msec/100 miles to 1 msec/100 miles. (The specific value measured depends not only on the speed of light along the media used — these might typically include a mix of wires, coaxial cable, microwave relay, and optical fibers — but also on switching and processing delays in the telephone exchanges passed.) In the next subsection, on CATV systems, we shall use 7.1 μsec/mile (4.4 μsec/km) as a typical propagation-delay value. This is obviously a much better number and is attainable because of the simplicity and much shorter span of the cable access network. In Chapter 9 following, on local area networks, we shall use 5 μsec/km as a typical propagation delay value.

The synchronization time component of the walk time can also be quite variable. In the examples discussed here, with signals implicitly transmitted at

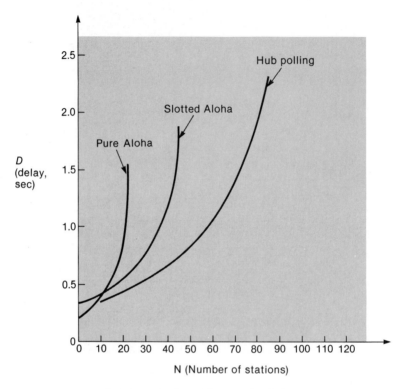

Figure 8-16 Inbound delays: hub polling and Aloha compared, 2000-mile system, $\lambda = 2$ packets/min/station, 4800-bps line capacity

high frequencies, this time is principally the startup time of the modems used, which is required to recover phase and timing. This time could range to as much as a few hundred msec for higher-bit-rate modems used to transmit data over the telephone network. In the token-passing local area network scheme discussed in the next chapter, in which modems are not used, the comparable synchronization time required to synchronize on an arriving bit stream will be measured in bits of latency delay. This will vary from a minimum value of 1 bit to as much as 10 bits and more. The token-passing system performance will, of course, reflect this latency delay.

Finally, time-delay calculations for both the random-access and the polling strategies depend on the line speed or capacity in bits/sec. Recall that the message transmission time m was added to the access delay to obtain the inbound delay. As the line speed increases the value of m decreases correspondingly.

Random-access schemes then tend to provide comparatively better performance in the low-activity range, since there they are limited by message-transmission time. Some of these observations will be borne out by examples both here and in the next chapter.

Note from Figs. 8-15 and 8-16 that the Aloha schemes are limited much more in their throughput capability than is polling. Since the maximum throughput of the pure Aloha scheme is 0.18 of capacity, this puts a corresponding limit on both the number of stations that can access a network and the activity per station. The slotted-Aloha scheme doubles the throughput capability to 0.368 of capacity. The polling schemes can operate up to line capacity, however, albeit with possibly long time delays; this is also borne out by Figs. 8-15 and 8-16. In the case of Fig. 8-15, for example, the $N = 120$ stations point of operation corresponds to a total line utilization of $\rho = 0.5$. The system could thus support even more stations if the resulting delays were tolerable. In the case of Fig. 8-16 the effective ρ for hub polling at $N = 90$ stations is 0.75.

8-3-1 Metropolitan-area Networks: CATV Systems

It was noted in passing, earlier in this chapter, that CATV systems with appropriate modification lend themselves to two-way interactive communications. A number of experimental two-way systems have been developed. One prominent example, the Warner Cable's QUBE system, uses polling [McGA]. Cox Cable's INDAX system uses CSMA/CD technology [ELLI]. CATV systems generally extend between 10 and 100 miles; they have had a particularly significant impact in metropolitan areas. When adapted for two-way communications, they are considered as prime candidates for use in metropolitan area communications.

CATV systems are commonly designed with a tree-type topology. A simple system, such as the one shown in Fig. 8-17, would consist of a main trunk cable feeding out from a headend. Trunk amplifiers installed along the trunk maintain the signal level and compensate for cable transmission characteristics. Multiple branches are connected to the trunk through bridger amplifiers. Branches may in turn split into sub-branches, to which subscriber drops are connected by passive taps. Ten thousand subscribers might typically be connected along this one trunk. Larger systems might have the headend connected to multiple hubs, each in turn controlling a number of trunks of the type shown in Fig. 8-17. The number of subscribers might then total more than 300,000 [McGA].

Cable systems provide bandwidths of between 300 and 450 MHz, with some versions going as high as 600 MHz. These are split into 6 MHz bands or channels

[ELLI] M. L. Ellis et al., "INDAX: An Operational Interactive Cabletext System," *IEEE J. on Selected Areas in Comm.*, vol. SAC-1, no. 2, Feb. 1983, 285-294.

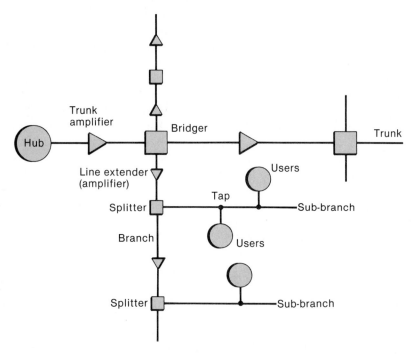

Figure 8–17 Single-trunk CATV layout

to carry TV signals. In the case of two-way data transmission, the frequency range 5–30 MHz or higher, still split into 6-MHz channels, is used to carry signals upstream (to the headend), and a number of 6-MHz downstream channels from the remaining bandwidth are allocated to downstream data traffic. Each 6-MHz channel can in turn handle multiple data channels, depending on the bit rate and modulation technique used. The INDAX system, for example, uses FM to provide 20 28-kbps channels, requiring 300 kHz of bandwidth each, in a 6-MHz band [ELLI]. Other modulation techniques, possible with higher signal-to-noise ratios, would allow higher bit rates to be used. These could go as high as 1 bps/Hz of bandwidth or even higher [SCHW 1980a].

Experimental interactive CATV systems have all been designed on the basis of communications to the headend and back, although proposals have been made for distributed two-way communications using packet-switching technology of the type discussed in the previous chapters of this book [SAADT]. If communications is always either upstream or downstream, to the headend and back, it is natural to think in terms of controlled or random access as the access strategies to be used. In the example we discuss here, we focus on roll-call polling, with the system-access controller located at the headend. This is the technique used in the QUBE system, as already noted [McGA]. Later, after describing the CSMA/CD technique, we consider briefly the possibility of applying that procedure to the same problem. As noted earlier, this has been the technique used in the INDAX system [ELLI].

In this example, a typical user is assumed to be located 6.25 miles from the headend. (The furthermost user is 12.5 miles from the headend.) Propagation delay is taken to be 7.1 μsec/mile (4.4 μsec/km). The one-way propagation delay is then $\tau/2 = 88.8$ μsec, and the total propagation delay for roll-call polling with N stations is approximately 88.8 N, $N \gg 1$. (See Eq. 8–7.) We use 256 kbps as the transmission capacity in this example. We assume that the normal response mode of HDLC is used to carry out the polling. Because of the potentially large number of subscribers to be polled in the CATV environment, however, we use the extended-address mode of operation. The address field is then 2 bytes, so that a total of 2^{16} or about 65,000 users can be accommodated on any one poll. (Separate polls can be carried out on different frequency bands, so that more subscribers can be accommodated if necessary.) The RR-frame used for polling is then 56-bits long (Figs. 8–4 and 8–6). Frames transmitted from user stations are I-frames and are all taken to be 128 bits long (56 bits of overhead plus 72 bits of data). Two cases are considered, one in which a user transmits a data frame every 15 sec on the average; the other, one frame every 30 sec. These numbers are consistent with those adopted as traffic assumptions for interactive CATV systems [McGA], [ELLI].

The resultant scan times t_c and average inbound delays D are shown tabulated in Table 8–4 as a function of the number of active subscribers, N, connected. Synchronization time has been neglected here. Note that with a user traffic load of $\lambda = 1$ frame/15 sec on the average, 5000 subscribers polled sequentially in a single cycle result in an average scan time of 1.84 sec. For the relatively high transmission capacity (256 kbps) and short frame length (128 bits) used in this example, the transmission time per frame, m, is very small compared with the scan time. Wait time is therefore negligible as well, and the inbound delay D is about one half the scan time, or 0.92 sec for these numbers. If one goes to an aggregate of 10,000 users polled in sequence, the scan time increases to 4.6 sec, and the average delay per station now increases to 2.3 sec. It is clear that this is about the threshold of operation of this system with these numbers. The effective utilization is still only $\rho = 0.33$ for $N = 10,000$ subscribers, so the system is only lightly loaded. The relatively long scan time is due primarily to the time required to transmit the poll message. Assuming a reduced traffic load of 1 frame/30sec/user, one can increase the number of subscribers polled somewhat, but not as much as one might like because of the relatively long time required to poll all the users. These numbers change considerably if

TABLE 8–4 Roll-call Polling, CATV System, 256 kbps*				
a. $\lambda = 1$ frame/15 sec	N	$\bar{t}_c(sec)$	$D(sec)$	ρ
	1000	0.32	0.16	0.03
	5000	1.84	0.92	0.167
	10000	4.6	2.3	0.33
	15000	9.2	4.6	0.5
b. $\lambda = 1$ frame/30 sec				
	1000	0.31	0.16	0.017
	5000	1.6	0.82	0.083
	10000	3.72	1.86	0.167
	15000	6.1	3.1	0.25

* 12.5 miles, distance to headend; 56-bit polling frame; 128-bit data frame

much higher bit rates are used. This reduces the time required to transmit the polling message and reduces the scan time correspondingly. Details are left to the reader.

Could a pure Aloha scheme be used instead? A simple calculation indicates that the *maximum possible* number of users that could be accommodated would be $N = 5500$ subscribers for $\lambda = 1$ frame/15 sec/station, and double that, 11,000 subscribers, for $\lambda = 1$ frame/30 sec/station. One must reduce these maximum numbers, based on $S_{max} = 0.18$, considerably to ensure stable operation of such a system. Pure Aloha is thus not appropriate for a head-end-oriented two-way CATV system. (The inbound delay is at most a few msec in these cases, so delay is obviously no problem. So far as delay is concerned the pure Aloha scheme outperforms the roll-call scheme by orders of magnitude in this example!) Pure Aloha could, however, be used in a scheme where small subsets of the total subscriber population access the system in random-access fashion. This turns out to be the case in a distributed, rather than head-end oriented, CATV system [SAADT] or in frequency reuse schemes where the CATV network is broken into relatively small regions to reduce the chance of collisions [MAXE].

[MAXE] N.F. Maxemchuk and A.N. Netravali, "A Multifrequency Multiaccess System for Local Access," ICC, June 1983.

8-4 Random Access Using CSMA/CD

As noted a number of times in this chapter, the basic Aloha random-access strategy provides a throughput of at most $1/e = 0.368$ of the channel capacity, and then only if stations transmit at precisely synchronized times. Is it possible to improve on the limits imposed by the Aloha schemes?

A large number of multiaccess strategies have been proposed and their performance analyses published in the literature. Some fall into the controlled-access category described here, while others are modifications of the random-access strategies. A complete categorization of the various schemes and a discussion of their performance characteristics appears in [TOBA]. Another survey of various multiaccess techniques is given by [KURO]. A particularly interesting class of multiaccess protocols that has attracted a lot of attention is one that uses adaptive tree-search or probing techniques to limit conflict among users that access a common channel. This class is described in detail in [HAYE, Chap. 9].

We spend the remainder of this chapter discussing the CSMA/CD protocol. It is based on, and provides throughput improvement over, the pure-Aloha technique. The Ethernet protocol, which has been adopted as one of the access standards in local area networks, incorporates the CSMA/CD technique. Ethernet implementations of local area networks are widespread. Details of the Ethernet standard, including packet format and a number of implementation issues, are provided in the next chapter. At this point we describe only CSMA/CD operation.

The basic concept of the CSMA/CD protocol is quite simple. All stations *listen* for transmissions on the line. A station that wishes to transmit does so only if it detects the channel is idle. This procedure is called *carrier sensing* (CS), and the access strategy using it is termed a CSMA (carrier sense multiaccess) scheme. It is apparent that collisions may still occur since stations are physically displaced from one another, and two or more stations may sense that the channel is idle and start transmitting, causing a collision. Once stations detect a collision (*collision detection*, or CD) they transmit a special *jam signal* notifying all other stations to that effect and abort their transmissions. The carrier sense (CSMA) feature provides throughput improvement over pure Aloha; collision detection, with a

---

[TOBA] F. A. Tobagi, "Multiaccess Protocols in Packet Communication Systems," *IEEE Trans. on Comm.*, vol. COM-28, no. 4, April 1980, 468-488; reprinted in [GREE].

[KURO] J. Kurose, M. Schwartz, and Y. Yemini, "Multiple-access Protocols and Time-constrained Communication," *Computing Surveys*, vol. 16, no. 1, March 1984, 43-70.

message aborted rather than transmitted to completion, provides still more improvement in throughput.

A number of CSMA techniques have been proposed and analyzed. They differ as to how they handle transmission once the line is sensed busy. In a p-persistent scheme, for example, a station that senses a busy channel transmits with a probability p once the channel goes idle. With a probability $(1 - p)$ it defers transmission for the propagation delay interval τ. A 1-persistent scheme is one in which the station attempts transmission as soon as the channel is sensed idle. In a nonpersistent scheme the station reschedules transmission to another time following a prescribed retransmission delay distribution, senses the channel at that time, and repeats the process. A number of other variants have been proposed as well [TOBA]. All of these schemes rely on stations being able to sense the end of a transmission soon after it is completed. They thus require that the end-to-end propagation delay τ be small compared with the message transmission length m. This condition is usually represented by having the parameter $a \equiv \tau/m \ll 1$. If the propagation delay becomes too long (a approaching 1 or greater), the schemes degenerate to the simple Aloha strategy. They are thus applicable primarily to local area networks or to larger networks operated at relatively low bit rates. (This increases the message length m.)

The CSMA/CD protocol, which follows the 1-persistent rule and adds the collision-detection feature to further improve the performance, is as already noted the protocol used in the Ethernet scheme [METC]. If a collision is detected and transmission aborted, a retransmission attempt is made after a random time interval, as in the Aloha schemes. This random retransmission interval is doubled each time a collision is detected, up to some maximum value, at which point the station gives up and notifies upper levels of transmission failure. This doubling of the interval is referred to as a binary backoff procedure and is found to improve system performance. (This was alluded to in our earlier discussion of the Aloha techniques, where it was noted that increasing the retransmission delay reduces the collision probability and eliminates a possible instability.)

The time-delay characteristic of CSMA/CD has been obtained by Lam [LAM 1980] using discrete-time analysis. Rather than reproduce his analysis here, which is quite detailed algebraically, we show how one obtains his expression for the maximum throughput through a plausibility argument. This points out the dependence of the throughput on the critical parameter $a \equiv \tau/m$. Lam's analysis is repeated in part in [HAYE, Chap. 8].

[METC] R. M. Metcalfe and D. R. Boggs, "Ethernet: Distributed Packet Switching for Local Computer Networks," *Comm. ACM*, vol. 19, no. 7, July 1976, 395–404.

[LAM 1980] S. S. Lam, "A Carrier Sense Multiple Access Protocol for Local Networks," *Computer Networks*, vol. 4, no. 1, Jan. 1980, 21–32.

Consider the bus topology of Fig. 8–18 (the same topology shown earlier in Fig. 8–2(b)). As noted there, stations are connected through passive taps to a two-way bus (cable or wire). Focus attention on the two stations furthermost apart, indicated by A and B in Fig. 8–18. We calculate the average time required to launch a message successfully on the bus. The reciprocal of this time is then the desired maximum throughput. By focusing on stations A and B, we get the maximum or worst-case time required to transmit a message and hence the worst-case maximum throughput. In keeping with the notation of Chapter 4 we call the time to successfully transmit a message the virtual transmission time t_v. This time has three components, as shown in Fig. 8–19. It is made up of the time m required to transmit a message (assumed to be a fixed time in this simple analysis), plus a time τ required to sense completion of transmission, plus multiples of 2τ-units of time to resolve collisions, once detected.

Why the time τ, the end-to-end propagation delay (Fig. 8–18), for sensing the end of carrier? Say that station A at one end of the bus is in the process of transmitting its message. Once A completes transmission, it still takes station B, at the other end of the bus, τ sec to sense the end of the transmission and begin its own transmission if it so desires. There is thus an unavoidable waste of τ sec due to the sensing algorithm. This effectively increases m by the value τ and reduces the maximum throughput accordingly.

Say that a collision between A's and B's transmissions takes place. In the worst case, it takes A and B 2τ sec to detect the collision and immediately turn off their transmissions. (To simplify this plausibility argument we neglect the added jam time during which a station notifies other stations that it has detected a collision.) This is shown in Fig. 8–20: A starts transmitting a message at some initial time. Just before A's message arrives at B, B decides to transmit. It senses the channel, thinks it is idle, and starts its own message going. There is an obvious collision, which is not detected by A until τ sec later. The total time to detect the collision is thus 2τ units of time, as shown in Fig. 8–20.

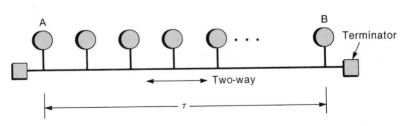

Figure 8–18 CSMA/CD model

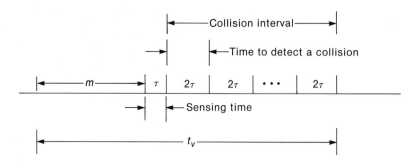

Figure 8–19 Calculation of virtual transmission time, CSMA/CD

Given a collision occurring, say that it takes $2\tau J$ units of time to resolve the collision. J represents the average number of retransmissions, given that a collision has occurred. It is thus comparable to the parameter $E = G/S - 1$, which we introduced in studying the Aloha systems (Eq. 8–13). The virtual transmission time is then given by Fig. 8–19:

$$t_v = m + \tau + 2\tau J$$
$$= m[1 + a(1 + 2J)] \qquad a \equiv \tau/m \qquad (8-19)$$

Note that the parameter $a = \tau/m$ arises quite naturally through this argument.

To continue with the analysis we must find a value for J. This depends on the retransmission strategy. For the Ethernet scheme, for example, one would

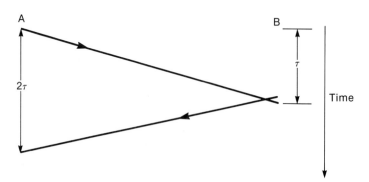

Figure 8–20 Worst-case detection of collision

have to carry out an analysis that involves the binary backoff strategy. Lam's simulation studies indicate that this is not necessary, that in fact the following simple model of retransmission provides accurate results ([LAM 1980]; see also [HAYE, Chap. 8]): *Assume* that the length of a collision interval (Fig. 8–19) is geometrically distributed in units of 2τ, with parameter v. Specifically, the interval is *one* unit (2τ sec) long, with probability v; two units long, with probability $v(1 - v)$; three units long, with probability $v(1 - v)^2$, and so forth. v is thus the probability of success at the end of an interval; $(1 - v)$ is the probability of a collision. In terms of this model, the average number of retransmissions is $J = 1/v$, for

$$J = \sum_{k=1}^{\infty} kv(1 - v)^{k-1} = 1/v \qquad (8-20)$$

This argument shifts the burden of finding J to that of finding v.

The probability v is now found using the following argument. Say that n stations ($n \gg 1$) are involved in possible transmissions. Let the probability that any one station wants to transmit in a 2τ-sec interval be p. The probability that exactly *one* station transmits and is successful is then

$$v = np(1 - p)^{n-1} \qquad (8-21)$$

It is left to the reader to show that the value $p = 1/n$ maximizes this expression and hence provides the greatest chance of success. Using this value for p and letting $n \gg 1$, as assumed, we find in the limit that

$$v_{\max} = \left(1 - \frac{1}{n}\right)^{n-1} \rightarrow e^{-1} \qquad n \rightarrow \infty \qquad (8-22)$$

The value of v to use in Eq. (8–20) is thus e^{-1}, and Eq. (8–19) becomes

$$t_v = m[1 + a(1 + 2e)] \qquad a \equiv \tau/m \qquad (8-19a)$$

Note that this model for retransmissions is reminiscent of slotted Aloha and does in fact lead to the slotted Aloha throughput value of e^{-1}.

The maximum message throughput, $\lambda_{\max}$, in units of messages/time, using the same arguments as in Chapter 4, is $1/t_v$. (This represents the maximum number of "equivalent" messages per unit time one could transmit over the common channel by concatenating them, one after the other.) Letting λ be the average number of messages/time transmitted over the channel, from *all* users combined, and normalizing to the transmission capacity $1/m$, in messages/time, one finds, from Eq. (8–19a), that

$$\rho \equiv \lambda m < \frac{1}{1 + a(1 + 2e)} = \frac{1}{1 + 6.44a} \qquad a \equiv \tau/m \qquad (8-23)$$

using the numerical value of e.

As an example, say that $a = 0.1$. This implies that the length of a message is 10 times the end-to-end propagation delay. For this value of a, we have

$$\rho < 0.6 \qquad\qquad (8-24)$$

Note that this is a considerable improvement over the pure-Aloha value of 0.18 and the slotted-Aloha value of 0.368. If a is reduced still more (either by shortening the cable length or reducing the transmission capacity in bps to increase m), ρ_{max} increases correspondingly, approaching the maximum possible value of 1. Equation (8–23), for the maximum normalized throughput in CSMA/CD, agrees with that obtained using the more sophisticated analysis of Lam. However, he does use the argument developed here to find J and hence v.

We shall apply this simple calculation of the maximum throughput of CSMA/CD to some local area network examples in the next chapter. Before continuing with that application, however, it is appropriate to return to the 12.5-mile metropolitan-area network example introduced briefly earlier. Recall that we used a value of $\tau/2 = 88.8$ μsec as the propagation delay of a typical station under the polling discipline. For CSMA/CD we must use the full end-to-end delay, $\tau = 177$ μsec. The headend detects collisions and sends jamming signals back to all users. (If stations themselves are required to detect collisions and abort transmission, the headend must echo all signals back over a downstream channel. This doubles the effective line length and hence the propagation delay τ.) As noted, we must thus have $m \gg \tau$ for CSMA/CD to work. Say that $a \leq 0.1$ is desired. Then $m = 1.77$ msec. For the 128-bit frames used earlier, this implies that the channel bit rate $C \leq (128/1.77) = 72$ kbps. Clearly the 256-kbps figure used in the polling example is inadequate. As the system gets larger in extent, the bit rate must be reduced accordingly. For a system covering a 25-mile region, we have $\tau = 0.36$ msec using the propagation-delay value of 7.1 μsec/mile. For $a \leq 0.1$, $m \geq 3.6$ msec, and $C \leq (128/3.6) \times 10^3 = 36$ kbps. This is, of course, just half of the bit rate for the 12.5-mile example. Note that Cox Cable's INDAX system, which uses the CSMA/CD protocol for CATV networks, uses a transmission speed of 28 kbps [ELLI]. This value is consistent with the bit rates just quoted. If one wants to use the CSMA/CD technique over relatively long distances, there is no choice but to either decrease the bit rate or increase the message length. For many applications these bit rates are quite sufficient. They are obviously much higher than the bit rates available over telephone lines (2400–9600 bps) and would presumably be quite adequate for interactive data transfer. However, they might be deemed inadequate for long file transfers.

How many stations could be accommodated on a single CSMA/CD channel for the 12.5-mile network used in the polling example of Table 8–4? Letting $\rho = 0.6$ and using $m = 1.77$ msec (this corresponds to 128-bit data frames with a line capacity $C = 72$ kbps), the total frame-transmission rate over the channel can be no more than $0.6/m = 340$ frames/sec. If a typical user inputs 1 frame/ 15 sec on the average, this allows 5100 users at most to access this one channel. The corresponding time delay would be quite high, however, so one might want to back off to considerably fewer users/channel. (Time delays in the CSMA/CD case are discussed in the next chapter.) Reducing the user-input rate to 1 frame/30 sec, the number doubles to 10,000 users at most. These numbers are to be compared with those of Table 8 – 4. Note that the bit rate there, however, is 256 kbps.

This completes our comparative discussion of polling and random-access schemes in this chapter. We pick it up again in the next chapter, where we specialize the discussion to the case of local area networks (LANs) covering distances of a few kilometers at most. The focus there is on the token-passing scheme comparable to hub polling, and on Ethernet, using the CSMA/CD protocol.

Problems

8–1   Consider a primary station A that polls four secondary stations B, C, D, E. The normal response mode (NRM) of HDLC is used to carry out the polling function. Diagram a typical sequence of frames transmitted to and from Station A, carrying the sequence out over at least one full poll cycle. Include the case in which information frames are occasionally sent out from station A as well, as shown in Fig. 8–6.

8–2   Refer to Fig. 8–6. As indicated in the text, station A interrupts the poll cycle to send the frame labeled DI11 to station D. What would happen if the P bit were set in this frame?

8–3   Derive Eq. (8–7), the expression for the overall propagation delay in a roll-call polling system with N equally spaced stations.

8–4   Show that the overall propagation delay in a roll-call polling system with N equally spaced stations, with the controller placed at the center of the string of stations, is given by

$$\tau' = \frac{\tau}{2}\left(1 + \frac{N}{2}\right)$$

Compare with Eq. (8–7).

8–5   Consider a polling network that consists of 10 stations spaced 300 miles apart. Calculate the total walk time for both roll-call and hub polling if the line speed is 4800 bps, polling messages are 48-bits long, and the synchronization time per station is 20 msec. The propagation delay is 2 msec/100 miles. Compare two cases, one with the controller at one end of the string, the other with the controller in the middle of the string.

8-6   Calculate the access delays for the example of Problem 8-5 if frames are 1200 bits long on the average and are exponentially distributed in length. Take the two cases $\rho = 0$ and $\rho = 0.5$.

8-7   Carry out the detailed calculations for the two polling examples summarized in Tables 8-1 and 8-2. Check the entries obtained there, with particular emphasis on the comparison between hub and roll-call polling summarized in Table 8-2.

8-8   Consider the two polling examples summarized in Table 8-2—one for 10 stations spaced 10 miles apart and the other for 10 stations spaced 200 miles apart. Generalize the results shown there (see Problem 8-7) to examples of nonexponential frame-length statistics. Two such examples, for roll-call polling only and for $\rho = 0.5$, appear in Table 8-3. Carry out the calculations for hub polling as well, and take $\rho = 0.8$, as well as other values of ρ, if desired.

8-9   An alternative derivation of the pure Aloha throughput equation (8-11) proceeds as follows:

$$G = S + \text{the average number of retransmissions}$$

Show that the average number of retransmission is $G(1 - e^{-2G})$ and then solve for S, obtaining Eq. (8-11).

Hint: What is the probability that *at least* one collision will take place?

8-10   A number of terminals use the pure-Aloha random-access strategy to communicate with a remote controller over a 2400-bps common channel. Each terminal transmits a 200-bit message, on the average, once every 2 minutes. What is the maximum number of such terminals that may use the channel? Repeat for the slotted-Aloha strategy. What if message lengths are increased to 500 bits? What if terminals transmit messages once every 3 minutes, on the average? How do the results change if the line speed is increased to 4800 bps?

8-11   Calculate inbound time delays for hub polling, pure Aloha, and slotted Aloha for the example of Figs. 8-15 and 8-16. Messages are 1200-bits long in all cases. Check the curves obtained. Carry out a similar set of calculations for roll-call polling and compare with those for hub polling.

8-12   Refer to the CATV metropolitan-area network example discussed in the text. Carry out the inbound delay calculations, checking the results obtained in Table 8-4. Repeat the calculation if 512-kbps and 1.544-Mbps transmission rates are used instead. (Note that this increases the bandwidth requirement proportionally.)

8-13   Refer to the CATV example summarized in Table 8-4. (See Problem 8-12 as well.) Evaluate the possibility of using the pure Aloha scheme for a system of this type.

8-14 n stations each transmit with a binary probability p. (The probability of no transmission is then $1 - p$.) Show that the probability of exactly *one* transmission is given by Eq. (8-21). Prove that Eq. (8-21) is maximized if $p = 1/n$.

8-15   Investigate the possibility of using a CSMA/CD access technique for a CATV system with the furthermost user 50 miles from the headend. Consider the three cases: data frames 128-bits, 512-bits, and 1200-bits long. What is the maximum line capacity (transmission rate) that can be used in each case if $a \equiv \tau/m$ is to be less than 0.1? Take the propagation delay to be 8 μsec/mile. Find the maximum number of users that can be accommodated for three cases: (1) users input 1 data frame/15 sec, (2) 1 data frame/30 sec, (3) 1 data frame/2 min ($\rho = 0.6$ in all cases).

8–16 Four stations are connected on a double wire to a controller, as shown in Fig. P8-16. Each generates an average of 75 1000-bit packets/sec destined for the controller. The line transmission rate is 600 kbps. The propagation delay is 5 μsec/km.

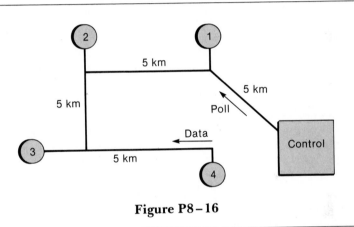

Figure P8–16

a. Roll-call polling is used. Calculate the average scan time if the polling message is 60 bits and each station requires 0.1 msec to synchronize to a poll. The stations are 5-km apart, as shown. Find the average scan time if the controller polls in a counterclockwise direction as shown.
b. Repeat part a. if hub polling is used.
c. Could a contention-type scheme such as slotted Aloha or CSMA/CD be used? Explain.

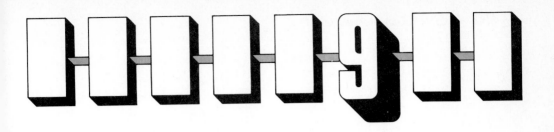

Local Area Networks

Previous chapters have concentrated primarily on geographically distributed —i.e., wide-area—networks. In the last chapter we mentioned metropolitan-area networks, those that cover a range of $10-100$ km, in conjunction with CATV systems. In this chapter we focus on *local area networks*, those that involve the interconnection of terminals, computers, workstations, and other intelligent systems within a building or a number of buildings that constitute a small campus. As noted in Chapter 8 in our discussion of various access mechanisms for packet-switched data networks, the systems on the local area network are commonly interconnected via the bus or the ring topology as shown in Figs. $9-1$(a) and $9-1$(b). The third topology shown in Fig. $9-1$, that of the star, is also used, but principally in conjunction with PBX (private branch exchange) communications. (The actual topology of the ring may turn out to be more like a star for reliability's sake, as will be noted later in this chapter, but the ring description is the one commonly used.)

The maximum extent of these networks end to end is at most several kilometers if they are to be operated at very high bit rates; this point was made in the previous chapter. Many access mechanisms have been proposed or implemented for local area networks. In this chapter we focus on three mechanisms that have been designated as standards: the CSMA/CD and token-passing access methods for the bus topology, and the token-passing ring. Other types of access methods, either proposed or implemented for special-purpose use, are described in the literature. (See for example [STAL] and [IEEE 1983a].)

[STAL] William Stallings, "Local Networks," *Computing Surveys*, vol. 16, no. 1, March 1984, 3–41.
[IEEE 1983a] "Local Area Networks," special issue, *IEEE J. on Selected Areas in Comm.*, vol. SAC-1, no. 5, Nov. 1983.

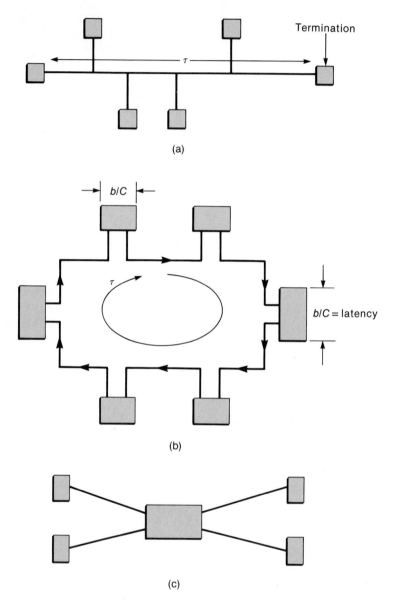

Figure 9–1 LAN topologies
a. Bus topology
b. Ring topology
c. Star topology

The local area network standards have been developed by the IEEE (the Institute of Electrical and Electronics Engineers) through a special committee called the IEEE 802 Committee. These standards cover the physical layer and a portion of the data link layer; they conform to the OSI Reference Model. The three standards noted in the last paragraph are labeled the IEEE Standard 802.3 for the CSMA/CD bus, the IEEE Standard 802.4 for the token-passing bus, and the IEEE Standard 802.5 for the token-passing ring. IEEE Standard 802.3 is almost identical to the Ethernet specification, which was developed jointly by Xerox, Intel, and Digital Equipment Corporation [ETHE]. We shall describe the CSMA/CD (Ethernet) and token-passing ring schemes in some detail, but first we compare their delay-throughput performance.

9-1 Comparative Performance, CSMA/CD and Token Ring

The performance of the CSMA/CD or Ethernet access method, pioneered by workers at Xerox [METC], and the token-passing ring, developed by workers at the Zurich Research Laboratories of IBM [BUX 1981a], can be compared using the equations developed in Chapter 8. (The basic token-passing scheme was first described by Farmer and Newhall of Canada [FARM].) Recall that we indicated in the last chapter that token passing is identical to hub polling except that the token-passing scheme has no central controller. All stations operate in a decentralized mode. All messages (frames) move around the ring and are actively repeated by each station through which they pass (Fig. 9-1b). A station reading its own address as a destination copies the frame, while passing it on, bit by bit, to the next station on the ring. The information contained in the frame will then be passed on to terminals, controllers, or other devices connected to the station. A circulating frame is removed from the ring by the station that transmitted it.

[ETHE] *The Ethernet, A Local Area Network, Data Link Layer and Physical Layer Specifications,* Digital Equipment Corp., Maynard, Mass; Intel Corp., Santa Clara, Calif.; Xerox Corp., Stamford, Conn.; Version 1.0, Sept. 30, 1980, and Version 2.0, Nov. 1982.

[METC] R. M. Metcalfe and D. R. Boggs, "Ethernet: Distributed Packet Switching for Local Computer Networks," *Comm. ACM,* vol. 19, no. 7, July 1976, 395–404.

[BUX 1981a] W. Bux et al., "A Reliable Token-Ring System for Local Area Communication," National Telecommunications Conference, New Orleans, 1981, pp. A.2.2.1–A.2.2.6.

[FARM] W. D. Farmer and E. E. Newhall, "An Experimental Distributed Switching System to Handle Bursty Computer Traffic," *ACM Symposium, Problems on Optimization of Data Communications,* Pine Mountain, Ga., Oct. 1963, 31–34.

The time required to repeat the frame at a station — i.e., the delay through the station — is termed the latency at the station (Fig. 9–1b). The *ring latency*, the equivalent of the total walk time L described in Chapter 8, is the sum of the propagation delay τ around the ring, plus the latency at each station.

Consider token passing now. Recall that receipt of the token represents permission to transmit. Say that a free token (a special type of frame) has been generated and is circulating around the ring. The first station with data to transmit seizes the token and proceeds to transmit. It is assumed to empty its buffer (another strategy, as noted in Chapter 8, is to transmit up to a fixed amount of data only) and then to pass the token on when this is completed. There are a number of ways of doing this. Each gives rise to a somewhat different model of the token-passing strategy. One procedure, similar to that of hub polling, is to issue the token as a part of the frame transmitted, in essence passing permission to transmit on to the next station as soon as the last bit of the frame has cleared the transmitting station. If frames are short, covering less than the once-around propagation delay τ, multiple tokens may exist on the ring.

Another strategy is to issue the token only when the transmitting station receives its message back again, having gone once around the ring. There are two versions possible here: Wait until the entire message has been received (and erased), or issue the new token when the previous token has been received. Both cases allow one token only on the ring at a time. Performance may be reduced somewhat in either case because of added delay. In the first case the frame (message) transmission time is effectively increased by the ring latency L. (Why is this so?) The second case provides an improvement over the first if the frames are shorter than the ring latency, in which case the effective service time at the station is the ring latency L. (Again, why is this so?)

It is apparent that the hub-polling access-delay equation (8–5) of the last chapter, if properly modified in each of these cases, can be used directly to determine the performance of the token-passing access strategy. Specifically, consider the average transfer time t_f required to transfer data from a source station to an arbitrary destination station along the ring. This is just the delay term (8–5) with frame transmission time m added, plus a factor of $L/2$ to account for propagation delay halfway around the ring to an average destination. The resultant equation for the transfer delay, for N stations along the ring, is given by

$$t_f = \frac{L(1 - \rho/N)}{2(1 - \rho)} + \frac{\lambda}{2}\frac{\overline{m^2}}{(1 - \rho)} + m + \frac{L}{2} \tag{9–1}$$

Here $\rho \equiv \lambda m$, λ is the total average traffic, in frames/sec, from all stations accessing the ring, m is the average frame length (data plus overhead), in units of time, and $\overline{m^2}$ is the second moment of the frame length distribution. In this chapter it is convenient to let λ represent the total traffic in the system, replacing

$N\lambda$ used in Chapter 8. (Note the comment made earlier that the service time or frame length here has to be appropriately modified in the case of single-token operation, the case to be considered in the examples following.)

Equation 9–1 appears in a paper by Bux in which he compares a number of local access methods [BUX 1981b]. His paper includes the CSMA/CD method as well, and we follow his approach in comparing the two access methods. The ring latency, or total walk time, L, is readily written in terms of the station latency. Let b be the station latency in bits. Dividing by the transmission capacity C gives the delay through the station (Fig. 9–1b). This is the delay between receiving a frame at a given station and transmitting it out on the ring. The ring latency is then just

$$L = \tau + Nb/C \tag{9–2}$$

for N stations actively connected to the ring. τ is, as previously noted, the propagation delay once around the ring. Figure 9–1(b) shows how Eq. (9–2) is obtained. The minimum station latency is $b = 1$ bit, and this is used in most of the examples discussed later.

Consider the CSMA/CD access method now. To compare with the token-passing ring, one needs an expression for the transfer delay t_f for this scheme. [BUX 1981b] contains the appropriate equation, as modified slightly from the original derivation by Lam [LAM 1980]. Recall from our discussion of CSMA/CD at the end of Chapter 8 that Lam used a discrete-time analysis based on slots 2τ units of time wide, where τ is the end-to-end delay along the bus. (The same parameter τ is used in both the CSMA/CD and the token-passing cases since it represents the propagation delay along the total length of the medium in both cases. See Figs. 9–1(a) and 9–1(b).) The equation for the normalized transfer time in the CSMA/CD case is given by (9–3):

$$
\begin{aligned}
t_f/m = \rho\, &\frac{[(\overline{m^2}/m^2) + (4e + 2)a + 5a^2 + 4e(2e - 1)a^2]}{2\{1 - \rho[1 + (2e + 1)a]\}} \\
&+ 1 + 2ea - \frac{(1 - e^{-2a\rho})\left(\dfrac{2}{\rho} + 2ae^{-1} - 6a\right)}{2[F_p(\lambda)e^{-\rho a - 1} - 1 + e^{-2\rho a}]} + \frac{a}{2}
\end{aligned}
\tag{9–3}
$$

This is obviously a rather complicated equation algebraically and so we will discuss it in limiting cases to help understand its significance. (The derivation of this equation is also quite tedious algebraically, which is one of the reasons why no attempt has been made here to reproduce the derivation.)

[BUX 1981b] W. Bux, "Local-Area Subnetworks: A Performance Comparison," *IEEE Trans. on Comm.*, vol. COM-29, no. 10, Oct. 1981, 1465–1473.

[LAM 1980] S. S. Lam, "A Carrier Sense Multiple Access Protocol for Local Networks," *Computer Networks*, vol. 4, no. 1, Jan. 1980, 21–32.

First some comments on the notation. The parameter a is the same one introduced at the end of Chapter 8 in our discussion of the limiting throughput of the CSMA/CD scheme: $a \equiv \tau/m$, τ the end-to-end propagation delay along the bus (Fig. 9–1a) and m the average frame length. This parameter plays a key role in all analyses of LAN access schemes that involve contention or some form of random access. The added term $a/2 = \tau/2m$ in Eq. (9–3) is the normalized average source-destination propagation time, comparable to the term $L/2$ in the transfer delay expression (9–1) for the token ring. $\rho = \lambda m$ and $\overline{m^2}$ were already discussed in connection with Eq. (9–1).

The function $F_p(\lambda)$ in the denominator of the next-to-the-last term of Eq. (9–3) is the Laplace transform of the frame-length distribution $f(t)$:

$$F_p(\lambda) = \int_0^\infty f(t)e^{-\lambda t}\, dt \tag{9–4}$$

Consider two special cases as examples. Say first that the frame length is a constant (deterministic) value m. Then $f(t) = \delta(t - m)$, with $\delta(t)$ the unit impulse function, $\overline{m^2}/m^2 = 1$, and

$$F_p(\lambda) = e^{-\rho} \qquad \overline{m^2}/m^2 = 1 \tag{9–5}$$

from Eq. (9–4). As the second example, let the frame length be exponentially distributed, with average length m. Then $\overline{m^2}/m^2 = 2$, and

$$F_p(\lambda) = 1/(1 + \rho) \qquad \overline{m^2}/m^2 = 2 \tag{9–6}$$

Other examples can obviously be calculated in a similar manner.

The two limiting cases of operation of CSMA/CD can be studied readily using Eq. (9–3). As the traffic intensity $\rho = \lambda m$ increases, note from the denominator of the first term of Eq. (9–3) that the average transfer delay increases in an unbounded way as ρ approaches a maximum value

$$\begin{aligned} \rho_{max} &= 1/[1 + (2e + 1)a] \\ &= 1/[1 + 6.44a] \end{aligned} \tag{9–7}$$

This is precisely the maximum value derived in Chapter 8 (Eq. 8–23). As a matter of fact the collision-resolution model used to carry out the transfer time analysis leading to Eq. (9–3) is the same as that used in the maximum-throughput calculation leading to Eq. (8–23) [LAM 1980].

At the other extreme, let $\rho \to 0$. It is left to the reader to show that the transfer time expression (9–3) approaches the minimum value

$$t_f|_{min} = m + \tau/2 \qquad \rho = 0 \tag{9–8}$$

(As a hint, note from Eq. (9–4) that $F_p(0) = 1$. The term in which this appears approaches $2ea$ as $\rho \to 0$ and cancels the $2ea$ term that precedes it.) This is exactly what we would expect from the CSMA/CD protocol. At very low traffic a station wishing to transmit can do so immediately, with no waiting, and with very little chance of a collision. Its average transfer time is then just the average service (transmission) time m plus one-half the end-to-end delay required, on the average, to reach a typical destination station. It is in this region of traffic that the CSMA/CD scheme outperforms controlled-access schemes such as token passing, as already noted in Chapter 8.

Equations (9–1) and (9–3) can now be used to compare the transfer-delay characteristics of the token-passing ring and CSMA/CD bus strategies. (The token-passing bus analysis is identical to that of the token-passing ring. The basic difference in the two cases is that in the token-passing bus a free token is addressed to a specific station. If stations pass control to their neighbors in one designated direction, it is apparent that Eq. (9–1) applies, but with τ now taken as the *round-trip* propagation delay of the bus.) Comparative performance curves for a number of typical local area network cases appear in Figs. 9–2 through 9–6, which are taken from [BUX 1981b]. Note the two significant differences between these examples and comparable ones studied in Chapter 8: The networks cover relatively short distances of 1 or 2 km, and they are operated at correspondingly higher bit rates than those discussed in Chapter 8. The transmission rates here range from 1 to 10 Mbps. These numbers are typical of cable-based LANs.

Figures 9–2, 9–3, 9–5, and 9–6 provide plots of average transfer time versus throughput, the performance tradeoff functions on which we have focused throughout this book. They differ slightly from the comparable curves of Chapter 8, however, in that they are shown normalized to the average *packet* length rather than the *frame* length (packet plus header). This is true for both abscissa (horizontal axis) and ordinate (vertical axis). If we take the average packet length to be $E(L_p)$ bits while the header length is L_h bits, the parameter m in Eqs. (9–1) and (9–3) is just $m = [E(L_p) + L_h]/C$, with C the cable transmission rate in bps. Figures 9–2 — 9–6 plot $t_f/E(L_p)/C$ versus $\lambda E(L_p)/C$ instead of our more usual t_f/m versus λm. Since $E(L_p) = 1000$ bits and $L_h = 24$ bits in these examples, the difference is slight in all cases.

Figure 9–2 provides performance results for the case of a 2-km LAN operated at 1 Mbps. The number of stations on the network has been chosen as $N = 50$, with a latency of 1 bit per station in the case of the token-passing ring. Letting the propagation delay along the cable be 5 μsec/km, we have $\tau = 10$ μsec. For a 1-Mbps transmission rate, the average frame length is $m = 1.024$ msec, and $a = \tau/m \doteq 0.01$. The CSMA/CD system should thus perform fairly well up to relatively high throughputs, as is demonstrated by the CSMA/CD curve of Fig. 9–2. Specifically, the maximum throughput normalized to $1/m$ (the capacity of the system in frames/sec) is $\rho_{\max} = 1/(1 + 6.44a) = 0.94$, from Eq. (9–7). This is fairly close to the maximum value shown in Fig. 9–2, with the throughput normalized to $E(L_p)/C$, as already noted. If we now consider the

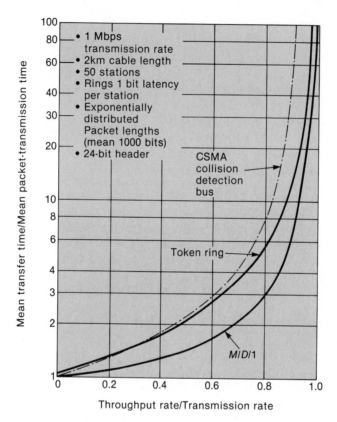

Figure 9–2 Transfer delay-throughput characteristics at 1 Mbps (from [BUX 1981b, Fig. 8], © 1981 IEEE, with permission)

token ring, the latency per station is $1/C = 1\,\mu sec$, so that for 50 stations we get 50 μsec. Adding to this the propagation delay τ, we have, as the overall walk time, $L = 60\,\mu sec$. From Eq. (9–1), the mean transfer delay at a throughput rate approaching zero is $m + L = 1.084$ msec, not much different from the transmission time $m = 1.024$ msec. This is apparent from Fig. 9–2. The token-ring and CSMA/CD schemes track one another fairly closely over most of the throughput range for this example. The token-ring performance begins to exceed that of the CSMA/CD bus at a normalized throughput of 0.4, but does not begin to differ substantially until the normalized throughput exceeds 0.8.

Although the focus here is on *normalized* delay, note that the absolute delays are quite small because of the relatively high transmission speed. We have already seen that the mean transfer delay at low traffic is very nearly the mean transmission time $m \doteq 1$ msec. At $\rho = 0.8$, the mean transfer delay is still only 8 msec for the CSMA/CD bus and 6 msec for the token ring.

When the cable speed is increased by an order of magnitude to $C = 10$ Mbps, the CSMA/CD scheme is affected tremendously. This is apparent from the curves of Fig. 9–3, with parameters chosen exactly the same as in Fig. 9–2 except for the increase in line speed. Since the average frame length m has now decreased to approximately 100 μsec, the parameter $a = \tau/m$ increases to 0.1, and the CSMA/CD scheme begins to fail at relative throughput rates approaching $1/(1 + 6.44a) = 0.6$. This is borne out by the CSMA/CD curve of Fig. 9–3.

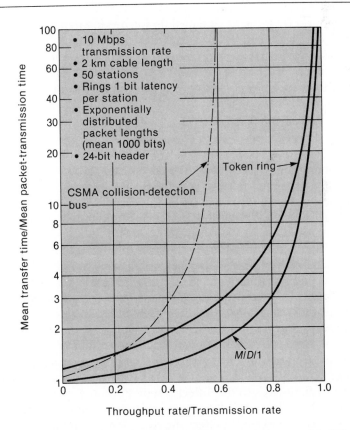

Figure 9–3 Transfer delay-throughput characteristics at 10 Mbps (from [BUX 1981b, Fig. 9], © 1981 IEEE, with permission)

Recall from our discussion of the CSMA/CD protocol in Chapter 8 that the carrier-sense and collision-detection mechanisms break down as the frame length m decreases, approaching the end-to-end delay τ. Relatively more collisions take place and the performance deteriorates for relatively low throughputs. The token-ring performance begins to exceed that of the CSMA/CD bus for a normalized throughput of about 0.2. The only way to improve the performance of the CSMA/CD scheme is to reduce the parameter $a = \tau/m$. For operation at $C = 10$ Mbps this means either using a shorter bus (say no more than 1 km) or a longer packet length.

The significance of a in the performance of a CSMA/CD scheme is further demonstrated by Fig. 9–4, which plots normalized transfer delay versus a for various values of $\rho = \lambda m$. Here the header length L_h has been taken equal to zero for simplicity, so that normalization is with respect to m. Equation (9–3) can then be used directly to obtain the curves of Fig. 9–4. Note that a must be limited to at most 0.02 for $\rho = 0.6$ or 0.05 for $\rho = 0.4$ to have the CSMA/CD strategy perform reasonably well on an average time-delay basis. As the parameter a decreases, collisions are more readily detected and the system begins to take on the characteristics of a centralized M/G/1 queue. This is borne out by the limiting form of Eq. (9–3) with $a = 0$. It is readily shown that one then gets

$$t_f = m + \frac{\lambda \overline{m^2}}{2(1-\rho)} \qquad a = 0 \qquad (9-9)$$

just the average delay (wait time plus transmission time) expression for the M/G/1 queue. (See Eq. (2–65).) As a check, for exponentially distributed frames we have $\overline{m^2} = 2m^2$. Then $t_f/m = 1 + \rho/(1-\rho) = 1/(1-\rho)$, the M/M/1 delay expression encountered many times in this book. For $\rho = 0.8$, $t_f/m = 5$; for $\rho = 0.6$, $t_f/m = 2.5$; for $\rho = 0.4$, $t_f/m = 1.67$. Note that these are precisely the values indicated for very small a in Fig. 9–4.

The M/G/1 result of Eq. (9–9) provides not only a limiting performance bound ($a \to 0$) for the CSMA/CD scheme, but serves to introduce a *best* performance bound for *all* multiaccess strategies. Visualize a common medium (bus, ring, or any other topology) with many stations connected. Let packets (and hence frames) be chosen to have the same deterministic (constant) length. They arrive with Poisson statistics. What is the very best strategy that one could adopt to transmit packets? For a given load (throughput) on the system, one would like to minimize the time delay. A little thought indicates that the "Maxwell demon" concept applies here: Say that the instant a packet arrives at a station a control message to that effect is sent at infinite speed (zero time) to a central controller. That controller replies with a scheduling number or a scheduling time for transmission. In effect the controller establishes a virtual single queue with packets transmitted first come–first served, in order of arrival. No other algorithm can do better with Poisson arrivals. This virtual queue behaves

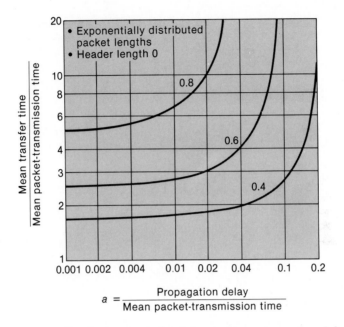

$$a = \frac{\text{Propagation delay}}{\text{Mean packet-transmission time}}$$

Figure 9–4 Normalized mean transfer delay versus propagation delay relative to mean packet-transmission time for the CSMA/CD bus (from [BUX 1981b, Fig. 13], © 1981 IEEE, with permission)

just like an M/D/1 queue, so that the bound on time delay is given by the M/D/1 result (from either Eq. (9–9) or Eq. (2–63)),

$$t_f|_{\min} = \frac{m}{(1-\rho)}\left(1 - \frac{\rho}{2}\right) \qquad (9-10)$$

This value of average transfer delay has been plotted in Figs. 9–2, and 9–3. Note that it does lie below all curves shown. It provides an indication of how much improvement is possible with other, more sophisticated scheduling algorithms.

Now consider Fig. 9–5. This shows the effect of the packet-length distribution on the transfer-delay characteristic. A 1-km 5-Mbps LAN has been chosen as an example. Figures 9–2 to 9–4 were all plotted for the exponentially distributed packet-length case. In Fig. 9–5 both the constant packet (deterministic) and larger variance cases are plotted. A hyperexponential distribution

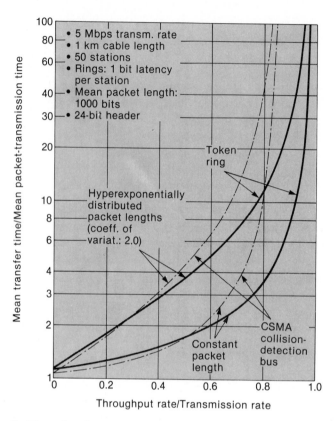

Figure 9–5 Transfer-throughput characteristic for constant and hyperexponentially distributed packet lengths (from [BUX 1981b, Fig. 10], © 1981 IEEE, with permission)

[KOBA] with a coefficient of variance $\sigma^2/m^2 = 2$ has been chosen as the larger variance case. (Recall that $\sigma^2/m^2 = 1$ for the exponential case.) The delay increases, of course, as the variance increases. This was first noted in Chapter 2 in discussing the M/G/1 queue. It is apparent here as well in both Eq. (9–1) and Eq. (9–3) through the dependence on $\overline{m^2}$. (The CSMA/CD result depends on the complete frame-length distribution as well through the Laplace transform $F_p(\lambda)$.) Note that the M/D/1 result again lies below both constant-packet-length curves, although for $\rho \le 0.8$, the constant-packet token-ring and M/D/1 delay

[KOBA] H. Kobayashi, *Modeling and Analysis: An Introduction to System Performance Evaluation Methodology*, Addison-Wesley, Reading, Mass., 1978.

results do not differ significantly. The results do depend significantly on the packet-length distribution, however, particularly in the range of operation $0.2 \le \rho \le 0.8$. Here delays differ by factors of 2 to 3 for the two frame-length distributions considered. This is apparent from Eqs. (9 – 1) and (9 – 3), in which the terms involving $\overline{m^2}$ are directly proportional to that quantity. On the other hand, we recall again that absolute delays are not that large. Thus, for a packet length of 1000 bits and $C = 5$ Mbps, $m = 200$ μsec. Even $t_f = 10m$ implies a transfer delay of only 2 msec at these high bit rates.

It would appear from these results that the token ring is to be preferred to CSMA/CD. As pointed out in the previous chapter, however, the CSMA/CD strategy is much simpler and was implemented much earlier commercially. It appears to be a good practical solution to the local area network problem for small networks that operate at relatively moderate bit rates with many highly bursty stations connected. As is apparent from the discussion thus far, CSMA/CD requires only passive taps on the bus; the token ring requires active repeaters (Fig. 9 – 1b).

The token-ring results of Figs. 9 – 2, 9 – 3, and 9 – 5 are based on a minimum station latency of 1 bit. Recall that this is the time required to read in (receive) from the line and write out (transmit) over the line (Fig. 9 – 1b). How sensitive are the performance results to this minimum latency choice? This obviously depends on the transmission speed and on the number of stations on the ring. Increased transmission speed reduces the latency in absolute time; the latency increases with number of stations.

Figure 9 – 6 shows the impact of latency on performance. Also shown is the variation with mode of operation. The multiple-token mode is the procedure mentioned at the beginning of this section in which the token is passed as soon as the last bit in a frame has completed transmission; the single-packet mode is the one in which the entire frame must first be received by the transmitter after a complete cycle around the ring before the token is passed on. In the single-token mode the transmitter waits only to receive the token. (This is the method adopted in the IEEE 802.5 standard.) It is apparent that the multiple-token mode is least sensitive to latency. In fact at high throughput the performance tends to become independent of latency. The presence of multiple tokens (multiple messages) on the ring poses questions of reliability, however, and leads to a much more complex system. It has thus not been adopted for use.

The other two schemes are quite sensitive to choice of latency. This point was noted in the previous chapter in our discussion of polling; it is apparent from Eqs. (9 – 1) and (9 – 2), on which Fig. 9 – 6 is based. The total walk time L varies with both the latency per station and the number of stations. In Fig. 9 – 6 the number of stations has been doubled from 50 to 100, as compared with the examples of the other figures. The increased delay due to walk time comes in, in two ways: the time to complete a scan and then the time to actually transfer a frame to a particular destination.

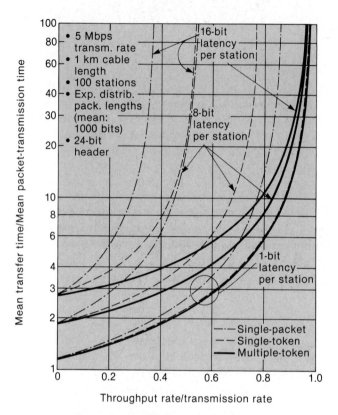

Figure 9 – 6 Transfer delay-throughput characteristic of token rings: impact of
station latency and operational mode (from [BUX 1981b, Fig. 11],
© 1981 IEEE, with permission)

The single-token mode, as seen from Fig. 9 – 6, is obviously better than the single-packet one. Recall that in the single-packet mode the transmission time is effectively increased by the walk time L to account for complete reception of the frame before releasing the token. Both m and $\overline{m^2}$ in Eq. (9 – 1) must be increased to account for this, further deteriorating the performance. (The effective traffic intensity $\rho = \lambda m$ must be increased as well.) In the single-token case, on the other hand, the service time m is replaced by the walk time L if frames are shorter than the ring latency. To qualify these comments relating to Fig. 9 – 6, consider the single-packet mode first. We replace m by $(m + L)$, giving us $\lambda m + \lambda L$ in place of ρ, and $\overline{m^2}$ by $\sigma^2 + (m + L)^2$. The maximum normalized throughput rate is then $\lambda m|_{\max} = 1/(1 + L/m)$. For the example of Fig. 9 – 6, with a 1-km cable,

5 Mbps transmission speed, and 1000-bit packets, $m = 200$ μsec, and $\tau = 5$ μsec. For 1-bit latency/station and 100 stations, $L = 25$ μsec, and $\lambda m|_{\text{max}} = 0.89$. For 8-bit latency, L increases to 165 μsec, $L/m = 0.825$, and $\lambda m|_{\text{max}} = 1/(1 + L/m) = 0.55$. Finally, for 16-bit latency per station, $L = 325$ μsec, $L/m = 1.6$, and $\lambda m|_{\text{max}} = 1/2.6 = 0.38$. These numbers are borne out by the curves of Fig. 9–6.

Now consider the single-token case. This was the one assumed in discussing token-ring performance in Figs. 9–2, 9–3, and 9–5. Recall that for this scheme the token is passed on to the next station when it is received back by the transmitting station. The minimum service time for the performance calculation is then the ring latency L. Frames longer than L in length are represented by their normal transmission time. This results in a modified frame-length distribution that must be used in the transfer-delay calculation. Say, for example, that the frame-length distribution $f(t)$ is exponential with average value m. (As noted earlier, this is very nearly applicable to the exponential distribution assumed for Figs. 9–2, 9–3, and 9–6. In these figures the *packet* lengths are exponential. Since the header of 24 bits is much shorter than the average packet length of 1000 bits, one can assume that both frame and packet are exponentially distributed.) We thus have

$$f(t) = \frac{1}{m} e^{-t/m} \, u(t) \tag{9–11}$$

with $u(t)$ the unit step function. This is sketched in Fig. 9–7(a). For the single-token analysis this must be modified to have frame lengths less than L-sec long set equal to L. The resultant frame-length distribution, shown in Fig. 9–7(b), then has the form

$$f'(t) = (1 - e^{-L/m}) \, \delta(t - L) + \frac{1}{m} e^{-t/m} u(t - L) \tag{9–12}$$

It is left for the reader to show that the average of this distribution, to be used in place of m in Eq. (9–1), is given by

$$E(t) = L + me^{-L/m} = m \left(\frac{L}{m} + e^{-L/m} \right) \tag{9–13}$$

The second moment $E(t^2)$, to be used in place of $\overline{m^2}$, is found similarly from Eq. (9–12).

It is apparent that this procedure results in an increased average service time $[E(t) > m]$ and hence reduces the maximum possible transmission rate. Specifically, the $(1 - \rho)$ terms in Eq. (9–1) are now given by $1 - \lambda E(t)$, and the maximum normalized transmission rate becomes

$$\rho_{\text{max}} = \lambda_{\text{max}}m = 1 \left/ \left(\frac{L}{m} + e^{-L/m} \right) \right. \tag{9–14}$$

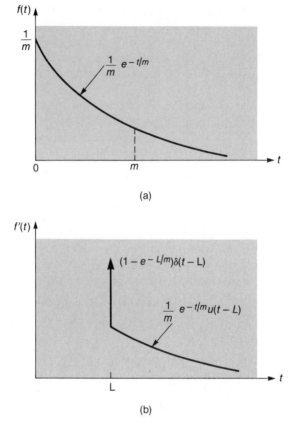

Figure 9–7 Frame-length distribution, single-token scheme
   a. Frame-length distribution
   b. Modified distribution

Consider the example of Fig. 9–6. We noted previously that $\tau = 5\,\mu\mathrm{sec}$ and $m = 200\,\mu\mathrm{sec}$. For 100 stations the ring latency (total walk time) is $L = 25\,\mu\mathrm{sec}$ and $L/m = 0.125$ for 1-bit latency per station, $L = 165\,\mu\mathrm{sec}$ and $L/m = 0.825$ for 8-bit latency per station, and $L = 325\,\mu\mathrm{sec}$, $L/m = 1.6$ for 16-bit latency per station. The respective values of $\rho_{\max}$ are given by $0.99, 0.79$, and 0.55, using Eq. (9–14). These results agree with those plotted in Fig. 9–6: For the 1-bit latency case the single-token scheme can operate out to almost the full transmission capacity, since the average frame length is eight times the total walk time. In the two other cases, with L nearly m or greater, the transmission capacity suffers. This shows the importance of keeping the latency per station as small as possible.

In the next section, we describe both the CSMA/CD and the token-passing ring protocols in more detail, focusing on some implementation questions and showing the frame formats used. We also show briefly how these LAN protocols fit into the OSI Reference Model.

9 – 2 IEEE 802 Local Area Network Standards

We noted at the beginning of this chapter that the IEEE has set up a series of standards for accessing local area networks. Included are IEEE standards 802.3 for a CSMA/CD bus, 802.4 for a token-passing bus, and 802.5 for a token ring. We focus in this section on the CSMA/CD bus protocol and the token ring, following on the discussion of their comparative performance in the preceding section.

It has already been noted that pioneering work on the development of the CSMA/CD method of randomly accessing a local area network was carried out by workers at Xerox [METC]. Their system, termed *Ethernet,* was later developed into a detailed specification through the joint endeavors of Digital Equipment Corporation, Intel, and Xerox [ETHE]. The IEEE 802.3 CSMA/CD local area network standard is almost identical [IEEE 1983c]. The token-ring standard, IEEE 802.5, is based on work carried out by workers at the IBM Research Laboratory in Zurich, Switzerland, and elsewhere [BUX 1981a].

The various access methods of the IEEE 802 family of local area network standards, which includes the 802.3 CSMA/CD access method, are designed to conform to the OSI Reference Model. They fit into the physical layer and a portion of the data link layer of that model (see Fig. 9 – 8). The access standard, whether CSMA/CD, token-passing bus, token-passing ring, or any other, specifies the formats and protocols used by a medium-access-control sublayer of the data link layer, as well as the physical layer. All access standards in turn communicate with higher OSI layers through a *logical link control* standard, IEEE 802.2, which constitutes the remainder of the data link layer. In the standard OSI view (Chapters 3 and 7) the logical link control uses the services of the medium-access-control sublayer to provide services to the network and other layers above. The logical link control carries out those data link control functions that are independent of medium and of medium-access control.

Two types of logical link control procedures are specified. In one, labeled *connection service,* error recovery based on that of the asynchronous balanced

[IEEE 1983c] *IEEE Standard 802.3, CSMA/CD Access Method and Physical Layer Specification,* IEEE Project 802, Local Area Network Standards, IEEE, New York, July 1983.

OSI layers   IEEE 802 LAN standards

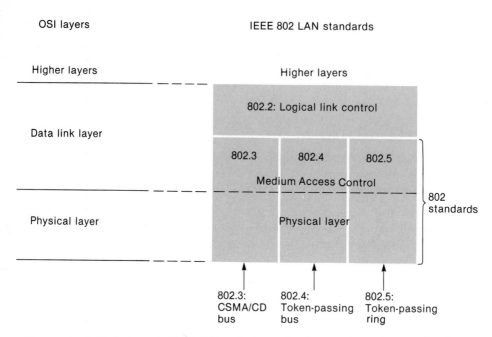

Figure 9–8 Relation of IEEE 802 LAN standards to OSI Reference Model

mode of HDLC (Chapter 4) is carried out. (As we shall see in the following subsections, the medium-access control, the sublayer below logical link control, provides a bit-error-detection capability that uses a frame check at the end of a frame. Errors detected are then passed up to the logical link control, or higher layers for action if necessary.) The transmitting and receiving stations on the LAN then "look" as if they are neighboring nodes in a network. The data link layer of the LAN presents an HDLC-like image to layers above.

The other procedure, called *connectionless service,* provides no error-recovery capability. It is up to the layers above to provide this recovery capability, including the retransmission of any messages that may be required. But note from Chapter 7 that the OSI transport protocol (TP) class 4 does in fact provide this capability. The logical link control connectionless service for LANs thus fits into the TP4 strategy. One application for the logical link control connectionless service might be local area networks interconnected by a wideband transport medium such as an optical-fiber transmission system. This is shown schematically in Fig. 9–9. The gateways shown there provide the necessary connection from a LAN to the optical-fiber system (or any other network that

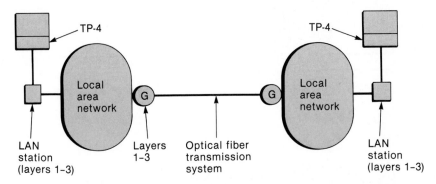

Figure 9–9 Interconnection of local area networks

interconnects the LANs). Layers 1 and 2 are the LAN layers shown in Fig. 9–8. Layer 3 could be the packet level of X.25. Another example involving the use of satellites to interconnect LANs appears in [DEAT].

9–2–1 Ethernet: CSMA/CD Local Area Network

We now describe how the CSMA/CD or Ethernet specifications (IEEE Standard 802.3) fit into the architectural picture of Fig. 9–8. In the next subsection we do the same for the token-passing ring.

Recall that the CSMA/CD protocol has the following features:

1. Listen for carrier; transmit only when there is no energy on the medium.

2. During transmission listen for collision; abort transmission and reschedule if collision is detected.

How are these access mechanisms implemented in Ethernet (or in the CSMA/CD standard)?

The physical layer is responsible for sensing traffic on the medium (coaxial cable in the case of Ethernet) and so indicating in a carrier-sense signal to medium-access control. This layer also compares the signal on the medium with the signal generated in case of a transmission and generates a collision detection signal if contention (interference) is indicated on the channel. These functions are carried out by the channel-access sublayer of the physical layer, shown in Fig.

[DEAT] George Deaton, "Multi-access Computer Nets: Some Design Decisions," *Data Comm.*, vol. 13, no. 14, Dec. 1984, 123–136.

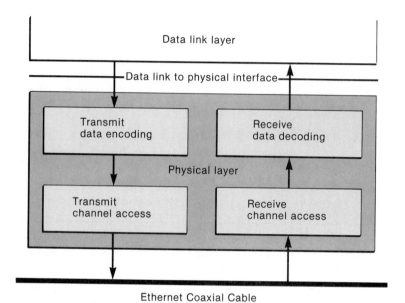

Figure 9–10 Physical layer functions, Ethernet (from [ETHE, Fig. 4–4] with permission)

9–10 [ETHE, Fig. 4–4]. The transmit channel-access component shown generates the collison-detect signal; both the receive channel-access and transmit channel-access components are used to generate a carrier-sense signal. Both of these signals are interpreted by the medium-access control sublayer of the data link layer.

   In addition to sensing the two signals—collision detect and carrier sense— the channel-access components transmit bits onto, and receive them from, the coaxial cable medium. The transmit data encoding component of the physical layer (Fig. 9–10) encodes the data bits into a zero-DC binary waveform appropriate for the coaxial medium, using Manchester encoding. An example of Manchester encoding appears in Fig. 9–11. In this encoding scheme the first half of the bit interval is used to transmit the logical complement of the bit during that interval; during the second half of the bit interval the uncomplemented value of the bit is transmitted. Thus 1's are represented by positive-going signal transitions; 0's, by negative-going transitions (see Fig. 9–11). The Manchester encoding/decoding functions are carried out by the transmit data encoding and receive data decoding components of the physical layer, shown in Fig. 9–10. These components also generate and remove a 64-bit pattern called

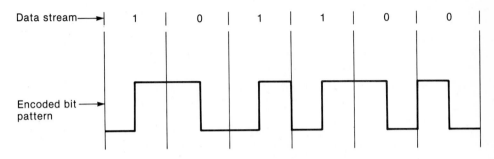

Figure 9-11 Manchester encoding

a preamble that precedes the actual data frame transmitted and that is used for synchronization purposes [ETHE].

As in the HDLC protocol described in Chapter 4, data are transmitted over the local area network in a frame that includes address and error-detection fields in addition to the data transmitted down from higher layers. The Ethernet frame format is depicted in Fig. 9-12. A frame is at most 1518 octets long. The first 12 octets provide destination and source addresses. The 2-octet type field is reserved for use by higher levels; it is not interpreted by the Ethernet protocol. The 4-octet frame check sequence at the end of the frame is used to detect errors. Unlike HDLC, however, error control is not exerted at the Ethernet layers; this function is relegated to the higher layers, as noted earlier. An indication of bit errors detected is sent up to the higher layers, however, for action at the logical link control sublayer or layers above.

The framing, addressing, and error-detection functions are carried out in the medium-access-control sublayer of the data link layer (Fig. 9-8). That sublayer is also assigned the task of acting on the carrier-sense and collision-detection signals generated by the channel-access components of the physical

Destination address	Source address	Type	Data	FCS
6 octets	6 octets	2 octets	46–1500 octets	4 octets

Figure 9-12 Frame format, Ethernet

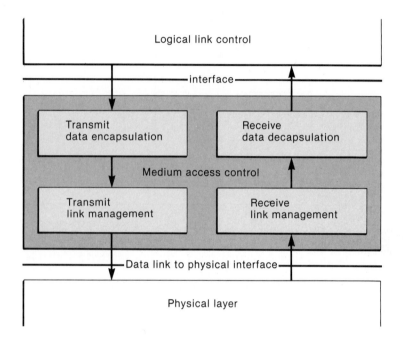

Figure 9–13 Medium access control, Ethernet

layer. Figure 9–13 shows how medium-access control is organized to carry out these functions [ETHE]. The link-management components carry out collision avoidance using the carrier-sense signal and contention resolution using the collision-detection signal. Framing, addressing, and error detection are handled by the data encapsulation components.

A typical Ethernet implementation, which follows the architecture just described, appears in Fig. 9–14 [ETHE, Fig. 4–1]. Note that in this implementation the addition of the preamble to the frame and the Manchester encoding of the resultant waveform are both carried out in the encode component located on the Ethernet controller board. The encoded-bit stream is sent out over the transceiver cable and is put onto the coaxial cable by the transceiver. That system also generates the carrier-sense and collision-detection signals, as already noted.

The Ethernet specification calls for a coaxial cable physical channel driven at 10 Mbps. From our discussion in Section 9–1 it is apparent that at this bit rate the Ethernet system cannot be too long. Specifically, the maximum total coaxial cable along the longest path between any two transceivers is specified to be 1500

Architecture

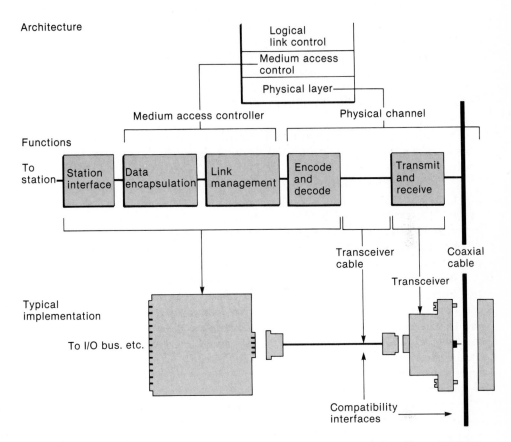

Figure 9-14 Ethernet architecture and typical implementation (from [ETHE, Fig. 4-1] with permission)

meters. The propagation velocity of the cable is taken to be 4.33 μsec/km, worst case. (This is then somewhat better than the value of 5 μsec/km assumed in Section 9-1.) This velocity produces an end-to-end propagation delay of 6.5 μsec. We shall use this figure as the end-to-end delay τ. τ will, however, be higher because of signal rise times and station signal detection delays. (This is noted later.) For the shortest frame length, 64 octets (Fig. 9-12), with another 9 octets added as the preamble, the lower limit of frame length is $m = 57.6$ μsec. This produces a value of $a = \tau/m = 0.11$, a relatively high value of the parameter a (Fig. 9-4). If the data field increases to 200 octets, the parameter a drops to

$a = 6.5/180 = 0.036$, a much more tolerable value. In addition, a minimum interframe spacing of 9.6 μsec is specified to provide appropriate interframe recovery time for controllers and for the physical channel [ETHE]. With this stipulation, a station that desires to transmit must defer for at least 9.6 μsec after sensing the end of a frame on the channel. This reduces the load on the cable somewhat and reduces the likelihood of collisions as well, particularly if frames are relatively short.

Stations on an Ethernet may not necessarily be connected on the same cable. Segments of cable may be connected through repeaters. Typical examples of configurations appear in Fig. 9–15. [ETHE, Fig. 7–16]. The specification puts limits on these configurations. It has already been noted that the maximum end-to-end path is 1500 meters. An individual segment (Fig. 9–15a) cannot exceed 500 meters in length. No more than two repeaters can be used in a path between two stations. The maximum number of stations on any Ethernet is 1024, with no more than 100 per segment. A point-to-point link between repeaters (Fig. 9–15c) cannot exceed 1000 meters.

Adding up delays along a worst-case path, and including the rise time of a propagated signal to detect it either as a carrier or a collision, the Ethernet specification finds the worst-case round-trip delay to be 45 μsec, or 450 bits at the 10 Mbps transmission rate [ETHE]. (Note that the one-way value of 22.5 μsec is much higher than the value $\tau = 6.5$ μsec quoted earlier.) This figure is used in determining the retransmission time in case of a collision. Specifically, once a collision is detected a jam signal of between 32 and 48 bits is transmitted to guarantee that all transmitting stations on the network detect the collision. The worst-case time for detecting a collision is then $450 + 48 = 498$ bits. A station that fails in a transmission attempt due to a collision cannot retransmit in less than this time. Retransmission delays are then measured as multiples of a parameter called the *slot time*, which is greater than 498 bits. The Ethernet specification defines the slot time to be 512 bits. This is precisely the parameter 2τ used in the worst-case CSMA/CD calculation of the last chapter, following Lam's analysis [LAM 1980] that led to Eq. (8–23) (Eq. (9–7) in this chapter) for the maximum throughput of a CSMA/CD access scheme.

Ethernet uses a truncated binary-backoff algorithm to govern retransmissions: The delay before retransmitting after detecting a collision is defined to be a random number r of slot times (512-bit intervals) distributed uniformly over the range 0 to 2^n, with n the nth retransmission attempt, until n reaches 10. The retry interval then remains at 0 to 2^{10} until n reaches 15. At this point the collision event is reported as an error to higher layers. If the binary-backoff portion of the algorithm is ignored, this indicates that the worst-case average frame length required over an Ethernet with $a = 0.1$ is 2560 bits or 320 octets. (See Eq. (9–7).) For smaller values of a, longer frame sizes are required.

9-2-2 Token-passing Ring

Token passing on a ring has also been standardized as an access method by the IEEE 802 committee; it appears as IEEE Standard 802.5 (Fig. 9-8) [IEEE 1984]. As noted earlier, the standard is based on work on the token ring technique developed by researchers at the IBM Research Laboratory in Zurich, Switzerland [BUX 1981a] and elsewhere.

The concept is quite simple; it was described earlier in this chapter and in the discussion of hub polling in Chapter 8. A station that receives permission to transmit via a special frame called a *token* does so. It then passes the token to the next station capable of receiving it. An example of a token-ring configuration was portrayed in Fig. 9-1(b). A more practical example is shown in Fig. 9-16 [BUX 1981a, Fig. 1]. A number of ring stations may be grouped together in star fashion, connected to a wiring or distribution panel, for ease of fault location and servicing. Distribution panels may be located at convenient places in a building. Bypass relays are used to disconnect inactive or malfunctioning stations. Stations B and F in Fig. 9-16 are shown disconnected from the ring by the bypass relays. A station is typically in either a *transmit* or a *repeat* state. In the transmit state it sends out its own frame after receiving permission to do so. In the repeat state it outputs a received frame bit by bit back onto the ring. It may copy the frame while repeating it if it recognizes itself as the destination address. It may also modify certain bits in the frame as it transmits them onto the ring. The electronic switch in Fig. 9-16 is a conceptual representation of either of these two states.

The format of a frame transmitted on a ring appears in Fig. 9-17(a). The token format appears in Fig. 9-17(b) [IEEE 1984]. The single-octet starting delimiter SD and ending delimiter ED denote the beginning and end of a frame. The 8-bit access control or AC field is used to implement the ring access protocol. The token bit T is set to zero in a token and one in a frame. The three P bits in the AC field are used to provide up to eight levels of priority in accessing the ring.

A station with information (a protocol data unit) to transmit, detecting a token with priority less than or equal to that of its waiting protocol data unit, may do so. It changes the token to a frame and starts transmitting the frame beginning with the SD field. The three R bits in the AC field are used to request that the next token be transmitted at the priority requested. These three bits would be set accordingly in the next token or frame repeated.

[IEEE 1984] *Draft E, IEEE Standard 802.5, Token Ring Access Method and Physical Layer Specifications,* IEEE Project 802, Local Area Network Standards, IEEE, New York, Aug. 1, 1984.

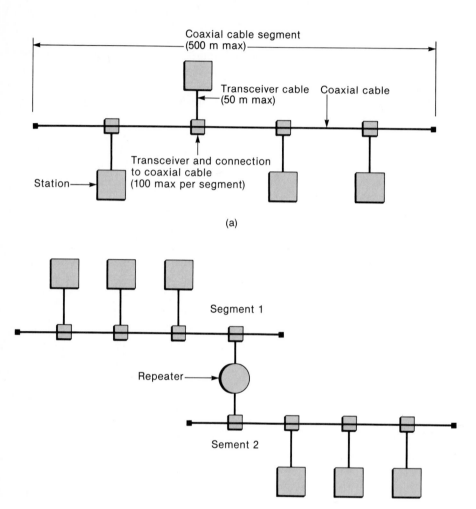

Figure 9-15 Ethernet configurations (from [ETHE, Fig. 7-1] with permission)
  a. Minimal configuration
  b. A typical medium-scale configuration
  c. A typical large-scale configuration

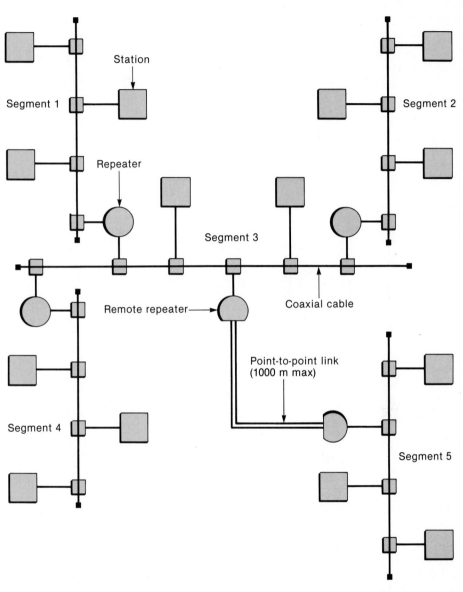

Station

Segment 1

Segment 2

Repeater

Segment 3

Coaxial cable

Remote repeater

Point-to-point link
(1000 m max)

Segment 4

Segment 5

(c)

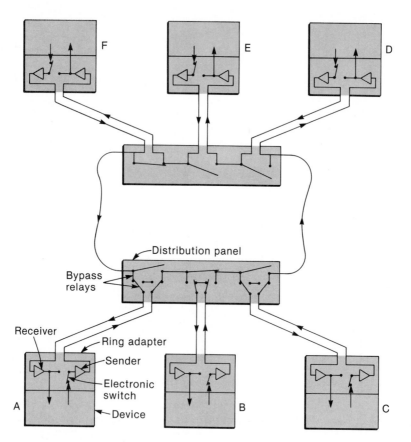

Figure 9–16 Token ring configuration (from [BUX 1981a, Fig. 1], © 1981 IEEE, with permission)

The other fields in the frame format are fairly straightforward. Destination and source addresses are either 2 or 6 octets long. Individual stations or groups of stations may be addressed. An all-1 destination address is used to broadcast information to all stations on the ring. The 4-octet frame-check sequence (FCS) field provides bit-error detection. As in the case of Ethernet, however, the token-ring access protocol does not attempt to correct any bit error detected. It signals the logical link control (Fig. 9–8) and other layers above that an error has been detected. Other types of error are signaled in this way as well.

The token-ring scheme, although decentralized, does require supervisory action to recover from possible problems with the token [BUX 1981a]. A token

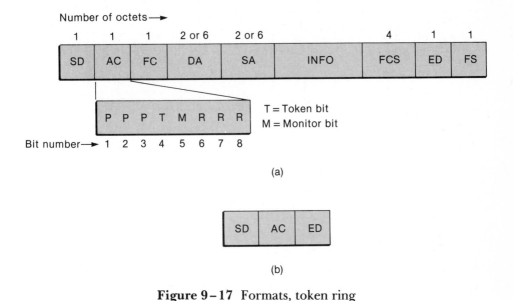

Figure 9–17 Formats, token ring
　　　　　a. Frame format
　　　　　b. Token format

may be lost, for example. (This may happen when the ring is initialized or because of corruption of one or more bits in the token itself; a fault in the starting delimiter field would cause this problem.) A busy token may continue to circulate indefinitely. (The *T* bit may be set to 1 by noise.) To recover from these and other possible problems one of the stations on the ring is designated as an *active monitor*. (Each station has the capability of serving as a monitor. Provision is made for each station to detect the possible failure of the active monitor and to have one of the other stations then take over this task [IEEE 1984].)

The active monitor uses timers and the *M* bit in the AC field of a token or frame (Fig. 9–17) to recover from token faults, as well as some frame faults. On receipt of each valid token or frame, the monitor resets a timer TVX. On expiration of this timer with no reset, special purge frames are transmitted continuously to signal all stations to switch to the repeat state and to clear the ring of possible garbage. The monitor then transmits a new token.

All frames and tokens as issued have the *M* bit in the AC field (Fig. 9–17) set to 0. On passing through the monitor station this bit is set to 1 for all frames and all tokens of priority greater than the lowest. All other stations repeat this bit as set. A token or frame that reaches the monitor with its *M* bit set to 1 is considered

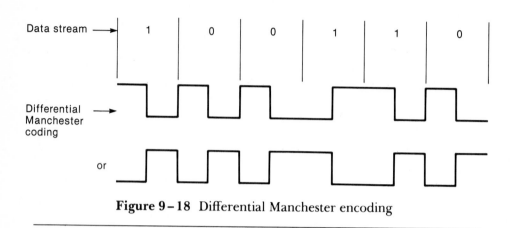

Figure 9–18 Differential Manchester encoding

invalid and is purged; a new token is eventually issued. The *M*-bit operation thus prevents tokens and frames from circulating indefinitely.

All aspects of the token-ring access protocol are carried out at the medium-access-control sublayer (Fig. 9–8). The physical layer receives bits one at a time from medium-access control, encodes them, and transmits them onto the medium. It performs the reverse operation as well, taking symbols from the medium, decoding them, and passing them up to medium-access control. The encoding procedure specified by the token ring standard is *Differential Manchester* coding. This method of encoding differs from Manchester coding, described previously (Fig. 9–11). In Differential Manchester encoding two polarities are used to carry binary information, and transitions still take place at the midpoint of a binary interval. In the case of a 1 bit, however, the initial half of the binary interval carries the same polarity as the second half of the previous interval. In the case of a 0, a transition takes place at both the beginning and the middle of the binary interval. An example appears in Fig. 9–18. Note that there are two possibilities, depending on the polarity at the end of the interval preceding the first one in the figure.

Problems

9–1 Refer to the discussion of the token ring.

 a. Show why multiple tokens may exist on the ring if the frames are short enough and permission to transmit is granted to the next station by a frame leaving the transmitting station.

b. Consider the case in which the token is passed on by a transmitting station only on receipt of its complete message. In calculating the delay performance of this scheme, why is the frame transmission time (the service time) increased by the ring latency?

c. The token is passed on by the transmitting station on receipt of its own token, embedded in the frame transmitted. Find the effective service time in these two cases: (1) the frame is shorter than the ring latency and (2) the frame is longer than the ring latency.

9-2 Verify Eqs. (9-5) and (9-6) using Eq. (9-4).

9-3 Refer to Eq. (9-3) for the CSMA/CD transfer time. Show that for very light traffic ($\rho \rightarrow 0$), the limiting (minimum) transfer time is given by Eq. (9-8). Show that with $a = 0$ one gets the M/G/1 result of Eq. (9-9).

9-4 Compare the transfer time-throughput performance of the token ring and the CSMA/CD bus for the following cases:

1. Two-km cable, 50 stations, 1-Mbps transmission rate, 1000-bit average frame length, exponentially distributed, 1-bit latency per station for the token ring case.

2. Same as case 1, but with 100 stations.

3. Same as case 1, but with a 10-Mbps transmission rate. Take the propagation delay to be 5 μsec/km.

Use the single-packet mode in the token ring case, for simplicity. Compare with Figs. 9-2 and 9-3. (Note that in these figures the *packet* length is distributed exponentially. Note also that the single-token mode has been assumed in carrying out the calculations for Figs. 9-2 and 9-3.)

9-5 Use Eq. (9-7) to verify the limiting values of a (delay increasing beyond bound) of Fig. 9-4.

9-6 Repeat case 1 of Problem 9-4 assuming the single-token mode for the token ring case. Compare the results for the two token-ring modes. Do these agree with comparable results of Fig. 9-6?

9-7 Determine the effect on the performance of the token-ring example of Problem 9-4, case 1, if the latency per station is increased to 8 bits. Is the result comparable to that shown in Fig. 9-6?

9-8 Repeat Problem 9-4, case 1, for 1000-bit fixed (deterministic) frame lengths. Compare with the results of Problem 9-4.

9-9 Use Eq. (9-12) (or Fig. 9-7) to calculate the mean and second moment of the single-token-mode frame-length distribution. Show that the mean is given by Eq. (9-13).

9-10 One hundred stations are distributed over a 4-km bus. A CSMA/CD protocol is used. The transmission rate is 5 Mbps and data frames are 1000 bits long, on the average. Calculate the maximum number of frames per second that each station can generate. Repeat if the bus length is reduced to 1 km. Repeat if the transmission capacity is doubled to 10 Mbps. Repeat if the frames are 10,000 bits long. Explain these results. Use 5 μsec/km as the bus propagation delay.

9–11 A coaxial cable bus 20-km long has 1000 stations connected to it: Electromagnetic energy propagates in both directions at a speed of 200,000 km/sec. The line capacity is 10 Mbps. Packets transmitted at each station are all 10 kbits long.

   a. A token (permission to transmit) is passed sequentially from station 1 to 2 to . . . to 1000, and then back to 1 again, starting the cycle over. (This is then called a token bus.) There are 8 bits of latency at each station. Each station generates 8 packets per 10 sec on the average. Calculate the average scan (cycle) time. What is the maximum rate of transmission of packets at each station?

   b. Pure random access (pure Aloha) is used instead, with stations transmitting at will. Calculate the maximum possible rate of transmission of packets at each station.

   c. Repeat b, if a CMSA/CD protocol is used.

   d. It is stated in the text that a performance bound on the best one can do with a common channel, as in this problem, is given by the time-delay–throughput characteristic of the M/D/1 queue. This implies queueing packets up and serving them (transmitting them over the bus) FCFS (FIFO).

      1. Justify this statement and explain why this procedure can never be achieved (hence the word *bound*).

      2. Calculate the maximum transmission rate of packets at each station in this case.

      3. Find the average time delay (access delay) at each station for this case if each station transmits 8 packets/10 sec, on the average (as in part a of this problem).

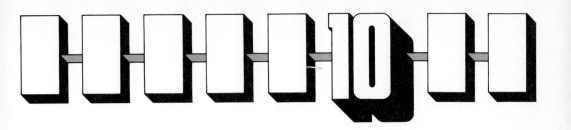

Introduction to Circuit Switching

Previous chapters have focused on packet-switched networks. In these networks data messages are blocked into shorter units called packets, which are then transmitted from source to destination along some routing path. Two techniques are used for this purpose: datagram or connectionless transmission, and virtual circuit transmission.

Circuit switching, or line switching as it is frequently called, will be discussed in this chapter and the one following. This kind of transmission requires a circuit to be set up from source to destination, as in the case of a virtual circuit, but then *dedicates* the actual physical circuit set up, end to end, to the users at either end. Recall that in virtual circuit transmission individual packets from multiple virtual circuits generally *share* the transmission facility. For both circuit switching and packet switching, the end-to-end physical circuit consists of a number of transmission links or circuits connected by intermediate circuit switches. The links could consist of time slots in a time-division multiplexed (TDM) system or frequency bands in a frequency-division multiplexed (FDM) system. Conceptually, then, the topological layout of a circuit-switching system looks no different from that of a packet-switched network.

Circuit switching is, of course, a much older technology than packet switching. It is the technology used worldwide over the years for telephony. Modern telephone systems and networks are used to transmit data, but voice has continued to constitute the bulk of their traffic. The fact that telephone networks as currently deployed are designed primarily to handle voice traffic is one of the reasons behind the development of specialized packet-switched networks devoted to data traffic over the past decade or so. As networks become integrated, handling voice, data, and other types of traffic, their characteristics may change to accommodate multiple types of traffic. Integrated or hybrid multiplexing strategies, used to combine or multiplex diverse types of traffic, with possibly different performance objectives, are considered in Chapter 12.

Aside from the fact that the worldwide telephone plant is ubiquitous and hence easily allows connections to be made between any two users almost anywhere in the world, under what conditions would one use packet switching for data transmission and under what conditions would one use circuit switching? The answer is not clear cut. Comparative costs and economics play a preeminent role [ROSN], [KUMM 1978]. The historical fact that most telecommunications worldwide is carried out using circuit-switching techniques cannot be denied; this fact obviously is crucial in determining the use of one technology or the other. However, it is apparent that bulk data—long files to be transferred, facsimile messages, digitized video transmission, and so on—can probably be transmitted cost effectively using circuit switching. On the other hand, bursty data—interactive terminal-to-computer transactions, for example—can be transmitted more efficiently using packet switching. The reason is obvious: It is very inefficient to dedicate a communications channel to a single pair of users, end to end, who transmit data only intermittently. One could let other users, who transmit bursty traffic as well, share the facilities, thus lowering the costs for all users. This is, of course, one of the reasons for the rapid deployment of packet-switching technology worldwide. On the other hand, a user with a continuous stream of data to transmit might be served better by a dedicated circuit-switched connection. The point at which one might want to move from one type of technology to another, however, is not clear. It depends not only on the size of the message (the number of bits) to be transmitted, but on the transmission capacity of the network links (very high-speed optical links, for example, could make even a 2-Mbit message appear like a short packet!), the response time requirements for the data, the time required to set up a connection, and, ultimately, the comparative costs, as noted earlier.

It also depends on how the circuit-switched network performance is evaluated. Most telephone circuit-switched networks operate on the basis of blocking

[ROSN] Roy D. Rosner, "Circuit and Packet Switching: A Cost and Performance Trade-off Study," *Computer Networks,* vol. 1, no. 1, June 1976, 7–26.

[KUMM 1978] K. Kummerle and H. Rudin, "Packet and Circuit Switching: Cost/Performance Boundaries," *Computer Networks,* vol. 2, no. 1, Feb. 1978, 3–17.

calls to allocate transmission resources fairly and efficiently. In some cases, however, calls may first be queued. Performance objectives are usually stated in terms of blocking probability, with consideration given as well to the time required to set up or connect a call. Packet-switched networks most commonly offer a time-delay performance objective, as we have already seen in this book; sometimes these networks use blocking to prevent congestion. It may thus be difficult to compare the use of packet- or circuit-switching techniques because the performance objectives may differ.

We shall be discussing the blocking performance of circuit-switching systems in detail later in this chapter and in the two chapters following, since blocking probability is a common measure of performance for these systems. However, to introduce the idea of circuit switching quantitatively, and to connect this technique smoothly with packet switching, described earlier, we shall first present a simple model of circuit-switched transmission using a *queued*-call, rather than a blocked-call, approach (see Section 10 – 1). This will enable us to introduce the basic concepts of circuit switching in a simple tutorial fashion. We will then be able to compare packet-switched and circuit-switched performance as well, coming to the obvious conclusion that *longer* messages should be circuit switched, while *shorter* messages should be packet switched. We shall carry out the comparison for a simple single-hop connection by comparing connect time in the case of the circuit-switched mode with time delay in the packet-switched case. (In either case, a user, not knowing the details of how his message is managed, is assumed to measure performance in terms of time delay. In one case the delay is due to the time required to set up a call; in the other, to the time to deliver the message.) This analysis, due to F. Closs [CLOSS], will be presented in Section 10 – 1. In Section 10 – 3, we introduce the elements of traffic analysis required for a quantitative study of circuit-switching systems. In the final two sections, we focus on modern digital-switching systems, describing the routing network within a switching system that enables a call coming in on one line (or trunk) to be connected to the appropriate outgoing trunk in an effectively nonblocked manner. We continue the analysis in the next two chapters by first describing call processing within a circuit switch in detail and then enlarging our scope to include routing in circuit-switched networks.

As has been the case thus far, where possible we attempt to be quantitative in our approach. Although the emphasis is on data transmission, the examples and numbers used in the chapters on circuit switching are of necessity drawn primarily from voice communications because of the overwhelming application to voice in these systems. It must be borne in mind, however, that the systems we

[CLOSS] F. Closs, "Message Delays and Trunk Utilization in Line-switched and Message-switched Networks," ISS '72, Tokyo, 1972.

describe are *digital* switching systems. Voice and data are thus essentially handled the same way as they move through a system. Voice and different types of data will differ as to their bandwidth requirements (or, equivalently, the bit rate required for real-time transmission), the length of a message (this is often called the *holding time* in the case of a voice call), and possibly their performance objectives. More will be said about these parameters when we discuss integrated multiplexing in Chapter 12.

10 – 1 Simple Model, Circuit Switching: Queued Mode

As noted previously, a circuit-switched network provides private connections, or circuits, end to end, for users who desire to communicate. Before data transmission can take place, the circuit must first be set up in a connection phase. After transmission concludes, a disconnect phase takes place. These modes of operation are similar to those involving in setting up and tearing down a virtual circuit, described in Chapter 5. They again require the transmission of appropriate control messages (or signals), first from the user who desires to establish a connection to the circuit switch (station or exchange) to which he or she is connected, then successively along the appropriate routing path through the network to the destination user. That user must in turn signal acknowledgement of the connection setup request. Provision must be made for signaling aborted setup attempts (for example, a "busy" signal in a telephone network), as well as other untoward events. Finally, the disconnect protocol and messages corresponding to this protocol must be defined as well. In normal telephone service nine categories of signals are required. These will be spelled out in detail in our discussion of call processing in Chapter 11.

In this simplified discussion we limit ourselves to the signaling messages shown moving through the network in Fig. 10–1. As indicated there, a data source or station initiates the call-setup phase by sending a "send request" signal to the switch to which it is connected. This signal is assumed to carry source-destination addresses and other information needed to set up the connection. In normal telephone practice, an "off-hook" signal (a change of state of the call-initiating telephone or data set) plus dial digits or a set of frequency tones carry this information. The terminal-to-switch path over which this signal is transmitted is often called the *local loop* in telephone plants and is so indicated in Fig. 10–1. After processing at the switch (including determining the appropriate outgoing channel or trunk—the switch-routing function), a "connect" message is forwarded to the next switch on the path to the destination. This is only done when an interswitch trunk is available for the connection. In the so-called forward-

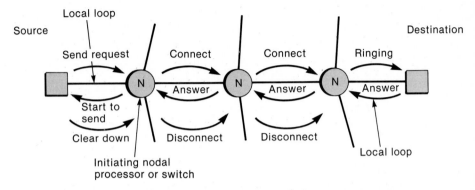

Figure 10-1 Signaling messages in a circuit-switched network

setup mode this trunk is "seized" for the call in question and held until the call is finally disconnected or upon issuance of an abort signal. This is the mode of operation we shall stress. (In the backward-setup mode, all trunks along the complete path must be available before they are assigned to the call in question.)

On reaching the final switch in the network (the one to which the destination is connected), with a complete path now set up, a "ringing" signal is delivered to the destination. The destination station, if ready to accept the call, replies with an "answer" or response message. This percolates back to the source, indicating that transmission can begin. The time from the initiation of the "send-request" message to the receipt of the "start-to-send" message is the call setup or connection time. In addition to this circuit-connection or establishment phase of circuit-switched communications, there is a disconnect phase as well, as already noted. At the completion of transmission the source station sends a "cleardown" signaling message to the switch to which it is connected; this switch in turn initiates a "disconnect" message, releasing the individual channels or trunks along the complete circuit. These messages are also indicated in Fig. 10-1.

There are different ways of sending the various signaling messages. In the traditional telephone mode, referred to as *inband signaling*, signaling messages are sent over the same trunks as the information messages or calls themselves. In recent years telephone networks have begun to introduce *common-channel signaling*, in which control messages or signals are carried over separate signaling channels. In the United States in particular this has led to the use of a separate packet-switched network called the CCS (common-channel signaling) network for transmitting call setup and routing packets, as well as other supervisory and control information. Common channel signaling will be described in Chapter 12. (CCS does not necessarily require a separate network. Signals could just as well be sent using dedicated signaling slots of a TDM frame.) We shall model the older inband signaling method in this introductory tutorial treatment. Closs provides a model for common-channel signaling as well.

Given this set of signals required to set up and tear down a complete connection, end to end, in a circuit-switched network, we can proceed to model and analyze the circuit-switched mode of operation, in particular to calculate the time required to set up a call. We follow the approach in [CLOSS], as already noted. The setup or connect time is a function of traffic in the network, length of control and data messages, signaling and data transmission rates, and the number of channels (trunks) available for communication. This exercise, though rather simplistic and artificial, is of interest because it provides a quantitative overview of the basic aspects of circuit switching. It will also enable us in the next section to compare the performance of the circuit-switched and packet-switched techniques, albeit in a very rudimentary way.

For this purpose, we model the circuit-switched network as one that *queues* call requests until a trunk or channel becomes free. In most operating networks calls are generally *blocked* if resources are not available for them. (The blocking model will be studied in detail in later sections.) To simplify the analysis, we assume that the network is fully connected: All nodes (switches, exchanges) are connected to one another. (Alternatively, we determine the call setup time, or connect time, for the case in which the route involves only one hop between source and destination nodes.) This enables us to avoid questions of routing at this point. The model to be analyzed then becomes that of Fig. 10–2. The two switches, nodes A and B, are connected by N channels or trunks, each of capacity C_L bits/sec. This is precisely the transmission rate of the local loop as well, as shown in Fig. 10–2, since the circuit made available for real-time transmission, end to end, is of the same capacity throughout. The data to be transferred from source to destination on its own private channel, once set up, will thus be transmitted at a rate of C_L bps throughout.

Recall that we indicated we would model the "inband" method of signaling and that, for simplicity at this point, we assume a queued rather than a blocking system. Send requests arriving at node A from stations connected to it are queued, after processing at the node, until one of the N channels becomes free (see Fig. 10–2). Once a channel is free, the channel is seized and a connect message is sent to node B, as shown in Fig. 10–1. The components of the call setup time, T_C, from the time of transmission of the send-request message to the receipt of the start to send message, are then indicated in Fig. 10–3. We have neglected the ringing and answer messages between node B and the destination station, although these could readily be added, since in the next section we will be comparing connect time using this model with packet-switched time delay to

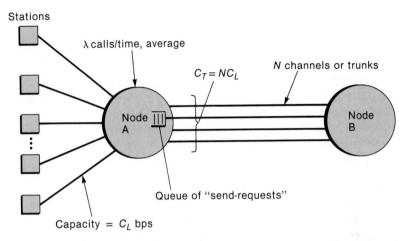

Figure 10-2 Pair of nodes in a fully connected circuit-switched network; queued model, inband signaling

node B only. The times indicated are the times required to send each of the signaling messages, as shown; processing in A and B is taken to require an average length of time $E(T_p)$ at each; and $E(W)$ is the average wait time on the queue at node A until one of the N channels becomes free (Fig. 10-2). (That processing at each node is a random quantity will become clear when we describe the call-processing function in detail in the next chapter.)

For simplicity we now take each of the signaling messages (other than the connect messages) to be of the same length and hence to require the same transmission time T_S. The connect-message transmission time is taken to be T_I.

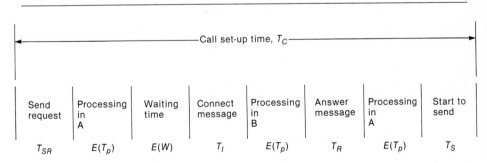

Figure 10-3 Components of call set-up time, Fig. 10-2 example

With this simplification, the connect time T_C, from Fig. 10–3, is given by the expression

$$T_C = 3T_S + T_I + 3E(T_p) + E(W) \qquad (10-1)$$

How does one calculate the average wait time $E(W)$? From Fig. 10–2 it is apparent that the model is that of an N-server queue, each transmission channel or trunk corresponding to one server. To carry the analysis any further in order to find the wait time $E(W)$, we must assume that calls or requests for a connection arrive at a Poisson rate and that the length of service is exponential. This gives rise to an M/M/N queueing model. (In Chapter 2 we discussed the M/M/N/N queueing system, which has zero waiting room, or blocking. Here the queue is taken to be infinite, and the M/M/N model arises.) The Poisson arrival assumption is a fairly valid one and is adopted in most analyses, as we have indicated throughout this book. The assumption of exponential service time is grossly inadequate, as will become apparent in the discussion that follows. We really have no choice, however, since using general service-time statistics complicates matters too much. All we can say is that the exponential service-time assumption is usually a conservative one, so that we at least err on the worst-case side.

If we do assume exponential service-time statistics, with average service time $1/\mu$ (or service rate μ, in units of calls or setup requests served per unit of time), and Poisson call arrivals at node A (see Fig. 10–2) of average rate λ, one gets the M/M/N model of Fig. 10–4. This is precisely a birth-death process (see Chapter 2) with the characteristics

$$
\begin{aligned}
\lambda_n &= \lambda && \text{all } n \\
\mu_n &= n\mu && n \leq N \\
&= N\mu && n \geq N
\end{aligned}
\qquad (10-2)
$$

Here n is the state of the queue (the number of calls or setup requests in the queueing system, including those in service).

From our discussion in Chapter 2, we know that the probabilities of states p_n are given by

$$p_n/p_{n-1} = \lambda_{n-1}/\mu_n \qquad (10-3)$$

so that

$$
\begin{aligned}
p_n &= \frac{(N\rho)^n}{n!} p_0 && n \leq N \\[2ex]
&= N^N \frac{\rho^n}{N!} p_0 && n \geq N
\end{aligned}
\qquad (10-4)
$$

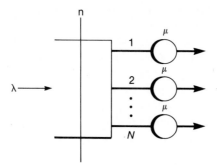

Figure 10–4 M/M/N model, queue of Fig. 10–2

We have defined ρ as the utilization parameter per trunk or channel

$$\rho \equiv \lambda/\mu N < 1 \qquad (10-5)$$

(Note that with N trunks the system has N times the capacity of that of the M/M/1 case.)

The unknown parameter p_0 is readily found, as usual, by requiring the sum of p_n over all possible states to be 1. It is apparent that p_0 is given by

$$p_0 = \left[\sum_{n=0}^{N-1} \frac{(N\rho)^n}{n!} + \frac{1}{1-\rho} \frac{(N\rho)^N}{N!} \right]^{-1} \qquad (10-6)$$

The average wait time $E(W)$ can be found directly by applying Eq. (10–4) to Eq. (10–6). Before doing this, however, we choose to focus on a particular performance parameter of the system that occurs quite often in studies of queued circuit-switching systems of the type being considered here. The wait time $E(W)$ will also be shown shortly to be directly expressible in terms of this function. The function of interest is the probability $P(\text{delay})$ that a message will be delayed. This is, of course, just the probability that there are at least N messages already in the M/M/N queueing system and hence in service. $P(\text{delay})$ is thus given by

$$P(\text{delay}) = \sum_{n=N}^{\infty} p_n = \frac{(\rho N)^N p_0}{(1-\rho)N!} \qquad (10-7)$$

after some simple manipulation. This equation for delay has found considerable use over the years in telephone-trunk design for queued or delay-type systems, in which calls are held rather than blocked or given a busy signal. For this reason Eq. (10–7) has been extensively tabulated. It is often called the *Erlang-C for-*

mula, or Erlang's formula of the second kind. (Recall that in Chapter 2 we derived the Erlang-B formula, or Erlang's formula of the first kind, which arises in blocking systems. That formula will be used later in this chapter when we discuss blocking systems.) Typical curves appear in the book by Collins and Pedersen [COLL], among others.

These tabulations and curves are useful in determining the number of trunks needed to obtain a specified probability of delay. Note that $\rho N = \lambda/\mu$ appears in both Eq. (10–6) and Eq. (10–7) as the controlling parameter. To distinguish between the per-trunk utilization ρ and the total traffic intensity ρN, the parameter A is often used to represent $\rho N = \lambda/\mu$. The probability of delay is then tabulated as the function $E_{2,N}(A)$:

$$P(\text{delay}) \equiv E_{2,N}(A) \qquad (10–8)$$

The units of $A = \lambda/\mu$ are measured in *Erlangs*. A system handling 2 messages/sec, each message lasting two sec on the average, has a traffic intensity of 4 Erlangs. Obviously more than four trunks would be needed to cope with this traffic. The Erlang unit of traffic was introduced briefly in Chapter 2, in discussing the Erlang-B formula, which describes the probability of blocking in a zero waiting room M/M/N/N queueing system. Erlang units are used extensively in circuit-switched traffic engineering studies, and we shall refer to these units repeatedly in the discussions following. Note that the parameter $A = \lambda/\mu$, in Erlangs, represents the *total load* on the system, while $\rho = A/N$ represents the per-trunk utilization.

As a typical calculation, assume that $\lambda/\mu = 0.8$ and take the two cases of $N = 1$ and $N = 5$. The $N = 1$ case is just the M/M/1 case, and it is easily shown that $E_{2,1}(A) = A = \rho = 0.8$, the probability that the one server is occupied. For the five-trunk case it is found that $E_{2,5}(0.8) = 0.0018$. The use of five trunks thus reduces the probability of delay from 0.8 to 0.0018, a factor of about 400.

Using the definition of the Erlang formula of the second kind, $E_{2,N}(A)$, as the probability of delay, one can readily show that a number of parameters of interest may be expressed in terms of this formula. This is the reason for its extensive use and tabulation. In particular, consider the mean number of messages $E(m)$ waiting for service. This must be the difference between the average number $E(n)$ in the system and the average number in service. The mean number is thus given by

$$E(m) = \sum_{n=N+1}^{\infty} (n - N)p(n)$$

$$= \left(\frac{\rho}{1 - \rho}\right) E_{2,N}(A) \qquad (10–9)$$

[COLL] A. A. Collins and R. D. Pedersen, *Telecommunications: A Time for Innovation*, Merle Collins Foundation, Dallas, 1973.

after some manipulation, and comparing with Eq. (10–7). As a special case, for $N = 1$, we get the M/M/1 result $\rho^2/(1 - \rho)$. It is left to the reader to show similarly that the average number of messages in the system is

$$
\begin{aligned}
E(n) &= A + E(m) \\
&= N\rho + E(m)
\end{aligned}
\tag{10–10}
$$

just the N-server extension of the M/M/1 result. (It is thus apparent that $A = N\rho = \lambda/\mu$ represents the average number of messages in service.) Finally, using Little's formula, we can find the time delays of interest. The average system delay, or time delay for short, is

$$
E(T) = \frac{E(n)}{\lambda} = \frac{1}{\mu} + E(W)
\tag{10–11}
$$

exactly as written in Chapter 2. The average wait time $E(W)$, the parameter of interest in this discussion, is in turn just

$$
E(W) = \frac{E(m)}{\lambda} = \frac{1}{\mu} \frac{E_{2,N}(A)}{N - A}
\tag{10–12}
$$

from Eq. (10–9), and recalling that $\rho = \lambda/N\mu = A/N$.

As an example of the application of these equations consider the previous example. Let $A = \lambda/\mu = 0.8$. As noted, for $N = 1$, $E_{2,1}(A) = A = 0.8$. The average wait time is $E(W) = \frac{1}{\mu}\left(\frac{0.8}{0.2}\right) = 4\left(\frac{1}{\mu}\right)$, and the average time delay is $E(T) = 5\left(\frac{1}{\mu}\right)$. If $\frac{1}{\mu} = 0.1$ sec, for example, $E(W) = 0.4$ sec, and $E(T) = 0.5$ sec. Now let five trunks be used. Recall that $E_{2,5}(0.8) = 0.0018$. Then the average wait time using Eq. (10–12) is $E(W) = 0.0004(1/\mu)$ sec! For $1/\mu = 0.1$ sec, this is just 40 μsec and clearly negligible. The average time delay $E(T)$ is essentially just the average transmission time $1/\mu = 0.1$ sec. The five-trunk system clearly results in vast improvements in the wait time. On the other hand, assume that a total output trunk (line) of capacity $C = 9600$ bps is available and that it is to be decided whether to allocate the trunk as five separate trunks of 1920 bps each or to leave it as one wider-band trunk. Assume that $1/\mu = 0.1$ sec comes from using a narrow-band line of 1920 bps. (Then the messages are 192 bits long on the average.) Five trunks of 1920 bps each result in the average wait time of 40 μsec, and an average time delay due to transmission of 0.1 sec. If one 9600-bps trunk is used instead, however, this becomes an M/M/1 queue with $\rho = 0.8/5 = 0.16$. The average message-transmission time is $0.1/5 = 0.02$ sec, and the average delay time is $E(T) = 0.02/0.84$, clearly much less than the delay time for the case of five trunks. Thus a higher-capacity line is preferable to subdividing it into smaller capacity lines. This point was also made in our discussion of Chapter 2.

The question of whether to subdivide a wideband line into narrower ones is appropriate for data communications, but not really for real-time circuit-switched voice communications. In this latter application a *fixed* bandwidth is needed for each voice call: 4 kHz for analog signaling, the bandwidth required to transmit 32 kbps or 64 kbps in the case of digital voice. Thus one cannot effect this subdivision in this application. (Packet voice, in which digitized voice segments are transmitted as packets over a packet-switched network, does allow this type of division.)

After this brief digression into the use of the Erlang-C formula, we now return to our original problem, that of the calculation of connect or call-setup time T_C for the queued circuit-switched model of Fig. 10–2. From Eq. (10–1) we need an expression for the average wait time $E(W)$. Using the M/M/N model, the wait time is given by Eq. (10–12). The only problem now is to determine the average service time $1/\mu$ that appears in that formula. From the earlier discussion leading to this model, this should be just the average time that one of the N trunks of Fig. 10–2 is held. The term *average holding time* is often used for this quantity, as already noted. This time is obviously related to the length of the call that is being set up. In our forward-setup mode of operation, however, the trunk is assigned to the user once his or her send-request message reaches the head of the queue in Fig. 10–2. The holding time thus begins with the transmission of the first connect message in Fig. 10–1 and concludes once the final disconnect message is received at the destination exchange. More precisely, the length of time a trunk is held in our model is the time required to initially signal and process the call request, the time required to actually transact the call, and then the time to disconnect the trunk end to end. For the simple single-hop example of Fig. 10–2, the components of the holding time appear as portrayed in Fig. 10–5. The different components of the holding time shown there are self-explanatory. Note, however, that to pinpoint the particular application we have labeled the average holding time T_H rather than the ubiquitous term $1/\mu$, used up to now. We have also used the symbol T_M to represent the average call or data-message length. The data transmitted in this case could be voice or data; as noted earlier, the system does not distinguish between the two once they are in digital form. We thus use the term "data" as a more general representation of the actual information message to be transmitted.

From Fig. 10–5, again assuming all signaling messages to be of the same time duration T_S, we have for the average holding time T_H in this example

$$T_H = 4T_S + T_I + 4E(T_p) + T_M \qquad (10–13)$$

This number is to be used in the average wait-time formula (10–12). Note, however, the problem we have already mentioned. Equation (10–12) assumes that the service time (holding time in this case) is distributed exponentially. It is apparent that the holding time in our case is not. As a matter of fact, from Fig.

10–5, the holding time T_H is the sum of a number of random variables (data message length and nodal processing times) plus some constant terms (the signaling messages). As noted earlier, however, we simply disregard this apparent problem (we really have no choice!) and proceed to take the holding time as distributed exponentially. Our discussions in Chapter 2 indicated that the exponential assumption is robust and conservative. One would thus expect the results obtained using this model to be good first-order ones, appropriate for our purposes.

With this caveat we write as the formula for the average waiting time on the queue of Fig. 10–2

$$E(W) = \frac{E_{2,N}(A)}{N - A} T_H \qquad (10-14)$$

with $A = \lambda T_H$ the load on the system in Erlangs, and the average holding time T_H given by Eq. (10–13). Using this expression for the wait time, one can calculate the call setup or connect time T_C from Eq. (10–1).

Consider two examples. Take $N = 10$ individual channels in both cases and neglect nodal processing time. (Processing time depends on the particular types of data messages transmitted, the hardware and software organization of the nodal switch, and the speed of the nodal processors, among other factors. Processing time is quite application dependent, and thus to calculate it is not a simple task. A typical calculation is carried out in the next chapter for the voice-call telephony environment and for a specific organization of the nodal switch.) In the first example assume that signaling and connect messages require the same time T_I to be transmitted, and in turn require one tenth the transmission time of the message. Hence $T_I = 0.1T_M$. As an example, say that the data messages are 100 characters long. The signaling messages are then 10 characters long. Since all messages are transmitted at the local loop speed of C_L bps, these convert directly to the same relative transmission times. For this example

$$T_H = 5T_I + T_M = 1.5T_M$$

and

$$T_C = 4T_I + E(W) = E(W) + 0.4T_M$$

from Eq. (10–13) and Eq. (10–1), respectively. Using Eq. (10–12) to calculate the average wait time we get the average connection-time curve of Fig. 10–6, designated by $T_{\text{sig}} = T_I$. In the second example we again take the connect message-transmission time $T_I = 0.1T_M$, but now assume that the other signaling messages are $0.1T_I$. This might be an appropriate model in some situations since the connect message must carry address information and possibly other control information, while the other signaling messages might be shorter. The connection time for this example is also shown in Fig. 10–6, designated as the curve

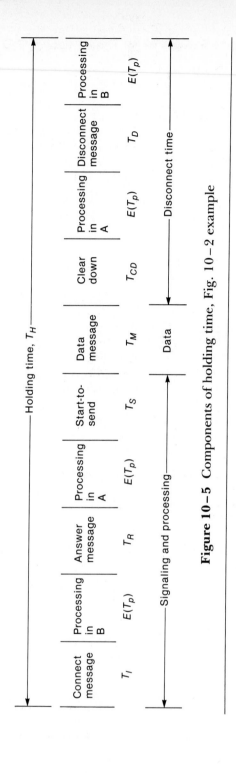

Figure 10–5 Components of holding time, Fig. 10–2 example

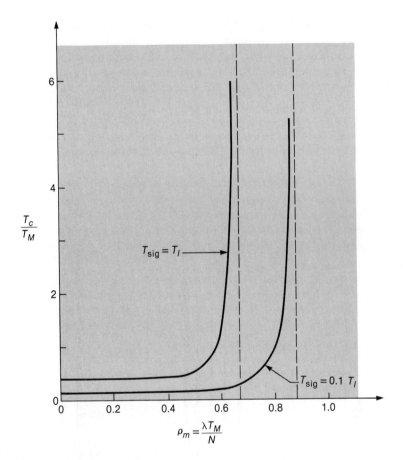

Figure 10–6 Normalized connection time, $N = 10$ channels, $T_I = 0.1T_M$

$T_{\text{sig}} = 0.1T_I$. Both curves are shown plotted as a function of a parameter ρ_M, defined as the per-channel data utilization. Thus we take $\rho_M \equiv \lambda T_M/N$. This contrasts with the per-channel traffic utilization $\rho \equiv \lambda T_H/N = A/N$, and with $A = \lambda T_H$ the overall utilization, in Erlangs. Note from Eq. (10–14) applied to this case that the maximum overall utilization is $A = 10$ Erlangs, corresponding to a maximum per-channel traffic utilization of $\rho = 1$. Since the total link capacity is being used for both signal transmission and data transmission, the effective capacity available for data message transmission is reduced. In the first example, with all signaling messages requiring the same transmission time, one tenth that of the average data message, the maximum per-channel data utilization is

$1/1.5 = 0.67$. In the second example, with $T_H = 1.14T_M$, the maximum value of ρ_M is $1/1.14 = 0.88$ and is so indicated in Fig. 10–6.

Note that in Example 1 the call-setup time has a minimum value of $0.4T_M$, that required to transmit the necessary signaling information. As the parameter ρ_M increases, due, for example, to more data stations or terminals accessing node A or to a greater usage by each terminal, the connection-setup time relative to message-transmission time begins to rise due to wait time for an available channel. In Example 2 the signaling messages are shorter and hence the call-setup times are correspondingly shorter. The system allows a greater utilization as well.

Some specific numbers are of interest. Say that data messages are 100 10-bit characters long on the average. Say that the channel transmission speed is $C_L = 110$ bps. (This is a typical teletypewriter speed.) Then the average message-transmission time is $1000/110 = 9.1$ sec long. For Example 1 the minimum connect time is 3.6 sec. However, if $\rho_M = 0.6$, $T_C/T_M = 1.34$, and the connect time is 12.2 sec. This corresponds to $\lambda = 0.66$ call/sec or 40 calls/minute. For 40 terminals connected in, this implies that each terminal initiates a call once a minute on the average. The message is 9 sec long, but the terminal must wait 12 sec on the average before transmission can begin. These numbers are pessimistic because of the exponential holding-time assumption noted earlier, but they point out one of the problems associated with circuit-switched data transmission: a call-setup time that may be inordinately long.

Now consider Example 2, with the signaling messages reduced in length. For the same per-channel data utilization $\rho_M = 0.6$, the call-setup time is $T_C = 0.19T_M = 1.7$ sec, clearly an improvement over the previous system. It is thus apparent that reduced signaling-message lengths are important in providing improved performance for line-switched networks.

As another typical example say that messages are 1000 8-bit characters long on the average. High-speed 4800-bps synchronous data terminals are used for communication. Then the data-message transmission time is 1.7 sec on the average. For Example 1 again, the minimum call-setup time with negligible traffic in the network is $0.4T_M = 0.7$ sec. For $\rho_M = 0.6$ again, however, the minimum setup time rises to $1.34T_M = 2.3$ sec. In this case $\lambda = 3.5$ messages/sec may be accepted, on the average, at the processing node. For Example 2, the minimum call-setup time is $0.13T_M = 0.22$ sec, while at $\rho_M = 0.6$ the average connect time is $T_C = 0.19T_M = 0.32$ sec. This figure would appear to be acceptable.

It is the possible occurrence of relatively long connection-setup times for short messages (in which the overhead due to signaling messages is proportionately high) in circuit switching that leads to the proposed use of packet switching for these cases. In such networks a multiplicity of users share the facility, and delay times due to queueing can be kept relatively low.

In the next section we model a packet-switched network in the same form as that of Fig. 10-2, enabling us to compare the relative performance of circuit switching and packet switching as data message lengths vary. Recall, however, that our circuit-switched model is rather simplistic and of the delay or queued type only. In later sections we focus on more realistic models of the circuit-switched process, particularly the blocked-calls type.

10-2 Circuit and Packet Switching Compared: Simple Model

In this section we again follow the approach of F. Closs [CLOSS] in carrying out a simplified comparison of circuit switching and packet switching. The circuit-switched model and analysis thereof were presented in the last section for the case of inband signaling. Closs has extended this work to include a model of out-of-band, or common-channel, signaling, with a number of the N trunks of Fig. 10-2 set aside specifically for signaling and the remaining trunks dedicated to circuit-switched data transmission. This work is not presented here. The reader is referred to his work for details of the analysis [CLOSS].

Our model of the packet-switched network, made fully comparable to that of Fig. 10-2 for the queued model of the circuit-switched network, appears in Fig. 10-7. The packets to be transmitted are precisely the data messages of the last section—i.e., the words message and packet are used synonymously. We calculate the packet-response time for this simple model and use it as a basis for comparison with the call setup or connect time in the circuit-switched case.

Specifically, to make the analysis here fully comparable to that of the circuit-switched case, assume the same number of terminals or stations as before (i.e., the same number connected to each node). The total arrival rate of messages at either node destined for the other node is thus again λ messages/time, assumed to be Poisson distributed. Terminals are connected to the nodes by relatively low-speed local loops (see Fig. 10-7), while the nodes themselves represent part of the packet-switched network, with high-speed lines or links of capacity C_T bps interconnecting the nodes. We assume full-duplex transmission, the internodal links thus being capable of carrying messages in either direction at the same rate.

We assume that terminals are connected to their appropriate node by a fixed local connection; thus connection time to the appropriate source node is not considered here. We will also neglect packet-transmission time over the relatively low-speed local loop, just as in the circuit-switched case, since this time is presumably the same for both types of switching. We consider network re-

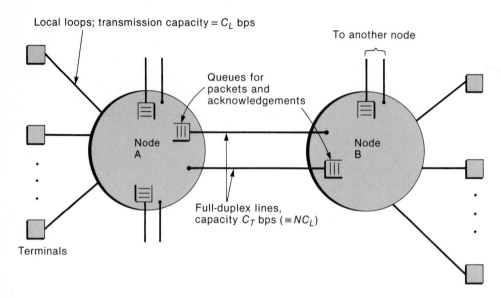

Local loops; transmission capacity $= C_L$ bps

To another node

Queues for packets and acknowledgements

Node A

Node B

Full-duplex lines, capacity C_T bps ($\equiv NC_L$)

Terminals

Figure 10–7 Pair of nodes, packet-switched network

sponse time over the packet-switched network only, for comparison with the circuit-switched case.

We model the packet-switching process as follows to determine the response time: Packets arriving at a node (for example, node A of Fig. 10–7) are queued up, first come–first served, for transmission over the high-speed internodal trunk. Once transmitted, at the rate of C_T bps, they are received at the destination node (in this example, node B of Fig. 10–7). We neglect propagation delay, just as in the circuit-switched case. Error checking is carried out at the receiving node and a positive acknowledgement is transmitted in the reverse direction, back to the sending (source) node. We again neglect the nodal-processing time required to check the packet and form the acknowledgement. The response time we would like to calculate for comparison with the circuit-switched connection time is measured from the time the packet arrives at the sending node to the time the acknowledgement is received there. Since processing and propagation delays are neglected here the latter time is just the time at which transmission of the acknowledgement is completed at the receiving node.

The calculation of this response time depends on the acknowledgement protocol as well as on the model used to analyze the packet and acknowledgement queueing delays. We shall in fact compare two different protocols in this simplified analysis; both were described in earlier chapters on packet switching.

One approach is to generate fixed-length acknowledgement packets and transmit these separately over the high-speed lines. Another approach is to embed the acknowledgements in data packets traveling in the reverse direction. (Recall that the HDLC protocol, like others, uses both techniques.) In the first case traffic on the line is increased because of the added packets; in the second case the traffic is kept down but the packet lengths are increased somewhat to accommodate acknowledgement information. In some networks that use separate acknowledgement packets, these packets are given priority over the (usually) longer data packets to reduce their queueing delay. We ignore this possibility here to keep the analysis simple.

Consider the first type of acknowledgement procedure. In this case each packet received generates a separate acknowledgement (ack) of fixed length, L_I bits. (We use the same notation as in the circuit-switched case since we will be assuming, in our comparison of packet switching and circuit switching, that acks have the same length as connect messages in the circuit-switched case.) The ack is then queued up, together with regular packets, at the receiving node and waits for transmission back to the sending node. Node A and node B in Fig. 10-7 are each assumed to generate packets for transmission to the other node; a schematic representation of packet arrivals at one of the outbound queues in either node takes the form of Fig. 10-8. Packets from A destined to B must queue up with acks from A to B, acknowledging correct receipt of prior packets from B to A. Similarly, acks from B to A, acknowledging receipt of prior packets from A to B, must wait together with packets from B to A. We assume the packets have the following characteristic: They each have a fixed header and control character portion L_I bits long, plus an exponentially distributed data portion m_c bits long on the average. Packet overhead is thus assumed to be the same size as acknowledgement messages. This model enables us to use M/G/1 analysis.

The node-to-node message response time T_D for this model of the packet-transmission and -acknowledgement process is readily obtained from M/G/1 analysis. Assume that the packets arrive at a node at a Poisson rate, with average arrival rate λ. Since each message generates an acknowledgement at the neighboring receiving node, λ acks/time, on the average, appear at the same node as well. This is shown in Fig. 10-8. We assume that these acknowledgement are Poisson distributed as well. Obviously, this cannot be true since the Poisson assumption requires independent arrivals, while the acknowledgement is clearly dependent on a packet transmission. However, since the acks in a given queue are determined by random packet arrivals and random wait times in another queue, it would appear that the Poisson assumption is reasonable in this case. The output buffer in Fig. 10-8 may thus be modeled as an M/G/1 queue, with an effective arrival rate of 2λ "equivalent" packets/time, half of which on the average are true packets, the other half of which are acks. Since all data is read out at the high-level rate of C_T bps, the average packet length in sec is

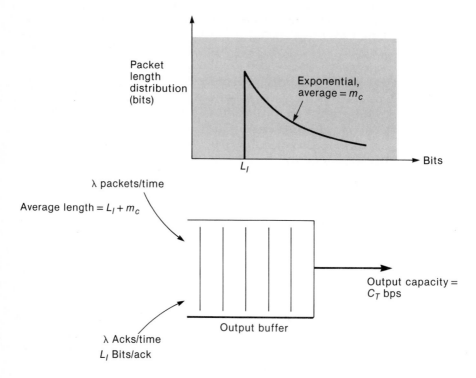

Figure 10–8 Transmission of separate acknowledgement packets

$(L_I + m_c)/C_T = t_h + t_m$, with t_m the data portion in sec and t_h the header portion. The average ack length in sec is also t_h. The average "equivalent" service time for the M/G/1 queue of Fig. 10–8 is then

$$E(\tau) = 0.5(t_h + t_m) + 0.5t_h \qquad (10-15)$$

since we assume the two types of input are equally likely to arrive and are served first come–first served. The effective ρ for the queue is then

$$\begin{aligned} \rho &= 2\lambda E(\tau) = \lambda(t_m + 2t_h) \\ &= \rho_M(1 + 2t_h/t_m) \end{aligned} \qquad (10-16)$$

with $\rho_M = \lambda t_m$ the effective *data* utilization factor. This ρ_M is exactly the same parameter defined previously in our discussion of circuit switching as the per-channel data utilization. There we wrote $\rho_M = \lambda T_M/N$, with $T_M = m_c/C_L$ the data message length in seconds, as referred to the local-loop-transmission capacity. Since $t_m = m_c/C_T$, and we assume that $C_T = NC_L$ in order to compare circuit switching and packet switching, it is apparent that $t_m = T_M/N$, and the two ρ_M's are the same.

Recall from Eq. (2–65) that the average wait time for an M/G/1 queue depends on the second moment $E(\tau^2)$ of the message lengths as well as on ρ. $E(\tau^2)$ is easily found for the model of Fig. 10–8. Specifically, we have

$$E(\tau^2) = 0.5[t_h^2 + (t_m + t_h)^2 + t_m^2] \qquad (10-17)$$

averaging over the two equally likely message types in the queue of Fig. 10–8. The first term in Eq. (10–17) is just the second moment of the fixed-length acks; the second term is the first moment squared of the packets; and the third term is the variance of the packet-length distribution. (The sum of the last two terms then gives the second moment of the packet-length distribution shown in Fig. 10–8.) Noting again that the equivalent packet arrival rate is 2λ, we then have as the average wait time for the queue of Fig. 10–8

$$E(W) = \frac{\lambda}{2} [t_h^2 + (t_m + t_h)^2 + t_m^2]/(1 - \rho) \qquad (10-18)$$

This may be put in a somewhat different normalized form involving the data utilization ρ_M by factoring out t_m. We then get

$$E(W) = \rho_M[t_m + t_h + t_h^2/t_m]/(1 - \rho) \qquad (10-18a)$$

with ρ given by Eq. (10–16).

Finally, the packet response time for the node-to-node packet-switched model of Figs. 10–7 and 10–8 is readily obtained by noting that it is due to the queueing delay of *packets* at node A (the transmitting node) and the queueing delay of *acks* at node B (the receiving node). Recall again that we are neglecting propagation delays and nodal processing delays. The response time then corresponds to two wait times (one at each node) plus the average transmission time of a packet and the transmission time of an ack. Calling the response time T_D we get, for this model,

$$T_D = t_m + 2t_h + 2E(W) \qquad (10-19)$$

with $E(W)$ given by Eq. (10–18a).

To compare specifically with the connection time T_C for the circuit-switched case, we must normalize to the equivalent loop data-transmission time T_M, as was done in Fig. 10–6. Letting $C_T = NC_L$ as previously and defining $k = t_h/t_m = L_I/m_c$ as the ratio of control-packet length to data-packet length, we get

$$T_D/T_M = \frac{1}{N}\left[1 + 2k + \frac{2\rho_M(1 + k + k^2)}{1 - \rho_M(1 + 2k)}\right] \qquad (10-19a)$$

This has been plotted in Fig. 10–9 as a function of ρ_M for the same case as the circuit-switched curves of Fig. 10–6: $N = 10$ and $k = t_h/t_m = 0.1$. The high-level capacity is thus 10 times the local loop capacity, and acks are one tenth the length of the average data packet. Note that for this example the comparison depends critically on the length of the signaling messages in the circuit-switched case. Thus if the signaling messages are the same length as the connect messages, ρ_M for the circuit-switched case has a maximum value of 0.67 and the connection time increases rapidly in the vicinity of this utilization factor. The maximum packet-switched data utilization is, on the other hand, $1/(1 + 2k) = 0.83$. If the

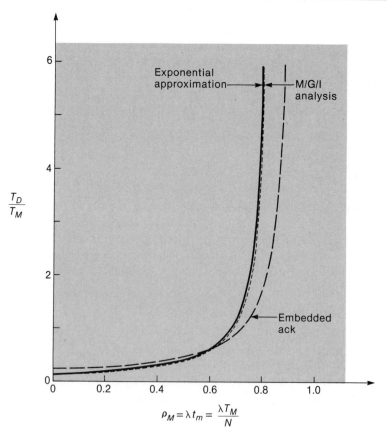

Figure 10–9 Normalized round-trip response time, packet-switched case, ack length $t_h = 0.1 t_m$, $t_m = 0.1 T_M$ ($N = 10$)

circuit-switched signaling messages are one tenth the length of the connect messages, however, the maximum utilization increases to 0.88 in the circuit-switched case.

Also shown in Fig. 10-9 are two curves for two other models of the node-to-node packet-switching process. One curve represents the time response for the embedded-ack technique noted earlier. In this case there are no separate ack messages transmitted. Instead, ack characters are included in each packet transmitted. As noted in our discussion of HDLC in Chapter 4, these need only provide the number of the previous packet transmitted in the reverse direction that is being acknowledged. The effective packet length is thus increased by very little. (In the separate-ack case the acknowledgement packet must carry synchronization, identification, and error-checking information. In many cases this packet is of the same form as the data packet with the data field left out.) The effective packet-arrival rate at a queue is now λ, instead of 2λ as in the separate-ack case. Figure 10-10 represents this case. The transmission time at the receiving end is now increased, however, since transmission of an entire packet rather than an ack packet is required for acknowledgement. It is left for the reader to show that the normalized response time in this case is given by

$$T_D/T_M = \frac{2}{N}\left[1 + k' + \frac{\rho_M(1 + k' + k'^2/2)}{1 - \rho_M(1 + k')}\right] \qquad (10-20)$$

Here $\rho_M = \lambda t_m$ again, and $k' = t'_h/t_m \geq k$, with t'_h the length of the header and control portion of the packet in this case. We take $k' \geq k$ to account for the enlarged packet length required to incorporate ack characters.

Equation (10-20) has been plotted in Fig. 10-9 for the special case $k' = k = 0.1$. We have thus ignored the additional packet length required to accommodate the acknowledgement process. This might correspond to a network where the average packet length is 100 characters and 10 header and control characters are used in addition. (Recall that in the HDLC protocol overhead consisted of 6 characters.) For an acknowledgement procedure requiring anywhere from 0.5 character (for example, 4 bits) to 3 characters this is probably a good approximation. Note from the curves of Fig. 10-9, and from a comparison of Eq. (10-19a) and Eq. (10-20), that the embedded-ack time response is poorer at low utilization and clearly better at the higher levels. It is poorer for small ρ_M because this procedure requires at the minimum the transmission of two full packets (one in each direction) to provide the round-trip response time. This is shown by the minimum value of $T_D = 2T_M(1 + k')/N = 2t_m(1 + k')$, from Eq. (10-20). The separate-ack procedure requires at minimum the transmission of one full packet (at the sending end) and one ack packet at the receiving end. Its response time is then $t_m + 2t_h$, as is apparent from Eq. (10-19) as well. At the higher values of ρ_M the additional ack messages add to the packet flow, increasing the response time considerably.

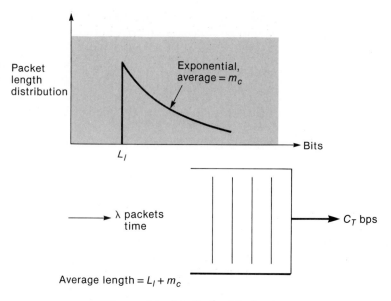

Figure 10–10 Embedded acks

Also shown in Fig. 10–9 is a curve labeled "exponential approximation." Here both data- and ack-packet lengths have been assumed to be distributed exponentially, one with the average value $t_m + t_h$, the other with the average value t_h. This thus provides a conservative upper bound on the model of Fig. 10–8, since the variance of both packet types in Fig. 10–8 is less than if they were both distributed exponentially. Note that for the example taken, however, the two curves are almost identical. It is left for the reader to show, using the wait-time expression (2–65) for the M/G/1 queue as applied to this model of two exponential arrivals, that the normalized response time in this case is given by

$$T_D/T_M = \frac{1}{N}\left[1 + 2k + \frac{2\rho_M(1 + 2k + 2k^2)}{1 - \rho_M(1 + 2k)}\right] \qquad (10-21)$$

It is this equation that has been plotted in Fig. 10–9 and labeled "exponential approximation." It is apparent that the only difference between this model and the one of Fig. 10–8 (plotted as the M/G/1 curve in Fig. 10–9) is in the second-moment portion $E(\tau^2)$ in the wait-time equation. This is seen by comparing Eq. (10–21) and Eq. (10–19a). Since $k = 0.1$ in Fig. 10–9 the two curves differ by very little. This analysis again validates the use of the exponential model in many situations.

Using the analysis of both the circuit-switched technique and the packet-switched mode of operation for the simple single-hop network of Figs. 10-2 and 10-7, respectively, we are now in a position to compare these two switching techniques for various cases other than the example already cited of $N = 10$ circuit-switched trunks. In particular, we would like to see how comparative performance varies with data-message length. We shall use the equations derived in this section for this purpose. Note again, however, that the equations for the circuit-switched model are of restricted form because of the emphasis on queued rather than blocked operation and because of the various simplifications made.

We noted earlier, in introducing circuit switching, that this procedure would seem to be the preferred technique for long data messages, while packet switching should appear preferable for shorter messages. This supposition is based on the observation that the minimum connect time for circuit switching requires a sequence of signaling and inquiry messages to be transmitted. For relatively short messages this invariant time may equal or exceed the message transmission time itself and be judged excessive by a network user. This is actually the case with short data messages transmitted over the common switched telephone networks. Connect times of up to 10 sec are quite normal in relatively busy periods. A 1000-character message transmitted at a 1200-bps rate would require about 7-sec transmission time, and a 5-15-sec connect time in this case might frustrate a user. This is one important reason for the move to common-channel signaling and other methods of speeding up the call setup time in telephone networks. As the networks move to all-digital transmission and switching capability (most circuit-switched networks currently use a combination of analog and digital technology) with very high-speed optical fiber transmission, both connection and transmission times will be expected to drop significantly.

In the case of a long message, the connect time in the circuit-switched case becomes small compared with the message transmission time. In the packet-switched case, this case leads to large network transmission times, and the response time increases correspondingly.* As indicated at the beginning of this chapter, the transmission of a half hour or more of digital or computer tape

---

* Note again that we are ignoring the local loop-transmission time required to get the message into the network in the first place. This is common to both modes of transmission. It is, however, the *only* transmission time required in the circuit-switched case since there is no storing and forwarding of messages. We also neglect here the possibility of dividing long messages into smaller packets. This provides an added advantage to packet switching in multihop networks.

output might better be done over a circuit-switched facility than over a packet-switched network.

The actual choice of circuit switching versus packet switching depends on many factors, as already noted, and it is clear that the dominant consideration will be economic in many cases. It is useful to quantify the time-delay comparison of the two schemes, however, and to see if the small-long message length considerations noted earlier do hold true as expected.

We shall in fact find that, generally speaking, shorter message lengths *are* more rapidly transmitted over a packet-switched channel. The result depends critically, however, on the utilization factor (higher utilizations tend to favor the circuit-switched case because of queueing delays in the packet-switched case); on the ratio N of high-level to low-level transmission speeds (or number of channels in the circuit-switched case); and on the control (signaling, connect, ack) message lengths.

Recall again that we are assuming a queued or nonblocking circuit-switched facility in this introductory treatment. In practice most circuit switching is of the blocking type, as noted several times. In addition, we are considering node-to-node switching or a fully connected topology. This tends to favor the circuit-switched results since the connect messages setting up the call need traverse only one link.

To compare the two switching techniques as a function of data message length we rewrite the equations for T_C (connect time) and T_D (response time) to focus on varying message length, m_c, in bits (or characters, or bytes, as the case may be). We keep the control- and signaling-message lengths, data utilization ρ_M, and line capacities fixed. As m_c increases in length, therefore, the message arrival rate λ must decrease to keep ρ_M fixed.

Consider first the circuit-switched case. Assume that all signaling messages are of the same length in bits (or characters, as the case may be) and are given by L_S. Connect messages are L_I bits long. (Then $T_S = L_S/C_L$, and $T_I = L_I/C_L$.) From Eq. (10–1) we then get, again neglecting nodal processing time,

$$\begin{aligned} T_C &= E(W) + T_I + 3T_S \\ &= E(W) + (L_I + 3L_S)/C_L \end{aligned} \tag{10–22}$$

The average waiting time $E(W)$ is again given by

$$E(W) = \frac{E_{2,N}(A)T_H}{N - A} \tag{10–14}$$

with the holding time T_H, obtained from Eq. (10–13) with processing time neglected, given by

$$\begin{aligned} T_H &= T_M + T_I + 4T_S \\ &= \frac{1}{C_L}(m_c + L_I + 4L_S) \end{aligned} \tag{10–23}$$

Here

$$A = \lambda T_H = N\rho_M \left(1 + \frac{L_I + 4L_S}{m_c} \right) \qquad (10-24)$$

The response time for the packet-switched case is given by

$$T_D = \frac{1}{C_T} \left[m_c + 2L_I \frac{2\rho_M(m_c + L_I + L_I^2/m_c)}{1 - \rho_M \left(1 + \frac{2L_I}{m_c} \right)} \right] \qquad (10-25)$$

This comes from Eq. (10–19a) with data and ack message lengths written out explicitly. We have assumed the separate-ack message procedure of Fig. 10–8.

These time-delay and connection-time equations may now be evaluated as a function of data message length, m_c, for some specific cases.

Example 1. The local loop-line capacity is 2.4 kbps and the network-level transmission capacity is 120 kbps. (Then $N = C_T/C_L = 50$.) Connect and ack messages are 160-bits (20-char) long, while circuit-switched signaling messages are 16-bits (2 char) long. The resultant curves of connect and response time for two values of ρ_M — 0.5 and 0.6 — are displayed in Fig. 10–11. Note that the circuit-switched curves bottom out to a very broad minimum connect time of 87 msec. This minimum is essentially just the time $(L_I + 3L_S)/C_L$ required to transmit the signaling and connect messages. Over most of this range, then, the waiting time $E(W)$ is negligible. (The number of individual lines, or servers, $N = 50$, is so large that $E_{2,50}(A) < 10^{-2}$ over the range from $10^3 < m_c < 10^4$ bits. For $m_c < 1000$ bits, the overall utilization A begins to approach the saturable value of 50 Erlangs, driving the connect time up for small m_c. For large m_c, greater than 10^4 bits, the limit of Fig. 10–11, the holding time begins to increase and will eventually drive the connect time up again.)

The packet-switched curves do drop below the circuit-switched curves in this case. For $\rho_M = 0.5$, for example, the packet-switched time delay is less than the equivalent circuit-switched connect time over the data-length range from 350–3000 bits. At smaller data-message (or packet) lengths the message arrival rate λ becomes so large to keep $\rho_M = \lambda t_m = \lambda m_c/C_T$ fixed, that the short data messages and ack packets combined saturate the system. Large queueing delays in this case are encountered at small values of m_c because of the way curves are plotted. For large m_c, on the other hand, the transmission time over the network link dominates and causes the time delay to increase. (Recall again that we are considering network-level time delay only. As already noted, both circuit switching and packet switching have an added local loop-transmission time required to get the message into the network node in the first place. This clearly dominates the network-level transmission time — in the example of Fig. 10–11 it is 50 times the network-level transmission time. Since this delay time is com-

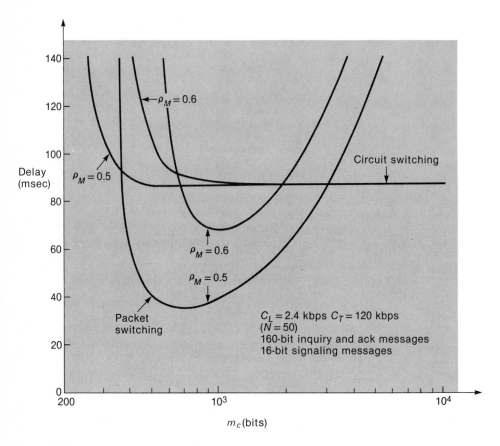

Figure 10–11 Comparison of packet switching and circuit switching,
Example 1

mon to both switching techniques it is left out here. The circuit-switched
method is thus not directly penalized for transmission time, as is the packet-
switched method. Transmission time in the circuit-switched case is manifested
indirectly in the call-setup time or connect time, through increased holding
time, which in turn increases the wait time component of the connect time.)

From Fig. 10–11 it is apparent that the packet-switched time delay is
extremely sensitive to the data utilization parameter ρ_M. For $\rho_M = 0.6$ the
packet-switched response time is below the circuit-switched connect time only
over the range from 650 to 1800 bits and rises rapidly on either end. For

$\rho_M = 0.7$ the packet-switched time delays are already outside the range of the figure and are much worse than the circuit-switched minimum connect time. For example, for $m_c = 1000$ bits, $T_D = 193$ msec, and for $m_c = 2000$ bits, $T_D = 154$ msec. Similarly, for $\rho_M = 0.8$, the minimum packet-switched time delay is 280 msec at a message length of 2560 bits and rises sharply about this value.

Example 2. For the second example we choose a much slower system, one that might be encountered with low-speed teletype transmission. We take the local loop-transmission speed to be $C_L = 11$ char/sec. (For 10-bit teletype characters this corresponds to 110 bits/sec.) The high-level network line capacity is 110 char/sec. (For 10-bit characters it is 1100 bits/sec, a nominal rate of data transmission over the analog-switched telephone network.) Connect and ack messages are assumed to be 10-characters long, while signaling messages are assumed to be 5 characters long. The resultant curves for $\rho_M = 0.5$ are plotted in Fig. 10–12. Note that connect and response times for this case are several seconds long as contrasted to the millisecond times of Fig. 10–11. The packet-switched system has a smaller time delay for packets less than 80 characters long. Its time delay increases essentially linearly with packet length above that value. The circuit-switching system again demonstrates a fairly constant connect time over the range of messages from 80 to 1000 characters in length.

It is apparent from these two examples that packet switching may provide a smaller response time in the smaller message (packet) ranges, while circuit switching does appear better for long message lengths. (Note again that these results have been obtained for a circuit-switched model with inband signaling channels. F. Closs has shown that the use of common signaling channels, assuming the same nonblocking model, provides even better connect-time results. In particular, an optimum assignment of high-level trunk capacity to the signaling channel exists that will minimize the delay time [CLOSS].)

It is possible to establish more general results than the two examples considered thus far by focusing on the systems for which packet switching provides a response-time improvement over circuit switching. From Eq. (10–25) we can find an approximate expression for the minimum packet-switched delay time. This is readily shown to be given by

$$T_D|_{\min} = \frac{L_I}{C_T} f(\rho_M) \tag{10–26}$$

with

$$f(\rho_M) = \left[\frac{4}{\left(\dfrac{1}{\rho_M} - 1\right)}\right]^2 \left[1 + \frac{1}{2}\left(\frac{1}{\rho_M} - 1\right) + \frac{\left(\dfrac{1}{\rho_M} - 1\right)^2}{8}\right] \tag{10–27}$$

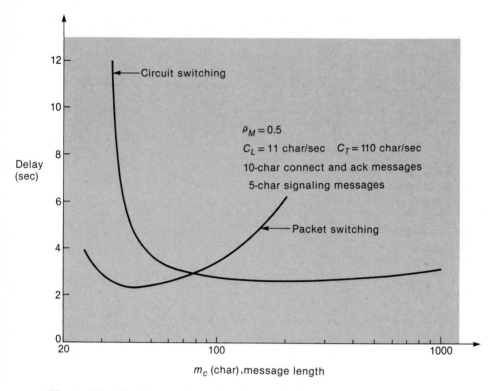

Figure 10-12 within the image shows axis labels: Delay (sec) on the vertical axis; and annotations:

$\rho_M = 0.5$
$C_L = 11$ char/sec $C_T = 110$ char/sec
10-char connect and ack messages
5-char signaling messages

Circuit switching
Packet switching

m_c (char), message length

Figure 10–12   Comparison of packet switching and circuit switching, Example 2

As $\rho_M \rightarrow 1, f(\rho_M) \rightarrow 16/(1 - \rho_M)^2$, indicating the extremely rapid deterioration of response time with utilization, as already noted.

As a check, take $\rho_M = 0.5$. Then $f(\rho_M) = 26$, and $T_D|_{\min} \doteq 26L_I/C_T$. For the case of Fig. 10–11 this gives 35 msec, clearly in agreement with the figure. For $\rho_M = 0.6, f(\rho_M) = 50$. The corresponding time delay for the example of Fig. 10–11 is 67 msec, clearly in agreement with the figure. For $\rho_M = 0.8, f(\rho_M) = 290$, showing the rapid increase with ρ_M.

The corresponding message or packet length at which the minimum time delay occurs is found to be

$$m_c|_{\min T_D} \doteq \left(\frac{4\rho_M}{1 - \rho_M} \right) L_I \qquad (10-28)$$

Note that this minimum-delay packet length increases with ρ_M, as is apparent from Fig. 10–11. It is also directly proportional to the ack packet length. As an example, say that $\rho_M = 0.5$. Then this packet length for minimum time delay is $4L_I$. For the case of Fig. 10–11, with $L_I = 160$ bits, this gives 640 bits, in agreement with the figure. Similarly, for $\rho_M = 0.6$, Eq. (10–28) applied to Fig. 10–11 gives $6L_I = 960$ bits, again in agreement with the figure. For Example 2, we had $L_I = 10$ characters, and $C_T = 110$ char/sec. Then $T_D|_{\min}$ from Eq. (10–26) is $26\,L_I/C_T = 2.4$ sec for $\rho_M = 0.5$, in agreement with Fig. 10–12. The message or packet length at which this minimum time delay occurs is $4L_I = 40$ characters, again in agreement with Fig. 10–12.

Now assume that the minimum connect time for the circuit-switched case is just

$$T_C|_{\min} = (L_I + 3L_S)/C_L \qquad (10-29)$$

from Eq. (10–22). We thus ignore the wait time and consider only the time required to transmit connect and signaling messages. This is close to the results obtained in Figs. 10–11 and 10–12 but clearly provides an advantage to circuit switching. From Eqs. (10–26) and (10–29) we find that packet switching will produce the smaller equivalent time delay if

$$\frac{f(\rho_M)L_I}{N} < L_I + 3L_S$$

or

$$C_T/C_L = N > f(\rho_M) \Big/ \left(1 + \frac{3L_S}{L_I}\right) \qquad (10-30)$$

Thus aside from a dependence on the relative length of signaling and connect messages, higher-capacity trunks will favor packet switching, *provided* the utilization parameter ρ_M is not too great. This is, of course, due to the inverse dependence of T_D on the network transmission speed C_T (see Eq. (10–25)). The circuit-switched call set-up time T_C, on the other hand, depends inversely on the local loop speed C_L. Keeping C_L fixed and increasing C_T reduces the packet-switched time delay T_D accordingly.

In Table 10–1 we tabulate the results of applying Eq. (10–30) to several cases of different ratios of L_S/L_I and different utilization parameters ρ_M. Obviously as we increase the signaling-message length we penalize the circuit-switched system more and deteriorate its performance relative to the packet-switched scheme. Note that Fig. 10–12 agrees with the results of Table 10–1. In that example we assumed $L_S = 0.5L_I$ and had $N = 10$. The minimum time delays are about the same. Had we chosen a higher transmission speed, say $C_T = 220$ char/sec, the packet-switched curve would have shifted down by a

TABLE 10–1	Minimum Trunk/Loop Transmission Speeds for Which Packet-switched Time Delay is Less than Circuit-switched Connect Time		
	ρ_M	L_S/L_I	$N = C_T/C_L$
	0.5	0	26
		0.1	20
		0.5	10
		1	7
	0.8	0	290
		0.1	223
		0.5	116
		1	76

factor of two (see Eq. (10–25)), with the circuit-switched curve remaining the same. Similarly, in Fig. 10–11, doubling the trunk capacity to $C_T = 240$ kbps would reduce all packet-switched time delays to 50 percent of the values shown, providing still greater improvement. In that figure we had $L_S = 0.1L_I$ and $N = 50$. From Table 10–1, C_T/C_L should be greater than 223 for $\rho_m = 0.8$ to have $T_D|_{min} < T_C$ for this case. For $C_T/C_L = 50$ and $\rho_M = 0.8$, then, circuit switching is clearly considerably better than packet switching, as noted earlier.

Whether these results, obtained for a very simplified model of the circuit-switching process, hold under more realistic conditions is not clear. They do substantiate in more quantitative fashion our intuitive notion that longer data messages may be transmitted more efficiently using circuit-switching technology. They also enable us to assess more quantitatively the effect of other parameters such as local-loop and network-transmission rates, the number of trunks or channels on a network link, and the load on the network. In the following sections and the next two chapters we develop more realistic models of circuit-switching technology, focusing primarily on the blocking model, the one most often used currently in that field.

10–3 Elements of Traffic Engineering

In the previous sections we provided simple calculations of time delay in packet switching and call-setup time in circuit switching in order to develop a comparative measure of performance for these two types of message-handling schemes.

As noted several times, however, the queued-call model for circuit switching is rarely the valid one. In most current circuit-switching systems, calls are blocked if resources (generally trunks or communications channels) are not available to support them. With modern computer-controlled switches, calls can be queued as well. The most sophisticated systems may use a combination of blocking and queueing. In this introductory treatment, however, we focus principally on the classical blocking mode of operation of circuit-switching systems. To lay the mathematical groundwork and to discuss the blocking process in switching systems more concretely, we devote this section to a brief overview of aspects of traffic theory particularly pertinent to classical circuit-switched performance evaluation. We rely on and extend our introductory treatment of queueing in Chapter 2. We do this in the context of a simple generic model for a switching system that, with the proper interpretation, can be made to provide the outgoing-link blocking performance of any of a number of switching systems. Much of the material in this section is taken from a book by R. Syski, one of the classical references on studies of traffic in telephone systems. Readers who desire more information are referred to this book [SYSK].

In Chapter 2 we discussed briefly the $M/M/N/N$ queueing system — the $M/M/N$ queue with no waiting room — as one example of a birth-death process. Because it has no waiting room it is called a blocking system: If N calls are already in progress, any additional calls arriving are blocked. The blocking probability in this case is given by the Erlang-B probability derived in Chapter 2 (Eq. 2–55). We shall see shortly that this is a special case of a more general probability distribution called the *Engset* distribution. To obtain the Engset distribution we focus attention on the generic model noted earlier of a switch or an exchange in a circuit-switched network (see Fig. 10–13). Recall from the previous sections that by circuit switching we mean a private (dedicated) connection set up end to end between two users. In the context of Fig. 10–13 this implies seizing one of the N output trunks when a call arrives on one of the M input lines. If all N outputs are busy (already taken), however, the incoming call is blocked. In a pure blocking system of the type we will be discussing, the call is then assumed lost. (In real life the user who sets up the call would probably try again. We do not model that characteristic in this introductory treatment.) Obviously, for blocking to occur we must have $M > N$; i.e., the number of inputs is greater than the number of output lines or trunks. In this case concentration is being used: As in packet-switched networks, resource sharing is used to reduce the cost. One relies on the statistical nature of calls to ensure that the probability of blocking is below some tolerable threshold.

[SYSK] R. Syski, *Introduction to Congestion Theory in Telephone Systems,* Oliver and Boyd, Edinburgh, 1960.

Input   Output

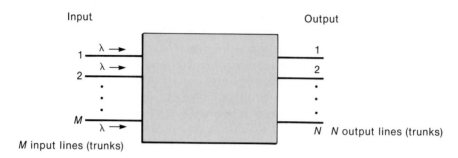

Figure 10–13 Generic model, circuit-switched exchange, single-output trunk group

The simple model of Fig. 10–13 is appropriate to a variety of switches used in practice, as already noted. For example, the M inputs could all be trunks coming from another office or exchange; the N outputs are similarly directed to another exchange in the network. A circuit-switched office of this type, with relatively little locally generated traffic, is called a *tandem* office, switch, or exchange. If many of the input lines are connected to local customers or users, the term *local* office or exchange is often used. A *private branch exchange,* or PBX, is a privately owned switch, generally of relatively small size, that is connected in turn, via output trunks, to the public network. The model of Fig. 10–13 shows one set of output trunks only. As in the case of packet-switched networks, circuit-switched offices serving as nodes in a network commonly have multiple output groups or links, each generally directed to a different office (although several groups may be connected to one office as well). Figure 10–14 provides an example of a network with each of these different switching-office types indicated. Also shown is a *remote concentrator,* a remote extension of a local switch that is often used to concentrate or multiplex remote users via one transmission facility to the local office to which it is directed. Note that such a concentrator performs the same function as a PBX, with the exception that it belongs to the public network rather than being privately owned.

In the case of the PBX the M_2 input sources shown (telephones or data sets or other digital sources) are concentrated into the N_2 trunks that connect the system to the local office. The remote concentrator similarly concentrates M_1 possible calls onto N_1 trunks. The generic model of Fig. 10–13 applied to local office A of Fig. 10–14 is somewhat more complex. In the case of calls routed to tandem office A, the N_3 trunks of Fig. 10–14 correspond to the N trunks of Fig. 10–13. The M potential inputs of Fig. 10–13 correspond to the composite of M calls from tandem B, the PBX, the remote concentrator, and the M_3 units

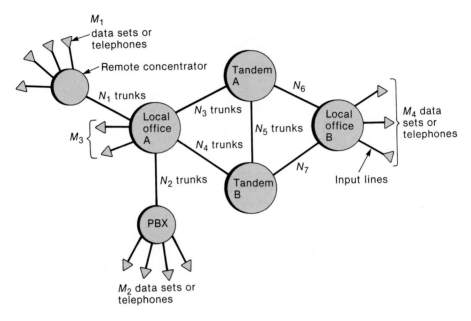

Figure 10–14 A circuit-switched network

connected directly to local office A, that are to be routed to tandem office A. In the case of a tandem switch such as tandem A in Fig. 10–14, the N output trunks of Fig. 10–13 correspond to any of the trunk groups or links (N_3, N_5, N_6) shown connected to tandem office A, while the M inputs correspond to those trunks from the other offices to which tandem A is connected that are to be switched or routed to that trunk group. All trunks in Fig. 10–14 are assumed to operate full-duplex—i.e., calls can be routed in both directions. Once a call is set up each of the two parties at either end is free to talk independently of the other. This is often called a "four-wire" connection. (In practice, separate time slots or frequencies, or physically separate transmission facilities, might be used to obtain the two-way capability needed.) Figure 10–13 thus models one of the *output* trunk groups of a switch and is used either to design the trunk group size (the number of trunks needed to accommodate the incoming calls expected) or to analyze the link or trunk-group blocking performance of the system once it is designed. This is similar to the queueing models, one for each outgoing line, that we used in analyzing packet-switched networks in earlier chapters.

In this section we study the blocking performance of the model of Fig. 10–13, given the traffic-arrival rate and circuit-holding time of the input lines or trunks shown. We ignore the processing and internal switching actually required at any one of the switches of Fig. 10–14 that directs an incoming call to the appropriate outgoing trunk group. In the next section we move into the switch itself, studying the methods developed for internally switching calls from one incoming trunk group, or from an incoming line, to an appropriate outgoing group. In the next chapter we discuss call processing within a switch. (Note again that the word "switch" here is sometimes used to refer to the entire switching office or exchange, or node in a network. It is sometimes used to refer to the actual switch within an exchange that directs calls from one trunk group or incoming line to another.)

Where then does the Erlang blocking model of Chapter 2 fit in? We shall show that this model is just the special case of Fig. 10–13 when the number of input lines M gets very large. It is the simplest model to use when studying the blocking performance at one of the outputs (trunk sides) of the switching offices (local or tandem) of Fig. 10–14.

Now consider the model of Fig. 10–13. Each of the M inputs (the sources) either is idle for an exponential length of average $1/\lambda$ or generates a call with exponential holding time of average length $1/\mu$. Each input as it arrives is assigned one of the outgoing trunks. Calls are turned away, or blocked, when all N trunks are occupied. The *composite* switch of Fig. 10–13 behaves statistically like a birth-death process, with arrival (birth) rate λ_n and departure rate μ_n when n calls are in progress. The state of the system of Fig. 10–13 is thus denoted by n. Specifically, it is apparent that the two state-variable rates are given, respectively, by

$$\lambda_n = (M - n)\lambda \quad \begin{cases} 0 \le n \le N \\ N \le M \end{cases} \qquad (10\text{–}31)$$

and

$$\mu_n = n\mu \quad 1 \le n \le N \qquad (10\text{–}32)$$

This birth-death process is shown pictured in Fig. 10–15. The state-probability balance equation, from Chapter 2 (Eq. 2–39), is again given by

$$\mu_{n+1}p_{n+1} = \lambda_n p_n \quad n \ge 0 \qquad (10\text{–}33)$$

with p_n the probability there are n calls in progress in the model of Fig. 10–13.

Solving Eq. (10–33) iteratively, as was done in Chapter 2, one obtains the following equation for p_n:

$$p_n/p_0 = \prod_{k=0}^{n-1} \lambda_k \bigg/ \prod_{k=1}^{n} \mu_k$$
$$= \left(\frac{\lambda}{\mu}\right)^n \binom{M}{n} \qquad (10\text{–}34)$$

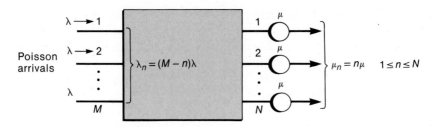

Figure 10–15 Birth-death model, Fig. 10–13, Engset distribution, n calls in progress

after inserting Eq. (10–31) and Eq. (10–32) for λ_k and μ_k, respectively, and simplifying the resultant expressions.

$$\binom{M}{n} = \frac{M!}{(M-n)!n!}$$

is the usual notation for the number of combinations of M objects taken n at a time.

Applying the probability normalization condition $\sum_{n=0}^{n} p_n = 1$ to Eq. (10–34) to determine the unknown probability p_0, one gets the *Engset distribution* for the probability p_n of the number of calls in process:

$$p_n = \left(\frac{\lambda}{\mu}\right)^n \binom{M}{n} \bigg/ \sum_{n=0}^{N} \left(\frac{\lambda}{\mu}\right)^n \binom{M}{n} \qquad 0 \le n \le N \qquad (10-35)$$

If we take as the probability of blocking P_B the probability that the system is fully occupied (recall again that the number of input lines M must then be greater than the number of output lines N), we get

$$P_B = p_N = \left(\frac{\lambda}{\mu}\right)^N \binom{M}{N} \bigg/ \sum_{n=0}^{N} \left(\frac{\lambda}{\mu}\right)^n \binom{M}{n} \qquad (10-36)$$

This is also called the *time congestion,* to distinguish it from another parameter called the *call congestion,* or the probability that a call is lost [SYSK, pp. 16, 199]. For birth-death processes where the arrival rate is dependent on the state of the system (Fig. 10–13 or its equivalent provide one example), these two measures of congestion are not necessarily the same, although the difference is usually small. When the arrival rate is state independent (the Erlang distribution, with Poisson arrival rates, is an example), the two measures are identical.

The two measures are distinguished in the following way: Consider a system with N resources (N trunks or channels in this case). Measure the fraction of time in a given observational interval that these resources are all busy, or utilized. For example, this might be the number of minutes (or seconds) in a given hour that all trunks are occupied. This fraction provides an estimate of the probability that all N resources are busy. In the context of the example under study here, we call this probability the *system blocking probability* P_B, or *time congestion*.

As the other possible measure of congestion, count the total number of calls arriving in a long interval of time and mark those that are lost due to a lack of system resources. (A call is lost if, on arrival, it finds all N outgoing trunks occupied.) The ratio of calls lost to the total number of calls offered in an observational interval provides an estimate of the *probability of loss* P_L, defined to be the *call congestion*. Thus, 100 calls lost out of 10,000 offered indicates $P_L = 0.01$.

To relate these two quantities we follow Syski's approach [SYSK, pp. 199, 220]. Let $p_N(a)$ be the conditional probability that a call arrives when the *system* is blocked (i.e., all N channels are occupied). Let $p(a)$ be the unconditional probability of arrival of a call. The probability of call arrival $p(a)$ multiplied by the probability P_L that a call, on arrival, finds the system blocked must equal the probability P_B that the system is blocked times the probability $p_N(a)$ that a call arrives when the system is blocked. We thus have the equality

$$P_L p(a) = P_B p_N(a) \tag{10-37}$$

or

$$P_L = \frac{p_N(a)}{p(a)} P_B \tag{10-38}$$

If the conditional probability $p_N(a)$ is independent of the state of system blocking — i.e., $p_N(a) = p(a)$ — the two measures of congestion are the same.

Take the birth-death process of Fig. 10-15 as an example. In this case the conditional probability of arrival $p_N(a)$ when the system is blocked must be proportional to $\lambda_N \Delta t$, with λ_N the call-arrival rate with the system in its blocking state $n = N$, as given by Eq. (10-31). Similarly, the unconditional probability of arrival $p(a)$ must be proportional to the quantity $\lambda_T \Delta t$, with λ_T the arrival rate averaged over all states:

$$\lambda_T = \sum_{n=0}^{N} \lambda_n p_n \tag{10-39}$$

From Eq. (10-38), then, for this example,

$$P_L = \frac{\lambda_N}{\lambda_T} P_B \leq P_B \tag{10-40}$$

since λ_N is less than λ_T as defined by Eq. (10-39). (Why is this so in this example?)

The blocking probability P_B is given by Eq. (10–36). Multiplying this expression by λ_N/λ_T, as obtained from Eqs. (10–31), (10–39), and (10–35), one finds, after considerable manipulation, that the time congestion P_L for the model of Fig. 10–15 is given by

$$P_L = \binom{M-1}{N}\left(\frac{\lambda}{\mu}\right)^N \Big/ \sum_{n=0}^{N} \binom{M-1}{n}\left(\frac{\lambda}{\mu}\right)^n \tag{10–41}$$

Note by comparison with Eq. (10–36) for P_B that these two expressions are very similar. In fact, they are related very simply by the identity

$$P_L(M) = P_B(M-1) \tag{10–42}$$

In particular, as $M \to \infty$ (i.e., $M \gg 1$), $P_L \to P_B$. This, as we show next, gives the Erlang-B blocking probability.

From Chapter 2 we recall again that the Erlang distribution, with the Erlang-B blocking probability for the blocking state obtained from this, was derived as the equilibrium-state probability solution to the M/M/N/N queueing system, or the M/M/N system with no waiting room. This was the system modeled by a birth-death process with birth rate $\lambda_n = \lambda$, independent of the state, and $\mu_n = n\mu$, $1 \le n \le N$. This is the special case of the model of Fig. 10–13 when the arrival rate λ_n is a constant λ, independent of state, corresponding to a Poisson arrival process. This model is portrayed schematically in Fig. 10–16. Intuitively, this model should arise when the number of inputs or sources M is much larger than the number of outputs N. For then, from Eq. (10–31), λ_n is very nearly the constant value $M\lambda$ no matter what the state n. The reason is obvious: With a very large number of inputs, the fact that a few (at most N) are already active does not change the overall probability of arrival appreciably.

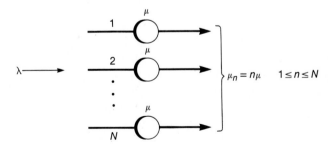

Figure 10–16 Birth-death model, Erlang distribution

To be precise, let $M \to \infty$ and $\lambda \to 0$ in the previous (Engset) model so that $M\lambda = \lambda'$, a constant. This gives rise to the Poisson arrival model. Using balance equations in this case, it is easy to show, as was done in Chapter 2, that for this case

$$p_n = A^n/n! \left/ \sum_{n=0}^{N} A^n/n! \right. \qquad A \equiv \lambda'/\mu \qquad (10-43)$$

Alternatively, one can derive this expression directly, though somewhat tediously, from Eq. (10–35). This is precisely the *Erlang* distribution obtained in Chapter 2 (Eq. 2–54). We have introduced the Erlang definition $A \equiv \lambda'/\mu$, as in the previous sections, because of its widespread use in telephony and general circuit-switched technology. In future work we drop the prime from λ' and use the symbol λ to represent the *total* (Poisson) arrival rate at the output link of a switching system such as that of Fig. 10–13. This is the notation shown in Fig. 10–16.

For this case, with Poisson, state-independent arrivals, the two measures of congestion, P_B and P_L, are identical, as already noted, or, as is apparent from either Eq. (10–40) ($\lambda_N = \lambda_T = \lambda$) or Eq. (10–42), with $M \to \infty$. The blocking probability, from Eq. (10–43), is just the Erlang-B formula discussed in Chapter 2.

$$P_B = p_N = A^N/N! \left/ \sum_{n=0}^{N} A^n/n! \right., \qquad A = \lambda/\mu \qquad (10-44)$$

This probability has been widely tabulated because of its widespread use. (See, for example, [COLL].)

Consider some examples of the applicability of Eq. (10–44). First say that $A = 12$ Erlangs. (This corresponds, as an example, to a very small telephone system with a call-arrival rate of $\lambda = 4$ calls/minute and an average holding time $1/\mu = 3$ minutes. Alternatively, this might correspond to a data circuit-switching system with $1/\mu = 10$ sec the average length of a data message and $\lambda = 1.2$ data messages/sec to be transmitted over one of the outgoing trunks.) How many trunks or channels are needed to accommodate this traffic load? The actual number depends on the blocking probability derived, but a little thought would indicate that as a rule of thumb, the number of trunks should be greater than the load in Erlangs. For recall that the utilization per trunk is A/N and that this number should generally be less than 1 for noncongested conditions. This is borne out by the entries of Table 10–2 ([COLL, p. 96]). Note from Table 10–2 that as N drops below A numerically (or as A/N exceeds 1) the blocking probability rises rapidly.

TABLE 10-2	Blocking Probability versus Number of Outputs, $A = 12$ Erlangs	
$P_B(\%)$	N (Number of Outputs)	A/N
1	20	0.6
4	17	0.7
8	15	0.8
19	12	1
30	7	1.7

Some useful approximations that give N as a linear function of A are available for relatively small A, $5 < A < 50$. These appear in [HILL, p. 97] and are repeated in Eqs. (10–45a) and (10–45b).

$$P_B = 1\% \qquad N \doteq 5.5 + 1.17A \qquad 5 < A < 50 \qquad (10\text{–}45a)$$

$$P_B = 0.1\% \qquad N \doteq 7.8 + 1.28A \qquad 5 < A < 50 \qquad (10\text{–}45b)$$

Large switching systems are, of course, equipped to handle much larger loads. As an example, the No. 4 ESS, the largest digital electronic switching system in the AT&T family of switches, achieved a load rating of 550,000 call attempts/hour in its final design [BSTJ 1977, p. 1021]. At an average call holding time of 3 minutes, this provides a load of 25,000 Erlangs. The number of trunk terminations used is 107,520. Dividing by two since two trunks (or two time slots) must be allocated to each call, this number is still more than twice the load in Erlangs. If the call-holding time is taken as 4 minutes, the Erlang load becomes 33,000 Erlangs.

There is some possible confusion, however, with respect to the Erlang rating. The term Erlang is also used to indicate the probability that a line or trunk is busy or occupied. Multiplying by the number of lines or trunks, one gets another Erlang measure by which switches are rated. For example, the No. 4 ESS is rated at 47,000 Erlangs, in addition to its design load of 550,000 call attempts/hour. This second rating divided by the number of trunks provides a measure of the probability that a trunk is busy. Both Erlang measures are used, leading to possible confusion. We shall see shortly that they are closely related. In this book, where possible, we shall use the term probability of line occupancy

[HILL] M. J. Hills, *Telecommunications Switching Principles*, MIT Press, Cambridge and London, 1979.

[BSTJ 1977] "No. 4 ESS," special issue, *BSTJ*, vol. 56, no. 7, Sept. 1977.

instead of Erlang and restrict our usage of Erlang where possible to our original definition — i.e., offered load, in calls/time, multiplied by the call holding time. (We shall have to deviate from this rule when describing real systems such as the No. 4 ESS, since these switches are rated in Erlang load in the second sense.)

The unit CCS is sometimes used in place of Erlangs to represent network load. A CCS is, by definition, 100 call-sec/hour. One call attempt/hour, with a 100-sec holding time per call, thus corresponds to 1 CCS. A simple calculation indicates that 1 Erlang corresponds to 36 CCS. Twelve Erlangs thus represent 432 CCS.

Returning to the Engset distribution, Eq. (10–35), another special case corresponds to the one for which the number of inputs M in Fig. 10–13 is equal to the number of outputs N. There is obviously no blocking in this case, since each call arriving finds a trunk available. From Eq. (10–35) the probability that n calls are in process is given by

$$p_n = \binom{N}{n}\left(\frac{\lambda}{\mu}\right)^n \bigg/ \sum_{n=0}^{N} \binom{N}{n}\left(\frac{\lambda}{\mu}\right)^n \qquad (10-46)$$

The denominator of Eq. (10–46) is just the series expansion of $\left(1 + \dfrac{\lambda}{\mu}\right)^N$ (the binomial expansion). Replacing the denominator with this closed-form expression and then simplifying further by letting the expression $\lambda/(\lambda + \mu)$ be represented by the parameter a, one finds easily that the probability of state p_n may be written in the alternative form

$$p_n = \binom{N}{n} a^n (1-a)^{N-n} \qquad a \equiv \lambda/(\lambda + \mu) \qquad (10-47)$$

This is recognizable as the *binomial distribution.*

The binomial distribution that arises in this special case of the switch model of Fig. 10–13 when $N = M$ is readily interpreted physically. Consider one of the N outputs (trunk or channel). It is either busy (occupied) or idle. If there are N sources and N outputs, as in this case, the sources may be looked at as having dedicated channels. A channel is then a two-state birth-death system, moving between an idle and a busy state. This is portrayed in Fig. 10–17. With a Poisson rate λ the channel goes from idle to busy (the call on the input corresponding to this channel arrives). With a Poisson rate μ the channel goes from busy to idle (a call is completed). For this isolated two-state system, the probability that the system is busy is just $\lambda/(\lambda + \mu) = a$; the probability that the system is idle is $1 - a = \mu/(\lambda + \mu)$. For N such binary (Bernoulli) systems, the probability that n channels are busy is then just the binomial probability $p_n = \binom{N}{n} a^n (1-a)^{N-n}$, precisely the form of Eq. (10–47).

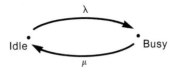

Figure 10-17 Two-state representation of one of the output ch. ·ls, Fig. 10-13, $N = M$

The parameter $a \equiv \lambda/(\lambda + \mu) = \rho/(1 + \rho)$, $\rho = \lambda/\mu$, the probability that a channel is busy, is just the quantity mentioned previously in discussing the second Erlang measure. Thus, a switch with a probability $a = 0.7$ that a line (or trunk or channel) is busy is commonly referred to as having a 0.7 Erlang rating. For 10,000 lines, the second load measure described earlier would then be 7000 Erlangs. This can be confusing, as noted. The last relation discussed provides a simple connection between the two definitions of Erlang. The parameter $\rho = \lambda/\mu$ is precisely the Erlang measure, in Erlangs/line, that we have consistently used thus far; we shall continue to use it here. The parameter a, to which the second definition of Erlang is applied, is $\rho/(1 + \rho)$, as noted. $a = 0.7$ thus corresponds to $\rho = a/(1 - a) = 2.3$ Erlangs/line in our original definition. One thus goes from one definition to the other using these simple relations.

It is apparent that the generic model of Fig. 10-13 can represent an interesting set of cases that arise frequently in circuit-switching systems. This was noted earlier. When one has $M > N$, the concentration case, the Engset distribution of Eq. (10-35) provides the probabilities of state from which various statistics of interest may be found. If $N = M$ one gets, as a special case, the binomial distribution Eq. (10-47). At the other extreme, if $M \gg N$, or, the alternative, calls arrive at the outputs of Fig. 10-13 at a Poisson rate, as shown in Fig. 10-16, one gets the Erlang distribution of Eq. (10-43). These distributions will be encountered later when we study the performance of circuit-switching systems.

The generic model of Fig. 10-13 is useful in modeling a variety of circumstances that occur in circuit-switching systems, as already noted. It can be used to model the concentration process directly. In a switching system with incoming calls (either locally generated calls or tandem calls—those arriving from another node or office in the network) switched to an appropriate set of output trunks, the probability of state at any one of the output trunk groups is given by the Engset distribution (or one of its derivatives) if blocking in the internal switch of the exchange may be neglected, and calls arriving at the output trunk group retain their Poisson character. This too was noted earlier.

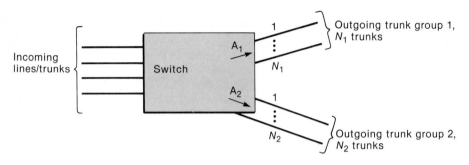

Figure 10–18 Simple model of a switching exchange

An example appears in Fig. 10–18. Incoming lines and/or trunks are shown at the left. Calls from these inputs are directed by the switch to the two outgoing trunk groups or links shown. If the effective Erlang load at each output is A_1 and A_2, respectively, as shown, with call arrivals at each modeled as a Poisson stream, the Erlang distribution of Eq. (10–43) applied to each link or trunk group provides the appropriate statistics at each point. In particular, Eq. (10–44) provides the blocking probability at each link. Alternatively, as already pointed out, if a design blocking probability is specified at each output link, and the respective loads A_1 and A_2 are known, the size of the trunk groups N_1 and N_2 can be ascertained. These link probabilities may in turn be used to calculate end-to-end blocking probabilities over a complete path in a network such as that of Fig. 10–14.

A more realistic situation, not considered here, may have a number of lines handling *both* incoming and outgoing calls, with some incoming calls directed right back onto the same trunk group or set of lines. (An obvious example occurs in a local exchange with a connection to be made between two end users both accessing that exchange.) Other, more complex cases for a variety of call-handling situations may be distinguished as well. The traffic analysis for a variety of access modes is carried out in a paper by Nesenbergs and Linfield [NESE].

10–4 Digital Switching Networks

We have alluded a number of times to the switching of calls at a given office or exchange from some input trunk group or set of lines to a trunk at a prescribed output group. Examples appear in Fig. 10–14. The same process of directing

[NESE] M. Nesenbergs and R. F. Linfield, "Three Typical Blocking Aspects of Access Area Teletraffic," *IEEE Trans. on Comm.*, vol. COM-28, no. 9, Sept. 1980, 1662–1667.

incoming messages that arrive on a given link to the appropriate outgoing link was encountered in the packet-switched environment discussed in previous chapters. The process was described in detail in Chapter 6 in our description of routing in networks.

No specific switch is required in a packet-switched node since all packets are *queued* for outbound transmission. In the case of circuit switching a dedicated connection between input and output lines or trunks at an exchange must be made, however, and a switch is necessary. Historically, so-called space-division switches have been generally used. More commonly nowadays, time-division switches are being implemented. In large switches both time and space switching are used. We shall discuss both techniques in this section and show how they are analogous to one another. We begin by focusing on the process of space-division switching, then show how the same process can be carried out through time-division techniques. The words switch and switching network are used interchangeably to describe the device that carries out the switching.

Consider first the case of switching any one of N inputs to one of N outputs. An example, for $N = 6$, appears in Fig. 10–19. The switch in this case corresponds to an $N \times N$ (square) array. Semiconductor switches or metallic contacts located at each of the crosspoints where input and output wires cross may be enabled, allowing any one of the input wires to be uniquely connected to any one of the outputs and thus establishing the connection. For this square array a connection between input and output is *always* possible (providing the output to which a connection is to be made is not already connected, i.e., busy). A switch of this type is called a *nonblocking* switch. A measure of its complexity is the number of crosspoints needed, generally N^2, or $N^2 - N$, if the inputs and

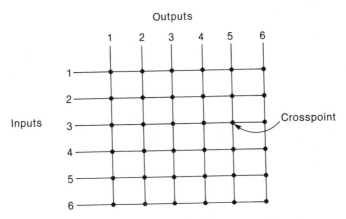

Figure 10–19 Square switching array, $N = 6$

outputs refer to the same set of terminals to be connected together. (In this latter case, the terminal connected to input line i is also connected to output line i, $1 \leq i \leq N$. A terminal can then both generate and receive a call.)

More generally, an $N \times K$ matrix may be used. (The notation used here is different from that of Fig. 10–13, but there should be no confusion since the calculation of blocking probability due to lack of an *output trunk*, as carried out in the previous section with reference to Fig. 10–13, will be consistently decoupled from the blocking properties of the switch, under consideration here.) It is apparent that if $K \geq N$ the switch is again nonblocking. For $K < N$, however, blocking by the switch is possible. An example, for $N = 8$ and $K = 4$, appears in Fig. 10–20. Four connections—1 to 2, 2 to 1, 3 to 3, and 4 to 4—are shown established. (Note that this implies that the outputs are different from the inputs.) Input lines 5–8 are thus blocked by the switch: Connections cannot be made to any of the output lines.

As the number of users or lines connected increases, the size and complexity of the switch increase correspondingly. As just noted, the complexity of a space switch is generally measured by the number of crosspoints required. For example, if 100,000 trunks are to be interconnected (this is roughly the number of trunk terminations in the AT&T No. 4 ESS tandem switch [BSTJ 1977]) and a square array is used, $N^2 = 10^{10}$ crosspoints are required, obviously a tremendously large number, even in this era of VLSI. Can one reduce this number and

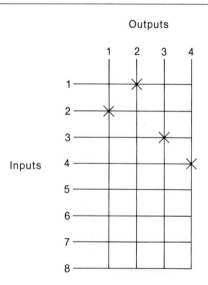

Figure 10–20 Switching array, 8×4

yet retain either a nonblocking switch or one with a very small probability of switch blocking? The common way is to go to multiple switching stages. Consider a three-stage space-division example. (Time-division switches will be considered shortly.)

Let the number of input lines or inlets N equal the number of outputs for simplicity. Group the N inlets (outlets) into N/n switch arrays, each such array comprising an $n \times k$ rectangular matrix. Figure $10-21$(a) shows an $n \times k$ switch

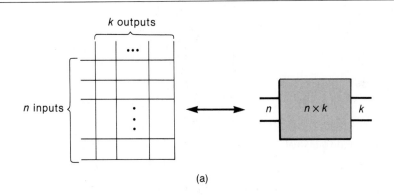

(a)

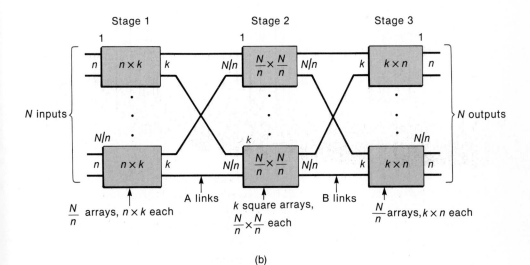

(b)

Figure 10–21 A three-stage switching network
   a. Typical $n \times k$ switch and its symbolic representation
   b. Three-stage switch

array and the simplified symbolic representation for it. Using this symbolic representation, Fig. 10–21(b) portrays the complete three-stage switch. There are N/n input and N/n output arrays, comprising stages 1 and 3, respectively. The second stage is made up of k square arrays, as shown, each with N/n inputs and outputs. Each of the output lines of a stage-1 array is connected to a different stage-2 array, in the same position as its array location in stage 1. Thus each of the output lines of the first array in stage 1 is connected to input line 1 of its corresponding array in stage 2, each output line of the second array in stage 1 is connected to its corresponding input line 2 in stage 2, and so forth. Similar connections are made between stage-2 and stage-3 arrays.

The total number of crosspoints in this case is

$$
\begin{aligned}
C &= 2 \left(\frac{N}{n} \right) (nk) + k \left(\frac{N}{n} \right)^2 \\
&= 2Nk + k \left(\frac{N}{n} \right)^2
\end{aligned}
\qquad (10-48)
$$

Through proper choice of the parameters n and k the complexity of the switch can be reduced to a value considerably less than N^2, the crosspoint count of the single-stage square array.

A multiple-stage switch of this type (with extensions to four, five, or more stages) may exhibit nonblocking or blocking behavior, as in the single-stage case. We first consider the nonblocking case and show that the switch complexity (as measured by the number of crosspoints) can be reduced below the single-stage case. By allowing some blocking, however, even more reduction in size (and hence cost) is obtained. Most large modern switches operate as blocking networks, but with very small blocking probabilities. They are thus often called *"essentially nonblocking"* switches. (The switch blocking probability, as will be shown in the next section, is typically much less than the link-blocking probability calculated in the previous section. Circuit-switched blocking behavior may then, to a good approximation, be modeled by the link-blocking probability calculated by the Engset or Erlang formulas.)

C. Clos of Bell Laboratories, in a seminal paper published in the *Bell System Technical Journal* in 1953, derived the requirement for a three-stage nonblocking switch [CLOS]. This is given simply, in the notation of Fig. 10–21, by the condition $k = 2n - 1$. This relation is easily demonstrated as follows: Say that a call coming in at a particular stage-1 array is to be connected to an outlet in a given stage-3 array. Say $(n - 1)$ inputs in the same stage 1 array are already

[CLOS] C. Clos, "A study of Nonblocking Switching Networks," *BSTJ*, vol. 32, no. 2, March 1953, 406–424.

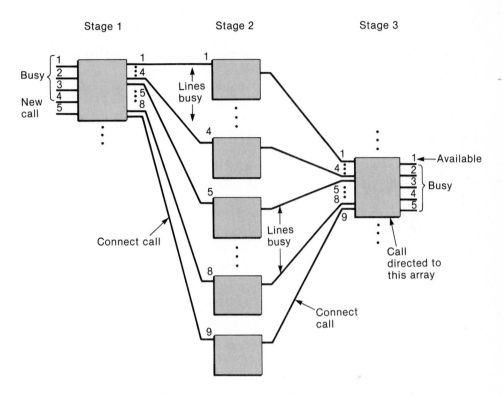

Figure 10-22 Example of a nonblocking network, $n = 5$, $k = 9$

connected (busy), as are $(n - 1)$ outputs in the desired stage-3 array. In the worst case let the $(n - 1)$ busy inputs and $(n - 1)$ busy outputs use different arrays in stage 2. To make a connection the desired call must be routed through an *unused* array in stage 2. Otherwise blocking will occur. (Recall, from Fig. 10-21, that each of the k lines connected from an array in stages 1 and 3 to stage 2 uses a different array in stage 2.) For an array in stage 2 to be available, then, we must have

$$k = (n - 1) + (n - 1) + 1 = 2n - 1 \qquad (10-49)$$

This condition is thus sufficient to guarantee an available path through the switch.

An example, with $n = 5$ and $k = 9$, appears in Fig. 10-22. A call arrives on input line 5 of the array shown in stage 1. All other inputs to this array are busy at this time. They are assumed to occupy lines 1-4 at the output of the array, as shown. (Any other four output lines could be used just as well.) This call is to be directed to the array indicated in stage 3. Lines 2–5 at the output of this array are shown as busy at the time. They must therefore be connected to four of the arrays in stage 2. These are shown as arrays 5–8 in Fig. 10–22. (In this worst-case analysis, with arrays 1–4 of stage 2 connected to the array of stage 1, any four of the five remaining arrays in stage 2 could have been chosen.) It makes no difference to which arrays in stage 1 these calls are connected. The fact that these calls are in progress eliminates the use of their lines for the new call to be set up. It is apparent that the new call can still be connected through to the output, however, using the remaining stage 2 array, number 9 as shown. Had there been only eight stage-2 arrays ($k = 2n - 2$), this would not have been possible.

For the nonblocking rule of Eq. (10–49), the number of crosspoints from Eq. (10–48) becomes

$$C|_{\text{nonblocking}} = 2N(2n - 1) + (2n - 1)\left(\frac{N}{n}\right)^2 \qquad (10\text{--}50)$$

This function exhibits a minimum with respect to n, the number of inputs in each of the stage-1 arrays. Specifically, it is readily shown that for $N \gg 1$, the optimum n is approximately

$$n_{\text{opt.}} \doteq \sqrt{N/2} \qquad (10\text{--}51)$$

with the corresponding number of crosspoints given by

$$C|_{\text{nonblocking opt.}} \doteq 4\sqrt{2}\, N^{3/2} \qquad N \gg 1 \qquad (10\text{--}52)$$

This is obviously an improvement over the N^2 behavior noted earlier. For example, take $N = 100,000$. Then the number of crosspoints required for the nonblocking three-stage switch is about 1.7×10^8, a considerable reduction from $N^2 = 10^{10}$, the number required for the square single-stage switch. For $N = 10,000$, Eq. (10–52) gives 5.6×10^6, compared with 10^8 for the single-stage case.

There is one problem, however, with the optimum value of n of Eq. (10–51). It does not guarantee that N/n is an integer, an obvious requirement from Fig. 10–21. As an example, for $N = 100,000$, $n = 222$ from Eq. (10–51). But N/n is not an integer in this case. The actual choice of n turns out to be noncritical, however. Thus, for this same example, if $n = 200$, $N/n = 500$, $k = 2n - 1 = 399$, and $C = 1.8 \times 10^8$, not much of a change from the minimum value 1.7×10^8. Similarly, for $n = 250$, $N/n = 400$, and $C = 1.8 \times 10^8$ also.

For a system this size ($N = 10^5$), even $C = 10^8$ (and the equivalent switch cost and complexity for time-division systems) is too large a number. The numbers can be reduced by allowing some blocking, as noted earlier. For this purpose, however, one must carry out a probabilistic analysis of the switch to determine the switch-blocking probability as a function of the switch parameters. This we shall do after we study time-division switch structures.

First note, however, that the nonblocking condition for the three-stage switch is readily extended to larger numbers of cascaded (tandem) stages by using building blocks of three-stage switches. For example, let each of the square $\frac{N}{n} \times \frac{N}{n}$ arrays in stage 2 of the three-stage switch of Fig. 10-21 actually be a three-stage nonblocking switch itself. The resultant switching network is a five-stage one and obviously has the same nonblocking property [INOS, p. 94]. Further extensions may be made as well.

10-4-1 Time-division Switching

Our introductory discussion of switching networks has focused thus far on space-division switching. These switches are equally applicable to analog and digital message transmission. We now turn our attention to the more modern time-division switches, which are applicable to digital switching only. These switches are completely analogous to the space-division ones, with nonblocking and blocking analyses carried out precisely the same way. As already noted, large modern switching systems most frequently consist of cascaded (tandem) combinations of the two types of switches.

A number of recent books describe digital switches, with particular emphasis on time-division switches, in much more detail than we do here. Reference will be made to these books from time to time. They include the works by H. Inose [INOS] and Collins and Pedersen [COLL] already cited, a textbook by John C. Bellamy [BELL], and a book edited by John C. McDonald [McDO]. This last work is a particularly up-to-date compendium of tutorial chapters on various aspects of digital switching. It includes discussions of hardware and software issues in detail, as well as descriptions of a number of practical switching systems.

To carry out time switching, all calls or messages to be switched must first be slotted into recurrent time samples, with a group of successive samples appearing on one physical line constituting a frame. The most common example in telephony involves the pulse-code modulation (PCM) transmission of voice sam-

[INOS] H. Inose, *An Introduction to Digital Integrated Communication Systems,* University of Tokyo Press, Tokyo, 1979.

[BELL] John C. Bellamy, *Digital Telephony,* John Wiley & Sons, New York, 1982.

[McDO] John C. McDonald, ed., *Fundamentals of Digital Switching,* Plenum Press, New York, 1983.

ples [SCHW 1980a, Chap. 3]. Each voice signal to be converted to digital format is sampled at a rate of 8000 samples/sec, or once every 125 μsec. Frames are thus 125-μsec long. These samples are in turn normally digitized (quantized) to 8 bits each, so that a typical PCM voice channel requires 64 kbps of transmission capacity. Individual voice channels are in turn combined or multiplexed into one time stream. A recent CCITT Recommendation replaces PCM transmission with adaptive delta modulation at 32 kbps, almost doubling the number of voice channels that can be transmitted.

Two multiplexing formats are used worldwide. The North American standard, developed originally by AT&T in the United States and adopted by Canada and Japan as well, combines 24 8-bit voice channels into one time stream operating at 1.544 Mbps. (One framing bit is added at the beginning of each frame. The aggregate bit rate is thus $(24 \times 8 + 1) = 193$ bits per 125-μsec frame.) This is called the T1 system. A typical T1 frame appears in Fig. 10–23 [SCHW 1980a, pp. 138–140]. Each 8-bit word in a channel is also referred to as a time slot in the time-multiplexed signal stream. Although the T1 format was originally developed for telephony, with voice transmission as its primary objective, the 24-channel frame can carry any other kind of signal as well. Fifty-six-kbps data signals fall naturally, into this format, for example. (The eighth or least significant bit in each T1 voice channel is used every six frames for signaling purposes. In the case of 56-kbps data, then, one would simply ignore the eighth bit.) Details of the interfacing of T1 trunks with digital switching systems appear in [McDO]. The international standard, used by countries outside North America and Japan, multiplexes or combines 30 64-kbps voice channels plus two signaling channels for an aggregate bit rate of 2.048 Mbps. This system, too, can also be used for data (nonvoice) transmission.

Higher multiplexing rates are possible as well. For example, the AT&T No. 4 ESS combines five T1 streams for an aggregate of 120 8-bit channels. Eight additional time slots are added, so that a frame to be switched consists of 128 time slots [BSTJ 1977, p. 1024]. See [McDO] for details of both digital and analog interfaces with digital switching systems.

Assuming that message or signal segments each occupy a time slot in a frame, how is time switching carried out? The simplest procedure is the obvious one. Each frame, as it arrives over an incoming port at the switch, is written into a memory. Switching is then accomplished by simply reading out the individual words in any desired (switched) order. Such a device is called a *time-slot interchanger* (TSI) [McDO, Chap. 5], [INOS, Chaps. 3 and 5]. An example appears in Fig. 10–24. The frame consists of five time slots or channels, of which only two,

[SCHW 1980a] M. Schwartz, *Information Transmission, Modulation, and Noise*, 3d ed., McGraw-Hill, New York, 1980.

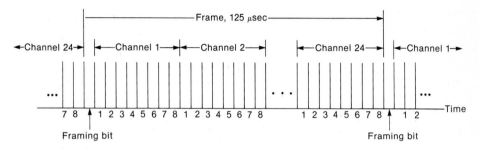

Figure 10–23 Frame format, T1 PCM system

X and Y, are assumed to be active and communicating with one another. On the input side, user X's data occupies channel 1; user Y's data occupies channel 3. After storing each frame in memory, Y's word is read out in, or transferred to, X's time slot; X's word is read out in Y's time slot. The dashed lines in Fig. 10–24 indicate that there is a minimum of a single-frame delay introduced in interchanging the data words. More complex operations are possible as well.

Just as space switches have a crosspoint cost limit, time-slot interchangers have limitations on the number of channels per frame that may be multiplexed.

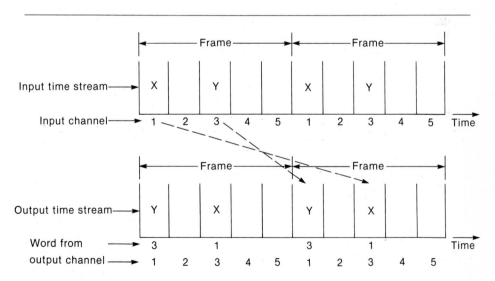

Figure 10–24 Example of time slot interchanging or switching

Multiplexing more channels requires more memory and becomes more costly as the number of channels to be stored increases. A more critical limit is determined by the access time required to read into and out of memory [BELL, p. 246]. Let t_c be the memory cycle time in microseconds required to both write into and read out a channel word (sample) from memory. Take a 125-μsec frame time as the most common example. The maximum number of channels that can be supported is then [BELL], [TAWA]

$$N = 125/2t_c \qquad\qquad (10-53)$$

For 500-nsec logic, this says that the maximum number is 125. For 50-nsec logic, this increases to 1250, and so on. Recall that the AT&T No. 4 ESS uses 120-channel time interchangers, with 128 time slots, in its switching network [BSTJ 1977]. A switching system handling just this number of trunks (recall that the complete No. 4 ESS handles on the order of 100,000 trunks) would require $N^2 = 14,400$ crosspoints if a single-stage space switch were used. If a three-stage nonblocking space switch were used instead, $C \doteq 7400$ crosspoints would be needed. The advantage of using time-switched technology is thus apparent. Note that the memory required is $8N$ bits, with 8 bits assumed per word. For $N = 120-128$ channels the memory required is only on the order of 1 Kbit. (Additional memory is required to control the memory addressing. Since each address is $\log_2 N$-bits long, this additional memory requires $N \log_2 N$ bits. For $N = 128$ channels, this only adds another $7N$ or 840 bits.)

Since memory cycle time limits the number of channels that may be switched using time-slot interchangers, larger switching systems have incorporated tandem time and space switching in their switching networks. Examples are the AT&T No. 4 ESS [BSTJ 1977] and the NTT (Japanese) DTS-11 [TAWA].

Consider in particular a time-slot interchanger followed by a space switch, followed in turn by an output TSI. The combination is designated a T-S-T switching network. The operation of this composite switch is completely analogous to the three-stage space switch of Fig. 10–21. (That switching network may be labeled an S-S-S switch.) To see this, consider the problem of switching N channels (time slots). Instead of doing this with one very fast TSI, group these N channels into N/n TSIs, each containing n time slots (channels) at its input. Let the number of output channels in a frame time be $k > n$. (Note that we are using the same notation as in the all-space-division case to preserve the analogy.) A typical frame at the input and output of a TSI appears in Fig. 10–25. Connect

[TAWA] K. Tawara et al., "Speech Path System for DTS-11 Digital Toll Switching System," *Review of the Electrical Comm. Laboratories*, NTT, Japan, vol. 30, no. 5, Sept. 1982, 771.

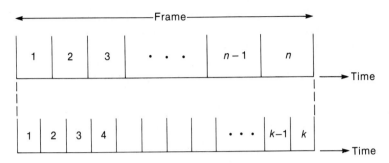

Figure 10–25 TSI frame with n and k ($k > n$) time slots at input and output, respectively

these N/n TSIs to a *single* $\dfrac{N}{n} \times \dfrac{N}{n}$ space switch as shown in Fig. 10–26. The N/n outputs of the space switch are in turn each connected to a time slot interchanger in a third T stage, with k time slots/frame at its input and n slots at its output.

  The crosspoint settings of the space switch are *changed* each of the k time intervals corresponding to the k time slots of the TSIs. The resultant space switch is termed a *time-multiplexed switch*. A simple analysis now indicates that if $k = 2n - 1$, one again has a *nonblocking* switching network, just as in the all-space network of Fig. 10–21. The T-S-T switch of Fig. 10–26 is thus completely analogous to the S-S-S switch of Fig. 10–21, as already noted. To demonstrate the nonblocking relationship, let $(n - 1)$ of the n input time slots in one

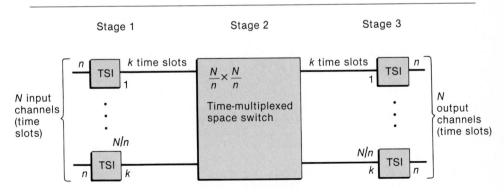

Figure 10–26 T-S-T switching network

of the input TSIs of Fig. 10–26 be busy. They in turn use $(n - 1)$ of the k different connections in time of the space switch. Let a call arrive on the remaining time slot of the same input TSI and let it be destined for one of the N/n output TSIs. Say that $(n - 1)$ output time slots of this output TSI are already occupied, or busy with calls. They also correspond to $(n - 1)$ of the k different connections in time of the space switch. In a worst-case analysis let the two sets of $(n - 1)$ space-switch crosspoint connections (those corresponding to the input TSI and those corresponding to the output TSI) be disjoint. These conditions are displayed in Fig. 10–27. In part (a) of the figure the k-slot time frame at the output of the stage-1 TSI, with $(n - 1)$ slots occupied (busy), is sketched schematically. Part (b) of Fig. 10–27 shows the k-slot time frame at the input of the stage-3 TSI, with its time axis aligned with that of the input (stage 1) TSI. $(n - 1)$ time slots are also assumed to be busy with calls, but are shown occupying a different or disjoint set than those of the input TSI. For a connection to be made between the two TSIs there must be at least one remaining free time slot in each of the two TSIs occurring at the same time. One such slot is shown appearing at the end of the frame. Since the space switch connecting the TSIs is square, any two TSIs can be connected through it. The two time slots must occur at the same time, however, as shown in Fig. 10–27. Under the conditions of Fig. 10–27, the new call can be connected, using the free slot occurring at the same time — the square switch allows the two TSIs to be connected. As previously then, the

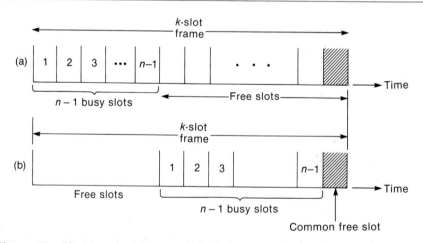

Figure 10–27 Nonblocking condition, time slot interchange, T-S-T switch
a. Input TSI, stage 1
b. Output TSI, stage 3

condition for this to happen, i.e., the call not to be blocked, is

$$k = (n - 1) + (n - 1) + 1 = 2n - 1$$

time slots, ensuring a nonblocking T-S-T switch.

The $k \dfrac{N}{n} \times \dfrac{N}{n}$ space arrays of Fig. 10–21 have thus been replaced by one $\dfrac{N}{n} \times \dfrac{N}{n}$ space array whose connections are independently changed or switched k times during a frame.

Consider some examples. First let $N = 960$ input channels be connected to 960 output channels. For a single-stage space switch this would require on the order of 10^6 crosspoints. In the T-S-T case, let $n = 120$ input time slots be used in the input TSIs. $N/n = 8$ input TSIs are thus required, with the same number at the output. For nonblocking operation $k = 2n - 1 = 239$ different connections of the space switch are required. (This might pose some problems because of the arbitrariness of the number! It might be better to use $k = 2n = 240$ if possible.) Since $N/n = 8$, an 8×8 space switch is called for. This is obviously a tremendous reduction in size over the single-stage 960×960 space switch. If a three-stage space switch were used, the number of crosspoints required would be $C \doteq 168,000$ crosspoints.

Consider another, much larger, example. Let $N = 96,000$ channels or trunks. Again let $n = 120$ time slots. Then $k = 239$ again to ensure nonblocking behavior. An 800×800 space switch is now needed, switching at the 239-slot rate. This T-S-T switch, combining 800 input and 800 output TSIs plus the 800×800 time-switched space switch, is completely comparable to a three-stage all-space (S-S-S) switch that requires on the order of 1.7×10^8 crosspoints.

Further reduction in switch complexity and cost is possible using an "essentially nonblocking" configuration. Many larger switching systems use this approach. The AT&T No. 4 ESS, a six-stage switch of the type T-S-S-S-S-T, is one example. In the next subsection we discuss multistage switches of the blocking type.

10–4–2 Blocking Probability Analysis of Multistage Switches: Lee Approximation

The determination of blocking probability in a multistage switch is inherently complex. Given a specific switching network, with the number of inputs and outputs as well as the probability of line (trunk) occupancy specified, the problem is to calculate the probability of not finding a free path (route) through the switch between a given input-output pair. This task is quite difficult: There are many possible paths to consider in a typical large switch, leading to combinatorial problems. More significantly, however, dependencies between blocking probabilities on different links along the path make the problem almost intractable. Approximations are thus used commonly for the calculations.

The simplest approximation, due to C. Y. Lee [LEE], assumes that the probabilities of finding individual links along a path busy are *independent*. If these probabilities are known, it then becomes a straightforward although possibly complex task to calculate the overall blocking probability. The problem is similar to that encountered in network reliability or survivability studies where, given the probability of failure of a link (equivalent to busy here), one calculates the overall reliability of a network [WILK]. Another example is that of calculating the end-to-end blocking probability in a circuit-switched network if the individual-link blocking probabilities are known. (See, for example, [LINP].) C. Y. Lee's approach in calculating the switch blocking probability is to use probability graphs in carrying out the calculations.

Consider a three-stage network as an example. As noted in our discussion of T-S-T nonblocking switch networks, there is a complete analogy between the T-S-T case and the all-space-division S-S-S network. The analogy carries over to the blocking switch case as well [COLL, pp. 22–39]. We thus need only consider one of the networks, either T-S-T or S-S-S. With appropriate identification of parameters in the final expression for blocking probability, one can use the same expression for both networks. The same analogy extends to other multistage configurations as well [COLL]. We focus here on the S-S-S case for simplicity. Figure 10–28 shows a portion of a three-stage space switch, indicating two typical channels that are to be connected together, one at the input and the other at the output. For a blocking switch we must, of course, have $k < 2n - 1$. In addition, let $k > n$. This ensures that the input and output arrays, or the TSIs in the T-S-T case, are inherently nonblocking. This would generally be the case for a tandem switch used to route calls from one set of incoming trunks to another set of outgoing trunks.

The case of $k < n$, or concentration, is precisely that discussed in the traffic analysis of the previous section. The input arrays themselves introduce blocking in this case. The overall blocking probability of the concentration of the input array *and* the arrays of the second (middle) stage becomes much more complicated to calculate. If the utilization per input channel is low, the blocking proba-

[LEE] C. Y. Lee, "Analysis of Switching Networks," *BSTJ*, vol. 34, no. 6, Nov. 1955, 1287–1315.

[WILK] R. S. Wilkov, "Analysis and Design of Reliable Computer Networks," *IEEE Trans. on Comm.*, vol. COM-20, no. 3, part II, June 1972, 660–678.

[LINP] P. M. Lin, B. J. Leon, C. R. Stewart, "Analysis of Circuit-switched Networks Employing Originating Office Control with Spill Forward," *IEEE Trans. on Comm.*, vol. COM-26, no. 6, June 1978, 754–765.

bility of the input array can be neglected, and the resultant system modeled the way we are proceeding. This procedure thus provides an approximate model for an end-office (local) switch or a PBX, with local users concentrated and switched over a relatively small number of trunks. Bellamy has used this model, including the assumption of low utilization per input, to provide some comparative results for end-office switches or PBXs with concentration [BELL, Table 5.2]. A more accurate approach for this case of concentration would follow the model of Nesenbergs and Linfield [NESE].

Returning now to the case of $k > n$ in Fig. 10–28, let the probability that a typical input channel is busy be a, just the binomial parameter $\lambda/(\lambda + \mu)$ introduced in the last section (see Eq. (10–47)) in discussing the probabilities of state of the structures of Figs. 10–13 and 10–15 for the case $M = N$. In saying that the parameter a is the same for each input channel, we are implicitly assuming an homogeneous system. This means then that the probability that a typical output channel in Fig. 10–28 is busy is a as well. We further assume that the incoming traffic is distributed uniformly over the k interstage links. The probability that an interstage link is busy is thus $p = an/k$.

Note that in the T-S-T case, the k interstage links correspond to k output slot times. It is apparent that in the homogeneous traffic case the traffic carried by the n input slots may also be taken as distributed uniformly over the k output slots, providing the same probability p that a TSI output slot (comparable to an interstage link in Fig. 10–28) is busy (occupied).

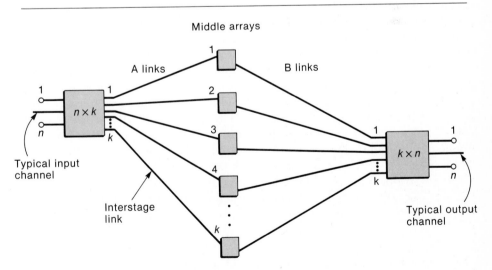

Figure 10–28 Calculation of blocking probability, three-stage switch

A complete path from input to output through one of the k nonblocking middle arrays in the S-S-S case of Fig. 10–28 traverses two links, as shown. Alternatively, in the T-S-T case, a connection from input TSI to output TSI requires matching time slots at the input and output of the middle space array. With the assumption of *independent* probabilities on each of the links in the switch network, the probability of blocking is the probability that no free path from input channel to output channel is available. (In the T-S-T case this corresponds to the probability that two time slots, one at an input TSI and the other at an output TSI, do not match up.) It is apparent that this probability is given by

$$P_B = [1 - (1 - p)^2]^k \qquad p = an/k \qquad (10-54)$$

Note that Eq. (10–54) does not equal zero for $k = 2n - 1$, the Clos nonblocking requirement, indicating the effect of the approximation.

As an example of the application of Eq. (10–54), let $n = 120$ and $k = 128$. These numbers appear in the No. 4 ESS system, although modified somewhat (see [BSTJ 1977] and our later discussion). Let $a = 0.7$ be the inlet channel utilization. Then the blocking probability for this case is

$$P_B = (0.882)^{128} \doteq 10^{-7}$$

For $a = 0.9$ the blocking probability rises to 0.042, a considerable change. This indicates the sensitivity of the result to input utilization.

What does one gain by going to this blocking configuration? In the S-S-S case it reduces the number of crosspoints needed, as noted earlier. By going to the T-S-T configuration one trades additional crosspoints for switching in time. As an example, consider the case of $N = 96,000$ total input lines used earlier. The optimum nonblocking three-stage space switch would require 1.7×10^8 crosspoints. The corresponding value of n, the number of input channels per input array, is then $n = 219$, clearly an impractical number. The choice of $n = 200$ with $k = 399$ would be much better, except that the optimum number of crosspoints is still much too large. For $n = 120$ and $k = 128$, the example just worked out, one gets $C = 1.06 \times 10^8$, a reduction by almost a factor of two but clearly still too high. Now consider the T-S-T version. With $n = 120$ and $k = 128$, the center space switch is of size 800×800, a clear reduction in crosspoints, at the added cost of requiring 800 TSIs at both the input and the output, with both the TSIs and the space switch operating at $k = 128$ slots or switch changes/frame time. Further reduction in the effective number of crosspoints or equivalently, the ability to operate at higher utilization/input channel, may be obtained by going to multiple space stages. As already noted, the No. 4 ESS switch is of the T-S-S-S-S-T type [BSTJ 1977].

The blocking probability of these more complex multistage switches may be calculated approximately by using the Lee graph approach. In particular, the analysis of the T-S-S-S-T switch using this approach is quite straightforward and is an extension of the three-stage analysis that leads to Eq. (10–54).

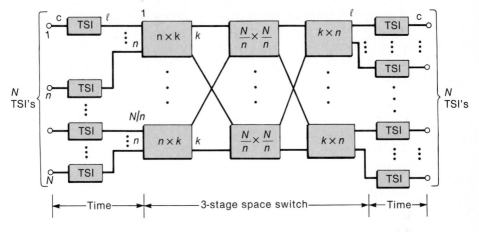

c = input time slots or channels
ℓ = output time slots or channels

Figure 10–29 A five-stage time-space switch (based on [COLL, Fig. 2.16])

Consider the T-S-S-S-T switch of Fig. 10–29 as an example [COLL, p. 36]. Note that the notation used in our previous T-S-T discussion has been changed to accommodate this larger system. (We adopt here the notation of Collins and Pedersen. See [COLL, p. 36].) Thus c represents the number of input time slots or channels in the stage-1 TSIs, with ℓ the corresponding number of output time slots. Alternatively, each input port handles c trunks (channels). There are a total of N TSIs in the first, as well as in the last, stages of the switch; the total number of trunks at input or output is thus $J = Nc$.

It is left to the reader to show, using the Lee method of analysis, that the approximate expression for blocking probability is given by

$$P_B = \{1 - q_1^2\,[1 - (1 - q_2^2)^k]\}^\ell \qquad (10\text{–}55)$$

where $q_1 = 1 - a/\alpha$, $q_2 = 1 - a/\alpha\beta$, and $\alpha \equiv \ell/c$ is the time stage expansion ratio (comparable to k/n in the T-S-T case), $\beta = k/n$ is the space stage expansion ratio, and a is again the probability that an input channel is busy.

Collins and Pedersen have carried out a number of typical design examples for this switch architecture as well as some others [COLL, Chap. 3]. In particular, consider a large switch example where the switch has a total of $J = Nc = 116{,}736$ input channels or trunks. [COLL, design (3), p. 50]. $N = 256$ ports are available, with $c = 456$ input time slots (channels)/port, and $\ell = 512$ time slots at the output of each TSI. The first-stage space switches are constrained to be 16×16 switches (i.e., $n = k = 16$). It is then left to the reader to show that

$C = 12,288$ crosspoints are required. For an input channel (time-slot) loading of $a = 0.9$, the (approximate) blocking probability of the switch is readily calculated to be $P_B = 6.4 \times 10^{-5}$. Since the time slots are each 8 bits long, recurring every $125\,\mu sec$, the TSI access time is 122 nsec. The space switch, which changes its configuration every slot interval, must switch within less than one bit interval, or within 30.5 nsec. It is also left as a problem to the reader to compare this design with one that uses the T-S-T configuration.

10–4–3 Improved Approximate Analysis of Blocking Switch

The problem with the Lee approximation is that in some cases it may not be very accurate. A more accurate, albeit more complex, approximation of the blocking probability of a multistage switch was provided in a classic paper in 1950 by C. Jacobaeus of Ericsson in Sweden. A calculation for a three-stage switch using this approach appears in [COLL, pp. 17–22]. This calculation focuses on the *time-congestion* analysis of the switch.

In the material following we derive a similar result for the *call congestion* (probability of loss). The analysis is due to M. Karnaugh [KARN]. Although portions of Karnaugh's derivation are similar to the Jacobaeus approach, the overall method of analysis is somewhat simpler conceptually. The resultant expression for the probability of loss turns out to be very close to the one based on Jacobaeus's work, as derived in [COLL]. This is to be expected, as noted in the last section when we discussed the distinction between time and call congestion.

In the derivation by Karnaugh the most general type of asymmetrical three-stage switch is considered, with each stage being allowed to be dissimilar and with the number of inputs and outputs differing as well. In the analysis considered here we specialize to the case of the symmetrical three-stage switch of Fig. 10–21 (or its T-S-T equivalent, as shown in Fig. 10–26).

Following Karnaugh, we begin by defining the probability of loss P_L, the call congestion, as previously: the fraction of offered calls over a long interval of time that is blocked and hence lost. Under equilibrium conditions, with the call-arrival rate independent of time, P_L must also be given by the ratio of the mean rate at which calls are lost to the rate at which they are offered. But as was the case with the Engset distribution, discussed in the last section, the call-arrival rate is a function of the state of the switch — i.e., the number of calls in progress in various parts of the switch. For a given state we can determine the call-arrival

[KARN] M. Karnaugh, "Loss of Point-to-Point Traffic in Three-Stage Circuit Switches," *IBM J. Research and Development*, vol. 18, no. 3, May 1974, 204–216; addendum, Sept. 1974, 465.

rate and then average over the states of the system. Specifically, let σ represent the state of the switch. Let $\lambda(\sigma)$ be the mean call-arrival rate at state σ, $P(L|\sigma)$ the conditional loss probability at this state, and $P(\sigma)$ the probability of being in state σ. Using these definitions, and summing over all the states, it is apparent that the desired loss probability is given by

$$P_L = \sum_\sigma P(\sigma)\lambda(\sigma)P(L|\sigma)/\sum_\sigma P(\sigma)\lambda(\sigma) \qquad (10-56)$$

The problem is now to determine the parameters of Eq. (10–56). To do this we must first determine the way in which a route or path is set up between an input channel and an output channel. In the case of the all-space switch of Fig. 10–21 this corresponds to finding a middle square array with input and output links going to the required input (stage 1) and output arrays (stage 3), respectively, that are both free. In the case of the T-S-T switch of Fig. 10–26 this corresponds to searching for two time slots, one at the output of the incoming TSI in question and the other at the input to the desired outgoing TSI, that are both free at the same time.

Some type of route- or path-finding algorithm must be used to search for a free path between the given input-output channel pair that are to be connected together. Since the path must go through one of the k middle switch arrays of Fig. 10–21 (or, equivalently, use one of the k time slots of Fig. 10–26), there are k possible paths that must be searched in this three-stage example. To simplify the analysis we assume that a *random route-hunting* algorithm is used. (This contrasts with a fixed or sequential path-search algorithm.) Then it is apparent that the average calling rate between any input-output channel pair in the switch must be the same—i.e., homogeneous conditions prevail throughout the switch. In particular, the loss probability must be identical between any input-output pair, and it suffices to calculate this probability for *any* such pair.

Consider any one of the input arrays and any one of the output arrays of Fig. 10–21. (Alternatively, focus on one of the input and one of the output TSIs of Fig. 10–26 in the T-S-T case.) Let $0 \le x \le n$ and $0 \le y \le n$ be the number of calls in progress in the input and output arrays chosen, respectively. Then the loss probability is given equally well by the expression

$$P_L = \sum_{x,y} P(x,y)\lambda(x,y)P(L|x,y)/\sum_{x,y} P(x,y)\lambda(x,y) \qquad (10-57)$$

with the pair (x,y) replacing the much larger state space σ.

In keeping with the random route-hunting assumption we now make the following simplifying assumptions:

1. x and y are independent—i.e., the number of calls in progress in the input and output arrays chosen are independent of one another. This assumption is found to be one of the major sources of error in this approach [KARN].

For large switch networks, however, one would expect the effect of this assumption to be small. With this assumption,

$$P(x,y) = P(x)P(y) \qquad (10-58)$$

2. To find $P(L|x,y)$, the conditional loss probability given x and y, one uses the "Jacobaeus assumption." This is described as follows. Call the k interstage links that connect the input (stage-1) arrays with the middle (stage-2) arrays the "A links." Call the k links that connect stage 2 with the output arrays the "B links." These are so indicated in Figs. 10–21 and 10–28. With x calls in progress at the input array in question and y calls in progress at the corresponding output array, there are $(k - x)$ idle A links and $(k - y)$ idle B links. A matched pair must be available if a path is to be set up. If no path is available a call is blocked. Under the Jacobaeus assumption the idle A links and B links are distributed independently, at random, over the k possible paths through the switch. (In an analogous manner, in the T-S-T case, there are $k - x$ time slots idle in the input or primary part of the switch; there are $k - y$ time slots idle in the output or second part of the switch. These idle slots are assumed to be distributed independently and randomly over the k time slots that comprise the total number of possible "paths" or choices.)

Blocking will occur if none of the $k - y$ idle links (slots in the T-S-T case) in the secondary part of the switch match the $k - x$ idle links (slots) in the primary part. An example with $k = 5$ appears in Fig. 10–30. In part (a) of this figure the circles represent links: Darkened circles represent the occupied (busy) links; open circles, the free ones. Part (b) shows the analogous picture in terms of time slots. In this example, with $x = 3$ busy links (slots) randomly selected from the k possible in the primary path, what is the chance that none of the $k - y = 3$ free links (slots) in the secondary will match a free link (slot) in the primary? This represents the loss probability under the conditions (state of the system) shown. Note that posed in this fashion the calculation of the loss probability becomes a problem in combinatorics.

More generally, say that there are $(k - x)$ free circles (links or slots) of the total of k available in the primary. There are $\binom{k}{k-x}$ ways in which these can occur. (For $k = 5$ and $k - x = 2$, as in Fig. 10–30, there are 10 such combinations.) But y circles (links or slots) in the secondary are occupied (busy). Then $\binom{y}{k-x}$ represents the number of ways in which the $(k - x)$ free circles in the primary cover the y busy circles of the secondary and hence are not available for completing the connection of the call. (In the example of Fig. 10–30, $\binom{y}{k-x} = 1$. There is thus *only* one way in which the $y = 2$ busy secondary

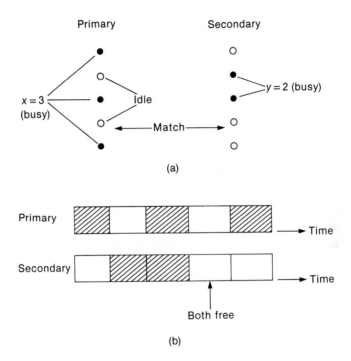

Figure 10–30 Example of busy and free links (slots), $k = 5$
a. Links
b. Time slots

circles and the $k - x = 2$ free primary circles match one another, resulting in loss. This is apparent from Fig. 10–30.)

The ratio of $\binom{y}{k-x}$ to $\binom{k}{k-x}$ gives the probability of loss. But this is just $P(L|x,y)$, since it is conditioned on the state (x,y). We thus have

$$P(L|x,y) = \binom{y}{k-x} \bigg/ \binom{k}{k-x} = \binom{x}{k-y} \bigg/ \binom{k}{k-y} \qquad (10-59)$$

$$= x!\,y!/k!(x+y-k)!$$

In the example of Fig. 10–30, $P(L|x,y) = 0.1$. This expression for $P(L|x,y)$ can now be introduced into Eq. (10–57) as part of the calculation of the loss probability P_L.

To complete the calculation, we must find expressions for $\lambda(x,y)$, $P(x)$, and $P(y)$. $P(x)$ and $P(y)$ are readily found. They represent, respectively, the probability that there are x and y calls present at the input and output of the switch. But with $k > n$, as already noted, the first and last stages of the switch are nonblocking. They can thus each be represented by the model of Fig. 10–13, with $M = N = n$ inputs (or outputs) in this case. As noted in the last section, this gives rise to the binomial distribution as the probability of state of this configuration. Specifically, with the parameter λ representing the average (Poisson) call-arrival rate per input channel and $1/\mu$ the average holding time of a call, both $P(x)$ and $P(y)$ are given by the binomial distribution of Eq. (10–47).

In this case, then,

$$P(x) = \binom{n}{x} a^x (1 - a)^{n-x} \tag{10–60}$$

and

$$P(y) = \binom{n}{y} a^y (1 - a)^{n-y} \tag{10–61}$$

with $a = \lambda/(\lambda + \mu)$, as in Eq. (10–47). These results are based on the independence of the input (x) and output (y) states assumed earlier (Eq. 10–58).

The one remaining function in Eq. (10–57) to be found is the state-dependent arrival rate $\lambda(x,y)$. This is also readily written once the assumption is made that the input state x and the output state y are independent. For $\lambda(x,y)$ must then be of the form $\lambda(x)\lambda(y)$. Recalling our discussion of the Engset distribution in the previous section, it is apparent that with x calls of n possible (Fig. 10–28) present at the input, the input arrival rate $\lambda(x)$ must in particular be given by the form $\lambda_x(n - x)$, λ_x a constant. By symmetry, the same dependence must hold at the output, with y replacing x. We thus write

$$\lambda(x,y) = \gamma(n - x)(n - y) \tag{10–62}$$

γ a constant.

Inserting Eqs. (10–59)—(10–62) into Eq. (10–57), it is readily found that the denominator separates and is easily evaluated by using the identity $E(x) = na$ for the binomial distribution. The summation in the numerator can be evaluated as well, and the following closed-form expression for the loss probability P_L obtained:

$$P_L = \frac{[(n - 1)!]^2 [2 - a]^{2(n-1)-k} a^k}{k! [2(n - 1) - k]!} \tag{10–63}$$

This is the desired formula for the loss probability of a three-stage switch (time-space or all space), in terms of the per-line (trunk or channel) probability of occupancy a, the input lines n and the first-stage output lines k.

An analogous calculation of the blocking probability or time congestion P_B, as originally carried out by Jacobaeus and derived in the references cited— [INOS, pp. 236, 237], [COLL, pp. 17–21]—produces the expression

$$P_B = \frac{(n!)^2 (2-a)^{2n-k} a^k}{k! [2n-k]!} \tag{10-64}$$

The loss probability or call congestion P_L and the blocking probability or time congestion P_B for the three-stage switch, under the assumptions of independence set down above, thus turn out to be related precisely as they were in the calculation of the Engset blocking probability in the last section:

$$P_L(n) = P_B(n-1) \tag{10-65}$$

How does Eq. (10–64) for the blocking probability (or the equivalent loss-probability equation (10–63)) compare with Lee's approximation for the three-stage switch, Eq. (10–54)? Note that with n inputs, at most n of the k outputs of the first stage can be busy. The Lee approximation does not account for this [COLL, pp. 23, 24]. For $k = n$ this input-call limitation is taken into account automatically, and the two expressions for blocking probability turn out to be the same. (The Lee approximation and the loss probability of Eq. (10–63) are the same for $k = n - 1$.) For $k > n$ the Jacobaeus formula consistently produces a smaller value than does the Lee formula. Proof of this plus a number of examples appear next.

That the Jacobaeus formula is still an approximation of the blocking probability is apparent by setting $k = 2n - 1$ in Eq. (10–64). This is of course the condition for a nonblocking switch. At this value of k, however, Eq. (10–64) produces a very small but non-zero value. Simulation studies indicate that the Jacobaeus (or Karnaugh) formula provides a better approximation of the switch blocking probability than does the Lee formula. (See [KARN] and references therein.)

Table 10–3 provides values of the loss probability P_L obtained using Eq. (10–63) for a number of switch designs and for various values of the probability of input-line call occupancy a [McDO, Table 9–5]. These relatively high values of a are characteristic numbers for trunk loading in tandem switches, as will become apparent in our discussion of some typical switch architectures in the next section.

The essentially nonblocking character of those designs becomes apparent from Table 10–3. At $a = 0.7$, usually cited as the design load for tandem switches with loading due primarily to transit (trunk) calls, all the designs have extremely low probabilities of loss (or blocking). This is due to the large values shown for both n and k. It is well known that even small increases in resources (lines and trunks in this case) provide dramatic improvement in performance when the applied loads are statistical in character and the resources are large.

TABLE 10–3	Loss Probability, Various Switch Designs, Jacobaeus (Karnaugh), Eq. (10–63)			

n	k	$a = 0.7$	$a = 0.8$	$a = 0.9$	$a = 1.0$
105	128	2.8×10^{-11}	3.3×10^{-9}	1.1×10^{-5}	0.0038
120	128	3.5×10^{-8}	1.3×10^{-4}	0.03	0.51
240	256	1.3×10^{-15}	1.8×10^{-8}	8.9×10^{-4}	0.3
480	512	3.3×10^{-30}	5.1×10^{-16}	1.1×10^{-6}	0.1
384	384	1.1×10^{-16}	1.1×10^{-7}	0.017	0.99
384	386	3.1×10^{-17}	4.7×10^{-8}	0.011	0.98

This is one example of the statistical law of large numbers. We have already seen this phenomenon in our brief discussion of the Engset and Erlang blocking formulas. It is obviously due to the Poisson or independent character of our call-arrival model but is generally assumed to be a valid representation of most real-life systems: As n and k increase together, arriving calls have a greater chance of finding a nonblocked path, even though more calls may be arriving at the overall switch, on the average, and the loss (or blocking) probability decreases rapidly. Thus, keeping the ratio of k/n fixed, P_L (or P_B) drops rapidly as n (and hence k) increases. A 240 × 256 switch has far better performance than a 120 × 128 switch. The 480 × 512 design shows the same order-of-magnitude improvement, for $a \le 0.9$. Increasing the expansion ratio, k/n, provides a dramatic improvement in performance. Note in particular, from Table 10–3, the vast reduction in loss probability in going from a 120 × 128 to a 105 × 128 design. As will be noted in the next section, these are precisely the design numbers adopted for the time portions of the AT&T No. 4 ESS switch.

Note that even the square switch design of 384 × 384 provides very good performance until a is close to unity. Increasing k slightly, to 386, results in a 2 to 3 times improvement in performance for $a = 0.7$ or 0.8. In all cases shown there is a dramatic increase in loss (or blocking) probability as a approaches 1. It is obvious that overloading even large switches must inevitably lead to congestion. Congestion-control mechanisms must thus be invoked. (Overload-control schemes for digital switching machines, used to reduce call-processing load, will be discussed in the next chapter.)

The exponential variation of P_L and P_B with a, n, and k, as apparent from Table 10–3, is readily shown by a study of Eqs. (10–63) and (10–64). For the special case of $n = k$ in the P_B equation and $(n-1) = k$ in the P_L equation (i.e., the square switch case), we get

$$P_B = (2-a)^k a^k = [1 - (1-a)^2]^k$$

just the Lee approximation for a three-stage switch [Eq. (10–54)], as already noted. Note the exponential decrease in performance with k for $a < 1$. Similarly, for the other values of n and k, the factorial expressions in Eqs. (10–63) and (10–64) vary exponentially with their parameters, leading to the dramatic changes in P_L and P_B with n and k, as already noted.

The exponential behavior of Eq. (10–64) (and Eq. (10–63) as well), in the more general case of $k > n$, is readily noted by invoking the Stirling approximation to the factorial:

$$n! \doteq \sqrt{2\pi} \, e^n \, n^{n+1/2}, \qquad n > 2 \qquad (10-66)$$

Using this approximation in Eq. (10–64), and defining $\beta = k/n$ as the expansion ratio of the $n \times k$ switch (time or space), one gets the following form of the Jacobaeus equation:

$$P_B = \frac{1}{\sqrt{\beta(2-\beta)}} \left(\frac{a}{\beta}\right)^k \left(\frac{2-a}{2-\beta}\right)^{2n-k} \qquad \beta \equiv k/n \qquad (10-67)$$

As an even simpler approximation, for β not too large, one can write

$$P_B \doteq \left(\frac{a}{\beta}\right)^k \left(\frac{2-a}{2-\beta}\right)^{2n-k} \qquad \beta = k/n \qquad (10-67a)$$

These equations lend themselves readily to computation and show the exponential variation of P_B with k as well.

Equations (10–67) and (10–67a) also indicate that the Jacobaeus approximation of the blocking probability should be smaller than that found using the Lee approximation. For comparing Eq. (10–54), the Lee approximation, with Eq. (10–67a), the Jacobaeus approximation, it is apparent that the leading term is the same in both cases. Since $(2n - k) \leq k$ for switch designs with expansion

---

TABLE 10–4 Blocking Probability Comparison, Lee and Jacobaeus Approximations, $a = 0.7$

n	k	P_B, Lee	P_B, Jacobaeus
64	64	0.002	0.002
64	68	0.0002	0.00016
64	72	1.5×10^{-5}	6.2×10^{-6}
64	76	8×10^{-7}	1.5×10^{-7}
120	128	10^{-7}	3.5×10^{-8}
240	256	10^{-14}	1.3×10^{-15}

($k \geq n$), the ones we have been discussing (i.e., those with no concentration or internal blocking in the $n \times k$ switches themselves), the second term in Eq. (10–67a) is smaller than the corresponding term in Eq. (10–54). Some numerical comparisons of the Lee and Jacobaeus approximations appear in Table 10–4. They show that the blocking probability found using the Jacobaeus approximation is consistently smaller than that found using the Lee approximation. Other, smaller switch examples appear in [BELL, Table 5.4].

10–5 Examples of Digital Switching Systems

Thus far in this chapter we have focused on some of the basic aspects of traffic theory appropriate to circuit switching and have provided an introduction to switching networks that contain both space and time stages. In this section we pause to illustrate some of the ideas presented in terms of some concrete examples. The two examples chosen are the AT&T No. 4 ESS, a very large digital tandem switch under centralized control, and the Italtel UT 10/3, a small- to medium-size exchange for combined local and tandem use that has a more distributed control architecture. Many other examples of digital switching systems are currently available, of course, and have been described in the literature. The two systems, which we shall describe quite briefly, are chosen for illustrative purposes only.

A selection of papers that describe many digital systems worldwide appears in a volume edited by Amos Joel [JOEL]. This book provides a list of digital switching systems in service or under development at the time of publication, as well as explanatory notes and comments, and a stylized set of diagrams that give a uniform presentation of the switching-network architecture of many of the systems described. A number of systems are described as well in the book edited by John C. McDonald, cited previously [McDO]. Systems are also described in two special issues of the *IEEE Transactions on Communications*, one for July 1979 [IEEE 1979] and one for June 1982 [IEEE 1982].

[JOEL] Amos E. Joel, Jr., ed., *Electronic Switching: Digital Central Office Systems of the World*, IEEE Press, New York, 1982.

[IEEE 1979] "Digital Switching," special issue, *IEEE Trans. on Comm.*, vol. COM-27, no. 7, July 1979.

[IEEE 1982] "Communication Software," special issue, *IEEE Trans. on Comm.*, vol. COM-30, no. 6, June 1982.

10-5-1 AT&T No. 4 ESS

As already noted a number of times, the AT&T No. 4 ESS is a very large digital switch, the largest manufactured by AT&T, and is designed for toll (tandem) applications. It was installed in 1976 and has received extensive modifications since then [BSTJ 1977], [BSTJ 1981], [BRUC]. Its system-capacity objectives were to handle 350,000 busy-hour call attempts (sometimes abbreviated as BHCA), to handle 110,000 trunks, and to carry a network load of 28,000 Erlangs. [BSTJ 1977, p. 1021]. The load in Erlangs refers to the traffic load at the switching network and translates to a per-trunk utilization or occupancy of $2 \times 28,000/110,000 \doteq 0.5$. One call uses two switching network terminations. The actual capacity levels achieved were 550,000 call attempts/hour at peak; 107,520 trunk terminations; and a network load of 47,200 Erlangs (trunk occupancy of $2 \times 47,200/107,520 \doteq 0.88$).

A rudimentary block diagram of the system, showing major subsystems only, appears in Fig. 10-31. As indicated there, system control is centralized using one large control-processing system (the AT&T 1A processor). Digital trunks arrive at the T1 rate (24 64-bps voice channels for a total bit rate of 1.544 Mbps), with frames repeating at 125-μsec intervals, as shown in Fig. 10-23. Analog trunks are converted to the same T1 (DS-1) format. Groups of five such T1 streams are then combined to form the DS-120 format. This format multiplexes 120 data time slots into 128 time slots per 125-μsec frame. The DS-120 signals are then fed into the switching network, which consists of a set of time-slot interchangers and a time-multiplexed space switch.

The switching network is of the T-S-S-S-S-T type, with the initial T-S stages and the final S-T stages grouped together and a double time-multiplexed space-stage network (S-S) appearing between the two. A block diagram of this basic architecture appears in Fig. 10-32. Note that the system is modularized, allowing the exchange to grow to its maximum capacity of 107,520 trunk terminations as desired.

First consider the TSI frames, which incorporate a time stage and a space stage. We focus here on the full 128-module system. A typical module of the 128 possible is sketched in Fig. 10-33. Seven DS-120 inputs, for a total of $120 \times 7 = 840$ data channels (trunks), are time and space switched in this module. The buffers indicated, one for each DS-120 input, are required because of

[BSTJ 1981] "No. 4 Electronic Switching System: System Evaluation," *BSTJ*, vol. 60, no. 6, part 2, July-Aug. 1981.

[BRUC] R. A. Bruce, P. K. Giloth, E. A. Siegel, Jr., "No. 4 ESS—Evolution of a Digital Switching System," special issue, "Digital Switching," *IEEE Trans. on Comm.*, vol. COM-27, no. 7, July 1979, 1001-1011; reprinted in [JOEL].

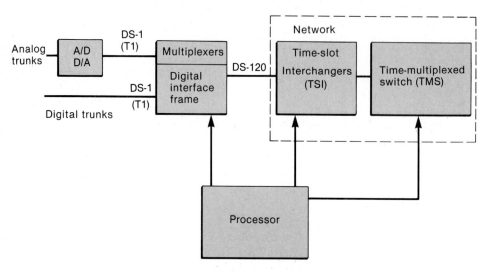

Figure 10–31 AT&T No. 4 ESS, basic block diagram

possible phase differences and time delays among the arriving trunks. They serve to ensure that the seven streams entering the decorrelator all do so in phase and time synchronism.

Note that the seven inputs with 128 time slots per frame each are distributed to eight time-slot interchangers where the first stage of switching (time) is carried out. The decorrelator is used to ensure that all seven input-data streams are spread uniformly out over the eight inputs to the TSIs. The effect is to reduce the trunk occupancy at the input to each TSI to seven eighths of its original, or to reduce the effective number of inputs to $120 \times 7/8 = 105$ input channels. These in turn are spread out over 128 channels at the TSI output.

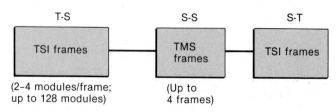

Figure 10–32 Switching network, No. 4 ESS

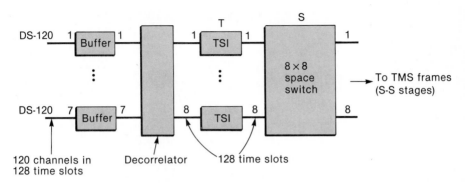

Figure 10-33 Details of a TSI module, No. 4 ESS

Each TSI thus appears as an $n \times k = 105 \times 128$ switch. From Table 10-3 of the previous section, the loss probability of a switch with these numbers, were it to be incorporated as part of a three-stage T-S-T switch, would be 10^{-11} at a per-channel loading of 0.7, rising to a loss probability of 10^{-5} at a per-channel loading of 0.9. Were the distribution of seven inputs among eight TSIs not to be done, the system would be of the 120×128 type, with considerably poorer performance, as seen from Table 10-3. Since the No. 4 ESS switching network is of the T-S-S-S-S-T type, not T-S-T, the blocking probability should be somewhat higher than the figures of Table 10-3. (See Problem 10-23 for blocking calculations of a five-stage switch.)

Performance specifications of the No. 4 ESS switching network call for a blocking probability of 0.005 at a per-trunk occupancy of $a = 0.7$, rising to 0.1 at $a = 0.9$. Simulations, using the original 1976 design, showed a performance better than these specifications. The blocking probability measured was below 0.005 at $a = 0.7$, rising to 0.01 at $a = 0.9$ and to 0.1 at $a = 0.95$ [BSTJ 1977, p. 1033].

The full No. 4 ESS has 128 modules of the type in Fig. 10-33. Since each terminates $120 \times 7 = 840$ trunks, the total capacity is $840 \times 128 = 107,520$ trunks, just the figure quoted earlier. The single-stage space switch of Fig. 10-33 provides eight outputs at each time slot. One hundred twenty-eight such modules then provide $128 \times 8 = 1024$ outputs, with 128 time slots each. These 1024 space outputs are in turn connected to the next S-S stages, organized as four 256×256 TMS (time-multiplexed switch) frames. The connections are arranged so that each 8×8 switch (one per TSI module) has two connections to each of the four TMS frames. An example for the first of the 128 8×8 space switches appears in Fig. 10-34. [McDO, Chap. 5]. As shown in that figure, each

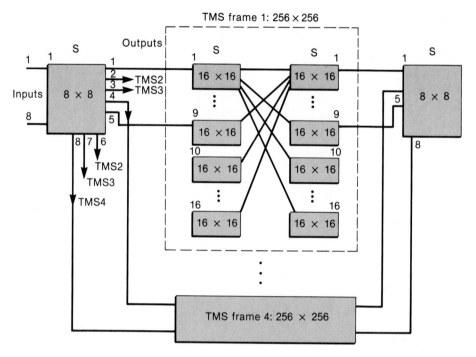

Figure 10–34 TMS frame, No. 4 ESS (partial connections only)

of the two space stages in the TMS frame consists of 16 16 × 16 space switches. This provides the desired 256 × 256 S-S configuration per TMS frame. Only partial connections are shown between the two middle space stages in the TMS frame.

The 1024 outputs of the four TMS frames are in turn connected to 128 TSI modules to complete the full network. These modules each contain a recorrelator to concentrate the 128 time slots appearing at the output of eight TSIs to seven DSI-120 streams with 120 data slots each. The picture is thus the mirror image of that shown in Fig. 10–33.

The total crosspoint count for a full 107,520-trunk No. 4 ESS is readily calculated to be 49,152, instead of the number 10^{10} that would be required with a single-space stage. A single TMS frame alone has $16^2 \times 32 = 8192$ crosspoints instead of the $256 \times 256 = 65,536$ that would be required with a single 256×256 space stage. Note also that 8 × 8 and 16 × 16 space switches are required. These are considerably simpler and less costly to build than larger switches. The use of 128 time slots, a power of 2, as the switching rate in the 125-μsec frame, enables clock signals to be derived more easily from a master clock. This is one of the primary reasons for the choice of powers of 2 throughout.

10-5-2 Italtel UT 10/3 Switch System

The Italian digital switching system, the Italtel UT 10/3, is designed in modular fashion for small- to medium-size exchanges. It can serve as a local switch, handling, with 7:1 concentration, over 20,000 subscriber lines and 2500 trunks at its maximum size; as a transit (tandem) switch connecting up to 4096 trunks; or a combination of the two (local/tandem switch). Its call-handling capacity is rated at over 120,000 busy-hour call attempts. As a local switch, with 14,000 subscribers and 4:1 concentration, its per-line occupancy is rated at $a = 0.145$, for a total of 2000 Erlangs. As a tandem switch, it is designed to have a per-trunk occupancy of $a = 0.7$, for a total rating of 2800 Erlangs [GALI], [MONT], [PROT].

The system-control and processing units are organized in a three-level distributed/hierarchical fashion. The system contains up to 16 fully interconnected modules, each with a number of line-interface units. All call handling and processing, including call routing, is carried out by the interface units and modules. Non-real-time administrative and maintenance functions, including operator control of the system, are handled by a service computer to which all interface modules are connected.

A block diagram of an interface module appears in Fig. 10-35. One such module can serve as a small stand-alone exchange. Note that it contains three basic units: a set of interface units that connect to lines and trunks, a time-stage unit (time-slot interchanger) that switches 256×256 time slots, and a module processor that provides control for the entire interface module. Each interface unit serves up to 256 network terminations divided into groups. A group is composed of 8, 16, or 32 lines or trunks and is controlled by a microprocessor. The interface unit provides analog-to-digital (A/D) or D/A conversion for analog subscriber lines or trunks; provides the appropriate interfacing between the network voice and/or data channels and the exchange internal time slots; detects and sends line signals; and carries out preprocessing of signals. A single

[GALI] R. Galimberti, G. Perucca, P. Semprini, "Proteo System: An Overview," ISS '81, Montreal, Sept. 1981.

[MONT] S. Dal Monte and J. Israel, "Proteo System-UT10/3: A Combined Local and Full Exchange," ISS '81, Montreal, Sept. 1981.

[PROT] *Proteo UT 10/3 Electronic Digital Exchange,* Application Notes, Italtel Co., Milan, Italy.

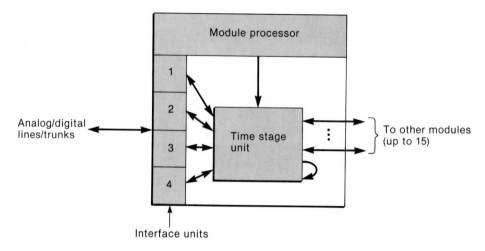

Figure 10–35 Italtel UT 10/3 interface module (from [GALI, Fig. 3], 1981
IEEE, with permission)

interface unit providing 256 time slots or channels would be used for a tandem
(trunk) system. This would interface directly with the time-stage unit, with no
concentration. Digital input would be provided by up to eight PCM streams
running at 2.048 Mbps each. (This is the CCITT international PCM recommendation, which provides 32 64-kbps channels. In the telephony case only 30 of
the channels are used to transmit 64-kbps voice/data signals; the remaining two
channels are used for signaling and other purposes [SCHW 1980a, p. 159].)

In the case of a local-exchange application, up to four line-interface units for
normal traffic, or seven such units for low-traffic cases, would be used. This
provides for up to 1024 network terminations, with 4:1 concentration into 256
time slots, in the former case, or 1792 ports, with 7:1 concentration, in the latter
case. Sixteen such interconnected modules, each providing 7:1 line concentration, produce the maximum-size local exchange capable of supporting 16 ×
1792 = 28,672 ports mentioned earlier.

The time-stage unit of Fig. 10–35 provides the 256 × 256 time switching
already noted. In addition, as indicated and as shown in Fig. 10–35, each
time-slot unit is fully connected to all other time slot units (one per module) in
the system. This provides space switching in addition. (Data in each time slot at a
module is sent out simultaneously to all modules, including itself. Only the
destination module, for which the data in the particular time slot is intended,
accepts the slot information, thus providing the desired switching function.)

The switching network for this sytem, with more than one module used, is thus of the T-S-T type. For only one module it is obviously of the T type.

The module processor supervises local lines and trunks, carries out digit analysis using messages sent to it by the interface unit processors, routes calls on the basis of the digit analysis, and carries out some maintenance, as well as other functions. Each module processor has available to it information regarding the allocation of lines and trunks to all the other interface modules. (However, it does not know the status of these lines and trunks.) Each module processor is fully duplicated to maintain reliability. All module processors are interconnected via intermodule communication controllers and are also connected to the higher-level exchange (service) computer that provides overall supervision, non-real-time administration and maintenance, mass storage, and so on. System configuration changes are all handled by operator control through the exchange computer. Each module processor is also connected via an intermodule serial bus controller to the interface units, and via a switch memory controller to the time-stage (switching) unit.

A more detailed block diagram of a line interface unit appears in Fig. 10-36 [PROT]. The unit shown in the example interfaces with analog subscriber lines. A trunk-interface unit would obviously not need the analog circuitry shown. The ALC blocks represent analog line cards. Each ALC interfaces with eight analog subscriber lines as shown. The ALC provides *battery* feed, *o*verload voltage protection, *r*inging, *s*upervision, *c*oding, *h*ybrid, and *t*est functions. (These seven functions, provided in most telephone systems, are grouped together under the acronym BORSCHT. See [BELL].) A/D and D/A operations are carried out by the analog line cards (ALCs) as well.

The DLC units represent digital line cards. Each interfaces with two ALCs, as shown. The DLC contains an 8-bit microprocessor designed to handle 16 subscriber lines or eight trunks. It carries out preprocessing of line signals. These signals could be off-hook signals or dial pulses in the case of lines; they could be connect (seizure) signals for call setup in the case of trunks. (They are discussed in more detail in the next chapter.) The line signals are detected, controlled, coded into message format (denoting whether the signal is an off-hook signal, a dial pulse, and so on), and transmitted as control messages to the module processor via the control shown.

The control buffer, on the transmit side, gathers control information from the DLCs for transmission to the module processor. On the receive side serial data received from the module processor is buffered and then distributed to the appropriate DLC.

The PCM buffer shown in Fig. 10-36 carries out serial-to-parallel conversion on transmission to the switching subsystem (the time-stage unit). Transmission on each outgoing line is at a 2.048-Mbps rate. An error-correction code is added during this process. Parallel-to-serial conversion is carried out in the reverse direction.

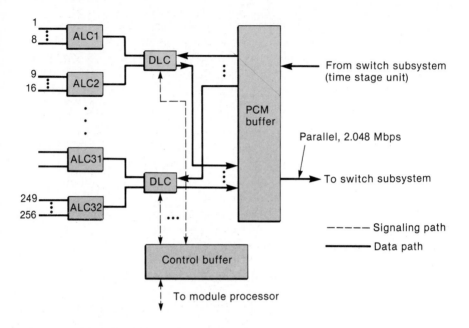

Figure 10−36   Line interface unit: subscriber line example (from [PROT]. Reprinted by permission of Italtel. Proteo UT 10/3 is a product of the LINEA UT Family of Electronic Switching Systems)

Figure 10−37 shows the module switching subsystem, containing the time-stage unit, that carries out the concentration and time-slot interchange operations and provides the connections with the other modules for the space-switching function. Also shown are connections to the module processor.

The time-slot compressor (TSC) concentrates all voice/data channels from the PCM buffers into 256 outgoing channels. Each channel is loaded into a RAM, in sequential order. The RAM is then read out, on command from the switch memory controller of the module processors, thus providing the necessary time switching and compression.

The high-speed junctor transmission (HSJTX) buffer receives the outgoing channels from the TSC buffer and distributes them, via 16 high-speed transmission junctors, to *all* modules, including the one in which the buffer is located. Information generated by the module processor for distribution to all modules is received by the transmission junctor buffer as well. Each transmission junctor (HSJTX)—the word usually used to denote the connecting circuit between switching stages in a switch — is hardwired to a companion receiving junctor (HSJRX) at another module, including one connected to its own receiving junctor. This provides the complete interconnection among all modules, allowing 256 channels to be distributed simultaneously on separate paths.

On the receiving side, each of the 256 channels (time slots) coming in from another module is stored in a RAM in the corresponding receiving junctor (HSJRX). The switch memory controller of the module processor selects (enables) one of the HSJRX memories in each time slot, thus completing the space connection for calls in progress that are directed to this module. These 256 channels, in sequence, are time multiplexed and sent on to the time-slot expander: The 256 time slots are loaded into memory banks, one for each interface unit, each with a 256-word capacity. By reading these out appropriately, connection is made to the appropriate port in the desired interface unit. This

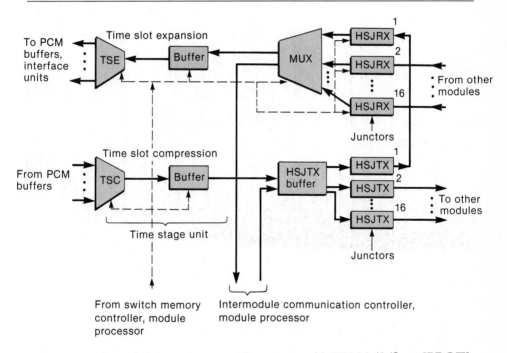

Figure 10-37 Switching subsystem (time stage unit), UT 10/3 (from [PROT]. Reprinted by permission of Italtel. Proteo UT 10/3 is a product of the LINEA UT Family of Electronic Switching Systems)

provides the required time-switching function, completing the T-S-T switching arrangement noted earlier.

In the next chapter, which describes call processing in digital circuit-switched systems, we return to the Italtel system. Among other system examples, we describe the software structure of the Italtel module processor.

Problems

10-1 The M/M/N queue model of Fig. 10–4 arises when one studies the wait-time characteristics of the queued model of circuit switching portrayed in Fig. 10–2. Show that the probability p_n that n calls (customers) are in service is given by Eq. (10–4), with p_0 given by Eq. (10–6).

Note: Problems 10–2 through 10–8 refer to the queued circuit-switched model of Fig. 10–2.

10-2 Show that the probability a call is delayed is given by the Erlang-C formula or Erlang's formula of the second kind $E_{2,N}(A)$, defined by Eq. (10–7). A, in Erlangs, is $N\rho = \lambda/\mu$. Calculate and plot $E_{2,N}(A)$ versus A for $N = 1, 2, 3, 5$, and more values if desired.

10-3 Prove that the average number of calls waiting for service is given by Eq. (10–9), while the average number in the queueing system is given by Eq. (10–10).

10-4 Show that the average wait time on a queue is given by Eq. (10–12).

10-5 $A = 0.6$ Erlang. Calculate the average wait time $E(W)$ and the average time delay $E(T)$ for $N = 1$ and $N = 3$.

10-6 The total capacity available is 56 kbps. Messages are 1000-bits long. Calculate the wait time $E(W)$ and the time delay $E(T)$ for the three separate cases $N = 1, 2$, and 5 trunks used. Do this for two situations: (1) The per-trunk utilization is $\rho = 0.6$; (2) A is fixed at 0.6 Erlang. Compare and discuss the significance of your results. *Hint:* As N increases, with the total transmission capacity fixed, what is happening to the speed at each individual trunk? Given a choice, is it better to use more or fewer trunks? (Note that varying the speed of individual trunks may be possible for data transmission but is generally not possible for real-time voice.)

10-7 Show that the cell setup time T_C and the average holding time T_H are given by Eq. (10–1) and Eq. (10–13), respectively. Show to your satisfaction that Figs. 10–3 and 10–5 are correct. Extend these formulas to the *two-hop* case, assuming that inband signaling messages must be queued at the second (intermediate) node as well. Assume that wait times at each node are independent.

10-8 Calculate and plot the normalized call setup time T_C/T_M (T_M is the average call or message length) versus $\rho_M \equiv \lambda T_M/N$, for the two cases $T_S = T_I$ and $T_S = 0.1T_I$. $N = 5$ trunks are used in both cases. λ is the average call-arrival rate, in calls/

time, Poisson distributed. Assume that the holding time is distributed exponentially. $T_I = 0.1T_M$.

10–9 Using the fully connected packet-switched network model of Fig. 10–7, calculate and plot the packet average time delay versus ρ for the two cases: (1) separate acknowledgements (acks); (2) embedded acks. Take the ack length to be one tenth the packet length.

10–10 Refer to Figs. 10–2 and 10–7. $C_T = 48$ kbps, $C_L = 9600$ bps. $\rho_M = \lambda T_M/N = 0.6$. Plot and compare the call-setup time and packet-time delay, respectively, for the circuit-switched (queued) and packet-switched models as the message length (call or packet in either case) varies from small to large values. Take all control messages to be 80 bits long. The packet header is 88-bits long, including 8 bits for an acknowledgement field. Use the embedded-ack analysis in the packet case.

10–11 Show that the normalized response time for the packet-switched model of Fig. 10–7 is given by Eq. (10–20) if acknowledgements are embedded in information packets. Show that the response time for the exponential approximation case is given by Eq. (10–21).

10–12 Derive the Engset distribution given by Eq. (10–35).

    a. For this distribution show that call and time congestion are related by $P_L(M) = P_B(M - 1)$.

    b. Let $M = N$. Show that p_n in Eq. (10–35) becomes the binomial probability

$$p_n = \binom{N}{n} a^n (1 - a)^{N-n} \qquad a = \lambda/(\lambda + u)$$

10–13 Derive the Erlang distribution (Eq. 10–43) in two ways:

    a. from M/M/N/N analysis

    b. from Eq. (10–35) for the Engset distribution.

10–14 Plot the Erlang-B blocking probability (Eq. 10–44) versus A for $N = 1, 5, 10, 20$. Superimpose $e^{-A}A^N/N!$ and compare. (This is sometimes suggested as an approximation for small P_B.)

10–15 a. A circuit-switching system has a call arrival rate $\lambda = 1$ call/3 sec. The average call-holding time is $1/\mu = 30$ sec. Find the minimum number of trunks N needed to have the blocking probability not exceed 1 percent.

    b. Compare with the approximation $N \doteq 5.5 + 1.17A$.

    c. Use the approximation $P_B \doteq e^{-A} A^N/N!$ and compare results.

    Note: A useful approximation to $N!$, $N \gg 1$ is the Stirling approximation, $N! \doteq \sqrt{2\pi} \, e^{-N} N^{N+1/2}$.

10–16 A useful recursive relation for calculating the Erlang-B blocking probability $P_B(N)$ defined by Eq. (10–44) is given as follows:

$$\frac{1}{P_B(N)} = 1 + \frac{N}{AP_B(N - 1)}, \quad P_B(0) = 1.$$

a. Derive this relation.

b. Use this recursive relation to calculate and plot $P_B(N)$ for $1 \leq N \leq 20$ and $1 \leq A \leq 20$ Erlangs.

10–17 Calls arrive at a circuit-switched exchange at the rate of 2 calls/sec. Calculate the load in Erlangs for the following call-holding times: 10 sec, 100 sec, 5 minutes. What is the load in CCS in each case? Repeat for the following two call-arrival rates: 1 call/2 sec; 50,000 calls/hour.

10–18 Consider a nonblocking three-stage network. Show that

$$n_{\text{opt.}} \doteq \sqrt{N/2}$$
$$C|_{\text{opt.}} \doteq 4\sqrt{2}\, N^{3/2} \qquad N \gg 1$$

10–19 Design a nonblocking three-stage space switch for $N = 10^4$ inputs to be connected to $N = 10^4$ outputs. Compare the resultant C with $C_{\text{opt.}}$.

10–20 Sketch a five-stage nonblocking space switch.

10–21 Repeat Problem 10–19 if a T-S-T switch is used.

10–22 Design a T-S-T switch of size $N = 2048$ inputs (outputs) if the maximum acceptable blocking probability is 0.002 (Lee approach) and the input utilization is 0.9/line. What would the number of crosspoints be for a comparable S-S-S switch? Compare with a nonblocking S-S-S switch.

10–23 Refer to the five-stage switch of Fig. 10–29 [COLL, p. 36]. Show, using the Lee method of analysis, that the approximate expression for blocking probability is given by Eq. (10–55).

10–24 Check the following T-S-S-S-T switch design, which is based on the configuration and notation of Fig. 10–29 [COLL, design 3, p. 50]. The switch has $J = N_C = 116{,}736$ input channels or trunks. $N = 256$ ports or TSIs, $c = 456$ input channels/port, $\ell = 512$ output time slots/port. 16×16 space switches ($n = k = 16$) are used as the first and third stages of the three-stage space switch. Show that $C = 12{,}288$ crosspoints are required. For an input-channel loading of $a = 0.9$, show that the blocking probability is $P_B \doteq 6.4 \times 10^{-5}$ (use the Lee approximation). The time slots are each 8-bits long recurring every 125 μsec. Show that the switching time of the space switch is 30.5 nsec.

10–25 Compare the five-stage design of Problem 10–24 with a T-S-T (single-space switch) design of your own choosing.

10–26 A 2×2 switch has calls arriving at a rate λ on each of the two input lines (channels) if idle. The average call holding time is $1/\mu$.

a. Find the probability p_j that the output lines are in state j, with j the number of calls in progress.

b. $\lambda/\mu = 1$. Calculate p_j for each of the states.

c. Consider *one* of the input lines. What is the probability that a call is in progress? Explain.

10–27 a. A 2:1 concentrator has two input channels leading to one output trunk. The call arrival rate is λ on an idle input channel. The average holding time is $1/\mu$.

Find and compare the blocking probability P_B and the loss probability P_L. Which of these two probabilities corresponds to the probability that an arriving call is blocked? Explain.

b. Calculate and compare the blocking probability of a 2:1 concentrator in these two cases:

1. There are two input channels and one output channel.
2. There are four input and two output channels.

$\lambda/\mu = 0.1$ in both cases.

c. Repeat b. for the loss probabilities.

10-28 A T-S-T switch is designed for a total of $N = 1024$ input channels, with $n = 32$ input time slots at each TSI (time-slot interchanger). The probability that an input channel (time slot) is occupied is 0.7.

a. Design a switch, using the Lee approximation, so that the blocking probability ≤ 0.001. Sketch the switch. How many crosspoints are needed?
b. Repeat a. but for a nonblocking design. (Again sketch the switch.)
c. Repeat a. and b. for comparable S-S-S designs. Compare the number of crosspoints in the four cases. Sketch the two S-S-S designs.
d. Another T-S-T design has 32 time slots at both the input and the output of each TSI ($n = k = 32$). Find the blocking probability. Sketch the equivalent S-S-S switch. What is its blocking probability? Compare the number of crosspoints in the two cases, T-S-T and S-S-S.
e. Find the nonblocking S-S-S design, with $N = 1024$ input channels, that minimizes the number of crosspoints. Compare with the S-S-S designs in c. and d.

10-29 **a.** Complete the calculation of loss probability P_L for a three-stage switch, following Karnaugh's approach, leading to Eq. (10-63).
b. Calculate P_L for $n = 120$, $k = 128$, and $a = 0.7, 0.8, 0.9$. Compare with Lee's approximation.

10-30 Consider a three-stage switch. k possible paths are available through stage 2 from stages 1 to 3. Let the probability that a path is busy be p_b. A random path-search algorithm is used. Show that the average number of paths to be searched either for a successful path to be found, *or*, after searching through all k paths, for switch blocking to be declared, is given by

$$E(N_p) = (1 - p_b^k)/(1 - p_b)$$

What approximations were used?
Calculate this number for a tandem switch with $n = 120$, $k = 128$, for the three cases $a = 0.7, 0.8, 0.9$; a is the probability that an input channel (trunk) is busy.

10-31 Consider a three-stage all-space (S-S-S) switch. The total number of inputs is $N = 512$. Each input group or array has $n = 16$, $k = 20$. The input utilization per line is $a = 0.7$.

a. Calculate the blocking probability using the Lee and Karnaugh-Jacobaeus approaches and compare.

    b. Two of the 16 inputs become permanently associated with a call (for example, they are permanently dedicated to a computer through a dialup port — see [BELL], for instance). The other inputs remain with the same call loading. Repeat a.

    c. Two of the 20 outputs of the input (stage-1) array become disabled. Repeat a.

    d. Refer to Problem 10–30. Calculate $E(N_p)$, the number of paths to be searched, in the three cases of a., b., and c. above.

Note: Problems 10-32 through 10-34 are longer ones and may be used as class projects. They involve further reading of the literature and extensions of some of the material discussed in this chapter.

10–32 Carry out a detailed comparison of the Lee and Jacobaeus (Karnaugh) three-stage switch-blocking approximations. See the references in the paper by Karnaugh [KARN] to other possible approaches to calculating the blocking probability. In particular, see the paper by K. Kummerle referred to there. Can you use these other approaches to develop better approximations?

10–33 Read the paper by Nesenbergs and Linfield [NESE] referred to in this chapter. Reproduce some of the calculations provided in the paper to familiarize yourself with the approach. Use the method described there to carry out a blocking probability calculation of a three-stage switch with $k < n$. Compare your results with those in [BELL, p. 230], with input blocking neglected.

10–34 Using simulation and/or analysis, compare some switch path-searching algorithms. (The results for random search are given in Problem 10–30.) See the comment in [BELL, p. 236], on call packing for one example. See a similar comment by V. Benes [BENE, p. 1387]. Think up some algorithms of your own.

[BENE] V. E. Benes, "Programming and Control Problems Arising from Optimal Routing in Telephone Networks," *BSTJ*, vol. 45, no. 9, Nov. 1966, 1373–1438.

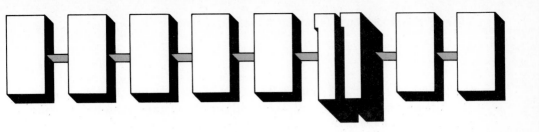

Call Processing in Digital Circuit-Switching Systems

In this chapter we focus on call handling in modern stored-program (computer-controlled) digital switching systems. We do this by example first, describing the software structure and call-processing task organization in a number of representative switching systems. We then provide a generic model of a somewhat idealized call-processing scenario to which we can apply simple queueing analysis. This enables us to determine such numbers as the call-handling capability of a switch (or, conversely, the requirements on the computer processor to provide a desired call-handling capacity), the average dial-tone delay as seen by the user (customer), and the call-setup delay through the switch. The discerning reader will note that the call-setup delay or call-processing time through the switch is precisely the quantity we neglected in carrying out our calculations of call-connect time in the idealized (queued-call) model of a circuit-switching system at the beginning of Chapter 10.

All switching systems incorporate overload- or congestion-control mechanisms to prevent systems from deadlocking, to allow call processing to proceed at times of intense demand and to allow the processors to continue to carry out necessary administrative tasks in the face of call-overload conditions. At the end of this chapter we describe some of the controls used, showing their connection to the control mechanisms used for packet-switched networks (which were described in Chapter 5). In the case of circuit switching the controls normally operate to delay the dial tone, block calls, or carry out a combination of these procedures. We carry out simplified analyses of several control mechanisms to obtain some qualitative tradeoff tables and curves, relating dial-tone delay and probability of blocking of calls to the tightness of the control.

11–1 Software Organization and Call Processing

Stored-program switching systems are designed to carry out a number of important processing tasks in addition to those relating specifically to call processing. Some of these are run in real time, while others represent non-real-time activities. Some must be handled immediately; others may be deferred. It is thus common practice to institute priority levels for various task categories, exactly as is done in most general-purpose computer systems. The particular organization of the priority classes, as well as the organization of the software structure based on these classes, vary from one system to another. Software organization depends as well on the control structure imposed, which ranges from large centralized switches, to multiprocessor hierarchical systems, to completely distributed processor systems. This is again similar to the case in modern general-purpose computer systems.

Generally, though, one can recognize three major types of tasks to be carried out by a computer-controlled digital switching system, whether of the centralized or the distributed type:

1. Provision must be made to handle faults that require immediate action. Fault interrupts are used for this purpose.

2. Obviously call-processing tasks must be accommodated. These are of two types — real-time tasks such as dial-pulse counting or sending that involve send or receive buffers, and other, less time-critical tasks that involve handling of calls.

3. Administrative programs, testing, and diagnostics must all be run, although these may be deferrable for a period of time.

As already noted, we shall focus principally on call processing in this chapter, but will mention other task categories through example. As a typical case, consider the organization of software tasks in the Japanese D-10 centralized analog switching system [TAKE]. Most centralized systems follow a similar organization. A three-group interrupt-level hierarchy is used here, with or-

---

[TAKE] S. Takemura, H. Kawashima, H. Nakajima, *Software Design for Electronic Switching Systems*, Peter Pereginus Ltd., Stevenage, U.K., 1979 (English edition edited by M.T. Hill).

dered priority levels included in the groups. In order of decreasing priority the hierarchy is as follows:

1. Fault-interrupt level

2. Clock-interrupt level

   a. High level, for dial-pulse counting and sending (activated every 4 msec)

   b. Low level, for network and trunk operations

3. Base levels, with three ordered classes

   a. Call processing

   b. Administrative programs

   c. Routine tests and diagnostics

Preemptive priority is used for the interrupts at levels 1 and 2. Nonpreemptive priority is used within groups. Thus call-processing tasks receive nonpreemptive priority over administrative programs. Within call processing itself there may be priorities. (One of our models of call processing, discussed later in this chapter, incorporates two priority levels among the call-processing tasks.) The 4-msec clock interrupt for dial-pulse reception enables dial pulses to be recognized immediately and to be processed in real time. In the case of distributed switching systems (the Italtel UT 10/3, described in the last chapter, and the AT&T No. 5 ESS, described briefly later in this chapter, are two of many such examples), line handling is usually carried out by dedicated microprocessors. Dial-pulse recognition, shown under the high clock-interrupt level here, would be handled by the line-interface units in this case.

In the rest of this chapter we ignore the fault and clock-interrupt levels, as well as other possible priority levels in the software organization of a digital switching processor, and focus principally on call-processing tasks. The type and number of these tasks will differ from switch to switch, but a basic set can be structured that is required in setting up and then clearing a call. To understand this structure we first consider the sequence of signals required in a network to set up and then clear a call. We focus for clarity on the simplest scenario involving the handling of an ordinary telephone call, often labeled a POTS ("plain, ordinary, telephone service"!) call.

Nine distinct signals are required. (The reader is asked to contrast these signals with the somewhat simpler set used in describing circuit switching in Section 10–1.) To standardize on notation here, we use the signals defined in United States telephone practice [AT&T]. (A number of signals other than

---

[AT&T] *Notes on the Network, Bell System Practices*, AT&T Co. Standard, Sec. 781-030-100, Issue 2, Dec. 1980, Sec. 5, Signaling, Figs. 1 and 2.

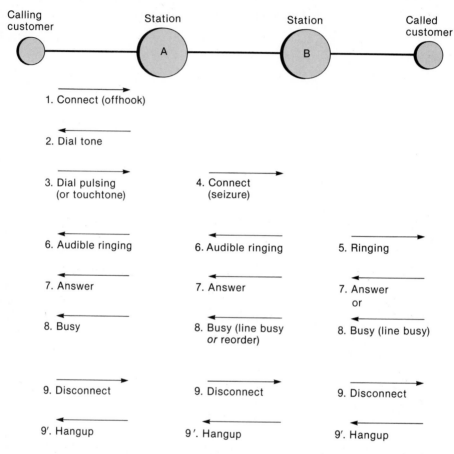

Figure 11–1 Sequence of signals transmitted, ordinary telephone call

those discussed here are also defined in [AT&T].) These are encapsulated in Fig. 11–1, which shows a calling party, two stations representing the network, and the called party at the other end. Note that the signals are very similar to, although they obviously carry different names from, those we encountered in Chapter 5 in setting up a virtual circuit in the packet-switched case.

A description of these signals follows:

1. A calling party goes "off hook." A *connect* signal is sent over the calling party's line to the station (exchange or office) to which it is connected.

2. The station replies with a steady *dial tone,* indicating that the equipment is ready for dialing.

3. The calling customer then starts transmitting the address of the called party, either by *dial pulsing* or by *touchtone signaling.*

4. The station, after determining the appropriate outgoing trunk, seizes the trunk and transmits a *connect (seizure)* signal to the next station. This signal requests service and holds the connection until the call is completed. (This is called *forward* signaling. For additional stations along the path to the called party, connect or seizure signals coming in on a trunk in turn generate new connect or seizure signals after appropriate processing.)

5. The destination station, to which the called customer is connected, transmits a *ringing* signal to the called party.

6. The station also starts an *audible ringing* signal going, back to the calling party, to indicate that ringing has begun.

7. The called party, answering with an off-hook signal, starts an *answer* message going, back to the calling party.

8. Alternatively, a *busy* signal is sent back to the calling party. Two types of busy signal are provided:

   a. *line busy* tone, indicating that the *called line* is busy, and

   b. *reorder* tone, indicating that the network is busy.

   This latter case could be due to no available paths, no available trunks, blockage in the equipment, or incomplete registration in the equipment. The reorder tone is set at twice the frequency of the line busy tone.

9. Completion of the call is signaled by "on hook" at either end. If the calling party completes, a *disconnect* signal is sent to the other party. If the called party goes on hook, a *hangup* signal is returned to the other party.

   Charging of the call starts with the answer signal.

   A typical local/tandem exchange (one that provides both local and transmit functions) in a network must be capable of generating and interpreting this sequence of signals. Four basic operations at an exchange can be distinguished in the processing of ordinary calls characterized by these signals [DUNC]:

[DUNC] T. Duncan and W. H. Huen, "Software Structure of No. 5 ESS—A Distributed Telephone Switching System," *IEEE Trans. on Comm.,* vol. COM-30, no. 6, June 1982, 1379–1385.

1. Scanning inputs, at the periphery of the system, for supervisory signal transitions and digits. The inputs can be lines, trunks, or a combination of the two. The signal transitions in the case of lines are on-hook or off-hook; in the case of trunks, the transitions would be changes in states, representing the arrival of a seizure signal. The digits can be dial tones, touchtone signals, or multifrequency signals in the case of trunks. Digits must obviously be detected in real time as they occur, as already noted.

2. Applying output signals to the periphery. These signals include dial tone (call origination detected), ringing, interoffice trunk signaling (the connect or seizure signal is one example), and call-progress tones to the originator. (This last category includes audible ringing and busy signals.)

3. Routing calls through the switch.

4. Allocating and establishing "talking paths" through the network, as well as deallocating and clearing calls.

A typical POTS call-setup scenario based on these four categories of operations might include the following software routines:

1. Scan periphery, request creation of terminal I/O if a call is detected (off-hook signal detected).

2. Apply dial tone to periphery and wait for digits.

3. Read digits and collect them from periphery; remove dial tone on detection of first digit.

4. Set up talking path through switch and create terminal process to control terminating phone or trunk.

5. Determine whether busy or on-hook at termination.

6. Apply ringing tone if on-hook.

7. Send audible tone to originator.

8. Detect off-hook at termination side; remove ringing tone at termination and at origination.

11 – 1 – 1   Example: Italtel UT 10/3*

How are these functions carried out in a real system? Consider the Italtel UT 10/3 local/tandem switching system as the first example. Recall from Fig. 10–35 that the UT 10/3 is a modular, hierarchically distributed switching

---

* Material in this subsection is based on [PROT].

system. Interface processors, under the control of a module processor, scan the inputs and apply output signals to the periphery. All other call-processing functions, such as reading, processing, and interpreting dial digits, routing calls, and setting up talking paths, are handled by the module processor.

The module processor software structure, for input/output management only, is shown schematically in Fig. 11–2 [PROT]. The structure is grouped into the operating-system level (for input/output management) and the application-system level. Input/output management consists of four functions as indicated: the module monitor, input control, process selection, and output control. Control is passed from one control function to another by the module monitor.

The *input control,* using an established priority sequence, selects the oldest message from the highest priority level that has data pending. The message is passed to *process selection control,* which determines the application process to be invoked. Two generic types of application process appear in Fig. 11–2: call handling and administration. (The administration process, controlled by requests from the supervisory exchange computer (see Chapter 10), handles administrative tasks relating to the exchange and the network. Here we consider only call handling.) On completion of processing, the application process produces output messages, placing them in a common buffer. Control is then returned to process selection, and from there back to the module monitor. The module monitor now passes control to *output control,* whose function is to take messages stored in the common buffer and queue them to appropriate buffers for output to external devices and internal queues.

Now consider the call-handling process in particular. This process consists of a number of tasks, each performed by an appropriate software module. The tasks are grouped into six categories, as shown in Fig. 11–2. Tasks carried out under *outgoing port selection,* for example, include digit analysis, call routing, outgoing module determination (of the maximum of 16 possible), and outgoing trunk hunting. *Speech resources management* includes tasks such as establishing speech paths, managing speech path resources, and addressing switching memory components. *Trunk* and *subscriber signaling* tasks are self-explanatory. *Charge-metering* management must take into account the type of call, the day, the time of day, and so forth. Finally, under the category *others* are grouped a number of tasks, including those involving the allocation and release of common resources such as ports and multifrequency transceivers.

A simple example of call handling shows the use of some of these software modules and the interplay between the module processor (described here) and

[PROT] *Proteo UT 10/3 Electronic Digital Exchange,* Application Notes, Italtel Co., Milan, Italy.

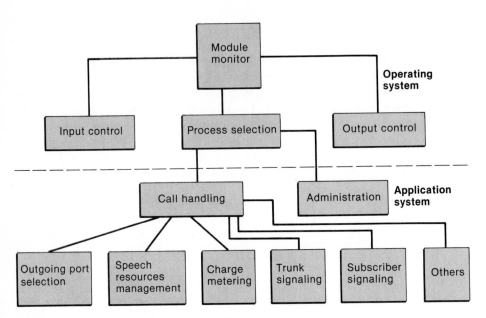

Figure 11–2 Italtel UT 10/3 module processor software structure: input/output management (from [PROT]. Reprinted by permission of Italtel. Proteo UT 10/3 is a product of the LINEA UT Family of Electronic Switching Systems)

the interface processors under its control [PROT]. Consider a dialed call for simplicity. Reference is made to Figs. 10–36 and 10–37 for details of the line-interface unit and the time-stage unit (time switch), respectively. First consider the generation of a seizure signal (off-hook) by a subscriber. The off-hook signal is detected by the line-interface unit and a message to that effect is sent to the module processor. This system in turn assigns a time slot to the subscriber in both the time-slot expansion (TSE) and time-slot compression (TSC) stages of the time-stage unit (Fig. 10–37), and then notifies the digital line card (DLC) at the line interface unit (Fig. 10–36) to that effect. Dial tone is then sent to the subscriber by the DLC. (The DLC provides modulation of a 425-Hz tone generated at the TSE and available to *all* outgoing lines.)

When the subscriber starts dialing, the DLC stops sending dial tone. Digits are transmitted to the module processor as received. This processor analyzes the digits and identifies the outgoing module to which the called subscriber or outgoing trunk, as the case may be, is connected. A message is then sent to that module. The outgoing module, on receiving the message, locates an appropriate outgoing port (subscriber or trunk). The interface unit for this port is then notified. If the called subscriber (or trunk) is available (i.e., free), the outgoing module assigns it a time slot in both the TSE and TSC stages (Fig. 10 – 37). In the case of a subscriber, a ringing current message is sent to the outgoing-line interface. The DLC at the outgoing-line interface, on receipt of the message, then sends the ringing current. At the same time, the DLC on the incoming (initiating) module is notified, and it sends a ringing control tone to the calling subscriber. In case an outgoing trunk is involved, a seizure (connect) signal must be sent.

When the called subscriber answers, the incoming and outgoing modules stop sending the ringing-control tone and current, respectively, and establish the voice path.

11–1–2 AT&T No. 5 ESS

The AT&T No. 5 ESS (electronic switching system) is also a modular, hierarchically distributed digital switching system, but both its hardware and its software organization are quite different from those of the Italtel UT 10/3.

The basic architecture of No. 5 ESS appears in Fig. 11 – 3 [ANDR 1981], [DAVI], [BOSC], [BAUM]. Four elements are included, as indicated: a number of interface modules, a time-multiplexed switch, a message switch, and a central processing unit (CPU). Most of the call-processing functions are handled by individual interface modules in decentralized fashion. A module provides an A/D and D/A function for analog lines and trunks, using an appropriate peripheral unit. It does time switching (512 time slots in and out) through the time-slot interchanger (TSI) and carries out most of the call-processing functions using an interface module processor. The central processor directs intermodule routing — i.e., sets up the path through the time-multiplexed switch — when necessary. It also provides system maintenance and controls external data links to remote operations support systems, as indicated in Fig. 11 – 3. Finally, the message switch controls message switching between interface modules, as well as between the interface modules and the central processor. HDLC format is used for these messages. Pairs of fiber-optic links, each member of a pair running at 32.768 Mbps, are used to provide data and speech path communications between interface modules, as shown.

[ANDR 1981] F. T. Andrews, Jr., and W. B. Smith, "No. 5 ESS—Overview," ISS '81, Montreal, 1981.

[DAVI] J. H. Davis et al., "No. 5 ESS System Architecture," ISS '81, Montreal, 1981.

[BOSC] H. L. Bosco et al., "No. 5 ESS—Hardware Design," ISS '81, Montreal, 1981.

[BAUM] S. M. Baumann et al., "No. 5 ESS Software Design," ISS '81, Montreal, 1981.

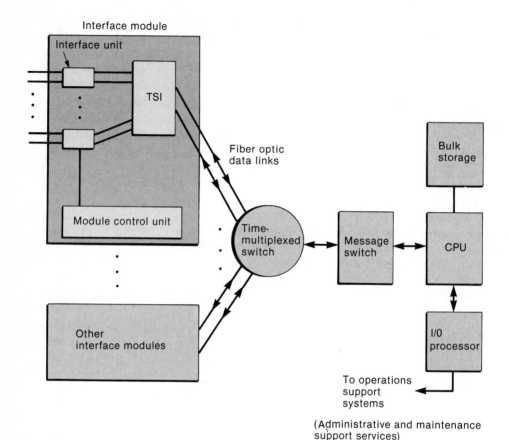

Figure 11–3 Architecture, AT&T No. 5 ESS (from [DAVI], © 1981 IEEE, with permission)

An interface module can handle up to 512 analog voice trunks (converted to 64-kbps per trunk, this produces a 32-Mbps output), 20 T1 inputs for a total of 480 digital trunks, or 4096 subscriber lines, using 8 : 1 concentration. Different line-interface units are used for each case, as in the Italtel system. The message switch in turn can support up to 127 interface modules. Using this modular approach, No. 5 ESS systems can be designed to serve as small local offices of less than 2000-line capacity; as larger offices of up to 50,000 lines handling 100,000 calls per busy hour with planned extension to 100,000 lines; and as toll offices of varying sizes. Remote switching modules can be added under the control of the main office [ANDR 1981].

A single interface module that switches calls through the time-slot inter-changer would serve for the smallest offices. Larger offices use the time-multi-plexed switch as well, providing an overall T-S-T switching function.

Now consider call processing in a typical No. 5 ESS. Four basic categories of software processes are required to handle call processing [DUNC]. They consist of a peripheral I/O process and a terminal-handling process associated with each active terminal, both types located in the interface module processors, and a routing- and terminal-allocation process and a switching-path-allocation process, both residing in the central processor. In addition, administrative, fault-re-covery, program-update, and other processes are associated with a typical ex-change [DUNC]. Figure 11-4, adapted from Duncan and Huen [DUNC, Fig. 6], indicates the location of the four basic processes in the central processor and in two interface module processors that communicate. As shown in Fig. 11-4, the peripheral I/O process provides the interface between the periphery (lines or trunks) and the terminal-handling process. The terminal-handling process carries out most of the call processing, using the characteristics and call features assigned to a particular terminal. The switching-path-allocation process in the central processor sets up the path through the switching network, using infor-mation from the routing and terminal-allocation process.

The steps involved in setting up a POTS call between two terminals A and B serve to clarify the use of these four processes [DUNC, pp. 1382, 1383]. The steps in our discussion are numbered and correspond to numbers in Fig. 11-4 to clarify the procedure. As in the UT 10/3 example, control passes from one process to another, each process carrying out a number of well-defined sequen-tial tasks.

1. The peripheral I/O scans the lines/trunks, and on detecting that a call is arriving, requests creation of a terminal-handling process (labeled A in Fig. 11-4).

2. Terminal process A applies dial tone via the peripheral I/O process and then waits for the digits.

3. On receiving the first digit, terminal process A removes dial tone and collects the digits from the peripheral I/O process.

4. Terminal process A sends a message to the routing and terminal-allocation process to locate the called party.

5. The routing and terminal-allocation process, after locating the called party, notifies the switch-allocation process to set up the speech path through the time-multiplexed switch. The routing-terminal allocation process creates a terminal-handling process B in the called interface module, to handle the called line (telephone) or trunk.

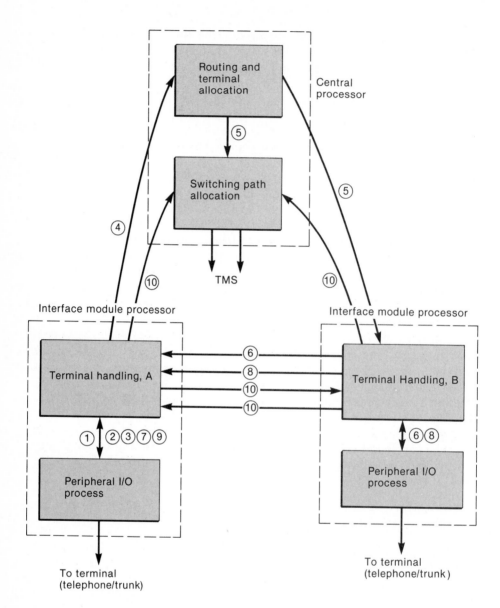

Figure 11–4 Call-processing software structure, AT&T No. 5 ESS (after [DUNC, Fig. 6], © 1982 IEEE, with permission)

6. Terminal process B communicates with terminal process A: It sends either a busy signal or a "setup complete" message, depending on conditions at the called interface. If the telephone at the called interface is on-hook, terminal process B applies a ringing tone.

7. Terminal process A, on receiving the "setup complete" message from B, applies audible ringing to the calling terminal.

8. When the called terminal goes off-hook, terminal process B removes the ringing and sends an answer message to A.

9. Terminal process A suspends audible ringing.

At this point the calling and called parties can begin their two-way conversation. Note how the various steps in setting up the connection between the two parties correspond to the sequence of signals described in the earlier discussion, as portrayed in Fig. 11–1.

10. When either terminal goes on-hook, the terminal process involved sends an appropriate release signal to the other side and notifies the switching-path-allocation process to release the speech path between the two interface modules.

11–2 Analysis of Call-processing Procedure

As indicated earlier and as apparent from the examples of call processing described in the last section, each switching system has its own way of carrying out procedures of call setup and call disconnect. Particular implementations obviously depend on the switch architecture, including the degree of decentralization and the particular assignment of tasks to individual processors in a system. The Italtel UT 10/3 assigns more tasks to the module processors and fewer to the line interface unit than does the AT&T No. 5 ESS. The UT 10/3 uses a number of peer module processors, however, while the No. 5 ESS has one central processing unit. The software implementations vary as well in their definitions of specific tasks to be carried out as part of call processing. Determination of the call-handling capability of a particular system, as well as the calculation of time-delay parameters of interest (such as dial-tone delay and call-setup time) are thus highly system dependent. Nonetheless, it is of interest to carry out a detailed quantitative analysis of a system — first, to solidify our knowledge of the call-handling procedure at a switch; second, to show how one would model call processing in a particular case; third, to actually develop some typical numbers and compare different ways of handling calls in a given system.

To be specific, we introduce here the model of a hypothetical local switch consisting of two modules, set up a queueing model appropriate to processing calls, and calculate both dial-tone delay and call-setup time. We compare two queueing disciplines: FIFO (FCFS) and nonpreemptive priority. All calls in this switch are assumed to be of the POTS type. They arrive at the line interface of one module and must be switched to a trunk associated with the second, outgoing module. Line scanning, as well as the delivery of dial tone and ringing signals, is assumed to be handled at the appropriate interface. Decisions as to when to deliver dial tone, routing determination, and all other aspects of call processing are assumed to be handled at the module processor. A model of the hypothetical system appears in Fig. 11–5. (Both modules handle lines and trunks as shown.) We shall assume that a fixed time delay of 5 msec is required to process messages at the interfaces, either directed from the line (trunk) to the module processor or in the reverse direction. (Details of processing at the line interface could be modeled as well if desired. One of the models of overload control analyzed in the next section does take into account some processing at the interface level.) The key objective here is to analyze a model of call processing as carried out at the level of the two module processors of Fig. 11–5; hence details of processing at the interface level are ignored. To carry out this objective we first define a specific set of elementary tasks into which to divide the process of setting up a locally generated POTS call. These tasks turn out to be 24 in number, with 14 carried out by the calling-module processor and 10 by the called-module processor. The various tasks involved and messages flowing, either to invoke a task or as the result of carrying out a task, are portrayed in Fig. 11–6. Tasks are drawn as rectangular boxes, and messages are labeled as moving into and out of the boxes. (Note that this division of the call-processing scenario into 24 tasks is purely arbitrary and it is introduced for illustrative

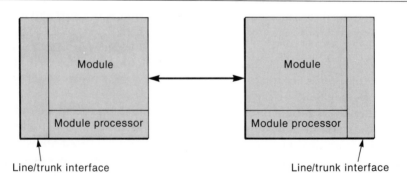

Figure 11–5 Hypothetical switch to be analyzed

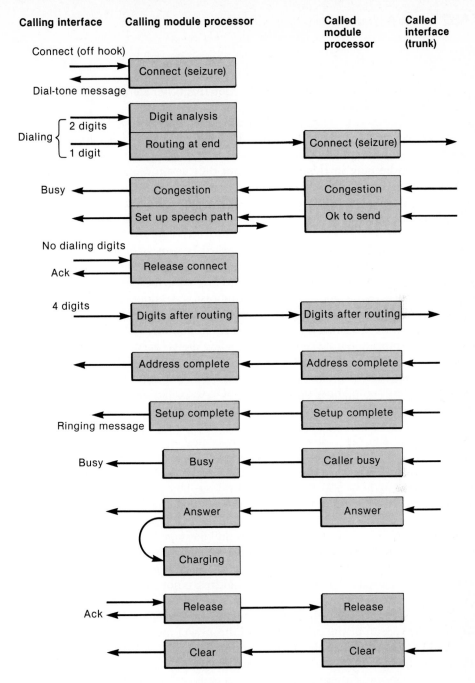

Figure 11–6 Elementary tasks and messages, hypothetical system

purposes only. Note from the previous section that other elementary task structures are possible as well.)

Most of the tasks and messages shown are self-explanatory. They follow the
sequence of signals of Fig. 11–1 and the calling-processing scenarios of the
previous section fairly closely. A few, however, require explanation. We neglect
area code addressing in this example; thus only seven digits are required to
completely set up a call. The first three digits, indicating the address of the called
party's exchange, are required to select an appropriate outgoing trunk at the
called-module interface. The first two digits each invoke a task called *digit
analysis* at the calling-module processor. The third digit invokes a *routing* task.
On completion of this task a connect message will be sent to the outgoing (called)
module, which will in turn cause a "connect" task to be invoked at that processor. This task, on completion, causes a connect (seizure) message to be sent to the
outgoing (called) interface. There are two possible responses to this message:
"congestion" (i.e., network resources not available) or "ok to send." Each response from the called interface causes the corresponding task to be generated,
as indicated. Since the time required for dialing digits (particularly by a human
caller) is normally much greater than that for processing the various tasks
(seconds compared with msec), a "resources busy" message would be received,
and the corresponding busy signal delivered to the caller, before the additional
four digits were dialed.

Any system must be designed to handle all types of error conditions. This
was obviously the case with the packet-switching architectures described in
earlier chapters. Circuit-switching systems are no exception. In the event that
the three dialing digits are not correctly received or are not completed, provision must be made to abort the call and release any processes assigned. The
release connect task carries out this function. Note then that the *digits after routing*
task, invoked for each of the four additional digits following the third routing
digit, will be processed only if an "ok to send" message has first been received.
The *address complete* message and tasks of the same name indicate that all seven
digits have been correctly received and that the address is valid. The *release* task
shown could be used to handle calls whose address is not complete, although this
possibility is not explicitly indicated in Fig. 11–6.

As noted in passing in the previous section, metering of the call (charging)
begins after receipt by the calling party of the answer message. This is indicated
in Fig. 11–6. Finally, the release task is shown being invoked by a message from
the calling interface. This message is presumed to be initiated by the calling
party in response to either of the two busy signals—congestion of resources or
called party busy. The release task is required to release all resources associated
with the call. This includes the speech path as well as software processes. The
clear task is the one invoked on completion of the call. In the example of Fig.
11–6 it is shown being initiated by the called party. This task also results in the

release of all resources. Timeouts and tasks that might be required to eliminate possible ambiguities or a number of possible error conditions other than those just noted are not indicated in Fig. 11-6; they will not be considered in our idealized and simplified analysis.

Using the set of tasks indicated in Fig. 11-6 we are in a position to carry out a quantitative analysis of the POTS call-handling scenario. We must first indicate how often each of the tasks is invoked, the time required to process a task, the number of messages generated, and where these messages are sent as a result of processing the task. This requires an estimate of the types of calls to be handled in a typical exchange and the tasks required for each, and some further assumptions as to the structure of this hypothetical exchange.

Different types of calls arise because of the different conditions encountered during call processing. Calls may be aborted before dialing begins or during the initial (three-digit) dialing phase; they may be aborted due to incomplete dialing (after routing and before completion of the seven-digit address); they may encounter congestion in the initiating exchange or somewhere else in the network; the called subscriber may be busy; there may be no answer; finally, the call may be completed successfully. Each of these different modes of operation results in the use of processing resources: Each requires work from the system and must be considered in evaluating system performance, even when the call is not carried through to completion. Each mode may require a different set of tasks to be carried out.

A typical call mix for European countries in 1978 has been described by Broux and Verbeck [BROU]. We use this mix in somewhat modified form, combining all network busy calls — whether due to blocking at the initiating exchange or to congestion elsewhere in the network — into one *network busy* call category. Six call categories in all can then be distinguished. These categories and the percentage of calls falling into each one are indicated in Table 11-1. Each of these call types uses a different set of the elementary tasks of Fig. 11-6. Table 11-2 indicates the number of times a task is invoked for each of the six different types of calls. The table also indicates the average number of times a given task is invoked in an "average" call, which is found by weighting the call type by its frequency of occurrence. Note that the digit-analysis task and the digits-after-routing task are normally invoked two times (for two successive digits) and four times, respectively, in processing a call. All other tasks are invoked either once or not at all.

As an example, consider the *set up speech path* task. This is invoked at the calling-module processor after an "ok to send" message has been received from

---

[BROU] A. Broux and M. Verbeck, "Metaconta 10 CN Exchanges: A New Generation of Switching Systems," *Electrical Comm.,* vol. 53, no. 1, 1978, 2-8.

TABLE 11-1 Typical Call Mix, European Countries, 1978

Call Type	Percent of Calls
1. Calls without dialing	10
2. Calls with incomplete dialing (after routing digit)	5
3. Network busy	5
4. Called subscriber busy	15
5. No answer	5
6. Answer	60

(after [BROU])

TABLE 11-2 Tasks Associated with Each Call Type

Calling-module Processor Tasks	Type of Call						Weight of Task w_i
	1	2	3	4	5	6	
Connect	1	1	1	1	1	1	1
Digit analysis	0	2	2	2	2	2	1.8
Routing at end	0	1	1	1	1	1	0.9
Congestion	0	0	1	0	0	0	0.05
Set up speech path	0	1	0	1	1	1	0.85
Release connect	1	0	1	0	0	0	0.15
Digits after routing	0	0	0	4	4	4	3.2
Address complete	0	0	0	1	1	1	0.8
Setup complete	0	0	0	0	1	1	0.65
Busy	0	0	0	1	0	0	0.15
Answer	0	0	0	0	0	1	0.6
Release	0	1	0	1	1	1	0.85
Clear	0	0	0	0	0	1	0.6
Charging	0	0	0	0	0	1	0.6
						Average per call	12.2
Called-module Processor Tasks							
Connect (seizure)	0	1	1	1	1	1	0.9
Congestion	0	0	1	0	0	0	0.05
Ok to send	0	1	0	1	1	1	0.85
Digits after routing	0	0	0	4	4	4	3.2
Address complete	0	0	0	1	1	1	0.8
Setup complete	0	0	0	0	1	1	0.65
Caller busy	0	0	0	1	0	0	0.15
Answer	0	0	0	0	0	1	0.6
Release	0	0	0	1	1	1	0.8
Clear	0	0	0	0	0	1	0.6
						Average per call	8.6

the called-module processor (Fig. 11–6). This implies that the routing digit has been received and that a connection can be made (i.e., there is no congestion). Eighty-five percent of all calls fall into this category (Table 11–1); thus on the average, this task will be invoked 0.85 times in an "average" call. The *release connect* task, as another example, will be invoked whenever the call must be aborted due to lack of dialing digits or to congestion. Table 11–1 indicates that 15 percent of all calls fall into this category. Averaging the tasks over the various types of calls, one finds that 12.2 tasks on the average are invoked at the calling-module processor in processing an "average" call; 8.6 tasks are on the average invoked at the called-module processor.

How does one now use this task structure in determining both the call-processing capability of the switch and various delays encountered in carrying out the processing? For this purpose we need a model of the task-processing procedure at a given module processor, one that indicates how the different tasks are invoked on initiation of a call. We assume that two input buffers are available at each module processor: one for receiving messages from its line-interface processor and the other for storing messages from the other processor. (Each module serves both as a called and a calling module.) Messages reaching the head of a queue cause the appropriate task-application process to be invoked. The task is run to completion, at the end of which time the output is placed in an output buffer. This output, on being dequeued, causes a series of messages to be transmitted. We assume that a message is directed to one of five different destinations in this simple model: the line interfaces, the other module processor, the switching network controller, a duplicate module processor required for reliability, and a higher-level administrative computer that handles call charging, station maintenance, and other administrative tasks. A queueing model based on this descriptive model appears in Fig. 11–7. For illustrative purposes the time required to dequeue messages at the input buffers is shown as 1.5 msec. Serving the output buffer is assumed to require 1 msec. Processing of the different output messages varies from message to message and so is not indicated explicitly in Fig. 11–7. We do assume, however, that once the processing of a task begins, it is carried to completion, concluding with the processing and transmission of the output messages shown in the figure.

Using this model and incorporating the particular service (priority) discipline used in handling queued messages, if any, one can calculate performance parameters of interest. For example, to determine dial-tone delay, one would calculate the time required for a connect (off-hook) message (Fig. 11–6) to wait in the input buffer and then be dequeued, add to this the time required to process the connect task called for by the connect message and to generate the output messages resulting from invocation of this task, and finally add the time required to receive a message from, as well as send a message to, the appropriate line interface. We shall carry out this particular calculation in detail later.

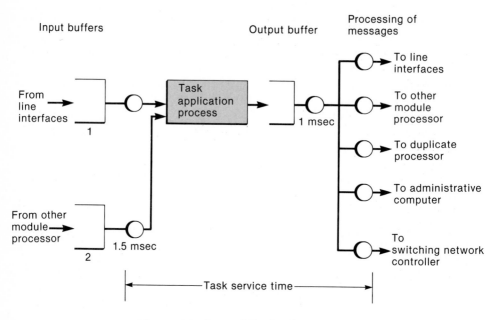

Figure 11–7 Model of call processing

To do the calculations we need to know the time required to process each task, as well as the message-processing (service) times shown in Fig. 11–7. For this purpose we will use some arbitrary numbers. Before proceeding with the detailed calculations, however, we focus on the calculation of the wait times in the input buffers. Note that the wait times are due to *task* messages waiting in each buffer. A call will cause a number of tasks to be invoked, in succession, as already noted. Say that the calls arrive at the line interface of a module at a Poisson rate of λ calls/sec. We now make the basic assumption that each call generates a fixed number of tasks at random times after the call arrival and that the task-arrival process is itself Poisson. In essence we are modeling the call-arrival process as due to the "average" call arriving as described in conjunction with our earlier discussion of Tables 11–1 and 11–2. A little thought will indicate that the independent arrival, or decoupling, of tasks is a fairly valid assumption as well, since the tasks are normally well dispersed in time. Thus, dialing digits, each of which generates a different task, will normally take times on the order of seconds to be carried out, while task processing may be measured in milliseconds.

To be specific, then, let a call generate f_1 tasks in input buffer 1, the one connected to the line interface (Fig. 11–7), and f_2 tasks in input buffer 2, the one connected to the other module processor. The task-arrival rate at input buffer 1 is then $\lambda_1 = f_1 \lambda$, while the rate at buffer 2 is $\lambda_2 = \lambda f_2$. From Table 11–2, as an example, we have $f_1 = 12.2$ and $f_2 = 8.6$. The total number of tasks generated per call in this example is $f = f_1 + f_2 = 20.8$.

To proceed we must specify the service discipline used. We compare in the analyses following two service strategies: the first is first come—first served (FCFS) or first in–first out (FIFO), independent of buffer; the second provides nonpreemptive priority to tasks in input buffer 1 (the one connected to the line interfaces). This second strategy will serve to reduce dial tone delay at the cost of increased call-setup time, as will be seen later. Within each buffer we maintain a FCFS (FIFO) discipline.

Consider the no-priority case first. The two buffers may now be collapsed into one equivalent one, with a total Poisson arrival rate of $(f_1 + f_2)\lambda = f\lambda$ tasks/sec. Each task requires a different time for its processing to be carried to completion, from dequeueing at the input buffer to final processing of the last output message generated. This task service time is indicated in Fig. 11–7. We thus have a model of Poisson arrivals at a queue, with varying service rates, precisely that of the M/G/1 queue discussed much earlier, in Chapter 2! Specifically, let m be the average service time of the composite task arrival stream and $E(m^2)$ the second moment of the service time, averaging over all values of the task service time of Fig. 11–7. We are then left with the equivalent M/G/1 queue of Fig. 11–8. From Eq. (2–65), we then have immediately as the average task wait time in either of the two input buffers

$$E(W) = \frac{\lambda f E(m^2)}{2(1 - \rho)}$$

$$\rho = \lambda f m$$
(11–1)

Equation (11–1) will be used in calculating average delays of interest in the nonpriority case in the material following.

Consider now the other service discipline noted earlier: nonpreemptive priority for tasks waiting in input buffer 1 of Fig. 11–7. From Eqs. (2–85) and (2–86) we can also immediately write, as the average wait times in input buffers 1 and 2, respectively,

$$E(W_1) = E(T_0)/(1 - \rho_1)$$
(11–2)

and

$$E(W_2) = E(W_1)/(1 - \rho)$$
(11–3)

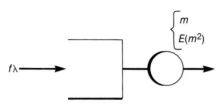

Figure 11–8 Equivalent M/G/1 model, call processing, no-priority case

Here $E(T_0) = \dfrac{\lambda f}{2} E(m^2)$ and $\rho_1 = f_1 \lambda m_1$, with m_1 the average service time of tasks in input buffer 1.

We now apply these equations using an illustrative example. In Table 11–3 we have repeated the 24 tasks of Fig. 11–6, and for each task we have indicated a typical (albeit arbitrary) value of processing time,* as well as the various output messages produced at the completion of the task and the time required to process each message. We have also indicated the input buffer at which the task would appear (obtained from Fig. 11–6). The column labeled *task service time* provides the sum of all the times for a given task, including 1 msec for dequeuing from the output buffer (Fig. 11–7), which is not indicated in the table. Blank entries under output messages indicate that messages are not transmitted to that destination.

Using the numbers in Table 11–3 one can calculate the different parameters in Eqs. (11–1) to (11–3). Thus let w_i be the weight associated with task i, as given in Table 11–2. Let x_i be the task service time, obtained from Table 11–3. Then it is apparent that the average task service time with no priority is

$$m = \sum_i w_i x_i / \sum_i w_i = \sum_i w_i x_i / f = 9.9 \text{ msec} \qquad (11-4)$$

using the entries of Tables 11–2 and 11–3. Similarly, using the line-interface buffer entries in Table 11–3 only, we have $f_1 = 12.2$ tasks per call on the average, as already noted, while $m_1 = \Sigma_i w_i x_i / f_1 = 9.5$ msec/priority-1 task. The corresponding priority-2 (other module-processor buffer) averages are $f_2 = 8.6$ tasks per call, again noted earlier, and $m_2 = 10.5$ msec per task. (Recall again that since each module serves both as a calling and a called module, f_1 can be

---

* These figures are obviously program- and processor-dependent; they depend on the number of instructions and instruction cycle times for each task. They would have to be calculated or measured for the particular processor used.

obtained by summing the weighted tasks in the first, line interface, column of Table 11–3; f_2 is obtained by summing the weighted tasks in the second, other module, column. The results agree with those previously found from Table 11–2, as they should.) As a check, we have

$$\frac{f_1 m_1 + f_2 m_2}{f} = m = 9.9 \text{ msec}$$

The second moment needed in Eqs. (11–1) to (11–3) is determined in a similar manner. Thus we have

$$E(m^2) = \frac{\Sigma_i w_i x_i^2}{\Sigma_i w_i} = 122 \text{ msec}^2 \tag{11–5}$$

again summing over all the entries of Table 11–3 for the nonpriority case. As a check, we have

$$E(m_1^2) = \frac{\Sigma_i w_i x_i^2}{f_1} = 111 \text{ msec}^2$$

summing over priority-1 entries only, while

$$E(m_2^2) = \frac{\Sigma_i w_i x_i^2}{f_2} = 138 \text{ msec}^2$$

summing over priority-2 entries. Then

$$E(m^2) = \frac{f_1 E(m_1^2) + f_2 E(m_2^2)}{f} = 122$$

as expected.

From Eq. (11–1) we then get, for the nonpriority case, with $f\lambda = \rho/m = \rho/9.9$,

$$E(W) = 6.16 \, \rho/(1 - \rho) \text{ msec} \tag{11–6}$$

Say now that we must keep the module-processor utilization ρ to within a maximum of 0.6 to allow the system to be used for tasks other than call processing, or to keep time delays within reasonable values. For the numbers we have chosen, we then have

$$\rho = 20.8(9.9 \times 10^{-3})\lambda \leq 0.6$$

or $\lambda \leq 2.9$ calls/sec. This corresponds to 10,500 busy-hour call attempts per module processor. For two such modules, the example we have chosen here, the maximum call-handling rate would be double this figure.

More generally, say that we have a processing system that handles the same mix of calls, with the same set of 24 tasks that could be invoked in any call-pro-

TABLE 11–3 Calculation of Task Service Time*

Calling-module Processor	Dequeue from Input Buffer		Applic. Process	Processing Time, Output Message					x_i Task Svc. Time
	1: Line Interface	2: Other Module		Line Interface	Other Processor	Duplicate Processor	Admin. Processor	Network	
Connect	1.5		10	1.5		8			22
Digit analysis (2)	1.5		4			3			6.5
Routing at end	1.5		5	1.5	1.5	3			12
Congestion		1.5	4	1.5	1.5	2			11.5
Set up speech path		1.5	4	1.5	1.5			5	14.5
Release connect	1.5		5	3		2			12.5
Digits after routing (4)	1.5		3		1.5				7
Address complete		1.5	3	1.5		2			9
Setup complete		1.5	3	1.5		2			9
Busy		1.5	3	1.5		2			9
Answer		1.5	3	1.5		2			9
Release	1.5		5	3		2		3	15.5
Clear		1.5	3.5	1.5		2		3	12.5
Charging	1.5		2				1.5		6

Called-module
Processor

Connect (seizure)	1.5	1.5	10	1.5		8	22
Congestion	1.5		4		1.5	2	10
Ok to send	1.5	1.5	3		1.5	2	9
Digits after routing (4)			3	1.5			7
Address complete	1.5		3		1.5	2	9
Setup complete	1.5		3		1.5	2	9
Caller busy	1.5		3		1.5	2	9
Answer	1.5		3		1.5	2	9
Release	1.5	1.5	5			2	9.5
Clear	1.5		2		1.5	2	8

* All table entries in msec

cessing scenario. Again let the average number of tasks spawned by an "average" call be $f = 20.8$. We can use this simple analysis to determine the average processing time m required of the processor to handle a specified number of call attempts in a given period of time. This provides us with the information necessary to choose the processor required. As an example, say that the call-handling capacity required per module is $\lambda = 29$ calls/sec, or 105,000 BHCA. This is 10 times the capacity found for the numbers arising from Table 11–3. For the same processor utilization $\rho = 0.6$, this implies that all tasks must be processed in one tenth the time required previously, with $m = 0.99$ msec/task on the average. The processor used must operate at 10 times the speed of the one required to produce the task-processing times of Table 11–3. All entries in Table 11–3 would thus be reduced by a factor of 10.

It has been noted a number of times that one of the primary purposes of carrying out the modeling and wait-time calculations based on the queueing model of Fig. 11–7 is to calculate a number of significant performance parameters. Specifically, given the models of Figs. 11–7 and 11–8, and Eqs. (11–1) to (11–3) for the wait time in both the FIFO and nonpreemptive priority cases, obtained from the models, one can calculate such parameters of interest as average dial-tone delay and average call-setup time.

Consider dial-tone delay first. From Fig. 11–6 for the hypothetical system under study here, it is apparent that this parameter depends on the time required to process the connect (seizure) task. More specifically, dial-tone delay is the time from the user going off-hook to the time dial tone is delivered to the user. In our model this time includes a fixed 5 msec-delay for the calling-interface processor to transmit a connect (off-hook) message to the calling-module processor, the wait time spent waiting in the input buffer (Fig. 11–7), 22-msec task service time (Table 11–3) to process the connect task to completion, plus another 5 msec for the line-interface processor to receive and process the dial-tone message and deliver dial tone to the calling party. For the FIFO case, the average dial-tone delay, in msec, is just

$$\text{dial-tone delay} = E(W) + 32 \qquad (11-7)$$

with $E(W)$ given by Eq. (11–1) or Eq. (11–6).

For the nonpreemptive priority case we get

$$\text{dial-tone delay} = E(W_1) + 32 \qquad (11-8)$$

with the delay again measured in msec, and the wait time $E(W_1)$ in the higher-priority input buffer 1 given by Eq. (11–2). Equation (11–5) for the second moment of the task service time must be used in Eq. (11–2). The resultant dial-tone delays, for a number of values of processor utilization ρ, appear in Table 11–4. Note, as expected, that dial-tone delay is improved with the priority discipline because of the reduction in the average wait time. With the num-

TABLE 11-4   Time Delays from Model of Fig. 11-7*

ρ	$E(W)$	$E(W_1)$	$E(W_2)$	Dial-tone Delay		Call-setup Time	
				Priority	No Priority	Priority	No Priority
0	—	—	—	32	32	44	44
0.5	6.16	4.3	8.6	36.3	38.2	56.9	56.3
0.7	14.4	7.1	23.6	39.1	46.4	74.7	72.8
0.9	55.4	11.2	112	43.2	87.4	167.2	154.8

* All delays in msec

bers chosen here this delay is obviously quite small, even for $\rho = 0.9$. Other choices of task service times in Table 11-3 would have resulted in different values for dial-tone delay. Some other examples appear in the problems at the end of this chapter.

Now consider the call-setup time. Here we calculate one component of this time only, the time required for call setup to be processed at *this* (hypothetical) digital office only. In practice one would have to repeat the calculation for every office along the path taken by the call-setup (seizure) message from originating office to terminating office. The call-setup time calculated here is essentially the call-processing time T_p encountered in our first (queued) model of circuit switching in Chapter 10, which we neglected in our calculations there. In the model under discussion here, this time is the time from receipt of a routing digit (Fig. 11-6) to the time a seizure signal is forwarded on an appropriate outgoing trunk. Note from Figs. 11-6 and 11-7 that two wait times are involved—one at input buffer 1 of the calling-module processor, with the routing-digit message from the line-interface processor waiting to activate a "routing-at-end" task; the other at input buffer 2 of the called-module processor, with the message delivered to that processor on completion of the "routing-at-end" task waiting to activate a connect (seizure) task. From Table 11-3 the "routing-at-end" task requires 12 msec to be completed, while the connect (seizure) task again requires 22 msec. Adding 5 msec at either end to account for time to receive and deliver messages to the line interface, one gets the following equations for the call-setup time, in msec, in *this* office only.

For the FIFO (no-priority) case,

$$\text{call-setup time} = 2E(W) + 44 \qquad (11-9)$$

For the nonpreemptive priority case,

$$\text{call-setup time} = E(W_1) + E(W_2) + 44 \qquad (11-10)$$

Values for this parameter also appear in Table 11 – 4 for various values of the processor utilization ρ. Note that for this parameter the effect of the priority discipline is to increase the call-setup time somewhat, although not by very much. Other choices of task service times would produce different results. Some examples are again left to the reader as problems at the end of this chapter.

Other service disciplines could be tried in a similar manner, to see the impact on both dial-tone delay and call-setup time. For example, one might want to *reverse* the priority chosen here and give higher priority to those tasks, such as those involved in the call-setup scenario, that involve processing at both module processors. This would reduce the call-setup time at the expense of increasing dial-tone delay. Details are left to the reader. The main point is that with the queueing model of Fig. 11 – 7 and the various task service times provided in Table 11 – 3, one could try out different service disciplines to see the effect on performance parameters of interest.

11 – 3 Overload Controls for Circuit-switching Machines

It was noted in the introductory paragraphs of this chapter that overload- or congestion-control mechanisms are often used in switching systems to prevent systems from deadlocking, to keep call processing viable at times of extremely heavy load, and to allow the control processors to continue to carry out necessary administrative tasks under overload conditions. Controls are necessary for circuit-switching systems just as they are necessary for packet-switched networks: Mechanisms must be introduced to protect finite resources in the face of high demand.

A large number of control mechanisms have been implemented or proposed for digital switching mechanisms [ITC 10], [ITC 9]. Some of them tend to be very similar to the control mechanisms for packet switching discussed in Chapter 5. As was the case there, two problems arise in designing an overload control mechanism: First, how does one measure the onset of congestion? Second, how does one control congestion to keep it from inundating a system? As in

---

[ITC 10] Session 5.2, "Overload Control," ITC-10, Tenth International Teletraffic Congress, Montreal, June 1983. The papers by P. Tran-Gia, B. T. Doshi and H. Heffes, F. C. Schoute, and L. J. Forys all discuss overload control. Another paper, by M. Eisenberg, given at Session 1.3 of the same conference, focuses on overload control.

[ITC 9] ITC-9, Ninth International Teletraffic Congress, June 1979. See the papers by A. Briccoli; J. A. G. Higuera and C. D. Berzosa; F. C. Schoute; P. Somoza and A. Guerreo; and K. Wilding and T. Karlstedt.

the packet-switched case, estimates of call-arrival rate and call-processing or task queues that exceed specified thresholds, among other measures, can be used to detect the onset of congestion. Controls used or proposed include window mechanisms, as in the packet-switched case, and mechanisms for controlling the rate of calls to be handled by the system processors. These controls will either block user traffic (i.e., produce a busy signal) or delay the delivery of a dial tone.

In this section we sample both approaches. We do this by first describing an idealized queueing model of call processing and then analyzing two controls based on this model: one that blocks calls and another that introduces dial-tone delay to keep congestion down. Note that these two mechanisms provide the same time-delay – throughput tradeoff encountered in Chapter 5 in our discussion of congestion controls for packet switching.

We then describe and compare analytically two overload controls appropriate to a system like the AT&T No. 5 ESS switch, discussed earlier in this chapter. Both controls are of the window type. One control delays the delivery of dial tone; the other blocks calls (produces a busy signal) after collecting the dial digits.

11-3-1 Idealized Models of Overload Control*

To study two different control mechanisms — one that blocks calls and another that delays the dial tone — we first introduce a simple queueing model of call processing in the unprotected (uncontrolled) state. Consider the infinite $M/M/1$ queue with feedback shown in Fig. 11-9. Calls are assumed to arrive at a single processor at a Poisson rate of λ calls/sec on the average. Calls are broken into tasks, as in the previous section, and it is these *tasks* that are served in the model of Fig. 11-9. Task-processing times are assumed to be distributed exponentially, with an average value of $1/\mu$.

A typical call completes and leaves the processing system with probability $(1 - \alpha)$. With probability α, additional tasks must be carried out, and the call is put back at the end of the processing queue. It is apparent that the average number of times a call cycles through the system is $1/(1 - \alpha)$, which then represents the average number of tasks/call. It is thus comparable to the parameter f introduced in our queueing model of the last section. As a check, consider the queue-throughput parameter γ shown in Fig. 11-9. Since $\gamma = \lambda + \alpha\gamma$, from flow conservation, we have

$$\gamma = \lambda/(1 - \alpha)$$

---

* Material in this subsection is based on [WALL].

[WALL] B. Wallstrom, "A Feedback Queue with Overload Control," ITC-10, Tenth International Teletraffic Congress, Montreal, June 1983, Session 1.3, paper 4.

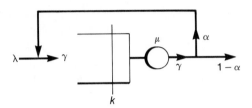

Figure 11–9 Idealized queueing model, call processing

γ thus represents the *task* throughput rate of the system and is comparable to $f\lambda$, the task-arrival rate of the previous section.

This idealized model is not directly comparable to the one used in the previous section to model a more realistic (albeit hypothetical) system. The model there implicitly assumed that tasks distributed randomly through all stages of the call-processing cycle were present in the queues (Fig. 11–7). That model was used to calculate the waiting time $E(W)$ (in the no-priority case) of a *single task*. This, plus the service time of a particular task, could be used to calculate the time for a *single pass* of that task through the processing system. By combining a sequence of tasks one could calculate the time required to process that group. Note that there was no feedback in that model, as is the case here. Here a given call is considered to spawn $1/(1 - \alpha)$ tasks on the average, and the time to get through the system is a measure of the calling-processing time, from the time the call enters the system to the time it leaves. The model in this section cannot possibly account for the call mix described in the previous section nor for the many types of tasks that can be encountered in handling different types of calls (Fig. 11–6). As an idealized model, it might be used to attempt to quantify a portion of the call-processing scenario. It might be used, for example, to describe in idealized fashion the tasks required to provide dial tone to a user, or some other cohesive subset of the call-processing procedure. More importantly for our purposes, however, the model serves the tutorial use of introducing simple models for control mechanisms that introduce either dial-tone delay or blocking of calls (a busy signal) in order to relieve congestion. These models can readily be analyzed, as shall be seen, and tradeoffs between delay and throughput can be quantified. In the next subsection, in our description of a more realistic model, the same tradeoffs occur.

Returning to the feedback queueing model of Fig. 11–9, it is left to the reader to show that the system behaves just like a simple M/M/1 queue with arrival rate $\gamma = \lambda/(1 - \alpha)$ and service rate $1/\mu$. (As a hint, one can invoke the open queueing network ideas of Chapter 5 and prove that even with feedback the product-form M/M/1 solution is retained.) Specifically, then, with k the state of the system—i.e., the number of tasks on queue, including any in service—one can immediately write as the probability distribution of the states

$$p_k = (1 - \rho)\rho^k \qquad \rho = \gamma/\mu = \lambda/\mu(1 - \alpha) \qquad (11-11)$$

The average number of tasks on queue is

$$E(k) = \rho/(1 - \rho) \qquad (11-12)$$

the average time for a single task to get through the queue is

$$\overline{T}_T = E(k)/\gamma = 1/\mu(1 - \rho) \qquad (11-13)$$

and the average time to process $1/(1 - \alpha)$ tasks, i.e., the call-processing delay for this model, is

$$\overline{T}_c = \overline{T}_T/(1 - \alpha) = 1/\mu(1 - \alpha)(1 - \rho) \qquad (11-14)$$

(As a check, from Little's formula, $\overline{T}_c = E(k)/\lambda$, producing Eq. (11–14) a different way.)

This provides the "unprotected" queue analysis. We now introduce the two overload control models described earlier. The first is a *call-blocking model*. This model simply consists of a *finite queue*, holding at most N tasks which blocks calls if N tasks are present. The model appears in Fig. 11–10. It is left to the reader to again show that the probability distribution of this finite-buffer queueing system is given by the finite M/M/1 distribution

$$p_k = \rho^k(1 - \rho)/[1 - \rho^{N+1}] \qquad \rho = \lambda/\mu(1 - \alpha) \qquad (11-15)$$

$$0 \le k \le N$$

The blocking probability is just

$$P_B = P_N = \rho^N(1 - \rho)/[1 - \rho^{N+1}] \qquad (11-16)$$

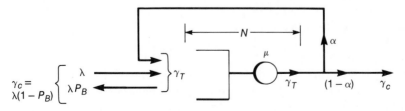

Figure 11–10 Call-blocking model

Since calls are blocked when the task queue is full, two throughputs must be distinguished. The net *call* throughput, in calls handled per unit time and labeled γ_c in Fig. 11–10, is just

$$\gamma_c = \lambda(1 - P_B) \tag{11–17}$$

while the *task* throughput, in tasks handled per unit time, is

$$\gamma_T = \gamma_c/(1 - \alpha) = \lambda(1 - P_B)/(1 - \alpha) \tag{11–18}$$

This parameter is also labeled as such in Fig. 11–10.

The call-processing delay $\overline{T}_c$, or the time to process $1/(1 - \alpha)$ tasks on the average, varies with the control parameter N as well. As N is reduced, $\overline{T}_c$ is reduced accordingly, but at the price of reducing the call throughput γ_c. Both functions, γ_c and $\overline{T}_c$, are found readily from Eqs. (11–15) to (11–17). Thus

$$\overline{T}_c = E(k)/\gamma_c = E(k)/\lambda(1 - P_B) \tag{11–19}$$

with $E(k)$, the average number of customers (tasks) in a finite $M/M/1$ queue, given by

$$E(k) = \sum_{k=0}^{N} kp_k = \frac{\rho}{(1 - \rho)(1 - \rho^{N+1})}[1 - (N + 1)\rho^N + N\rho^{N+1}] \tag{11–20}$$

and P_B given by Eq. (11–16). From these equations, one finds readily that

$$\overline{T}_c = \frac{\rho}{\lambda(1 - \rho)}\left[\frac{1 - (N + 1)\rho^N + N\rho^{N+1}}{1 - \rho^N}\right] \tag{11–21}$$

The first term, in front of the brackets, is just the uncontrolled, infinite-queue result, written previously as Eq. (11–14).

Some special cases are of interest. First let $\rho = 1$. As in our congestion-control analysis of packet-switched networks in Chapter 5, this value of utilization represents the beginning of the onset of heavy congestion. (The finite-queue model of Fig. 11–10 enables us to handle values of $\rho \geq 1$, as already noted a number of times in this book, beginning with Chapter 2.) It is left to the reader to show, either by setting $\rho = 1 - \epsilon$ and letting $\epsilon \to 0$ or by using L'Hôpital's rule in Eq. (11–21) that at this value of ρ, the average time delay required to process the $1/(1 - \alpha)$ tasks is just

$$\overline{T}_c|_{\rho=1} = (N + 1)/2\mu(1 - \alpha) \tag{11–22}$$

This shows the limit in delay introduced by the control. Since in this model a typical call requires $1/\mu(1 - \alpha)$ units of time to be processed, at most $(N + 1)/2$ such units of time are required at $\rho = 1$. This contrasts with $\overline{T}_c$ in the (idealized) unprotected infinite-buffer model of Fig. 11–9. (In practice, of course, there is no such thing as an infinite buffer. With a more realistic model one might show an unstable situation occurring without control, which could lead to deadlock.) The effect of the control is thus quite apparent; it agrees with the notion of limiting the time delay, encountered previously in Chapter 5. The tightest control is represented, of course, by letting $N = 1$, i.e., limiting the number of tasks in queue to 1 at most. The price paid is the usual reduction in call through-put (discussed in Chapter 5).

Before moving on to the time-delay–throughput tradeoff, it is of interest, as in Chapter 5, to consider the limiting case of traffic intensity. Letting $\rho \to \infty$, to check the effectiveness of the control at this extreme value of traffic, one finds readily from Eq. (11–21) that $\overline{T}_c$ approaches

$$\overline{T}_c|_{\rho \to \infty} = N/\mu(1 - \alpha) \qquad (11–23)$$

For large N this is about double the time delay experienced at $\rho = 1$. A composite curve showing the normalized time delay $\mu(1 - \alpha)\overline{T}_c$ as a function of ρ, in the controlled case, appears in Fig. 11–11. (Note that at values of $\rho \leq 0.6$ or so, for large enough N, the bracketed term of Eq. (11–21) is close to 1, and the delay incurred is almost that of the uncontrolled case, given by the leading term of Eq. (11–21), the familiar infinite M/M/1 queue time delay.)

Now consider the effect of this call-blocking control on the throughput. As the task buffer size N is reduced calls will be blocked more often, reducing

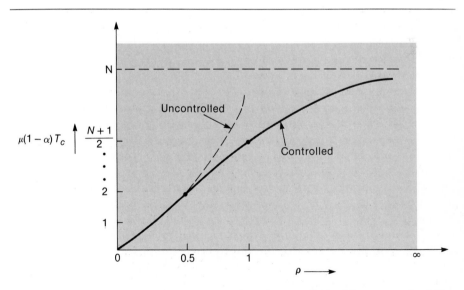

Figure 11–11 Normalized call-processing time, call-blocking control of Fig. 11–10

throughput. To demonstrate this, consider Eq. (11 – 16) for the blocking probability. Again take the case $\rho = 1$, the point at which heavy congestion begins to be encountered. It is left to the reader to show that at this point the blocking probability is given by

$$P_B|_{\rho=1} = 1/(N+1) \qquad (11-24)$$

The relative throughput is $(1 - P_B) = N/(N+1)$. Thus, for the tightest control, with $N = 1$, $P_B = 0.5$, 50 percent of all calls are blocked, and the throughput is 50 percent of the load at that point. As N increases, the blocking probability decreases and the relative throughput increases, but call-processsing time $(\overline{T_c})$ goes up as well, as shown by Eq. (11 – 22). This is precisely the same tradeoff mechanism we encountered in Chapter 5, in our discussion of congestion controls. For $N = 2$, only 33 percent of the calls are blocked, 67 percent get through, and the delay is $(N + 1)/2 = 1.5$ times the minimum processing time.

We can be more precise about the throughput, for all values of ρ, by normalizing γ_c, the call throughput, to $\mu(1 - \alpha)$, the *maximum* number of calls per unit time that can be handled by the model of Fig. 11 – 10. (Recall that $1/\mu$ is the average *task* processing time and $1/\mu(1 - \alpha)$ the average *call* processing time, by definition.) Doing this, we find, from Eq. (11 – 17), that the normalized call throughput, as expected (see Chapter 2), is given by

$$\gamma_c/\mu(1 - \alpha) = \rho(1 - P_B) \qquad (11-25)$$

For $\rho = 1$, this is just $N/(N+1)$, as noted previously. For $\rho \gg 1$, it is readily shown, from Eq. (11 – 16), that $P_B \rightarrow 1 - 1/\rho$, and $(1 - P_B) \rightarrow 1/\rho$. The blocking probability becomes very high and most calls are turned away, but the task queue is always full at its maximum value of N, and the call throughput rate is at its maximum possible value of $\mu(1 - \alpha)$ calls per unit time. Figures 11 – 12 and 11 – 13, for the blocking probability P_B and the normalized throughput rate $\gamma_c/\mu(1 - \alpha)$, respectively, summarize these elementary notions. (Note again that this was exactly the physical operation described in Chapter 2 in the discussion of the blocking characteristics of the finite M/M/1 queue.)

We discuss this simple model for call-blocking control further after we describe the second control model based on the unprotected call-processing model of Fig. 11 – 9. The second model incorporates a call-processing *delay* mechanism. This could be interpreted as a scheme that delays dial tone rather than blocking a call. Consider specifically the queueing model of Fig. 11 – 14 [WALL]. The single queue of the unprotected call-processing model of Fig. 11 – 9 has been replaced by *two* queues: a finite *server queue* holding at most M *tasks* (including one in service) and an *external queue* that keeps *calls* waiting until their task processing can begin when they are let into the server queue.

If M tasks are waiting in the server queue (including the one in service), additional calls arriving are kept waiting in the external queue. If fewer than M

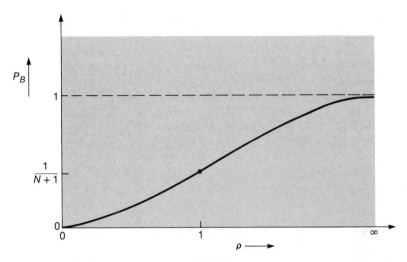

Figure 11–12 Blocking probability characteristics, queue of Fig. 11–10

tasks appear in the server queue, calls are immediately steered to the server queue, without incurring a delay in the external queue. Passing from one queue to the other is assumed to take zero time in this model. The external queue can be visualized as one holding calls that are waiting to receive dial tone, and delays in that queue will henceforth be called dial tone delays for simplicity.

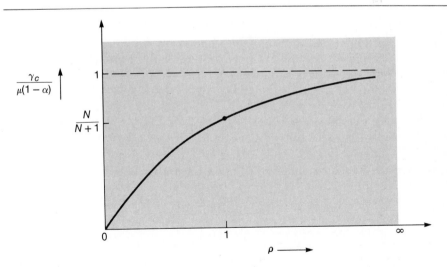

Figure 11–13 Relative throughput, call-blocking control of Fig. 11–10

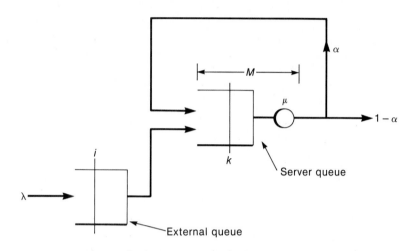

Figure 11–14 Model, delayed call-processing control

To analyze this system, let i represent the number of calls waiting in the external queue, and let k again represent the number in the finite server queue, including the one in service. Hence $0 \leq k \leq M$, $0 \leq i < \infty$. We use the symbol M for the control here to distinguish it from the call-blocking control discussed previously. (Note that there is no blocking in this control scheme because of the infinite external queue assumed.) To study the statistical properties of this two-queue system, let $P(i,k)$ represent the (two-dimensional) probability that k tasks are waiting in the server queue, while i calls are waiting in the external queue. It is apparent that for $0 \leq k \leq M$, i must be zero, while for $i \geq 1$, $k = M$. Thus the only $P(i,k)$ terms defined are

$$P(0,k) \qquad 0 \leq k \leq M$$

and

$$P(i,M) \qquad 0 \leq i < \infty$$

One may write balance equations to find these terms. Four such equations are needed; they follow in order:

(1) $$\lambda P(0,0) = \mu' P(0,1) \tag{11-26}$$

(2) $$(\lambda + \mu') P(0,k) = \lambda P(0,k-1) + \mu' P(0,k+1) \qquad 1 \leq k \leq M-1 \tag{11-27}$$

Here the parameter $\mu' = \mu(1 - \alpha)$ has been introduced to simplify notation.

These first two equations are precisely those of a finite M/M/1 queue. Continuing, we have as the third balance equation

(3) $$(\lambda + \mu')P(0,M) = \lambda P(0, M - 1) + \mu'P(1,M)$$ (11-28)

The last term in this equation represents the rate of transition from the external queue to the server queue. Finally, the balance equation for the external queue, precisely that of the M/M/1 queue, is given by

(4) $(\lambda + \mu')P(i,M) = \lambda P(i - 1,M) + \mu'P(i + 1,M)$ $i = i, 2, 3, \ldots$ (11-29)

It is left to the reader to show that the solution to this set of equations is of the product form (Chapter 5):

$$P(0,k) = (1 - \rho)\rho^k \qquad 0 \le k \le M$$ (11-30)

and

$$P(i,M) = (1 - \rho)\rho^{i+M} \qquad 0 \le i < \infty$$ (11-31)

Here $\rho \equiv \lambda/\mu(1 - \alpha)$ as previously. Note that the complete system behaves like a single M/M/1 queue with $\ell = (i + k)$ customers. As such, the average time spent in the complete system is

$$E(T) = 1/\mu(1 - \alpha)(1 - \rho)$$ (11-32)

independent of the control M. The purpose of the control is to transfer delay from one queue to the other. Thus as M is reduced, corresponding to a tightening of control, the call-processing delay decreases, at the expense of increasing the external-queue or dial-tone delay. The first delay, the call-processing delay, is obviously related to the average number of tasks in the server queue. From Eqs. (11-30) and (11-31), it is given by

$$E(k) = \sum_{k=0}^{M} \sum_{i=0}^{\infty} kP(i,k)$$

$$= \sum_{k=0}^{M} kP(0,k) + M \sum_{i=1}^{\infty} P(i,M)$$ (11-33)

$$= \frac{\beta}{(1 - \rho)} [1 - \rho^M]$$

after some manipulation of terms and simplification. Since the single M/M/1 queue has as the average number on queue $E(\ell) = \rho/(1 - \rho)$ and since $i + k = \ell$, we have immediately for $E(i)$, using Eq. (11-33),

$$E(i) = \rho^{M+1}/(1 - \rho)$$ (11-34)

It is left to the reader to show, as a check, that this result may be obtained directly from Eq. (11-31) and the definition of $E(i)$.

The variation of the average number in each of the two queues with change of the control parameter M is readily demonstrated through an example. Let $\rho = 0.9$, a high congestion point. (Remember again that unlike the previous finite queue control, we must maintain $\rho < 1$ with this model to ensure stability.) Then the average number on both queues together is always $\rho/(1-\rho) = 9$, independent of M. Figure 11–15 shows the corresponding average number on each of the two queues plotted as a function of M.

The corresponding time delay in each of the two queues is found using Little's formula. Specifically, we have, as the average *call* delay due to waiting in the server queue (the call-processing delay),

$$E(T_s) = E(k)/\lambda \qquad (11-35)$$

while the average delay in the external queue, dial-tone delay, is given by

$$E(T_e) = E(i)/\lambda = E(T) - E(T_s) \qquad (11-36)$$

(Why does one divide by λ? What does one get if $E(k)$ is divided by $\gamma = \lambda/(1-\alpha)$?) Table 11–5 provides the resultant values for the various normalized time delays (again normalized to the average call-processing time $1/\mu\,(1-\alpha)$) for the same example as that of Fig. 11–15, $\rho = 0.9$. Note the tradeoff mentioned earlier, which is also apparent from Fig. 11–15. As the control parameter M (the number on the server queue) is decreased, tightening

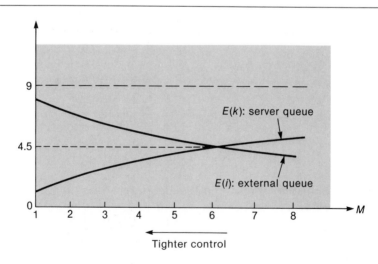

Figure 11–15 Effect of control of Fig. 11–14 on average queue size, $\rho = \lambda/\mu = 0.9$

TABLE 11–5   Effect of Delayed Processing (External Queue) Control, from Fig. 11–14

Control M	E(k)	Normalized Delays		
		$\mu(1-\alpha)E(T_e)$	$\mu(1-\alpha)E(T)$	$\mu(1-\alpha)E(T_s)$
1	0.9	9	10	1
2	1.7	8.1	10	1.9
3	2.4	7.3	10	2.7
4	3.1	6.6	10	3.4
5	3.7	5.9	10	4.1
6	4.2	5.3	10	4.7
7	4.7	4.8	10	5.2

the control, the delay shifts from call-processing delay to external-queue ("dial-tone") delay.

A comparison of the two types of control mechanisms studied in this section via the idealized model appears in Table 11–6. We have chosen $\rho = 1$ for the call-blocking control case and $\rho = 0.9$ for the delayed-call control case to simplify the calculations. We have included, in addition to the delays and blocking probability, the average number $E(k)$ in the serving queue in either case. This may be interpreted as providing a measure of the load on the processor, though a direct comparison is difficult because of the difference in the performance characteristics of the two controls. Thus the call-blocking control introduces blocking to reduce congestion, as measured by the call-processing delay. Yet at the same time this reduces the call throughput and forces users (customers) to retry. The delayed call-processing (delayed dial tone) control reduces call-processing delay by forcing newly arriving calls to wait on an external queue. In practice this would be implemented by delaying the delivery of dial tone. This strategy will be discussed in more detail in the next section.

A study of Table 11–6 indicates that, aside from the basic problem of not really being able to compare the effect of blocking probability on users in one case with dial-tone delay in the other, the two control mechanisms can be made to provide the same performance so far as call-processing delay and processor load are concerned. Consider the blocking-control case with $P_B = 0.1$ and a relative throughput of 0.9, for example. The normalized call-processing delay is 5 from Table 11–6. If one chooses a control of $M = 7$ for the queued control, the normalized call-processing delay is 5.2. The processor loads, measured in terms of the average number of tasks in the server queue, are 4.5 and 4.7, from Table 11–6. The numbers are quite comparable. On the other hand, the blocking control does block 10 percent of the calls in this model but with no dial-tone delay, while the queued control introduces a relative dial-tone delay of 4.8 with no blocking of calls.

We shall not pursue this model at this point, leaving further examples and calculations to the reader. As noted earlier, the model was introduced because of its relative simplicity and because it enables us to focus directly on the two control mechanisms. In the next section we model two similar controls, using a more realistic model of a circuit-switched processing system. We find there that by proper choice of the control parameter in either case the two controls can again be made to provide about the same performance.

11–3–2 Overload Controls for Hierarchical Distributed Systems

To complete this discussion of overload controls, we have selected two window-type controls described in the literature that are applicable to systems such as the AT&T No. 5 ESS, discussed earlier in this chapter, that have a two-level processor hierarchy.

One could impose an overload control at each level, just as is done in some of the packet-switched networks described in Chapter 5. The control we describe here is designed to protect the higher-level central processor. The lower-level line/trunk interface processors could be protected independently by a mechanism such as that described and analyzed by M. Eisenberg [EISE]. We follow, in greatly simplified form, the work of Doshi and Heffes [DOSH 1982], [DOSH 1983]. They have studied, in addition to window-type controls, rate-based controls that limit the rate at which calls are sent to the control processor. They have also compared FIFO and LIFO (last in–first out) service disciplines. We describe the FIFO models and analyses only.

The LIFO algorithm has been modified to reward customers that have continued to wait beyond a specified threshold by putting them at the head of the queue at that time. This algorithm has been found, through analysis, simula-

[EISE] M. Eisenberg, "A Strict Priority Queueing System with Overload Control," ITC-10, Tenth International Teletraffic Congress, Montreal, June 1983, Session 1.3.

[DOSH 1982] B. T. Doshi and H. Heffes, "Comparison of Control Schemes for a Class of Distributed Systems," 21st IEEE Conf. on Decision and Control, Orlando, Fla., Dec. 1982.

[DOSH 1983] B. T. Doshi and H. Heffes, "Analysis of Overload Control Schemes for a Class of Distributed Switching Machines," ITC-10, Tenth International Teletraffic Congress, Montreal, June 1983, Session 5.2, paper 2.

TABLE 11–6 Comparison of Two Controls, Call-processing Model, Fig. 11–9

Control (N or M)	Processor Load $E(k)$		Normalized Call-processing Delay		Queued External Delay	Blocked P_B
	Queued ($\rho = 0.9$)	Blocked ($\rho = 1$)	Queued $\mu(1-\alpha)E(\overline{T_s})$	Blocked $\mu(1-\alpha)\overline{T_c}$	$\mu(1-\alpha)E(T_e)$	
1	0.9	0.5	1	1	9	0.5
3	2.4	1.5	2.7	2	7.3	0.25
5	3.7	2.5	4.1	3	5.9	0.17
7	4.7	3.5	5.2	4	4.8	0.125
9		4.5		5		0.1

tion, and measurement in the field, to provide superior overall performance [FORY]. It accounts for, and tends to correct for, customer behavior under overload. It is found that as dial-tone delay increases in overload, customers may dial prematurely or may abandon the call completely and try again. With the FIFO discipline this can result in an unstable situation, leading to deadlock as load increases persist (precisely the phenomenon noted in Chapter 5). The LIFO discipline, with its reward to patient customers, eliminates the instability. It has been adopted for use in the AT&T No. 1 ESS, the AT&T large analog-stored program-control system [FORY]. To simplify the presentation here, we consider only the FIFO strategies.

The model of the switching system to be controlled appears in Fig. 11–16. Note that the model is similar to that of the AT&T No. 5 ESS, described earlier in this chapter. (See Fig. 11–3 in particular.) M multiple peripheral controllers (PCs) are shown connected to one another and a control processor (CP) through a message switch (MS). The PCs are assumed to do most of the call handling. Specifically, in this model the PC detects off-hook, provides dial tone, and collects digits. It then transmits a message to the CP via the MS, requesting that a route be set up through the network switch. The CP determines the appropriate outbound PC and its destination line/trunk, sends messages via the MS to both PCs to that effect, and sets up the connection through the network switch. Details of this call-processing procedure appear in our description of the AT&T No. 5 ESS exchange in Subsection 11–1–2.

The basic congestion problem arising is that the CP bottlenecks if too many calls are waiting to be processed. The CP then cannot carry out the necessary non-call-processing work (the central control, administration, and maintenance work noted at the beginning of this chapter). A number of control strategies can be adopted. We focus on window-based schemes in this subsection, as already noted. These schemes fall into two categories. In the first, the PC can delay giving dial tone, can block customers, or both; this is called a pre-PC control scheme. The second strategy blocks customers by giving them a busy signal *after* the dial digits have been collected; it is called a post-PC scheme. Note that the delayed dial tone and the blocking strategies are precisely those modeled in a very idealized fashion in the previous section.

The general queueing model of the system to be controlled, based on the system model of Fig. 11–16, appears in Fig. 11–17. It focuses on one PC only, with the model of each PC assumed to be of the same form [DOSH 1983, Fig. 2]. Note the external queue appearing in the PC portion of the model. This is similar to the external queue of Fig. 11–14, used to buffer calls waiting for dial

[FORY] L. J. Forys, "Performance Analysis of a New Overload Strategy," ITC-10, Tenth International Teletraffic Congress, Montreal, June 1983, Session 5.2, paper 4.

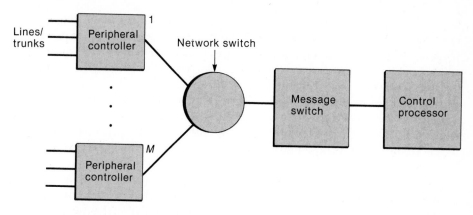

Figure 11–16 Model of hierarchical digital switching system

tone. The model used to represent the dialing of digits is of the infinite server $M/M/\infty$ type (Chapter 2), with dialing time assumed exponential and of average value $1/\mu_D$ units of time. This implies that any number of calls may be in the dialing stage. Dialing is followed by a processing stage in the PC, with messages being formed for delivery via the message switch (MS) to the central processor (CP). Processing (service) time is taken to be exponential, with an average of

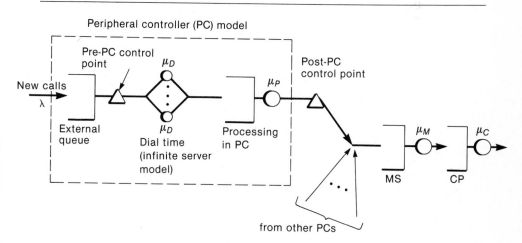

Figure 11–17 Queueing model, digital switching system (after [DOSH 1983, Fig. 2], with permission)

$1/\mu_P$ time units. The message switch and central processor, operating in tandem, accept messages and process calls (the call-setup stage), respectively, from all PCs, with exponential service rates of average values $1/\mu_M$ and $1/\mu_C$, respectively.

Pre-PC control strategy   How are the window controls now exerted? Consider the pre-PC window-control strategy first. In this scheme a maximum of K calls for each PC can be in either the PC dialing stage, in the PC processing stage, or in the MS or CP for processing. If K calls at a PC *are* in any one of these stages, new calls arriving are held in the external queue and dial tone is delayed. A model of this window-control scheme for a single PC (the other PCs have the same control applied) appears in Fig. 11–18. Note that the resultant closed queueing network, focusing on a single PC only, is similar to that proposed for the sliding-window control in Chapter 5 (see Fig. 5–11). As a call completes processing, another one may be admitted to the system. Here, similar to the (artificial) $(M+1)$ queue of Fig. 5–11, a "token" store is indicated. As a call is completed, its "token" is conceptually placed in the store. A call that attempts to enter the controlled system from the external (dial-tone delay) queue must first receive a token from the store. If the store is empty, i.e., K calls are in progress for this PC, new calls attempting to enter are held in the external store.

It is clear that, by modeling the control of each PC in an M PC system by the same circulating token model, one obtains a combined open-closed queueing network of the type discussed in Chapter 5. Some of the methods noted there for

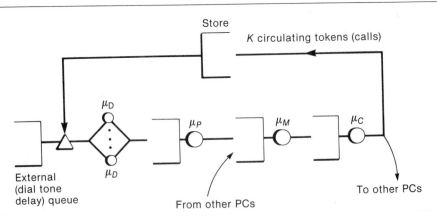

Figure 11–18   Model, pre-PC window control (after [DOSH 1982, Fig. 3(a)], © 1982 IEEE, with permission)

solving such networks could be used. Instead, we resort, as in Chapter 5, to simplifications and approximations where possible.

First, assume that in a period of congestion a call is *always* waiting in the external queue for processing to begin (i.e., dial tone to be delivered). As a call completes processing and leaves the system, another immediately takes its place. K calls per PC must thus always be in progress somewhere in the processing stages. This enables us to decouple the circulating-token, closed-network model of Fig. 11–18 from the external queue. We can calculate the throughput (calls processed per unit time) of the closed network. By flow conservation this must be the throughput of the external queue; it is obviously a function of the window size K. It also represents the *capacity* (maximum throughput) of the external queue since it is the limiting call-handling rate imposed by the window control. With this call-handling capacity known, the external queue statistics, from which dial-tone delay is calculated, may readily be found for any call-arrival rate (load on the system) less than the capacity, using our queueing analysis of Chapter 2. We can in fact determine the tradeoffs among call-handling capacity, call-processing time, and dial-tone delay, all as a function of the window control K, for this model.

This method of decomposing a complex queueing network into smaller, more manageable problems is often called the *method of decomposition*. A very similar solution approach has been used by Kobayashi to solve a computer system performance model arising in the analysis of an interactive virtual storage system with multiprogramming [KOBA, pp. 178–187]. A related, much more detailed discussion appears in [SAUE, Sec. 6.3], where the token model introduced in Fig. 11–18 is used explicitly.

We now proceed with the details of the analysis. Under the assumption noted previously that there are always calls waiting for a dial tone in the external queue, the model of Fig. 11–18 simplifies to the one shown in Fig. 11–19(a). The parameter $\lambda_i^T(K)$ indicated is the throughput of the ith PC noted earlier that must be found to enable decomposition to be used. The parameter λ_{cp}^T is the total throughput, or call-handling capacity of the complete system, obtained by adding all the $\lambda_i^T(K)$'s. Note that this closed queueing network model for the ith PC, together with the $(M-1)$ other identical networks resulting when all M PC controls are modeled, gives rise to a composite closed queueing network with M chains, all coupled through the common MS and CP queues. (Implicit in the modeling is the assumption that all M PCs have identical traffic and processing characteristics.) This composite network model can be solved using standard

[KOBA] H. Kobayashi, *Modeling and Analysis: An Introduction to System Performance Evaluation Methodology,* Addison-Wesley, Reading, Mass., 1978.

[SAUE] C. H. Sauer and K. M. Chandy, *Computer Systems Performance Modeling,* Prentice-Hall, Englewood Cliffs, N.J., 1981.

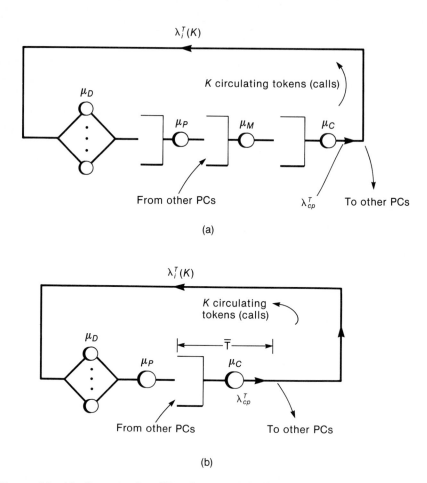

$\lambda_i^T(K)$

K circulating tokens (calls)

From other PCs λ_{cp}^T To other PCs

(a)

$\lambda_i^T(K)$

K circulating tokens (calls)

From other PCs   To other PCs

(b)

Figure 11–19 Steps in simplification, model of Fig. 11–18
   a. Simplification, model of Fig. 11–18, using decomposition
   b. Final simplified model

methods for solving queueing networks, as already noted. However, we shall resort to further simplification and an approximation procedure to obtain results much more readily.

Specifically, consider some typical numbers [DOSH 1983]. Let the average message switch-processing time be $\mu_M^{-1} = 11$ msec, the average control-processor service time be $\mu_C^{-1} = 23$ msec, and the PC-processing time be $\mu_P^{-1} = 52$ msec. The maximum call throughput of the complete system for these numbers is clearly limited by $\mu_C = 1000/23$ calls/sec, or 157,000 calls/hour. The throughput λ_{cp}^T shown in Fig. 11–19(a) can never exceed this value and can only approach it as the window control $K \to \infty$. But for $M \gg 1$ (we shall use $M = 20$ PCs in a complete example to follow), $\lambda_i^T(K) = \lambda_{cp}^T/M \ll \mu_C$. In addition, for these numbers, $\lambda_i^T(K) \ll \mu_P = 1000/52$ calls/sec, or 72,000 calls/hour, the call-handling capacity of a single PC. Hence the call-handling capacity is limited by the CP, not by the PCs: The PC-processing queue in Fig. 11–19(a) is almost always empty, and that queue may be dropped from the figure. As another simplification, we also neglect MS processing, since $\mu_M > \mu_C$. (This latter approximation is, of course, not as valid as the previous one but still leads to reasonably correct results, as we shall indicate later in comparing our calculations with those in [DOSH 1983]. The approximation results in a greatly simplified analysis.) The final single-PC model to be analyzed is then that of Fig. 11–19(b).

This model can, of course, also be analyzed using closed queueing network solution techniques. Instead we resort to one final approximation that reduces the computational requirements enormously. We have used the symbol $\overline{T}$ in Fig. 11–19(b) to represent the average time delay through the CP (queueing plus service time); it represents the CP portion of call-processing time. We can find $\overline{T}$ in terms of the other parameters of Fig. 11–19(b) using Little's formula. Specifically, we have

$$K = (\mu_D^{-1} + \mu_P^{-1} + \overline{T})\lambda_i^T(K) \qquad (11-37)$$

We need another expression for $\overline{T}$ to be able to calculate the PC throughput $\lambda_i^T(K)$. This is where we introduce another approximation. (Lest the reader become frightened at all these approximations, let him note that analysis calls for bold, courageous steps! Hopefully one can justify them by alternative approaches or by simulation if necessary.) Focus on the model in Fig. 11–19(b) of the CP system itself. This is just a single queue with service rate (capacity) μ_C and load (arrival rate) λ_{cp}^T. With product-form conditions in the original closed queueing network of Fig. 11–19(a) assumed to hold (i.e., all service times in that figure are exponential, with FIFO service used throughout), the CP queue statistics are of the M/M/1 type. If $MK \gg 1$, the queue can be modeled quite accurately as an infinite M/M/1 queue, and we write immediately for the time delay $\overline{T}$

$$\overline{T} = \frac{1}{\mu_C - \lambda_{cp}^T} \qquad (11-38)$$

(This last assumption is necessary; without it, even with a product-form solution the time delay is not necessarily of the form of Eq. (11–38). Recall from Eq. (5–50) that the λ_i appearing there for the ith queue is not a true throughput but a *relative* arrival rate. The normalization constant $g(N,M)$, in terms of which a time delay such as $\overline{T}$ would be found, must be evaluated by summing over all states of the composite closed network. One gets the M/M/1 solution only if the constraint of N customers in the closed network is relaxed. This is precisely what happens if we let $MK \gg 1$. Alternatively, a physical interpretation states that with a large number of queueing chains serving the CP queue, i.e., $M \gg 1$, the CP call-arrival statistics with average value λ_{cp}^T should approach the Poisson distribution.)

Defining $\tau \equiv \mu_D^{-1} + \mu_P^{-1}$, $\rho \equiv \lambda_{cp}^T/\mu_C$, and replacing $\overline{T}$ in Eq. (11–37) by Eq. (11–38), one gets, as the operative equation for the complete system of Fig. 11–19(b),

$$\frac{1}{(1-\rho)} + \mu_C\tau = KM/\rho \tag{11–39}$$

This is a quadratic equation in $\rho = \lambda_{cp}^T/\mu_C$ and may be solved, for μ_C, τ, and M fixed, for different values of the window-control parameter K. The example cited earlier, with numbers taken from [DOSH 1983], is a useful one to discuss. Recall that in that example we had $\mu_C^{-1} = 23$ msec, $\mu_P^{-1} = 52$ msec, and $M = 20$ PCs. Taking the average digit-dialing time μ_D^{-1} to be 2.9 sec (the number cited in the same reference), one finds that $\mu_C\tau = 129$. (Other examples, including the sensitivity of the analytic results to the choice of the digit-dialing time μ_D^{-1}, appear among the problems at the end of this chapter.) Using Eq. (11–39) to calculate ρ for different values of K, and from this to find λ_{cp}^T and $\overline{T}$, one gets the results shown in Table 11–7. Note, as expected, the limit introduced on the maximum throughput λ_{cp}^T by the window. In particular, as the window size K increases, the maximum throughput increases, approaching the call-handling capacity of the system, $\mu_C = 157,000$ calls/hour on the average, as defined for this system. The usual time-delay–throughput tradeoff appears as well: As K increases, increasing the throughput λ_{cp}^T, the CP processing or response time $\overline{T}$ increases as well. Table 11–7 also includes a column labeled "call-setup time." This time, $\overline{T} + \mu_P^{-1}$, would be the average time (neglecting the MS time delay) required to process a call through the switching system *after* dialing has been completed.

What value of window size K should be used? Note, from Table 11–7, that not only does $\overline{T}$ increase with K, but that the CP utilization ρ does as well. It was this parameter that, both at the beginning of this chapter and in our analysis of call processing for the hypothetical two-module switching system, we indicated we wanted to control. We noted that some processing capability should be left for administrative (non-call-handling) tasks, even in the case of heavy load. To see how administrative load enters into the limitation on the call-handling load, we use a simple model due to Doshi and Heffes [DOSH 1983] for calculating the effect of such a load. First note that in calculating the CP call-carrying capacity μ_C one must take into account CP processing overload in addition to the handling of call-processing tasks. Modeling this overhead as a fixed fraction OH of

TABLE 11–7   Performance of Window-based Pre-PC Control, Figs. 11–17 through 11–19*

K	ρ	λ_{cp}^T (calls/hr)	CP Response Time $\overline{T}$, msec	Call-setup Time, $\overline{T} + \mu_P^{-1}$, msec
3	0.46	72,000	42.6	95
4	0.61	95,000	59	111
5	0.75	117,000	92	144
6	0.88	138,000	192	244
7	0.95	149,000	460	512

* $M = 20$ PCs, $\mu_C = 157,000$ calls/hr, $\mu_P^{-1} = 52$ msec, $\mu_D^{-1} = 2.9$ sec

the CP processor utilization, and letting x_{cp} be the average time required to process a call at the CP, we have

$$\mu_C^{-1} = x_{cp}/(1 - OH) \qquad (11-40)$$

As an example, with $OH = 0.08$ (8 percent of the processor work is devoted to overhead tasks other than pure call-processing tasks), and $\mu_C^{-1} = 23$ msec, the number used in calculations thus far, $x_{cp} = 21.2$ msec is the average time actually devoted, per call, to handling call processing in the CP. Now say that the administrative load corresponds to a fraction AL of the CP processor utilization. This simple model, as in the case of the introduction of the overhead OH, ignores details of how administrative processing is done, representing this load as a constant fraction AL of the utilization. Then the call-carrying capacity of the CP is reduced to a value

$$\mu_C' = \frac{1 - OH - AL}{x_{cp}} < \mu_C \qquad (11-41)$$

Specifically, let $OH = 0.08$ and $x_{cp} = 21.2$ msec as previously. With these values we had $\mu_C = 1000/23$ calls/sec, or 157,000 calls/hour. For various values of administrative load AL, one can calculate the residual call-carrying capacity μ_C'. But this is precisely the value of λ_{cp}^T we desire, for by limiting λ_{cp}^T to that value using the window control, we allow the CP processor to continue to carry on administrative tasks, even in the face of severe congestion. We thus have, for the example just used,

$$\lambda_{cp}^T = \frac{(0.92 - AL)1000}{21.2} \text{ calls/sec}$$

$$\qquad (11-42)$$

or

$$170,000 \, (0.92 - AL) \text{ calls/hour}$$

As an example, say that 25 percent of the CP processor time is to be devoted to administrative work. From Eq. (11–42), one gets $\lambda_{cp}^T = 114{,}000$ calls/hour as the limit to be placed on the call throughput. From Table 11–7, in turn, one cannot quite get this value with the pre-PC window control. The closest controlled value appears to be 117,000 calls/hour on the average, with a window setting (calls/PC) of $K = 5$. Working backwards from Eq. (11–42), one then finds that with this value for the control, $AL = 0.23$, or 23 percent of the processor work could be devoted to administrative (non-call-handling) tasks. If the window control is increased to $K = 7$, allowing 149,000 calls/hour to be processed on the average, AL is reduced to 0.04. This shows the sensitivity of both call throughput and fractional administrative load to the window control.

Consider now the impact of this control on the dial-tone delay. Recall that with the pre-PC window control under discussion here, calls are held in an external queue prior to receiving dial tone (Fig. 11–18). This results in an admission delay, which is precisely the dial-tone delay. Our decomposition approach, described earlier in words, can be represented pictorially in terms of the diagram of Fig. 11–20(a). This figure indicates that the maximum throughput of the PC is $\lambda_i^T(K) = \lambda_{cp}^T/M$ calls per unit time. As noted earlier, this value of throughput is defined to be the maximum possible when calls are always waiting to be processed in the external queue. So far as the external queue is concerned, this corresponds to *its* call-handling capacity, so that that queue can be represented in the equivalent (decomposition) form of Fig. 11–20(b). For an infinite external queue the actual call-arrival rate (load) must be less than $\lambda_i^T(K)$. For a finite queue this is the maximum rate at which calls will be served. Calls arriving at a higher rate will find themselves occasionally blocked as the external queue saturates. The model thus allows a combination of dial-tone delay (to a maximum value) and the blocking (busy or reorder) signal to be analyzed. We focus here on the infinite queue, dial-tone delay case only.

The analysis of the queueing system of Fig. 11–20(b) depends on the statistical distribution of the service time, whose average value is $\lambda_i^T(K)^{-1}$. With Poisson arrivals we get an M/G/1 queue. We know from Chapter 2 that M/M/1 analysis serves as a conservative estimate for the M/G/1 case if the variance of the service time is not too large. An M/M/1 model also avoids the problem of calculating the variance. We thus adopt the M/M/1 model as an approximation. In this case, as shown in Chapter 2, the average *wait time* in the external queue (the average dial-tone delay) may be written immediately as

$$E(W) = \frac{\rho/\lambda_i^T(K)}{1 - \rho} \qquad \rho \equiv \lambda/\lambda_i^T(K) \tag{11–43}$$

As an example, take $\lambda_{cp}^T = 117{,}000$ calls/hour for the entire system, just the case of a window control of $K = 5$ for the parameters of Table 11–7. We then have $\lambda_i^T(K) = 117{,}000/20 = 5850$ calls/hour or 1.63 calls/sec as the maximum

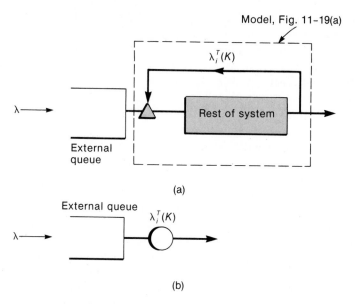

Figure 11–20   Decomposition approach in calculation of dial-tone (admission) delay
a. Initial model
b. Equivalent model

call-handling rate of each PC. The average dial-tone delay obviously depends on the average load. Take 104,000 calls/hour as the load on the entire system. The load per PC is then 5200 calls/hour. This gives us $\rho = 5200/5850 = 0.89$ as the utilization of the PC external-queue system. From Eq. (11–43), for these numbers, we get

$$E(W) = 5 \text{ sec}$$

This assumes FIFO service. Doshi and Heffes get a 4.6-sec average dial tone delay for the same numbers using the LIFO strategy [DOSH 1983]. If one increases the window size to $K = 7$, the capacity of the PC system increases to $\lambda_i^T(K) = 149{,}000/20 = 7450$ calls/hour (Table 11–7), or 2.06 calls/sec on the average. For this number, and the same load as before, we have $\rho = 0.7$, and Eq. (11–43) gives

$$E(W) = 1.1 \text{ sec}$$

Doshi and Heffes get 0.09 sec for the comparable LIFO case. (Note that our results appear quite reasonable in comparison to those of Doshi and Heffes, justifying the various approximations made.) The problem, as already noted and as apparent from Table 11−7, is that the CP response time and the related call-setup time go up quite rapidly as K increases.

Post-PC control strategy We return now to Fig. 11−17 and focus on an analysis of the post-PC window-control strategy. The results will then be compared, for the same example, with those of the pre-PC strategy just discussed. Some approximations will again have to be made to obtain a simple yet substantially valid model for analysis. In this control scheme all calls receive a dial tone. Calls will be blocked after their final digits have been collected if K calls are being processed in either the message switch (MS) or the central processor (CP). Blocking in practice means delivering a reorder (busy) signal to the subscriber in question. As in the blocked versus delayed-control models of the idealized control strategies of the previous section, we again have a blocking control to compare with the dial-tone delay control just considered.

Putting a window (token) control on each PC at the post-PC control point shown in Fig. 11−17, one gets the closed queueing network model of Fig. 11−21. As previously, λ_{cp}^T is the net throughput, in calls processed per unit time, through the message switch and the central processor. $\overline{T}$ is again the

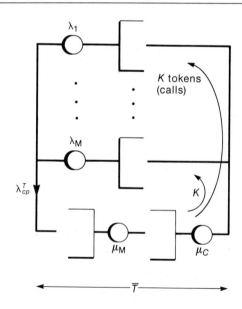

Figure 11−21 Model of post-PC window control, Fig. 11−17

call-processing time through the two systems. The λ_i's shown, $i = 1, 2, \ldots ,$ M, are the loads applied at each of the PCs. Since calls will now be blocked the net throughput through each PC will be less than the applied load. The queue in each of the λ_i branches is an artificial queue introduced to model the window control. Details are left to the reader. (Compare with Fig. 5–11 for the sliding-window control, for example.)

The case of particular interest is the one where the load at each PC is the same, i.e., $\lambda_i = \lambda$, all i. This is the case just analyzed for the pre-PC control. First, let $\mu_M = 0$ again to simplify the analysis. With $\lambda_i = \lambda$, all i, Fig. 11–21 can be redrawn as Fig. 11–22. By symmetry, we have $\lambda_i^T(K) = \lambda_{cp}^T/M \equiv \lambda^T$, as previously. We use the symbol λ^T to simplify the notation. Let $E(n)$ be the average number of calls in each of the PC branches, while $E(m)$ is the average number in the CP (bottom) branch. Since MK calls circulate throughout this closed system

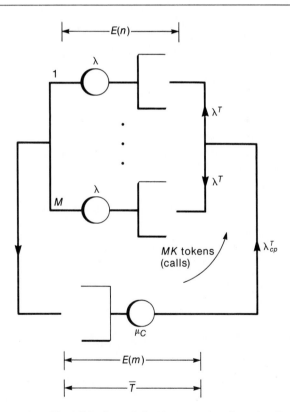

Figure 11–22 Simplified model, Fig. 11–21, $\lambda_i = \lambda$, all i, $\mu_M = 0$

with the window control, we must have $E(m)$ and $E(n)$ related by the expression

$$E(m) = M[K - E(n)] \tag{11-44}$$

In addition, by Little's formula, we have

$$\overline{T} = E(m)/\lambda_{cp}^T = M[K - E(n)]/\lambda_{cp}^T \tag{11-45}$$

It now remains to find $E(n)$ and λ_{cp}^T for a given PC λ and window size K to find the corresponding CP response time $\overline{T}$. This is where we introduce some plausible approximations.

Focus on an individual PC queue. What does it see, by symmetry? Since all M PCs access the common CP system, the work or service rate available to an individual PC must be μ_C less the net call-handling throughput of the other $(M-1)$ PCs. Call this net throughput γ. We must then have

$$\gamma = (M-1)\lambda^T = \frac{(M-1)}{M} \lambda_{cp}^T \tag{11-46}$$

The equivalent closed two-queue model involving one PC appears in Fig. 11–23. But note that for consistency the PC throughput λ^T must satisfy the throughput equation

$$\lambda^T = \frac{\gamma}{(M-1)} = \lambda(1 - p_0) \tag{11-47}$$

This is the single-server relation introduced in earlier chapters; it states that the throughput equals the service rate (λ) times the probability that the server is not idle. We now use two approximations. We first approximate the upper queue of Fig. 11–23 by a finite M/M/1/K model. This enables us to find an expression for p_0, the probability that the queue is empty. This is just

$$p_0 = \frac{1-\rho}{1-\rho^{K+1}} \qquad \rho = \frac{\mu_C - \gamma}{\lambda} \tag{11-48}$$

from our finite M/M/1 analysis of Chapter 2. Now take as the total load on the system the capacity μ_C of the CP unit. Specifically, let

$$\lambda = \mu_C/M \tag{11-49}$$

We then have

$$\rho = \frac{\mu_C - \gamma}{\lambda} = M(1 - \gamma/\mu_C) = M(1 - a) \tag{11-50}$$

with $a \equiv \gamma/\mu_C$. But from Eqs. (11–47), (11–48), and (11–49)

$$a = \gamma/\mu_C = \left(\frac{M-1}{M}\right)\rho(1 - \rho^K)/(1 - \rho^{K+1}) \tag{11-51}$$

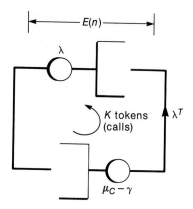

Figure 11-23 Reduced model for analysis

Combining Eqs. (11-50) and (11-51) one gets a single equation that can be solved for ρ (or a), for given values of K and M. This also gives us corresponding values of λ_{cp}^T, using the preceding equations. Details are left to the reader.

Returning to the approximation of the upper queue of Fig. 11-23 by a finite M/M/1/K queue, for that queue we have already seen in Chapter 2 that

$$p_n = (1 - \rho)\rho^n/(1 - \rho^{K+1}) \tag{11-52}$$

and

$$E(n) = \sum_{n=0}^{K} np_n = \frac{\rho}{(1 - \rho)}\left[\frac{1 - (K + 1)\rho^K + K\rho^{K+1}}{(1 - \rho^{K+1})}\right] \tag{11-53}$$

Given ρ we calculate $E(n)$, and, from Eq. (11-45), we get $\overline{T}$.

This analysis has been used to calculate the CP throughput λ_{cp}^T and the CP response time $\overline{T}$ for the same example used previously in the pre-PC window control analysis. Recall that we used $\mu_C = 157{,}000$ calls/hour as the CP call-handling capacity and assumed $M = 20$ PCs. The results for the post-PC control, for three values of the window control parameter, appear in Table 11-8. Note that the range of window sizes possible here is smaller than that for the pre-PC control scheme (see Table 11-7). In particular, the value of $K = 1$, with only one call allowed in the CP per PC, already provides a rather high CP utilization.

The results of Table 11-8 have been plotted in Fig. 11-24, together with the results of the pre-PC analysis that appeared earlier in Table 11-7. Note how closely they track one another. This indicates that either control could be used in the high throughput range. The pre-PC control does result in a wider range

K	a	λ_{cp}^T (calls/hr)	$\overline{T}$ (msec)	P_B
TABLE 11–8	Performance of Window-based Post-PC Control, Figs. 11–17 and 11–21*			
1	0.776	127,000	104	0.19
2	0.864	142,000	218	0.095
3	0.912	150,000	408	0.045

* Load $M\lambda = \mu_C = 157{,}000$ calls/hour, $M = 20$ PCs

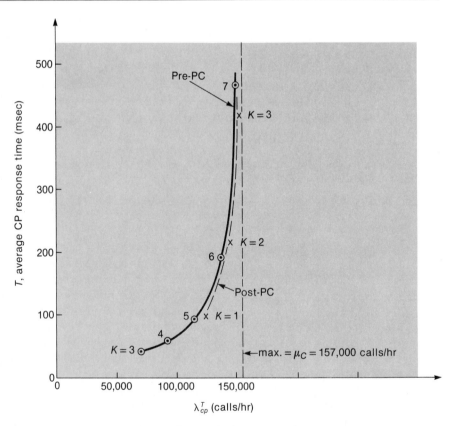

Figure 11–24 Comparison of two window-control strategies, $M = 20$ PCs

of control, however, and can be pushed down to lower throughput values, enabling more administrative processing to be carried out. For example, in the post-PC case, with $K = 1$ and $\lambda_{cp}^T = 127,000$ calls/hour, Eq. (11–42) indicates that $AL = 0.17$. Using the post-PC control, then, at most 17 percent of the processor load can be diverted to administrative tasks in a period of congestion. If $K = 2$ is used instead, this drops to $AL = 0.08$. The two controls also operate differently, as has been noted a number of times. The pre-PC control introduces dial-tone delay; the post-PC control blocks *after* receipt of dialed digits. Table 11–8 also provides values for the blocking probability for the three values of window size K. Note that 19 percent of the calls would be blocked for $K = 1$, while this drops to 9.5 percent for $K = 2$. This appears to be quite tolerable in periods of heavy congestion. As in the case of the two related controls described in the idealized model analysis of the previous section, more cannot be said concerning the comparison of the two types of control. One blocks users; the other delays dial tone. The two performance criteria, blocking probability and dial-tone delay, cannot be compared further in any objective fashion.

Interestingly, the results of the two analyses, plotted in Fig. 11–24, agree quite closely with the results of Doshi and Heffes [DOSH 1983, Fig. 5], validating the various approximations made in the analyses here.

Problems

11–1   Figure 11–4, based on a figure in [DUNC], indicates the software calls required to handle a POTS call in the AT&T No. 5 ESS. Prepare a similar figure, or a flow chart, indicating how a POTS call would be handled in the Italtel UT 10/3.

11–2   Figures 11–5 to 11–7 develop a call-processing model for a hypothetical switch. Develop a similar model for the AT&T No. 5 ESS, using the call-processing scenario described in the text. It may be advisable to refer to some of the original papers on this switch (cited in the text) to help with the modeling.

11–3   The call-processing model of Figs. 11–5 to 11–7 is based on an arbitrary set of 24 tasks described in the text. Try developing a similar model using a somewhat different task structure. The model itself is applicable to a local exchange. How would this change in the case of a tandem switch? How would a combination local/tandem switch be modeled?

11–4   Refer to the hypothetical call-processing model of Figs. 11–5 to 11–7. For this model, with the numbers provided in the text, calculate and plot a. dial-tone delay and b. call-setup time in this switch, as a function of processor utilization ρ. Do this for two cases: (1) no priority and (2) nonpreemptive priority to call-processing tasks arriving at the link interface buffer. Compare the two cases.

11–5   Determine the sensitivity of the results of Problem 11–4 to the choice of some of the parameters: Change some of the task-processing times, add or delete tasks if possible, change (add or delete) messages sent at the end of task execution. How do these affect the dial-tone delay and call-setup time?

11–6   Refer to Problem 11–4. Try another priority discipline; for example, choose priority by *task* rather than by the buffer at which the task appears. Is it appropriate, for example, to provide higher priority to shorter tasks? Use two priorities again, for simplicity. Determine the effect on dial-tone delay and call-setup time. Another priority discipline that could be tried is suggested in the text.

11–7   A vehicular telephone system requires charging at a receiving mobile system. Modify Fig. 11–6 to take this into account. (Here the *called* processor is charged as well.) Charging the receiving party adds 0.6 task/call. Modify Table 11–2 appropriately. Do the same for Table 11–3, using an appropriate number for processing time. What effect does this have on delay calculations?

11–8   Change the call mix of Table 11–1 and determine the impact on the call-processing calculations of the hypothetical system described in the text.

11–9   Refer to Fig. 11–9, an idealized feedback queueing model for call processing. Show that the probability of state is given by Eq. (11–11). Satisfy yourself that Eq. (11–14) is the call-processing delay for this model.

11–10 The processor in the model of Fig. 11–9 experiences overload when $\rho = \lambda/\mu(1 - \alpha) \to 1$. One possible overload control blocks calls when the number of tasks on queue reaches a specified threshold N. The model of Fig. 11–10 results.

   a. Show that the probability distribution of the finite M/M/1 queue of Fig. 11–10 is given by Eq. (11–15).
   b. Show that the average call throughput and average call-processing delay are given, respectively, by Eq. (11–17) and Eq. (11–21).
   c. Show that the call-processing delay at $\rho = 1$ is given by Eq. (11–22), while the call throughput is $\gamma_e = \lambda N/(N + 1)$. Explain the control mechanism in terms of these values. (Note that these results are very similar to those of the sliding-window control, discussed in Chapter 5.)

11–11 Refer to Fig. 11–14, the model of an idealized overload control that delays call processing. Show that the state probabilities for this system are given by Eqs. (11–30) and (11–31). Verify the control curves of Fig. 11–15.

11–12 Check the entries in Table 11–6, comparing the two controls for the call-processing model of Fig. 11–9. Extend the table to cover cases other than those shown there.

11–13 Refer to Fig. 11–18, the model of the pre-PC overload window control discussed in the text. Simplifying this model to the one shown in Fig. 11–19(b), show that Eq. (11–37) results. How is Eq. (11–38) obtained? Finally, derive Eq. (11–39), which describes the performance of the system.

11–14 a. Use Eq. (11–39) to check the entries of Table 11–7, describing the performance of the pre-PC overload window control of Figs. 11–18 and 11–19.

b. Change the average digit dialing time to $\mu_D^{-1} = 5$ sec. Calculate the performance now and compare with Table 11–7. How sensitive are the results to the digit dialing time? Reduce the dialing time to 2 sec. What is the result?

c. Change some of the other parameters used in obtaining Table 11–7 and determine the effect on system performance. In particular, try different values of M (but with $M > 10$ in all cases—why?). Plot the call setup time versus λ_{cp}^T, the calls/hour, or traffic in the system, as the window size, K, varies.

d. Determine the average dial-tone delay, given by the M/M/1 approximation Eq. (11–43), for some of the cases calculated in earlier parts of this problem. Explain the results obtained.

11–15 Show that the model of Fig. 11–21 represents the post-PC window control of Fig. 11–17.

11–16 a. Refer to the models of Figs. 11–21 to 11–23. Show how Fig. 11–23 is obtained.

b. Derive Eqs. (11–51) and (11–53), used in the analysis.

c. Check the entries of Table 11–8.

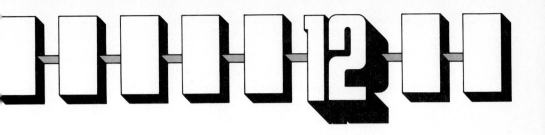

The Evolution toward Integrated Networks

Previous chapters have covered essentially two areas: packet-switched networks and circuit switching. Historically, circuit-switched networks were developed to provide a voice-communication capability. Packet-switched networks, developed in the 1970s, were designed to handle interactive data traffic. Future networks will be required to handle a variety of types of traffic: interactive data, longer file transfers, digital voice, facsimile, video, remote control information, and so forth. Studies are being made of ways of integrating networks to handle this heterogeneous traffic mix efficiently and cost effectively. Networks of the future may combine aspects of both packet switching and circuit switching. In fact, packet-switched networks are evolving to handle other types of traffic besides interactive data, while more and more circuit-switched networks are being designed to handle data traffic in addition to voice and to provide a packet-switched capability as well.

In this chapter we introduce the concept of integrated networks. As done elsewhere in this book, we provide both a qualitative discussion and, where possible, a quantitative analysis. We do this by discussing first the evolution of the telephone network, which was originally designed to handle voice calls, to a

627

network geared to handling a variety of types of traffic. We then focus on integration in a network and provide some simple analytical models for studying the integration of circuit-switched calls and packets in one network.

Recall that in the chapters on circuit switching we first described the switching network at an exchange. We then analyzed the process of call setup and modeled overload controls designed to reduce congestion in a circuit-switched exchange. In Section 12–1, devoted to routing in circuit-switched networks, we leave the exchange and take a more systemwide view, moving onto the network itself. We first describe briefly the classical method of hierarchical routing of calls as carried out in most of the world's telephone networks. We then move on to the newer nonhierarchical methods of routing that are being introduced in various networks. The nonhierarchical approach is similar to that of routing in packet-switched networks, studied in detail in Chapter 6, with all nodes in the network considered peer nodes. Nonhierarchical routing is being introduced as a result of studies showing that it will result in substantial cost savings as compared with the more traditional method of hierarchical routing in circuit-switched networks.

Used in conjunction with alternate-path routing, in which calls blocked on one path are allowed to try additional paths, nonhierarchical routing turns out to result in possible instabilities, or at least in deteriorated performance at overload (applied loads higher than those for which a network was designed). Overload controls are thus needed to eliminate this problem, and one particular method, called trunk reservation, is described in this chapter.

Following the discussion of nonhierarchical routing in Section 12–1 we move on to a brief discussion in Section 12–2 of the CCITT No. 7 signaling network, which is used to carry call-routing information. In implementation this is a packet-switched network based on the X.25 layer architecture. The AT&T implementation in the United States, the CCS (common-channel signaling) network, has been designed not only to carry call-routing information but to also handle many other features appropriate to modern digital networks.

Following the description of the CCITT No. 7 signaling system, we turn to the final topic in this book: integrated networks, those capable of carrying both packet- and circuit-switched data of all types. In Section 12–3 we describe some of the current standards efforts in CCITT that relate to *integrated services digital networks* (ISDNs). We then introduce the subject of analysis of network integration by focusing on the simplest example possible: a single link that carries both circuit-switched and packet-switched traffic. Several ways of handling the combined traffic are considered, and because of the complexity of even this simple example, approximate techniques of analysis are introduced. This last section, Section 12–4, should provide an introduction to the subject of analysis of integration in networks.

12–1 Routing in Circuit-switched Networks

12–1–1 Hierarchical Routing*

As just noted, the classical method of routing in telephone networks world-wide is hierarchical routing. Telephone switching offices or exchanges are classified according to their level in a hierarchy. The North American hierarchy, for example, consists of the five basic classes shown in Fig. 12–1. Offices below home on one or more of the offices above. There are 10 regions in the United States, each with a regional center, and 2 regions in Canada.

Alternate routing is used in accordance with the hierarchy shown in Fig. 12–1: A first-choice route is attempted for a given call; if this route is not available, one or more alternate routes will be tried in order, moving up the hierarchy. The simplest example, involving a single alternate route, is represented by the triangle of Fig. 12–2. The two class-4 toll centers shown there are conected by a so-called direct high-usage trunk group. Toll calls to be routed between these two toll centers would first attempt the direct trunk group. (As the name indicates, a trunk group consists of a specified number of trunks or channels, each capable of handling one call.) Those calls not able to be carried by this high-usage group would overflow onto the alternate route shown, routed through a higher-level primary center. This is shown designated as a final route. Calls not able to be carried over this path would be blocked, and a network busy signal returned to the source of the call (see Chapter 11). The alternate route also consists of a specified number of trunks.

In practice, many alternate routes might be made available, depending on the telephone company, the point at which a call enters the hierarchy, and so forth. A typical routing pattern utilizing the full hierarchy, one of many possible routing plans, appears in Fig. 12–3 [AT&T, Fig. 2, Appendix 2, as modified]. Thirteen possible routes connecting end offices A and B are shown, each successive one moving further up the hierarchy. The first path tried is the direct high-usage trunk group that connects end offices A and B. If all trunks on this route are occupied, the call attempt overflows onto route 2 via a class 4-office to which both end offices are connected. If the second link in this path, which connects the class 4-office and end-office B, is completely filled, the call may overflow onto route 3. This procedure continues until the thirteenth and final route is attempted. Note that if any of the links moving directly up and down the

---

* Material in this subsection is based on [AT&T, Sec. 3, "The Switching Plan"].
[AT&T] *Notes on the Network, Bell System Practices*, AT&T Co. Standard, Sec. 781-030-100, issue 2, Dec. 1980.

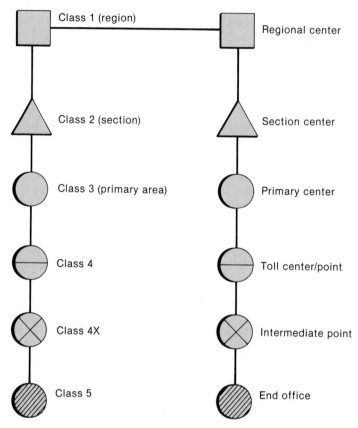

Figure 12–1 North American telephone office hierarchy

ladder are completely filled (no trunks are available), the call will finally be blocked. A link of this type represents a final trunk group for the call; these links are designated by solid lines. Note the number of routing triangles of the type of Fig. 12–2 that appear in Fig. 12–3. Routes 1 and 2 represent such a triangle, as do routes 3 and 4, and so forth. End offices are not all interconnected in a given network. Calls may thus be frequently routed via toll or higher-level centers, giving rise to the triangle of Fig. 12–2.

12–1–2 Nonhierarchical Routing

As noted at the beginning of this chapter, studies have shown that nonhierarchical routing, with offices (nodes) in a network treated as peers so far as routing is concerned, can provide cost improvement over hierarchical routing.

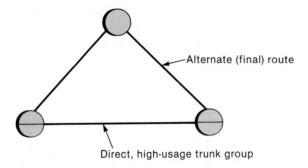

Figure 12–2 Alternate routing, routing triangle

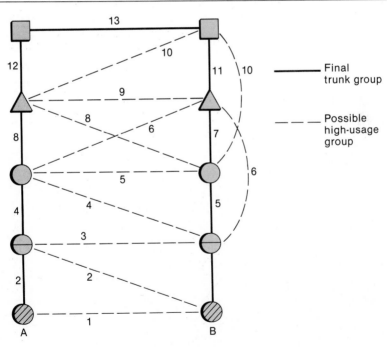

Figure 12–3 Typical hierarchical routing pattern (from [AT&T, Fig. 2, App. 2, as modified], "Notes on the Network," Bellcore, with permission)

Major networks are thus either replacing or planning to replace hierarchical routing techniques with nonhierarchical ones. Instead of looking like the ladder of Fig. 12–1, the network configuration then becomes a mesh of the type discussed in earlier chapters on packet switching. An example appears in Fig. 12–4.

A primary reason for the cost effectiveness of nonhierarchical routing, particularly in a relatively large country, is that traffic patterns tend to change in different offices and different parts of the network as a function of the time of day or season of the year. Hierarchical routing is essentially static and must be designed for peak traffic in the network. Some network capacity will thus remain idle during nonpeak intervals. Adaptive routing, implicit in nonhierarchical routing, can be adjusted to changing traffic patterns with time as traffic at various offices peaks at different hours. A given load can thus be accommodated by fewer resources. Alternately, more traffic can be accommodated for the same set of resources if adaptive, nonhierarchical routing is used. More dynamic routing of this type, as contrasted with the static routing of the hierarchical network, requires rather sophisticated computer-controlled switching machines and improved methods of signaling. The advent of computer-controlled digital switches, described in the last two chapters, and the CCITT No. 7 signaling networks, described later, have made nonhierarchical routing possible.

However, a problem arises with this kind of routing. If alternate routing is still used—i.e., when the first-choice path is not available, alternative paths, defined in a nonhierarchical sense, are tried (Fig. 12–4)—instabilities may be

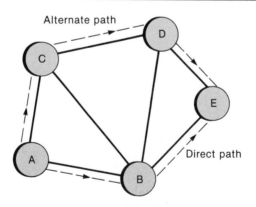

Figure 12–4 Nonhierarchical routing

encountered if the load increases beyond the design value. Controls thus have to be introduced. This will be discussed later.

How are routes determined in nonhierarchical routing? A variety of approaches have been adopted, just as was the case with packet switching (see Chapter 6). The European portion of the U.S. government Autovon network uses originating office control, in which alternate routes are selected as required by the exchange (office) initiating the call [LINP]. AT&T uses a similar, decentralized nonhierarchical routing strategy, in which each node in the network has a prescribed series of alternate paths to be tried. The number and sequence of paths varies with the day and time of day [ASH]. Bell Canada uses a centralized dynamic approach. The routes in its procedure change adaptively as traffic conditions change. A central routing processor receives information periodically from switches in the network as to the number of idle trunks they have available on each of their outgoing links, as well as information as to the load on their call-processing units. Routes are then selected and distributed back to the switches [CAME]. In both the AT&T and the Bell Canada cases, routes are chosen to be no more than two links long. There is thus only one tandem switch to be chosen for a given alternate path. The Canadian routing strategy, carried out on a trial basis first for the metropolitan area of Toronto, is based on earlier work of Szybicki [SZYB]. An early paper that describes advantages to be gained by the use of nonhierarchical routing is by C. Grandjean [GRAN].

As noted previously, studies indicate that nonhierarchical routing can provide a 15 percent cost reduction over hierarchical routing [ASH]. Instabilities may develop, however, if alternate routing is used. To demonstrate the problem arising, and then later to indicate how an appropriate control can alleviate this problem, we use a simple fully connected, symmetrical network model with homogeneous traffic generated at each node to represent the nonhierarchical

[LINP] P. M. Lin, B. J. Leon, C. R. Stewart, "Analysis of Circuit-switched Networks Employing Originating Office Control with Spill-Forward," *IEEE Trans. on Comm.*, vol. COM-26, no. 6, June 1978, 754–765.

[ASH] G. R. Ash, R. H. Cardwell, R. P. Murray, "Design and Optimization of Networks with Dynamic Routing," *BSTJ*, vol. 60, no. 8, Oct. 1981, 1787–1820.

[CAME] W. H. Cameron et al., "Dynamic Routing for Intercity Telephone Networks," ITC-10, Tenth International Teletraffic Congress, Montreal, June 1983, Session 3.2, paper 3.

[SZYB] E. Szybicki, M. E. Lavigne, "The Introduction of an Advanced Routing System into Local Digital Networks and Its Impact on the Networks' Economy, Reliability, and Grade of Service," ISS '79, Paris, 1979.

[GRAN] C. Grandjean, "Call Routing Strategies in Telecommunication Networks," ITC-5, Fifth International Teletraffic Congress, New York, June 1967, 261–269.

routing strategy. The model and analysis are due to Krupp [KRUP]. The problem of instability in nonhierarchical networks was first pointed out by Nakagome and Mori [NAKA].

Say that there are N nodes in the fully connected, symmetrical network to be analyzed. Figure 12–5 shows two examples, for $N = 3$ and $N = 4$ nodes. The direct path from source to destination is then, of course, just one link long. Alternate paths are two links long, as shown in Fig. 12–5, and there are always $M = N - 2$ such alternate paths available. Let A be the external offered load in Erlangs per node pair in either direction. (Either side may generate a call in a circuit-switched connection; the load is the same in either case.) For the fully connected, symmetrical network under discussion here, A is then the direct-path applied load on each link as well, as shown in Fig. 12–5. The *total* external load applied to the network is $L = \frac{1}{2}N(N - 1)A$.

With alternate path routing used, the total load actually offered to a link consists of the external offered load, A, plus the alternate-routed load overflowing from other direct paths. Call this total offered load a, in Erlangs. It is this load that must be used in carrying out the analysis and, in particular, in calculating the call-blocking probability and carried load as a function of offered load, the performance parameters of interest. To carry out the calculations, the following simplifying assumptions must be made:

1. The link blocking probabilities are independent.

2. Both arriving and overflow calls are Poisson.

3. The Erlang-B blocking probability formula holds.

Poisson call-arrival statistics are, of course, standard and have been assumed throughout this book, including the previous chapters on circuit-switching technology. The overflow calls, those blocked from a direct path and overflowing onto an alternate path, are clearly not Poisson. (Why is this so?) We use the Poisson assumption anyway, for simplicity. Simulation studies accompanying the analysis summarized here indicate that the analytical results obtained using this assumption are quite accurate and provide meaningful results [KRUP], [AKIN 1983].

[KRUP] R. S. Krupp, "Stabilization of Alternate Routing Networks," ICC '82, Philadelphia, June 1982, 31.2.1–31.2.5.

[NAKA] Y. Nakagome and H. Mori, "Flexible Routing in the Global Communication Network," ITC-7, Seventh International Teletraffic Congress, Stockholm, June 1973, paper 426.

[AKIN 1983] J. M. Akinpelu, "The Overload Performance of Engineered Networks with Hierarchical and Nonhierarchical Routing," ITC-10, Tenth International Teletraffic Congress, Montreal, June 1983, Session 3.2, paper 4.

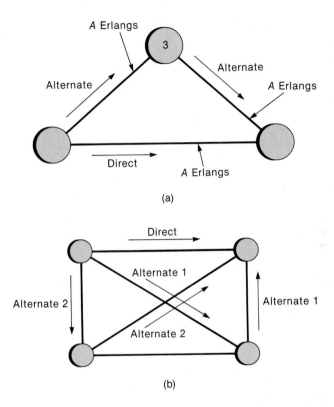

Figure 12-5 Fully connected, symmetrical networks
a. $N = 3$ nodes; one alternate path
b. $N = 4$ nodes; two alternate paths

Under assumptions 2 and 3 just stated, the Erlang blocking probability on a link of n trunks, with total offered load a, is given by the standard formula

$$p = a^n/n! \Big/ \sum_{i=0}^{n} a^i/i! \qquad (12-1)$$

Given the *link*-blocking probability p, the *call*-blocking probability z is easily calculated. This is the probability that a call blocked at a direct path finds no alternate path available. Recall that for our fully connected, symmetrical model each of the $M = N - 2$ alternate routes is two links long (Fig. 12-5). Using the assumption that link-blocking probabilities are independent and each equal to p, the call-blocking probability is just

$$z = p[1 - (1 - p)^2]^M = p[1 - q^2]^M \qquad (12-2)$$
$$q = 1 - p$$

The carried portion of the externally offered load is then given by

$$C = A(1 - z) \qquad (12-3)$$

This is the link throughput and is, as noted, a performance parameter of critical interest.

The only problem now is that we do not know the total load a offered to a link, which is needed to calculate the link-blocking probability p (Eq. 12–1). It turns out that this cannot be calculated explicitly but that a relation connecting it to the external offered load A can be found, enabling the problem to be solved. To find this relation, consider the total load K actually carried on a link. This must be, as noted earlier, the sum of the nonblocked portion of the actual (external) offered load plus the carried portion of the load overflowing from other links.

The carried portion of the load overflowing from other links has two equal components in this fully connected, symmetrical network model. This is shown by example in the three- and four-node networks of Fig. 12–6, which correspond to the same two network examples of Fig. 12–5. Note that in the three-node example of Fig. 12–6(a), link 1–2 carries, in addition to the load directly routed from 1 to 2, traffic flowing from node 3 to node 2 that is blocked from the direct path 3–2 and overflows onto the two-link path 3–1–2. It also carries the overflow traffic blocked from the direct path, node 1 to node 3. The same three components of traffic flow in the reverse direction over the link, from node 2 to node 1. By symmetry each of the two other links carries the same traffic. Similar results apply to each of the links in the four-node network of Fig. 12–6(b), as can be seen by inspection. Summing the total carried load on a typical link in the general case of N fully connected nodes, with $M = N - 2$ alternate paths, one then has

$$K = A(1 - p) + 2 Ap(1 - z/p) \qquad (12-4)$$

The first term in Eq. 12–4 is the nonblocked portion of the actual (external) offered load A. The second term is the sum of the two equal overflow components of the carried load in a typical link. (Ap is the portion of the external offered load that overflows onto an alternate path; $(1 - z/p)$ represents the probability that this overflow load is not blocked.)

But the total load carried on a link must also equal the nonblocked portion of the total load a offered to a link. Thus, we must also have

$$K = a(1 - p) \qquad (12-5)$$

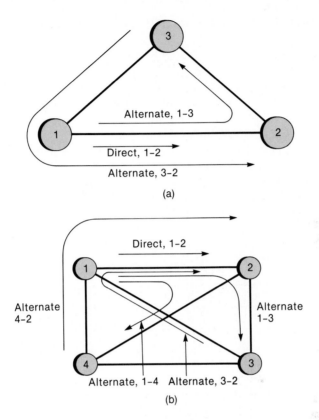

Figure 12–6 Total traffic flowing over a typical link with alternate routing used
a. $N = 3$ nodes
b. $N = 4$ nodes

Equating Eqs. (12–4) and (12–5), one readily finds that

$$A = \frac{a(1 - p)}{1 + p - 2z} \qquad (12-6)$$

This represents the desired relation between the actual (external) offered load A and the (fictitious) total offered load a. Given a, one finds p and z from Eq. (12–1) and Eq. (12–2), respectively, and then A, using Eq. (12–6). The carried portion of the externally offered load is then readily found from Eq. (12–3). One can then plot curves of carried load C versus the offered load A for various values of n (trunks/link group, Eq. (12–1)) and M (the number of alternate routes).

A typical set of curves calculated by Krupp appear in Fig. 12–7 [KRUP, Fig. 1]. These are plots of carried load per trunk, C/n, versus the offered load/trunk, A/n, for the case of $n = 100$ trunks/group. (Recall that a group represents the total number of trunks allotted to a link.) Note the dramatic appearance of a cusp, representing an unstable situation, as the number of alternative paths increases from $M = 1$ to $M = 2$. Even the case of one alternative route ($M = 1$) shows a sharp drop in carried load as the offered load increases beyond a normalized offered load of 0.875 Erlang/trunk in this case. Simulation curves obtained by Krupp, and reproduced in Fig. 12–8 [KRUP, Fig. 1a], are similar in form. They do not exhibit the instability as dramatically as do the analytical curves of Fig. 12–7, since the two branches of each curve in

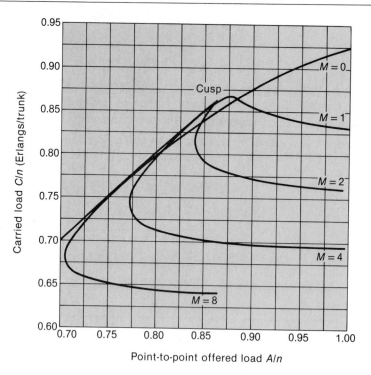

Figure 12–7 Analytic results, general overload performance of network; M alternate routes and $n = 100$ trunks/group (from [KRUP, Fig. 1], © 1982 IEEE, with permission)

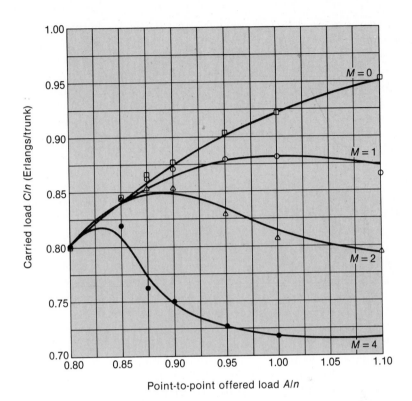

Figure 12–8 Simulation results, general overload performance of network; M alternate routes and $n = 100$ trunks/group (from [KRUP, Fig. 1(a)], © 1982 IEEE, with permission)

Fig. 12–7 appear to be averaged in the simulation results. They do show the severe drop-off in carried load as the offered load increases beyond a certain point. The performance deteriorates as more alternate routes are allowed to be used. The $M = 0$ curve, that for nonalternate routing, does not, of course, exhibit the instability at overload. It does provide somewhat poorer performance at lower loads.

Why the possibility of an instability if alternate routing is used? This is because an alternately routed call uses *two* trunks (one on each link in the alternate path), rather than the one trunk required in the direct path. For small carried load this presents no problem and in fact improves the performance over the case with no alternate routing. (Note that the curves with $M = 2$ and $M = 4$

alternate paths in Fig. 12–7 lie above the $M = 0$ (the no-alternate path case) at offered loads smaller than the ones at which instability sets in.) For higher loads alternately routed calls take up double the resources, blocking directly routed calls in both links used and causing them to overflow more often, which in turn blocks direct calls even more, setting in motion a possibly unstable situation.

Do these results carry over to more general, more realistic nonsymmetrical nonhierarchical networks? Akinpelu has extended the mathematical model with alternate routing to networks of this more general type and has found, both through analysis and through simulation models verifying the analysis, that the instabilities due to alternate routing do persist [AKIN 1983], [AKIN 1984]. Studies of network models representing subsets of engineered network models (i.e., those designed using real topologies and real data) do not exhibit the instabilities. They do indicate that, with alternate routing used, the nonhierarchical networks show a drop in carried load once the load for which they have been designed is exceeded [AKIN 1984]. A drop of this type is not exhibited by the older hierarchical networks. The nonhierarchical networks provide significant cost improvement (or, equivalently, can carry more load) over hierarchical networks up to a specified design load, but deteriorate beyond this point if alternate routing as described here is used.

There are thus two options. One can abandon alternate routing and use direct routing only. This is then similar to the various methods of shortest-path routing for packet networks described in Chapter 6. Alternately, one can introduce controls on the alternately routed traffic to prevent it from blocking direct calls and hence destabilizing the network. A particular class of controls, using trunk reservation for direct or first-routed calls, has been studied extensively and will be adopted in nonhierarchical networks as they are deployed [KRUP], [AKIN 1983], [AKIN 1984]. This scheme is analogous to the input-buffer-limiting congestion-control strategy summarized in Chapter 5. In this scheme a specified number of trunks in a group are reserved for direct-routed traffic. By proper choice of this number the instability due to alternately routed traffic can be eliminated and the falloff in carried load beyond a specified offered load eliminated. We shall carry out the analysis of this procedure in Subsection 12–1–3.

Other methods of control are possible as well. In one scheme alternate routing is allowed up to a specified threshold and only direct routing allowed above this point. This corresponds to reserving all trunks for direct-routed traffic above this load. In another scheme overflow traffic may try an alternate path with a probability θ. With a probability $(1 - \theta)$ it is blocked. θ may be made to vary with load, if desired. By proper adjustment of θ one can almost match the

[AKIN 1984] J. M. Akinpelu, "The Overload Performance of Engineered Networks with Nonhierarchical and Hierarchical Routing," *AT&T Bell Labs. Technical J.*, vol. 63, no. 7, Sept. 1984, 1261–1281.

effect of trunk reservation without the need to attempt an alternate path. Combinations of these strategies may be used as well. Details appear in [YUM]. Yum has also studied the comparative cost of providing alternate routing with and without controls and has shown how it does produce a cost penalty compared with nonalternate routing [YUM].

12–1–3 Control of Alternately Routed Traffic: Trunk Reservation for First-routed Traffic*

As shown in Subsection 12–1–2, the use of alternate routing in nonhierarchical circuit-switched networks improves performance to some extent at light loads, but can result in instability at overload. Trunk reservation has been proposed to alleviate this problem. In this control strategy, first-routed traffic is allowed to access all n trunks of a link trunk group; alternately routed traffic — calls overflowing from other attempted routes — is not allowed to access more than $m < n$ trunks of the group. If all m are occupied, alternately routed calls are blocked from this link. $(n - m)$ trunks are thus always reserved for first-routed traffic only.

Note that this type of reservation scheme is very similar to that used in input-buffer limiting in packet-switched networks (see Chapter 5). Recall that in our earlier discussion transit traffic (comparable to first-routed traffic here) was allowed to occupy all N buffers of a node; input traffic, newly generated at the node, was allowed to occupy at most N_I of the N buffers. It was shown that, without this control, instability, leading to deadlock and zero throughput, ensues when the offered load is large enough. An analogous phenomenon is encountered here. The object is to choose the control parameter $m < n$ so as to eliminate the potential instability at overload while at the same time keeping the throughput at lower loads relatively high.

The analysis of the trunk reservation scheme is relatively straightforward. We again model the call arrival-departure process, as in Chapters 2 and 10, as an n-server birth-death process. Using the same assumptions and notation as in Subsection 12–1–2, say that the total traffic offered to a link, both first-routed and alternately routed, is a Erlangs. For $j \leq m$ calls active in a trunk group containing $n > m$ trunks, we then have as the probability of this state

$$p_j = \frac{a^j}{j!} p_0 \qquad 0 \leq j \leq m \qquad (12\text{–}7a)$$

[YUM] T-K Yum, *Routing in Nonhierarchical Networks*, Ph.D. dissertation, Columbia University, May 1985.
* References for this subsection include [KRUP] and [AKIN 1983].

This is the same Erlang formulation for a multiserver system first encountered in Chapter 2 and then again in Chapter 10. For $j > m$ calls, however, the first-routed traffic only is present. The offered load of this traffic is $A < a$ Erlangs. More specifically, we write A in the form $a(1 - r)$, with r a parameter to be found. It is then left to the reader to show, in a very straightforward manner, that for $m < j \le n$, we have as the probability of state

$$p_j = A^{j-m} p_m / j(j - 1) \cdots (m + 1)$$
$$= \frac{a^j(1 - r)^{j-m}}{j!} p_0 \qquad\qquad m < j \le n \qquad (12-7b)$$
$$A \equiv a(1 - r)$$

The zero-state probability, p_0, is found as usual by summing over all possible states:

$$\frac{1}{p_0} = \sum_{j=0}^{m} \frac{a^j}{j!} + \sum_{j=m+1}^{n} \frac{a^j(1 - r)^{j-m}}{j!} \qquad (12-8)$$

Two probabilities of interest can now be found from the expressions for p_j. The first is the probability that a first-routed call will be blocked; this is just the probability p_n. Defining this to be a parameter p as in Subsection 12–1–2, we have, from Eq. (12–7b),

$$p \equiv p_n = \frac{a^n(1 - r)^{n-m}}{n!} p_0 \qquad (12-9)$$

For $m = n$, i.e., the case in which no trunks are reserved, p becomes the Erlang-B probability of Eq. (12–1).

The second probability to be found is the probability $\hat{q}$ that no more than $(m - 1)$ trunks are busy. This is then the probability that an alternate routed call overflowing onto this link finds a trunk free and is not blocked. This is obviously given by

$$\hat{q} = \sum_{j=0}^{m-1} \frac{a^j}{j!} p_0 \qquad (12-10)$$

Consider now the average number of trunks occupied. This is just

$$K = \sum_{j=0}^{n} j p_j$$
$$= a(1 - r)(1 - p) + ar\hat{q} \qquad (12-11)$$
$$= A(1 - p) + ar\hat{q}$$

substituting Eqs. (12–9) and (12–10) into the expression for K and simplifying. Details are left to the reader. Consider the interpretation of the third line of Eq. (12–11). The first term—$A(1 - p)$—is the portion of the trunks occupied due to the first-routed traffic that is not blocked. It is obviously also the carried load due to first-routed traffic. The second term — $ar\hat{q}$ — must then be the portion of the trunks occupied due to alternately routed (overflow) traffic. It also represents the portion of the carried load due to overflow traffic.

The quantity K, the average number of trunks occupied, may also be interpreted as the carried load on the link, or the average normalized throughput, in Erlangs. This equivalent relation between average number of trunks (servers) occupied and average normalized throughput was first pointed out in Chapter 2, in Eqs. (2–56) and (2–57). K thus has the same interpretation as in Subsection 12–1–2, and as such may be calculated another way. As was done in Eq. (12–4), K can be written directly as the sum of first-routed carried load and alternately routed carried load:

$$K = A(1 - p) + 2\,Ap[1 - (1 - \hat{q}^2)^M] \qquad (12-12)$$

The second term has again been written for $M \le (N - 2)$ alternate routes. Note again how it is obtained: There are two contributions to the overflow traffic on each link in a fully connected, symmetrical network. With a probability p the first-routed traffic overflows. The overflow traffic offered to a link is thus Ap Erlangs. The probability that this traffic will be blocked on a given two-link path is $1 - \hat{q}^2$. The probability it will *not* be blocked and will be accepted as a call is $1 - (1 - \hat{q}^2)^M$. Equation (12–12) follows directly.

Equating (12–11) to (12–12) one can obtain a relation connecting A and a, as was done in Subsection 12–1–2. The relation is written more simply if one first writes an expression for the call-blocking probability. Labeling this quantity $\hat{z}$ to show its relation to the blocking probability z found in the uncontrolled case discussed previously, one has

$$\hat{z} = p(1 - \hat{q}^2)^M \qquad (12-13)$$

(Note that this is the probability that a call is first blocked, with probability p, on a first-routed link, and then, with probability $(1 - \hat{q}^2)^M$, cannot find a free alternate path among the M such paths potentially available.) Using Eq. (12–13), one then obtains the following expression relating A and a:

$$A = \frac{a(1 - p) + ar[\hat{q} - (1 - p)]}{1 + p - 2\hat{z}} \qquad (12-14)$$

This is to be compared with Eq. (12–6) for the equivalent relation with no trunks reserved. As a check, let $\hat{q} = q = 1 - p$, $\hat{z} = z$. Equation (12–14) does in fact become Eq. (12–6).

An equivalent relation between A and a is given by

$$A = \frac{ar\hat{q}}{2(p - \hat{z})} \qquad (12-15)$$

Details are left to the reader.

How does one use these expressions? Choose a value of a. Pick an initial arbitrary value for A. The value of r is then chosen to satisfy the identity $A = a(1 - r)$. Calculate r, $\hat{q}$, and $\hat{z}$, giving rise to another value of A, using Eq. (12–14) or Eq. (12–15). Repeat the process, iterating until the procedure converges.

Given A, a, and $\hat{z}$, one can then calculate the carried load per link, C, as was done previously:

$$C = A(1 - \hat{z}) \qquad (12-16)$$

An example of the use of this analysis of the trunk reservation-control scheme applied to a symmetrical network appears in Fig. 12–9, taken from the paper by

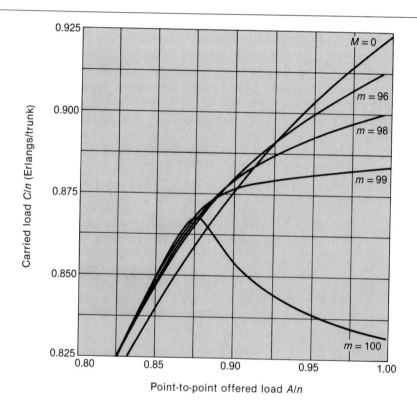

Figure 12–9 Analytic results, general overload performance with m unreserved trunks/group; $M = 1$ alternate route and $n = 100$ trunks/group (from [KRUP, Fig. 2], © 1982 IEEE, with permission)

Krupp [KRUP, Fig. 2]. This example shows the result of applying trunk reservation to a network with $M = 1$ alternate path. This corresponds to one of the examples of Fig. 12-7, with $n = 100$ trunks/trunk group. The curve labeled $m = 100$ in Fig. 12-9 is the one with no trunk reservation and corresponds to a portion of the $M = 1$ curve of Fig. 12-7. (The scale is different in the two curves.) Note that as m decreases, with the number of trunks $(n - m)$ reserved for first-routed traffic increasing, the downturn in carried load is eliminated and the carried load at higher offered loads is increased. At lower offered loads, however, increasing the number of reserved trunks reduces the carried load. A choice thus has to be made between improving the performance at overload while not deteriorating the performance at lower loads too much. Similar results are obtained for an example with $M = 4$ alternate routes [KRUP, Figs. 3, 4].

This analysis for trunk reservation in the case of symmetrical nonhierarchical networks carries over directly to more general, asymmetrical networks [AKIN 1983]. Similar results are found for much larger, more complex networks [AKIN 1983], [AKIN 1984]. The drop in carried load as the offered load increases, due to alternate routing, is found to disappear through the use of trunk reservation control. Similar results have been obtained in other simulation studies [HAEN]. Without controls, the performance of a dynamic nonhierarchical routing strategy becomes poorer than that of hierarchical routing as overload increases. With trunk reservation control the performance of both schemes is improved in the overload region, with the performance of both found to be about the same.

12-2 Common-channel Signaling for Circuit-switched Networks

In the previous section we discussed the routing of calls in a circuit-switched network, with emphasis on nonhierarchical routing. We indicated there that a specified direct or first-routed path is tried first. If no trunks are available, alternate routes are then attempted in a prescribed order. It is clear that to actually carry out the routing procedure signaling messages must be sent from exchange to exchange along the path specified. A number of these control messages were described in Chapter 11, including the connect (seizure), answer, busy, and disconnect messages (see Fig. 11-1).

[HAEN] D. G. Haenschke, D. A. Kettler, E. Oberer, "DNHR: A New SPC/CCIS Network Management Challenge," ITC-10, Tenth International Teletraffic Congress, Montreal, June 1983, Session 3.2, paper 5.

How are these messages sent in a circuit-switched network? There are two possibilities: The messages may be sent on the same channels or trunks used to carry the actual call or information transmitted (voice, data, and other types of traffic), or they may use separate channels or networks. The first procedure is called in-band signaling; the second, out-of-band or *common-channel signaling* (CCS). These two methods were mentioned briefly in Section 10–1 in our introduction to circuit switching. We then carried out an analysis of in-band signaling for the sake of simplicity.

With the advent of digital switching, common-channel signaling has rapidly become the preferred way of handling the connection of calls in circuit-switched networks. The objectives of CCS are to provide significant improvements in call-connect time (see Chapters 10 and 11) and considerably increased signaling capacity. These are made possible by modern digital techniques, the widespread use of computer-controlled digital systems (as described in Chapter 11) and by the availability of wide-band digital transmission facilities.

A separate signaling network may be used for common-channel signaling, or signaling may be done using the same physical facilities of the circuit-switched network, occupying separate channels (commonly time slots) set aside for this purpose. Since signaling messages themselves are packets or short blocks of data, common-channel signaling has developed using packet-switched technology. A telephone network that adopts CCS as its method of handling the setting up and tearing down of circuit-switched calls thus uses both technologies discussed in this book: circuit switching for the calls themselves; packet switching for the control messages required to handle the connection and disconnection of calls. A separate packet-switched network may thus be established to handle control messages for the circuit-switched network.

A common-channel signaling system developed by the CCITT and adopted as a recommendation in 1981 is called the CCITT Common Channel Signaling (CCS) System No. 7, or simply System No. 7. It is rapidly becoming recognized as the international standard [CCS 7], [ANSI]. This signaling system, designed using the concepts of packet switching and tailored to conform with the OSI model, has been developed for use with both national and international traffic, for local and long-distance networks, for interexchange signaling, and for various types of channels, including both terrestrial and satellite channels.

The introduction of a packet-switched network used to carry call-signaling messages makes other data-handling services possible as well. These services

---

[CCS 7] *CCITT Specification for Signaling System No. 7*, CCITT, Geneva, 1984.

[ANSI] *American National Standard Specification of Signaling System No. 7*, American National Standards Inst., Inc., Issue 1, draft document, Jan. 1985. (This document is based on the 1984 CCITT specification as modified for use within the United States.)

include the simultaneous transmission of various types of traffic, data, voice, and video, among others. They make available the possibility of special types of call services (such as ''800'' call features and services, call forwarding, and database administration). They also make it possible to transmit data in either packet-switched or circuit-switched mode. Some of these possibilities will be discussed in Subsection 12–2–3.

To carry out the routing and signaling function, messages must obviously be sent via the packet-switched signaling network from one exchange in a circuit-switched network to another. Signal transfer points (STPs), either located at an exchange or geographically separate, are designated to provide the generation and transfer of signaling messages. A portion of such a network is shown in Fig. 12–10. One signaling channel is shown associated with two exchanges A and B. Messages involving the use of trunks (channels) that connect these two exchanges are in this case transmitted directly between them. However, signals relating to A and B may also be transferred via an intermediate STP, as shown. This mode of operation is called a *nonassociated* one. Both modes of operation have been defined in System No. 7. Signal-routing paths are strictly predetermined in a given network. In practice, then, one path *or* the other would be used to carry signals between two exchanges.

An example of an existing packet-switched network used to carry call-signaling messages between exchanges in a circuit-switched network is the AT&T CCS network. This network was established in 1976. It originally used an earlier CCITT signaling system, the CCITT signaling system No. 6—called, in the AT&T terminology, Common Channel Interoffice Signaling. A diagram of a subset of this network in the United States, used to handle the routing of

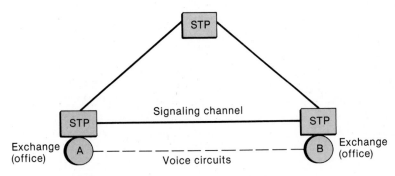

Figure 12–10 Portion of a CCS network

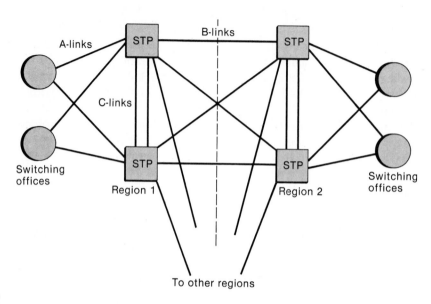

Figure 12–11 AT&T CCS network (from [FRER, Fig. 1], reprinted with permission from *BSTJ.* © 1982 AT&T)

long-distance AT&T calls and a great deal of other traffic, appears in Fig. 12–11 [BSTJ 1978], [BSTJ 1982], [FRER, Fig. 1].

The network is fully connected for reliability. Each of the 10 regions in the United States (at the top of the hierarchy of Fig. 12–1) has two interconnected duplicated STPs. The regional STPs are augmented by area STPs. Switching offices throughout the country home on the STPs in their region, as shown in Fig. 12–11. The A-links designate the links that provide access to the network from a switching office. The STPs in the network are interconnected by so-called B-links, while duplicate STPs in a region are connected by C-links. In the original 1976 network, links were of 2400- or 4800-bps transmission capacity. The newer System No. 7 is designed for 56-kbps and 64-kbps transmission capacities, which speed up the call-connect process, allow faster transmission of signaling messages, and enable signaling capacity to be increased dramatically.

[BSTJ 1978] "Common Channel Interoffice Signaling," special issue, *BSTJ*, vol. 57, no. 2, Feb. 1978.

[BSTJ 1982] "Stored Program Controlled Network," special issue, *BSTJ*, vol. 61, no. 7, part 3, Sept. 1982.

[FRER] R. F. Frerking and M. A. McGrew, "Routing of Direct Signaling Messages in the CCIS Network," special issue, *BSTJ*, vol. 61, no. 7, part 3, Sept. 1982, 1599–1609.

The CCITT System No. 7 protocol is designed to conform with the OSI Reference Model, as already noted. It has two parts, a *user part* and a *message transfer part*. The user part comes in several varieties, each one corresponding to higher-layer protocols that enable user functions, possibly on dissimilar machines, to communicate with one another. Examples of user parts include a telephone user part for basic telephone service, a data user part for circuit-switched data service, an operations and maintenance user part, and an ISDN (integrated service digital network) user part for providing combined voice, data, and video services. The user parts make use of the network delivery services provided by the *message transfer part* [CCS 7], [ANSI], [HLAW]. The message transfer part provides a connectionless (datagram-type) but sequenced transport service. Note how this organization conforms to the OSI Reference Model. System No. 7 is a layered architecture, but the layers are referred to as levels. The message transfer part comprises levels 1–3, which correspond to OSI layers 1 and 2 and the lower part of layer 3. A comparison between the OSI and System No. 7 architectures appears in Fig. 12–12 [ANSI, Overview, Fig. 1]. The message transfer part is labeled MTP in this figure. The functional block labeled SCCP (signaling connection control point) provides the conversion from the message transfer part to the network service (both connectionless and connection oriented) specified by the OSI model. Note that the higher layers of the OSI model connect directly to the SCCP.

The System No. 7 layered model also allows network application processes, residing within layer 3 of the OSI Reference Model, to access the SCCP and the message transfer part through a layered model equivalent to that of the OSI model, but residing wholly within the network layer. These are indicated as the application service part (N.ASP) and are numbered System No. 7 levels 4–4, 4–5, and 4–6 in Fig. 12–12. A network application is an entity that uses the services of the application service part (N.ASP), the SCCP, and the message transfer part. An example of such an application, shown in Fig. 12–12, is the transaction capabilities application part, which is used for non-call-connection services such as "800" service or calling card service [BSTJ 1982].

It is apparent from Fig. 12–12 that different user parts enter the model at different points in the hierarchy. The ISDN user part, for example, deals solely with services within the network and hence appears in the network layer. User parts residing within the network layer, but requiring presentation, session, and transport features similar to those provided by the overall OSI model, obtain these from the N.ASP levels. Finally, user parts lying outside the network itself, such as operations and maintenance (OA&M in Fig. 12–12), appear just like any

---

[HLAW] F. Hlawa and A. Stoll, "Common Channel Signaling Based on CCITT System No. 7," *Telephony*, Feb. 9, 1981.

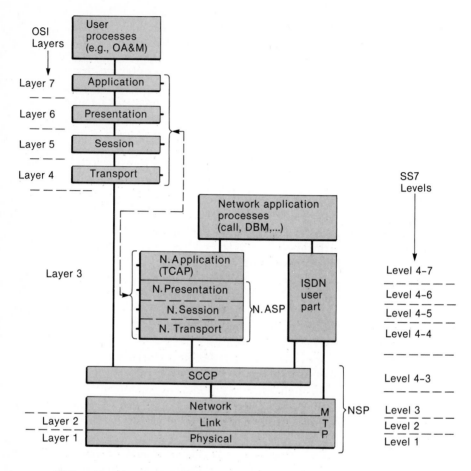

MTP = Message Transfer Part
SCCP = Signaling Connection Control Part
OA&M = Operations Administration and Maintenance
TCAP = Transaction Capabilities Application Part
ASP = Application Service Part
NSP = Network Service Part
DBM = Database Management

Figure 12–12 Signaling system No. 7 layered reference model (from [ANSI, Overview, Fig. 1])

other user process in the OSI model and require the services of the OSI application layer.

No matter where the particular user part appears in the layered reference model, it ultimately requires the services of the message transfer part, as previously noted and as indicated in Fig. 12–12. Each user part—whether the telephone user part, operations and maintenance, ISDN, or any of the others defined—has standard messages and procedures defined for a particular service. For example, in the telephone user part, operating procedures have to be established for interexchange signaling. These signals are exactly the call setup, disconnect, answer, and other signals described earlier and discussed in detail in Chapter 11. Their use and handling are the function of the user part, and, in this case, would appear at level 4 (Fig. 12–12) as well as higher levels, if necessary. The message transfer part, at levels 1–3, provides the network delivery service.

Level 1 of the message transfer part deals with the physical layer, as is apparent from Fig. 12–12; level 2 deals with the data link control. The System No. 7 data link control, covering messages transmitted between adjacent STPs, is based on HDLC, as will be seen shortly. The frame format is similar to that of HDLC, but the various signaling messages transmitted differ. Finally, level 3 of the message transfer part is precisely the level concerned with message distribution and with control of the CCS network.

12–2–1 Message Transfer Part, Signaling Link Level

Now consider the link-level protocol. Note again that this protocol governs communication between adjacent STPs over the common signaling channel (Fig. 12–10), just as in the link-layer protocols discussed in Chapter 4. The blocks of data transmitted over the link, corresponding to frames in the HDLC nomenclature, are called *signal units* here. Three types of signal units are distinguished: a message signal unit, a status signal unit, and a filler signal unit. The first type carries a *message*, corresponding to the information part or *packet* of the HDLC frame, plus a message transfer part corresponding to the HDLC frame header. The *filler* signal unit is used to transmit ack and nak information when messages are not available. The message field carries information as to the signal being transmitted, appropriate addresses, and a variety of other variable-length fields. By expanding the message field size one can use the message signal units for purposes other than common-channel signaling. The CCS network then becomes a full-fledged packet-switched network and can be used to carry out many of the stored program-control functions alluded to earlier [BSTJ 1982]. The format of the message signal unit appears in Fig. 12–13 [HLAW]. The format of the filler signal unit is similar, except that it has no message (*SIF*) field and no *SI*- and *NI*-fields preceding it. (These fields as well as the other fields of

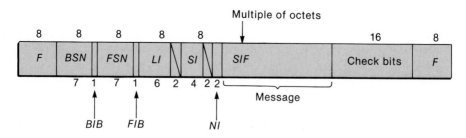

Figure 12–13   Format, message signal unit, System No. 7 (numbers shown are in bits)

the signal unit format are described later in this subsection.) The CCS *status* signal unit, used to initiate transmission on a link or to recover from loss of transmission, also has no *SI*- and *NI*-fields and incorporates a 1- or 2-byte control field in place of the message (*SIF*) field of the message signal unit.

In normal operation, message signal units are transmitted back and forth between adjacent STP nodes. They play the role of I-frames in the HDLC protocol. When message signal units are not available for transmission, however, filler signal units are transmitted instead. Note the appearance, in the message signal unit format of Fig. 12–13, of the 8-bit *F* (flag) characters at the beginning and end of the signal unit. These are precisely the *F*-characters, 01111110, used in the HDLC frame format (see Chapter 4). The 16-bit check field uses exactly the same error-detecting check polynomial as in HDLC. The rest of the message signal unit format, however, differs from that of the HDLC I-frame described in Chapter 4. Note, for example, that there is no explicit address field identifying either the transmitting (sending) STP or the receiving STP nodes. The 7-bit backward sequence number (*BSN*) and forward sequence number (*FSN*) fields shown in Fig. 12–13 play the roles of $N(R)$ and $N(S)$, respectively, in HDLC. Sequence numbering from 0 to 127 is thus used in this signaling system.

Explicit positive or negative acknowledgement of message signal units is carried out using the sequence number fields and the backward indicator (*BIB*) and forward indicator (*FIB*) bits associated with each. As in HDLC, the *FSN*-field is incremented each time a new message signal unit is transmitted. The backward sequence number (*BSN*), assigned by the receiving system to signal units it transmits, refers to a message signal unit it has received from its neighboring node. Negative acknowledgement associated with this *BSN* is indicated by *inverting* the *BIB* bit. This inversion is maintained in all subsequent signal units transmitted in the same direction until another faulty signal unit is received and a negative acknowledgement again indicated by inverting the bit. The forward indicator bit (*FIB*) is used by the transmitter to indicate retransmission of a message signal unit: It is inverted when a negatively acknowledged message signal unit is retransmitted. All signal units following maintain the same bit status until another negative acknowledgement is received. A go-back-*N* retransmission protocol is used in System No. 7. All signal units are stored at the sending side until positively acknowledged. On receipt of a negative acknowledgement the signal unit in question and all signal units following are retransmitted. Filler signal units are normally transmitted whenever no new message signal units are waiting to be transmitted. As already noted, these signal units carry the same *BSN* and *FSN* fields, with associated backward and forward indicator bits, as the message signal units. The *FSN*, however, is not incremented. It carries the sequence number of the last message signal unit transmitted. Filler signal units thus serve to positively or negatively acknowledge signal units received in the reverse direction. An example of an error-free signal unit exchange is shown in Fig. 12–14. *BIB* and *FIB* bits are not indicated since they maintain the same value throughout error-free transmission. The first number shown represents the *BSN* and the second the *FSN*, in agreement with Fig. 12–13. The letter *M* is used to represent a message signal unit, while *F* represents the filler signal unit. Since either message or filler signal units are always

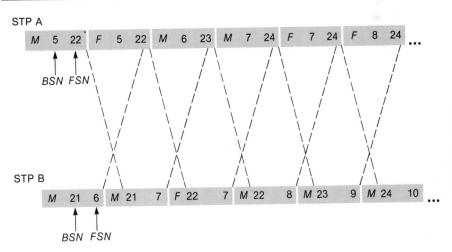

Figure 12–14 Error-free signal unit exchange, System No. 7; M = message, F = filler

present on the line, as shown in Fig. 12–14, there is no need to transmit two F-flags. Only the F-flag at the beginning of a signal unit need be transmitted.

The go-back-N retransmission procedure is modified when longer-delay propagation links are encountered to speed up the error-recovery process. A method called *preventive cyclic retransmission* is used on all satellite links and on intercontinental links with a one-way propagation delay of 15 msec or more. In this method, previously transmitted but still unacknowledged message signal units are retransmitted cyclically, without waiting for acknowledgement, when no new messages are available to be sent. A simple performance analysis for preventive cyclic retransmission, as well as the basic go-back-N method, will be presented later.

The message (SIF) field in Fig. 12–13 varies in multiples of octets (bytes), as shown there. In international applications its maximum size is 61 octets. In national applications the message size may be larger. Messages in United States networks, for example, may range up to 256 octets in length [ANSI, Q.703]. The length indicator field, LI (Fig. 12–13), is used to distinguish between the three types of signal units, as well as to indicate message length where applicable. Thus a value of 0 indicates a filler signal unit; 1 or 2 corresponds to the status signal unit; and 3–63 provide for various lengths of the message signal unit.

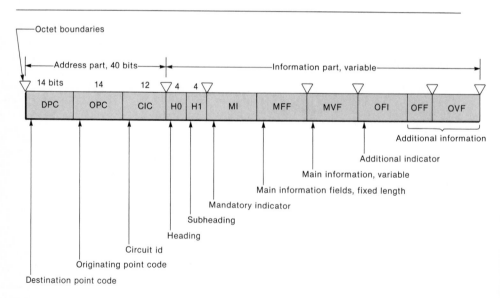

Figure 12–15 Message (SIF) format, telephone user part, international telephone traffic (based on [HLAW, Fig. 4])

The number 63, in particular, indicates the message field is longer than 61 octets. The *NI* (national indicator) field is used to distinguish between national and international messages. Finally, the service indicator (*SI*) field indicates the user part (service) involved in the transmission of the message. As noted previously, examples of a user part include a telephone user part, a data user part, and an ISDN user part (Fig. 12–12).

Consider as an example the telephone user part, defined to provide basic telephone service. Messages under the control of this part are associated with the various signals involved in setting up and releasing circuits for a call, as noted earlier and as described in detail in Chapter 11. The message length in this part is generally 10 octets or less, with its format structured as shown in Fig. 12–15 [HLAW, Fig. 4]. The address part, 40 bits in length, is shown as it would be used for international telephone traffic. Fourteen bits are each used to identify the originating and destination exchange addresses; 12 bits identify a speech circuit for these two exchanges. The rest of the message, which is variable in length, is called the information part. The first 8 bits identify the signal. Some signals, such as a release or release acknowledgement, require no additional information and thus terminate at this point. Others, such as the dialing-digit messages, carry additional information and thus require added information fields.

12-2-2 Signaling System Performance

Using the signal unit formats of Figs. 12–13 and 12–15, as well as our discussion of the data link level involving exchange of signal units between adjacent STPs, it is possible to determine the time-delay–throughput performance of System No. 7 under various link loads [ANSI, Q.706]. This analysis can in turn be related to the call-handling (trunk load) capacity of a CCS link. Only the error-free condition case is considered here. Analytic results for the system in the case of disturbances, with error retransmission required, appear in [ANSI, Q.706].

Consider the basic transmission scheme first, without the use of preventive cyclic retransmission. Message signal units, possibly varying in length, are assumed to arrive at a Poisson rate of λ_1 messages per unit time and to queue up for transmission over the CCS link. (We neglect message-processing time and focus on queueing time only.) If messages are not available to be transmitted, filler signal units of fixed length T_f, in units of time, are transmitted instead (Fig. 12–14). Although the arrival of the filler signal units does not, strictly speaking, constitute a Poisson process (why is this so?), we ignore this potential problem and assume at this point that they do arrive at an average Poisson rate of λ_2 units/time. We shall see shortly that the message-signal-unit delay equation obtained using this assumption appears reasonable and intuitively correct. The contribution to the link utilization by the filler signal units is then just $\rho_2 = \lambda_2 T_f$.

A message signal unit arriving during the transmission of a filler signal unit waits for completion of that signal unit and is then transmitted immediately. It is thus apparent that single-server nonpreemptive priority queueing analysis with two classes of customers is called for to determine the waiting time of the message signal units. Equations (2–83) and (2–85) can be applied directly to this problem. The only, rather subtle, point to be made here is that since there are no gaps in transmission (Fig. 12–14), the link operates at full utilization. Its traffic intensity or utilization ρ must be set equal to 1.

Let the average message signal unit length, in units of time, be T_m. The second moment of the length is taken as some value $E(\tau_1^2)$. The portion of the link utilization due to message signal units is given as usual, by $\rho_1 = \lambda_1 T_m$. Combining message and filler signal unit utilizations, we then have

$$\rho = 1 = \rho_1 + \rho_2$$

$$\rho_1 = \lambda_1 T_m \qquad \rho_2 = \lambda_2 T_f$$

(12–17)

From Eqs. (2–83) and (2–85) applied to this problem, we have as the average wait time of the message signal units, the performance parameter of interest here,

$$E(W_1) = E(T_0)/(1 - \rho_1)$$

(12–18)

with the residual service time $E(T_0)$ given by

$$E(T_0) = \tfrac{1}{2}\lambda_1 E(\tau_1^2) + \tfrac{1}{2}\lambda_2 T_f^2$$

$$= \frac{\lambda_1 E(\tau_1^2)}{2} + \frac{\rho_2 T_f}{2}$$

(12–19)

Use has been made here of the fact that the filler signal units are of fixed length T_f, as already noted.

Writing $\rho_2 = 1 - \rho_1$ from Eq. (12–17) and applying Eq. (12–19) to Eq. (12–18), we get, as the average wait time for the message signal units at one link of a CCS system,

$$E(W_1) = \tfrac{1}{2}T_f + \tfrac{1}{2}\lambda_1 \frac{E(\tau_1^2)}{(1 - \rho_1)}$$

(12–20)

This is the equation appearing in [ANSI, Q.706]. Note the physical interpretation here. With very little traffic in the system (ρ_1 and λ_1 small), arriving message units must wait for a signal filler unit to complete transmission. On the average this is one half of the filler transmission length, in units of time. As the load begins to increase, the normal $M/G/1$ wait time, due to other message signal units waiting for transmission, is added and begins to predominate.

Now consider some typical calculations. We must provide some numbers for the filler length T_f and the message length parameters T_m and $E(\tau_1^2)$. We first note that since there are no gaps in the transmission of signal units (Fig. 12–14), there is no need to transmit both beginning and end F-fields (Fig. 12–13) of a signal unit. One F-field suffices; it serves to demark consecutive signal units. This then reduces the length of a typical signal unit by 1 octet. Since the filler signal unit carries one F-field, the BSN, the FSN, the LI, and the 2-octet check bit fields (Fig. 12–13), it is 6 octets or 48 bits long. Assume that an average message signal unit is 15 octets or 120 bits long. This provides 8 octets or 64 bits for the message (SIF) field (Figs. 12–13 and 12–15). For a 56-kbps transmission link, we then get, for the average message signal unit length, $T_m = 2.15$ msec, while the filler signal unit length is $T_f = 0.85$ msec.

Consider first the special case in which all message signal units are the same length T_m. This then means that M/D/1 analysis (Chapter 2) applies. Using $E(\tau_1^2) = T_m^2$ in this case (the variance of the message length is zero), and adding the message signal unit length T_m to the wait time to find the average link time delay $E(T)$, one gets, from Eq. (12–20),

$$E(T) = \tfrac{1}{2}T_f + \frac{T_m}{(1 - \rho_1)}(1 - \tfrac{1}{2}\rho_1) \qquad \rho_1 = \lambda_1 T_m \qquad (12\text{–}21)$$

Taking first the case of very little message traffic on the link for the 56-kbps example just described, we have, as the minimum average message unit time delay at each CCS link,

$$E(T) = \tfrac{1}{2}T_f + T_m = 2.58 \text{ msec} \qquad \rho_1 = 0 \qquad (12\text{–}22)$$

For $\rho_1 = 0.5$ this delay rises to 3.65 msec. For $\rho_1 = 0.9$ it becomes 12.3 msec. These numbers are clearly quite small and even if summed over two or three cascaded links can lead to fast call-setup times. (Compare the numbers, for example, with those obtained in the examples at the beginning of Chapter 10. Note, however, as discussed in Chapter 11, that processing time and queueing delays at the switching nodes may increase these times markedly.)

How do these numbers relate to the *call*-handling capacity of a system? In particular, what values of ρ_1 might one expect for various call-traffic conditions? Let the Erlang load at an exchange using an STP node be A Erlangs. Say that the average call is 300 sec- (5 min-) long. Then the number of calls/sec accessing the STP node is $\lambda_c = A/300$ calls/sec. Each call generates a series of message signal units (see Chapter 11). As a typical example, say that there are six such message units. (These might include a single call-setup message, an end-of-pulsing or end-of-dial-digits message, an answer signal, and a release-ack signal. In other cases additional dial digit messages or other control messages might be needed.) The average number of CCS message signal units transmitted per second is then

$$\lambda_1 = 6A/300 \text{ signal units/sec} \qquad (12\text{–}23)$$

Say that $A = 10,000$ Erlangs, as an example. Then $\lambda_1 = 200$ message signal units/sec. For the value of $T_m = 2.15$ msec just calculated, $\rho_1 = \lambda_1 T_m = 0.43$. If $A = 20,000$ Erlangs, the message signal unit utilization doubles to $\rho_1 = 0.86$. It is thus apparent that with these numbers the CCS link is capable of handling 10,000 to 20,000 Erlangs of traffic. If the average call length is reduced, the load can obviously increase.

How sensitive are these time-delay results to the fixed-message-length (M/D/1) assumption? If $E(\tau_1^2)$ increases by 20 percent to $1.2\, T_m^2$ (this is a model used in the CCS Recommendation performance calculation), the average time delay $E(T)$ increases from 3.65 to 3.87 msec for $\rho_1 = 0.5$, a clearly insignificant change. For $\rho_1 = 0.9$ the time delay increases from 12.3 msec to 14.2 msec, again a tolerable change. Larger variations in message lengths lead of course to further increases in the average delay.

Preventive cyclic maintenance, used, as noted earlier, to speed up the error recovery process on long propagation paths, tends to increase the average message delay under error-free conditions [ANSI, Q, 706, Fig. 2]. To see why this happens, consider a very long propagation path. Messages waiting for acknowledgement will then be continually retransmitted cyclically, even with a relatively moderate traffic load. A message signal unit will thus almost always be in the process of being transmitted: this could be one being transmitted for the first time or one waiting to be acknowledged. Filler signal units will still be used to fill gaps in transmission, but will only infrequently be required. A new message arriving will thus always have to at least wait for the completion of transmission of a message signal unit. Since message signal units are longer than filler signal units, this will increase the link waiting time for message signal units.

The waiting-time analysis in this case of preventive cyclic retransmission can be carried out in the following manner. There are three types of signal units transmitted over a CCS link in this case: 1) arriving message signal units, 2) message signal units waiting for acknowledgement that are retransmitted cyclically if no new messages arrive, and finally, 3) the filler signal units transmitted if neither one of the first two message types is available for transmission. This is again a case of nonpreemptive priority transmission, but with *three* classes of customers in the order just listed. Letting the respective average arrival rates be λ_1, λ_2, and λ_3, in order of priority, the residual service time $E(T_0)$ is now given by

$$
\begin{aligned}
E(T_0) &= \tfrac{1}{2}(\lambda_1 + \lambda_2)E(\tau_1^2) + \tfrac{1}{2}\lambda_3 T_f^2 \\
&= \tfrac{1}{2}(\lambda_1 + \lambda_2)E(\tau_1^2) + \tfrac{1}{2}\rho_3 T_f
\end{aligned}
\tag{12-24}
$$

with $\rho_3 = \lambda_3 T_f$ now defined as the filler signal unit link utilization.

But now consider the form of Eq. (12–24) and interpret it in terms of the situation under study. With very little traffic on the link a newly arriving message signal unit waits either for another message signal unit to complete transmission or for a filler signal unit to complete transmission. In the latter case the average time for completion is $T_f/2$. What is the probability that a filler signal will have been transmitted? Let the acknowledgement round-trip propagation delay between CCS nodes be T_L units of time. If this interval of time is small a waiting message unit is quickly acknowledged, there is little chance a new message will arrive, and filler units are more likely to be transmitted. Alternatively, if the propagation delay is long, waiting message units are more likely to be retransmitted, new message units are more likely to arrive, and filler units are more likely to be transmitted.

The probability that no message units will arrive in T_L units of time so that filler signal units are continuously transmitted in this time is the Poisson probability of zero arrivals, $e^{-\lambda_1 T_L}$. This enables us to write

$$\rho_3 = e^{-\lambda_1 T_L} \tag{12–25}$$

Since $\rho_1 + \rho_2 = (\lambda_1 + \lambda_2)T_m = 1 - \rho_3$, we then have, from Eq. (12–24),

$$E(T_0) = \frac{(1 - \rho_3)E(\tau_1^2)}{2T_m} + \rho_3 \frac{T_f}{2} \tag{12–24a}$$

The average wait time in this case is then

$$E(W_1) = \frac{(1 - e^{-\lambda_1 T_L})E(\tau_1^2)}{2T_m(1 - \rho_1)} + \frac{e^{-\lambda_1 T_L} T_f}{2(1 - \rho_1)} \tag{12–26}$$

Since the round-trip propagation delay must be greater than 15 msec for preventive cyclic retransmission to be invoked, and $T_m = 2$ msec is much smaller than this, it is apparent that, even with relatively light loading, message signal units are being transmitted continually over the link. As a good approximation, then,

$$E(W_1) \doteq \frac{E(\tau_1^2)}{2T_m(1 - \rho_1)} \tag{12–26a}$$

with $e^{-\lambda_1 T_L} \doteq 0$, or $\lambda_1 T_L = \rho_1 T_L/T_m > 2$.

As a special case, let the message signal units be of fixed length T_m. Then the average wait time is

$$E(W_1) \doteq \frac{T_m}{2(1 - \rho_1)} \qquad \rho_1 T_L/T_m > 2 \tag{12–26b}$$

and the average time delay at a CCS node, not including processing time, is

$$E(T) \doteq \frac{T_m}{(1 - \rho_1)}\left[\frac{3}{2} - \rho_1\right] \tag{12–27}$$

Consider the same example we used earlier in discussing the basic transmission method. We had $T_m = 2.15$ msec for a 56-kbps link. With $T_L = 30$ msec, $T_L/T_m = 14$, so that $\rho_1 T_L/T_m > 2$ corresponds to $\rho_1 > 0.14$. Say that $\rho_1 = 0.5$, as an example. Then the average time delay using the preventive cyclic retransmission method is $E(T) = 2 \, T_m = 4.3$ msec. Using the basic transmission method the delay was $E(T) = 3.5$ msec. Thus, as expected, the preventive cyclic maintenance method does increase the CCS link delays somewhat.

More detailed results, including the effect of errors on the link, appear in [ANSI, Q.706, Figs. 2 – 7].

12 – 2 – 3 High-level Features

It is apparent that the introduction of a packet-switched signaling system such as CCS allows that system to be used for the transmission of data other than the signaling and control information that sets up and takes down circuits in a typical telephone environment. This point was made at the beginning of this section and was reinforced by noting that the user parts incorporated in System No. 7 include not only the telephone user part for basic telephone service, but a data user part, an ISDN user part, and a variety of network application processes. Included here are such features as database access and management and a variety of transaction-type services not necessarily related to call connection (Fig. 12 – 12).

One example of a set of such services are those provided by AT&T Communications in the United States. These services include calling card service, which permits a customer with a telephone credit card to dial calls without operator assistance, and "800" service, through which calling parties can make calls without charge to themselves. These two services and others like them make use of the AT&T Stored Program Control Network, a network of stored program control switching systems interconnected by the common-channel signaling network. These switching systems include the AT&T No. 4 ESS and No. 5 ESS digital switching systems, described in Chapters 10 and 11. By appropriate design of the software in the computers around which those switching systems are built, a variety of non-call-connection features can be incorporated. Interconnecting the computers through the CCS network makes the Stored Program Control Network possible.

Calling-card service features for the AT&T system are described in a series of papers in [BSTJ 1982]. In particular, an overview appears in [BASI]. An overview of 800 service appears in [SHEI]. Other applications using the Stored Program Control Network appear in [RAAC].

[BASI] R. G. Basinger et al., "Calling Card Service — Overall Description and Operational Characteristics," *BSTJ*, vol. 61, no. 7, part 3, Sept. 1982, 1655 – 1674.

[SHEI] D. Sheinbein and R. P. Weber, "800 Service Using SPC Network Capability," *BSTJ*, vol. 61, no. 7, part 3, Sept. 1982, 1737 – 1744.

[RAAC] G. A. Raack, E. G. Sable, R. J. Stewart, "Customer Control of Network Services," *IEEE Comm. Mag.*, vol. 22, no. 10, Oct. 1984, 8 – 14.

12–3 Integrated Services Digital Networks

Chapters 3 – 9 of this book focused on packet-switched data networks; the layers of the OSI Reference Model were used as a mechanism to integrate the discussion. Chapters 10 and 11 and the previous sections of this chapter focused on circuit-switched networks, as exemplified by telephone networks worldwide. Although still used principally to carry voice, the use of telephone networks for data transmission has increased rapidly, particularly with the advent of the digital switching systems discussed in Chapters 10 and 11.

It is apparent from the previous sections that packet-switched and circuit-switching technology, two seemingly different modes of handling traffic, are moving closer together. Thus the move toward nonhierarchical routing in the public carrier circuit-switched networks makes these networks look topologically like the packet-switched networks discussed in earlier chapters. The development of packet-switched CCS systems that handle control signaling as well as data transmission for the public circuit-switched networks is another indication of the move toward integration of these two ways of handling data.

Future communication networks are expected to handle a variety of data traffic types, covering a range of applications as diverse as very low bit-rate control and alarm channels for the home and business, interactive information services, electronic mail, digital voice, facsimile, file transfers, and wideband digital video services among many others. Some of the communication services required might be carried out using packet-switched technology, some circuit-switched, and some hybrid techniques yet to be developed. Integration of these services and traffic types corresponding to them, for transmission over one network or over a number of networks in parallel, might be carried out at the customer's premises or by the network operator. A single system (terminal or intelligent workstation) with a number of data inputs coming in to it might be used, or the outputs of a number of stations might be multiplexed together for transmission over a network or series of networks.

These networks of the future have been termed *integrated services digital networks* (ISDNs). To some extent small portions of this concept already exist. Modern digital switching machines of the type discussed in Chapters 10 and 11 have packet-switched modular add-ons that enable packet-switched and circuit-switched connections to be handled by the same machine. The No. 5 ESS system described in Chapter 11 is an example of such a system. In addition to the circuit-switched capability discussed in Chapter 11, various voice-digital access formats defined by the CCITT are supported in No. 5 ESS. Packet-switched interface units are available for connection to X.25 packet-switched networks, as are connections to the CCS network. Details appear in [JOHN].

Many modern PBXs (private branch exchanges) integrate voice and data in one system as well. Local area networks are also being designed to handle a variety of traffic types.

Three elements are generally recognized as being required for an ISDN [ANDR 1984]: All-digital channels end to end are used; the network handles a multiplicity of services with possibly differing bandwidths using interleaved bit streams; there are standard interfaces for user access. The standard interface concept, in particular, has been attacked by CCITT. Study groups of CCITT are working on the development of interfaces that will be compatible with existing 64-kbps digital voice channels and that will incorporate signaling (control) channels as well. An example of such an interface, the first one developed as a CCITT recommendation, is the "2B + D" narrowband interface. This consists of two 64-kbps "B channels" for information transfer and a 16-kbps "D channel" for signaling and other uses. The three channels, totaling 144 kbps of transmission capability, are interleaved. A B channel could be used for digital voice or circuit-switched data; the D channel could be used for carrying packet-switched data as well as control packets. Wider band interfaces, based on an $nB + D$ structure, with the D channel considered a 64-kbps channel, are also being developed as CCITT recommendations. $n = 23$ makes them compatible with the 1.544-Mbps T1 standard; $n = 30$ makes them compatible with the worldwide 2.048-Mbps digital transmission standard [SCHW 1980a]. Other interface recommendations are designed to handle even higher bandwidth services such as video and high-speed facsimile. A layered architecture approach is being used to allow conformity with the OSI model [KOST], [DUC]. In the ISDN world, communications traffic, both between users and for control purposes with networks, should be able to move freely among packet-switched networks, circuit-switched networks, and control networks such as CCS System No. 7 [DUC].

[JOHN] J. W. Johnson et al., "Integrated Digital Services of the 5 ESS System," ISS '84, Florence, Italy, May 1984, Paper 14A3.

[ANDR 1984] F. T. Andrews, Jr., "ISDN '83," special issue on ISDNs, *IEEE Comm. Mag.*, vol. 22, no. 1, Jan. 1984, 6–10.

[SCHW 1980a] M. Schwartz, *Information Transmission, Modulation, and Noise,* 3d ed., McGraw-Hill, New York, 1980.

[KOST] D. J. Kostas, "Transition to ISDN—An Overview," special issue on ISDNs, *IEEE Comm. Mag.*, vol. 22, no. 1, Jan. 1984, 11–17.

[DUC] N. Q. Duc and E. K. Chew, "ISDN Protocol Architecture," *IEEE Comm. Mag.*, vol. 23, no. 3, March 1985, 15–22.

It is clear that within the confines of the CCITT ISDN-interface recommendations, as well as other proposals for integrating different types of traffic that might be suggested, there are various ways to multiplex or combine traffic. At the link level, different types of traffic might each be assigned a separate channel of specified bandwidth; they might compete for a group of channels in FCFS or priority order; they might share channels dynamically; some types of traffic might be assigned channels on a circuit-switched (dedicated) basis, while others might use them on a packet-switched basis. Within a network or series of interconnected networks, different types of traffic might be handled differently. Circuit-switched and packet-switched traffic could be routed differently, even between the same source-destination addresses, or they could be routed on identical paths. They might share the same network facilities; they could use different resources.

Since the field of integrated services digital networks is quite broad and is still in its early stages of development, it cannot be covered in any depth in this book. Instead, we choose to focus on one class of problems involving integration of disparate traffic types that has received extensive attention in recent years. This is the area of integrated or hybrid multiplexing: the study of various ways of combining two or more traffic types for transmission over a common digital-transmission facility. Examples include the multiplexing of circuit-switched traffic such as voice together with packet-switched data onto a common transmission link, and the combining of circuit-switched traffic streams of differing bandwidth (voice, files of data, and video, among others) onto a common transmission facility. A pictorial representation of such an integrated multiplexer appears in Fig. 12–16 [SCHW 1983, Fig. 2]. A number of circuit-switched inputs, of possibly differing bandwidths, b_1 to b_K, with different holding times, $\mu_{c_1}^{-1}$ through $\mu_{c_K}^{-1}$, and different arrival rates, λ_{c1} through λ_{cK}, are to be combined with packets from a common packet queue onto a common time-division multiplexing (TDM) link of capacity C bps. The TDM link is shown as having a frame structure, with each frame of F bits divided into a frame portion of F_c bits and a packet portion of $F - F_c$ bits. Other configurations may be visualized as well.

A number of analytic studies have appeared in the literature that model and analyze various combining strategies using the basic architecture of Fig. 12–16. We shall discuss a few of these. Models of multiplexing or combining schemes such as that of Fig. 12–16 can be applied to a number of areas under study in the ISDN field. They can be used to model the combined output of one integrated or multimedia workstation carrying voice, data, graphics, and video at the interface to a network. They can be used to model the multiplexing of traffic

[SCHW 1983] M. Schwartz and B. Kraimeche, "An Analytic Control Model for an Integrated Node," IEEE Infocom '83, San Diego, Calif., April 1983.

from a variety of users onto a wideband digital channel for transmission into a network. They can be used to study various ways of combining integrated traffic along the transmission links of an integrated network.

In the remainder of this chapter we analyze combining strategies that are based on the architecture of Fig. 12–16. We make no attempt to be exhaustive, but rather focus on a few simple examples of combining schemes. Our emphasis will be on ways of modeling these techniques and then on methods for analyzing the models developed. Comparisons will be made where possible. The material draws on and extends the queueing analysis of Chapter 2 and chapters following.

12–3–1 A Mathematical Prelude: Moment-generating Functions

Moment-generating functions or z-transform techniques are used extensively in the study of queueing systems. We will require their use in the analysis of the combining or hybrid multiplexing techniques to follow. It is thus appro-

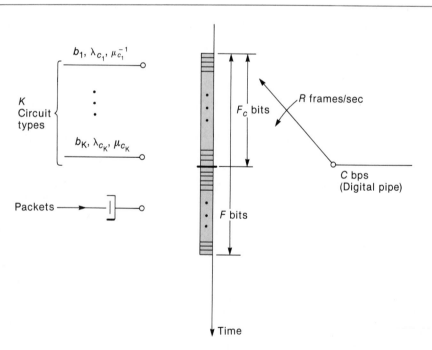

Figure 12–16 TDM frame configuration of an integrated channel

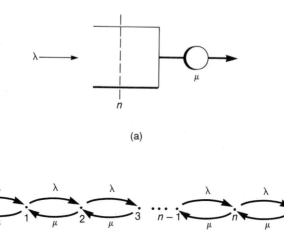

Figure 12-17 M/M/1 queue revisited
a. M/M/1 queue model
b. State diagram, M/M/1 queue

priate to introduce their use in queueing analysis here, extending the queueing analysis of Chapter 2. Here we consider the use of these functions only in M/M/1 and M/M/2 analysis. Other applications appear in [KLEI 1975a]. The use of these functions in the study of M/G/1 queues is summarized in [SCHW 1977].

Consider the M/M/1 queue shown in Fig. 12-17. Recall from our discussion of Chapter 2 and chapters following that in this simplest of queueing systems the arrival process is Poisson with average arrival rate λ, and the service-length distribution is exponential with average departure rate μ. The state diagram for the infinite queue at equilibrium then takes the form of Fig. 12-17(b). This is a special case of the more general birth-death process introduced in Chapter 2. Using the state diagram of Fig. 12-17(b) one can write the

[KLEI 1975a] L. Kleinrock, *Queueing Systems, Volume 1: Theory,* John Wiley & Sons, New York, 1975.
[SCHW 1977] M. Schwartz, *Computer-Communication Network Design and Analysis,* Prentice-Hall, Englewood Cliffs, N.J., 1977.

balance equation first introduced in Chapter 2 (Eq. 2–15), which governs the equilibrium statistics of the M/M/1 queue:

$$(\lambda + \mu)p_n = \lambda p_{n-1} + \mu p_{n+1} \qquad n \geq 1 \qquad (12-28)$$

p_n is the probability that the system is in state n. The balance equation for state $n = 0$ and $n = 1$ is similarly given by

$$\lambda p_0 = \mu p_1 \qquad (12-29)$$

Equation (12–28) was solved in Chapter 2 by noting that it in fact reduced to a much simpler balance equation

$$\lambda p_n = \mu p_{n+1} \qquad (12-30)$$

One then finds p_n recursively in terms of lower values of n, coming up with the solution

$$p_n = p_0 \rho^n \qquad \rho = \lambda/\mu \qquad (12-31)$$

The probability p_0 that the queue is empty is then found by the fundamental rule of probability measure that the sum of the probabilities over all states of the system must equal 1.

In more complex queueing systems that involve infinite queues, including those to be discussed in the rest of this section and Section 12–4 following, the balance equations are not readily solved by such recursive techniques. One therefore resorts to the moment-generating function or transform technique that we introduce here. Equation (12–28) is an example of a *difference equation* that occurs in discrete-parameter processes. The use of transform techniques transforms a difference equation to an algebraic equation that is often solved more readily using the techniques of algebra. Readers familiar with z-transforms applied to discrete-time systems or Laplace transforms applied to continuous-time systems will recall the same approach being used there [OPPE].

The moment-generating function or z-transform used in probability theory is defined as

$$G(z) = \sum_{n=0}^{\infty} p_n z^n \qquad (12-32)$$

Note that this function must be applied to the *infinite*-state system. Finite-state systems (for example, the finite M/M/1/N queue with blocking or the window-type flow-control schemes of Chapter 5) are usually solved by solving the finite

---

[OPPE] A. V. Oppenheim and A. S. Willsky, with I. T. Young, *Signals and Systems,* Prentice-Hall, Englewood Cliffs, N.J., 1983.

set of equations that define their operations. Some properties of this function are in order:

(1)
$$G(1) = \sum_{n=0}^{\infty} p_n = 1 \qquad (12-33)$$

(2)
$$\left.\frac{dG(z)}{dz}\right|_{z=1} = \sum_{n=0}^{\infty} np_n = E(n) \qquad (12-34)$$

(3)
$$\left.\frac{d^2G(z)}{dz^2}\right|_{z=1} = E(n^2) - E(n) \qquad (12-35)$$

(4)
$$\sum_{n=1}^{\infty} p_{n-1}z^n = zG(z) \qquad (12-36)$$

(5)
$$\sum_{n=0}^{\infty} p_{n+1}z^n = z^{-1}[G(z) - p_0] \qquad (12-37)$$

Properties (2) and (3), which indicate that by differentiating $G(z)$ with respect to z one can generate the moments of p_n, show why this function is called the moment-generating function. (It is clear that continued differentiation generates further moments.) Properties 4 and 5 show that difference terms convert to corresponding algebraic terms in z. The derivation of these properties, using the defining relation (12–32) for $G(z)$, is left to the reader.

Now consider the solution to Eq. (12–28), the governing equation of the M/M/1 queue, by the use of moment-generating functions. Multiply each term in that equation by z^n and sum over all values of n from 1 to ∞. It is left to the reader to show that, using Eqs. (12–36) and (12–37), one obtains the algebraic equation

$$(\lambda + \mu)[G(z) - p_0] = \lambda z G(z) + \mu\{z^{-1}[G(z) - p_0] - p_1\} \qquad (12-38)$$

Collecting all terms in $G(z)$ on the left-hand side and all other terms on the right-hand side, using the so-called boundary equation (12–29) to cancel the term involving p_1, and then simplifying the resultant equation, one finds the solution for $G(z)$ to be given by

$$G(z) = \frac{p_0}{1 - \rho z} \qquad \rho \equiv \lambda/\mu \qquad (12-39)$$

It is of interest to note that in the reduction of Eq. (12–38) to obtain Eq. (12–39), a factor of $(z - 1)$ common to the left- and right-hand sides of the resultant equation appears. The factor occurs commonly in the application of moment-generating functions to the analysis of queueing systems; it has been canceled in writing Eq. (12–39).

An additional equation must be used to find the unknown probability p_0. It is apparent that this must be the probability sum equation (12–33). Setting $G(1) = 1$, one finds that $p_0 = (1 - \rho)$. (Note that this was exactly what was found in going from Eq. (2–17) to Eq. (2–18) in Chapter 2.) The final solution for $G(z)$ for the M/M/1 queue is then given by

$$G(z) = \frac{(1 - \rho)}{(1 - \rho z)} \qquad (12\text{–}39\text{a})$$

As a check, it is left to the reader to show, by differentiating Eq. (12–39a), that

$$E(n) = \left. \frac{dG(z)}{dz} \right|_{z=1} = \frac{\rho}{(1 - \rho)} \qquad (12\text{–}40)$$

This agrees, of course, with the result obtained in Chapter 2 (Eq. 2–27), as used throughout this book.

How does one now find the desired probability p_n that the system is in state n? The object is to expand $G(z)$ in a power series in z and pick off the coefficient of z^n. (Recall the defining relation (12–32) for $G(z)$.) This is readily done in this case. (It turns out to be much more difficult and tedious algebraically in more complex queueing systems. More commonly, one differentiates $G(z)$ to find moments of the desired distribution.) Specifically, it is apparent that for $|\rho z| < 1$ or $\rho = \lambda/\mu < 1$, the denominator of Eq. (12–39a) may be expanded in an infinite series, providing the following infinite-series expansion for $G(z)$:

$$G(z) = \sum_{n=0}^{\infty} (1 - \rho)(\rho z)^n \qquad \rho z < 1 \qquad (12\text{–}39\text{b})$$

The desired probability p_n is the coefficient of z^n and is given by

$$p_n = (1 - \rho)\rho^n \qquad (12\text{–}41)$$

just as found in Chapter 2. (See Eq. (2–18).)

It is apparent that this method of finding the probability distribution p_n is much more lengthy (albeit quite straightforward) than the direct way used in Chapter 2. However, the method will be shown to be particularly useful in the material following, in our analysis of some simple models for the integration of voice and data. These models lead to two-dimensional birth-death processes that, in the infinite queue case, can only be solved by transform methods.

Now consider the M/M/2 queue as a second example. Recall that this queue was also used as an example in Chapter 2. The queue model and its state diagram appear in Fig. 12–18. Note that the basic difference between this queueing system and the M/M/1 queue is that for two or more customers present the rate of service is 2μ instead of μ. (Under the assumption of exponential service statistics there is a probability $2\mu\Delta t$ that a service completion will occur in the interval $(t, t + \Delta t)$ if the system is in any of the states $n \geq 2$.) One can

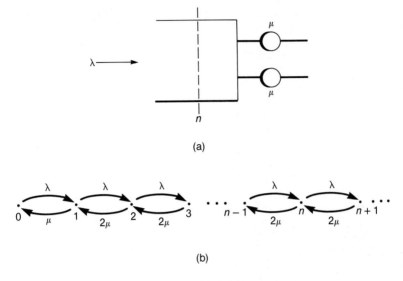

(a)

(b)

Figure 12–18 M/M/2 queue
a. Model
b. State diagram

---

set up the balance equations for this system using Fig. 12–18(b) and then solve them by introducing the moment-generating function, just as was done previously in the case of the M/M/1 queue. We choose instead to take the simpler approach of finding the moment-generating function from the state probabilities themselves. The more general case, starting from the balance equations representing Fig. 12–18(b), is left as an exercise for the reader.

From Eq. (2–41) we have as the equilibrium state probabilities for the M/M/2 queue

$$p_n/p_0 = \left(\frac{\lambda}{\mu}\right)(\lambda/2\mu)^{n-1} \qquad n \geq 1 \qquad (12-42)$$

(From Fig. 12–18(b) it is apparent, from recursive considerations, that $p_1/p_0 = \lambda/\mu$, $p_2/p_1 = \lambda/2\mu$, . . . , $p_n/p_{n-1} = \lambda/2\mu$, and so forth.) Using the parameter ρ to represent λ/μ rather than $\lambda/2\mu$ as was done in Chapter 2 (the present usage is somewhat more convenient at this point), one has as an alternate form for p_n/p_0 for the M/M/2 queue

$$p_n/p_0 = \rho \left(\frac{\rho}{2}\right)^{n-1} \qquad n \geq 1 \qquad (12-42a)$$

Since we must have $\sum_{n=0}^{\infty} p_n = 1$, it is readily shown that the unknown zero state probability p_0 is given by

$$p_0 = (2 - \rho)/(2 + \rho) \tag{12-43}$$

Now let us introduce the moment-generating function $G(z)$. Using the defining relation (12–32), one has, from Eq. (12–42a),

$$
\begin{aligned}
G(z) &= \sum_{n=0}^{\infty} p_n z^n \\
&= p_0 \left[1 + \rho \sum_{n=1}^{\infty} (\tfrac{1}{2}\rho z)^j \right] \tag{12-44} \\
&= p_0 \left[\frac{2 + \rho z}{2 - \rho z} \right]
\end{aligned}
$$

after carrying out the indicated sum and simplifying. Note that p_0 can be immediately found from Eq. (12–44) if desired by setting $G(1) = 1$. The resultant value agrees, of course, with Eq. (12–43).

Differentiating Eq. (12–44), one finds quite readily that the average number of customers in the queue, including those in service, is given by

$$E(n) = \left. \frac{dG(z)}{dz} \right|_{z=1} = \frac{4\rho}{(2 - \rho)(2 + \rho)} \tag{12-46}$$

The expressions for the moment-generating function for both the M/M/1 and M/M/2 queues, Eqs. (12–39) and (12–44), will be used later in validating results for a hybrid multiplexing scheme.

12–3–2 Models for Integrated Voice and Data

In this introduction to integrated communication systems we focus on the multiplexing of two types of traffic only. One type, epitomized by voice, as one example, requires circuit-switched service. The other type consists of packet-switched traffic. The circuit-switched traffic is blocked if transmission channels (trunks) are not available. The packet-switched traffic may be queued while waiting for service.

A limited number of studies of the multiplexing of heterogeneous users of blocked, circuit-switched service as exemplified by Fig. 12–16 have been car-

[GIMP] L. A. Gimpelson, "Analysis of Mixtures of Wide- and Narrow-band Traffic," *IEEE Trans. on Comm. Tech.*, vol. 13, no. 3, Sept. 1965, 258–266.

[INOS] H. Inose, *An Introduction to Digital Integrated Communication Systems*, University of Tokyo Press, Tokyo, 1979.

ried out [GIMP], [INOS]. These users differ in their bandwidth (bit rate) requirements, their holding or service times, and their arrival rates, as shown in Fig. 12–16. Studies have also been made of systems that integrate wideband circuit-switched traffic with narrowband queued traffic [YAMA]. Extensions of this analysis to more general classes of service involving both queueing and blocking, and various service disciplines applied to traffic with differing bandwidth requirements, have been made by Kraimeche and Schwartz [KRAI 1984], [KRAI 1985]. This work has also incorporated studies of the integration of packet-switched traffic with heterogeneous circuit-switched users [SCHW 1983]. These studies are obviously applicable to systems integrating interactive packet-switched data, circuit-switched voice, circuit-switched wideband video, and other types of traffic, as already noted. In the models described here we consider the much simpler case in which all circuit-switched users are of the same bandwidth. They are thus grouped together as one class. The packet-switched data traffic is similarly assumed to be queued at a common queue served by the transmission facility. With no packet-switched data present we have precisely the Erlang-B model discussed in Chapters 2 and 10. With circuit-traffic absent the analysis is precisely that carried out in Chapter 2. The new feature studied here is the effect when both classes of traffic compete for the same transmission facility.

To be more specific consider a digital transmission link such as that of Fig. 12–16, which connects nodes in an integrated network. An example might be a T1 link operating at 1.544 Mbps, as mentioned in Chapter 10. The TDM frame structure to be used in the analysis following, a special case of Fig. 12–16, consists of N slots each b bits wide; this is depicted in Fig. 12–19. In the T1 example, there are 24 slots, each 8-bits wide, repeating at 125-μsec intervals. (An additional bit per frame is added for synchronization purposes.) In the heterogeneous traffic case of Fig. 12–16 different users would be assigned different numbers of slots, depending on their bandwidth requirements. In our analyses here all users require the same bandwidth, one slot of b bits per frame. A circuit-switched user keeps its slot assignment once made, from frame to frame, as long as required. The user is blocked if no assignment is available. Packet-switched traffic, on the other hand, is queued for transmission. The N

[YAMA] T. Yamaguchi and M. Akiyama, "An Integrated Hybrid Traffic Switching System Mixing Preemptive Wideband and Waitable Narrowband Calls," *Electronics and Communications in Japan*, vol. 53-A, no. 5, 1970, 43–52.

[KRAI 1984] B. Kraimeche and M. Schwartz, "Circuit Access Control Strategies in Integrated Digital Networks," Infocom '84, San Francisco, Calif., April 1984.

[KRAI 1985] B. Kraimeche and M. Schwartz, "Analysis of Traffic Access Control Strategies in Integrated Service Networks," *IEEE Trans. on Comm.*, vol. COM-33, no. 9, Oct. 1985, pp. 1085–1093.

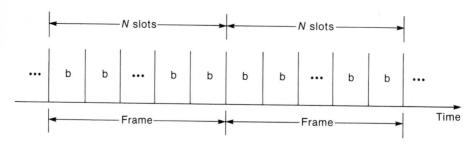

Figure 12-19 TDM frame structure

slots of Fig. 12-19 then represent the N trunks or channels available for transmission over a link, as described in previous chapters.

Two methods of analysis are available to study the integration of traffic onto a TDM frame of the type of Fig. 12-19. One is discrete-time analysis, in which the slot structure is explicitly accounted for in the analysis. The other method is the continuous-time analysis used throughout this book. This latter model ignores the discrete slots of the TDM frame and focuses on the concept of "channels" only. It is a valid model for analysis if the frame length is small compared with the service time required to transmit a circuit-switched call or a data packet. Consider the T1 case as an example. One slot of 8 bits every 125 μsec corresponds to 64-kbps transmission capacity. Say that a data packet 1200-bits long is to be transmitted at this rate. One hundred fifty frames would be required to transmit this packet. Clearly the 125-μsec spacing between 8-bit segments of the packet is small compared with the 18.75 msec (150 frames) required to transmit the packet. The packet thus appears to be handled by a single channel operating at the 64-kbps rate. The continuous-time model is an even better approximation for circuit-switched voice. Calls may last anywhere from 1 minute to 20 minutes, clearly much longer than the frame length. The discrete nature of the TDM frame can thus be neglected, and continuous-time analysis used. In essence, with continuous-time analysis we assume N channels operating in *parallel,* just the model adopted earlier in this book. This leads to the continuous model of Fig. 12-20, a special case of the more general model of Fig. 12-16, as already noted.

Two types of traffic are shown arriving at a transmission link modeled by N channels in parallel. A controller is used to schedule the traffic on arrival. In the general case of Fig. 12-16, a controller would know each user's frame bandwidth requirement and would decide whether to immediately transmit an arriving message, schedule it for later transmission, or block it immediately. In the model of Fig. 12–20, with two users only, we have arbitrarily taken the special case in which one class is blocked if a time slot is not immediately available, while the other class is queued for transmission as shown. Specifically, the class-1 traffic shown, with average arrival rate λ_1 and average holding time (service time) $1/\mu_1$, is blocked if it cannot be handled on arrival. Blocking may be due to lack of a channel or because of a decision made by the controller. This type of traffic obviously represents circuit-switched voice or any other traffic that requires only one of the N channels. Class 2-traffic has an arrival rate λ_2 and an average service (transmission) time $1/\mu_2$; this traffic is queued if slots are not available. (Note that we use the word *call* to represent the unit of class-1 traffic and the word *packet* to represent class-2 traffic. This has, of course, been customary usage in the circuit-switched and packet-switched fields, respectively, as noted throughout this book. The distinction is in a sense meaningless in connection with the analysis here. The two classes of traffic under consideration both require the same bandwidth for transmission. They are distinguished only by the way in which they are handled on arrival — one group is blocked, the other is queued. The queued calls could just as well be "circuit-switched" calls unless they may be preempted by the blockable calls, in which case the distinction becomes moot. To be concrete in our discussion, however, we shall frequently refer to the class-1 calls as *voice calls* and the class-2 arrivals as *data packets*.)

A variety of strategies might be envisioned for controlling the use of the output channels by the two types of calls. They could be handled in FIFO order; priority could be given to the voice calls; or a data packet might preempt a voice call. A specified number N_1 of the N channels might be allocated to the class-1 calls; the class-2 calls would be allocated the remaining $N_2 = N - N_1$ channels (as in Fig. 12–16) plus, on a temporary basis, any of the N_1 channels not occupied.

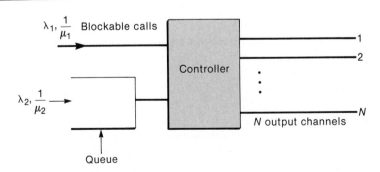

Figure 12–20 Continuous time model, integration of circuit-switched and packet-switched traffic

Other control policies could be invented as well. In the following subsections we analyze two simple schemes: the FIFO strategy (Subsection 12–3–3) and one in which the voice (class-1) calls can preempt class-2 calls if necessary (Subsection 12–3–4). The analysis is carried out for the case of $N = 1$ channel only. Reference is made to the literature for the (more practical) case of larger numbers of channels.

In the last section of this book (Section 12–4) we describe and analyze in detail the strategy just noted in which $N_1 < N$ slots (channels) per frame are reserved for class-1 (voice) calls. $N_2 = N - N_1$ slots are allocated to class-2 (data) packets. However, unused class-1 slots in a given frame may be used, for that frame only, by class-2 packets. A scheme of this type has been termed a *movable-boundary* strategy.

12–3–3 Integration Using the FIFO Discipline

In the first in–first out mode of operation the controller of Fig. 12–20 assigns a channel to either user in order of arrival. The analysis for the general case of N channels has been carried out by Bhat and Fischer using moment-generating functions [BHAT]. Here we study only the case of $N = 1$ channel. Closed-form approximations for larger systems ($N > 1$) for the parameters of interest — the blocking probability of class-1 (voice) calls and the average time delay of class-2 packets — appear in Bhat and Fischer's work. Some of this work is discussed at the end of this subsection.

As has been the case previously in this book, Poisson arrivals and exponential service times are assumed throughout the analysis. With $N = 1$ channel available for either type of traffic arrival, it is apparent that a class-1 (voice) arrival will always be blocked unless the channel is free. Class-2 arrivals, however, will be buffered whenever the channel is occupied. To find the performance of the integrated system, the blocking probability of class-1 calls and the time delay incurred by the class-2 packets just noted, one must find the probabilities of state of the system.

The state space of the integrated system is in general a two-dimensional one, defined by the number i of class-1 calls in progress and by the number j of packets in the system. For $N = 1$ transmission channel, $i = 0$ or 1; $j = 0, 1, 2, \ldots$, with an infinite buffer assumed. This gives rise to the two-dimensional state space depicted in Fig. 12–21. The two-dimensional state probabilities to be found are p_{ij}. With the Poisson arrival and exponential service-time assumptions just noted, it is apparent that transitions between states follow the pattern shown in Fig. 12–21, with transition rate parameters indicated. Thus the tran-

[BHAT] U. N. Bhat and M. J. Fischer, "Multichannel Queueing Systems with Heterogeneous Classes of Arrivals," *Naval Res. Logist. Quarterly*, vol. 23, no. 2, 1976, 271–283.

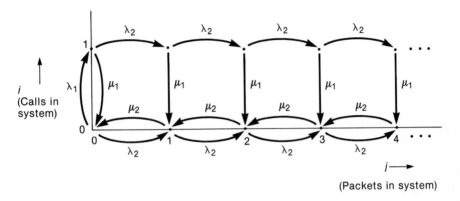

Figure 12-21 State diagram, integrated system, FIFO discipline, $N = 1$ channel

sition out of the free-channel state $(0,0)$ can occur at a rate λ_1 or λ_2, whichever traffic class arrives first. Once the system is in state $i = 1$, with a class-1 call in progress, a transition out of that state can either take place down to $i = 0$, on completion of the call, with rate μ_1, or, if in state j, up to state $j + 1$ on arrival of a class-2 packet. If the system is in state $i = 0$, $j \geq 1$, class-1 calls are blocked, and the system moves up along the $i = 0$ axis.

Two sets of balance equations can be written, one for $i = 0$ and the other for $i = 1$. For $i = 0$, we have

$$(\lambda_1 + \lambda_2)p_{00} = \mu_1 p_{10} + \mu_2 p_{01} \tag{12-47}$$

and

$$(\lambda_2 + \mu_2)p_{0j} = \lambda_2 p_{0j-1} + \mu_1 p_{1j} + \mu_2 p_{0j+1} \qquad j \geq 1 \tag{12-48}$$

Similarly, for $i = 1$, it is apparent that one gets

$$(\mu_1 + \lambda_2)p_{10} = \lambda_1 p_{00} \tag{12-49}$$

and

$$(\mu_1 + \lambda_2)p_{1j} = \lambda_2 p_{1j-1} \qquad j \geq 1 \tag{12-50}$$

These must be solved to obtain p_{ij}.

Equation (12-49) immediately provides an expression for p_{10} in terms of p_{00}. Using Eq. (12-49) in Eq. (12-47), one can also obtain a simple expression for p_{01} in terms of p_{00}. Defining $\rho_1 = \lambda_1/\mu_1$, $\rho_2 = \lambda_2/\mu_2$, as has been the case throughout this book, and letting $\alpha \equiv \mu_2/\mu_1$ be the ratio of the two rates of transition (or equivalently the ratio of class-1 to class-2 service times), we get

$$p_{01} = \rho_2[1 + \rho_1/(1 + \alpha\rho_2)]p_{00} \qquad (12-51)$$

Equation (12–50) is easily solved recursively to obtain

$$p_{1j} = \left(\frac{\alpha\rho_2}{\alpha\rho_2 + 1}\right)^j p_{10} = \left(\frac{\alpha\rho_2}{\alpha\rho_2 + 1}\right)^j \left(\frac{\rho_1}{1 + \alpha\rho_2}\right) p_{00} \qquad (12-52)$$

using Eq. (12–49).

The only remaining set of probabilities to be found is p_{0j}. This cannot be found in a straightforward way from Eq. (12–48) because of the presence of the coupling term $\mu_1 p_{1j}$, which connects $i = 0$ with $i = 1$. The approach we use is to introduce the moment-generating function. In particular, define $G_0(z)$ as the moment-generating function of the probabilities p_{0j}:

$$G_0(z) \equiv \sum_{j=0}^{\infty} p_{0j}z^j \qquad (12-53)$$

Following the approach of Subsection 12–3–1, we now make use of identities similar to Eqs. (12–36) and (12–37) to convert the difference equation (12–48) to an algebraic equation in $G_0(z)$. Specifically, we need the following identities, the derivation of which is left to the reader:

$$\sum_{j=1}^{\infty} p_{0j}z^j = G_0(z) - p_{00} \qquad (12-54)$$

and

$$\sum_{j=1}^{\infty} p_{0j+1}z^j = z^{-1}[G_0(z) - p_{00} - zp_{01}] \qquad (12-55)$$

In addition, it is left to the reader to show, using Eq. (12–52), that

$$\sum_{j=1}^{\infty} p_{1j}z^j = \frac{\alpha\rho_1\rho_2 z}{(\alpha\rho_2 + 1)[1 - \alpha\rho_2(z - 1)]}p_{00} \qquad (12-56)$$

Multiplying each term in Eq. (12–48) by z^j, summing over all $j \geq 1$, and using Eqs. (12–53)—(12–56), one gets the following equation for $G_0(z)$:

$$G_0(z)(1 - \rho_2 z)(z - 1) = p_{00}\left[zA(z) + (z - 1) - \frac{z\rho_1\rho_2}{(1 + \alpha\rho_2)}\right] \qquad (12-57)$$

The function $A(z)$ is given by

$$A(z) = \rho_1\rho_2 z/(\alpha\rho_2 + 1)[1 - \alpha\rho_2(z - 1)] \qquad (12-58)$$

As a check, note that with $\rho_1 = 0$—i.e., the class-1 calls are absent—one gets precisely Eq. (12–39), the moment-generating function for the M/M/1 system.

A simple trial indicates that $z = 1$ is a root of the polynomial on the right-hand side of Eq. (12–57). $(z - 1)$ can thus be factored out to cancel the $(z - 1)$ term on the left-hand side. The final expression for $G_0(z)$ is then given by

$$G_0(z) = p_{00} \left[1 + \frac{z\rho_1\rho_2}{1 - \alpha\rho_2(z - 1)} \right] / (1 - \rho_2 z) \qquad (12\text{–}59)$$

The only remaining unknown is p_{00}, the probability that the system is idle. To find p_{00} we again make use of the fact that all probabilities must sum to one. Thus we have

$$\sum_{i,j} p_{ij} = \sum_{j=0}^{\infty} p_{0j} + \sum_{j=0}^{\infty} p_{1j} = 1 \qquad (12\text{–}60)$$

From the defining relation (12–53) for $G_0(z)$, we have

$$\sum_{j=0}^{\infty} p_{0j} = G_0(1) = p_{00}(1 + \rho_1\rho_2)/(1 - \rho_2) \qquad (12\text{–}61)$$

using Eq. (12–59). From Eq. (12–56) or Eq. (12–52), we also have

$$\sum_{j=0}^{\infty} p_{1j} = \rho_1 p_{00} \qquad (12\text{–}62)$$

Inserting Eqs. (12–61) and (12–62) into Eq. (12–60), we then find that

$$p_{00} = (1 - \rho_2)/(1 + \rho_1) \qquad (12\text{–}63)$$

Note that with $\rho_1 = 0$, one again gets the M/M/1 result $(1 - \rho_2)$.

We are now in a position to calculate the performance of this single-channel integrated system. In particular, the blocking probability of the class-1 calls is just

$$P_B = 1 - p_{00} = (\rho_1 + \rho_2)/(1 + \rho_1) \qquad (12\text{–}64)$$

If we set $\rho_2 = 0$ — i.e., the case for which the class-2 packets are absent — we get $P_B = \rho_1/(1 + \rho_1)$, the Erlang-B blocking probability of a single-channel blocking (loss) system. (See Eq. (2–55).) The effect of the queued packets competing with the class-1 calls for the common channel is thus to increase the blocking probability.

To calculate the average time delay of class-2 (queued) traffic in the system, we first calculate the average number of packets in the system. This is just

$$\begin{aligned}
E(j) &= \sum_{i,j} j p_{ij} = \sum_{j=0}^{\infty} j p_{0j} + \sum_{j=0}^{\infty} j p_{1j} \\
&= G_0'(1) + \sum_{j=0}^{\infty} j p_{1j}
\end{aligned} \qquad (12\text{–}65)$$

with $G_0'(1) \equiv \dfrac{dG_0(z)}{dz}\bigg|_{z=1}$. From Eq. (12–59), with p_{00} given by Eq. (12–63), one finds after some calculation that

$$G_0'(1) = \frac{\rho_2}{(1-\rho_2)} + \frac{\alpha\rho_1\rho_2^2}{(1+\rho_1)} \qquad (12\text{–}66)$$

Using Eq. (12–52), or differentiating Eq. (12–56) with respect to z and setting $z = 1$, one finds that

$$\sum_{j=1}^{\infty} jp_{1j} = \alpha\rho_1\rho_2 p_{00} = \alpha\rho_1\rho_2\left(\frac{1-\rho_2}{1+\rho_1}\right) \qquad (12\text{–}67)$$

The average number of packets in the system is thus

$$E(j) = \frac{\rho_2}{(1-\rho_2)} + \frac{\alpha\rho_1\rho_2}{(1+\rho_1)} \qquad (12\text{–}68)$$

Note that the first term is just the M/M/1 result, obtained when $\rho_1 \to 0$. The second term represents the increase in the number of packets on queue due to the competition with class-1 calls for the use of the channel. The equilibrium condition for the class-2 queue is $\rho_2 < 1$, independent of ρ_1, since the service discipline is FIFO and class-1 calls are blocked if the channel is not available.

Using Little's formula one finds immediately that the time delay $E(T)$ through the system, in normalized form, is given by

$$\mu_2 E(T) = E(j)/\rho_2 = \frac{1}{1-\rho_2} + \frac{\alpha\rho_1}{(1+\rho_1)} \qquad (12\text{–}69)$$

In practice, if the class-1 calls are voice calls, $\alpha \equiv \mu_2/\mu_1 = 1/\mu_1/1/\mu_2 \gg 1$. (Recall that voice calls may last many minutes, while packet-transmission times may be on the order of msec.) The time delay thus increases enormously because of the α that appears in the numerator of the second term in Eq. (12–69). As an example [WEIN], let $1/\mu_1 = 100$ sec and $1/\mu_2 = 10$ msec. Then $\alpha = 10^4$. Say that $\rho_1 = 0.1$ and $\rho_2 = 0.4$ to set the blocking probability at 0.45. (From Eq. (12–64), the only way to reduce P_B substantially is to keep *all* traffic low. Recall that this is a one-channel system.) Then $\mu_2 E(T) = 1.7 + 990 = 992$, as contrasted with the M/M/1 result $\mu_2 E(T) = 1.7$. The actual time delay for this example is $E(T) = 9.9$ sec, substantially greater than the value $E(T) = 17$ msec when the data packets have a separate channel for themselves.

[WEIN] C. J. Weinstein, M. L. Malpass, M. J. Fischer, "Data Traffic Performance of an Integrated Circuit- and Packet-Switched Multiplex Structure," *IEEE Trans. on Comm.*, vol. COM-28, no. 6, June 1980, 873–877.

Then why integrate voice and data if they interfere so severely with one another? Note again that this is a *one-channel* example, chosen because of its relative analytic simplicity. The use of many channels ($N \gg 1$), required to keep the blocking probability low for reasonable class-1 usage, would improve the picture considerably. In addition, as will be shown later when we discuss the movable-boundary scheme, substantial improvement in performance is obtained when a number of channels are shared, rather than segregated into two separate usage categories.

It was noted at the beginning of this subsection that Bhat and Fischer had developed closed-form approximations for the blocking probability and average time delay for this FIFO-discipline integrated system [BHAT]. They indicate that this is necessary because numerical difficulties arise in solving the equations obtained for the case of a large number of channels ($N \gg 1$). The form of the approximation for the blocking probability for the class-1 calls is particularly simple. They note that the blocking probability appears relatively independent of the parameter $\alpha \equiv \mu_2/\mu_1$, over wide ranges of α and for various values of N, ρ_1, and ρ_2. They thus propose as the approximation the function obtained for the special case $\alpha = 1$.

Note that in the case of $N = 1$ channel we found the result for the blocking probability, Eq. (12–64), to be independent of α. This is to be expected for this case when Poisson arrival class-1 calls are turned away if the single-server channel is occupied by either of the two types of traffic [BHAT]. It is not the case exactly when $N > 1$. The approximation proposed for the blocking probability for any number of channels is expressed in terms of the Erlang-B formula, discussed a number of times in this book. Let this formula be written in the form $E(\rho,N)$, with ρ the total load in Erlangs and N the number of channels. It is then defined as

$$E(\rho,N) = \rho^N/N! \left/ \sum_{j=0}^{N} \rho^j/j! \right. \tag{12–70}$$

It is easy to show, by writing the terms in the denominator out and dividing through by the numerator, that the Erlang-B formula obeys the following recursive form:

$$\begin{aligned} E(\rho,N) &= \frac{1}{1 + [N/\rho E(\rho,N-1)]} \\ &= \frac{\rho E(\rho,N-1)}{N + \rho E(\rho,N-1)} \end{aligned} \tag{12–71}$$

This is a very useful recursive relation to calculate $E(\rho,N)$, particularly when N is large. The approximation proposed by Bhat and Fischer for the blocking probability of the class-1 calls in this integrated FIFO system is given by

$$P_B = \frac{\rho E(\rho, N-1)}{N - \rho_2 + \rho E(\rho, N-1)} \tag{12-72}$$

with $\rho = \rho_1 + \rho_2$.

This expression is exact for $\alpha = \mu_2/\mu_1 = 1$. Note also that when $\rho_2 = 0$—i.e., the class-2 packets disappear—one gets precisely Eq. (12–71). It is left to the reader to show that when $N = 1$ channel, this formula agrees with Eq. (12–64). As noted previously, Bhat and Fischer have checked this approximation for various examples and have found it to be quite accurate [BHAT]. The approximate expression for the class-2 time delay is somewhat more complex and so will not be reproduced here. The reader is referred to [BHAT] for details.

In the next subsection we analyze a control strategy that provides preemptive priority for the class-1 (voice) calls. The results are as to be expected: The blocking probability of the class-1 calls is reduced, but at the price of increased time delay for the class-2 packets. This is the classical tradeoff in performance. But the specific improvement obtained in blocking probability and the added cost in time delay can only be determined by analysis. This we do for the case of $N = 1$ channel only. We refer to results from the literature for performance comparisons of larger system.

12–3–4 Integration with Preemptive Priority

In this subsection, we continue with the control model of Fig. 12–20, focusing now on the integration strategy in which class-1 calls are given preemptive priority over class-2 packets. More precisely, an arriving class-1 call finding all N channels occupied will preempt a class-2 customer (packet) if one is receiving service. That packet will then be queued, waiting in a FIFO manner for a channel (server) to become free in order to complete its service. The class-1 call is blocked if, on arrival, all channels are busy serving calls of the same class. The analysis following proceeds in the same manner as that of Subsection 12–3–3. Again we restrict ourselves to the case of $N = 1$ channel for simplicity. We then compare the FIFO and preemptive-priority disciplines for this special case. Reference is made to the literature for comparative results for larger systems.

The basic paper describing this strategy is that by Fischer [FISC 1977], who sets up the balance equations and then uses the generating-function method to study the N-channel case. Earlier, related work appears in papers by Segal

[FISC 1977] M. J. Fischer, "A Queueing Analysis of an Integrated Telecommunications System with Priorities," *INFOR*, vol. 15, no. 3, Oct. 1977, 277–288.

[SEGA] and Halfin and Segal [HALF]. Fischer also analyzes a control strategy in which the class-2 packets are given preemptive priority over the class-1 calls. The reader is referred to these papers for details of the work.

With the class-1 calls in the model of Fig. 12–20 given preemptive priority over class-2 packets, it is clear that they experience no interference from class-2 traffic. Their blocking probability is then just governed by the Erlang-B formula, so that we immediately have

$$P_B = \rho_1^N/N! \bigg/ \sum_{j=0}^{N} \rho_1^j/j! = E(\rho_1, N) \qquad (12-73)$$

using the notation of Subsection 12–3–3. The parameter $\rho_1 = \lambda_1/\mu_1$ is as usual the traffic intensity, with λ_1 the average (Poisson) arrival rate of the class-1 calls and $1/\mu_1$ the average holding (service) time. As a special case, when $N = 1$, we have

$$P_B = \rho_1/(1 + \rho_1) \qquad (12-74)$$

This expression is to be compared with Eq. (12–64) showing the effect of the interfering data packets when FIFO service is used. Equation (12–73) is, more generally, to be compared with the approximate FIFO blocking probability expression of Eq. (12–72). As expected, the preemptive-priority strategy decreases the class-1 blocking probability.

The calculation of the average time delay of the class-2 packets is much more involved, requiring two-dimensional balance equations to be set up and moment-generating functions used to solve them. The two-dimensional state space for the preemptive-priority integrated system appears in Fig. 12–22. Comparing this diagram with that of Fig. 12–21 for the FIFO strategy, it is apparent that the only difference is the set of upward transitions, with rate λ_1, from the $i = 0$ to the $i = 1$ points, for all $j \geq 1$. The reason for these transitions is clear: A class-1 call, arriving with probability $\lambda_1 \Delta t$ in the infinitesimal interval $(t, t + \Delta t)$, immediately preempts a class-2 packet in service, causing the system to move from state $(0, j)$ to state $(1, j)$.

To find the desired state probabilities p_{ij}, $i = 0, 1; j = 0, 1, 2, \ldots$, from which the average time delay of the packets can be found, we proceed, as previously, to set up the balance equations corresponding to the state space of Fig. 12–22. These are then solved by using moment-generating functions.

[SEGA] M. Segal, "A Preemptive Priority Model with Two Classes of Customers," ACM/IEEE Second Symposium on Problems in the Optimization of Data Communications Systems, Palo Alto, Calif., Oct. 1971, 168–174.

[HALF] S. Halfin and M. Segal, "A Priority Queueing Model for a Mixture of Two Types of Customers," *SIAM J. Appl. Math.*, vol. 23, no. 3, Nov. 1972, 369–379.

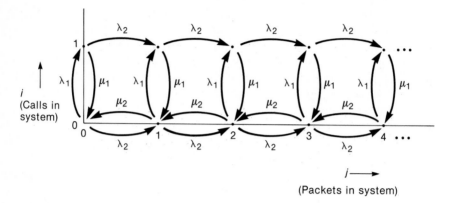

Figure 12–22 State diagram, integrated system, preemptive priority to class-1 calls, $N = 1$ channel

Comparing Figs. 12–21 and 12–22, it is apparent that the equations relating p_{00}, p_{01}, and p_{10} (the boundary conditions) are the same in the two cases. We thus have, exactly as written in Eqs. (12–47) and (12–49),

$$(\lambda_1 + \lambda_2)p_{00} = \mu_1 p_{10} + \mu_2 p_{01} \tag{12-75}$$

and

$$(\mu_1 + \lambda_2)p_{10} = \lambda_1 p_{00} \tag{12-76}$$

Just as in the FIFO case, we can solve these for p_{10} and p_{01} to obtain

$$p_{10} = \frac{\lambda_1}{\mu_1 + \lambda_2} p_{00} = \frac{\rho_1 p_{00}}{1 + \alpha\rho_2} \tag{12-77}$$

and

$$p_{01} = \rho_2 p_{00}[1 + \rho_1/(1 + \alpha\rho_2)] \tag{12-78}$$

Here $\rho_1 = \lambda_1/\mu_1$, $\rho_2 = \lambda_2/\mu_2$, and $\alpha \equiv \mu_2/\mu_1$, as defined previously.

The general balance equations for $j \geq 1$ differ from the FIFO case because of the λ_1 transition at all values of $j \geq 1$ (see Fig. 12–22). Specifically, we have

$$(\lambda_1 + \lambda_2 + \mu_2)p_{0j} = \lambda_2 p_{0j-1} + \mu_1 p_{1j} + \mu_2 p_{0j+1} \qquad j \geq 1 \tag{12-79}$$

and

$$(\lambda_2 + \mu_1)p_{1j} = \lambda_1 p_{0j} + \lambda_2 p_{1j-1} \qquad j \geq 1 \tag{12-80}$$

Equation (12–80) in particular now differs significantly from Eq. (12–50) for the FIFO case because of the added coupling term $\lambda_1 p_{0j}$ on the right-hand side. The probabilities p_{ij} can no longer be found directly in terms of p_{10} and hence p_{00}. Instead one must resort to the use of *two* moment-generating functions to convert Eqs. (12–79) and (12–80) to two algebraic equations with two unknowns that can then be solved simultaneously.

Specifically, define the two functions

$$G_i(z) \equiv \sum_{j=0}^{\infty} p_{ij} z^j \qquad i = 0, 1 \tag{12–81}$$

Note that these must satisfy the probability measure sum

$$G_0(1) + G_1(1) = \sum_{i,j} p_{ij} = 1 \tag{12–82}$$

This last equality will again be used to find p_{00}. Multiplying Eqs. (12–79) and (12–80) by z^j, summing over all values of j, $j \geq 1$, and using Eq. (12–81) to identify the appropriate functions obtained, one finds, as the two algebraic equations,

$$\begin{aligned}(\lambda_1 + \lambda_2 + \mu_2)[G_0(z) - p_{00}] &= \lambda_2 z G_0(z) + \mu_1[G_1(z) - p_{10}] \\ &+ \frac{\mu_2}{z}[G_0(z) - p_{00} - z p_{01}]\end{aligned} \tag{12–83}$$

and

$$(\lambda_2 + \mu_1)[G_1(z) - p_{10}] = \lambda_1[G_0(z) - p_{00}] + \lambda_2 z G_1(z) \tag{12–84}$$

Details are left to the reader.

Using Eqs. (12–77) and (12–78) to replace terms in p_{10} and p_{01} by p_{00}, then solving Eqs. (12–83) and (12–84) simultaneously for $G_1(z)$ and $G_0(z)$, one finds these two quantities to be given by

$$G_1(z) = \frac{\rho_1 G_0(z)}{1 + (1 - z)\alpha\rho_2} \tag{12–85}$$

and

$$G_0(z) = \frac{p_{00}[1 + (1 - z)\alpha\rho_2]}{[\alpha z^2 \rho_2^2 - z(\rho_2 + \rho_1\rho_2 + \alpha\rho_2^2 + \alpha\rho_2) + 1 + \alpha\rho_2]} \tag{12–86}$$

The reader is again asked to verify these results. (Note that a common factor of $(z - 1)$ has been cancelled out to obtain the results.) To find the remaining unknown quantity p_{00} use is made of Eq. (12–82). It is readily shown that p_{00} is given by

$$p_{00} = \frac{1 - \rho_2(1 + \rho_1)}{1 + \rho_1} = \frac{1}{1 + \rho_1} - \rho_2 \qquad (12-87)$$

Note that for p_{00} to be a positive number, a necessary condition for equilibrium in a queueing system of the type under consideration, as noted in Chapter 2 [COX], we must have

$$\rho_2 < 1/(1 + \rho_1) \qquad (12-88)$$

The maximum class-2 utilization of the system is thus reduced by the presence of the higher-priority class-1 traffic.

As a check, note from Eq. (12–74) that $1/(1 + \rho_1) = 1 - P_B$. Equation (12–88) can thus be rewritten

$$\rho_2 < 1 - P_B \qquad (12-88a)$$

or

$$\lambda_2 < \mu_2(1 - P_B) \qquad (12-88b)$$

The interpretation of Eq. (12–88) is thus very simple: $(1 - P_B)$ is the fraction of the channel capacity not used by the (preemptive) class-1 traffic, which in turn represents the average capacity made available to the class-2 traffic. Equilibrium conditions prevail for the infinite-queue data system if the average arrival rate λ_2 is less than the average capacity of the class-2 portion of the system.

The expression (12–86) for $G_0(z)$ is also readily checked. Let $\rho_1 = 0$ so that the class-1 traffic disappears. It is then readily shown that the term in the numerator of Eq. (12–86) that multiplies p_{00} is a factor of the denominator. Canceling this factor and simplifying, one finds, after substituting in for p_{00} with $\rho_1 = 0$, that

$$G_0(z) = \frac{1 - \rho_2}{1 - \rho_2 z} \qquad (12-89)$$

Comparing with Eq. (12–39a) it is clear that this is the expression for the moment-generating function for the M/M/1 queue, just the system remaining when the class-1 traffic in the system under study is removed.

We now use Eqs. (12–85) and (12–86) to find the performance parameters of interest. We have already written an expression for the blocking probability of the class-1 traffic by noting that that traffic is unaffected by class-2 traffic. We can check this expression by using $G_1(z)$ and $G_0(z)$. Specifically, we have as the

[COX] D. R. Cox and H. D. Miller, *The Theory of Stochastic Processes*, Methuen, London, 1965.

blocking probability in this single-server preemptive priority system the probability that a class-1 call is already in service in the system. This is given by the sum of all probabilities along the $i = 1$ line in Fig. 12–22. More precisely,

$$P_B = \sum_{j=0}^{\infty} p_{1j} = G_1(1) \tag{12-90}$$

from the definition (12–81) of $G_1(z)$. Using Eqs. (12–86) and (12–87) in Eq. (12–85) to find $G_1(1)$, we find

$$P_B = \rho_1/(1 + \rho_1) \tag{12-90a}$$

just the single-channel Erlang-B probability of Eq. (12–74). As a further check, recall that $G_0(1) + G_1(1) = 1$. Then $P_B = 1 - G_0(1)$, and Eq. (12–90a) is obtained again.

The average time delay of the class-2 data packets is found by invoking Little's formula and using moment-generating functions. As in the previous FIFO case, the average number of packets in the queue of Fig. 12–20, including any in service, is given by

$$E(j) = \sum_{j=0}^{\infty} j(p_{0j} + p_{1j}) \tag{12-91}$$

(See Eq. (12–65).)

From the definition (12–81) of $G_0(z)$ and $G_1(z)$, it is apparent that the sum in Eq. (12–91) is given, in terms of moment-generating functions, by

$$E(j) = G_0'(1) + G_1'(1) \tag{12-91a}$$

with $G_0'(z)$ again the short-hand notation for $dG_0(z)/dz$. Using Little's theorem, we then have

$$\mu_2 E(T) = \mu_2 E(j)/\lambda_2 = E(j)/\rho_2 \tag{12-92}$$

Differentiating the expressions (12–85) and (12–86) for $G_1(z)$ and $G_0(z)$, respectively, setting $z = 1$, and then using Eqs. (12–91a) and (12–92), one finds, after much manipulation and simplification, that

$$\mu_2 E(T) = \frac{(1 + \rho_1)^2 + \alpha\rho_1}{[1 - \rho_2(1 + \rho_1)][1 + \rho_1]} \tag{12-92a}$$

This is the desired expression for the time delay of the class-2 traffic for the class-1 preemptive-priority strategy. As a check, with $\rho_1 = 0$, one again gets the M/M/1 delay expression $1/(1 - \rho_2)$.

Equation (12–92a) is to be compared with the comparable equation (12–69) for the FIFO case. Note first that the maximum class-2 load in the preemptive-priority case is given by $\rho_2 < 1/1(1 + \rho_1)$, as noted earlier and as apparent from Eq. (12–92a). For the FIFO case we had $\rho_2 < 1$. Preemptive priority for the class-1 traffic thus reduces the maximum class-2 load. The use of this priority discipline must also result in increased time delay for the class-2 packets as the maximum load value is approached. Consider as an example the numbers used earlier in discussing the FIFO analysis results. We took the case where $1/\mu_1 = 100$ sec, $1/\mu_2 = 10$ msec, $\alpha = 10^4$, $\rho_1 = 0.1$, and $\rho_2 = 0.4$. The blocking probability in the FIFO case was $P_B = 0.45$. Here it is 0.09. The normalized time delay in the FIFO case was $\mu_2 E(T) = 992$. Here it is $\mu_2 E(T) = 1600$. A 5-to-1 reduction in blocking probability results in a 60 percent increase in time delay. This is the tradeoff between these two integrated control strategies that we alluded to earlier. The increase in time delay here is due primarily to the reduced channel capacity available to the class-2 packets, as just noted. To show this take $\alpha \gg 1$, $\rho_1 \ll 1$, and $\rho_2 < 0.5$. Then the time delay in the FIFO case is dominated by the second term in Eq. (12–69), while the delay in the preemptive-priority (PP) case is dominated by the second term in the numerator of Eq. (12–92a). Taking the ratio of these expressions we have

$$\frac{E(T)_{PP}}{E(T)_{FIFO}} \doteq 1/(1 - \rho_2), \qquad \alpha \gg 1, \qquad \rho_1 \ll 1 \qquad (12-93)$$

The comparable ratios of the blocking probabilities are similarly given by

$$\frac{P_{BPP}}{P_{BFIFO}} \doteq \frac{\rho_1}{\rho_2} \qquad \rho_1 \ll 1 \qquad (12-94)$$

Equations (12–94) and (12–93) clearly show the reduction in blocking probability and increase in time delay due to the factor ρ_1 in the first case and $1/(1 - \rho_2)$ in the second case.

Some numerical results for larger systems are presented in the papers cited. [FISC] plots curves showing the improvement in blocking probability introduced by moving from the FIFO discipline to the preemptive-priority case for $N = 3$ channels. Curves of average time delay for various values of the parameters are also plotted for preemptive priority with $N = 3$ channels. [BHAT] has tables of delay and blocking probability for the FIFO case for $N = 3$ and $N = 5$ channels.

12–4 Movable-boundary Strategy

We indicated in Subsection 12–3–2 that another method of controlling service (transmission) of the blockable class-1 calls and queued class-2 data packets (Fig. 12–20) is to divide a frame of N slots (Fig. 12–19) into two sections. One section, containing N_1 slots, is allocated to the class-1 calls. The other section, with $N_2 = N - N_1$ slots, is reserved for class-2 data packets. The class-2 packets may occupy any of the N_1 class-1 slots not currently in use. An arriving class-1 call, however, is allowed to preempt a class-2 packet occupying one of its allocated slots if necessary for it to receive service. This control strategy has been termed the *movable-boundary* scheme.

The movable-boundary strategy was first proposed by Kummerle [KUMM 1974], Zafiropoulo [ZAFI], and their colleagues at IBM Zurich. The frame structure suggested is indicated in Fig. 12–23. The method clearly offers a performance advantage for the class-2 users over a fixed-boundary strategy, in which the same reserved slot allocations are made. The blocking performance of class-1 calls remains the same in either case. Data packets can be expected to have their queueing delay reduced with the movable-boundary strategy since advantage is taken of gaps in the class-1 slot assignment to transmit more packets than would otherwise be the case. More efficient use is thus made of the communications resource, the transmission link. Detailed descriptions of the time-division multiplexing frame structure for such a scheme appear in [COVI] and [ROSS]. Implementation questions are discussed in [PORT] and [RUDI 1978].

Although substantial improvement in time-delay performance for the class-2 packets may be expected, care must be taken with the movable-boundary scheme. Specifically, if an attempt is made to utilize the class-1 slots released on the average to increase the class-2 throughput above what would normally be handled by the N_2 slots reserved per frame (Fig. 12–23), extraordinarily long data queues may result. This is called the *overload* region of operation of the class-2 packets. This phenomenon occurs for the case in which the class-1 holding times are long compared with the packet lengths; it is due to excess (overload) packets arriving during the long class-1 holding time. This result has been verified by simulation [WEIN]. We demonstrate this result analytically in Subsection 12–4–1 for the special case of $N = 2$ slots, with $N_1 = N_2 = 1$ slot allo-

[KUMM 1974] K. Kummerle, "Multiplexer Performance for Integrated Line- and Packet-switched Traffic," ICCC, Stockholm, 1974, 508–515.

[ZAFI] P. Zafiropoulo, "Flexible Multiplexing for Networks Supporting Line-switched and Packet-switched Data Traffic," ICCC, Stockholm, 1974, 517–523.

[COVI] G. Coviello and P. Vena, "Integration of Circuit/Packet Switching by a SENET (Slotted Envelope Network) Concept," National Telecommunications Conference, New Orleans, Dec. 1975, 42–12—42–17.

[ROSS] M. Ross et al., "Design Approaches and Performance Criteria for Integrated Voice/Data Switching," *Proc. IEEE*, vol. 65, no. 9, Sept. 1977, 1283–1295.

[PORT] E. Port et al., "A Network Architecture for the Integration of Circuit and Packet Switching," ICCC, Toronto, Aug. 1976, 505–514.

[RUDI 1978] H. Rudin, "Studies in the Integration of Circuit and Packet Switching," ICC, Toronto, June 1978, 20.2.1–20.2.7.

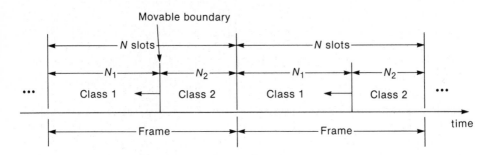

Figure 12–23 Movable-boundary strategy

cated to each of the two user classes. A high data-traffic approximation introduced later in Subsection 12–4–3 verifies this result as well [GAVE]. The implication of this observation is that the data-traffic *time delay* can be *reduced* using the movable-boundary strategy, but that this scheme should not be relied on to increase the data throughput. Flow control of the data traffic is thus required to ensure that the data-traffic load is constrained properly [WEIN].

It can be shown that the maximum class-2 (data) utilization for the general N-slot movable boundary scheme, assuming an infinite data queue, is given by

$$\rho_2 < N - E(i) = N - \rho_1(1 - P_B) = N_2 + [N_1 - \rho_1(1 - P_B)] \quad (12\text{–}95)$$

with P_B the Erlang-B blocking probability of the class-1 traffic. This agrees with intuition, since it suggests that the data queue is at equilibrium only for $\rho_2/(N - E(i)) < 1$, with $N - E(i)$ the average number of channels (servers) made available to the data queue. Equation (12–95) will be verified in the special case of the $N = 2$ movable-boundary system analyzed in Subsection 12–4–1.

Equation (12–95) may be written alternatively as

$$\rho \equiv \rho_2 + \rho_1(1 - P_B) < N \quad (12\text{–}96)$$

The parameter ρ represents the total utilization of the system of Fig. 12–20. This obviously agrees with intuition as well: $\rho_1(1 - P_B)$ is the class-1 utilization; ρ_2 is the class-2 utilization. The total utilization ρ must be less than N, the total number of channels (servers) to ensure that overall system is in equilibrium. A more general form of Eq. (12–96) is shown in [KRAI 1985] to be the condition

[GAVE] D. P. Gaver and P. P. Lehoczky, "Channels that Cooperatively Service a Data Stream and Voice Messages," *IEEE Trans. on Comm.*, vol. COM-30, no. 5, May 1982, 1153–1162.

for equilibrium for a movable-boundary scheme in which class-1 and class-2 traffic each require different numbers of slots per frame.

The condition for equilibrium of the data queue, Eq. (12-95), appears to indicate that the data throughput may exceed the data-system-dedicated capacity of N_2 slots/frame. This is correct. However, as just noted, inordinately large queues are found to appear in the overload region $\rho_2 > N_2$, or $\lambda_2 > \mu_2 N_2$. This will be demonstrated in our analysis of the movable-boundary system for the special case $N = 2$, $N_1 = N_2 = 1$. It must be emphasized again that flow-control techniques should be used with a movable-boundary scheme to ensure that the overload region is not penentrated.

The general analysis of the movable-boundary scheme becomes quite complex algebraically. Approximation methods are thus appropriate for analyzing this integrated multiplexing strategy analytically. We propose such a method for the underload region ($\rho_2 < N_2$) in Subsection 12-4-2, comparing it with the result of the exact analysis for the $N = 2$ slot case. We also discuss a fluid-flow approximation for the overload region ($\rho_2 > N_2$) [GAVE], again validating the method by comparing it with the exact results for $N = 1$ and $N = 2$ slot schemes.

It is worth noting at this point that the $N = 1$ slot/frame example of the preemptive-priority integration strategy discussed in the previous section is the same as the movable boundary scheme for this special case. The average class-2 (data) time-delay expression obtained, Eq. (12-92a), shows the strong dependence on class-1 holding time that we have noted. In particular, as noted in the previous section, if $\alpha = \mu_2/\mu_1 \gg 1$, the usual case for voice and data packets, large packet-time delays may be expected. It is this linear dependence on α that drives the packet-queueing delays up.

12-4-1 Continuous-time Analysis of Movable-boundary Scheme

Before proceeding with the continuous-time analysis for the special case of $N = 2$ slots/frame, it is worthwhile reviewing the movable-boundary strategy as a possible way to integrate the two classes of traffic of Fig. 12-20. As indicated earlier, both class-1 calls and class-2 packets are assumed to arrive at Poisson rates of λ_1 calls/time and λ_2 packets/time, respectively. Both are assumed to be distributed exponentially in length, with average values $1/\mu_1$ and $1/\mu_2$ (in units of time), respectively. We take the frame length to be relatively short so that continuous-time analysis can be used. The class-1 (voice) calls are assigned N_1 of the N channels (slots) per frame available. $N_2 = N - N_1$ slots are allocated to the data packets, assumed to be queued up in first come-first served (FCFS) or first in-first out (FIFO) order (Fig. 12-23). Data packets may use voice channels if available, but preemptive priority of voice over data is enforced. Thus a data packet using a voice channel is immediately put back onto the data queue when a voice call arrives. The voice calls under these conditions are thus not affected by the data and see an N_1-channel blocking system. The voice traffic-blocking probability is precisely the Erlang-B expression, given in this case by

$$P_B = \rho_1^{N_1}/N_1! / \sum_{i=0}^{N_1} \rho_1^i/i! \qquad (12-97)$$

This decoupling of the voice analysis from the data was also possible in analyzing the preemptive-priority strategy of the last section (see Eq. (12-73)). There no slots were specifically reserved for the data packets, so that we implicitly had $N_1 = N$, $N_2 = 0$.

We now take the special case of the movable-boundary scheme with $N = 2$ channels and with $N_1 = N_2 = 1$ channel allocated to each of the two traffic classes. Letting $i = 0$ or 1 be the states of the (blockable) voice traffic and $j = 0, 1, 2, \ldots$ again be the states of the queued data traffic, one gets the two-dimensional state diagram of Fig. 12-24 for the movable-boundary strategy in this case. It is to be compared with Figs. 12-21 and 12-22 for the FIFO and priority cases. Note that for $i = 1$, the case in which a voice call is present, the data system looks like an M/M/1 system. For $i = 0$, with no voice calls present, the data traffic is allowed to use up to two channels. The system in this case looks like an M/M/2 system. The vertical transitions between $i = 0$ and $i = 1$ are obviously due either to a voice call arriving, at Poisson rate λ_1, which switches the data system from a two-server to a one-server system or to a voice call being completed, with rate μ_1, which switches the data system from a one-server to a two-server system.

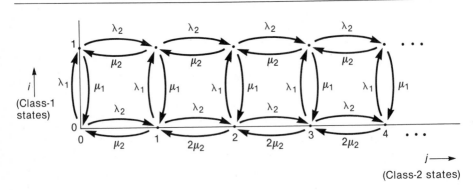

Figure 12-24 Movable-boundary state-transition diagram, $N = 2$ channels, $N_1 = N_2 = 1$ channel

Five balance equations are required in this case. Two are general in form and cover the evolution of p_{1j} and p_{0j}, respectively. Using the state diagram of Fig. 12–24, these are written relatively easily as

$$(\mu_1 + \mu_2 + \lambda_2)p_{1j} = \lambda_1 p_{0j} + \mu_2 p_{1j+1} + \lambda_2 p_{1j-1} \qquad j \geq 1 \qquad (12–98)$$

and

$$(\lambda_1 + \lambda_2 + 2\mu_2)p_{0j} = 2\mu_2 p_{0j+1} + \mu_1 p_{1j} + \lambda_2 p_{0j-1} \qquad j \geq 2 \qquad (12–99)$$

These are similar, of course, to those written previously for the FIFO and preemptive-priority strategies of Subsections 12–3–3 and 12–3–4. (Note that the more general case, with N_1 channels reserved for class-1 calls, would require $N_1 + 1$ such equations, one for each value of i, the state of the class-1 subsystem.)

It is apparent from the restrictions on j, the state of the class-2 system in Eqs. (12–98) and (12–99), that three additional boundary equations are required to cover the cases $i = 1, j = 0$, and $i = 0, j = 0, 1$. These three equations, representing the balance equations at states (1,0), (0,0) and (0,1), respectively, are also written by inspection from Fig. 12–24:

$$(\lambda_2 + \mu_1)p_{10} = \lambda_1 p_{00} + \mu_2 p_{11} \qquad (12–100)$$

$$(\lambda_1 + \lambda_2)p_{00} = \mu_2 p_{01} + \mu_1 p_{10} \qquad (12–101)$$

$$(\lambda_1 + \lambda_2 + \mu_2)p_{01} = 2\mu_2 p_{02} + \mu_1 p_{11} + \lambda_2 p_{00} \qquad (12–102)$$

Note that these three equations involve five unknown state probabilities: p_{00}, p_{01}, p_{10}, p_{11}, and p_{02}. Two additional equations are needed to completely determine these probabilities. One will be given by the probability measure $\Sigma_{i,j} p_{ij} = 1$. The other will be seen later to come from a specification on the roots of the denominator polynomial of one of the moment-generating functions.

We proceed as previously by defining two moment-generating functions, one for each value of i:

$$G_i(z) \equiv \sum_{j=0}^{\infty} p_{ij} z^j \qquad i = 0, 1 \qquad (12–103)$$

Multiplying Eq. (12–98) term by term by z^j and summing over all values of $j \geq 1$, the equation is again converted into an algebraic equation that involves the two generating functions $G_0(z)$ and $G_1(z)$. Details are left to the reader. Equation (12–100) is used to cancel the terms involving p_{10}, p_{00}, and p_{11} that appear in the equation. After some simplification, one is left with the following equation:

$$G_1(z)[z\mu_1 + (z - 1)(\mu_2 - \lambda_2 z)] = \lambda_1 z G_0(z) + \mu_2(z - 1)p_{10} \qquad (12–104)$$

Repeating the same procedure for Eq. (12–99), one gets

$$G_0(z)[z\lambda_1 + (z-1)(2\mu_2 - \lambda_2 z)] = \mu_1 z G_1(z) + \mu_2(z-1)(2p_{00} + zp_{01}) \quad (12-105)$$

Use has been made here of Eq. (12–101) to simplify the equation.

Note from both Eq. (12–104) and Eq. (12–105) that one finds that

$$G_1(1) = \rho_1 G_0(1) \qquad \rho_1 \equiv \lambda_1/\mu_1 \qquad\qquad (12-106)$$

Since we also have

$$G_0(1) + G_1(1) = \sum_{i,j} p_{ij} = 1 \qquad\qquad (12-107)$$

one finds that

$$G_0(1) = \sum_{j=0}^{\infty} p_{0j} = 1/(1+\rho_1) \qquad\qquad (12-108)$$

and

$$G_1(1) = \sum_{j=0}^{\infty} p_{1j} = \rho_1/(1+\rho_1) \qquad\qquad (12-109)$$

But the probability that class-1 (voice) calls are blocked is the probability that the system is in state $i = 1$. This is just $G_1(1)$. We thus have

$$P_B = G_1(1) = \rho_1/(1+\rho_1) \qquad\qquad (12-109a)$$

precisely as expected for a blocking system with $N_1 = 1$ server. (See Eq. (12–74) or Eq. (12–90a), for example.) This agrees, of course, with the statement that the blocking probability of the general N-channel movable-boundary scheme, with N_1 slots (channels) reserved for class-1 calls, is given by Eq. (12–97).

An additional equation relating the five unknown probabilities p_{00}, p_{01}, p_{02}, p_{10}, and p_{11} may now be obtained with $G_0(1)$ and $G_1(1)$ specified. In particular, adding Eq. (12–104) to Eq. (12–105), canceling the common factor $(z-1)$, and then writing ρ_2 for λ_2/μ_2, one finds that

$$G_0(z)(2 - \rho_2 z) + G_1(z)(1 - \rho_2 z) = 2p_{00} + (p_{01}z + p_{10}) \quad (12-110)$$

Letting $z = 1$ and using Eqs. (12–108) and (12–109), we have immediately that

$$2p_{00} + p_{01} + p_{10} = 1 - \rho_2 + 1/(1+\rho_1) \qquad\qquad (12-111)$$

This equation plus Eq. (12–101) enable us to solve independently for p_{01} and p_{10} in terms of p_{00}.

Some simplifying notation is in order at this point. Recall that in Eq. (12–96) we defined the overall utilization ρ. Specializing to the case under consideration here, with $N = 2$ slots, $N_1 = N_2 = 1$ slots assigned per user class, one has

$$\rho \equiv \rho_2 + \rho_1(1 - P_B) = \rho_2 + \rho_1/(1 + \rho_1) < 2 \qquad (12-112)$$

Let the parameter a represent the quantity $(2 - \rho)$. This is then a measure of the average capacity remaining with total utilization ρ. From Eq. (12–112) we have

$$a \equiv 2 - \rho = (1 - \rho_2) + 1/(1 + \rho_1) \qquad (12-113)$$

Equation (12–111) can then be written in simplified form as

$$2p_{00} + p_{01} + p_{10} = a \qquad (12-111a)$$

Using this equation and Eq. (12–101), with the parameters $\rho_1 = \lambda_1/\mu_1$, $\rho_2 = \lambda_2/\mu_2$, and $\alpha = \mu_2/\mu_1$ introduced to simplify the notation as previously, one obtains the following two equations for p_{01} and p_{10}:

$$p_{01} = \frac{(2 + \rho_1 + \alpha\rho_2)p_{00} - a}{(\alpha - 1)} \qquad (12-114)$$

$$p_{10} = \frac{\alpha a - (\rho_1 + \alpha\rho_2 + 2\alpha)p_{00}}{(\alpha - 1)} \qquad (12-115)$$

The additional equation required to obtain p_{00} will be discussed shortly.

Returning to the two moment-generating functions $G_0(z)$ and $G_1(z)$, we can solve Eqs. (12–104) and (12–105) simultaneously to find explicit expressions for each. These can be written in a variety of ways. One useful set has the following form (details are left as an exercise):

$$G_0(z) = \frac{z(2p_{00} + p_{01}z + p_{10}) + \alpha(z - 1)(1 - \rho_2 z)(2p_{00} + p_{01}z)}{\alpha(2 - \rho_2 z)(z - 1)(1 - \rho_2 z) + z(2 - \rho_2 z) + \rho_1 z(1 - \rho_2 z)} \qquad (12-116)$$

$$G_1(z) = \frac{\rho_1 z G_0(z) + \alpha(z - 1)p_{10}}{\alpha(z - 1)(1 - \rho_2 z) + z} \qquad (12-117)$$

Using these two equations one can calculate all statistical quantities of interest.

Before proceeding with the calculations, however, it is of interest to check these expressions for $G_0(z)$ and $G_1(z)$. Note from Fig. 12–24 that there are two limiting cases (already alluded to) of the movable-boundary strategy as defined. If the class-1 traffic ρ_1 is very small (alternatively, λ_1, the class-1 arrival rate, is small compared with the service rate μ_1), the system stays in the $i = 0$ state most of the time. The data queue sees effectively two servers and should thus behave like an M/M/2 system. At the other extreme, with $\lambda_1 \gg \mu_1$, or $\rho_1 \to \infty$, the dedicated class-1 (voice) channel is occupied with class-1 calls all of the time, and the data queue sees one server (its dedicated channel) only. It should behave like an M/M/1 system in this case.

Consider the case $\rho_1 \to \infty$ first. Referring to Eq. (12–116), it is apparent that $G_0(z) \to 0$. Then all the probabilities along the $i = 0$ axis must be zero. In particular, $p_{00} = p_{01} = 0$. From Eq. (12–116) we then have in this limiting case

$$\rho_1 G_0(z) \to p_{10}/(1 - \rho_2 z) \qquad \rho_1 \to \infty \qquad (12\text{–}118)$$

Substituting this into Eq. (12–117), one finds

$$G_1(z) = p_{10}/(1 - \rho_2 z) \qquad \rho_1 \to \infty \qquad (12\text{–}119)$$

This is precisely the expression found earlier (Eq. 12–39) for the M/M/1 moment-generating function. The probability p_{10} must equal $(1 - \rho_2)$ in this limiting case to have $G_1(1) = 1$. Our surmise, based on Fig. 12–24 and the movable-boundary model it represents, is thus validated for the case $\rho_1 \to \infty$.

Now consider the case $\rho_1 = 0$. It is left to the reader to show that we now get $p_{10} = 0$, $p_{01} = \rho_2 p_{00}$, $G_1(z) = 0$, and

$$G_0(z) = p_{00}(2 + \rho_2 z)/(2 - \rho_2 z) \qquad \rho_1 = 0 \qquad (12\text{–}120)$$

This agrees of course with the M/M/2 result of Eq. (12–44), checking our second surmise. (As an aside, it is always important to carry out limiting checks of this type to ensure correctness of the analysis. It is quite easy, and often almost unavoidable, to make trivial mistakes in carrying out the algebraic manipulations, as the reader will most surely find in checking the analysis here.)

Returning to Eqs. (12–116) and (12–117), we complete the analysis by calculating the average delay $E(T)$ of the class-2 (data) traffic. Specifically, from the moment-generating property of $G_0(z)$ and $G_1(z)$, one first finds the average number of packets on the data queue. It is left to the reader to show that this is given by

$$\begin{aligned} E(j) &= G_0'(1) + G_1'(1) \\ &= G_0'(1)(1 + \rho_1) + \alpha p_{10} - \alpha(1 - \rho_2)\rho_1/(1 + \rho_1) \end{aligned} \qquad (12\text{–}121)$$

Here $G_0'(z)$ and $G_1'(z)$ again represent derivatives with respect to z. Use has been made of Eq. (12–117) in calculating $G_1'(1)$.

Calculating $G_0'(1)$ from Eq. (12–116) and simplifying, one finds $E(j)$ to be given by

$$E(j) = \frac{1}{a(1 + \rho_1)}\left[\alpha p_{10} + (1 + \rho_1)(\rho_2 + p_{01}) - \frac{\alpha\rho_1}{(1 + \rho_1)}(1 - \rho_2)\right] \qquad (12\text{–}121a)$$

Note the term a which equals $(2 - \rho)$ in the denominator. This is a good check on the result since one expects $E(j)$ to behave as $1/a$ from queueing considerations. As a further check, consider first the case $\rho_1 \to 0$. It is apparent that

$$E(j) \to \frac{\rho_2 + p_{01}}{2 - \rho_2} = \frac{4\rho_2}{(2 - \rho_2)(2 + \rho_2)} \qquad \rho_1 = 0 \qquad (12\text{–}121b)$$

which is the M/M/2 result. For $\rho_1 \to \infty$, one finds

$$E(j) \to \rho_2/(1 - \rho_2) \qquad \rho_1 \to \infty \qquad (12-121c)$$

which is the M/M/1 result, again validating expression (12-121a).

Using Little's formula, the normalized time delay is immediately written as

$$\mu_2 E(T) = E(j)/\rho_2 = \frac{1}{a} \left\{ \frac{\alpha}{\rho_2(1+\rho_1)} \left[p_{10} - \left(\frac{\rho_1}{1+\rho_1}\right)(1-\rho_2) \right] + \frac{p_{01}}{\rho_2} + 1 \right\}$$
$$(12-122)$$

This expression is still not complete, since p_{10} and p_{01}, and hence p_{00} (using Eqs. (12-114) and (12-115)), must still be found. This requires one additional equation in these probabilities. To find this equation we return to Eq. (12-116) for $G_0(z)$. Note that the denominator of $G_0(z)$ is a third-order polynomial in z. From the defining series (12-103) for the moment-generating function, it is apparent that the series must converge for $|z| \leq 1$. Real roots of the denominator polynomial that are less than 1 must be canceled by the same root appearing in the numerator if $G_0(z)$ is to be a valid moment-generating function. In particular, for the example under study here, the third-order polynomial at the denominator turns out to have one root less than 1. A comparable root must thus appear in the numerator, providing the necessary additional equation for p_{00}.

Specifically, let

$$G_0(z) = N_0(z)/D_0(z) \qquad (12-123)$$

Let $z = z_0$ be a root of the denominator $D_0(z)$. We thus have, from Eq. (12-116),

$$D_0(z_0) = 0 = \alpha(2 - \rho_2 z_0)(z_0 - 1)(1 - \rho_2 z_0) + z_0(2 - \rho_2 z_0) + \rho_1 z_0(1 - \rho_2 z_0)$$
$$(12-124)$$

This cubic equation must be solved for the value $z_0 < 1$. The numerator polynomial $N_0(z)$ at this value of z must then be

$$N_0(z_0) = 0 = z_0(2p_{00} + p_{01} z_0 + p_{10}) + \alpha(z_0 - 1)(1 - \rho_2 z_0)(2p_{00} + p_{01} z_0)$$
$$(12-125)$$

This equation plus Eqs. (12-114) and (12-115) provide the three equations from which p_{00} may be found.

It is obvious that in general this procedure can only be carried out numerically for various values of ρ_1, ρ_2, and α. It is possible in this movable-boundary example with $N = 2$ channels, however, to find closed-form approximate solutions for p_{00}, p_{01}, and p_{10}, and hence for the normalized time delay $\mu_2 E(T)$ under the condition $\alpha \gg 1$. Since $\alpha = \mu_2/\mu_1$ can be as large as 10^4 under realistic voice/data traffic conditions, this restriction on the analysis is a valid one. It turns out that the calculation of the root $z_0 < 1$, using Eq. (12-124), must be carried out separately for the two data traffic conditions $\rho_2 < 1$ and $\rho_2 > 1$. The first corresponds to the underload region and the second to the overload region, described earlier. We present only the results of the calculations, leaving the details to the reader as an exercise.

Specifically, we find the normalized time delay in the two data-traffic regions to be given by the following two expressions:

(1) $\rho_2 < 1, \alpha \gg 1$

$$\mu_2 E(T) = \frac{4}{a(1 + \rho_1)(2 + \rho_2)} + \frac{\rho_1}{a(1 + \rho_1)} \tag{12-122a}$$

$$a \equiv 2 - \rho = (1 - \rho_2) + 1/(1 + \rho_1)$$

(2) $\rho_2 > 1, \alpha \gg 1$

$$\mu_2 E(T) = \frac{1}{a}\left[1 + \frac{a}{2 + \rho_2} + \frac{\alpha(\rho_2 - 1)\rho_1}{\rho_2(1 + \rho_1)^2}\right] \tag{12-122b}$$

$$a = 2 - \rho = 2 - \rho_2 - \rho_1/(1 + \rho_1)$$

First consider the underload result, Eq. (12–122a). Note that this expression turns out to be independent of α, the ratio of the class-1 holding time to the class-2 packet length, so long as $\alpha \gg 1$! It is left to the reader to show, as a check on Eq. (12–122a), that if $\rho_1 = 0$, one gets the M/M/2 average time delay; for the case $\rho_1 \to \infty$, one gets the M/M/1 delay; and, finally, as $\rho_2 \to 0$—i.e., the class-1 (data) traffic is negligible—$E(T) \to 1/\mu_2$, just the class-1 service time, as expected.

The time delay in the overload region, given by Eq. (12–122b), however, is proportional to α. This agrees with the result obtained previously for the movable-boundary scheme with $N = 1$ slot. Refer to Eq. (12–92a) for the time delay in that case. (Recall that the movable-boundary strategy is identical with the preemptive-priority strategy in that case.) Since $\alpha \gg 1$ has been assumed in obtaining this result, it is apparent that the time delay increases rapidly as one attempts to penetrate the $\rho_2 > 1$ region by driving more data traffic into the system. The relative increase in time delay is proportional to $(\rho_2 - 1)\alpha$, as is apparent from Eq. (12–122b). This is the same observation made earlier in this section when we discussed the movable-boundary strategy in general: It represents a useful scheme for reducing time delay with normal data loads applied to the system ($\rho_2 < N_2$ in general) but should not be used to obtain additional data throughput.

Interestingly, both Eq. (12–122a) and Eq. (12–122b), although explicitly derived for the cases $\rho_2 < 1$ and $\rho_2 > 1$, respectively, do converge to the same value for $\rho_2 = 1$:

$$\mu_2 E(T) = \frac{4}{3} + \rho_1 \qquad \rho_2 = 1 \tag{12-122c}$$

Note that this agrees with the M/M/2 result, with $\rho_1 = 0$. However, the time delay goes out of bound as $\rho_1 \rightarrow \infty$, as would be expected since $\rho_2 = 1$ represents a nonequilibrium situation in that case.

Equations (12–122a) and (12–122b) can be used to evaluate the time-delay performance of the class-2 (data) portion movable-boundary scheme and to show how this scheme compares with a fixed-boundary strategy, in which the class-1 and class-2 traffic are restricted to the N_1 and N_2 channel assignments. Instead of comparing time delays, however, it is more useful to compare average wait time on the queue. Using Eq. (12–122a), the normalized wait time for the movable-boundary scheme in the underload region, with $N = 2$ slots (channels) and $N_1 = N_2 = 1$ slots, is given by

$$
\begin{aligned}
\mu_2 E(W) &= \mu_2 E(T) - 1 \\
&= \frac{\rho_2}{a}\left[1 - \frac{2}{(1 + \rho_1)(2 + \rho_2)}\right] \qquad (12\text{–}126) \\
&= \frac{\rho_2(\rho_2 + 2\rho_1 + \rho_1\rho_2)}{a(1 + \rho_1)(2 + \rho_2)} \qquad \rho_2 \le 1
\end{aligned}
$$

with $a = (1 - \rho_2) + 1/(1 + \rho_1)$.

Equation (12–126) is plotted in Fig. 12–25 for various values of class-1 traffic ρ_1 (or, equivalently, class-1 blocking probability $P_B = \rho_1/(1 + \rho_1)$), and is compared there with the average wait time for an equivalent fixed-boundary scheme. Since $N_2 = 1$ channel only is available in that case, the average normalized wait time for the fixed-boundary scheme is just the M/M/1 wait time $\mu_2 E(W) = \rho_2/(1 - \rho_2)$. Note the sizable improvement in wait time obtained by use of the movable-boundary strategy in the region $\rho_2 > 0.5$. The fixed-boundary wait time (and hence time delay as well) begins to increase rapidly as ρ_2 exceeds 0.5. This point has been reiterated in many places throughout this book. The movable-boundary scheme provides a dramatic improvement because its wait time (and time delay) in the underload region is determined by the *average number* of channels available, just $a = 2 - \rho = 2 - [\rho_2 + \rho_1/(1 + \rho_1)]$ in this case. Even for sizable class-1 traffic, with $\rho_1 = 0.8$, enough added channel capacity is made available to the class-2 traffic to reduce its delay considerably.

How does the movable-boundary scheme perform in the case of larger, more realistic systems? The algebra involved in extending the analysis of this paragraph to substantially larger systems becomes quite complex and tedious, and numerical problems may arise in calculating the roots of the polynomials encountered. Instead it is appropriate to introduce approximation techniques that can be used to evaluate the performance quite readily. We discuss two such approximations, one for the underload region $\rho_2 < N_2$ and the other for the overload region $\rho_2 > N_2$, in the next two subsections. These approximations are validated by comparison with the results just presented.

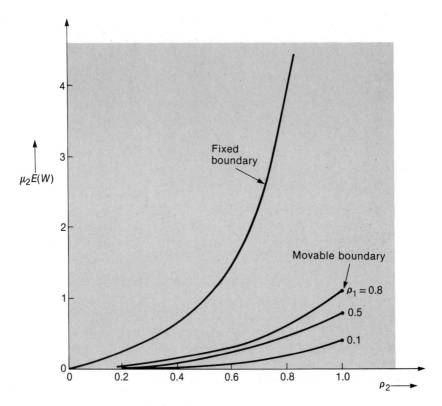

Figure 12–25 Movable-boundary scheme compared with fixed boundary, $N = 2$, $N_1 = N_2 = 1$

12–4–2 Movable-boundary Scheme: Approximate Analysis, Underload Region

As just indicated, in order to proceed with the analysis of the movable-boundary scheme for large numbers of channels (slots in a frame), it is necessary to use approximation procedures. Here we introduce an ad hoc formula appropriate for the underload region of operation, $\rho_2 < N_2$, with N_2 the number of channels designated as class-2 (data) channels. In Subsection 12–4–3 we discuss an approximation method based on a fluid-flow approach that is appropriate for the overload region, $\rho_2 > N_2$. The approximation formula introduced here has no theoretical justification. It is chosen to have the right form and is shown to agree with the analytical result obtained in Subsection 12–4–1 for the case of $N = 2$ channels, with $N_2 = 1$ channel reserved for data. The approximate overload analysis of Subsection 12–4–3 does have theoretical justification, which is outlined there.

Both approximation methods are appropriate under the continuous-time model assumptions only. Recall that these assumptions require the frame length to be small compared with both the class-1 and class-2 service times and also require the service times to be distributed exponentially. Occhiogrosso and his coworkers [OCCH] give a simple approximation for the performance of the movable-boundary scheme in the underload region for the case of frames that are long compared with the class-2 service time.

We obtain our general continuous-time approximation for the wait time in the underload region by making some observations:

1. The condition for equilibrium of the data (class-2) queue is

$$\rho_2 < N - E(i) = N - \rho_1(1 - P_B) = N_2 + [N_1 - \rho_1(1 - P_B)] \quad (12\text{–}95)$$

as noted earlier. (Here P_B is the usual Erlang-B blocking probability. $E(i) = \rho_1(1 - P_B)$ is the average number of class-1 calls in the system, with ρ_1 the class-1 traffic utilization.) Alternatively, the total utilization ρ is bounded by N:

$$\rho \equiv \rho_2 + \rho_1(1 - P_B) < N \quad (12\text{–}96)$$

This implies that whatever equation we write for the wait or delay times must be inversely proportional to

$$a \equiv N - \rho = N - [\rho_2 + \rho_1(1 - P_B)] \quad (12\text{–}127)$$

This is obviously a generalization of expression (12–113) for the average capacity remaining with utilization ρ, which we introduced in the special case of $N = 2$ channels.

2. We again assume that the ratio α of the class-1 holding time to class-2 service is very large, $\alpha = \mu_2/\mu_1 \gg 1$. It is reasonable to assume then that the data queue reaches equilibrium for each of the $N_1 + 1$ class-1 (voice) states, $0 \le i \le N_1$.

If the class-1 system were always to operate in a state i, the number of servers (channels) available for the class-2 packets would be $N - i$, and the average wait

---

[OCCH] B. Occhiogrosso et al., "Performance Analysis of Integrated Switching Communications Systems," National Telecommunications Conference, Los Angeles, Dec. 1977, 12:4-1–12:4-11.

time would be given by

$$E(W|i) = \frac{1}{\mu_2} E_{2,N-i}(\rho_2)/(N - i - \rho_2) \qquad (12-128)$$

with $E_{2,m}(\rho)$ the Erlang-C distribution, or the Erlang distribution of the second kind, introduced in Chapter 10. Recall that this distribution was defined to be the probability that all m servers are occupied, and was given by [see Eqs. $(10-6)$—$(10-8)$]

$$E_{2,m}(\rho) \equiv \sum_{k=m}^{\infty} p_k = \rho^m/[\rho^m + (m - \rho)(m - 1)! \sum_{k=0}^{m-1} \rho^k/k!] \qquad (12-129)$$

As a check, with $m = 1$, this simplifies to the M/M/1 result

$$E_{2,1}(\rho) = \rho \qquad (12-130)$$

For $m = 2$, the M/M/2 result simplifies to

$$E_{2,2}(\rho) = \rho^2/(2 + \rho) \qquad (12-131)$$

However, the movable-boundary scheme does *not* operate in one class-1 state i. Instead the state of occupancy moves randomly, albeit slowly, among the $N_1 + 1$ possible values $0 \leq i \leq N_1$. Since the class-1 (voice) subsystem is taken to be a pure blocking system (Fig. $12-20$), the probability $p(i)$ that the system is in state i, the probability that i calls are active, is the Erlang probability of Chapter 2 (Eq. $2-54$),

$$p(i) = \rho_1^i/i \Big/ \sum_{j=0}^{N_1} \rho_1^j/j! \qquad (12-132)$$

It appears reasonable to assume, then, that the wait time of the class-2 packets is found by averaging over the Erlang-C distributions of the various states. We thus choose as our approximate expression for the normalized average class-2 wait in the general movable-boundary scheme, with N slots (channels) per frame available,

$$\mu_2 E(W) = \frac{1}{a} \sum_{i=0}^{N_1} p(i)E_{2,N-i}(\rho_2) \qquad (12-133)$$

Here

$$a = N - \rho, \text{ as in Eq. } (12-127)$$

$$p(i) \text{ is given by Eq. } (12-132)$$

and

$$E_{2,N-i}(\rho_2) \text{ is defined by Eq. } (12-129)$$

As an example, take the special case $N = 2$, $N_1 = 1 = N_2$ analyzed in Subsection 12-4-1. Then, from Eq. (12-133), $\mu_2 E(W)$ is approximated by

$$\mu_2 E(W) = \frac{1}{a} [p(0)E_{2,2}(\rho_2) + p(1)E_{2,1}(\rho_2)] \qquad (12-134)$$

Substituting Eq. (12-130) and Eq. (12-131) in for $E_{2,1}(\rho_2)$ and $E_{2,2}(\rho_2)$, respectively, and writing $p(0) = 1/(1 + \rho_1)$, $p(1) = \rho_1(1 + \rho_1)$, we find that

$$\begin{aligned}
\mu_2 E(W) &= \frac{\rho_2}{a(1 + \rho_1)} \left[\frac{\rho_2}{2 + \rho_2} + \rho_1 \right] \\
&= \frac{\rho_2(\rho_2 + 2\rho_1 + \rho_1\rho_2)}{a(1 + \rho_1)(2 + \rho_2)}
\end{aligned} \qquad (12-134a)$$

Comparing with Eq. (12-126), the expression obtained using the exact analysis, we note that the two results are identical! Whether the approximation Eq. (12-133) remains a good one (let alone equivalent to the exact expression!) for larger systems has not been determined. Based on the preceding arguments and on the result obtained for $N = 2$, $N_1 = 1$, it would appear that the approximation should remain valid.

As an example of the use of this approximation consider the movable-boundary scheme with $N = 5$, $N_1 = 2$, $N_2 = 3$. Take $\rho_1 = 0.6$ as the class-1 traffic utilization. The blocking probability is then $P_B = p(2) = 0.10$ from Eq. (12-132), just the Erlang-B blocking probability result. $p(0) = 0.56$ and $p(1) = 0.34$ from Eq. (12-132) as well. The average number of channels occupied by the class-1 (voice) traffic is $\rho_1(1 - P_B) = 0.54$, so that $4.46 = 5 - 0.54$ channels on the average are available for the class-2 packets. The average number of channels free in the system is then $a = 4.46 - \rho_2$. The approximation to the normalized class-2 wait time is then given by

$$\mu_2 E(W) = \frac{1}{a} [p(0)E_{2,5}(\rho_2) + p(1)E_{2,4}(\rho_2) + p(2)E_{2,3}(\rho_2)] \qquad (12-135)$$

with $E_{2,m}(\rho_2)$ found from Eq. (12-129).

This wait time has been tabulated in Table 12-1 for a number of values of $\rho_2 < N_2 = 3$. Also tabulated are the equivalent normalized wait times for a fixed-boundary scheme with $N_2 = 3$ channels. Note that the wait time in this case is just given by Eq. (12-128), with $(N - i) = 3$ here. The movable-boundary scheme again outperforms the fixed-boundary scheme by a considerable amount. Actually the impact of the increasing data load is only felt for $\rho_2 \geq 2$, or a normalized fixed-boundary load $\rho_2/N_2 = 2/3$ or more. It is apparent that the fixed-boundary delay rises rapidly beyond this point as ρ_2 approaches 3. The movable-boundary wait time is still quite small, however, since its increase is

TABLE 12-1   Comparison of Movable-boundary and Fixed-boundary Schemes*

ρ_2	$\mu_2 E(W)$	
	Movable Boundary	Fixed Boundary
1.0	0.006	0.05
1.5	0.019	0.16
2.0	0.06	0.5
2.5	0.26	1.4

* $N = 5$ slots, $N_1 = 2$ slots, $N_2 = 3$ slots, $\rho_1 = 0.6$, $P_B = 0.10$

mediated primarily by $a = 4.46 - \rho_2$ in this example. At $\rho_2 = 3$, the maximum data utilization in the underload region, the effect of decreasing a still has not been felt substantially.

12-4-3 Movable-boundary Scheme: Fluid-flow Approximate Analysis, Overload Region

In this subsection we continue the approximate analysis of the movable-boundary scheme geared to larger systems, but now consider the overload region $\rho_2 > N_2$. The approximation introduced here is based on a *fluid-flow approach*, useful in its own right in studies of queueing systems. The approach adopted follows that of [GAVE]. An alternate approximation method using a fluid-flow approach appears in [LEON]. The fluid-flow technique allows one to demonstrate the extraordinarily high buildups of data queues possible during periods of intense voice activity. Recall that this occurs when voice calls are present for relatively long periods of time, not allowing data packets to use voice channels that, on the average, one might expect to be made available to them. This again demonstrates the need to introduce data-flow control, as noted earlier in this section.

The basic idea behind the fluid-flow approach is to model the data portion of the movable-boundary system as a deterministic system. Say that there is no flow control exerted on the data and that the data-arrival rate, λ_2, is temporarily

[LEON] A. Leon-Garcia, R. H. Kwong, G. F. Williams, "Performance Evaluation Methods for an Integrated Voice-Data Link," *IEEE Trans. on Comm.*, vol. COM-30, no. 8, Aug. 1982, 1848–1858.

greater than its departure rate, based on the number of channels (slots) made available to the data. The data queue then starts building up and quickly assumes such large values that individual (discrete) packet arrivals and departures are masked by the arrival-departure process. The queue-length distribution can be looked at as being that of a *continuous* process, with arriving flow λ_2 packets/sec and departing flow $(N - i)\mu_2$ packets/sec. Here as previously the integer i represents the number of voice calls in the system. The net rate at which the data queue builds up with i voice calls in the system is then

$$r_i = \lambda_2 - (N - i)\mu_2 \qquad (12-136)$$

If $r_i > 0$, or $\lambda_2 > (N - i)\mu_2$, the data queue will tend to increase linearly at this rate if the queue length is large enough to allow the continuous process approximation to be a valid one. If $r_i < 0$, the queue will tend to decrease.

As an example, say that $N = 15$ channels. $N_1 = 10$ channels are reserved for voice calls (class-1 traffic), and $N_2 = 5$ channels are allocated to class-2 data packets. Let the voice traffic be such that, *on the average*, 5 of the voice channels are free and available for data packets. The total number available, on the average, is thus 10 channels. The usual equilibrium analysis indicates that if $\lambda_2 < 10 \mu_2$, the data queue will not build up indefinitely and the data queue statistics will reach stationary (equilibrium) values. In particular, let $\mu_2 = 100$ packets/sec. (A packet is then 10 msec long on the average.) Pick $\lambda_2 = 900$ packets/sec as the (overload) packet-arrival rate. (In a flow-controlled situation, one would choose $\lambda_2 < 500$ packets/sec on the average.) So long as there are fewer than $i = 6$ voice calls in the system, $r_i = \lambda_2 - (N - i)\mu_2 < 0$, the data queue tends to empty, and no problems arise.

Consider an interval of time, however, in which $i = 7$ voice calls are present. λ_2 remains at 900 packets/sec. Then $r_i = (900 - 800)$ packets/sec is the rate at which the data queue tends to build up. If the average voice-holding time is $1/\mu_1 = 100$ sec, the queue length will increase by an average of 10,000 packets in this time. It is clear, then, that large excursions in the data queue length are possible because of the large ratio of call-holding time to packet-service time ($\alpha = \mu_2/\mu_1 = 10^4$ in this example) when the data packet-arrival rate λ_2 exceeds the value $N_2\mu_2$, its normally assigned rate. This is the point made earlier in noting that the movable-boundary scheme should be used to reduce packet-time delays and not to increase the packet throughput. These large excursions in queue length indicate why the deterministic fluid-flow model with a continuum of values for the number of packets on the data queue is a valid one.

Figure 12–26 is taken from simulations carried out by C. J. Weinstein and coworkers at Lincoln Laboratory [WEIN, Fig. 1]. It portrays a real-time plot of voice calls and the corresponding data queue length, as a function of time, for a movable-boundary scheme with a 15-slot frame structure — precisely the example just described. The numbers used $(N_1, N_2, \mu_1, \mu_2, \lambda_1, \lambda_2)$ are also those

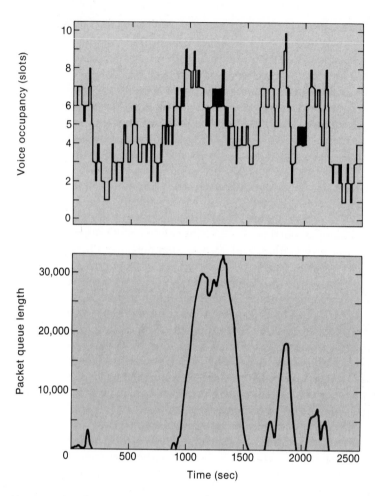

Figure 12–26 Simulation results, movable-boundary strategy, $N = 15$ slots/frame, $N_1 = 10$, $N_2 = 5$, $\lambda_1 = 0.05$ calls/sec, $\mu_2 = 100$ packets/sec, $\lambda_2 = 900$ packets/sec (from [WEIN, Fig. 1], © 1980 IEEE, with permission)

chosen in the preceding example. Note how the figure verifies the simple fluid-flow model just described. During those intervals of time in which 6 or fewer voice calls are present, the data queue is in underload, with very few packets waiting, on the scale of the figure. During overload, however, the packet queue tends to build up at the rate r_i, as the reader may verify. In particular, during the

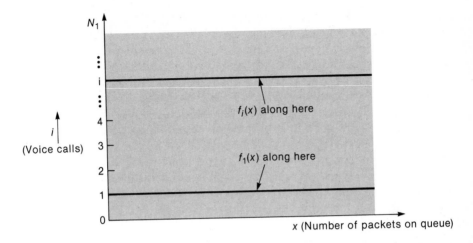

Figure 12-27 Continuous variable arising in fluid-flow approximation, movable-boundary scheme

time interval from 1000 to 1500 sec, the system is almost continuously in overload, and the queue builds up to more than 30,000 packets. On the average the voice calls in progress do hover about 6, so that the system is in equilibrium, but because of the long call-holding times, very large packet lengths ensue, as already mentioned.

Analysis is required to study the details of operation of this system in overload. For this purpose we use the fluid-flow approximation mentioned earlier [GAVE]. Since the packet queue length does behave almost like a continuous variable in the overload region (Fig. 12-26), we represent it in this approximation by the continuous variable $x(t)$. The objective then is to find the probability density function $f_x(t)$ or, at equilibrium, $f(x)$, the continuous variable counterpart of the discrete state probability p_j introduced earlier in this chapter.

More precisely, we shall find the set of equilibrium conditional density functions $f_i(x)$ conditioned on having i class-1 calls in the system. This is demonstrated schematically in Fig. 12-27. The approach will be similar to the one used in Chapter 2 in first finding the probabilities of state of the M/M/1 queue. Recall [Eq. (2-12)] that we were able to set up an equation relating the state probability $p_j(t + \Delta t)$ at time $t + \Delta t$ to its value Δt units of time earlier, at time t. We use a similar approach here. The equations generated are called forward

---

[COX] D. R. Cox and H. D. Miller, *The Theory of Stochastic Processes*, Methuen, London, 1965.

equations [COX]. Before developing these equations, however, we digress to discuss the fluid-flow rate parameters that are required for the analysis.

It must be noted that by the very nature of the fluid-flow approximation, the data queue must be large. We shall in fact allow the continuous variable x to approach zero. It is apparent that discrepancies between this approximation and the exact discrete approach will increase as we decrease x. In particular, boundary conditions for the data queue statistics—i.e., the probabilities p_{i0} that the queue is empty—must be found separately. The approach used will be discussed later.

We proceed as follows: Say that there are i class-1 (voice) calls in the system, $0 \le i \le N_1$. Then $(N - i)$ channels are available for data packets, and these will be handling $(N - i)\mu_2$ packets/sec. With a packet-arrival rate of λ_2 packets/sec, the packet queue length will be changing at the rate

$$\left.\frac{dx}{dt}\right|_i \equiv r_i = \lambda_2 - (N - i)\mu_2$$
$$= \mu_2[\rho_2 - (N - i)] \qquad 0 \le i \le N_1 \tag{12–136a}$$

If the arrival rate λ_2 is greater than $\mu_2(N - i)$, the queue will be increasing at the rate r_i; if the arrival rate λ_2 is smaller than $\mu_2(N - 1)$, the queue will be decreasing (emptying) at the rate r_i. For very large queues, for which the fluid approximation is valid, the queue size will increase (or decrease) linearly with time at the rate r_i, so long as i calls remain connected. As the number of voice (class-1) calls changes, the rate parameter r_i changes accordingly.

As a check on the validity of this fluid flow approach, it is apparent that the average rate of change of queue length must be negative; if it were positive the queue would tend to keep increasing, leading to an unstable situation. For a stable queue (i.e., one in statistical equilibrium) we must thus have

$$\sum_{i=0}^{N_i} r_i p_i < 0 \tag{12–137}$$

with p_i the probability that there are i calls in the system. But under the assumptions made this is just the Erlang probability

$$p_i = \rho_1^i/i! \left/ \sum_{j=0}^{N_1} \rho_1^j/j! \right. \tag{12–138}$$

It is left to the reader to show, inserting Eqs. (12–138) and (12–136) into Eq. (12–137), that for equilibrium conditions to prevail one must have

$$\rho_2 < N - \rho_1(1 - P_B) \tag{12–139}$$

This is precisely the same as Eq. (12–95), postulated earlier as the condition for equilibrium in the controlled system of Fig. 12–20. Alternatively, as shown in Eq. (12–96), the total utilization ρ is given by

$$\rho \equiv \rho_2 + \rho_1(1 - P_B) < N \qquad (12-140)$$

The simple fluid-flow approach thus yields a result in agreement with the result obtained more generally by using continuous-time queueing theory arguments similar to those described earlier in this section.

We now proceed to set up the forward equations for finding the desired data queue conditional-probability density functions $f_i(x)$, $0 \leq i \leq N_1$. As in the discrete probability counterpart of Chapter 2, we first consider the time-varying density function

$$f(x,i,t)$$

This represents the probability density function that there are x packets in the system at time t, conditioned on having i class-1 (voice) calls present. We obtain the forward equation representing the evolution of $f(x,i,t)$ with time by relating its value at time $t + \Delta t$ to its value at time t, an infinitesimal time interval Δt earlier.

Specifically, it is left to the reader to show that the following forward equation does in fact represent the evolution of the density function with time:

$$\begin{aligned}
f(x,i,t + \Delta t) = {} & f(x - r_i\Delta t,i,t)[1 - (\lambda_1 + i\mu_1)\Delta t] \\
& + f(x - r_{i-1}\Delta t,i - 1,t)\lambda_1\Delta t \\
& + f(x - r_{i+1}\Delta t,i + 1,t)(i + 1)\mu_1\Delta t
\end{aligned} \qquad (12-141)$$

The equation is valid only for $1 \leq i < N_1$ and must be modified appropriately at the boundary points $i = 0$ and N_1. Expanding $f(x,i,t + \Delta t)$ in a two-variable Taylor series with respect to both x and t, canceling terms, dividing through by Δt, letting $\Delta t \to 0$, and then taking the stationary time (equilibrium) case $\dfrac{\partial f}{\partial t}(x,i,t) = 0$, one gets the following set of first-order differential equations:

$$\begin{aligned}
r_i \frac{df_i(x)}{dx} = {} & -(\lambda_1 + i\mu_1)f_i(x) + \lambda_1 f_{i-1}(x) \\
& + (i + 1)\mu_1 f_{i+1}(x) \qquad 1 \leq i < N_1
\end{aligned} \qquad (12-142)$$

At the extreme values of i, one gets in addition the two differential equations

$$r_0 \frac{df_0(x)}{dx} = -\lambda_1 f_0(x) + \mu_1 f_1(x) \qquad (12-143)$$

and

$$r_{N_1} \frac{df_{N_1}(x)}{dx} = -N_1\mu_1 f_{N_1}(x) + \lambda_1 f_{N_1-1}(x) \qquad (12-144)$$

Dividing each equation through by the appropriate rate parameter, the set of differential equations may be written in the compact matrix-vector form

$$\frac{d\mathbf{f}(x)}{dx} = -A\mathbf{f}(x) \qquad (12-145)$$

with $\mathbf{f}(x)$ the column vector of the $(N+1)$ $f_i(x)$'s and A an $(N_1+1) \times (N_1+1)$ matrix containing the coefficients modifying the $f_i(x)$ terms in Eqs. $(12-142)$–$(12-145)$.

As an example, consider the simplest movable boundary case $N = N_1 = 1$, $N_2 = 0$, analyzed exactly earlier using moment-generating functions. For this case, only Eqs. $(12-143)$ and $(12-144)$, with $N_1 = 1$, are needed. One then has

$$A = \begin{bmatrix} \dfrac{\lambda_1}{r_0} & -\dfrac{\mu_1}{r_0} \\[2ex] -\dfrac{\lambda_1}{r_1} & \dfrac{\mu_1}{r_1} \end{bmatrix} \qquad (12-146)$$

The formal solution to Eq. $(12-145)$ may be written directly, in vector form, as

$$\mathbf{f}(x) = \mathbf{c}e^{-Ax} \qquad (12-147)$$

with $\mathbf{c}$ a set of constants to be determined [NOBL]. Alternatively, each conditional density function, the group representing the components of $\mathbf{f}(x)$, may be written as the weighted sum of a set of exponentials in the eigenvalues of A [NOBL]. Thus we have

$$f_i(x) = \sum_k c_{ik}e^{-\lambda_k x} \qquad (12-148)$$

with the λ_k's the eigenvalues of the matrix A.

Since the $f_i(x)$'s are probability density functions defined over all values of x (Fig. $12-27$), it is clear that only positive eigenvalues must be included in the summation. The matrix A has (N_1+1) eigenvalues. It may be shown that the positive eigenvalues represent that subset for which the rate parameter $r_i > 0$ [GAVE]. From Eq. $(12-136)$ it is clear that the $N+1$ rate parameters may be written in ascending order.

$$r_0 < r_1 < r_2 < \cdots < r_m < 0 < r_{m+1} < \cdots < r_{N_1} \qquad (12-149)$$

m represents the largest integer i for which $r_i = \mu_2[\rho_2 - (N-i)]$ is still negative. There are then $N_1 - m$ positive rate parameters, and hence $N_1 - m$ positive

[NOBL] B. Noble and J. W. Daniel, *Applied Linear Algebra*, Prentice-Hall, Englewood Cliffs, N.J., 1977.

eigenvalues to be included in the calculation of $f_i(x)$. The $N_1 - m$ values of i for which the rates r_i are positive represent states for which the data queue length is increasing. They are thus called "up" states [GAVE].

We carry out the analysis for the $N = N_1 = 1$, $N_2 = 0$ case to show the approach used and also to compare the fluid-flow approximation result with the one obtained using the exact analysis. For this example, the condition of equilibrium from Eq. (12–139), agreeing with the result of the earlier exact analysis, is

$$\rho_2 < 1 - \frac{\rho_1}{1 + \rho_1}$$

or

$$\rho \equiv \rho_2 + \rho_1/(1 + \rho_1) < 1$$

For $N = N_1 = 1$, $N_2 = 0$, the A matrix is given by Eq. (12–146). It is left to the reader to show that for this matrix the two eigenvalues are $\lambda = 0$ and

$$\lambda = \frac{1}{\alpha} \frac{1 - \rho_2(1 + \rho_1)}{\rho_2(1 - \rho_2)} \tag{12–150}$$

Here $\alpha \equiv \mu_2/\mu_1$ as in the preceding analysis. As a check, $r_0 = \mu_2(\rho_2 - 1) < 0$, since $\rho_2 < 1$ is always true here. Also, $r_1 = \lambda_2 = \mu_2\rho_2 > 0$. There is thus only one "up" state, the one for which the single channel available is occupied by a voice call, with the data queue increasing linearly with time at the arrival rate λ_2.

From Eq. (12–148) the two density functions are given by

$$f_0(x) = c_0 e^{-\lambda x} \qquad x > 0 \tag{12–151}$$

and

$$f_1(x) = c_1 e^{-\lambda x} \qquad x > 0 \tag{12–152}$$

with c_0 and c_1 two undetermined constants yet to be found.

We noted previously that the data queue conditional density functions $f_i(x)$ are to be interpreted for $x > 0$ only. They do not provide values for the probability that $x = 0$. For $x = 0$, the empty queue state, or the boundary in this problem, the conditional probabilities p_{i0} must be found separately. In particular, from the physics of the fluid-flow approach, we must have

$$p_{i0} = 0, \qquad i > m \tag{12–153}$$

That is, in the "up" state the queue under the fluid-flow approximation can never be zero since its rate of increase is positive. There are thus $(m + 1)$ unknown probabilities on the boundaries to be found as well. We demonstrate the calculations involved in the special case of $N = N_1 = 1, N_2 = 0$. First note that, in general, for any class-1 (voice) state i, with i calls in progress, we must have the condition

$$p_i = \int_0^\infty f_i(x)dx + p_{i0} \tag{12-154}$$

Since probability p_i is the Erlang probability, it provides N_1 equations to be used in finding the unknown constants. For the example under consideration here, with $N = N_1 = 1$, $N_2 = 0$, the discrete probability $p_{10} = 0$, since this represents an "up" state: i.e., in this case, the one transmission channel is occupied and the data queue must be increasing under overload conditions. We thus have

$$\int_0^\infty f_1(x)dx = c_1/\lambda = p_1 = \rho_1/(1 + \rho_1) \tag{12-155}$$

using Eq. (12–152) to evaluate the integral. This provides an equation for the unknown constant c_1. Thus

$$c_1 = \lambda p_1 = \frac{\rho_1}{\alpha(1 + \rho_1)} \frac{[1 - \rho_2(1 + \rho_1)]}{\rho_2(1 - \rho_2)} \tag{12-156}$$

introducing the expression (12–150) for the eigenvalue λ.

Unfortunately the constant c_0 in the expression for $f_0(x)$ cannot be found uniquely in the same manner. The boundary probability p_{00} is *not* zero since the system is in a "down" state ($r_0 < 0$) when no class-1 call is present (see Fig. 12–28). We have one equation for both c_0 and p_{00}:

$$p_0 = \frac{1}{1 + \rho_1} = \int_0^\infty f_0(x)dx + p_{00}$$
$$= \frac{c_0}{\lambda} + p_{00} \tag{12-157}$$

We need an additional independent equation to find c_0 and p_{00} uniquely.

This equation is found by noting that c_0 and c_1, the undetermined constants introduced in Eqs. (12–147), (12–151), and (12–152), must represent components of the eigenvector $\mathbf{c}$ associated with the eigenvalue λ. (In general, the coefficients c_{ik} in Eq. (12–148) represent a subset of the elements of the eigenvector $\mathbf{c}_i$ associated with the eigenvalue λ_i.) To demonstrate this, recall that $\mathbf{f}(x)$ is a solution to the matrix-vector equation

$$\frac{d\mathbf{f}(x)}{dx} = -A\mathbf{f}(x)$$

Writing

$$\mathbf{f}(x) = \mathbf{c}e^{-\lambda x} \quad \mathbf{c} = \begin{bmatrix} c_0 \\ c_1 \end{bmatrix}$$

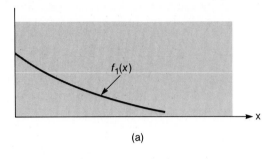

(a)

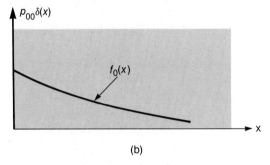

(b)

Figure 12-28 Fluid flow approximation, movable-boundary scheme
a. $i = 1$, voice call present ($p_{10} = 0$)
b. $i = 0$, no voice call present

with

$$\frac{d\mathbf{f}(x)}{dx} = -\lambda \mathbf{c} e^{-\lambda x}$$

we must have

$$A\mathbf{c} = \lambda \mathbf{c}$$

Hence

$$(A - \lambda I)\mathbf{c} = 0$$

and $\mathbf{c}$ must be an eigenvector associated with the eigenvalue λ [NOBL]. It is left for the reader to show that in the example under discussion this then provides the necessary additional equation

$$c_0 = p_2 c_1 / (1 - p_2) \qquad (12-158)$$

Simplification in the algebra results if we introduce the traffic intensity parameter $\rho \equiv \rho_2 + \rho_1/(1 + \rho_1)$. Using this to rewrite c_1 in Eq. (12–156), one gets

$$c_1 = \frac{1}{\alpha} \frac{\rho_1(1 - \rho)}{\rho_2(1 - \rho_2)} \qquad (12-159)$$

From Eq. (12–158) one then has

$$c_0 = \frac{1}{\alpha} \frac{\rho_1(1 - \rho)}{(1 - \rho_2)^2} \qquad (12-160)$$

Finally, from Eq. (12–157), one has

$$p_{00} = \frac{1}{1 + \rho_2} - c_0/\lambda$$
$$= \frac{1 - \rho}{1 + \rho_1} \qquad (12-161)$$

after some manipulation.

The mean data queue length, the performance parameter of interest, follows directly. This is $E(x)$, comparable to $E(j)$ in the exact analysis. $E(x)$ is given by

$$E(x) = \int_0^\infty x[f_0(x) + f_1(x)]dx$$
$$= \frac{\alpha \rho_1 \rho_2}{(1 - \rho)(1 + \rho_1)^2} \qquad (12-162)$$

after introducing Eqs. (12–151) and (12–152), using the known values of λ, c_0, and ρ, and carrying out the integration. (Recall that $\rho \equiv \rho_2 + \rho_1/(1 + \rho_1)$.)

The average normalized time delay using the fluid-flow approximation is then given by

$$\mu_2 E(T)|_{\text{fluid}} = E(x)/\rho_2 = \frac{\alpha \rho_1}{(1 - \rho)(1 + \rho_1)^2} \qquad (12-163)$$

The exact continuous-time analysis expression for the time delay for this case was given previously by Eq. (12–92a). Rewriting that equation by replacing ρ_2 with $\rho - \rho_1/(1 + \rho_1)$, one finds that

$$\mu_2 E(T) = \left[\frac{\alpha \rho_1}{(1 + \rho_1)^2} + 1 \right] \Big/ (1 - \rho)$$
$$= \mu_2 E(T)|_{\text{fluid}} + 1/(1 - \rho) \qquad (12-164)$$

The difference between the exact and the approximate (fluid-flow) expression is the error term $1/(1 - \rho)$. For large values of $\alpha \equiv \mu_2/\mu_1$ the difference is clearly negligible. As an example, take $\alpha = 10^4$ and $\rho_1 = 0.1$. The voice-blocking probability is then $P_B = \rho_1/(1 + \rho_1) \doteq 0.09$. The first term in the numerator of Eq. (12–164) is about 800, clearly much larger than 1. Even if $\alpha = 100$ and $\rho_1 = 0.5$, the first term is on the order of 20, again much larger than 1. The fluid-flow approximation is thus an excellent one in this case. The method provides a good, relatively simple way to study the effect of overload in systems such as the movable-boundary one.

As the second application of the fluid-flow approximation to a movable-boundary system, consider the $N = 2$, $N_1 = 1 = N_2$ example worked out in detail earlier. In Subsection 12–4–2 we showed that the underload-region time-delay approximation was exact for this example. In the overload region the fluid-flow approach can be shown to provide the following expression for the normalized time delay:

$$\mu_2 E(T)|_{\text{fluid}} = \frac{\alpha(\rho_2 - 1)\rho_1}{a\rho_2(1 + \rho_1)^2} \qquad \rho_2 > 1 \qquad (12–165)$$

$$a \equiv 2 - \rho$$

The derivation of this expression is left as an exercise (see Problem 12–23).

Comparing with Eq. (12–122b), the time-delay expression in the overload region found using the moment-generating function approach, it is apparent that the fluid-flow approximation is identical with the third term in that expression. Clearly the fluid-flow approximation is wrong at the point $\rho_2 = 1$, where it predicts zero delay! This is because the fluid-flow approach predicts complete emptying of the data queue at this value of the traffic utilization. As ρ_2 increases, however, the third term in Eq. (12–122b) begins to dominate if $\alpha \gg 1$, and the approximate value, as given by Eq. (12–165), becomes more accurate. Note particularly that the approximation captures the linear increase with α, the phenomenon described earlier. The two time-delay expressions, (12–122b) from the exact analysis and (12–165) from the fluid-flow approximation, are compared in Table 12–2 for the special case of $\rho_1 = 1$ or $P_B = 0.5$. Note that as both α and $(\rho_2 - 1)$ increase the approximation becomes more accurate. Note in particular, with $\alpha = 10^4$, how large the time delays become as ρ_2 increases. This is the phenomenon described earlier in this chapter; it indicates the need for flow control if the movable-boundary scheme is used.

Other examples may be worked out using the fluid-flow approximation to determine the performance of the movable-boundary scheme in the data-overflow region. A simple generalization of the example just discussed is the case in which there are N slots/frame, with $N_1 = 1$ slot dedicated to class-1 (voice) users, and $N_2 = N - 1$ slots dedicated to class-2 (data) users. The approximate class-1 average time delay, as an extension of Eq. (12–165), is then found to be given by

$$\mu_2 E(T)|_{\text{fluid}} = \frac{\alpha(\rho_2 - N_2)\rho_1}{a\rho_2(1 + \rho_1)^2} \qquad \rho_2 > N_2 \qquad (12–166)$$

TABLE 12–2 Overload Region, Movable Boundary Scheme, $N = 2$, $N_1 = N_2 = 1$: Exact Analysis and Fluid-flow Approximation Compared

	$\mu_2 E(T)$			
	$\rho_2 \rightarrow 1.1$	1.2	1.3	1.4
$\alpha \downarrow 10$	3.39	5.03	8.19	17.4 ← exact
	0.57	1.39	2.88	7.14 ← approx.
100	8.5	17.5	34.1	81.7
	5.7	13.9	28.8	71.4
1000	59.6	143	294	724
	56.8	139	288	714
10^4	571	1393	2890	7153
	568	1389	2884	7143

with $a \equiv N - \rho$, as introduced originally in Eq. (12–127). The derivation of Eq. (12–166) is left as an exercise for the reader (see Problem 12–24).

All of the analysis in this section, as has been the case throughout this book, has been predicated on a Poisson arrival model for each of the two classes of traffic. This is generally accepted to be a reasonable model for both voice and packets. Voice has the property, however, that there are gaps or intervals of silence occurring as much as 60 percent of the time that a person is speaking. Use has been made of this phenomenon in practice by attempting to fill the gaps with additional voice calls, enabling a given set of trunks to carry as much as twice the voice traffic for which it is designed. The technique was originally called a *time assigned speech interpolation* (TASI) scheme. In the digital world it is called a *digital speech interpolation* (DSI) scheme. It is obviously of interest to see whether DSI techniques can be used to improve the overall system performance. The fluid-flow approach has been extended by O'Reilly specifically to study this problem [OREI]. Related work appears in [SRIR], [WILL], and [FISC

[OREI] P. O'Reilly, *A Fluid-Flow Approach to Performance Analysis of Integrated Voice-Data Systems with DSI*, GTE Laboratories, Waltham, Mass., April 1985.

[SRIR] K. Sriram, P. K. Varshney, J. G. Shanthikumar, "Discrete-Time Analysis of Integrated Voice/Data Multiplexers with and without Speech Activity Detectors," *IEEE J. on Selected Areas in Comm.*, Vol. SAC-1, no. 6, Dec. 1983, 1124–1132.

[WILL] G. F. Williams and A. Leon-Garcia, "Performance Analysis of Integrated Voice and Data Hybrid-Switched Links," *IEEE Trans. on Comm.*, vol. COM-32, no. 6, June 1984, 695–706.

[FISC 1979] M. J. Fischer, "Data Performance in a System where Data Packets Are Transmitted during Voice Silent Periods—Single Channel Case," *IEEE Trans. on Comm.*, vol. COM-27, no. 9, Sept. 1979, 1371–1375.

1979]. ([SRIR] uses a discrete-time model for the system, while the others use continuous-time models.) O'Reilly has used as a model of speech alternating, exponentially distributed intervals of silence and speech activity (the latter interval is called a *talkspurt*). This alternating exponential model is a common one for DSI analysis. O'Reilly again shows a significant improvement over a comparable fixed-boundary scheme. Simulation and comparison with results obtained by [SRIR] using the discrete-time approach indicate that the fluid-flow model is quite accurate so long as the ratio of voice talkspurt length to data packet length (comparable to our α) is greater than 30.

Problems

12–1 Carry out the analysis of alternate routing of circuit-switched traffic for a fully connected, symmetrical network model, and show that the offered load A and the total load a offered to a link are related by Eq. (12–6). Take the example of $n = 5$ trunks/link group and $M = 1$ alternate route. Calculate and plot the resultant carried load C versus offered load A. Compare with the case of nonalternate routing ($M = 0$). Repeat for $n = 10$ trunks/link group. *Note:* It may be helpful in carrying out the calculations to calculate the Erlang-B blocking probability using the recursive form given by Eq. (12–71) in the text. (See also Problem 10–16.) Try larger examples ($n > 10$ and $M > 1$) if time permits.

12–2 Refer to the discussion in the text on trunk reservation for control of alternately routed traffic. Prove that the probability that j calls are active is given by Eqs. (12–7a), (12–7b), and (12–8). Show that the average number of trunks occupied is given by Eq. (12–11). Show that the first-routed offered load A and the total offered traffic a, at a link, are related by Eqs. (12–14) and (12–15).

12–3 Consider the following two cases of an alternate routing procedure with $M = 1$ alternate route to be attempted: $n = 5$ and $n = 10$ trunks/link group. Calculations for these two cases were carried out in Problem 12–1. Introduce a trunk reservation control in each case and compare the resultant carried-load versus offered-load curves with those obtained in Problem 12–1. Let the control parameter m vary by steps of 1 from the uncontrolled case $m = n$. Discuss your results. As in Problem 12–1, try to extend the calculations to larger systems, time permitting. It may again be helpful to use the recursive form of the Erlang-B blocking probability given by Eq. (12–71) in the text.

12–4 Refer to the discussion on the performance analysis of common-channel signaling in the text. The average wait time for message signal units at a CCS link is shown to be given by Eq. (12–20). Calculate and plot the average message unit time delay, $E(T)$, as a function of normalized message signal unit load ρ, for the two cases of 4800-bps and 56-kbps links. Compare the results. Message signal units are all 120-bits long, while filler signal units are 48-bits long. Repeat if message signal units vary randomly in length, with $E(\tau_1^2) = 1.4\,T_m^2$. Compare with the calculations for $E(\tau_1^2) = 1.2\,T_m^2$ discussed in the text.

12-5 Given a discrete random variable x with probabilities $p_j = p(x = j)$, $j = 0, 1, 2, \ldots$, the moment-generating function $G_x(z)$ is defined as

$$G_x(z) \equiv E(z^x) = \sum_{j=0}^{\infty} p_j z^j$$

a. Prove the following properties of $G_x(z)$:

1. $G_x(1) = 1$

2. $\left.\dfrac{dG_x(z)}{dz}\right|_{z=1} = E(x)$

3. $\left.\dfrac{dG_x(z)}{dz^2}\right|_{z=1} = E(x^2) - E(x)$

4. Given $y = \sum_{i=1}^{n} x_i$, with the x's independent discrete random variables,

$$G_y(z) \equiv E(z^y) = \prod_{i=1}^{n} G_{x_i}(z)$$

b. Using the definition and properties of a., show that the moment-generating functions, mean value $E(x)$ and variance σ_x^2, of each of the following discrete probability distributions are given by the functions indicated:

1. Poisson distribution

$$p_j = \frac{\lambda^j e^{-\lambda}}{j!} \qquad j = 0, 1, 2, \ldots$$

$$G_x(z) = e^{-\lambda(1-z)} \qquad E(x) = \sigma_x^2 = \lambda$$

2. Geometric distribution

$$p_j = pq^{j-1} \qquad q = 1 - p \qquad j = 1, 2, \ldots$$

$$G_x(z) \equiv \frac{pz}{1 - qz} \qquad E(x) = \frac{1}{p} \qquad \sigma_x^2 = \frac{q}{p^2}$$

3. Bernoulli distribution (x has the values 0 and 1 only)

$$p_0 = q \qquad p_1 = p = 1 - q$$
$$G_x(z) = q + pz \qquad E(x) = p \qquad \sigma_x^2 = pq$$

4. Binomial distribution

$$p_j = \binom{n}{j} p^j q^{n-j} \qquad q = 1 - p \qquad 0 \le j \le n$$

$$G_x(z) = (q + pz)^n \qquad E(x) = np \qquad \sigma_x^2 = npq$$

Hint: Note that x may be written $x = \sum_{j=1}^{n} x_j$, the x's independent Bernoulli variables.

12-6 Applying the method of moment-generating functions to the solution of Eq. (12–28), the governing equation of the M/M/1 queue, show that one obtains Eq. (12–39a). Find $E(n)$, the average queue occupancy, by differentiating Eq.

(12–39a), checking the well-known result, Eq. (12–40) (Eq. (2–27) in Chapter 2). Find the second moment $E(n^2)$, and from this the variance σ^2, as well.

12–7   Find the moment-generating function for the M/M/2 queue by setting up the balance equations representing this queueing system (see Fig. 12–18(b)) and converting them to an equivalent algebraic equation involving the moment-generating function. (*Hint:* Use the same approach applied in the text to the M/M/1 queue.) The result should agree with Eqs. (12–43) and (12–44). Use this to find the first moment $E(n)$, the second moment $E(n^2)$, and the variance σ^2.

12–8   Refer to the analysis of the FIFO discipline for integrating voice and data over an N-channel (slot) TDM frame, described in the text. Focus on the simplest case of $N = 1$ channel.

   a. Carry through the details of the analysis, arriving at Eq. (12–59) for the moment-generating function $G_0(z)$ and Eq. (12–63) for the probability that the system is idle. The intermediate identities and Eqs. (12–54) and (12–55) are to be proved as part of the analysis.

   b. Calculate the average time delay of class-2 (queued) traffic, using Eqs. (12–59) and (12–63). In particular, prove the relation (12–65) and show that the average number of packets (class-2 traffic) in the system is given by Eq. (12–68). From this, verify that the average time delay for this system is given by Eq. (12–69).

   c. Calculate the average packet-time delay for this system for the cases $1/\mu_1 = 0.1$ sec, 1 sec, 10 sec, and 5 minutes. The average packet length is in all cases $1/\mu_2 = 10$ msec. Take $\rho_1 = 0.1$ and $\rho_2 = 0.4$ to keep $P_B = 0.5$. Compare with the M/M/1 result in which the data packets have a channel to themselves. Try some other examples with $1/\mu_2$, ρ_1, and ρ_2 varying as well.

12–9   Verify that the two-dimensional state diagram of Fig. 12–22 does in fact represent the preemptive priority integrated system for $N = 1$ channel.

12–10 Draw a two-dimensional state diagram representing the operation of an $N = 2$ channel FIFO system. This extends the diagram of Fig. 12–21 for the $N = 1$ channel case. Compare with Fig. 12–24 for the $N = 2$ channel movable-boundary scheme. This latter scheme provides a dedicated channel to the class-2, queued data, traffic, with the other class-1 channel made available to the data on a preemptive-priority basis. Draw the comparable state diagram for an $N = 2$ channel preemptive-priority scheme, with *both* channels dedicated to class-1 traffic, and available, with preemption possible, to class-2 traffic. Compare with the FIFO and movable-boundary strategies.

12–11 Show how one obtains Eqs. (12–83) and (12–84) as the two equations whose solution provides the moment-generating functions for the preemptive-priority system with $N = 1$ channel. Solve these equations, together with the equations relating the boundary probabilities, to obtain Eqs. (12–85)—(12–87). Show that Eq. (12–86) reduces to the M/M/1 result (Eq. 12–89) if $\rho_1 = 0$.

12–12 Show that the average time delay of class-2 queued packets in the single-channel ($N = 1$) preemptive-priority system is given by Eq. (12–92a). Calculate the time delay and the class-1 blocking probability for the cases calculated in part c. of Problem 12–8 for the FIFO scheme. Compare with those results.

12–13 Consider the movable-boundary strategy discussed in this chapter. Take the special case of $N = 2$ channels, $N_1 = N_2 = 1$ channel each (Fig. 12–23). Show that Fig. 12–24 represents the state diagram for this system, and that Eqs. (12–98)—(12–102) are the appropriate balance equations describing this system. Draw state diagrams for the system with $N = 3$ channels for the two cases, $N_1 = 1$ and $N_1 = 2$ channels, and write down the balance equations for each of these two cases.

12–14 Consider the movable-boundary example discussed in the text with $N = 2$ channels, and $N_1 = N_2 = 1$ channel dedicated to the class-1 and class-2 traffic, respectively. Carry out the details of the analysis described in the text, verifying Eqs. (12–114)—(12–117). Show, using these equations, that the average number of data packets on queue is given by Eqs. (12–121) and (12–121a).

12–15 Starting with Eq. (12–122) for the average time delay of the movable-boundary scheme with $N = 2$, $N_1 = N_2 = 1$, show that one obtains Eq. (12–122a) as the solution in the underload region, $\rho_2 < 1$, and Eq. (12–122b) in the overload region, $\rho_2 > 1$. Both solutions require the assumption that $\alpha \gg 1$. *Hint:* In obtaining Eq. (12–122a), refer to Eq. (12–116). One must find a root of the denominator polynomial that is less than 1. Since $\alpha \gg 1$, some thought indicates that the root z_0 must be given by $z_0 = 1 - \epsilon$, $\epsilon \ll 1$. Hence $(z - 1)$ in Eq. (12–116) is just $-\epsilon$, while all other values of z may be taken equal to 1. Solve explicitly for $\alpha\epsilon$ and show it is given by

$$\alpha\epsilon = \frac{1}{(1 - \rho_2)} + \frac{\rho_1}{(2 - \rho_2)} \qquad \rho_2 < 1$$

Use this result, plus Eqs. (12–114) and (12–115), in the numerator of Eq. (12–116) to find explicit expressions for p_{00}, p_{01}, and p_{10}. Recall that the numerator polynomial must satisfy $N(z_0) = 0$. In particular, show that

$$p_{10} = \frac{\rho_1}{1 + \rho_1} (1 - \rho_2) \qquad \rho_2 < 1$$

Verify the underload time delay result Eq. (12–122a), using Eq. (12–122). In the overload case, $\rho_2 > 1$, use a similar approach to find the root $z_0 < 1$.

12–16 Show that Eq. (12–122a) for the $N = 2$-channel movable-boundary time delay in the underload region reduces to the appropriate value in each of these special cases: $\rho_1 = 0$, $\rho_1 \to \infty$, $\rho_2 = 0$.

12–17 Consider the movable-boundary scheme with $N = 2$ channels, and $N_1 = N_2 = 1$ channel dedicated to class-1 (blocked) and class-2 (queued) traffic, respectively. Take $\rho_1 = 0.1$ to keep the blocking probability relatively low. Plot the normalized class-2 time delay $\mu_2 E(T)$ as a function of the class-2 utilization ρ_2 for the three cases $\alpha = \mu_2/\mu_1 = 10, 100, 10^4$. Cover the entire range of ρ_2, including the overload region. Discuss the effect of increasing α on the time delay.

12–18 Say that each channel in the $N = 2$-channel movable-boundary scheme corresponds to 64-kbps transmission. Each of the two slots in a frame is 8-bits long. The average class-2 packet length is $1/\mu_2 = 10$ msec.

   a. What is the overall bit rate of the system?
   b. What is the length of a frame, in μsec? How many frames, on the average, are required to transmit a packet? What is the average packet length, in bits? Is the continuous-time assumption a valid one in this case?
   c. Consider the following two cases of class-1 traffic:

   1. voice, with average holding time $1/\mu_1 = 5$ minutes.
   2. bulk file transfer, 10 Mbytes in length.

   In both cases let $\rho_1 = 0.1$. Calculate and plot the average class-2 packet-time delay as a function of ρ_2, the class-2 utilization, for both cases. Cover the entire range possible for ρ_2. What happens to the delays as ρ_2 increases beyond $\rho_2 = 1$? Compare the delays to those obtained using a fixed boundary scheme at $\rho_2 = 0.9$ ($N_1 = N_2 = 1$, but with the class-2 traffic not allowed to access the class-1 channel).

12–19 Use the approximation Eq. (12–133) to the average wait time of class-2 data packets in the underload region to calculate the average wait time as a function of ρ_2 for a system with $N = 6$ slots/frame, $N_1 = 3$ slots reserved for class-1 calls, $N_2 = 3$ slots reserved for class-2 packets. Take $\rho_1 = 0.6$. Compare these results with the equivalent wait time results for a fixed-boundary scheme. Compare with the following two movable-boundary cases:

   1. $N = 5$, $N_1 = 2$, $N_2 = 3$ (see Table 12–1 in the text).
   2. $N = 6$, $N_1 = 4$, $N_2 = 2$.

   Explain the results.

12–20 Refer to the movable-boundary example considered in Table 12–1 in the text. Compare the performance of that example with the following three $N = 5$-channel (slot) cases: $N_1 = 1, 3, 4$. In all three cases take $\rho_1 = 0.6$.

12–21 Consider the fluid-flow approximate analysis for the movable-boundary overload region discussed in the text.

   a. Verify the equilibrium condition (12–139), using Eqs. (12–136)—(12–138).
   b. Show that the evolution with time of the data queue conditional probability density function $f(x,i,t)$ is given by the forward equation (12–141).
   c. Derive Eqs. (12–142)—(12–144), the differential equations whose solution provides the data queue conditional-probability density functions at equilibrium. Show that they may be written in the matrix-vector form of Eq. (12–145). Provide explicit expressions for matrix A in the two cases $N = N_1 = 1$, $N_2 = 0$ (see Eq. (12–146)) and $N = 2$, $N_1 = N_2 = 1$.

12–22 Consider the special case of a movable-boundary scheme with $N = N_1 = 1$, $N_2 = 0$. This is the example for which the fluid-flow overload analysis is carried out in the text. Check the analysis, filling in the details left out, and prove the overload class-2 packet time delay for this example is given by Eq. (12–163). In particular, show that the non-zero eigenvalue is given by Eq. (12–150), that the two data queue conditional density functions are given by Eqs. (12–151) and (12–152), with the constants c_0 and c_1 given by Eqs. (12–160) and (12–159), respectively, and that the mean queue length is given by Eq. (12–162). Plot the packet time delay versus p_2 for $p_1 = 0.1$ and $\alpha = 10^4$, and compare with the exact time delay.

12–23 Use the fluid-flow technique to find an approximate expression for the average data queue time delay in the overload region of a movable-boundary scheme with $N = 2, N_1 = N_2 = 1$ slots (channels) per frame. Show that this is given by Eq. (12–165). Carry out all details of the analysis.

12–24 Consider the movable-boundary scheme with N slots/frame, one of which ($N_1 = 1$) is dedicated to class-1 (blocked) users. Show, using the fluid-flow technique, that the approximate class-2 average time delay is given by Eq. (12–166).

12–25 This problem is a project that could be used as a term project or could be carried out jointly by a group of students. Refer to the papers by Akinpelu on nonhierarchical circuit-switched routing discussed in this chapter ([AKIN 1983], [AKIN 1984]). [AKIN 1983] in particular develops equations for the performance of nonhierarchical alternately routed schemes in nonsymmetric networks with and without trunk reservation controls applied. Choose a reasonably sized nonsymmetric network (of at least 10 nodes, say) with a homogeneous traffic load, for simplicity, and carry out the analysis suggested, calculating carried load versus offered load. Vary the number of trunks per link as well as the number of trunks reserved. Determine the minimum number of reserved trunks required in each case to eliminate the potential deterioration in performance as the offered load increases to high (overload) values. Compare the results to those obtained by Akinpelu, as well as those obtained by Krupp for a symmetrical network [KRUP].

Can you think of *other* control strategies, possibly adaptive, to reduce the negative aspects of alternate routing at high load? Try to analyze these strategies, using symmetrical networks if necessary. (See [YUM] for examples.)

12–26 The following projects extend the discussion on the movable-boundary scheme in the text.

    a. Refer to the papers referenced at the end of this book on the incorporation of DSI (digital speech interpolation) techniques into a movable-boundary strategy. These include [OREI], [WILL], and [FISC 1979]. Try to extend the analysis of the simple movable-boundary examples provided in this chapter to include the effect of DSI. Carry out the analysis for larger systems if possible. Compare with the results of [SRIR] using a discrete-time model.

    b. Refer to the paper by Sriram et al. [SRIR]. Study this paper and others cited that use the discrete-time model for the movable-boundary scheme. Work out some simple examples and compare with results of the continuous-time model discussed in this chapter.

References

[ABRA] N. Abramson, "The Aloha System," in *Computer Networks*, N. Abramson and F. Kuo, eds., Prentice-Hall, Englewood Cliffs, N.J., 1973.

[AHO] A. V. Aho, J. E. Hopcroft, and J. D. Ullman, *The Design and Analysis of Computer Algorithms*, Addison-Wesley, Reading, Mass., 1974.

[AHUJ] V. Ahuja, "Routing and Flow Control in Systems Network Architecture," *IBM Syst. J.*, vol. 18, no. 2, 1979, 298–314.

[AKIN 1983] J. M. Akinpelu, "The Overload Performance of Engineered Networks with Hierarchical and Nonhierarchical Routing," ITC-10, Tenth International Teletraffic Congress, Montreal, June 1983, Session 3.2, paper 4.

[AKIN 1984] J. M. Akinpelu, "The Overload Performance of Engineered Networks with Nonhierarchical and Hierarchical Routing," *AT&T Bell Labs. Technical J.*, vol. 63, no. 7, Sept. 1984, 1261–1281.

[ANDR 1981] F. T. Andrews, Jr., and W. B. Smith, "No. 5 ESS—Overview," ISS '81, Montreal, 1981.

[ANDR 1984] F. T. Andrews, Jr., "ISDN '83," special issue on ISDNs, *IEEE Comm. Mag.*, vol. 22, no. 1, Jan. 1984, 6–10.

[ANSI] *American National Standard Specification of Signaling System No. 7*, American National Standards Inst., Inc., Issue 1, Draft document, Jan. 1985. (This document is based on the 1984 CCITT specification as modified for use within the United States.)

[ASH] G. R. Ash, R. H. Cardwell, R. P. Murray, "Design and Optimization of Networks with Dynamic Routing," *BSTJ*, vol. 60, no. 8, Oct. 1981, 1787–1820.

[ATKI] J. D. Atkins, "Path Control: The Transport Network of SNA," *IEEE Trans. on Comm.*, vol. COM-28, no. 4, April 1980, 527–538; reprinted in [GREE].

[AT&T] *Notes on the Network, Bell System Practices*, AT&T Co. Standard, Sec. 781-030-100, Issue 2, Dec. 1980, Sec. 5, Signaling, Figs. 1 and 2.

[BARA] P. Baran, "On Distributed Communication Networks," *IEEE Trans. on Comm. Syst.*, vol. CS-12, no. 1, March 1964, 1–9.

[BART] Paul D. Bartoli, "The Application Layer of the Reference Model of Open Systems Interconnection," *Proc. IEEE*, vol. 71, no. 12, Dec. 1983, 1404–1407.

[BASI] R. G. Basinger et al., "Calling Card Service—Overall Description and Operational Characteristics," *BSTJ*, vol. 61, no. 7, part 3, Sept. 1982, 1655–1674.

[BASK] F. Baskett et al., "Open, Closed, and Mixed Networks of Queues with Different Classes of Customers," *J. ACM,* vol. 22, no. 2, 1975, 248–260.

[BAUM] S. M. Baumann et al., "No. 5 ESS Software Design," ISS '81, Montreal, 1981.

[BELL] John C. Bellamy, *Digital Telephony,* John Wiley & Sons, New York, 1982.

[BENE] V. E. Benes, "Programming and Control Problems Arising from Optimal Routing in Telephone Networks," *BSTJ,* vol. 45, no. 9, Nov. 1966, 1373–1438.

[BERT] H. V. Bertine, "Physical Level Protocols," *IEEE Trans. on Comm.,* vol. COM-28, no. 4, April 1980, 433–444; reprinted in [GREE].

[BEST] M. Best, "Optimization of Nonlinear Performance Criteria Subject to Flow Constraints," 18th Midwest Symposium on Circuits and Systems, Concordia University, Quebec, Aug. 1975, 438–443.

[BHAT] U. N. Bhat and M. J. Fischer, "Multichannel Queueing Systems with Heterogeneous Classes of Arrivals," *Naval Res. Logist. Quarterly,* vol. 23, no. 2, 1976, 271–283.

[BOSC] H. L. Bosco et al., "No. 5 ESS—Hardware Design," ISS '81, Montreal, 1981.

[BROU] A. Broux and M. Verbeck, "Metaconta 10 CN Exchanges: A New Generation of Switching Systems," *Electrical Comm.,* vol. 53, no. 1, 1978, 2–8.

[BRUC] R. A. Bruce, P. K. Giloth, E. A. Siegel, Jr., "No. 4 ESS—Evolution of a Digital Switching System," special issue, "Digital Switching," *IEEE Trans. on Comm.,* vol. COM-27, no. 7, July 1979, 1001–1011; reprinted in [JOEL].

[BSTJ 1977] "No. 4 ESS," special issue, *BSTJ,* vol. 56, no. 7, Sept. 1977.

[BSTJ 1978] "Common Channel Interoffice Signaling," special issue, *BSTJ,* vol. 57, no. 2, Feb. 1978.

[BSTJ 1981] "No. 4 Electronic Switching System: System Evaluation," *BSTJ,* vol. 60, no. 6, part 2, July-Aug. 1981.

[BSTJ 1982] "Stored Program Controlled Network," special issue, *BSTJ,* vol. 61, no. 7, part 3, Sept. 1982.

[BURR] *Burroughs Network Architecture (BNA), Architectural Description, Reference Manual,* vol. 1, Burroughs Corp., Detroit, Mich., April 1981.

[BURT] H. O. Burton and D. D. Sullivan, "Errors and Error Control," *Proc. IEEE,* vol. 60, no. 11, Nov. 1972, 1293–1301.

[BUX 1980] W. Bux, K. Kummerle, and H. L. Truong, "Balanced HDLC Procedures: A Performance Analysis," *IEEE Trans. on Comm.,* vol. COM-28, no. 11, Nov. 1980, 1889–1898.

[BUX 1981a] W. Bux et al., "A Reliable Token-Ring System for Local Area Communication," National Telecommunications Conference, New Orleans, 1981, pp. A.2.2.1–A2.2.6.

[BUX 1981b] W. Bux, "Local-Area Subnetworks: A Performance Comparison," *IEEE Trans. on Comm.,* vol. COM-29, no. 10, Oct. 1981, 1465–1473.

[BUZE] J. P. Buzen, "Computational Algorithms for Closed Queueing Networks with Exponential Servers," *Comm. ACM,* vol. 16, No. 9, Sept. 1973, 527–531,

[CALL] Ross Callon, "Internetwork Protocol," *Proc. IEEE,* vol. 71, no. 12, Dec. 1983, 1388–1393.

[CAME] W. H. Cameron et al., "Dynamic Routing for Intercity Telephone Networks," ITC-10, Tenth International Teletraffic Congress, Montreal, June 1983, Session 3.2, paper 3.

[CANT] D. G. Cantor and M. Gerla, "Optimal Routing in a Packet-switched Computer Network," *IEEE Trans. on Computers,* vol. C-23, no. 10, Oct. 1974, 1062–1069.

[CARLA] A. B. Carleial and M. E. Helman, "Bistable Behavior of Aloha-type Systems," *IEEE Trans. on Comm.,* vol. COM-23, no. 4, April 1975, 401–410.

[CARLD] D. E. Carlson, "Bit-oriented Data Link Control Procedures," *IEEE Trans. on Comm.,* vol. COM-28, no. 4, April 1980, 455–467; reprinted in [GREE].

[CCITT] *CCITT Draft Recommendation X.200, Reference Model of Open Systems Interconnection for CCITT Applications,* Geneva, June 1983.

[CCS7] *CCITT Specification for Signaling System No. 7,* CCITT, Geneva, 1984.

[CEGR] T. Cegrell, "A Routing Procedure for the TIDAS Message-switching Network," *IEEE Trans. on Comm.* vol. COM-23, no. 6, June 1975, 575–585.

[CHAN] K. M. Chandy, U. Herzog, and L. S. Woo, "Parametric Analysis of Queueing Networks," *IBM J. Research and Development,* vol. 19, no. 1, Jan. 1975, 43–49.

[CHAP] A. Lyman Chapin, "Connections and Connectionless Data Transmission," *Proc. IEEE,* vol. 71, no. 12, Dec. 1983, 1365–1371.

[CLOS] C. Clos, "A Study of Nonblocking Switching Networks," *BSTJ,* vol. 32, no. 2, March 1953, 406–424.

[CLOSS] F. Closs, "Message Delays and Trunk Utilization in Line-switched Networks," ISS '72, Tokyo, 1972.

[COLL] A. A. Collins and R. D. Pedersen, *Telecommunications: A Time for Innovation,* Merle Collins Foundation, Dallas, 1973.

[CONA] J. W. Conard, "Services and Protocols of the Data Link Layer," *Proc. IEEE,* vol. 71, no. 12, Dec. 1983, 1378–1383.

[COVI] G. Coviello and P. Vena, "Integration of Circuit/Packet Switching by a SENET (Slotted Envelope Network) Concept," National Telecommunications Conference, New Orleans, Dec. 1975, 42–12—42–17.

[COX] D. R. Cox and H. D. Miller, *The Theory of Stochastic Processes,* Methuen, London, 1965.

[DANE] A. Danet et al., "The French Public Packet-switching Service: The Transpac Network," ICCC, Toronto, Aug. 1976, 251–260.

[DAVI] J. H. Davis et al., "No. 5 ESS System Architecture," ISS '81, Montreal, 1981.

[DAY] John D. Day and Hubert Zimmermann, "The OSI Reference Model," *Proc. IEEE,* vol. 71, no. 12, Dec. 1983, 1334–1341.

[DEAT] George Deaton, "Multi-access Computer Nets: Some Design Decisions," *Data Comm.,* vol. 13, no. 14, Dec. 1984, 123–136.

[DEC] *DECnet, Digital Network Architecture, Routing Layer Functional Specification,* Version 2.00, Digital Equipment Corp., Maynard, Mass., May 1, 1983.

[DIJK] E. W. Dijkstra, "A Note on Two Problems in Connection with Graphs," *Numer. Math.,* vol. 1, 1959, 269–271.

[DOSH 1982] B. T. Doshi and H. Heffes, "Comparison of Control Schemes for a Class of Distributed Systems," 21st IEEE Conf. on Decision and Control, Orlando, Fla., Dec. 1982.

[DOSH 1983] B. T. Doshi and H. Heffes, "Analysis of Overload Control Schemes for a Class of Distributed Switching Machines," ITC-10, Tenth International Teletraffic Congress, Montreal, June 1983, Session 5.2, paper 2.

[DUC] N. Q. Duc and E. K. Chew, "ISDN Protocol Architecture," *IEEE Comm. Mag.,* vol. 23, no. 3, March 1985, 15–22.

[DUNC] T. Duncan and W. H. Huen, "Software Structure of No. 5 ESS—A Distributed Telephone Switching System," *IEEE Trans. on Comm.,* vol. COM-30, No. 6, June 1982, 1379–1385.

[EISE] M. Eisenberg, "A Strict Priority Queueing System with Overload Control," ITC-10, Tenth International Teletraffic Congress, Montreal, June 1983, Session 1.3.

[ELLI] M. L. Ellis et al., "INDAX: An Operational Interactive Cabletext System," *IEEE J. on Selected Areas in Comm.,* vol. SAC-1, no. 2, Feb. 1983, 285–294.

[EMMO] Willard F. Emmons and A. S. Chandler, "OSI Session Layer: Services and Protocols," *Proc. IEEE,* vol, 71, no. 12, Dec, Dec. 1983, 1397–1400.

[ETHE] *The Ethernet, A Local Area Network, Data Link Layer and Physical Layer Specifications,* Digital Equipment Corp., Maynard, Mass.; Intel Corp., Santa Clara, Calif.; Xerox Corp., Stamford, Conn.; Version 1.0, Sept. 30, 1980, and Version 2.0, Nov. 1982.

[FARM] W. D. Farmer and E. E. Newhall, "An Experimental Distributed Switching System to Handle Bursty Computer Traffic, " ACM Symposium, Problems on Optimization of Data Communications, Pine Mountain, Ga, Oct. 1963, 31–34.

[FAYO] G. Fayolle et al., "Stability and Optimal Control of the Packet Switching Broadcast Channels," *J. ACM,* vol. 24, July 1977, 375–380.

[FISC 1977] M. J. Fischer, "A Queueing Analysis of an Integrated Telecommunications System with Priorities," *INFOR,* vol. 15, no. 3, Oct. 1977, 277–288.

[FISC 1979] M. J. Fischer, "Data Performance in a System where Data Packets Are Transmitted during Voice Silent Periods—Single Channel Case," *IEEE Trans. on Comm.,* vol. COM-27, no. 9, Sept. 1979, 1371–1375.

[FLET 1978] J. G. Fletcher and R.W. Watson, "Mechanisms for a Reliable Timer-based Protocol," *Computer Networks,* vol. 2, no. 4/5 1978, 271–290.

[FLET 1982] J. G. Fletcher, "An Arithmetic Checksum for Serial Transmission," *IEEE Trans. on Comm.,* vol. COM-30, no. 1, Jan. 1982, 247–252.

[FOLT 1980a] H. C. Folts, "Procedures for Circuit-switched Service in Synchronous Public Data Networks," *IEEE Trans. on Comm.,* vol. COM-28, no. 4, April 1980, 489–495; reprinted in revised form in [GREE].

[FOLT 1980b] H. C. Folts, "X.25 Transaction-Oriented Features—Datagram and Fast-Select," *IEEE Trans. on Comm.,* vol. COM-28, no. 4, April 1980, 496–499; reprinted in revised form in [GREE].

[FORD] L. R. Ford, Jr., and D. R. Fulkerson, *Flows in Networks,* Princeton University Press, Princeton, N.J., 1962.

[FORY] L. J. Forys, "Performance Analysis of a New Overload Strategy," ITC-10, Tenth International Teletraffic Congress, Montreal, June 1983, Session 5.2, paper 4.

[FRAN] H. Frank and I. T. Frisch, *Communication, Transmission, and Transportation Networks,* Addison-Wesley, Reading, Mass., 1971.

[FRAT] L. Fratta, M. Gerla, and L. Kleinrock, "The Flow Deviation Method: An Approach to Store-and-Forward Communication Network Design," *Networks,* vol. 3, no. 2, 1973, 97–133.

[FRER] R. F. Frerking and M. A. McGrew, "Routing of Direct-Signaling Messages in the CCIS Network," special issue, *BSTJ,* vol. 61, no. 7, part 3, Sept. 1982, 1599–1609.

[GALI] R. Galimberti, G. Perucca, P. Semprini, "Proteo System: An Overview," ISS '81, Montreal, Sept. 1981.

[GARL] L. L. Garlick, R. Rom, J. B. Postel, "Reliable Host-to-Host Protocols: Problems and Techniques," Fifth Data Communications Symposium, Snowbird, Utah, Sept. 1977, 4–58 — 4–65.

[GAVE] D. P. Gaver and P. P. Lehoczky, "Channels that Cooperatively Service a Data Stream and Voice Messages," *IEEE Trans. on Comm.*, vol. COM-30, no. 5, May 1982, 1153–1162.

[GERL] M. Gerla, *The Design of Store-and-Forward Networks for Computer Communications*, Ph.D. dissertation, Dept. of Computer Science, UCLA, 1973.

[GIES] A. Giessler et al., "Flow Control Based on Buffer Classes," *IEEE Trans. on Comm.*, vol. COM-29, no. 4, April 1981, 436–443.

[GIMP] L. A. Gimpelson, "Analysis of Mixtures of Wide- and Narrow-band Traffic," *IEEE Trans. on Comm. Tech.*, vol. 13, no. 3, Sept. 1965, 258–266.

[GORD] W. L. Gordon and G. F. Newell, "Closed Queueing Systems with Exponential Servers," *Operations Research*, vol. 15, no. 2, 1967, 254–265.

[GRAN] C. Grandjean, "Call Routing Strategies in Telecommunication Networks," ITC-5, Fifth International Teletraffic Congress, New York, June 1967, 261–269.

[GREE] P. Green, ed., *Computer Network Architectures and Protocols*, Plenum Press, New York, 1982.

[GROE] I. Groenback, *The TCP and ISO Transport Service—A Brief Description and Comparison*, NATO Technical Memorandum STC TM-726, SHAPE Technical Center, The Hague, Netherlands, Feb. 1984.

[GTET] *Functional Description of GTE Telenet Packet Switching Networks*, NFD-005.014, GTE Telenet Communications Corp, Vienna, Va., May 1982.

[GUNT] K. D. Günther, "Prevention of Deadlocks in Packet-switched Data Transport Systems," *IEEE Trans. on Comm.*, vol. COM-29, no. 4, April 1981, 512–524.

[HAEN] D. G. Haenschke, D. A. Kettler, E. Oberer, "DNHR: A New SPC/CCIS Network Management Challenge," ITC-10, Tenth International Teletraffic Congress, Montreal, June 1983, Session 3.2, paper 5.

[HAGO] J. Hagouel, *Issues in Routing for Large and Dynamic Networks*, PhD. dissertation, Columbia University, New York, 1983.

[HALF] S. Halfin and M. Segal, "A Priority Queueing Model for a Mixture of Two Types of Customers," *SIAM J. Appl. Math.*, vol. 23, no. 3, Nov. 1972, 369–379.

[HAYE] J. F. Hayes, *Modeling and Analysis of Computer Communications Networks*, Plenum Press, New York, 1984.

[HILL] M. J. Hills, *Telecommunications Switching Principles*, MIT Press, Cambridge and London, 1979.

[HITC] L. E. Hitchner, *A Comparative Study of the Computational Efficiency of Shortest Path Algorithms*, Univ. of California, Berkeley, Operations Research Center Report ORC 68-25, Nov. 1968.

[HLAW] F. Hlawa and A. Stoll, "Common Channel Signaling Based on CCITT System No. 7," *Telephony*, Feb. 9, 1981.

[HOBE] V. L. Hoberecht, "SNA Function Management," *IEEE Trans. on Comm.*, vol. COM-28, no. 4, April 1980; reprinted in [GREE].

[HOLL] Lloyd L. Hollis, "OSI Presentation Layer Activities," *Proc. IEEE,* vol. 71, no. 12, Dec. 1983, 1401–1403.

[HSIE] W. Hsieh and B. Kraimeche, "Performance Analysis of an End-to-End Flow Control Mechanism in a Packet-switched Network," *J. Telecomm. Networks,* vol. 2, 1983, 103–116.

[IBM] "Systems Network Architecture," special issue, *IBM Systems J.,* vol. 22, no. 4, 1983, 295–466.

[IEEE 1979] "Digital Switching," special issue, *IEEE Trans. on Comm.* vol. COM-27, no. 7, July 1979.

[IEEE 1982] "Communication Software," special issue, *IEEE Trans. on Comm.,* vol. COM-30, no. 6, June 1982.

[IEEE 1983a] "Local Area Networks," special issue, *IEEE J. on Selected Areas in Comm.,* vol. SAC-1, no. 5, Nov. 1983.

[IEEE 1983b] "Open Systems Interconnection (OSI)—New International Standards Architecture and Protocols for Distributed Information Systems," H. C. Folts and R. desJardins, eds., *Proc. IEEE,* vol. 71, no. 12, Dec. 1983.

[IEEE 1983c] *IEEE Standard 802.3, CSMA/CD Access Method and Physical Layer Specification,* IEEE Project 802, Local Area Network Standards, IEEE, New York, July 1983.

[IEEE 1984] *Draft E, IEEE Standard 802.5, Token Ring Access Method and Physical Layer Specifications,* IEEE Project 802, Local Area Network Standards, IEEE, New York, Aug. 1, 1984.

[INOS] H. Inose, *An Introduction to Digital Integrated Communication Systems,* University of Tokyo Press, Tokyo, 1979.

[IP] *Internet Protocol, Military Standard,* MIL-STD-1777, U.S. Department of Defense, May 20, 1983.

[IRLA] M. I. Irland, "Buffer Management in a Packet Switch," *IEEE Trans. on Comm.,* vol. COM-26, no. 3, March 1978, 328–337.

[ISO 1979] *Data Communication—High Level Data Link Control Procedures—Elements of Procedures, ISO International Standard 4335,* International Organization for Standardization, Geneva, 1979.

[ISO 1983] *ISO International Standard 7498, Information Processing Systems—Open Systems Interconnection—Basic Reference Model,* International Organization for Standardization, Geneva, Oct. 1983.

[ISO 1984a] *ISO International Standard 8072, Information Processing Systems—Open Systems Interconnection—Transport Service Definition,* International Organization for Standardization, Geneva, 1984.

[ISO 1984b] *ISO International Standard 8073, Information Processing Systems—Open Systems Interconnection—Transport Protocol Specification,* Geneva, 1984; appears in *Computer Comm. Rev.,* vol. 12, nos. 3 and 4, July/Oct. 1982.

[ITC 9] ITC-9, Ninth International Teletraffic Congress, June 1979.

[ITC 10] Session 5.2, "Overload Control," ITC-10, Tenth International Teletraffic Congress, Montreal, June 1983.

[JACK] J. R. Jackson, "Job Shop-like Queueing Systems," *Management Science,* vol. 10, no. 1, 1963, 131–142.

[JACO] I. M. Jacobs et al., "Packet Satellite Network Design Issues," National Telecommunications Conference, Nov. 1979, 45.2.1–45.2.12.

[JAFF] J. M. Jaffe and F. H. Moss, "A Responsive Distributed Routing Algorithm for Computer Networks," *IEEE Trans. on Comm.,* vol. COM-30, no. 7, July 1982, 1758–1762.

[JOEL] Amos E. Joel, Jr., ed., *Electronic Switching: Digital Central Office Systems of the World,* IEEE Press, New York, 1982.

[JOHN] J. W. Johnson et al., "Integrated Digital Services of the 5ESS System," ISS '84, Florence, Italy, May 1984, Paper 14A3.

[KAHN 1978] R. E. Kahn et al., "Advances in Packet Radio Technology," *Proc. IEEE,* vol. 66, no. 11, Nov. 1978, 1468–1496.

[KAHN 1979] R. E. Kahn, "The Introduction of Packet Satellite Communications," National Telecommunications Conference, Nov. 1979, 45.1.1–45.1.8.

[KARN] M. Karnaugh, "Loss of Point-to-Point Traffic in Three-Stage Circuit Switches," *IBM J. Research and Development,* vol. 18, no. 3, May 1974, 204–216; addendum, Sept. 1974, 465.

[KELL] F. P. Kelly, *Reversibility and Stochastic Networks,* John Wiley & Sons, Chichester, U.K., 1979.

[KLEI 1965] L. Kleinrock, "A Conservation Law for a Wide Class of Queueing Systems," *Naval Res. Logist. Quarterly,* vol. 12, 1965, 181–192.

[KLEI 1972] L. Kleinrock, *Communication Nets: Stochastic Message Flow and Delay,* McGraw-Hill, New York, 1964; reprinted, Dover Publications, 1972.

[KLEI 1975a] L. Kleinrock, *Queueing Systems. Volume 1: Theory,* John Wiley & Sons, New York, 1975.

[KLEI 1975b] L. Kleinrock and S. S. Lam, "Packet Switching in a Multiaccess Broadcast Channel: Performance Evaluation," *IEEE Trans. on Comm.,* vol. COM-23, no. 4, April 1975, 410–423.

[KLEI 1976] L. Kleinrock, *Queueing Systems. Volume 2: Computer Applications,* John Wiley & Sons, New York, 1976.

[KLEI 1977] L. Kleinrock and F. Kamoun, "Hierarchical Routing for Large Networks: Performance Evaluation and Optimization," *Computer Networks,* vol. 1, no. 3, 1977, 155–174.

[KLEI 1980] L. Kleinrock and F. Kamoun, "Optimal Clustering Structures for Hierarchical Topological Design of Large Computer Networks," *Networks,* vol. 10, no. 3, 1980, 221–248.

[KNIG] K. G. Knightson, "The Transport Layer Standardization," *Proc. IEEE,* vol. 71, no. 12, Dec. 1983, 1394–1396.

[KOBA] H. Kobayashi, *Modeling and Analysis: An Introduction to System Performance Evaluation Methodology,* Addison-Wesley, Reading, Mass., 1978.

[KOHN 1974] A. G. Konheim and B. Meister, "Waiting Lines and Times in a System with Polling," *J. ACM,* vol. 21, no. 3, July 1974, 470–490.

[KOHN 1980] A. G. Konheim, "A Queueing Analysis of Two ARQ Protocols," *IEEE Trans. on Comm.,* vol. COM-28, no. 7, July 1980, 1004–1014.

[KOST] D. J. Kostas, "Transition to ISDN — An Overview," special issue on ISDNs, *IEEE Comm. Mag.,* vol. 22, no. 1, Jan. 1984, 11–17.

[KRAI 1984] B. Kraimeche and M. Schwartz, "Circuit Access Control Strategies in Integrated Digital Networks," Infocom '84, San Francisco, Calif., April 1984.

[KRAI 1985] B. Kraimeche and M. Schwartz, "Analysis of Traffic Access Control Strategies in Integrated Service Networks," *IEEE Trans. on Comm.,* vol. COM-33, no. 10, Oct. 1985, 1085–1093.

[KRON] R. Kronz, S. Lee, and M. Sun. "Practical Design Tools for Large Packet-switched Networks," Infocom '83, San Diego, Calif., April 1983.

[KRUP] R. S. Krupp, "Stabilization of Alternate Routing Networks," ICC, Philadelphia, June 1982, 31.2.1–31.2.5.

[KUMM 1974] K. Kummerle, "Multiplexer Performance for Integrated Line-and Packet-switched Traffic," ICCC, Stockholm, 1974, 508–515.

[KUMM 1978] K. Kummerle and H. Rudin, "Packet and Circuit Switching: Cost/Performance Boundaries," *Computer Networks,* vol. 2, no. 1, Feb. 1978, 3–17.

[KURO] J. Kurose, M. Schwartz, Y. Yemini, "Multiple-access Protocols and Time-constrained Communication," *Computing Surveys,* vol. 16, no. 1, March 1984, 43–70.

[LAM 1974] S. S. Lam, *Packet Switching in a Multi-Access Broadcast Channel,* Ph.D. dissertation, Dept. of Computer Science, UCLA, April 1974.

[LAM 1975] S. S. Lam and L. Kleinrock, "Packet Switching in a Multiaccess Broadcast Channel: Dynamic Control Procedures," *IEEE Trans. on Comm.,* vol. COM-23, no. 9, Sept. 1975, 891–904.

[LAM 1977] S. S. Lam, "Queueing Networks with Population Constraints," *IBM J. Research and Development,* vol. 21, July 1977, 370–378.

[LAM 1979] S. S. Lam and M. Reiser, "Congestion Control of Store-and-Forward Networks by Input Buffer Limits: An Analysis," *IEEE Trans. on Comm.,* vol. COM-27, no. 1, Jan. 1979, 127–134.

[LAM 1980] S. S. Lam, "A Carrier Sense Multiple Access Protocol for Local Networks," *Computer Networks,* vol. 4, no. 1, Jan. 1980, 21–32.

[LANG] A. Langford, K. Naemura, R. Speth, "OSI Management and Job Transfer Services," *Proc. IEEE,* vol. 71, no. 12, Dec. 1983, 1420–1424.

[LAVE] S. S. Lavenberg and M. Reiser, "Stationary State Probabilities of Arrival Instants for Closed Queueing Networks with Multiple Types of Customers," *J. Appl. Prob.,* vol. 17, 1980, 1048–1061.

[LAZA 1983] A. A. Lazar, "Optimal Flow Control of a Class of Queueing Networks in Equilibrium," *IEEE Trans. on Automatic Control,* vol. AC-28, no. 8, Aug. 1983, 1001–1007.

[LAZA 1984] A. A. Lazar and T. G. Robertazzi, "The Geometry of Lattices for Markovian Queueing Networks," *Columbia Research Report,* July 1984.

[LEE] C. Y. Lee, "Analysis of Switching Networks," *BSTJ,* vol. 34, no. 6, Nov. 1955, 1287–1315.

[LEON] A. Leon-Garcia, R. H. Kwong, G. F. Williams, "Performance Evaluation Methods for an Integrated Voice-Data Link," *IEEE Trans. on Comm.,* vol. COM-30, no. 8, Aug. 1982, 1848–1858.

[LEWA] Douglas Lewan and H. Garret Long, "The OSI File Service," *Proc. IEEE,* vol. 71, no. 12, Dec. 1983, 1414–1419.

[LINI] Peter F. Linington, "Fundamentals of the Layer Service Definitions and Protocol Specifications," *Proc. IEEE,* vol. 71, no. 12, Dec. 1983, 1341–1345.

[LINP] P. M. Lin, B. J. Leon, C. R. Stewart, "Analysis of Circuit-switched Networks Employing Originating Office Control with Spill Forward," *IEEE Trans. on Comm.,* vol. COM-26, no. 6, June 1978, 754–765.

[LINS] Shu Lin and Daniel J. Costello, Jr., *Error Control Coding: Fundamentals and Applications*, Prentice-Hall, Englewood Cliffs, N.J., 1983.

[LITT] J. D. C. Little, "A Proof of the Queueing Formula $L = \lambda W$," *Operations Res.*, vol. 9, no. 3, 1961, 383–387.

[LOWE] Henry Lowe, "OSI Virtual Terminal Service," *Proc. IEEE*, vol. 71, no. 12, Dec. 1983, 1408–1413.

[MAXE] N. F. Maxemchuk and A. N. Netravali, "A Multifrequency Multiaccess System for Local Access," ICC, June 1983.

[McCL] F. McClelland, "Services and Protocols of the Physical Layer," *Proc. IEEE*, vol. 71, no. 12, Dec. 1983, 1372–1377.

[McDO] John C. McDonald, ed., *Fundamentals of Digital Switching*, Plenum Press, New York, 1983.

[McGA] T. P. McGarty and G. J. Clancy, Jr., "Cable-based Metro Area Networks," *IEEE J. on Selected Areas in Comm.*, vol. SAC-1, no. 5, Nov. 1983, 816–831.

[McQU 1974] J. M. McQuillan, *Adaptive Routing Algorithms for Distributed Computer Networks*, BBN Report 2831, May 1974 (also available as Ph.D. dissertation, Division of Engineering and Applied Physics, Harvard University, 1974).

[McQU 1978] J. M. McQuillan, G. Falk, and I. Richer, "A Review of the Development and Performance of the ARPAnet Routing Algorithm," *IEEE Trans. on Comm.*, vol. COM-26, no. 12, Dec. 1978, 1802–1811.

[McQU 1980] J. M. McQuillan et al., "The New Routing Algorithm for the ARPA-NET," *IEEE Trans. on Comm.*, vol. COM-28, no. 5, May 1980, 711–719.

[MERL] P. M. Merlin and A. Segall, "A Fail Safe Distributed Routing Protocol," *IEEE Trans. on Comm.*, vol. COM-27, no. 9, Sept. 1979, 1280–1287.

[METC] R. M. Metcalfe and D. R. Boggs, "Ethernet: Distributed Packet Switching for Local Computer Networks," *Comm. ACM*, vol. 19, no. 7, July 1976, 395–404.

[MONT] S. Dal Monte and J. Israel, "Proteo System-UT10/3: A Combined Local and Full Exchange," ISS '81, Montreal, Sept. 1981.

[NAKA] Y. Nakagome and H. Mori, "Flexible Routing in the Global Communication Network," ITC-7, Seventh International Teletraffic Congress, Stockholm, June 1973, paper 426.

[NBS 1983a] *Specification of a Transport Protocol for Computer Communications, Volume 1: Overview and Services*, Inst. for Computer Sciences and Technology, National Bureau of Standards, Gaithersburg, Md., Jan. 1983.

[NBS 1983b] *Specification of a Transport Protocol for Computer Communications, Volume 2: Class 2 Protocol*, Inst. for Computer Sciences and Technology, National Bureau of Standards, Gaithersburg, Md., Feb. 1983.

[NBS 1983c] *Specification of a Transport Protocol for Computer Communications, Volume 3: Class 4 Protocol*, Inst. for Computer Sciences and Technology, National Bureau of Standards, Gaithersburg, Md., Feb. 1983.

[NBS 1983d] *Specification of a Transport Protocol for Computer Communications, Volume 4: Service Specifications*, Inst. for Computer Sciences and Technology, National Bureau of Standards, Gaithersburg, Md., Jan. 1983.

[NBS 1983e] *Specification of a Transport Protocol for Computer Communications, Volume 5: Guidance for the Implementor*, Inst. for Computer Sciences and Technology, National Bureau of Standards, Gaithersburg, Md., Jan. 1983.

[NBS 1983f] *Specification of a Transport Protocol for Computer Communications, Volume 6: Guidance for Implementation Selection,* Inst. for Computer Sciences and Technology, National Bureau of Standards, Gaithersburg, Md., June 1983.

[NESE] M. Nesenbergs and R. F. Linfield, "Three Typical Blocking Aspects of Access Area Teletraffic," *IEEE Trans. on Comm.,* vol. COM-28, no. 9, Sept. 1980, 1662–1667.

[NEWP] C. B. Newport and P. Kaul, "Communication Processors for Telenet's Third Generation Packet-switching Network," Eascon 77, Arlington, Va. Sept. 1977.

[NOBL] B. Noble and J. W. Daniel, *Applied Linear Algebra,* Prentice-Hall, Englewood Cliffs, N.J., 1977.

[OCCH] B. Occhiogrosso et al., "Performance Analysis of Integrated Switching Communications Systems," National Telecommunications Conference, Los Angeles, Dec. 1977, 12:4-1–12:4-11.

[OPDE] A. Opderbeck, J. H. Hoffmeier, R. L. Spitzer, "Software Architecture for a Microprocessor-based Packet Network," National Computer Conference, Anaheim, Calif., June 1978.

[OPPE] A. V. Oppenheim and A. S. Willsky, with I. T. Young, *Signals and Systems,* Prentice-Hall, Englewood Cliffs, N.J., 1983.

[OREI] P. O'Reilly, *A Fluid-Flow Approach to Performance Analysis of Integrated Voice-Data Systems with DSI,* GTE Laboratories, Waltham, Mass., April 1985.

[PAPO] A. Papoulis, *Probability, Random Variables, and Stochastic Processes,* 2d ed., McGraw-Hill, New York, 1984.

[PENN] M. C. Pennotti and M. Schwartz, "Congestion Control in Store and Forward Tandem Links," *IEEE Trans. on Comm.,* vol. COM-23, no. 12, Dec. 1975, 1434–1443.

[PERL] R. Perlman, "Fault-tolerant Broadcast of Routing Information," *Computer Networks,* vol, 7. no. 6, Dec. 1983, 395–405.

[PORT] E. Port et al., "A Network Architecture for the Integration of Circuit and Packet Switching," ICCC, Toronto, Aug. 1976, 505–514.

[POST] J. B. Postel, "Internetwork Protocol Approaches," *IEEE Trans. on Comm.,* vol. COM-28, no. 4, April 1980, 604–611; reprinted in [GREE].

[PRICE] W. L. Price, "Data Network Simulation Experiments at the National Physical Laboratory, 1968-1976," *Computer Networks,* vol, 1, 1977, 199–208.

[PROT] *Proteo UT 10/3 Electronic Digital Exchange,* Application Notes, Italtel Co., Milan, Italy.

[RACC] G. A. Raack, E. G. Sable, R. J. Stewart, "Customer Control of Network Services," *IEEE Comm. Mag.,* vol. 22, no. 10, Oct. 1984, 8–14.

[RAJA] A. Rajaraman, "Routing in TYMNET," European Computing Conference, London, May 1978.

[RAUB] E. Raubold and J. Haenle, "A Method of Deadlock-free Resource Allocation and Flow Control in Packet Networks," ICC, Toronto, Aug. 1976.

[REIS 1980] M. Reiser and S. S. Lavenberg, "Mean-value Analysis of Closed Multichain Queueing Networks," *J. ACM,* vol. 27, no. 2, April 1980, 313–322.

[REIS 1981] M. Reiser, "Mean-value Analysis and Convolution Method for Queue-Dependent Servers in Closed Queueing Networks," *Performance Evaluation,* vol. 1, 1981, 7–18.

[REIS 1982] M. Reiser, "Performance Evaluation of Data Communication Systems," *Proc. IEEE,* vol, 70, no. 2, Feb. 1982, 171–195.

[RIND] J. Rinde, "Routing and Control in a Centrally Directed Network," National Computer Conference, *AFIPS Conf. Proc.*, vol. 38, 1977, 211–216.

[ROSE 1980] E. C. Rosen, "The Updating Protocol of ARPANET'S New Routing Algorithm," *Computer Networks*, vol. 4, no. 1, Feb. 1980, 11–19.

[ROSE 1981] E. C. Rosen, "Vulnerabilities of Network Control Protocols: An Example," *Computer Comm. Rev.*, July 1981, 11–16.

[ROSN] Roy D. Rosner, "Circuit and Packet Switching: A Cost and Performance Trade-off Study," *Computer Networks*, vol. 1, no. 1, June 1976, 7–26.

[ROSS] M. Ross et al., "Design Approaches and Performance Criteria for Integrated Voice/Data Switching," *Proc. IEEE*, vol. 65, no. 9, Sept. 1977, 1283–1295.

[RUDI 1976] H. Rudin, "On Routing and 'Delta-Routing': A Taxonomy and Performance Comparison of Techniques for Packet-switched Networks," *IEEE Trans. on Comm.*, vol. COM-24, no. 1, Jan. 1976, 43–59.

[RUDI 1978] H. Rudin, "Studies in the Integration of Circuit and Packet Switching," ICC, Toronto, June 1978, 20.2.1–20.2.7.

[RYBC] Antony Rybczynski, "X.25 Interface and End-to-End Virtual Circuit Service Characteristics," *IEEE Trans on Comm.*, vol. COM-28, no. 4, April 1980, 500–510; reprinted in revised form in [GREE].

[SAADS] S. Saad and M. Schwartz, "Input Buffer Limiting Mechanisms for Congestion Control," ICC, Seattle, June 1980.

[SAADT] T. N. Saadawi and M. Schwartz, "Distributed Switching for Data Transmission over Two-Way CATV," *IEEE J. on Selected Areas in Comm.*, vol. SAC-3, no. 2, March 1985, 323–329.

[SABN] K. Sabnani and M. Schwartz, "Performance Evaluation of Multidestination (Broadcast) Protocols for Satellite Transmission," National Telecommunications Conference, New Orleans, Nov. 1981.

[SAUE] C. H. Sauer and K. M. Chandy, *Computer Systems Performance Modeling*, Prentice-Hall, Englewood Cliffs, N.J., 1981.

[SCHU] Gary D. Schultz, "Anatomy of SNA," *Computerworld*, vol. 15, no. 11a, March 18, 1981, 35–38.

[SCHW 1976] M. Schwartz and C. K. Cheung, "The Gradient Projection Algorithm for Multiple Routing in Message-switched Networks," *IEEE Trans. on Comm.*, vol. COM-24, no. 4, April 1976, 449–456.

[SCHW 1977] M. Schwartz, *Computer-Communication Network Design and Analysis*, Prentice-Hall, Englewood Cliffs, N.J., 1977.

[SCHW 1979] M. Schwartz and S. Saad, "Analysis of Congestion Control Techniques in Computer Communication Networks," *Proc. Symp. on Flow Control in Computer Networks*, Versailles, Fr., Feb. 1979; J. L. Grange and M. Gien, eds., North-Holland Publishing Co., Amsterdam, 113–130.

[SCHW 1980a] M. Schwartz, *Information Transmission, Modulation, and Noise*, 3d ed., McGraw-Hill, New York, 1980.

[SCHW 1980b] M. Schwartz and T. E. Stern, "Routing Techniques Used in Computer Communication Networks," *IEEE Trans. on Comm.*, vol. COM-28, no. 4, April 1980, 539–552; reprinted in [GREE].

[SCHW 1980c] M. Schwartz, "Routing and Flow Control in Data Networks," NATO Advanced Study Inst.: New Concepts in Multi-user Communications, Norwich, U.K., Aug. 4–16, 1980; Sijthoof and Nordhoof, Neth.

[SCHW 1982a] M. Schwartz, "Performance Analysis of the SNA Virtual Route Pacing Control," *IEEE Trans. on Comm.*, vol. COM-30, no. 1, Jan 1982, 172–184.

[SCHW 1982b] M. Schwartz and T-K. Yum, "Distributed Routing in Computer Communication Networks," 21st IEEE Conference on Decision and Control, Orlando, Fla., Dec. 1982.

[SCHW 1983] M. Schwartz and B. Kraimeche, "An Analytic Control Model for an Integrated Node," IEEE Infocom '83, San Diego, Calif., April 1983.

[SEGA] M. Segal, "A Preemptive Priority Model with Two Classes of Customers," ACM/IEEE Second Symposium on Problems in the Optimization of Data Communications Systems, Palo Alto, Calif., Oct. 1971, 168–174.

[SHEI] D. Sheinbein and R. P. Weber, "800 Service Using SPC Network Capability," *BSTJ*, vol. 61, no. 7, part 3, Sept. 1982, 1737–1744.

[SHIE] D. R. Shier, "On Algorithms for Finding the *k* Shortest Paths in a Network," *Networks*, vol. 9, no. 3, 1979, 195–214.

[SIMO] J. M. Simon and A. Danet, "Controle des ressources et principes du routage dans le reseau Transpac," *Proc. Symp. on Flow Control in Computer Networks*, Versailles, Fr., Feb. 1979; J. L. Grange and M. Gien, eds., North-Holland, Amsterdam, 33–44.

[SITA] *A Pocket Guide to SITA*, Information and Public Relations Department, SITA, Neuilly-sur-Seine, Fr., July 1983.

[SNA 1980] *Systems Network Architecture Format and Protocol Reference Manual: Architectural Logic*, SC30-3112-2, IBM, Research Triangle Park, N.C., 1980.

[SNA 1982] *Systems Network Architecture, Technical Overview*, GC30-3073-0, IBM, Research Triangle Park, N.C., 1982.

[SPRO] D. E. Sproule and F. Mellor, "Routing, Flow, and Congestion Control in the Datapac Network," *IEEE Trans. on Comm.*, vol. COM-29, no. 4, April 1981, 386–391.

[SRIR] K. Sriram, P. K. Varshney, J. G. Shanthikumar, "Discrete-Time Analysis of Integrated Voice/Data Multiplexers with and without Speech Activity Detectors," *IEEE J. on Selected Areas in Comm.*, vol. SAC-1, no. 6, Dec. 1983, 1124–1132.

[STAL] William Stallings, "Local Networks," *Computing Surveys*, vol. 16, no. 1, March 1984, 3–41.

[STER] T. E. Stern, "An Improved Routing Algorithm for Distributed Computer Networks," IEEE International Symposium on Circuits and Systems, Workshop on Large-scale Networks and Systems, Houston, Tex., Apr. 1980.

[STUD] P. von Studnitz, "Transport Protocols: Their Performance and Status in International Standardization," *Computer Networks*, vol. 7, 1983, 27–35.

[SUNS 1977] C. A. Sunshine, "Efficiency of Interprocess Communication Protocols for Computer Networks," *IEEE Trans. on Comm.*, vol. COM-25, no. 2, Feb. 1977, 287–293.

[SUNS 1978] C. A. Sunshine and Y. K. Dalal, "Connection Management in Transport Protocols," *Computer Networks*, vol. 2, 1978, 454–473.

[SYSK] R. Syski, *Introduction to Congestion Theory in Telephone Systems*, Oliver and Boyd, Edinburgh, 1960.

[SZYB] E. Szybicki, M. E. Lavigne, "The Introduction of an Advanced Routing System into Local Digital Networks and Its Impact on the Networks' Economy, Reliability, and Grade of Service," ISS '79, Paris, 1979.

[TAJI] W. D. Tajibnapis, "A Correctness Proof of a Topology Information Maintenance Protocol for Distributed Computer Networks," *Comm. ACM*, vol. 20, July 1977, 477–485.

[TAKE] S. Takemura, H. Kawashima, H. Nakjima, *Software Design for Electronic Switching Systems,* Peter Pereginus Ltd., Stevenage, U.K., 1979 (English edition edited by M. T. Hill).

[TAWA] K. Tawara et al., "Speech Path System for DTS-11 Digital Toll Switching System," *Review of the Electrical Comm. Laboratories,* NTT, Japan, vol. 30, no. 5, Sept. 1982, 771.

[TCP] *Transmission Control Protocol, Military Standard,* MIL-STD-1778, U.S. Department of Defense, May 20, 1983.

[TOBA] F. A. Tobagi, "Multiaccess Protocols in Packet Communication Systems," *IEEE Trans. on Comm.,* vol. COM-28, no. 4, April 1980, 468–488; reprinted in [GREE].

[TOWS] D. Towsley and J. K. Wolf, "On the Statistical Analysis of Queue Lengths and Waiting Times for Statistical Multiplexers with ARQ Retransmission Schemes," *IEEE Trans. on Comm.,* vol. COM-27, no. 4, April 1979, 693–702.

[TYME] L. Tymes, "Routing and Flow Control in TYMNET," *IEEE Trans. on Comm.,* vol. COM-29, no. 4, April 1981, 393–398.

[WALL] B. Wallstrom, "A Feedback Queue with Overload Control," ITC-10, Tenth International Teletraffic Congress, Montreal, June 1983, Session 1.3., paper 4.

[WARE] Christine Ware, "The OSI Network Layer: Standards to Cope with the Real World," *Proc. IEEE,* vol. 71, no. 12, Dec. 1983, 1384–1387.

[WECK] S. Wecker, "DNA: The Digital Network Architecture," *IEEE Trans. on Comm.,* vol. COM-28, no. 4, April 1980, 510–526; reprinted in [GREE].

[WEIN] C. J. Weinstein, M. L. Malpass, M. J. Fischer, "Data Traffic Performance of an Integrated Circuit- and Packet-Switched Multiplex Structure," *IEEE Trans. on Comm.,* vol. COM-28, no. 6, June 1980, 873–877.

[WEIR] D. F. Weir, J. B. Holmblad, A. C. Rothberg, "An X.75-based Network Architecture," ICCC, Dec. 1980, Atlanta, Ga., 741–750.

[WILK] R. S. Wilkov, "Analysis and Design of Reliable Computer Networks," *IEEE Trans. on Comm.,* vol. COM-20, no. 3, part II, June 1972, 660–678.

[WILL] G. F. Williams and A. Leon-Garcia, "Performance Analysis of Integrated Voice and Data Hybrid-switched Links," *IEEE Trans. on Comm.,* vol. COM-32, no. 6, June 1984, 695–706.

[X.25] *Draft Revised Recommendation X.25, Interface between Data Terminals Operating in the Packet Mode in Public Data Networks,* Study Group VII, CCITT, Geneva, Feb. 1980; reprinted in *Computer Comm. Rev.,* vol. 10, nos. 1 and 2, Jan./April 1980, 56–129.

[YAMA] T. Yamaguchi and M. Akiyama, "An Integrated Hybrid Traffic Switching System Mixing Preemptive Wideband and Waitable Narrowband Calls," *Electronics and Communications in Japan,* vol. 53-A, no. 5, 1970, 43–52.

[YUM] T-K Yum, *Routing in Nonhierarchical Networks,* Ph.D. dissertation, Columbia University, May 1985.

[ZAFI] P. Zafiropoulo, "Flexible Multiplexing for Networks Supporting Line-switched and Packet-switched Data Traffic," ICCC, Stockholm, 1974, 517–523.

[ZIMM] H. Zimmermann, "OSI Reference Model—The ISO Model of Architecture for Open Systems Interconnection," *IEEE Trans. on Comm.,* vol. COM-28, no. 4, April 1980, 425–432; reprinted in [GREE].

Glossary

ABM Asynchronous balanced mode of HDLC.

Access delay A performance measure of polling systems. It is measured from the time a data packet arrives at one of the stations sharing the common channel to the time the packet begins transmission.

Ack Abbreviation for acknowledgement.

ADCCP Abbreviation for "Advanced Data Communication Control Procedure." American version of HDLC.

AK Acknowledgement transport protocol data unit, OSI transport layer.

ANSI Abbreviation for "American National Standards Institute."

Application layer The uppermost one in the seven-layer hierarchy. It is the layer that ensures that two application processes, cooperating to carry out the desired information processing across a network, understand one another.

ARM Asynchronous response mode of HDLC.

ARPAnet A computer network developed by the Advanced Research Projects Agency of the U.S. Department of Defense.

ARQ Abbreviation for "Automatic Repeat Request."

BAC Balanced asynchronous class of HDLC.

Bifurcated routing Multiple path routing.

Bind triangle Sequence of messages transmitted in SNA when setting up a session.

Binary backoff procedure A procedure used to adjust the retransmission time in the CSMA/CD random access technique. It doubles the retransmission interval each time a collision is detected.

Bit stuffing A technique in HDLC or SDLC to eliminate the possibility that data might have the same sequence as a flag sequence by inserting a zero at the transmitter any time five 1's appear.

BIU Basic information unit in SNA.

Blocking probability The probability that customers are turned away in a queue or other service system.

BLU Basic link unit in SNA.

BTU Basic transmission unit in SNA.

Call congestion The ratio of calls lost due to a lack of system resources to the total number of calls over a long interval of time.

Call set-up time The time needed to set up an end-to-end dedicated path in circuit-switching systems.

CC Connection confirm transport protocol data unit, OSI transport layer.

CCITT Abbreviation for "International Telegraph and Telephone Consultative Committee."

CDT Abbreviation for "Credit Allocation." This is used for flow control in the OSI transport layer.

Circuit switching A technology that establishes a dedicated path between any pair or group of users attempting to communicate.

Common application service elements Part of the protocol in the application layer that is common to all processes and interfaces with the presentation layer.

Common channel signaling (CCS) Out-of-band signaling for circuit-switched networks. Control messages are carried over separate signaling channels.

Connectionless data service Service at a given layer of the OSI Reference Model in which there is no connection setup phase.

COS Class of service, SNA.

CR Connection request transport protocol data unit, OSI transport layer.

CSMA/CD Carrier sense, multiple access with collision detection. It is a multiaccess technique in which stations listen before transmitting. A transmitting station detecting a collision aborts its transmission.

Data link layer The layer that ensures correct transmission of information between adjacent nodes in a network.

Datagram service Service at the network layer in which successive packets may be routed independently from end to end. There is no call setup phase. Datagrams may arrive out of order.

DC Disconnect confirm transport protocol data unit, OSI transport layer.

DCE Data circuit terminating equipment through which the DTE is connected to a network.

Deadlock A situation in which traffic ceases to flow and throughput drops to zero.

Delay time The sum of waiting time and service time in a queue.

Differential Manchester encoding A modified version of Manchester coding. In this scheme the phase of successive binary intervals is switched.

DNA DEC's digital network architecture.

DR Disconnect request transport protocol data unit, OSI transport layer.

DT Data transport protocol data unit, OSI transport layer.

DTE Data-terminal equipment (user systems).

ECMA Abbreviation for "European Computer Manufacturers Association."

Erlang A unit widely used in circuit-switching systems to measure the load on the system. The product of call arrival rate and call holding (service) time.

Erlang-B distribution Erlang distribution of the first kind, or Erlang loss formula.

ERN Explicit route number in SNA.

Essentially nonblocking switches Those switches with switch-blocking probability much less than link-blocking probability.

Explicit route The end-to-end physical path in SNA into which the virtual route is mapped.

FCFS (FIFO) A service discipline of queueing systems, based on the first come first served rule.

Filler signal unit Signal unit of CCS used to transmit ack and nak information when messages are not available.

FIPS Abbreviation for "Federal Information Processing Standard."

Flow control Used to mean both congestion control and the control, by a receiver, to prevent overflow of its buffer.

FMD Abbreviation for "Function Management Data."

Forwarding data base Routing table containing the least cost outgoing link for each destination in DNA.

Frame Unit of data of the data link layer.

Full duplex Two-way simultaneous communication with either user transmitting at will.

Gateway An interface that provides the necessary protocol translation between disparate networks.

Graceful close Method used to terminate a connection at the transport layer with no loss of data.

Half duplex Two-way communication with only one user transmitting at a time.

HDLC Abbreviation for "High-level data link control." The international standard for data link control developed by ISO.

Hierarchical routing Multiple level routing. Used both in packet switching and circuit switching.

Hub polling One of the polling techniques. Permission to transmit is passed sequentially from one designated user to another.

ILS (input buffer limiting scheme) A flow control scheme that blocks overload locally generated arrivals by limiting their number at a buffer.

Inband signaling Control messages are sent over the same trunks as the information messages or calls.

Incarnation number A unique name or number sent within a data unit to avoid duplicate data unit acceptance.

IP Abbreviation for "Internetwork Protocol" in the ISO activities, as well as Internet Protocol in ARPA protocol activities.

ISDN Abbreviation for "Integrated Services Digital Networks." All-digital network handling a multiplicity of services with standard interfaces for user access.

ISO Abbreviation for "International Organization for Standardization."

LAPB Abbreviation for "Balanced link access procedure." The data link level of X.25. Same as a subset of the asychronous balanced mode of HDLC.

Local area network (LAN) A network covering small geographic areas.

Local loop The signal path between terminal and switch.

Logical channel number Virtual circuit identified at the packet level of X.25.

Logical link A virtual circuit concept of DNA, existing at the End communications and Session control layer above.

Looping Problem encountered in distributed datagram routing in which packets return to a previously visited node.

LU Logical unit, access port for users in SNA.

Manchester encoding A way of encoding to get a zero-DC binary waveform. In this encoding scheme, half of the bit interval is transmitted with a positive signal and the other half is transmitted with a negative signal.

Message signal unit Signal unit of CCS that carries a message corresponding to the information part or packet of the HDLC frame plus a message transfer part corresponding to the HDLC frame header.

Modem Device that converts data signals to analog signals for transmission over the voice telephone network.

MTP Message transfer part of CCS.

NAU Network addressable unit. SNA term for LU, PU, and SSCP. Each unit in SNA has a unique address.

Network layer The layer that manages routing and flow control of packets. Refers to the OSI Reference Model.

NRM Normal response mode of HDLC.

OSI Abbreviation for "Open System Interconnection."

Pacing control SNA term for flow control.

Packets Small blocks of data into which messages are broken, for transmission across a network.

Packet switching A technology that transmits packets from source to destination.

Path control IBM SNA's network layer.

PBX Abbreviation for "Private Branch Exchange." A privately owned switch, generally of relatively small size, connected via output trunks to the public network.

PC   Pacing count in SNA.

PDU   Protocol data unit, the unit of data in the OSI Reference Model containing both protocol-control information and user data from the layer above.

Permuter table   Routing table in TYMNET.

Physical layer   The lowest layer in communication architectures providing a direct connection betweeen neighboring nodes in a network. This layer is responsible for bit integrity.

PIU   Path information unit in SNA.

Poisson process   One of the most common arrival processes in queueing theory which has the memoryless property of successive arrivals.

Polling   A controlled method of access to networks.

POTS   Abbreviation for "Plain, Ordinary Telephone Service."

Presentation layer   The layer in communication architectures responsible for ensuring the proper syntax of information being exchanged.

Primitives   Abstract representations of interactions across the service access points, indicating information is passed between the service user and service provider. There are four types of primitives in the OSI Reference Model — request, indication, response, and confirm.

Protocol   Agreement between two peer entities on the means of communication.

Pure ALOHA   A random access technique developed by the University of Hawaii in the early 1970s. In this scheme a user wishing to transmit does so at will. Collisions are resolved by retransmitting after a random period of time.

PU   Physical unit that manages the communication resources at a given node in SNA.

Push   Mechanism in TCP for ensuring immediate transmission of data.

REJ   Abbreviation for "Reject." A special control frame used in HDLC.

Remote concentrator   A remote extension of a local switch often used to concentrate or multiplex remote users via one transmission facility to the local switch to which directed.

RH   Request or response header (SNA term).

Ring latency   The time required to repeat the frame at a station on a ring.

RNR   Abbreviation for "Not Ready to Receive." A special control frame used in HDLC.

Roll-call polling   A technique in which every station is interrogated sequentially by a central system.

Routing data base   Distance table in DNA.

RR   Abbreviation for "Ready to Receive." A special control frame used in HDLC.

RU   Request or response unit, basic unit of data in SNA.

Scan time   The time between two successive polls to a station.

SCCP   Signaling connection part of CCS.

SDLC Abbreviation for "Synchronous Data Link Control." Data link control in IBM's SNA.

Session Connection between end users in SNA.

Session layer The layer in the OSI Reference Model that manages and controls the dialogue between the users of the service.

Shortest path A least cost path between a given source-destination pair.

Slotted ALOHA A random access technique extending pure ALOHA to the case in which messages may only be transmitted in slotted intervals of time.

SNA Abbreviation for IBM's Systems Network Architecture. This is a seven-layer communication architecture.

SSCP System services control point in SNA that manages all resources within an SNA domain.

Status signal unit Signal unit of CCS used to initiate transmission on a link or to recover from loss of transmission.

T1 system A digital communication system designed to handle 24 voice channels at 64 kbps each.

Tandem switch An intermediate switch connecting two other switching exchanges.

TCP Abbreviation for "Transmission Control Protocol." ARPAnet-developed transport layer protocol.

Telnet Terminal-remote host protocol developed for ARPAnet.

TGC Transmission group control in SNA.

TH Transmission header, SNA term.

Time congestion The fraction of time that resources (outgoing trunks) are busy.

Time division multiplexing (TDM) Access method that allocates each user a prescribed portion (a slot) of an outgoing time frame.

Time-multiplexed switch The space switch of which the cross point settings are changed each time slot.

TP Abbreviation for "Transport Protocol," OSI Reference Model.

TPDU Abbreviation for "Transport Protocol data unit."

Transition probabilities Probabilities of moving from one state to another.

Transport layer The layer that provides the appropriate service to the session layer. The layer that shields the session layer from the vagaries of underlying network mechanisms.

Trunk A communication channel between two switching systems.

TSAP Abbreviation for "Transport Service Access Point" in the OSI transport protocol layer.

TSDU Abbreviation for "Transport Service Data Unit" in the OSI transport protocol layer.

TSI Abbreviation for "Time Slot Interchanger." Information is transferred from one time slot at the input to another slot at the output.

ULP Abbreviation for "Upper Layer Protocol." Layer above TCP.

Virtual circuit A transmission path set up end to end with different user packets sharing the path. Packets are constrained to arrive at the destination in sequence.

Virtual route Virtual circuit in IBM's SNA.

Virtual route pacing control SNA congestion control at the path control level.

VRPRS Virtual route pacing response in SNA.

Walk time The time required to transfer permission to poll from one station to another.

Window control A credit or token scheme in which a limited number of messages or calls only are allowed into the system.

X.25 The CCITT three-layered interface architecture for packet switching connecting a DTE to a DCE.

Index